AQUATIC
OLIGOCHAETA
OF THE WORLD

Frontispiece.

Photomicrographs of Alluroides brinkhursti.

A. Anterior end, showing phyngeal glands, thick dorsal wall of pharynx with retractor muscles. Note the brain and thickened septa.

B. Longitudinal section through genital region—the median dorsal spermathecal pore is in IX, testis and sperm funnel in X, sperm sac almost occluding XI, atrium and prostate as well as ovary in XIII.

C. Oblique T.S. through atria, showing lumen on right, penis on left. Note longitudinal and transverse sections of connections between prostates and lining cells on right and left resp.

D. Longitudinal section through IX-XI. Note septum 10/11 with its muscular thickening recurved to form sperm sac (top centre, immediately above sperm funnel).

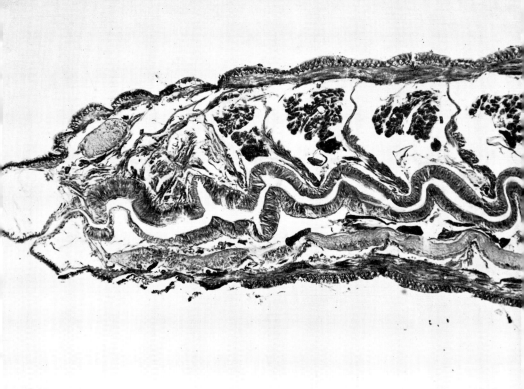

A

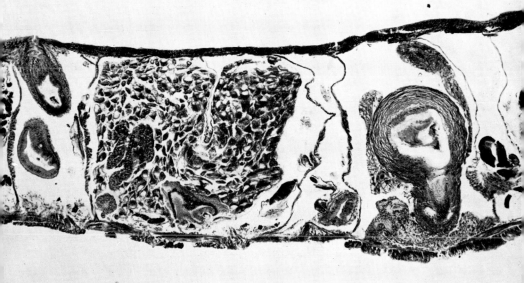

B

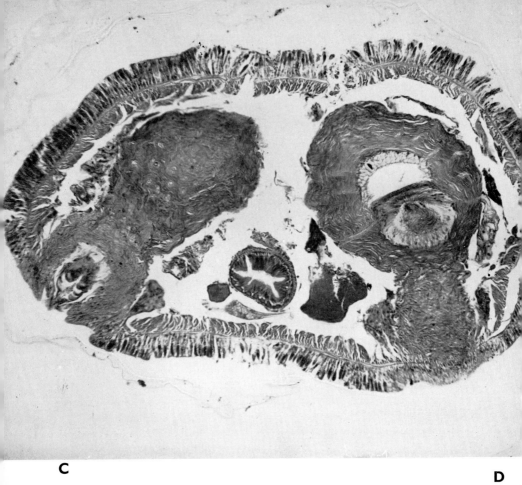

C

D

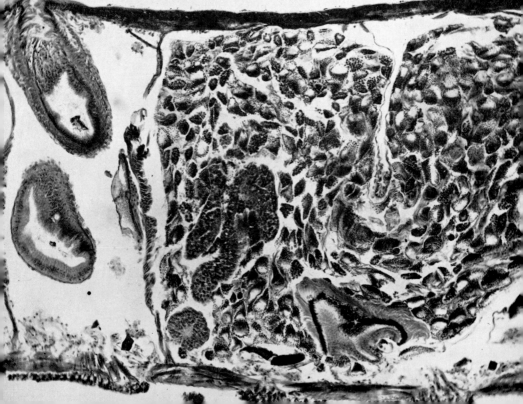

R. O. Brinkhurst
B. G. M. Jamieson

AQUATIC OLIGOCHAETA OF THE WORLD

WITH CONTRIBUTIONS BY

D. G. COOK
D. V. ANDERSON
J. VAN DER LAND

UNIVERSITY OF TORONTO PRESS

OLIVER AND BOYD
Tweeddale Court
Edinburgh EH1 1YL
A Division of Longman Group Limited

First published 1971
© R. O. Brinkhurst & B. G. M. Jamieson, 1971
All rights reserved
First published in Canada and the United States by
UNIVERSITY OF TORONTO PRESS
Toronto and Buffalo

ISBN 0 8020 1833 5
Microfiche ISBN 0 8020 0158 0

Printed in Great Britain by
Northumberland Press Limited
Gateshead

This work is dedicated to
H. B. NOEL HYNES
"The founder of the feast"

PREFACE

This book was conceived as a result of a conversation between the senior author and a representative of the publishers in a small cottage, formerly inhabited by the mortuary keeper, located behind the Zoology Department of Liverpool University. Up to that time the concept was limited to a review of the British Tubificidae, or perhaps the aquatic Oligochaeta of Britain, but, with the prospect of a publisher for a broader based systematic revision, a start was made on a study of the aquatic Oligochaeta of the world. Several colleagues joined the project, and many more have assisted us in the preparation of this work. Some, like Dr. A. M. Beeton of the University of Wisconsin—Milwaukee, Dr. Ruth Patrick of the Academy of Natural Sciences, Philadelphia, and Mr. R. E. Noble of the South African C.S.I.R., Pretoria, Mr. R. W. Sims, British Museum (Natural History) and Dr. M. Dzwillo of the Hamburg Museum opened doors which led to the exploration of unworked areas. Several former students and associates, such as C. R. Kennedy, J. Hiltunen, M. Johnson and K. E. Chua have unstintingly supplied unpublished data in manuscript form and helped the huge task of accumulating the literature. The extensive contributions by Dr. D. G. Cook are not limited to those appearing in this volume under his authorship.

The production of the manuscript and the extensive journeys which made possible the loan of many large collections were financed by a variety of sources. The Royal Society of London, The American Philosophical Society, The American Society of Limnology and Oceanography, The Nature Conservancy, The National Research Council of Canada, The U.S. Public Health Service (now Federal Water Pollution Control Administration, Department of the Interior), The University of Liverpool, The University of Queensland, The University of Sydney, The Australian Research Grants Committee and many other agencies supplied funds which supported the project in one way or another.

Half way through production of this volume, the senior author moved the project to the University of Toronto, which has generously supported the work through to its completion. The former Chairman, Dr. K. C. Fisher and his successor, Dr. D. A. Chant, were instrumental in securing a one-year appointment for Dr. Jamieson, who was released by the University of Queensland to take up this position, so that the final stages of preparation could be carried out with less of a geographical barrier between us. Dr. D. G. Cook has been supported in North America by a Fellowship from the Systematics and Ecology Programme of the M.B.L., Wood's Hole, through the generosity of its Director, Dr. M. R. Carriker.

The list of contributors to whom all of us owe a large debt of gratitude should include the names of all those individuals who have sent collections to us for identification. These have ranged from one or two specimens from out-of-the-way places to extensive collections of many thousands of specimens, all donated to us with the minimum of conditions imposed. Without these collections the work would have been greatly hampered, and the absence of a list of the donors reflects only the length that such a list would assume.

I am certain that my fellow authors would agree that, overriding all of these contributions has been the patience with which our wives and families have borne the preoccupation of their menfolk with what must have seemed at times to be the never-ending pursuit of an unattainable goal.

RALPH BRINKHURST
Toronto, December 1969.

CONTENTS

Note: The various chapters may be attributed to the following authors:

R.O.B.: 3, 4 (microdriles); 6-10, 12 (11 in part).
B.G.J.: 1, 3, 4 (megadriles); 14-15.
D.G.C.: 1 (microdriles); 8 (marine genera), 5, 11.
D.V.A.: 2.
J.v.d.L.: 13.

The senior author (R.O.B.) accepts responsibility for the arrangement of material (references, figures, chapter sequences) and for the format used for the synonomies, references etc.

Illustrations other than those completed by the authors were re-drawn from original sources by Miss Y. Bracke, Mrs. J. Vanstone, Mrs. Monika Shaffer and Mr. Nelson Campbell.

PART ONE **BIOLOGY**

INTRODUCTION

In recent years the trend of biological research has been directed into the biochemical laboratory and away from museums and the libraries that are the basis of taxonomical research, but this trend is reversing. The demands of the resurgent discipline of natural history in the modern quantitative garb we call ecology have made startlingly apparent our lack of knowledge concerning the identity of often the commonest organisms around us. These demands have become heightened by the increasing awareness of the place of man in the ecosystem, not just as a tourist, but as a participant, and often a destructive participant. The pollution of our environment has made us aware of the need to study the intricate mechanisms determining the distribution and abundance of animals, which in turn leads to the need to identify species with accuracy. Aquatic oligochaetes, particularly tubificids, become an all-too obvious feature of organically polluted waters, as the author became aware in studying the recovery of a river in the Midlands of Britain. This group of animals is clearly one which proves useful to biologists surveying inland and estuarine waters for signs of pollution of various types. In addition, these worms are perhaps the most important group concerned with the retrieval of organic matter (and hence energy) from the sediments of lakes and rivers, organic matter or energy that might be buried by successive layers of silt were it not for their activity. Our understanding of the dynamics of the circulation of energy in a lacustrine ecosystem may be crucial in decisions relating to expensive programmes of eutrophication control. If a large proportion of the energy entering the lake is utilized only until it becomes trapped in the sediments, then control of nutrient input will no doubt lead to a reclamation of the lake or the establishment of a less eutrophic condition (in those parts of the

3

world where this is a desirable goal). Conversely, if most of the energy remaining in the silt is returned to the system through the activities of oligochaetes and other organisms, probably mediated by bacteria, then input control will merely stabilize an existing level of production for a long period. The answers to several important questions may be provided by future studies of the largely unexplored communities of lake and river beds.

That these sort of questions are apparent to many limnologists becomes immediately obvious to anyone attempting a revision of the taxonomy of the aquatic organisms involved through the unceasing requests for identification of collections from pollution control agencies, fisheries laboratories, universities and museums across the world. This volume has been produced to meet this very obvious demand. In one respect, however, it will fail to meet the demand. The style and layout of a systematic treatise of this sort makes it difficult to use as a guide to the fauna of any particular region. For this purpose, local keys are of much greater value, but these require a broad general base in order to establish some uniformity in nomenclature, which is where this volume should contribute most to these practical problems. A good illustration of the frustration that will be felt by the non-specialist attempting to identify a collection with this work may be cited. In the Tubificidae many species may be identified from keys using only externally visible characters and such hard parts as penis sheaths that are readily detectable from the simplest temporary whole mounts. This means that keys are written which proceed direct to the identification of species, as the determination of the genus may require dissection or sectioning. This would not be appropriate or indeed possible in a key to the entire family, and so the subfamilies and genera are keyed-out in the traditional way in the present work. The keys presented here are often based on a few characters, as they are intended to be guides to the species descriptions rather than complete in themselves. Hence any identifications made by consulting the keys should be confirmed by referring to the text and illustrations. Regional keys to the faunas of Britain, North America and Africa have already been produced, and keys to other areas will appear in due course. These will, of course, have to be continually revised in the light of new records and descriptions, a task rendered easier by their production in a cheaper format than that of a world revision. The main purpose of this publication, then, is to act as a basis for such local reviews, as well as to focus more attention on this particular group of animals. It is to be hoped that a new impetus to oligochaete research may develop from this.

Although new studies will rapidly render sections of this work out of date, it is hoped that it will serve as a basic reference for these future studies. The most recent major review of the Oligochaeta in the English language was produced by J. Stephenson in 1930, but that text was primarily meant as a contribution to zoology in general, and the taxonomic section deals only with the oligochaetes down to the generic level. Much published work has appeared in the forty years that have elapsed since the preparation of that work. For example, the large and highly significant contribution by Professor S. Hrabě of Brno, Czechoslovakia, almost all post-dates that of Stephenson's work, as does much of the published work of his compatriot, L. Cernosvitov, whose contributions came to such an untimely end in Britain in 1946. Had he lived, I am sure that we would have benefited from the production of a general revision at a much earlier date. Several major reviews of families have been produced in recent years, particularly that on the Naididae by Christina Sperber published in 1948 (upon which the author relied very heavily for the chapter on that family herein) and the more recent study of the Aeolosomatidae and allied forms by Bunke, published in 1967. Apart from these synthetic studies and major reviews of the fauna of certain areas (the fauna of Brazil studied extensively by Professor E. Marcus, and Eveline du-Bois Reymond Marcus for example— cited as Marcus and Marcus E. respectively in references throughout the text) the most recent account of the known species of the Oligochaeta is that produced by Professor W. Michaelsen of Hamburg in 1900. This book was a synthesis of the very considerable activity in oligochaete systematics that occurred before the turn of the century, including the monographs by F. Vejdovsky and F. E. Beddard and the many contributions of the classical authorities such as J. d'Udekem, E. Claparède, G. Eisen and others. It also formed the working basis for Michaelsen's own subsequent manifold publications as well as those of W. B. Benham, K. Bretscher, E. Piguet, F. Smith and Stephenson himself, as well as the late work by Beddard. Despite the lack of illustrations and the great expansion in the number of species described since 1900, this book is still of primary significance to students of the Oligochaeta.

Forty years ago Stephenson suggested that the Oligochaeta was already too large a group for anyone to handle alone. That this prediction proved accurate may be seen by the number of people that have contributed directly or indirectly to this review of only part of the diversity exhibited by the group. For this reason some areas (such as the Enchytraeidae) have been given scant treatment as no one would relieve us of the task of reviewing those parts of the literature with which we were largely ignorant, and some subjects, such as physiology and regeneration studies, have been ignored for the same reason or because good syntheses are already available. The decision as to what to cover and what to leave out must always be a personal choice in the final analysis, and we feel confident that we have left plenty of scope for other texts in this area to augment what we have attempted. This volume deals with all of the families and subfamilies that are completely aquatic. The Enchytraeidae are reviewed to the generic level partly because of a lack of expertise and partly because many species are not truly aquatic. It is hoped to review the remaining oligochaete families at the generic level in a further volume, the few truly aquatic species therein being indicated.

Many will note a lack of evidence of the use of recently developed numerical methods, which, with the notable exception of studies by R. Sims and B. G. Jamieson, have not been a feature of recent publications in this area. Whilst not attempting to excuse this serious omission, it is, perhaps, worthwhile pointing out that, while we continue to operate on the basis of the law of priority, the first requirement in any such study is to untangle the confused nomenclatural history of the species concerned before proceeding to a consideration of higher systematics. If we have been successful, and if not too many errors have crept into the manuscript, then it is to be hoped that less time will need to be spent delving into the past and more attention can be focussed on morphogenetic studies, biochemical analyses and quantitative measures of kinship in the future. These advances in methodology may help to ease the drudgery of working over large collections, but, while we retain the Binominal System and the International Code required for its smooth operation, our systematics will still be rooted in basic taxonomic work at the species level, in which work there is no substitute for a broad experience of the group and a willingness to accept evidence from any source, anatomical, geographical, behavioural, physiological, biochemical or even ecological, for where is the species that has no ecological reality?

1

ANATOMY

GENERAL ORGANIZATION

Oligochaetes are typically segmented, bilaterally symmetrical, hermaphroditic annelids with a spacious coelom, a pre-oral prostomium, an anterior ventral mouth and a posterior anus. In cross section they are usually cylindrical and in length they range from less than 1 mm (some *Chaetogaster* species, Naididae) to about 400 mm (*Haplotaxis gordioides*). Each segment, except the peristomium, usually bears four discrete bundles of setae, two dorso-lateral and two ventro-lateral. The body wall is composed of epidermis, a layer of circular muscles and an inner layer of longitudinal muscle. In a few anterior segments at sexual maturity, the epidermis is thickened and contains a high proportion of secretory cells forming a dorso-lateral or annular clitellum. A ganglionated ventral nerve cord extends the length of the body and connects with a dorsal brain anteriorly. The mouth opens into the gut which extends throughout the body and is constricted by the septa between each segment. Closely applied to the gut is a dorsal blood vessel, and above the nerve cord a ventral vessel. These are connected in each segment, either directly by commisures, or indirectly via a gut plexus. Excretion is by means of basically segmental nephridia which are usually absent in some anterior segments and in the region of the genitalia. In some cases they may be reduced to one or a few in number. The genital system consists of one or two pairs of testes, rarely more, which are associated with pairs of male funnels and vasa deferentia. Posterior to the male system, one or two pairs of ovaries and small, ventral female funnels are present. Typically, ectodermal pouches which store sperm after copulation, the spermathecae, are found near the region of the gonads.

7

MICRODRILES* (D. G. COOK)

EXTERNAL ANATOMY

SIZE The extremes in length have already been mentioned. Diameter likewise ranges from between 50 μ to 4 to 5 mm. In general, however, the microdriles are typically vermiform and are usually about 30 to 60 times longer than they are wide. Again the number of body segments is subject to wide variation, ranging from about 10 in some Naididae to about 300 in some Haplotaxidae. In most families there is considerable intra-specific variation in size and segment number (Hrabě, 1938) even in those worms which have very few segments (as *Chaetogaster langi*, which has from 8 to 21 segments—Sperber, 1948). Thus little taxonomic significance can be placed on absolute size or segment number, but the general appearance, of the worm, or more objectively, the ratio of these characters, may be of diagnostic value. For example most Tubificidae, Lumbriculidae, Alluroididae, Phreodrilidae and many Naididae are typically 30 to 60 times longer than they are wide. However, *Tubifex ignotus* is consistently very thin and thread-like (about 150 times longer than wide) while most *Chaetogaster* species are short and plump. Some haplotaxids are very long and very thin.

PROSTOMIUM The prolobic or zygolobic prostomium is usually rounded and is approximately as long as it is broad. Its probable tactile function is enhanced in some species by the development of a narrow, proboscis-like extension as in *Stylaria* (Naididae) (Fig. 7. 11B, C) and many species of *Rhynchelmis* and *Kincaidiana hexatheca* (Lumbriculidae) (Fig. 5. 4I). A possible rudiment of a proboscis is also found in the marine tubificids *Peloscolex intermedius* and *Limnodriloides medioporus* in the form of a small, thin-walled papilla (Cook, 1969). On the other hand the prostomium may be very small and pointed as in the tubificid *Epirodrilus antipodum* (Cernosvitov, 1939) or completely retractable within the peristomium as in *Peloscolex variegatus* (Brinkhurst, 1962).

SEGMENTATION In external appearance the microdrile families range from smooth ascarid-like worms which show little of their segmentation as for example *Lamprodrilus baicalensis* (Lumbriculidae) to worms with well developed intersegmental furrows as in *Lamprodrilus koretneffi* (Michaelsen, 1901). As well as the basic segmental divisions, many species of Lumbriculidae and Tubificidae exhibit some form of superficial, secondary annulation. The degree to which this is apparent is, to some extent, dependent on the state of contraction of the individual. Nomura (1926) reported secondary annuli on *T. tubifex* while Dixon (1915) denied their presence. Such

---

* Microdriles, in this work consist of the families Lumbriculidae, Haplotaxidae, Tubificidae, Naididae, Phreodrilidae, Opistocystidae, Dorydrilidae and Alluroididae (plus Enchytraeidae where considered).

annulation, however, seems to be a constant feature in most species of *Rhynchelmis* and *Trichodrilus* (Lumbriculidae).

Many Naididae reproduce asexually by fission and form chains of individuals. This manifests itself externally from a series of more or less profound constrictions in the middle region of the body to true chains of individuals showing some degree of cephalization; e.g. absence of dorsal setae from anteriormost segments and pharyngeal rudiments.

PIGMENTATION   The microdriles are usually unpigmented and translucent but most families appear red by virtue of the haemoglobin content of the blood. This can be masked by the colour of the gut contents, or, in transmitted light, by the opacity of the chloragogen cells. Some species, however, do show some true pigmentation. *Branchiodrilus menoni* (Naididae) has bands of pigment located in large cells which can be found in the muscle layers of the body wall or around the blood vessels (Stephenson, 1912). In *Lumbriculus variegatus* there is dark green pigment beneath the anterior body wall (Cook, 1967a). Histological investigation has since shown that this pigment is located in the longitudinal muscle layer.

SENSORY HAIRS   In the older literature much attention was given to "sensory hairs" imparting a "dense furry" appearance to some microdriles (Vejdovsky, 1884; Goodrich, 1895). From personal observation these hairs have been found in live specimens of *Lumbriculus variegatus*, *Eclipidrilus lacustris*, *Stylodrilus heringianus* and some Naididae. These structures, which are, in the author's experience, destroyed by normal fixation techniques (formalin, Bolin's fixative and mercuric chloride) are delicate extensions of epidermal sensory cells and are most likely found in all microdriles. Marcus (1944) demonstrated their sensory nature in *Slavina* (see below).

PAPILLAE   In most species of *Peloscolex* (Tubificidae) the body wall has a characteristic opaque appearance which is imparted by the presence of heavily cuticularized papillae, with or without accumulated foreign particles adhering to them. In *P. benedeni* these are large leaf-shaped structures about 10 to 15 μ in height (Stephenson, 1922b) while in *P. gabriellae* they are absent, or are represented by rings of small granulae, or are fully developed (Brinkhurst, 1965b). The papillae are probably shed prior to breeding (Dahl, 1960; Brinkhurst, 1964a). Similar papillae are found very rarely in other microdriles.

BRANCHIAE   A conspicuous feature of the Opistocystidae, many Naididae, *Branchiura sowerbyi* (Tubificidae) and *Phreodrilus branchiatus* (Phreodrilidae) are extensions of the body wall which are respiratory in the function, the branchiae or gills. *Opistocysta* is characterized by the presence of one small dorso-median projection and two elongate lateral projections of the terminal segment (Cernosvitov, 1936) (Fig. 10. 1H-K). In the Naididae the gills are usually more or less complex foldings to digitiform extensions of the terminal segment (Fig. 7. 13A). *Dero sawayai* has two pairs of gills, one dorsal and one ventral (Marcus, 1943; Naidu, 1962), while *Dero botrytis* possess about 50 digitiform gills which are arranged in six rows running dorso-ventrally (Marcus, 1943) (Fig. 7. 15E). The gills of *Dero digitata* exhibit much intra-specific variation. They are located in a trumpet-shaped, ciliated branchial fossa which is open dorsally, and consist basically of four pairs of extensions some, or all of which may be elongate and digitiform (Chen, 1940). Of a different form are the branchiae of *Branchiodrilus* (Naididae) which possesses paired digitiform gills, dorso-lateral in position and enclosing the hair setae (Fig. 7. 12E). In *B. semperi* the gills are about four times the diameter of the body in length on the second segment but they become shorter posteriorly until they disappear at the fiftieth or sixtieth segment (Stephenson, 1912).

The only tubificid genus to be described which has gills is *Branchiura*. They are paired digitiform projections, one pair to each segment and are dorso-median and ventro-median in position (Fig. 8. 36e). In *B. sowerbyi* the gills are usually found on all posterior segments, the anteriormost being the shortest (Chen, 1940). Stephenson (1917) reported that his *Kawamuria japonica* (= *B. sowerbyi*) had no gills. Chen (1940) attributed this to the physical environment, in which opinion he is supported by Yamaguchi (1953) whose specimens of *B. sowerbyi* from different parts of Japan showed variation from 5 to 50 pairs of gills.

In his description of *Phreodrilus branchiatus*, Beddard (1884a) reported that this species has paired cylindrical gills about as long as the diameter of the body, placed dorso-laterally on about 13 of the posterior segments.

CLITELLUM AND GENITAL PORES A characteristic of sexually mature Oligochaeta is the possession of a thickened region in the body wall which is glandular in nature and which produces the cocoon at oviposition. In the microdriles this clitellum is one cell thick and appears as a more or less opaque, papillate region, occupying from 2 to 6 segments including those bearing the genital openings. In the Alluroididae it is saddle-shaped and confined to the dorsal part of segments. In general the genital pores are located in the region of the fifth to the thirteenth segments except in the case of the Opistocystidae whose male pores occur on the 22nd segment. In general, the male, female and spermathecal openings are paired, and are situated ventro-laterally in adjacent segments or segments in close proximity. In most of the microdriles there is a single pair of each, but in some instances there may be some replication of one or more type of genital pore. In some species of *Lamprodrilus* and *Lumbriculus* (Lumbriculidae), from two to four pairs of male pores may be found, and haplotaxids possess two pairs of both male and female pores. Spermathecae, which are very susceptible to intra-familial, intra-generic or even intra-specific variation in number, can have paired openings on from one to seven segments in Lumbriculidae, and one to four in Haplotaxidae. The number and sequential arrangement of the genitalia and their associated pores has great systematic importance and will be discussed in detail in the section on the morphology of the genitalia and in the chapter on phylogeny.

The male pores are usually the most conspicuous and in appearance range from simple, more or less circular openings, to pores situated on raised papillae or surrounded by concentric folds in the body wall (in *Lumbriculus*), or to the extreme case of *Stylodrilus heringianus* (Lumbriculidae) in which the pores are situated at the tips of a pair of long external, non-retractible, penes. A modification found in some Lumbriculidae, Tubificidae, Naididae and Phreodrilidae is the development of eversible pits in the body wall into which the true male pores open, creating secondary external pores. In *Slavina evelinae* (Naididae) and some species of Tubificidae belonging to the genera *Monopylephorus*, *Limnodriloides* and *Smithsonidrilus*, this process has continued, resulting in the development of a common median chamber into which the male pores open. Externally this appears as a simple, more or less open depression, to an elongate slit arranged transversely. Brinkhurst (1963, 1965a) has pointed out that this bursa may be evanescent and only present at full sexual maturity. In some species of *Eclipidrilus* and *Tatriella* (Lumbriculidae) the male genitalia of one side of the body have failed to develop and in these species the male pore appears as a single mid-ventral opening. Some species of *Lamprodrilus* (Lumbriculidae) appear to possess two pairs of male pores on one segment. The smaller of these, however, are the openings of accessory copulatory glands (Michaelsen, 1901). The glands have taken the form of longitudinally arranged strips which open in a mid-ventral furrow in *L. pygmaeus sulcatus* (Isossimoff, 1962). *Rhynchelmis* (Lumbriculidae) also possesses accessory copulatory glands derived from rudimentary male ducts which may open singly or paired in the region of the true male openings.

The spermathecal pores are likewise usually paired circular openings situated ventro-laterally although they may be dorso-lateral in *Lumbriculus variegatus*, some tubificid species, and in many haplotaxids and phreodrilids. In common with the male pores, the spermathecal openings may also be situated in a common median bursa (e.g. *Limnodriloides medioporus*, Tubificidae) or, by loss from one side of the body, be represented by a single mid-ventral pore (e.g. *Eclipidrilus lacustris*, Lumbriculidae; *Monopylephorus parvus*, Tubificidae). In conjunction with the unpaired spermathecal state, a peculiar phenomenon has occurred in some Alluroididae and *Phallodrilus monospermathecus* (Tubificidae) for which a selective advantage is difficult to trace. In these groups the spermathecal pore or pores are mid-dorsal. This state, and the situation found in some species with ventral, but unpaired, spermathecae, and paired male genitalia, would appear to make copulation more difficult than in the case of both being paired or unpaired. One explanation would be that peculiar mating behaviour resulting from this may more effectively ensure intra-specific copulation in co-existing species.

Possibly of systematic importance from the familial to the specific level is the location of the male and spermathecal pores relative to the ventral setae. In those tubificids for which data is available, the pores occur anterior to the setae, while in Lumbriculidae and Dorydrilidae they are always posterior to the setae. Naididae show both positions even within one genus. It is evident that more data is needed before an assessment of the importance of this feature as a taxonomic character can be made.

The female pores in the microdriles are very inconspicuous. When visible they appear as paired transverse slits located ventrally in the posterior inter-segmental furrow of the male pore segment, or in one or more segments immediately posterior to this in a similar location.

SETAE* Implanted in, and protruding from, the body wall of each seg-ment except the peristomium and a few terminal segments there are usually four bundles of setae, two dorso-lateral and two ventro-lateral. The setae, which are of ectodermal origin, are composed of chitin and a proteinoid part (Avel, 1959). Each setal bundle may contain from one up to about twenty setae which are not necessarily of similar morphology or length. In some enchytraeids it is well known that within a single bundle the lengths of the setae differ. Stephenson (1915) showed that in some Naididae the node is situated nearer to the distal end of the outer setae of a bundle than in the inner ones. In *Eclipidrilus lacustris* (Lumbriculidae) this author (Cook. 1967a) demonstrated a similar phenomenon and also showed that within each bundle, the inner setae (those nearest the median plane) are slightly longer than the outer. This type of dissimilarity within the bundles is probably a widespread phenomenon in the microdriles. In the genital region the ventral setae may be absent or modified to subserve a copulatory function. Either spermathecal or penial setae may be developed, in a few instances both are similarly modified in the same species.

Basically the setae are of two types: hair setae (which occur only in the dorsal bundles of Opistocystidae, Phreodrilidae, most Naididae and many Tubificidae) and sigmoid setae or crotchets (which may be found in both dorsal and ventral bundles of all microdriles). The latter consist of an S-shaped shaft which at some point has a thickened region, the node or nodulus (Fig. 7. 1P). Their distal ends may be pointed, rounded or bifid. The latter type, whose teeth exhibit varying degrees of development, may be ornamented especially in dorsal bundles, by a series of fine intermediate teeth (Fig. 7. 16I). The dorsal setae of many Naididae and Phreodrilidae are straighter than their ventrals (Fig. 7. 7B-D) and in *Peloscolex swirenkovi* (Tubificidae) the posterior dorsals also lose their sigmoid form, but by the excessive development of the shaft distal to the node (Fig. 8. 19E). These seem to represent a condition intermediate between sigmoid setae and true hair setae.

* The term seta(e) is employed here as it seems to dominate recent English-language publication. The senior author would prefer chaetae (viz. Oligochaeta) as being the more specific term, but acknowledges the majority view should prevail.

The absence of a node and a slender, elongate form characterize the hair setae. In both Naididae and Tubificidae they are either smooth or may possess fine serrations or lateral hairs surrounding, or localized to one side of, the shaft (Fig. 7. 8J; 7. 10C; 7. 11M; 7. 16A). Due to the diversity of setal types and the taxonomic importance of their number, distribution and detailed morphology, a brief outline of the setal forms characteristic of the various families will be given.

Lumbriculidae and Dorydrilidae (excluding *Lycodrilus*) have no hair setae and no modified genital setae. The sigmoids have well developed nodes and are pointed or sometimes bifid distally. The dorsal setae in some species of *Lamprodrilus* and *Trichodrilus* differ from the ventrals but within a singe bundle the setae are of the same form. In rare instances in *Lamprodrilus* the setae are wholly or partly absent (Isossimoff, 1962).

The Alluroididae similarly have no hairs, and possess two pointed or rounded setae in each bundle. However, the ventral setae of the male pore segment may be modified.

Haplotaxidae are characterized by one or two setae in each bundle. In *Haplotaxis gordioides* the ventral setae are large, strongly sigmoid or hooked and occur singly while the dorsals are small and may be absent from a number of posterior segments (Fig. 6. 2A, B, D, G, K, L, N, O). *H. glandularis* has one bifid and one single-pointed seta in each bundle (Yamaguchi, 1953).

Phreodrilidae possess hair setae and very small needle-like crotchets in the dorsal bundles, and single-pointed or bifid sigmoids ventrally, which may be present together in a single ventral bundle (Fig. 9. 1A-D). Setae of the spermathecal segment are often absent or modified.

The Tubificidae exhibit a wide diversity of setal form. In many species smooth or serrate hair setae are found in some or all of the dorsal bundles. These may be associated with bifid sigmoids similar to those found in the ventral bundles of the same species (*Aulodrilus pluriseta* for instance) but, more commonly, these sigmoid dorsal setae are further ornamented with a series of fine intermediate teeth. These are termed pectinate setae or, where the lateral or primary teeth are indistinguishable from the intermediates, palmate setae. In other species, or even genera in which hair setae are absent, the setae may be bifid with rudimentary upper teeth (many marine species) or with large distinct teeth differing in relative length (*Limnodrilus*). Penial or spermathecal setae may be modified, rarely both in the same species. Genital setae may have elongate teeth making a hollow-tipped distal end (*Peloscolex, Potamothrix, Psammoryctides*), or knobbed or hooked distal ends (*Rhyacodrilus, Bothrioneurum*) sickle-shaped (*R. falciformis*) or spoon-shaped (*Aulodrilus, Siolidrilus*).

The marine species *Adelodrilus anisosetosus* possesses one giant penial, together with a fan-like row of very thin, straight setae with a strongly reflexed hook distally, in each bundle (Cook, 1969).

The setae of the Naididae and Opistocystidae are similarly very diverse. In many genera the dorsal setae are absent on a few anterior segments and completely absent in the case of *Chaetogaster*. Hair setae are present in the dorsal bundles of all genera except *Amphichaeta, Paranais, Homochaeta, Uncinais* and *Ophidonais*. They are usually smooth, but in some genera may be serrate or even plumose (for example *Dero pectinata*). They are usually of similar length throughout the body, but in a few species, including *Ripistes parasitica*, the hairs are very long in certain regions of the body. Other regional dissimilarities may also exist, as for example in most species of *Nais* and many of *Dero* whose anterior ventral setae, which were formed in a budding zone, are generally longer and straighter than those in posterior segments. In the genera which have no hair setae, the dorsal crotchets are similar in form to the ventrals, but in most other cases the dorsals are modified to more or less straight shafted setae which are single pointed or bifid distally (Fig. 7. 5I and O). These, the needle setae of most naidid workers, often lack a node, or have only an inconspicuous one, and hence they closely approach hair setae in general form. The ventral bifid setae are thinner in relation to their length than those of other families (e.g. Fig. 7. 5E-G), and in many cases may be more strongly sigmoid or have more localized and sharper curves. In common with the Tubificidae, the Naididae often possess penial, and more rarely spermathecal setae. The penials are usually larger than the adjacent ventrals, and the distal ends may be similar to the latter or modified by the reduction of one of the teeth. An extreme modification in both the penial and spermathecal setae is found in some species of *Pristina*. In *P. amphibiotica* the penials have very long teeth which meet at the distal end and which are united by a thin webbing (Stephenson, 1926). In appearance these are similar to the hollow spermathecals of *Potamothrix heuscheri* (Tubificidae).

ANATOMY AND HISTOLOGY OF BODY WALL

The histology of the body wall has received little attention in the microdriles. It is of similar general morphology throughout the body length and consists of five layers. An outer, non-cellular cuticle overlies (and is probably secreted by) the epidermis which is composed of cuboidal to columnar cells of various types. Beneath the epidermis there are two layers of muscles; an outer circular layer and an inner layer of longitudinal muscles which are bounded internally by a thin peritoneal epithelium.

CUTICLE   The cuticle is colourless and usually very thin, as for example in *Tubifex tubifex* in which it is 3 μ thick (Dixon, 1915). In *Haplotaxis africanus*, however, it may be nearly as thick as the epidermis (Benham, 1909) and appears to be more comparable to that of the earthworms than to the relatively thin cuticles of the truly aquatic families. As some Haplotaxidae are terrestrial in habit it may be concluded tentatively that a relatively thick cuticle is an adaptation to this environment, a contention which is supported by the Enchytraeidae, many of which are also terrestrial and which have a robust cuticle.

EPIDERMIS The epidermis possesses two major types of cell. By far the most frequent are supporting cells which are more or less undifferentiated cuboidal to columnar epithelium which are said to have basal processes within the circular muscle layer (Atheston, 1899). Glandular cells, which contain globules of secretory material are usually randomly mixed with these supporting cells. However, Yamaguchi (1953) showed that, in *Haplotaxis glandularis*, these cell types vary in frequency in different regions of the body. For instance, the gland cells are predominant within each major secondary annulus but they are entirely absent in minor annuli. Also within the epidermal layer are small basal cells which probably serve a replacement function (Atheston, 1899) and various types of modified sensory cells which will be described elsewhere (see p. 24). Also to be discussed later (see p. 18) will be the specialized glands of the Haplotaxidae, and glands connected with sexual maturity (see pp. 31-41).

The clitellum, which is developed in the region of the gonads at sexual maturity, is one cell thick and is characterized by the presence of a very high proportion of glandular cells. In the Naididae these are up to ten times thicker than normal epithelial cells, are filled with secretory globules and have a basal nucleus (Sperber, 1948). Dixon (1915) reported that the clitellar epidermis of *Tubifex tubifex* was 40 to 45 μ thick while the normal epidermis was 6 to 8 μ thick. Similarly in *Eclipidrilus lacustris* (Lumbriculidae) the cells of the clitellum are much more elongate than the surrounding epidermis and contain numerous spherical particles of deeply-staining secretory material (Cook, 1967a) (Fig. 1.1A-B).

Another seemingly constant regional difference in the surface epithelium was reported by Berg and Ditlevsen (1899) who found that the cells in the intersegmental furrows were more squamous than those of the segmental epidermis.

CIRCULAR MUSCLE The fibres of the circular muscle, which are similar in histological structure to the longitudinal fibres, are arranged at right angles to the long axis of the body. The nuclei of the fibres are all located in the lateral line (Hesse, 1894). The circular muscle is usually a single layer under the epidermis. It is generally very thin, about half the thickness of the epidermis being usual (Fig. 1. 1A-B) but some exceptional cases of its variability within the Lumbriculidae have been reported. Michaelsen (1905) found that it was more than five times thicker than the epidermis and twice as thick as the longitudinal layer in *Lamprodrilus inflatus*. Isossimoff (1962) stated that while the normal thickness of the circular muscle in *Stylodrilus crassus crassus* was 40 μ, it was 150 μ thick in the fifth segment of *S. crassus crassior*.

In some Naididae and Tubificidae the circular muscle layer seems to be

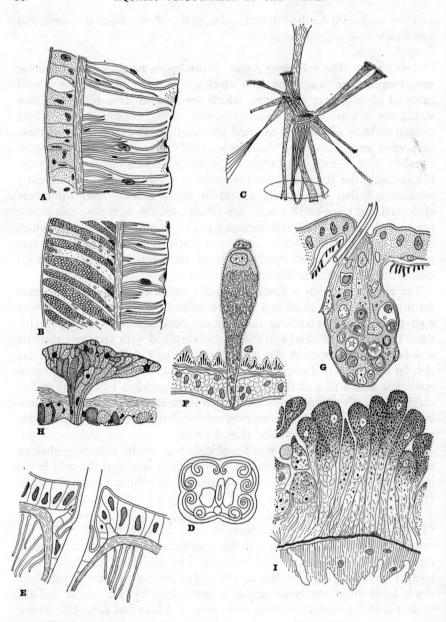

Fig. 1.1. Body wall and coelom.
Transverse section through body wall. A. *Eclipidrilus lacustris*, pre-genital segment.
  B. Same, clitellum.
Setal musculature. C. *Stylaria lacustris*.
Longitudinal muscle. D. *Rhynchelmis* sp.
Setal sac. E. *Lumbriculus variegatus*.
Epidermal glands. F. Unicellular gland of *Haplotaxis glandularis*. G. Dorsal setal gland
  of same. H. Timm's gland, *Haplotaxis*.
Chloragogen. I. *Potamothrix hammoniensis*.

more complex than a single thin band of muscle fibres. In *Monopylephorus irroratus* (as *acklandicus*) according to Chen (1940), some fibres from the circular muscle pass into the coelom forming a second circular layer inside of the longitudinal one and also contribute to the septal muscles. In *Nais*, *Chaetogaster* and *Lumbriculus*, De Bock (1901) found a layer of diagonal muscle in which the fibres cross each other at right angles forming a wide meshed network and which pursue a spiral course around the body. This author stated explicitly that the diagonal layer was absent in *Tubifex*.

LONGITUDINAL MUSCLE Muscle histology in microdriles was examined by Hesse (1894) and De Bock (1901), the latter account being somewhat marred by a peculiar terminology, which, even today, is a vexed question in many areas of oligochaete morphology (see pp. 33-41). The following outline, in which a muscle fibre is defined as a single, functionally contractile, unit, is based on the description of *Lumbriculus variegatus* by De Bock (1901). Each muscle fibre, which may be considered equivalent to a single muscle cell, is composed of an elongate and very tall contractile part (Fig. 1. 1A) and a nucleus which lies outside of the latter. The contractile elements are located at the periphery of the fibre and thus appear to be double in transverse section. In the longitudinal muscle layer the fibres are arranged parallel to the long axis of the body. Unlike the situation in the Lumbricidae each fibre abuts the circular muscle layer, and so they are not arranged in the familiar "caissons" or units. A peculiar exception to this arrangement (which otherwise appears to be characteristic of the microdrile families) is found in many species of *Rhynchelmis* (Lumbriculidae). In this genus the longitudinal muscle layer is broken up into eight bands, the margins of which curl inwards towards the centre of the body at the median, horizontal and two oblique planes (Fig. 1. 1D).

In Tubificidae and Naididae the longitudinal muscle fibres do not appear to be so tall as those in *Lumbriculus* (De Bock, 1901), so that the muscle layer is relatively thinner than in the Lumbriculidae.

PERITONEUM The peritoneum is a thin layer of squamous epithelium which separates the longitudinal muscle layer from the coelom. The only exception of which this author is aware, was reported in *Monopylephorus limosus* in which the peritoneal epithelium in the anterior part of the body consists of pear-shaped cells with their rounded ends directed towards the coelom (Nomura, 1915).

THE SETAL SACS Each seta is contained within a tubular invagination of the epidermis. Dixon (1915) stated that in *Tubifex tubifex* the cuticle extends into the outer part of the sac for some way but illustrated it as continuing proximal to the setal node. In Lumbricidae, Sajovic (1907) showed that the cuticle extended to the setal node, at which point the epidermis became thin and closely applied to the setal shaft. It would be expected that this arrangement, which allows for tractible movements of the setae, pertains to the microdriles. For *Lumbriculus variegatus* I have been able to show that the cuticle does extend for a little way past the node but is reflexed upon itself (Fig. 1. 1E). The seta figured is in the retracted condition.

Setae develop from follicles derived from the epidermis by proliferation (Beddard, 1895; Nomura, 1913; Dixon, 1915). The young follicle is syncitial and the seta develops, distal end first, within it (Nomura, 1913, 1915). At the base of the setal sac, a reserve or replacement follicle, often containing a setal rudiment, is usually present.

SETAL MUSCULATURE   The muscles associated with setal movement have been described in *Tubifex tubifex* (Dixon, 1915), *Monopylephorus limosus* (Nomura, 1915) and *Stylaria lacustris* (Liebermann, 1932). The two former authors both recognized inter-follicular or setal retractor muscles, and a number of parieto-vaginal or setal protractor muscles. Liebermann (1932) in a more exhaustive account of *Stylaria lacustris*, showed that in addition to one inter-follicular and three protractor muscles, each setal bundle was provided with two anterior flexor muscles, two posterior flexors and two muscles which rotated the whole setal bundle (Fig. 1. 1C).

GLANDS IN THE BODY WALL   The most important glands are those connected with the sexual organs and will be discussed under that heading. Apart from these, and the gland cells which seem to be present in the epidermis of all microdriles, three major types of glands have been found in some Haplotaxidae; unicellular glands, Timm's glands and septal glands.

Yamaguchi (1953) described large, club-shaped unicellular glands which projected into the coelom of *Haplotaxis glandularis*. These cells have a basal nucleus and contain granules of secretory matter (Fig. 1. 1F).

In *H. gordioides* a number of pear-shaped gland cells, which open to the exterior in the mid-ventral line beneath the nerve cord, were described by Timm (1883), Beddard (1895) and Hrabě (1931). These structures, known as Timm's glands, are penetrated by a nerve and have thus been attributed a sensory function, despite their glandular appearance (Fig. 1. 1H).

Yamaguchi (1953) in his description of *H. glandularis* described a pair of sub-spherical bodies, or setal glands, which projected into the body cavity and opened immediately dorsal to the dorsal setae (Fig. 1. 1G). Setal glands, although well documented in some megadriles, have not been described in other microdrile families except those which are associated with the genital setae (*vide infra*).

INTERNAL ANATOMY

COELOM

General. The coelom is a spacious cavity extending throughout the length of the body, in which the internal organs are suspended by septa and mesenteries. The septa divide the cavity, which is filled with coelomic fluid, into a number of discrete chambers or segments. In most cases the septa are located beneath the external intersegmental furrows but Issosimoff (1948, 1962) reported an exception to this in some Lumbriculidae. He showed that in *Lamprodrilus* the septa had moved forward in relation to both the intersegmental furrows and the fundamental metamery of the ventral nerve cord.

In *Lamprodrilus vermivorus,* Michaelsen (1905, 1926) found that the coelom of the pregenital segments was more or less completely occluded by muscular tissue.

Coelomic Fluid. The coelomic fluid combines many functions and is of great importance in the life of the animal. It acts as a hydrostatic skeleton upon which the muscular system may act to induce movement, and is also a vehicle for the free cells it contains, many of which are phagocytic and therefore an important defence system (Avel, 1959). The fluid is also an intermediate reservoir for excretory products and is the substrate on which the nephridia act.

Due to the small size of most microdriles the biochemistry of the coelomic fluid has not been studied. However, Bahl (1946) showed that in *Pheretima* (Megascolecidae) the coelomic fluid contains more calcium, sodium and chloride ions than the blood, but less protein. This author found no glucose, free amino acids or fats in the coelom and thus the fluid does not act as a transport system for food, at least in solution.

Coelomic Lining. The coelomic cavity is lined by the relatively un-specialized parietal peritoneum (see p. 17) and a visceral peritoneum, some of whose cells are modified as chloragogen cells. The chloragogen cells are found surrounding the gut and the dorsal blood vessels, beginning in the region of the fifth to the tenth segment.

Detailed descriptions of the chloragogen cells are to be found in Rosa (1898) and Liebmann (1931). In *Potamothrix hammoniensis* (Tubificidae), the latter author found the chloragogen cells to be elongate and pear-shaped, containing many yellow-coloured granules aggregated in the swollen, or distal end of the cell (Fig. 1. 1I). These inclusions, the chloragosomes, con-tain a purine and a lipid which imparts the yellow colour to the chloragogen (Avel, 1959).

Functionally the chloragogen has been compared to the vertebrate liver (Avel, 1959). It is thought to have a role in the metabolism of glycogen and protein, and also in excretion in that it contains a high concentration of ammonia and urea. The concentration of amino acid in the chloragogen of *Peloscolex multisetosus* is described in Chapter 4.

Free Coelomic Cells. Probably all oligochaetes possess a number of free cells within the coelom. Liebmann (1942) distinguished a number of types in the Lumbricidae and it is possible that detailed studies would reveal a similar variety of corpuscles in microdriles. From a taxonomic point of view, the very conspicuous coelomocytes found in some tubificid genera are important (*Bothrioneurum, Monopylephorus, Rhyacodrilus*). These granular, spherical to ovoid cells completely fill the coelom of some species. In *Monopylephorus limosus* the diameter of the coelomocytes is 5 to 6 μ (Nomura, 1915) and in *Bothrionurum iris* 10 to 19 μ (Aiyer, 1925). Coelomocytes also occur in some naidids (Sperber, 1948) and in some Phreodrilidae (Stout, 1958). The validity of using the presence or absence of coelomocytes as a generic character in Tubificidae is open to doubt as small numbers of these structures have been observed in genera in which they were thought to be absent (Cook, 1969).

SEPTA   The septa are composed of three layers: an inner layer of muscle fibres and two outer layers of peritoneal epithelium (Nomura, 1913). Typically a septum from the posterior part of the body is penetrated by the gut (to which the edges are closely applied), the ventral nerve cord, the dorsal and ventral blood vessels, and by a pair of nephridial tubules. The musculature is often rather complex, but consists mainly of fibres running dorso-ventrally, laterally and, from the ventral surface, circum-intestinally (Fig. 1. 2A) (Nomura, 1913; Pointner, 1911). The former author stated that the septal muscle fibres terminated in the circular muscle of the body wall. Haffner (1927) in describing the musculature associated with the blood vessels of *Lumbriculus variegatus* mentions that a band of muscle fibres from the dorsal body wall pass beneath the dorsal vessel, but dorsal to the gut, and that fibres having a similar origin in the dorsal part of the septa, run obliquely to the dorsal blood vessel (Fig. 1. 2B).

Probably in all families of microdriles the septa of the first two to five segments are incomplete or absent (Stephenson, 1930). In general the septa are thin, delicate structures but a number of pre-genital septa are often thickened by the development of the muscle layer. This has been reported in some Tubificidae (Dahl, 1960), Haplotaxidae (Brinkhurst, 1966), and Alluroididae (Brinkhurst, 1964b; Jamieson, 1968a).

The septa are modified or distorted in some regions of the body in most microdriles, the most important adaptation being the development of elongate pouches in the septa of the genital segments. These structures which may extend posteriorly for many segments, are variously known as sperm sacs, seminal vesicles and egg sacs (see pp. 31-41). A peculiar modification was noted by Nomura (1913) in *Limodrilus hoffmeisteri* (Tubificidae), in which the posterior septa of the fifth to the ninth segments possessed paired ventral pouches of unknown function.

Most microdriles possess groups of chromophilic cells in a few anterior segments, which are often closely applied to the septa and have thus been misnamed septal glands. These structures are associated with the digestive system and will be discussed under "Pharyngeal glands".

AILEMENTARY CANAL   The microdrile gut consists of an anterior,

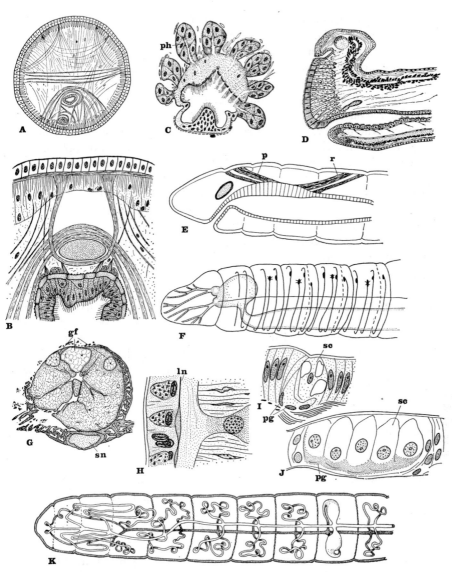

Fig. 1.2. Septa, alimentary canal, nervous system and vascular system.
Septal musculature. A. *Isochaeta virulenta*. B. *Lumbriculus variegatus*.
Pharynx. C. Transverse section *Aulophorus furcatus* ph. pharyngeal gland. D.
Longitudinal section *Aulophorus carteri* with protruded pharynx. E. Musculature of
Lumbriculus variegatus p. protractors, r. retractors.
Nervous system. F. *Lumbriculus variegatus*. G. Transverse section of nerve cord of
Branchiura sowerbyi GF. giant fibres; SN sub-neural blood vessel. H. Lateral line
of *Lumbriculus variegatus*. ln. lateral nerve. I, J. Sections through eye of *Stylaria
lacustris*. sc. sense cell. pg. pigment.
Anterior blood vessels. K. *Tubifex tubifex*.

ventrally opening mouth cavity composed of very thin epithelium, a pharynx, and a simple tubular hind gut which extends the length of the body and terminates in a posterior ventral anus. In some Haplotaxidae a muscular gizzard intervenes between the pharynx and the hind gut, and in the Naididae a stomach or intestinal dilation is similarly located.

A thickened pad of tall collumnar cells situated in the dorsal part of the anterior region of the gut constitutes the pharynx. It has been observed in all microdrile families and is capable of being protruded through the mouth (Fig. 1. 2D). Associated with the pharynx are groups of chromophilic cells which have been termed septal glands (Sperber, 1948; Brinkhurst, 1966) or simply chromophile cells (Stephenson, 1930), and whose structure and function have been misunderstood. Stephenson (1922a, 1930) denied that they had any connection with the digestive system in microdriles. Marcus (1943) demonstrated that in the Naididae discrete masses of chromophile cells were directly associated with the pharynx by narrow cell connections but thought the cells to be the periferal, basophilic parts of the pharyngeal epithelium (Fig. 1. 2C). Sperber (1948) was unable to support this view but confirmed that connection with the pharyngeal lumen existed. Similar penetration by these cells has been observed in Haplotaxidae (Brinkhurst, 1966) and personally in some Lumbriculidae and Tubificidae (Cook, unpublished data). Thus it is contended that the so-called septal glands or chromophilic cells of other families are, and should be termed, pharyngeal glands. In some Naididae (Marcus, 1943) and Lumbriculidae (Cook, 1967a), two sets of muscles from the dorsal body wall are attached to the pharynx: anterior protractor muscles and posterior retractors (Fig. 1. 2E).

The structure of the microdrile gut is summarized by Stephenson (1930) and detailed in certain groups or species by Smith (1900a), Nomura (1913, 1915), Dixon (1915), Mehra (1922) (Tubificidae); Goddard (1909a) (Phreodrilidae); Forbes (1890), Benham (1905, 1909), Brinkhurst (1966) (Haplotaxidae); Sperber (1948) (Naididae); Jamieson (1968a) (Alluroididae). In general the hind gut begins immediately after the pharynx and is a simple cylindrical tube, without a typhlosole. As typified by *Lumbriculus* this part of the gut is composed of ciliated collumnar lining cells, a very thin muscle layer, and a layer of chloragogen cells (see Fig. 1. 2B).

The only tubificids which do not conform to this pattern are *Limnodriloides* species which have a pair of blind digitiform gut diverticulae in the eighth or ninth segment directed anteriorly and attached to the anterior septum of their segment (Boldt, 1928) (Fig. 8. 24F), and *Tubifex pseudogaster* whose gut possesses a glandular thickening in the ninth segment (Dahl, 1960). Sperber (1948) found a differentiated stomach situated just posterior to the pharynx in all species of Naididae which she studied. This author found that the stomach, which may or may not be marked off from the rest of the gut by a dilation, is histologically distinct from other regions of the alimentary canal. Its epithelium is characterized by a strongly developed brush border which is present in addition to the cilia. Sperber also states that the stomach epithelium is thicker, especially ventrally, than in other parts of the gut.

In some Haplotaxidae, a gizzard intervenes between the pharynx and hind gut. Brinkhurst (1966) described and illustrated this structure in *Haplotaxis gordioides* in which it consists of a thick layer of circularly arranged muscle fibres surrounding a thickly cuticularized intestinal epithelium.

NERVOUS SYSTEM

General Anatomy. Accounts of the morphology of the nervous system of various microdriles are to be found in Eisen (1879, 1886), Goodrich (1895), Normua (1913, 1915, 1926), Dixon (1915), Chen (1940) (Tubificidae): Vejdovsky (1884), Brode (1898), Stephenson (1907a, 1907b), Southern (1909) (Naididae): Isossimoff (1926) (Lumbriculidae): Forbes (1890) (Haplotaxidae): Goddard (1909a) (Phreodrilidae). The general morphology of the brain and ventral nerve cord seem to be consistent throughout the Oligochaeta. The brain is a bilobed structure consisting of two pairs of ganglia located one above the other, which is situated dorsal to the buccal cavity in the first segment. A number of prostomial nerves are given off anteriorly and a pair of circum-pharyngeal connectives ventro-laterally which join the ventral nerve cord at the subpharyngeal ganglion in segment II. The ventral nerve cord is composed of two closely adjacent nerve trunks running the length of the body in the mid-ventral line, which are swollen or ganglionated in every segment. In *Lumbriculus variegatus*, Isossimoff (1926) found that in every segment except the first two, four pairs of lateral nerves arise from the ventral nerve cord which innervate the body wall. The first, third, and, in middle segments, the fourth pair of lateral nerves unite dorsally, forming complete nerve rings. The second pair terminate in ganglia near the dorsal setae (Fig. 1. 2F). There is probably, according to Isossimoff, a lateral nerve which unites with each segmental nerve in a series of small ganglia located in the lateral line (Fig. 1. 2H). In the general form of the nervous system this account of *Lumbriculus* agrees closely with that of Dixon (1915) for *Tubifex* although this author does not mention a lateral nerve. A so-called sympathetic or visceral nervous system is well documented in the Lumbricidae (Millott, 1943) consisting of segmental nerves passing to the gut from the body wall plexus near each septum, and a pair of small ganglia branching off the circum-pharyngeal connectives, which innervate the pharynx. Brode (1898) described a pharyngeal nerve in the naidid *Aulophorus vagus* and it would seem probable that this type of visceral innervation occurs in all Oligochaeta.

Histology. In transverse section the ventral nerve cord is seen to consist of two major nerve trunks lying side by side and three major giant nerve fibres running close to their dorsal side. Each nerve trunk is surrounded by a sheath, which fuses medially to form a vertical septum. The whole nerve cord is enveloped by a layer of longitudinal muscle fibres and a peritoneal covering (Keyl, 1913) (Fig. 1. 2G). The latter author found that in *Branchiura sowerbyi* (Tubificidae) the giant fibres are very large, especially in the posterior part of the body, and some may attain the same diameter as the whole of the nerve cord.

Sense Organs. In *Tubifex*, Atheston (1899) found isolated, or loose groups of sensory cells within the epidermis. These are flask-shaped cells innervated by nerve fibres entering the cells a little distance from their bases, and are found most abundantly in the prostomial and posterior regions of the body. Some of these cells possess one or a number of thin hair-like processes externally (see pp. 9-15). These sense cells are similar to those described by Marcus (1944) in *Slavina appendiculata* (Naididae) (Fig. 7. 8S).

In *Bothrioneurum* (Tubificidae) a peculiar sensory structure is present in the form of a ciliated pit situated asymmetrically on the dorsal surface of the prostomium. The cells composing it are taller than the surrounding epidermis and the muscle layers of the body wall immediately beneath the pit are absent (Beddard, 1901).

In some Naididae pigmented eye spots are present on the lateral surface of the first segment. The eyes in *Stylaria lacustris* which were described by Hesse (1902), consist of five or six large unpigmented cells situated beneath and medial to the row of sense cells (Fig. 1. 21, J). Each cell contains a number of vacuoles filled with a clear fluid, and a transparent hyaline ovoid body about the same size as the nucleus. Dehorne (1916) denied the presence of the latter structures ("phaosomes") but found that a mass of nerve cells situated anterior to the eyes connected each sense cell to a short nerve leading to the circum-pharyngeal connectives.

Also peculiar to the Naididae (most *Chaetogaster*) is an organ, the refractile body, situated medially and in the posterior part of the brain, which may function as a statocyst or balancing organ (Sperber, 1948). It is a lens-shaped vesicle whose contents are refractive and consist of brownish aggregates which apparently dissolve in some of the solvents employed in histological techniques (*op. cit.*).

VASCULAR SYSTEM An extensive literature exists on the anatomy of the blood system especially by the earlier workers who tended to study live material, in which it is a conspicuous feature.

The basic plan of the vascular system is fairly uniform throughout the microdriles. Dorsally a large pulsating blood vessel, usually lying close to the gut surface, runs the entire length of the body, divides anteriorly in the region of the brain, and joins with two ventral vessels in the first segment. The latter unite in one of the pre-genital segments (usually from III to VII) and runs the length of the body as a single vessel located just dorsal to the nerve cord. In each segment except the first, the dorsal and ventral vessels are connected either by circum-intestinal connectives, or indirectly in post-genital segments, through a blood vessel plexus around the gut which consists of a fine capillary network closely applied to the gut surface. In the genital segments the dorso-ventral connectives may be very elongate and run through the sperm and egg sacs for many segments posterior to their point of origin. Blood flows anteriorly in the dorsal vessel, posteriorly in the ventral vessel, dorsally through the alimentary plexus and ventrally through the connectives (Haffner, 1927).

In most families the blood vessels are composed of a very thin membrane surrounded by thin layers of circular and longitudinal muscle fibres and a peritoneal covering which may, in parts, consist of chloragogen cells.

The anatomical details of this basic plan vary between the major groups, thus notes on each family are presented below.

Tubificidae

Details of various species are to be found in Udekem (1853), Claparède (1862), Eisen (1886), Beddard (1892, 1895, 1896), Goodrich (1895), Benham (1903, 1907, 1915), Moore (1905), Southern (1909), Nomura, (1913, 1915, 1926, 1929), Dixon (1915), Meyer (1916), Mehra (1922), Cernosvitov (1939), Chen (1940), Yamoto (1940), Marcus (1942), Dahl (1960), Naidu (1965).

The major modifications in this family are the presence of hearts and secondary dorsal and ventral vessels. In *Tubifex*, Dixon (1915) showed that, anteriorly, the blood vessels followed the basic microdrile plan (Fig. 1. 2K) but that, in the fifth segment, a supra-intestinal vessel branched off the dorsal vessel which extended to behind the genitalia. It is in close contact with the intestinal plexus but in VIII, it is connected to a swollen pair of circum-intestinal vessels known as the hearts (Fig. 1. 3A). Stephenson (1930) states that this condition is also found in *Limnodrilus, Clitellio, Bothrioneurum, Branchiura,* and *Psammoryctides.* Beddard (1895) showed that in *Bothrioneurum* a sub-intestinal vessel was present in addition to the supra-intestinal and ventral vessels (Fig. 1. 3B). Goodrich (1895) found no secondary dorsal or ventral vessels in *Monopylephorus rubroniveus* but described the hearts in this species as filiform dorso-ventral vessels in some pre-genital segments, which contained a number of one-way valves. He also found similar valves in the dorsal vessel at the junction of these connectives (Fig. 1. 3C). In *Branchiura sowerbyi* a subneural blood vessel appears to be present (Keyl, 1913).

The blood vessels of the posterior segments in Tubificidae are similar to the condition described in *Tubifex* by Dixon (1915). In this species a pair of long coiled connectives lying freely in the coelom and a pair of small commisures closely applied to the gut, join the dorsal and ventral vessels. A number of small vessels also drain into the gut plexus.

In *Branchiura sowerbyi* the gills are supplied with blood by long loops of the dorso-ventral connectives (Fig. 1. 3D) (Stephenson, 1930).

Naididae

The structure of the naidid vascular system was summarized by Sperber (1948). In general the dorsal vessel is closely applied to the gut especially

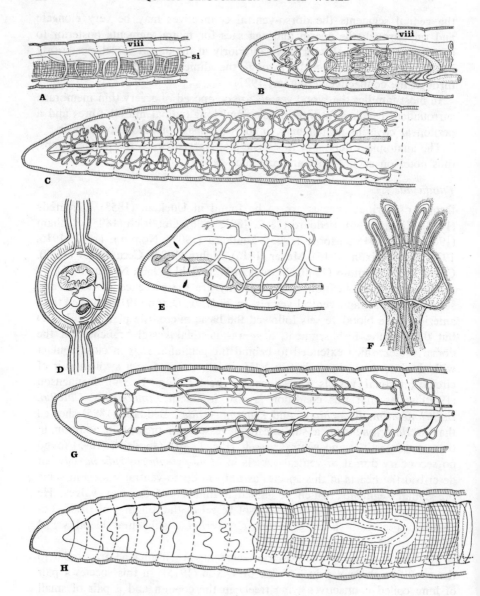

Fig. 1.3. Vascular system.
Anterior blood vessels. A. *Tubifex tubifex* si. supra-intestinal vessel. B. *Bothrioneurum vejdovskyanum*. C. *Monopylephorus rubroniveus*. E. *Ophidonais*. G. *Lumbriculus variegatus*. H. *Trichodrilus moravicus*.
Blood vessels to branchiae. D. *Branchiura sowerbyi*. F. *Dero*

in posterior segments, and is often displaced to one side of it so that it lies ventro-laterally. Dorso-ventral connectives are present in a few anterior segments only and these may anastomose to form a simple plexus (Fig. 1. 3E), as in some Lumbriculidae (see below). In those species with posterior gills, the dorsal vessel branches into two major trunks which send branches into each gill projection and drain into a semi-circular vessel joining the ventral vessel (Fig. 1. 3F). The intestinal plexus begins behind the genital region.

Lumbriculidae

In general this family follows the basic microdrile plan closely but some of its members are unique in possessing blind sac-like blood vessels which branch off the dorsal vessel in post-clitellar segments. The latter are often used as specific criteria and are fully covered in the systematic section.

Haffner (1927), in an exhaustive account of the vascular system in *Lumbriculus variegatus*, showed that the dorso-ventral connectives were long and coiled, and in the first few segments anastomosed forming a pharyngeal plexus (Fig. 1. 3G). In the genital region and up to about the twentieth segment, the connectives are simple tubular structures in the posterior part of each segment. In more posterior segments the only connection between the dorsal and ventral vessels is by the alimentary plexus. In this species a pair of blind lateral blood vessels which are usually branched, are present in the anterior part of each segment from about the tenth. In *Eclipidrilus lacustris*, Cook (1967a, 1967b) showed that the blind lateral vessels (two pairs in this species) possessed numerous small blind caecae which penetrated the musculature of the body wall (Fig. 1. 4A).

Details of the vascular system in other species are to be found in Claparède (1862), Eisen (1881, 1888, 1895), Yamaguchi (1936), Southern (1909), Hrabě (1961), Cook (1967a, 1967b). One peculiarity seems to exist in the anterior dorso-ventral connectives of some *Trichodrilus* species. Hrabě (1938) showed that in *T. moravicus* the six anteriormost of these vessels join the ventral vessel one segment posterior to their points of origin on the dorsal vessel (Fig. 1. 3H).

Dorydrilidae

Cook (1967a) demonstrated that in *Dorydrilus michaelseni* the vascular system is similar to that of the Lumbriculidae, but that the dorso-ventral connectives exist only in the second and third segments and are represented in one or two segments posterior to this only by short blind extensions of the dorsal vessel. Lateral blood vessels and commisures are absent from the fourth segment.

Haplotaxidae

In *Haplotaxis gordioides* the dorsal and ventral vessels are united in each segment by connectives which are very long and which pursue a characteristic sinuous course (Vejdovsky, 1884; Brinkhurst, 1966 (Fig. 1. 4B). Beddard maintained that these lateral vessels in *H. menkeanus* (=*H. gordioides?*) arise from the ventral vessel, pass around the coelom and rejoin the ventral vessel.

Phreodrilidae

The blood system of this family seems to be rather variable. Beddard (1891) found that *P. subterraneus* possessed two dorsal vessels anteriorly, and that the secondary or supra-intestinal vessel carried a pair of swollen lateral hearts (Fig. 1. 4C). Goddard (1909a), however, described *P. notabilis* as having simple dorsal and ventral vessels with connectives from about the fourth segment and no hearts (Fig. 1. 4D). Stout's (1958) account of *P. major* is similar to the above.

Alluroididae

Jamieson (1968a) summarized the existing knowledge of the vascular system of this family as, ". . . nothing is known of it beyond the existence of single dorsal and ventral vessels, the presence of meandering lateral loops in anterior segments as far posteriad as 12 (*S. transvaalensis*) and the absence of a subneural vessel (*A. pordagei, S. transvaalensis*). . . ."

NEPHRIDIAL SYSTEM Nephridia are the organs of excretion and osmoregulation. In the microdriles they consist basically of a ciliated funnel, the nephrostome, a short canal which penetrates a septum, and a looped or coiled nephridial tube which opens to the exterior by a small pore located near the ventral setae of the segment behind that bearing the funnel (Fig. 1. 4E). Following the classification system of Bahl (1947) this type is known as an open, exonephric, holonephridium. They are found in all microdrile genera except *Chaetogaster* (Naididae) in which the nephrostome is absent (closed exonephric holonephridium).

In the megadrile families which exhibit a great variety of nephridial types, these organs are fairly well known, are of great taxonomic importance and have been classified in various ways (summarized in Stephenson, 1930; Bahl, 1947).

A close study of the anatomy and distribution of nephridia may also provide valuable taxonomic information within the microdriles, but as reliable information is sparse, only a brief summary of pertinent literature and nephridial diversity within the various familes is given here.

For the most part nephridia are paired and occur in each segment except a few anterior and genital segments.

Tubificidae

Detailed accounts are to be found in Claparède (1862), Eisen (1886), Goodrich (1895), Beddard (1896), Smith (1900a), Southern (1909), Nomura (1913, 1925, 1929), Dixon (1915), Mehra (1922), Boveri-Boner (1920), Naidu (1965) and Jamieson (1968b).

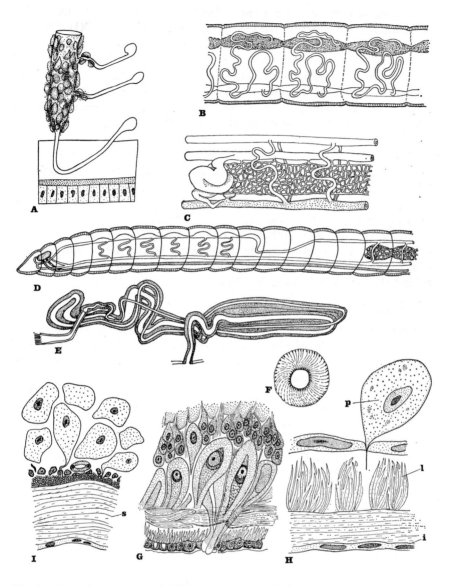

Fig. 1.4. Vascular system nephridial system and genital system.
Blood vessels. A. *Eclipidrilus lacustris* showing blind caecae of posterior lateral blood
 vessel embedded in body wall musculature. B. *Haplotaxis gordioides* commisural
 vessels. C. *Phreodrilus subterraneus*. D. *Phreodrilus notabilis*.
Nephridium. E. *Tubifex*.
Genital system. F. Transverse section of spermatophore of *Tubifex tubifex*. G. Ventral
 gland of *Nais paraguayensis*. H. Transverse section of the atrial wall of *Lumbriculus
 variegatus*, p. prostate cell; 1. longitudinal muscle; i. inner lining. I. Transverse
 section of atrial wall of *Eclipidrilus lacustris*, s. spiral muscles.

In general nephridia are present from about the sixth to the eighth segments but are absent in the genital region (usually X and XI). They are often absent or poorly developed on one side of the body.

In *Tubifex* the immediate post-septal part of the tube is thin-walled and joins, via a small ampulla containing a mass of granular matter, a thick-walled, wide tube which opens to the exterior by a small contractile bulb (Dixon, 1915). In *Macquaridrilus* this terminal structure is greatly enlarged and projects into the body cavity (Jamieson, 1968).

Naididae

Specific accounts are to be found in Boveri-Boner (1920), A. Dehorne (1923, 1925).

The closed nephridia of *Chaetogaster* have already been mentioned. They have an inverted U-shape and are present beginning in the sixth or seventh segment. The post-sepal tubule in *Stylaria lacustris* has a uniform histology except for a dilated ampulla immediately behind the septum (Boveri-Boner, 1920). Nephridia are apparently distributed in pairs beginning just behind the genitalia in most Naididae.

Lumbriculidae

Detailed accounts of various species are to be found in Claparède (1862), Smith (1900b), Beddard (1908), Boveri-Boner (1920), Mrazek (1926) and Hrabě (1960).

Nephridia may be absent in certain segments or on one side of the body. Smith (1900b) showed that *Eclipidrilus asymmetricus* possessed one nephridium in each segment behind the genitalia, which are located in alternate sides of the body in successive segments. In some species of *Trichodrilus* and *Stylodrilus* the nephridial tubules are produced into elongate loops which extend posteriad through a number of segments (Beddard, 1908; Southern, 1909). This possibly represents a link between the normal arrangement and that found in *Lumbriculus variegatus* in which the nephridia of each side of the body, and of successive segments, are joined through a system of united tubules which extend mid-ventrally throughout the length of the body from about the tenth segment (Boveri-Boner, 1920).

Haplotaxidae

Forbes (1890), Beddard (1891), Benham (1909), Smith (1918), Boveri-Boner (1920), Jackson (1931) and Brinkhurst (1966) provide useful information on the excretory system in this family.

Boveri-Boner (1920) found that in *Haplotaxis gordioides* the nephridia are large organs surrounded by a thick layer of peritoneal cells, which fill the coelomic cavity. They are found in a few pre-genital, and all post-genital segments. Jackson (1931) stated that they were present in all segments from the fifth, including the genital region, in *H. darlingensis* (=*H. africanus*).

Phreodrilidae

Benham (1907), Goddard (1909a) and Stout (1958) found that nephridia are tubificid-like in some species, and are often asymmetrically disposed. They are either paired or single, alternating or on one side of the body, in post-genital segments, but are absent or restricted pre-genitally. Goddard (1909b) found that in a worm "greatly resembling" *Phreodrilus mauiensis* the terminal chamber was distinct and flask-shaped, and that the cells of the proximal part of the nephridial duct were histologically similar to hirudinean or oligochaete muscle cells. In *P. niger* the nephridia which open on segment VI are said to extend posteriad to about the tenth segment, similar to those of some Lumbriculidae (Beddard, 1894a).

Alluroididae

Jamieson (1968a) summarized the existing knowledge of this family. Nephridia are paired from about IX to XI posteriad. In common with the rest of the microdriles they are open exonephric holonephridia, but do not possess a terminal bladder.

GENITALIA The Oligochaeta are hermaphroditic and their genital organs contain male, female and spermathecal elements.

The male genitalia consist of one to four pairs of testes which are associated with vasa deferentia and usually storage or intromittant organs termed the atria. The testes are located ventrally on the anterior septa of their segments, and in many cases this, and the posterior septa possess elongate pouches, the sperm sacs, in which the later stages of spermatogenesis occur and which serve as sperm resevoirs. A pair of male funnels are located on the posterior septum of each testicular segment and open into a pair of narrow, ciliated vasa deferentia which drain into paired or rarely single, atria. The latter occur in the segment immediately posterior to the testes segment (Tubificidae, Naididae, Phreodrilidae, Opistocystidae, Dorydrilidae) or in the same segment (Lumbriculidae); that is, the vasa deferentia penetrate one septum at the most. In Haplotaxidae, atria are not developed and in the Alluroididae the genitalia are more comparable to the megadrile system as in this family the testes and atria are separated by two segments devoid of gonads; thus the vasa deferentia penetrate three septa. The atrium is usually a muscular chamber which in some cases is known to be a compound organ of ectodermal and mesodermal origin (Hrabě, 1939a; Cook, 1967a). The atrium, which either terminates in a penis or opens to the exterior as a simple pore, is often associated with glandular cells which are termed, in microdriles, prostate glands. The ventral setae in the region of the atrial segment are modified in some species, in which case they are known as penial setae. These are often embedded in, or associated with, glandular tissue or discrete copulatory glands. Other accessory glands are present in some species and these often take the form of pads of tissue situated ventrally in the proximity of the atria.

The female system is very simple and consists of paired ovaries which are attached to the anterior septa of their segments, and paired, small, ciliated funnels which are situated ventrally on the posterior septa of the ovarian segments. Egg sacs, similar in structure and function to the sperm sacs, are usually present. The ovarian segments are invariably posterior to the testicular segments.

The spermathecae are ectodermal, usually pedunculate pouches which contain the sperm after copulation (e.g. 7. 4H and Fig. 7. 6L). They are usually paired structures which open ventrolaterally in the region of the gonads.

As the number, form and sequential arrangement of the parts of the genitalia are of fundamental systematic importance, and as the morphology of the male ducts varies between and within families, more detailed information on the genitalia at the familial level is given below.

Tubificidae

Detailed accounts of the anatomy of some tubificid species are given by Beddard (1892), Goodrich (1895), Benham (1903), Nomura (1913, 1915), Dixon (1915), Hrabě (1935, 1939b), Černosvitov (1939), Jamieson (1968b). General accounts are to be found in Brinkhurst (1963) and in the systematic section of the present volume. Basically the tubificid genital system consists of one pair of testes situated in the tenth segment associated with a pair of atria in XI, a pair of ovaries and female funnels in the atrial segment, and a pair of spermathecae in the testicular segment (Fig. 8. 34B). Sperm and egg sacs are usually developed from posterior, and often anterior, projections of the septa of the gonadial segments.

Generic criteria in Tubificidae are based on the morphology of the male genitalia and it would be superfluous here to detail all its possible modifications. In general terms a ciliated male funnel on septum 10/11 opens into a vas deferens which enters an atrium consisting of an inner lining and an outer muscular layer covered with peritoneal epithelium. In the species possessing prostate glands, the atrial lining, from which the prostate tissue is derived (Hrabě, 1939a) is usually thin, but in *Clitellio arenarius* which has no prostate, the atrial lining is very thick and may itself perform the function of prostate tissue. Prostate glands are composed of pear-shaped cells, the ends of which penetrate the atrial wall and which either cover the atria diffusely as for example in *Rhyacodrilus*, or, the more usual condition, are localized to one part of the atrial surface as a discrete organ. In *Tubifex* the prostate gland is a subspherical body joined to the atrium near the vas deferens by a narrow stalk formed from the elongate ends of the prostate cells. *Telmatodrilus* is peculiar in possessing a number of discrete prostate glands arranged along the length of its tubular atrium (Eisen, 1879; Brinkhurst, 1965b). The atria are very variable in form. In general they are more or less tubular structures communicating with the exterior as a simple pore or as a modified penis. The latter is formed from folds in the body wall, the cuticle being retained (Fig. 8. 22K). In some species the penial cuticle is thickened, forming a penis sheath, a structure which reaches its ultimate development in some *Limnodrilus* species in which the penis is a very thick cuticular tube. Another type of intromittant organ is found in some tubificids formed from a simple invagination of the body wall which is capable of being everted, termed a pseudopenis (Brinkhurst, 1965a). In *Bothrioneurum* and *Smithsonidrilus* a bulbous paratrium of obscure origin and function opens into a common chamber with a tubular atrium. It is contended by Cook (1969) that the secretory, storage and intromittant functions of the atrium have become separated in such entities and that the peculiar *Adelodrilus* with a prostate-bearing penial bulb, is close to the ancestral form of the two former genera. *Branchiura* possess an atrial diverticulum but of a different structure to the paratrium of *Bothrioneurum* and *Smithsonidrilus*. It is evident that morphogenetic studies on these genera are necessary to establish the homologies, and the terminology, of these structures. Penial setae are found in many species of this family (see p. 12; Setae), which are sometimes associated with glandular tissue (Smith, 1900a; Mehra, 1922) and are located on the atrial segment.

Spermathecae are ectodermal structures formed by the invagination of the body wall epithelium, and consist of a distinct narrow duct and a voluminous ampulla. In *Tubifex* the spermathecal wall is composed of an inner layer of cuboidal cells, a layer of scattered muscle fibres and an outer peritoneal layer (Dixon, 1915). In most cases the spermathecae are paired but in *Monoplephorus parvus* and *Phallodrilus monospermathecus* they have failed to develop on one side of the body, leaving single median structures opening ventrally in the former (Moore, 1905) and dorsally in the latter (Knollner, 1935). Spermathecae are entirely absent in *Bothrioneurum* and *Jolydrilus*, and may be absent in *T. tubifex*. Spermathecal setae are present in many tubificid species.

In a number of members of this family the sperm contained within the atria prior to copulation and within the spermathecae after copulation, are in compact oriented bundles, the spermatophores, which are often of characteristic shape and thus systematic importance. The spermatophores of *Tubifex* are elongate structures which, in cross section, consist of an inner central circular lumen or axis around which the sperm heads are arranged and an outer coat of sperm tails which radiate from the axis in a spiral fashion (Dixon, 1915) (Fig. 1. 4F). The structures attached to the body wall and containing sperm in *Bothrioneurum* are not homologies of these spermatophores and they are termed sperm bearers* in the systematic section.

---

* It might be better to call these spermatophores as this term is used in the megadriles and to call the organized sperm bundles in the spermathecae of tubificids by the old term, spermatozeugma.

Naididae

The number and sequential arrangement of the parts of the genitalia of the Naididae are identical to those of the Tubificidae but they are located anterior to the tubificid position: in the naidids the atria are mostly situated in the sixth segment. Sperber (1948) summarized the anatomy and histology of the genitalia and more detailed accounts of various species are to be found in Stephenson (1915), Mehra (1924, 1925), Stolte (1933) and Marcus (1943).

According to the account of Sperber (1948), the naidid male funnels are usually cup-shaped but are often deformed by pressure in the sperm sac and in some cases extend into this cavity. A pair of vasa deferentia of varying length and consisting of more or less cuboidal ciliated cells, enter a pair of atria at varying positions from the apical to the proximal end. In some species the epithelium of the vasa deferentia is thick and glandular (Fig. 7. 24H). The atria are usually more or less spherical to elongate bodies which communicate with the exterior by narrow discrete ejaculatory ducts (Fig. 7. 2B). True penes appear to be absent but, in *Slavina evalinae*, Marcus (1944) reported that the ejaculatory ducts open into a common eversible chamber which constitutes an eversible pseudopenis in the sense of Brinkhurst (1965a). The atrial wall is composed of an inner ectodermal lining, an outer layer of muscle fibres derived from the body wall layers according to Mehra (1924), and a peritoneal covering. In general prostate cells are pear-shaped with a vacuolate or granular cytoplasm which cover either the whole or part of the vasa deferentia, or the atrial wall. Sperber (1948) summarized the occurrence of prostate cells on normal or glandular vasa deferentia as, ". . . prostate cells are combined with a not (or only faintly) glandular sperm duct wall in most species of the first three genera (*Nais, Ophidonais, Uncinais*) and most *Pristina*, and with a strongly glandular wall in *Nais elinguis, Specaria, Stylaria lacustris* and probably *Pristina biserrata*, while the wall is glandular without prostate in *Pristina longiseta* and *evelinae*, and not or only slightly glandular without prostate in the rest, as far as can be deduced from the literature and my own investigations." This author also states that a thick atrial lining of glandular appearance is often associated with ". . . the occurrence of a prostate covering of varying density and extension." She observed that the ends of the cells penetrate the atrial muscle. Mehra (1924) thought the prostate cells in *Stylaria* and *Slavina* to be of mesodermal origin, while Hrabě (1939a) showed these cells to be ectodermal in some Lumbriculidae and Tubificidae and concluded that they were so in Naididae. It seems, however (see above), that in the Naididae two morphologically distinct types of prostate cell may be present; atrial prostate which may be derived from the atrial lining, and prostates associated with the vasa deferentia which are unlikely to be ectodermal due to their close connection with a mesodermal structure. Obviously this is another area in which developmental studies of selected species are necessary.

Penial setae are often developed in Naididae (see p. 14; Setae) which are usually situated on the atrial segment but in some species of *Pristina* may be located one or two segments behind the atria. Both penial, and where they occur, the spermathecal setae, are usually associated with glandular tissue (Fig. 7. 24H).

Associated with reproduction but of uncertain function, some naidids possess ventrally situated glandular tissue in some pre-genital segments. Marcus (1943) described these ventral glands, perhaps better termed the accessory copulatory glands, as containing ten to twenty large pear-shaped to elongate cells containing secretory material, the ends of which formed part of the body wall in the mid-ventral line (Fig. 1. 4G).

All Naididae except *Aulophorus tonkinensis*, *Pristina amphibiotica* and *P. idrensis*, possess spermathecae. They are similar in structure to those of Tubificidae and usually have discrete narrow ducts and more or less spherical ampullae. True spermatophores* have not been reported in this family (Sperber, 1948) but some degree of sperm orientation does occur in *Stylaria*, *Ophidonais* and *Piguetiella*.

Lumbriculidae

The genitalia of the lumbriculids are highly diverse and vary in number, position and arrangement. The atrial segment(s) is located between the seventh and the fourteenth segments. Basically two types of male system may be recognized in the family (Fig. 4. 2), both of which are unique in the Oligochaeta in that one pair of testes and their associated atria occur in the same segment (Cook, 1968). First, and probably the most primitive type, is that in which one pair of testes occur in the same segment as their respective atria; a condition which may be repeated in a number of segments as in some *Lamprodrilus* species and *Lumbriculus multiatriatus.*† From this, the monotesticular condition (meaning one pair of testes associated with each pair of atria), it is possible to derive the second or bitesticular state in which two pairs of testes are associated with one pair of atria occurring in the most posterior testis-bearing segment. This situation is further complicated by the fact that the bitesticular state can revert to the monotesticular condition by the loss of the anterior pair of testes and vasa deferentia found in some *Eclipidrilus* and *Rhynchelmis* species in which anterior testes are absent, but functionless, often small rudimentary male funnels and vasa deferential are present in addition to the functional ones (see Chapter 4).

Male funnels may be small and cup-shaped (*Trichodrilus*) or large convoluted structures (some *Rhynchelmis* species) which occur on the posterior septum of the testes segment. The vasa deferentia are tubular ciliated elongate structures composed of more or less cuboidal epithelium, which usually drain into the apical end of the atria. The atria are paired bodies, except in some *Eclipidrilus* species and *Tatriella* where a single median

* See footnote p. 33.
† Primitively two pairs of testes seem likely (see p. 169).

atrium is found. Each is composed of a thin lining epithelium (ectodermal), an outer muscle layer and a peritoneal covering. The pear-shaped prostate cells penetrate the peritoneum and muscle layer either singly or in bunches (Fig. 5. 5B). The atrial muscle may attain great thickness and complexity in some species and in two species studied in detail consists of two discrete muscle layers (Cook, 1967a). In *Lumbriculus variegatus* the muscle layers consist of an inner layer of fibres arranged at right angles to the long axis of the atrium (circular) and a layer of longitudinally arranged bundles of fibres (Fig. 1. 4H). In *Eclipidrilus lacustris* the inner muscle coat is composed of many layers of muscle fibres which are arranged spirally around the long axis of the atrium. The direction of the spiral alternates in successive layers, imparting a cross hatched appearance to the atrium in surface view (Fig. 1. 4I). A thin layer of longitudinal muscle fibres are found outside the spiral layers. In general form the atria may be small subspherical bodies to thick elongate cylinders or narrow, vas deferens-like structures as in some *Rhynchelmis* species.

The atria terminate in narrow ducts which open to the exterior either as simple pores, or via modified penial structures. The penis is formed from folds in the body wall which may or may not be contained within another fold, the penis sac. A second type of penis, found in *Lumbriculus*, is composed of the lining cells in the proximal part of the atrial duct which become very elongate and may be protruded through the atrial pore (see Dorydrilidae).

Penial and spermathecal setae are not found in the Lumbriculidae, but at full maturity the ventral setae of the genital segments may be lost.

Accessory copulatory glands are present in some Lumbriculidae. In some *Rhynchelmis* species these consist of rudimentary male ducts which have no connection with gonads but have a thick glandular covering (Hrabě, 1927). Altman (1936) described ventral glands composed of groups of elongate cells opening mid-ventrally in some pre-genital segments of *Rhynchelmis elrodi* which appear to be similar in structure to those of *Allonais* (Naididae) described by Marcus (1943). In some species of *Lamprodrilus* accessory copulatory glands are often highly developed. In *L. satyriscus* these are large glandular masses situated ventrally behind the atrial ducts and opening to the exterior on conical papillae enclosed with a body wall fold (Isossimoff, 1962) (Fig. 5. 4D).

The female system is peculiar* in the Lumbriculidae in that the ovaries and female funnels are located one, or in *Styloscolex* two segments behind the atrial segment(s). Usually one pair of ovaries is present, but some *Lumbriculus* and *Styloscolex* species possess two.

Spermathecae are present in all members of the family with the exception of *Bichaeta saguinea,* and about 20% of *Tatriella slovenica* specimens (Hrabě, 1939b). Their number (usually one to five pairs) and distribution

---

* Owing to the position of the atria (see p. 172).

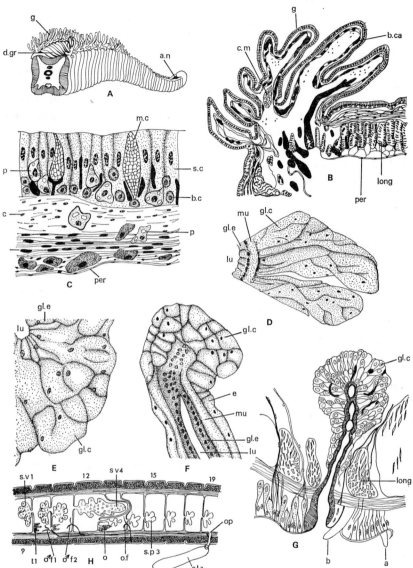

Fig. 1.5. Anatomy of the glossoscolecids.
A to C—*Alma nilotica*: A—posterior end; B—transverse section of body wall through gill; C—section through gill wall.
D to F—*Sparganophilus tamesis*, prostate-like glands: D—anterior gland, part of section through glandular part; E—posterior gland, same; F—posterior gland, section through transition between duct and glandular part.
G—*Sparganophilus smithi*, section through body wall and a prostate-like gland.
H—*Alma nilotica*, semidiagrammatic longitudinal section of the genital region.
a, seta *a*; an, anus; b, seta *b*; b.c, basal cell; b.ca, blood capillary; c, connective tissue; cla, clasper; c.m., mass of coelomocytes; d.gr, dorsal groove; e, "epidermis"; g, gill; gl.c, gland cells; gl.e, glandular epithelium; long, longitudinal musculature; lu, lumen; f1 and 2, sperm funnels; mu, muscular layer; m.c, mucous cell; o, ovary; o.f, oviducal funnel; op, communication between coelom of body and of clasper; p, phagocytic cell penetrating body wall; per, peritoneum; s.c, supporting cell; s.p 3, third septal pouch; s.v 1 and s.v 4, anterior and posterior seminal vesicles.

in relation to the male genital system are important generic and specific criteria; thus details will be found in the systematic section. In general they are usually paired structures with discrete narrow ducts and voluminous ampullae, occurring in front of, behind, or within, the atrial segment. A trend, noticeable in the male system to reduce or concentrate the genitalia also occurs in the spermathecal system; thus in some *Eclipidrilus* species the spermatheca is a single median structure. A peculiar phenomenon occurs in some *Rhynchelmis* species in which the spermathecal ampullae are in direct communication with the gut cavity (Mrazek, 1900), a feature found in many enchytraeids but only some tubificids of the genus *Rhyacodrilus*. Spermatophores are not developed in this family.

Dorydrilidae.

In the genus *Dorydrilus* the male genitalia consist of a pair of testes situated in the ninth segment which are associated with a pair of atria located in the tenth segment. A pair of relatively large male funnels occur on the anterior face of septum 9/10 and these open into a pair of narrow vasa deferentia which penetrate this septum, run along the inner surface of the ventral body wall, turn dorsally and join the paired atria at their anterio-dorsal apices. The atria are relatively large pear-shaped bodies which occupy most of their segment. They consist of an inner lining which is relatively thin apically but very thick proximally, and a thick outer muscle layer. Scattered diffuse prostate cells occur towards the apical end of the atrium. At sexual maturity the cells forming the proximal linings of the atria become very elongate and protrude through the male pores as compact, cylindrical, internal penes which appear to be capable of considerable extension. The outer ring of cells have thin cuticular membranes which envelop these internal penes (Fig. 10. 1D).

There is a single pair of ovaries and female funnels in the same segment as the atria.

One or two pairs of spermathecae with long discrete, often coiled ducts and more or less spherical ampullae are present. These are situated in the atrial or post-atrial segment, or in both of these. Modified genital setae and spermatophores are unknown in the Dorydrilidae. The genital anatomy of *Lycodrilus* is poorly understood.

Opistocystidae

The only accounts of this family are those of Cernosvitov (1936) and Cordero (1948).

The genitalia are dorydrilid in their sequential arrangement but are uniquely located far posterior of those of other microdrile families. One pair of testes and ovaries occur in XXI and XXII respectively (Cernosvitov, 1936); Cordero (1948) states, however, that they are in XV and XVI.* A pair of vasa deferentia drain into the apical end of the naidid-like atria which

* See Chapter 10.

are covered with a diffuse layer of prostate cells and open in the ovarian segment.

A pair of spermathecae with long discrete ducts open near the anterior septum of the post-atrial segment.

Haplotaxidae

Accounts of various species of this family are to be found in Forbes (1890), Beddard (1890, 1891), Michaelsen (1903a, 1905, 1907), Benham (1909), Smith (1918), Stephenson (1930), Yamaguchi (1953), Hrabě (1958), Cekanovskaya (1962) and Brinkhurst (1966).

The arrangement of the gonads and the absence of atria in this family suggest affinity with an ancestral oligochaete condition (see Chapter 4). Typically two pairs of testes and ovaries, located in X to XIII, are present. In some species the most posterior ovaries are lost, and in another the gonads are displaced one segment anteriad. The male ducts consist merely of two pairs of male funnels and vasa deferentia which open to the exterior more or less ventro-laterally, on the segment immediately posterior to their associated testes (Fig. 6. 2). Atria, penes and prostate glands are absent, but the terminal ends of the vasa deferentia are lined with cuticle (Brink-hurst, 1966), suggesting a possible atrial (ectodermal) rudiment (but *vide infra*). *Haplotaxis violaceus* is peculiar in that the anterior vasa deferentia penetrate two septa and open in the ovarian segment, just anterior to the posterior pair. Secretory tissue, in the form of copulatory or setal glands which are situated in one or more of the gonadal segments, has been described in a number of Haplotaxidae.

From one to four pairs of spermathecae, which open ventro-laterally to dorso-laterally near the anterior intersegmental furrow of their segments, are situated immediately in front of the gonadal segments (usually in VIII or IX, or between VI to IX) (Fig. 6. 1).

The female system is of the usual microdrile type except for *H. smithii* whose funnels are said to penetrate the posterior septum of their segment and open anterior to the ventral setae of the succeeding segment (Michaelsen, 1925).

The lack of atria and prostate glands in the Haplotaxidae, combined with the frequent occurrence of copulatory glands (50% of *Haplotaxis* species) or ventral setal glands (25%, excluding those with copulatory glands), is interesting from a phylogenetic viewpoint. It suggests that the microdrile atrium (with ectodermal lining cells) may have developed from a copulatory gland-like structure (ectodermal), which became associated with the vasa deferentia. This type of structure could have acquired a storage or propulsive function by expansion and the elaboration of musculature derived from the body wall, while retaining its original glandular function in the form of internal lining cells, or by their relocation outside of the storage chamber as external prostate glands (known to be ectodermal in some cases) (Hrabě, 1939a).

Phreodrilidae

Detailed accounts of various phreodrilid species are to be found in Beddard (1891, 1894a), Michaelsen (1903b), Benham (1904), Goddard (1909a, 1909b), Goddard and Malan (1913a, 1913b), Stephenson (1930), Jackson (1931), Stout (1958) and Brinkhurst (1965a).

The number and arrangement of the parts of the genitalia are consistent within the family but may be displaced one or two segments anteriad in some species. One pair of testes and ovaries occur in the eleventh and twelfth segments respectively, a pair of male ducts are located in the ovarian segment, and a pair of spermathecae are situated in the post-atrial, usually the thirteenth, segment. A second paid of spermathecae are found, in XIV, in *Phreodrilus nothofagi* (Stout, 1958), while *P. fusiformis* is said to have a single median spermatheca (Goddard, 1909a). The male ducts are relatively complex and the homologies of the parts, which were discussed by Brinkhurst (1965a) and whose interpretation is followed in this account, will be doubtful until morphogenetic studies are undertaken. Basically a a pair of narrow vasa deferentia join a pair of atria either basally or medially. The latter have a thin muscle wall and a thick layer of lining cells which possibly act as internal prostate tissue (as in the tubificid *Clitellio arenarius*). External prostate glands are entirely absent (Fig. 9. 1). The male ducts terminate either as well-developed penes or as eversible or protrusible pseudopenes (Brinkhurst, 1965a). The latter type are often associated with an elaboration of the proximal part of the duct. At its maximum development, seen in *P. subterraneus*, this part of the male duct is very elongate and coiled, and the lining is separated from the muscle layer. This same modification has been observed in *Monopylephorus* and *Branchiura sowerbyi* (Tubificidae) and others.

The spermathecae are paired and open through narrow ducts in the post-atrial segment. In a number of species a valve-like apparatus has been observed at the junction of the spermathecal duct and ampulla (Beddard, 1894; Michaelsen, 1903b; Goddard and Malan, 1913a). In many species the spermathecae open to the exterior as simple pores ventro-laterally. However, in the subgenus *Phreodrilus* the pores are dorsal and open into muscular vestibulae formed from an inversion of the body wall and in *P. (Insulodrilus)* the pores are on the anterior margin of their segment and are closely associated with the female pores.

Paired ovaries and female funnels are situated in the atrial segment.

Alluroididae

In some respects the genitalia of the Alluroididae are intermediate between those of the microdriles and the megadriles. Detailed accounts are to be found in Beddard (1894b, 1906), Brinkhurst (1964b) and Jamieson (1968a).

In the Alluroidinae a pair of testes situated in X are associated with a pair of male funnels and vasa deferentia which penetrate two segments devoid of gonads and open into a pair of atria (or in one case, directly to the exterior) on XIII. One pair of ovaries and female funnels are situated in the atrial segment, and one to three pairs of spermathecae which usually open dorsally, occur in some pre-testicular segments.

The Syngenodrilinae, containing the single species *Syngenodrilus lamuensis*, has paired testes in both X and XI associated with two pairs of vasa deferentia which open directly to the exterior in XIII. Three pairs of tubular prostate glands (in the megadrile sense, *vide infra*) open on XI to XIII independently of the male pores. A pair of ovaries occur in XIII and two pairs of spermathecae open on the seventh and eighth segments.

Variations in the form of the male ducts are difficult to interpret in some cases. In *Alluroides pordagei* the vasa deferentia join the apical end of a pair of elongate tubular atria which extend posteriad into about the sixteenth segment. The atrium is of usual microdrile construction, having lining cells, a muscular layer, and an outer layer of prostate cells, the processes of which penetrate the muscle layer (Brinkhurst, 1964b; Jamieson, 1968a). The atria of *Alluroides brinkhursti* are similarly constructed but are pear-shaped and have a very thick muscle layer. The vasa deferentia of *Standeria* join a pair of atria which are devoid of external prostate cells but in *Brinkhurstia* the vasa deferentia open into small chambers (probably equivalent to pseudopenes), independent of the narrow, tubular prostate-covered atrial structures (Jamieson, 1968a).

In *Brinkhurstia*, *Syngenodrilus* and also in some phreodrilids in which the vasa deferentia enter the atria basally, the storage and propulsive functions of the atria have probably been lost (or never developed), but the glandular function is retained. Following this interpretation of the male genitalia, and the possible origin of the atrium from a glandular structure (see Haplotaxidae, p. 39), the atrium, used in the microdrile sense as in this account, and the tubular prostate glands, in the sense of authorities on megadriles, are homologous.

GLOSSOSCOLECIDAE (B. G. M. JAMIESON)

EXTERNAL ANATOMY

The Glossoscolecidae are all "megadriles", varying in size from a few centimetres to some 63 cm in *Martiodrilus crassus*, 1·26 m in *Glossoscolex giganteus*, 1·94 m in *Microchaetus rappi* and 2·1 m (by 2·4 cm wide) in *Rhinodrilus fafner*. The number of segments is in the hundreds, and exceeds 600 in *Rhinodrilus fafner*. Some of the Glossoscolecinae are smaller than haplotaxids. Secondary annulation, giving the segments a biannulate, triannulate or even hexannulate appearance is common both in the Alminae and in the terrestrial glossoscolecids. In terrestrial forms, coincidence of secondary annulation with frequent absence of anterior setae renders segmental ennumeration difficult.

BODY FORM The body is always elongate and vermiform in the Glossoscolecidae but the cross section varies in shape. In terrestrial glossoscolecids it is approximately circular, but in the Alminae, which are aquatic and include *Alma*, *Callidrilus*, *Glyphidrilus*, *Drilocrius*, *Glyphidrilocrius* and *Criodrilus*, it is quadrangular (square or trapezoidal) except at the extreme anterior end or in the forebody. It appears that all Alminae are capable of depressing the dorsal surface, at least at the posterior end, to form a longitudinal groove or gutter and in *Alma* this has been shown to form a tubular "lung" in which air bubbles are trapped (Beadle, 1957); a respiratory function has also been ascribed to it in *Drilocrius* (Carter and Beadle, 1932) and in *Glyphidrilus* (Nair, 1938). In these aquatic forms the anus is dorsal and subterminal or is dorso-terminal and records of a posterior anus are probably a result of autotomy. The dorsal location of the anus is somewhat surprising in view of the respiratory function of the dorsal groove. The anus is also dorsal in the aquatic genera *Sparganophilus* and *Biwadrilus* but there, though a dorsal groove may form, the cross section is not quadrangular. A quadrangular or trapezoidal cross section is known elsewhere in the oligochaeta only in the lumbriculid *Rhynchelmis* and in the Lumbricids *Bimastos* and *Eiseniella*.

THE PROSTOMIUM In the aquatic glossoscolecids, the prostomium is zygolobous, i.e. is fused with the peristomium and neither impinges on nor is separated by any furrow from the latter. There are some doubtful records of the prolobous condition, in which the prostomium is demarcated from the peristomium by a furrow coinciding with the anterior margin of the latter, or of the epilobous condition, in which a dorsal "tongue" of the prostomium extends on to the peristomium but not as far as the posterior border of the latter. In the terrestrial forms it is generally prolobous but is so often withdrawn because of invagination of the first one to three segments that its form is unknown in many species. It is frequently prehensile, and in *Pontoscolex corethrurus* is extended in life as a tapering trunk-like appendage the tip of which probes the environment and periodically adheres by a sucker-like terminal disc to surrounding objects (personal observations). It is also described as prehensile in *Martiodrilus* (=*Thamnodrilus*) *crassus* (v. Pickford, 1940). The prostomium forms a very long, slender sensory proboscis, reminiscent of that of the naidid *Pristina*, in *Andiodrilus biolleyi*, a glossoscolecid living in epiphytic bromeliads, and is similarly elongated in *Onychochaeta*, *Diachaeta* (=*Hesperoscolex*) and *Periscolex* (Picada, 1913). It is nasute in *Alma nasuta* (q.v.) and is reputedly (personal correspondence) proboscis-like in other species of *Alma* in life.

PIGMENTATION Pigmentation of the body wall has taxonomic value, and should be recorded in life, but reference to it has been omitted in the taxonomic section on the Glossoscolecidae (Chapter 15) because of the unreliability of observations based on preserved material. *Biwadrilus*,

Sparganophilus and *Criodrilus* are pigmentless, with the exception of clitellar pigmentation in the last two genera, but some species, at least, of *Alma* and *Glyphidrilus* are pigmented. *Alma emini* shows a bright green pigmentation in life which the author has seen elsewhere only in the lumbricid *Allolobophora chlorotica*; and *Glyphidrilus gangeticus* in preservation is reddish.

BRANCHIAE The only megadriles known to possess gills are *Alma eubranchiata* and *A. nilotica*. The existence of gills in *A. schultzei* (Michaelsen, 1915) is refuted in the present study (p. 784). Their anatomy in *A. nilotica* has been described by Gresson (1927) and in more detail, by Khalaf El Duweini (1957; Fig. 15 A-C). In this species they are branched pinnate or plume-like extensions of the body wall surmounting the walls of the dorsal groove immediately above the dorsal setal couples, the row on each side extending through the posterior tenth to one-sixth of the body with the exception of the last six or seven segments. Their form and extent in *A. eubranchiata* (*v.* Michaelsen, 1910) are similar.

THE CLITELLUM AND COCOONS An extensive clitellum, exceeding ten segments in length, is characteristic of the Glossoscolecidae, and a shorter clitellum is uncommon. The clitellum only sporadically extends through more than ten segments in the other families (e.g. 12-13 segments in the ocnerodrile *Nematogenia lacuum*) and it never attains the extremes of length or the great variation in location which occur in the Glossoscolecidae. The location and length in terms of numbers of segments, reach their greatest variability in *Alma*. In this genus extreme limits recorded for the anterior border are segments 35* to 247 and for the posterior border are 49 to 295; the minimum number of segments occupied is 20 and the maximum is 69. (The most anterior location in *Alma* occurs in *A. togoensis* (35 or 36 to 49-63) and the most posterior in *A. eubranchiata* (225-247 to 268-295).) Stephenson (1930a) states that the clitellum in the Glossoscolecidae begins behind segment 14, so that the female pore is in front of it. While this is frequently true and is almost invariably the case in the Glossoscolecini, there are numerous exceptions. Thus in *Microchaetus* it may begin as far forward as segment 11 and normally begins in segment 13 or 14 (occupying in this genus 6 to 44 segments); and in *Criodrilus* it begins in or behind segment 14.

The structure of the clitellum of *Alma emini* has been described by Grove (1931; Fig. 1. 6D) and corresponds closely with that observed by the same author in *Diachaeta exul* (Fig. 1. 6C), in *Sparganophilus tamesis* by Eisen (1896a: 162) and by the writer (Fig. 1. 6A), in *Callidrilus ugandaensis* in the present study, by Nair (1938, Fig. 1.6B) in *Glyphidrilus annandalei*, and

---

* Arabic numerals are employed in this section and in Chapter 15 because they are more convenient than roman numerals for the frequent references which are made to high segmental counts in the Glossoscolecidae.

by Khalaf El Duweini (1951) (Fig. 1. 6E) in *Alma nilotica*. The clitellum of *Biwadrilus* is similar but has, in addition to the fine- and coarse-grained cells, club-shaped peripheral cells with fine or coarse granules (Nagase and Nomura, 1937). In *Criodrilus lacuum*, Benham (1887) observed only glandular cells with small spherical globules. In *A. nilotica* the clitellum at first sight appears to have three layers but there is in fact only one layer, the cells of which all extend to the cuticle but whose inner ends lie at different levels. The shortest (outermost) cells are normal epidermal supporting cells with a few sensory cells and a few mucin-secreting cells irregularly distributed amongst them. The cells which appear to make the "middle layer" are glandular cells which are fairly numerous and are irregularly distributed. They contain large granules and there is evidence that they secrete the cuticle and membrane of the cocoon. The apparent "third layer" is composed of cells appearing to form several tiers and arranged in groups which are separated from one another by thin lamellae of connective tissue. These contain fine granules of an albuminous secretion (Khalaf El Duweini, 1951).

The latter author, as did Grove (1931) for *A. emini* and Grove and Cowley (1927) for *Eisenia*, presents evidence that the fine-granule cells secrete the albuminous contents of the cocoon. Grove (1931) considered this relative abundance of mucin-secreting cells in the clitellum of *A. emini* to indicate secretion of a copulatory slime tube. Their paucity in the clitellum of *A. nilotica* corresponds with the absence of a slime tube in this species.

Correlated with the great length of the clitellum in many Glossoscolecidae is the unusual elongation of the cocoon and, presumably, the large number of developing embryos in such forms. The longest recorded are those of *Alma multisetosa* in which they attain a length of 130-155 mm and may contain 32 embryos and where the clitellum may occupy 57 segments (Grove, 1931). The cocoons reach a length of 110 mm in *A. nilotica* in which each produces 8-22 young (Khalaf El Duweini, 1951, from whose detailed account Fig. 15. 6G is redrawn); 70 mm in *Criodrilus lacuum* (v. Janda, 1926), in which they release from 2-8 young (Oerley, 1887); and they average 42 mm (by 12 mm) in *Glyphidrilus annandalei* (Fig. 1. 6F) in which each contains, on average, seven ova (Nair, 1938). Their elongate spindle-shape in these aquatic taxa contrasts with the more nearly ovoid form of cocoons of other families and appears to be characteristic (diagnostic?) of the Alminae as perhaps is the relatively large number of young released. The cocoons of *Sparganophilus tamesis* (=*S. eiseni*) are almost ovoid, though with attenuated extremities and release from 1-4 young (Cernosvitov, 1945; Harman, 1965, Fig. 1. 6H).

GENITAL PORES   The male pores are borne on conspicuous protuberant porophores in *Biwadrilus* (anteclitellar on segment 13); *Criodrilus* (intraclitellar on segment 15; Fig. 15. 11) and in *Drilocrius* (ante- or just intra-

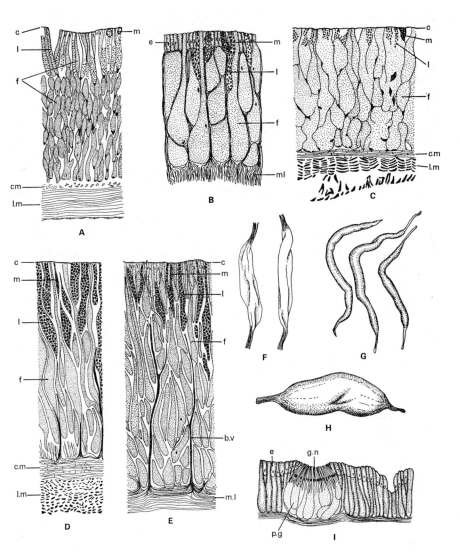

Fig. 1.6. Anatomy of glossoscolecids.
A to E—sections through the clitellum: A—*Sparganophilus tamesis* (original); B—*Glyphidrilus annandalei*; C—*Diachaeta exul*; D—*Alma emini*; E—*Alma nilotica*; F-H—cocoons; F—*Glyphidrilus annandalei*; G—*Alma nilotica*; H—*Sparganophilus tamesis*.
I—sagittal section of genital marking of *Glyphidrilus annandalei*.
b.v, blood vessel; c, cuticle; c.m, circular muscle; e, epidermal cell; f, fine-grained gland cell; g.n, nuclei of gland cells; 1, large-grained gland cell; m, mucin cell; m.l, muscular layer; p.g, papillary gland.

clitellar on segment 15 or 16; Fig. 15). In *Glyphidrilocrius* they are intraclitellar on segment 17 in deep depressions in wing-like tubercula (Fig. 15. 9). In *Drilocrius alfari* they are located on segment 16 near the bases of remarkable elongate laminae with the appearance of claspers (Fig. 15. 10A). Claspers are elsewhere seen only in the genus *Alma*, all species of which possess them (Fig. 15. 7 and 15. 8). In *Alma* they differ from those of *Drilocrius alfari* in bearing genital setae and in carrying the male pores. Their bases occupy 2 to 3 segments and are normally centred on segment 19 or intersegment 18/19. Their form and action is discussed on p. 769. In *Callidrilus* the male pores are intraclitellar on segment 17, shortly in front of the prostate pores, on a pair of small papillae behind which are longitudinal tubercula pubertatis (Fig. 15. 1). The glossoscolecid tendency to backward movement of the male pores is well demonstrated in *Glyphidrilus* in which the inconspicuous pores have been recorded from intersegment 19/20 to as far posteriorly as segment 30 but probably occur further posteriorly in *G. stuhlmanni*. In this genus they lie ventral to and at or near the posterior limits of a pair of tubercula pubertatis which are usually elevated as delicate "wings" or alae (Fig. 15. 3 and 15. 4). These are discussed on p. 747. The male pores of *Sparganophilus* are inconspicuous but (as in *Glyphidrilus*) can be located, if the body wall is partly macerated, by tracing the whitish sperm-filled vasa deferentia to the surface in external (or internal) examinations. They lie in 19 on or lateral to a pair of elongate and sometimes almost alate tubercula pubertatis (Fig. 15. 14) which are sometimes broken up into a series of papillae.

Turning to the terrestrial glossoscolecids, the male pores, in *Kynotus* are anteclitellar on segment 15 or 16 in level fields or, in erection, on claspers; intraclitellar on 15-16 between swollen lips in *Hormogaster;* intraclitellar a varying distance behind segment 16 in *Microchaetus*; and intraclitellar (only in *Opisthodrilus* postclitellar) in the Glossoscolecini and (always?) behind 16. In *Eudevoscolex*, alone in the Glossoscolecidae, there are two pairs of male pores, a condition known elsewhere in the megradriles only in the Moniligastridae and in the "octochaetid" *Hoplochaetella*. Elongate tubercula pubertatis occur in *Hormogaster*, and in the Glossoscolecini, and (though termed "copulatory walls" or "Pubertätswälle") in *Microchaetus* and *Tritogenia*.

Female pores are inconspicuous, on segment 13, or (*Glyphidrilus kukenthali*) 13 and 14, in all Glossoscolecidae. The spermathecal pores are rarely visible externally. Their distribution is discussed under spermathecae below and in Chapter 15.

GENITAL MARKINGS Papillae or other glandular elevations, other than the tubercula pubertatis and those bearing the genital pores, are termed "genital markings". These are absent from the Sparganophilinae and Biwadrilinae but occur in some Alminae. They are especially well-developed in *Glyphidrilus* where they take the form of oval papillae which, except in *G. tuberosus*, are differentiated into a peripheral rim and a central area which may be elevated or depressed (Fig. 15. 3 and 15. 4). Significantly, in view of its proposed relationship, similar rimmed genital markings are formed in *Callidrilus* (Fig. 15. 1). Nair has shown the markings in *Glyphidrilus annandalei* to be composed of elongate glandular cells with thin layers of connective tissue between them (Fig. 1. 6I). He considers it probable that they produce a secretion which facilitates adhesion of concopulants. Their distribution in the various species of *Glyphidrilus* is discussed on p. 748. Special genital markings are uncommon in *Alma, Drilocrius* and *Glyphidrilocrius*. In *Criodrilus* glandular fields of different type surround the genital setae (Fig. 15. 11). *Microchaetus* and *Tritogenia* also have glandular elevations which here form a conspicuous papillae around the genital setae; in addition *Microchaetus* has the "copulatory walls" (tubercula pubertatis) already referred to, and *Tritogenia* has paired "puberty cushions" (Pubertätspolster) extending through several clitellar segments. *Hormogaster* also has genital seta papillae in addition to tubercula pubertatis.

SETAE   The nature of oligochaete setae has been discussed on p. 12. Throughout the aquatic Glossoscolecidae (Sparganophilinae, Biwadrilinae and Alminae) there are four pairs of setae per segment, forming eight longitudinal rows (the lumbricine arrangement) and commencing on segment 2.

The setae are sigmoid and simple-pointed in all Gossoscolecidae but the genital setae of *Biwadrilus bathybates* are bifid and the terrestrial species *Diachaeta thomasi* and *Pontoscolex corethrurus*, and some species of *Periscolex*, have sporadic bifid body setae (Beddard, 1892; Michaelsen, 1918). In the terrestrial subfamilies, especially in the larger species, the setal arrangement is usually lumbricine but the setae of the anterior segments tend to be reduced and may be absent in part or the whole of the preclitellar region. Thus in *Kynotus michælseni* setae are absent from the first and second segments while in *K. darwini* they are absent from the first sixteen segments. Absence of setae from the third or more segments also occurs sporadically in *Microchaetus*, in *Tritogenia* and in the Glossoscolecini.

Departures from the simple lumbricine condition occur in some genera of the Glossoscolecini. In *Diachaeta thomasi* the setae are irregularly closely and widely paired in successive segments, throughout the body, while in *Pontoscolex corethrurus*, those at the posterior end of the body, are in quinqunx (p. 737). In *Periscolex* setae may be numerous in each segment (the perichaetine condition); this is clearly an independent development of this condition which is seen elsewhere in many megascolecids.

Where setae have the lumbricine arrangement it is customary in taxonomic descriptions to denote the four setae of a side by the symbols *a-d* consecutively, the ventralmost seta on each side being seta *a*. The ratio of intersetal distances is of taxonomic value at the specific or higher levels and is expressed by the formula $aa:ab:bc:cd:dd$, usually in conjunction with the ratio of the dorsal-median intersetal distance relative to the circumference ($dd:u$). The possible phylogenetic significance of the smaller size of this ratio in the Alminae and, apparently, the Hormogastrinae as compared with the Glossoscolecinae and Kynotinae is discussed in Chapter 4.

The normal somatic setae of the Glossoscolecidae, with the exception of those of *Biwadrilus*, which are unornamented, are commonly (always?) ornamented with transverse or spirally arranged scales, ridges, toothrows or striations. Genital setae are developed in all subfamilies excepting the Sparganophilinae. These are usually elongated setae which depart more or less noticeably from the sigmoid form of the somatic setae and are usually borne on glandular protuberances of the body wall or are associated with special internal glands but in *Alma* they are carried on the clasper and are very much smaller than the somatic setae (see p. 769 Fig. 15. 5). In *Biwadrilus bathybates* (Fig. 15. 12) they may be termed penial setae as they are located at the male apertures. Genital setae have not been reported for Almini, other than *Alma*, but in *Criodrilus lacuum* (Criodrilini) they occur on several clitellar segments. In this species (Fig. 15. 11) they sometimes display an ornamentation which includes the transverse ornamentation usual in somatic setae of the Glossoscolecidae and the four distal longitudinal grooves of the genital setae of the Lumbricidae and Hormogastrinae. Genital setae are widespread in the Kynotinae and Glossoscolecinae.

Discrete glands known as setal glands may be associated with some of the setae. In *Microchaetus* (Fig. 1. 9) and *Kynotus*, these are especially large, prostate-like structures, surrounding the genital setae. In *M. benhami* the central lumen is surrounded by columnar cells, a muscular sheath, and outside these, pear-shaped groups of gland cells which communicate with the lumen (Rosa, 1891), a morphology remarkably like that of the atria of the Alluroidinae. The anterior prostate-like glands of *Sparganophilus* (p. 64) may also be regarded as setal glands.

DORSAL PORES Intersegmental median unpaired dorsal pores, which allow escape of coelomic fluid to the exterior of the body, are present in many Glossoscolecini but are absent from the aquatic subfamilies and from the Microchaetini, Hormogastrinae and Kynotinae.

ANATOMY AND HISTOLOGY OF BODY WALL

The structure of the body wall in Oligochaetes has been discussed on p. 14 and it will suffice here to comment briefly on the limited literature on that of the Glossoscolecidae.

In *Biwadrilus bathybates* (Nagase and Nomura, 1937) the body wall has the typical layering of the oligochaeta, namely, an external cuticle (nearly 4 μ thick); an epidermis ("hypodermis"), with a single layer of cells, divisible into columnar, gland and supporting cells, this layer being nearly 60μ thick in the forebody, and internal to the epidermis, in sequence, circular and longitudinal muscle layers and the coelomic peritoneum. The nature of the secretion of the gland cells, which are apparently of only one type, is not specified. The supporting cells are strongly compressed between the columnar cells. At the bases of the epidermal cells are occasional bulbiform groups of cells (sensory buds) from which sensory hairs penetrate the cuticle. These buds are similar to those described by Vejdovsky (1884) and Collin (1888) for *Criodrilus lacuum*.

Within the Alminae, the epidermis of *Criodrilus lacuum* consists of narrow tubular cells with oval nuclei between which, basally, are small rounded cells with rounded nuclei. Goblet cells are present as narrow cells filled with granules, and with the protoplasm and nucleus at the inner ends. Their scarcity is presumed to be correlated with the aquatic existence. The capillary loops of the blood vessels pass between the cells of the epidermis (Benham, 1887). That of *Callidrilus ugandaensis* (personal observations) closely resembles the epidermis of *Lumbricus* (*v.* Cerfontaine, 1890), having numerous large-grained gland cells resembling the mucous cells and somewhat fewer cells with amorphous vacuolated contents which appear identical with the albuminous gland-cells of *Lumbricus*. Between these, there are elongated supporting cells, as in the latter genus, but basal cells were not certainly recognized. As in *Criodrilus* but unlike *Lumbricus*, blood capillaries extend into the epidermis. Khalaf El Duweini (1957), in describing the epithelium of the gills in *Alma nilotica*, states that it resembles that of the remainder of the body but has larger, more numerous goblet cells; no other types of gland cells are described or illustrated. Basal and supporting cells are present. Material of *Alma* available to the writer is unsuitably preserved for histological examinations.

There is no comparative literature on the body-wall musculature of the Glossoscolecidae and an investigation of its structure has not been undertaken in the present work. Nagase and Nomura (1937) state that in *Biwadrilus* the longitudinal musculature is composed of bundles of irregularly or sometimes pinnately arranged muscle fibres. Only in this genus, of the Glossoscolecidae, is the musculature associated on each side with a lateral line. This penetrates the musculature but its structure is not described. Khalaf El Duweini (1957) illustrates pinnately arranged muscle fibres for *Alma nilotica* (Fig. 1. 5B). That of *Criodrilus lacuum* is described by Rhode (1885); in this species the muscle fibres have dissociated and show no definite arrangement within their connective tissue caissons. A similar condition obtains in the lumbricid *Eisenia foetida* in front of the clitellum, though behind the clitellum the arrangement of pores is pinnate. Care is clearly necessary in comparative studies of musculature to ensure that sections from equivalent regions in several parts of the body are compared. Izoard (1952, 1958) reviews evidence for regarding the "muscle fibres" of the longitudinal muscle of earthworms as myofibrils in syncytial fibres. Stolte

(1935) and Avel (1959) give useful reviews of literature on oligochaete musculature.

INTERNAL ANATOMY

COELOM   The coelom (see p. 18) is spacious in the Glossoscolecidae. Septation is complete except in a few anterior segments. Coelomocytes of *Alma nilotica* and their function in post-reproductive phagocytosis of the claspers are described by Khalaf El Duweini (1950a).

SEPTATION   Thickening of the musculature in some septa is common in megadriles and has some taxonomic value but assessment of relative degrees of thickening is highly subjective and is partly dependent on the degree of contraction and methods of fixation employed. Reference to it has therefore been omitted from the systematics section on the Glossoscolecidae.

ALIMENTARY CANAL

General Anatomy. The anatomy of the alimentary canal in the two aquatic genera with the most species, *Glyphidrilus* and *Alma* is discussed on p. 770 (Fig. 15. 2, 15. 6), and information on other genera is given in the generic and specific accounts. It will suffice here to note some general features of the alimentary canal in the Glossoscolecidae.

Gizzards, both oesophageal or intestinal, are absent from *Sparganophilus* and *Biwadrilus* but all other glossoscolecids have some regional thickening of the alimentary musculature. This consists of a single oesophageal gizzard in the Kynotinae (in segment 5), the Glossoscolecini (in segment 6), the Microchaetini (in segment 7), and in *Glyphidrilus* (in segment 7 or 8 and sometimes extending into an adjacent segment). In *Callidrilus* strong gizzard-like thickening of the oesophagus occurs in segments 5 and 6 or in 7 to 8 or to 9. Muscular thickening is restricted to the anterior region of the intestine in *Alma* and *Glyphidrilocrius* being remarkably far posterior, in segments 38-40, in *Glyphidrilocrius. Drilocrius* is variable, usually possessing a rudimentary gizzard in segment 5 and/or 6 but sometimes lacking recognizable thickening of the oesophagus, and commonly with appreciable muscular thickening at the beginning of the intestine. *Criodrilus lacuum* has slight oesophageal thickening in the region of segments 5 to 7 and has anterior thickening of the intestine. *Hormogaster* has three, well-developed oesophageal gizzards in segments 6-8 and, according to Rosa (1889), some thickening at the beginning of the intestine.

Extremes for origin of the intestine are segment 9, in *Sparganophilus* and segment 38, in *Glyphidrilocrius ehrhardti*. A dorsal typhlosole is present in the Glossoscolecinae (including the Glossoscolecini and Microchaetini), in the Hormogastrinae, and in all Alminae but is absent from the Sparganophilinae, Biwadrilinae and Kynotinae.

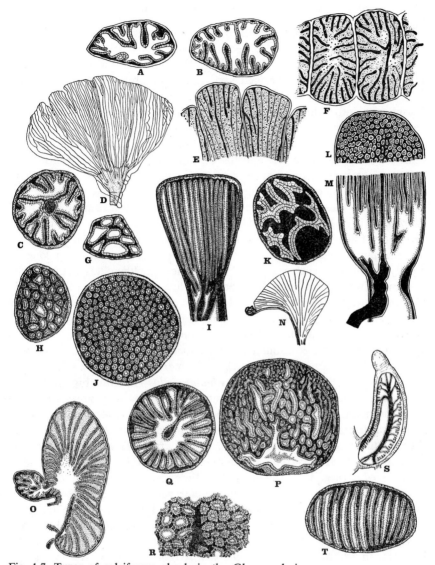

Fig. 1.7. Types of calciferous glands in the Glossoscolecinae.
Ridged sacs (Leistentaschen)—t.s., A—*Diachaeta thomasi*; B—*Periscolex fuhrmanni*.
Partitioned sacs (Fachkapseltaschen)—t.s., C—*Quimbaya* (=*Thamnodrilus*) *cameliae*.
Deeply ridged sacs (Saumleistentaschen)—*Inkadrilus aberratus*. D—lateral view; E—l.s. and F—t.s.
Tubule sacs (Schlauchtaschen)—G—*Onychochaeta windlei*, t.s. of a sac of segment 8; H—*Thamnodrilus* (=*Aptodrilus*) *fuhrmanni*, t.s. through apical region.
Panicled tubular sacs (Rispenschlauchtaschen)—I to K, *Thamnodrilus* (=*Aptodrilus*) *excelsus*; L to M, *Rhinodrilus paradoxus*; N, *Thamnodrilus gulielmi*.
I—l.s.; J and K—t.s. of the apical and basal regions, respectively; L—t.s. through the middle of a sac; M—l.s. through the basal half of a sac; N—l.s.
Composite tubular sacs (Kompositenschlauchtaschen)—O—*Martiodrilus* (=*Thamnodrilus*) *duodenarius*, approximately sagittal section; P—*Enantiodrilus borellii*, t.s., proximal to the middle. Q to S—composite tubular sacs of the honeycomb type (Wabentaschen)—Q to R—*Martiodrilus* (=*Thamnodrilus*) *agilis*, t.s. and tangential section of a sac; S—*Martiodrilus* (=*Thamnodrilus*) *crassus*, internal view of sac.
Lamellar sacs (Lamellentaschen)—T—*Andiorrhinus brunneus*, t.s. l.s., longitudinal section; t.s., transverse section of a sac.

Calciferous Glands. Michaelsen (1918, 1937) recognizes the following types of calciferous glands in the Glossoscolecini. An understanding of their structure is necessary for identification of the genera of this tribe. The simplest type is the "ridged sac" (Leistentaschen) seen in *Diachaeta* (Fig. 1. 7A) and *Periscolex* (Fig. 1. 7B), consisting of outpouchings of the oesophagus into which extend the transverse folds of the oesophageal lining,. the folds running lengthwise in the sacs. "Partioned sacs" (Fachkapseltaschen) resemble ridged sacs in which certain of the folds are sufficiently high to have met and fused in the axis of the sac, thus dividing it into a number of compartments around a central axis, as in *Quimbaya* (=*Thamnodrilus*) *cameliae* (Fig. 1. 7C). Intermediate between these two forms are the "deeply ridged sacs" (Saumleistentaschen), seen in *Inkadrilus aberratus* (Fig. 1. 7D-F), in which the ridges are higher and hemlike. Secondary folds may be given off from the primary folds of a partitioned sac and may unite to form secondary tubular chambers running longitudinally in the sac; such sacs are "tubule sacs" (Schlauchtaschen) and are seen in a simple form, though probably with some branching of lumina, in *Onychochaeta windlei* (Fig. 1. 7G) and, with considerable branching, in *Thamnodrilus* (=*Aptodrilus*) *fuhrmanni* (Fig. 1. 7H). Tubule sacs in which the basal cavity divides successively and dichotomously are termed "panicled" or "panicled-tubular sacs" (Rispenschlauchtaschen) and are well exemplified by *Thamnodrilus* (=*Aptodrilus*) *excelsus* (Fig. 1.7 I-K), *Rhinodrilus paradox* (Fig. 1. 7L-M), and *Thamnodrilus gulielmi* (Fig. 1. 7N, see Michaelsen, 1937). Alternatively, a number of tubes may arise separately from the main or central lumen, and not by successive branching, though individual tubes may be branched. Such "composite-tubular sacs" (Kompositenschlauchtaschen) distinguish *Martiodrilus* from *Thamnodrilus* and are exemplified by *Martiodrilus* (=*Thamnodrilus*) *duodenarius* (Fig. 1. 7O), *Enantiodrilus* (Fig. 1. 7P) and, in the Microchaetini, *Microchaetus microchaetus*. Variants of the composite-tubular sacs in which the central or main lumen is wide and extends throughout the length of the sac and gives off numerous short, blind tubes along its course like the chambers of a bovine reticulum, "honeycomb-sacs" (Wabentaschen) are seen, for instance, in *Martiodrilus* (=*Thamnodrilus*) *agilis* (Fig. 1. 7Q-R) and *M.* (=*T.*) *crassus* (Fig. 1. 7S). Lastly, calciferous glands which are divided, across the lumen, by lamellae are termed by Michaelsen "lamellar-sacs" (Lamellentaschen), and are seen in *Andiorrhinus* (Fig. 1. 7T) and in *Andiodrilus*.

Calciferous glands, at least of an extramural kind, are absent from the remainder of the Glossoscolecidae.

NERVOUS SYSTEM

The cerebral ganglia never occupy the prostomium in megadriles but in *Biwadrilus bathybates*, the nervous system of which is described and illustrated by Nagase and Nomura (1937), the medianly fused ganglia occupy

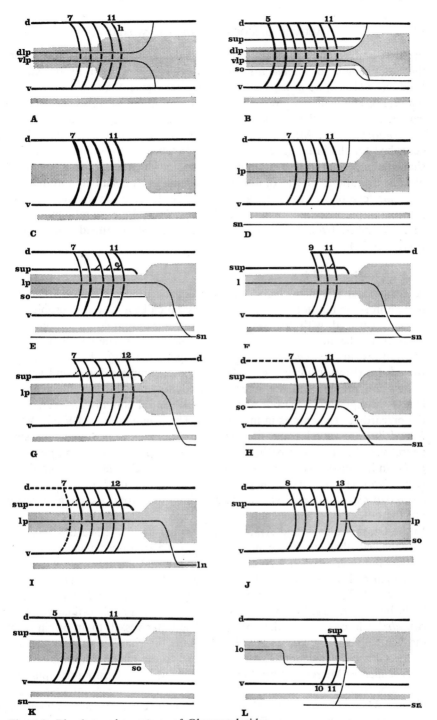

Fig. 1.8. Blood vascular systems of Glossoscolecidae.

Diagrams showing the major blood vessels. The thickness of vessels indicated does not reflect actual relative widths. With the exception of (A), the segment of origin of the intestine is not indicated.

A—*Sparganophilus*; B—*Biwadrilus*; C—*Microchaetus pentheri*; D—*Criodrilus lacuum*; E—*Glyphidrilus ceylonensis*; F—*Glyphidrilus gangeticus*; G—*Callidrilus*; H—*Drilocrius*; I—*Alma*; J—*Glyphidrilocrius*; K—*Hormogaster redii*; L—*Glossoscolecini* (*Pontoscolex corethrurus*).

c, supra-oesophageal connective; d, dorsal vessel; dlp, dorsal latero-parietal; h, heart; lo, latero-oesophageal; lp, lateroparietal; sn. subneural; so, subaesophageal; sup, supra-oesophageal; v, ventral vessel.

the first and second segments. McKey-Fender and MacNab (1953) report that the ganglia of *Criodrilus lacuum*, which have been illustrated by Vejdovsky (1884), lie at least partly in the first segment. So far as is known, the cerebral ganglia have moved further posteriorly in all other glossoscolecids, their furthest displacement being into the fourth segment in *Pontoscolex corethrurus* (v. Michaelsen, 1928). Auditory sense organs have been described for the latter species by Eisen (1896b). Each consists of a dome-shaped, epidermal cell containing a vacuole which encloses a spherical otolith; vibrations of the otolith are presumably transmitted through the investing cytoplasm to a basal nerve. The innervation of the sensory prostomial proboscis of *Andiodrilus biolleyi*, from the cerebral ganglia, has been described by Picado (1913). Sensory buds have already been mentioned (p. 24).

VASCULAR SYSTEM

A dorsal and a ventral blood vessel are present in all Glossoscolecidae though in some but not all of those species which have a supra-oesophageal vessel connected with the hearts, the dorsal vessel may be suppressed anterior to the hearts. The vascular system most resembles that of the microdriles in *Sparganophilus* (Sparganophilinae; Fig. 1. 8A) and in *Biwadrilus* (Biwadrilinae; Fig. 1. 8B). In these, complete dorsal and ventral vessels are present and are interconnected not only by the five pairs of hearts (in segments 7-11) but also by dorso-ventral commissural vessels in the precardiac segments as in microdriles. The commissurals of V and VI in *Biwadrilus* are valvular and may be regarded as additional hearts. A sub-neural vessel is absent in both genera but *Biwadrilus* differs from *Sparganophilus* in having developed a supra-oesophageal vessel as in some microdriles (e.g. *Tubifex*) and many megadriles. The hearts in *Biwadrilus*, as in *Hormogaster*, do not receive connectives from the supra-oesophageal vessel but the latter vessel sends connectives in each segment to the dorsal vessel. In the Glossoscolecini and Almini, on the other hand, the hearts are latero-oesophageal, that is, have connections with the supra-oesophageal as well as the dorsal vessel. Both *Sparganophilus* and *Biwadrilus* have a pair of latero-parietal vessels on each side running forwards from an origin in segment 14. The dorsal latero-parietal originates from the dorsal vessel in both, but the ventral latero-parietal originates from a subintestinal vessel in *Biwadrilus* in contrast, in the absence of the latter vessel, to an origin from the ventral vessel in *Sparganophilus*. Perhaps the simplest vascular system in the Glossoscolecidae, at least with regard to the major trunks, is seen in *Microchaetus* (Microchaetini; Fig. 1. 8C). In this tribe supra-oesophageal and subneural vessels and, so far as is known, latero-parietal vessels are absent. *Microchaetus pentheri* has dorso-ventral commissural vessels in segments 5 to 11, those of 7 to 11 being moniliform hearts (Fig. 1. 9B). Intrasegmental doubling, with intersegmental union, of the dorsal vessel often occurs in some of the heart segments in *Microchaetus* and *Tritogenia*, as is sporadically the case in other megadriles including *Hormogaster* and *Pontoscolex*. It seems probable that the extreme simplicity, in general plan, of the circulatory system of the Microchaetini is primitive and not a secondary reduction.

Within the Alminae the vascular system of *Criodrilus* (Criodrilini) is both the simplest and from other evidence (Chapter 4) the most primitive (Fig. 1. 8D). Though a subneural vessel and a pair of latero-parietal vessels are present, there is no supra-oesophageal vessel. The latero-parietal vessels join the dorsal vessel in segment 12. The hearts occupy segments 7-11 as in the preceding genera. Persistence of the subneural vessel to the anterior end of the body appears, from the evidence of other Alminae, to be a primitive feature.

The least modified condition of the vascular system in the Almini appears to be that seen in *Glyphidrilus ceylonensis* (Fig. 1. 8E). Details of the vascular system in *Glyphidrilus* are given on p. 750. A feature of the system in *G. ceylonensis* which is presumed to be primitive is the extension of the dorsal and subneural vessels to the anterior extremity, but whether the existence of supra-oesophageal connectives to the hearts (in 9-11) is derived, as considered by Gates (1958c), or is primitive, is debatable. Gates envisaged a "proto-*Glyphidrilus*" in which such connectives were totally absent, in the presence of a supra-oesophageal vessel. It seems possible, however, that primitively the oesophageal blood plexus connected by a commissure on each side with the dorsal blood vessel (or the bases of the hearts) in the absence of the dorsal median specialization of the plexus (which we term the supra-oesophageal vessel.) Before development of a supra-oesophageal vessel it would be necessary for the oesophageal plexus to communicate with the hearts or the dorsal vessel in each segment in which hearts coexisted with the plexus, but on development of the supra-oesophageal as a longitudinal collecting vessel, linking the plexus of separate segments, the number of connectives could be reduced. *Glyphidrilus ceylonensis*, in this view, still shows the primitive condition of no definite, or at most a weakly-developed, supra-oesophageal vessel, and complete dorsal (and subneural) vessels and accordingly the plexus connects by "supra-oesophageal" connectives with the hearts in a maximal number of segments, 9-11. Restriction of connectives to segments 10 and 11 in the closely related *G. annandalei* and in *birmanicus*, *papillatus* and *weberi*, may, then, be regarded as a reduction on development of a supra-oesophageal vessel. In all of these species the hearts occupy segments 7-11 and the hypothesis advanced here of reduction of connectives is supported by the fact that in *G. gangeticus* (Fig. 1. 8F) which is clearly a highly specialized species, only the hearts of segment 11 are latero-oesophageal. Evidence for the advanced rather than primitive nature of *G. gangeticus* is restriction of the hearts to three pairs, in 9-11 (rarely 8-11), anterior termination of the dorsal vessel at the first hearts and failure of the subneural vessel to persist in front of segment 16 or 17. Hearts are seen behind segment 11 for the first time in *G. stuhlmanni* in which they occupy 8-13; correlated with this is the extreme elongation, for the genus, of the oesophagus in this species. In *Callidrilus* (Fig. 1. 8G) the vascular system, though imperfectly known, appears to be on the same plan as *Glyphidrilus*. Hearts are present in 7-12 and the last two pairs or all of them are latero-oesophageal; the dorsal vessel ends anteriorly with the first pair.

Drilocrius (Fig. 1. 8H) has retained the apparently primitive location of hearts in segments 7-11. A supra-oesophageal vessel is present and some at least of the hearts are latero-oesophageal. The dorsal and subneural vessels may or may not be absent from the forebody. An apparent departure from *Callidrilus* and *Glyphidrilus* is the origin of the suboesophageal, rather than the latero-parietal vessels, from the subneural vessel but this requires confirmation. Latero-parietal vessels have been described for *D. alfari* but their origin is unknown. The vascular system of *Alma* (Fig. 1. 8I) is discussed in detail on p. 000. It will suffice here to note that the median subneural vessel has been replaced by paired latero-neurals which are continuous anteriorly with the latero-parietals. Hearts lie in segments 7 or 8 to the preovarian segment (i.e. to 12 or, in *A. eubranchiata*, in which interpolation of a supernumerary heart-containing segment has occurred, to 13). The dorsal vessel (always?) ends anteriorly with the first hearts; the supra-oesophageal vessel may be similarly restricted or may continue on to the pharynx. All hearts may be latero-oesophageal or only the last three pairs may possess supra-oesophageal connectives.

Glyphidrilocrius (Fig. 1. 8J) also lacks a subneural vessel and, again, from its specialized morphology, this appears to be a secondary condition. The oesophagus is enormously elongated and, as in the case of *Glyphidrilus stuhlmanni*, the hearts occupy 8-13, presumably following loss of those of 7 and addition of pairs in 12 and 13 to the primitive battery. The dorsal vessel ends anteriorly with the first hearts. If presence of supra-oesophageal connectives in several segments is primitive relative to restriction to a few segments, as proposed above, it is somewhat surprising that all the hearts have connectives in *Glyphidrilocrius* especially as the hearts of 12 and 13 are presumably new acquisitions. The answer may perhaps lie in the extraordinary development of the supra-oesophageal which exceeds the dorsal vessel in diameter. Origin of a pair of posterior latero-parietals from the suboesophageal vessels is unique and probably has no phylogenetic relation with origin of anterior latero-parietals from the median suboesophageal vessel in *Biwadrilus*.

The vascular system of *Hormogaster* (Fig. 1. 8K) has been described by Pitzorno (1899) and by Omodeo (1948). Complete dorsal, ventral and subneural vessels are present and, as in *Biwadrilus*, the hearts lie in segments 5 to 11, and lack connectives to the supra-oesophageal vessel. The latter also anastomoses extensively (in segments 9-13) with the dorsal vessel, a further parallel with *Biwadrilus*.

The greatest modification of the vascular system in the Glossoscolecidae occurs in the Glossoscolecini and is furthest advanced in *Pontoscolex* (Fig. 1. 8L) (Eisen, 1900; Gates, 1943). Here the only remaining hearts are those of segments 10 and 11 and they retain the supra-oesophageal connections but have lost the connections with the dorsal blood vessel. The dorsal vessel remains complete but the subneural has not been traced in front of 9/10 and is presumably depleted by a pair of large commissurals passing to the dorsal vessel in segment 12. The supra-oesophageal is very short (in 10-13 only?); probably a correlation with the presence of three pairs of large calciferous glands in 7-9. Latero-parietals have not been described but a pair of latero-oesophageals is present, uniting to form a sub-oesophageal vessel. In *Martiodrilus crassus* (*v.* Pickford, 1940) a less modified condition exists; the hearts of 10 and 11 are oesophageal, as in *Pontoscolex*, but hearts are present in 7-9, and appear to be dorso-ventral, their dorsal connections being solely with the dorsal vessel, though ventral connectives have not been verified. The dorsal vessel apparently terminates on the gizzard. In *Diachaeta* there are only two pairs of hearts, as in *Pontoscolex*, but they link the dorsal and ventral vessels (Beddard, 1890).

NEPHRIDIA

Nephridia are absent from the ten or more anteriormost segments in the aquatic subfamilies of the Glossoscolecidae in the adult state, as is sporadically the case in the Glossoscolecini. Despite this absence in the adult, Hague (1923) demonstrated nephridia as far forward as segment 3 in embryos of *Sparganophilus tamesis* (=*eiseni*). The structure of the nephridia of *Alma nilotica* has recently been described by Khalaf El Duweini (1965). Nair (1938) briefly describes those of *Glyphidrilus annandalei*, and those of *Biwadrilus* are described by Nagase and Nomura (1937). In all of the aquatic subfamilies they are exonephric stomate holonephridia (terminology of Bahl, e.g. 1947).

In *A. nilotica* they commence, in the adult, in one of segments 13 to 15. The funnel is preseptal. The post septal portion consists of a much lobed mass of hypertrophied peritoneal cells in which the nephridial tubule is embedded. The tubule is not branched and makes two long, slightly convoluted parallel loops. The wall of the first (ental) loop consists of finely granular syncytial cytoplasm; cilia line parts of the narrow lumen. This portion corresponds to the "narrow tube" of lumbricid nephridia. The first limb of the second loop corresponds to the "middle tube" and has, as in the Lumbricidae, a brownish, opaque appearance. This portion is also syncytial. Its wall is slightly thicker than that of the preceding portion and the diameter of its lumen slightly wider. Cilia form two longitudinal tracts throughout its length. The ectal limb of the second loop corresponds to the lumbricid "wide tube". It, like the latter, begins with a dilated portion, the equivalent of the "ampulla". This limb is again syncytial but is unciliated; the diameter of wall and lumen are hardly greater than in the ental limb. The "wide tube" is continuous to the nephropore. The portion in the body wall has some

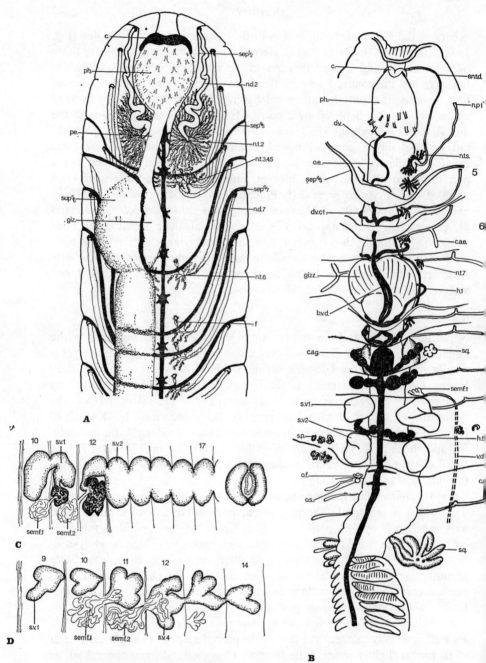

Fig. 1.9. Anatomy of Glossoscolecidae.

A, B—dissections to show excretory systems: A, *Pontoscolex corethrurus*; B, *Microchaetus pentheri*; C, D—seminal vesicles of *Alma*: C, *A. nasuta*; D, *A. eubranchiata cathuractae*.

b.d.v, bifid dorsal vessel; c, cerebral ganglia; ca.g, calciferous gland; cae, caecum of nephridium; d.v, single dorsal vessel; d.v.c, dorsoventral commissural vessel; ent.d, duct of enteronephric nephridium; f, nephridial funnel; gizz, gizzard; h 1, 2, first and last hearts; n.c, ventral nerve cord; n.d 2, 7, ducts of nephridia of the segments thus numbered; np 1, first nephropore; n.t 2-7, tubules of nephridia of the segments thus numbered; o, ovary; oe, oesophagus; o.f, oviducal funnel; o.s, ovisac; ph, pharynx; sem.f 1, 2, first and second seminal funnels; sep 1/2, 4/5, 5/6, respective septa; s.g, septal gland; sp, spermatheca; s.v 1, first seminal vesicle; s.v 2, 4, last seminal vesicles; v.d, vas deferens.

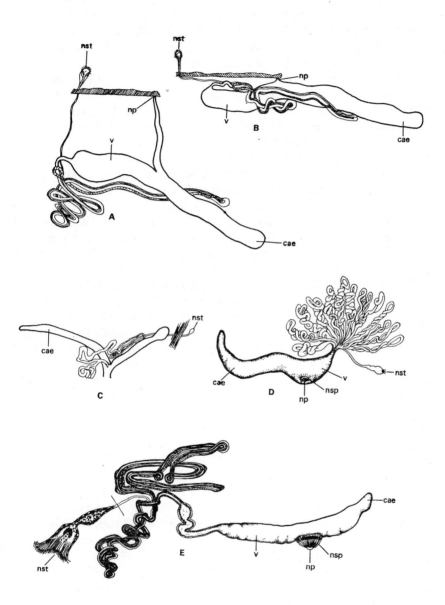

Fig. 1.10. Nephridia of Glossoscolecidae.
A, B—*Hormogaster praetiosa*: A, an anterior—B, a posterior nephridium; C—*Kynotus cingulatus*, posterior nephridium; D—*Microchaetus rappi*, nephridium; E—*Pontosco-lex corethrurus*, a nephridium from a clitellar segment.
cae, caecum, np, nephropore; n.sp, sphincter of nephropore; nst, nephrostome; v, terminal vesicle.

muscular and connective tissue fibres but there is no well-defined terminal sphincter. Khalaf's description thus indicates similarity of nephridial structure with that of the Lumbricidae and dissimilarity with the Glossoscolecinae as exemplified by *Pontoscolex corethrurus* (Fig. 1. 9A, 1. 10E) (Bahl, 1947) and by *Microchaetus* (Fig. 1. 9B and 1. 10D).

Similarities between the nephridia of *Microchaetus* and *Pontoscolex*, which have been confirmed in the present study; are their increasingly tufted form in the direction of the anterior end of the body, with numerous tightly coiled loops; their very large, caplike terminal sphincters; their long lateral caeca; and the thick anteriorly directed ducts of the anterior nephridia. Nephridial sphincters have also been reported for *Diachaeta*, *Onychochaeta*, *Meroscolex* and *Opisthodrilus* but are apparently absent from other genera of the Glossoscolecini. Nephridial caeca have been observed in *Martiodrilus crassus* in which each nephridium has multiple funnels (Pickford, 1940; Bahl, 1947). Caeca, but not sphincters, occur in *Hormogaster* and *Kynotus* (Fig. 1. 10A, B and C). The possible phylogenetic significance of the structure of the nephridia and the location of their pores in the family is discussed in Chapter 4.

The nephridia of the Glossoscolecinae always, so far as is known, retain their funnels, and though they may appear tufted through development of numerous coiled loops in the Glossoscolecini and in *Microchaetus*, are not, in these taxa, subdivided into meronephridia despite Bahl's loose application of this name to the conjoined loops in *Pontoscolex*. A simple form of true meronephry, i.e. development of more than one pair of discrete nephridia per segment, is, however, seen in *Tritogenia* which otherwise closely resembles *Microchaetus* though lacking nephridial caeca and sphincters. Enteronephry has been demonstrated in the present study for the anteriormost nephridia of *Microchaetus pentheri* (Fig. 1. 9B). Bahl (1947) discredited early observations of a similar condition in *Pontoscolex*.

GENITAL SYSTEM

The Gonads.   In the Glossoscolecidae the basic arrangement of the gonads is that of the Haplotaxidae, with paired testes in segments 10 and 11 and paired ovaries in 12 and 13. Proandry (testes in 10 only) or metandry (testes in 11 only) frequently supervene, however, and only very rarely (in *Enantiodrilus borgellii* (Glossoscolecini) and *Glyphidrilus kukenthali* (Almini)) does hologyny (ovaries in 12 and 13) occur. The remaining Glossoscolecidae, so far as is known, are metagynous, having lost the ovaries of segment 12. This strong tendency to loss of the anterior ovaries contrasts with a tendency in the Haplotaxidae to loss of the posterior (never the anterior) ovaries.

Functional hologyny, that is, occurrence of ovaries in 12 and 13, with gonoducts passing to the exterior from each segment, is known elsewhere in the megadriles only in the acanthodrile *Diplocardia sandersi* (*v.* Gates, 1958). Sporadic adult hologyny, as an intraspecific abnormality, was found to be common in several species of *Allolobophora* and *Lumbricus* examined by Woodward (1892, 1893) but Gates (1958) stresses that in cases of sporadic hologyny the animals remain functionally metagynous as oviducts are absent from segment 12. He draws attention to abnormalities of *Enantiodrilus borelli*, including presence of ova in the "testes" of segment 11 and concludes that it is parthenogenetic. Sperm in the spermathecae of *Diplocardia sandersi* indicated, on the other hand, that reproduction was biparental and the same is true of *Glyphidrilus kukenthali*. It seems possible that the hologynous condition of *G. kukenthali* is a retention of the primitive condition of the presumed haplotaxid ancestor or at least reflects a genetic liability with regard to the location of the ovaries in the Glossoscolecidae which permits reversion, even in biparental populations, to the primitive arrangement.

The ovaries vary greatly in form, even intragenerically, in the Glossoscolecidae. In *Sparganophilus* (personal observation) and in *Hormogaster* (*v.* Rosa, 1889) the ovary terminates, at the free edge, in a single chain of oocytes as in the Lumbricidae. In *Criodrilus* the ovary is flattened and lobed or paddle-shaped, with several irregular rows of ova, a notable contrast with the Lumbricidae, including the morphologically remarkably similar *Bimastos palustris*. In *Biwadrilus*, also, the ovaries are indented lobes. In *Callidrilus* and most species of *Alma* and *Glyphidrilus* each ovary has the form of a ribbon or lamina which is very extensive transversely to the long axis of the body and which is commonly ruffled (plicate); occasionally it is bushy. The laminate or ribbonlike form has not been reported for *Drilocrius* but transverse laminae with many chains of oocytes occur in *Glyphidrilocrius*. In *Kynotus* each ovary is much subdivided (Rosa, 1892) and, in the Glossoscolecini, those of *Martiodrilus crassus* are "many fingered glands" (Pickford, 1940). Each ovary of *Microchaetus saxatilis* (personal observation) is a subspherical fine-grained body in which individual oocytes are not externally distinguishable.

Testis Sacs. Testis sacs are chambers the walls of which are distinct from the normal coelomic peritoneum and which enclose the testes and, in this family, the sperm funnels. They are developed in *Microchaetus* and *Tritogenia* and frequently in the Glossoscolecini.

Seminal Vesicles and Septal Pouches. Seminal vesicles (sperm sacs) are evaginations, into neighbouring segments, of the septa which bound the testis segments. There are, therefore, maximally four pairs, in segments 9-12. The posterior pair often deflects septum 13/14 posteriorly for one to a few segments and in the Glossoscolecini may be very extensive. Thus, in *Diachaeta thomasi* the seminal vesicles extend as far back as segment 119 and from about 75 bear numerous digitiform processes (Beddard and Fedarb, 1889). In *Alma nasuta* (Alminae) diverticula of the seminal vesicles of segment 12 extend into segment 35 and in *A. eubranchiata catarrhactae* the vesicles of 12 extend into segment 14 (Fig. 1. 9C, D) (Michaelsen, 1935b).

Backward pouchlike evaginations of septa independent of those forming the seminal vesicles occur in *Callidrilus, Drilocrius, Glyphidrilocrius,* and in some species of *Glyphidrilus* and of *Alma.* Those of septum 13/14 may contain oocytes as in *Glyphidrilus annandalei* (Nair, 1938), *Drilocrius, Glyphidrilocrius,* and *Hormogaster,* and are then termed ovisacs; they are the only such pouches in the taxa mentioned. Those which do not contain oocytes (including any on septum 13/14) are termed septal pouches. The greatest number of these yet recorded occurs in *Alma eubranchiata,* in which they depend from the anterior septa of segments 13 or 14 to at least 28. Khalaf El Duweini (1954) showed those of *Alma nilotica,* in which they occupy segments 13-22, to contain coelomocytes, as did Cognetti (1911) for *Alma emini,* in which there are six to seven pairs occupying segments 13, 14-18, 20; ovisacs are apparently absent in *Alma.* The ovisacs of some species, e.g. *Martiodrilus crassus* (*v.* Pickford, 1940), some species of *Drilocrius* and *Microchaetus pentheri* (personal observations), are posterior evaginations of the female funnels rather than of the septum.

Sperm Funnels and Ducts. The sperm funnels are carried on the posterior walls of segment 10 and/or 11 and their anatomy in the family shows no noteworthy peculiarities. The sperm ducts, on the other hand, are peculiar in being deeply embedded in the musculature of the body wall throughout their courses from the funnels to the male pores. This feature is, so far as is known, constant in the Glossoscolecidae and apparently diagnoses them from all other families. Pool (1937) implies that the vasa deferentia are embedded in the body wall in the lumbricid *Allolobophora,* and de Ribaucourt (1901) reports this condition for *Allolobophora caliginosa,* but in specimens of *A. caliginosa trapezoides* examined by the writer the ducts, though lying below the level of the general peritoneum of the ventral body wall, are superficial relative to the coelomic surface when compared with the Glossoscolecid condition. In *Bimastos palustris* (*v.* Moore, 1895) the sperm ducts are not obvious unless gorged with spermatozoa, "being imbedded in the tissues of the body wall." This partial embedding of the vasa deferentia in the body wall in some Lumbricidae provides support for the view that the family is closely related to the Glossoscolecidae (see p. 188).

The vasa deferentia discharge at the male pores without the intervention of atria or other specializations, or association with prostate glands, in *Sparganophilus,* the Alminae (excluding *Callidrilus* and *Criodrilus*) and the Glossoscolecinae, including the Microchaetini but excluding many Glossoscolecini.

Glands and Bursae associated with the Male Ducts. Dilations or chambers at the ectal end of the vasa deferentia and protruding into the coelom occur in *Kynotus, Criodrilus,* commonly in *Glossoscolex* (together with some other Glossoscolecini) and in *Biwadrilus.* In *Kynotus cingulatus* (Benham, 1896) there is a large muscular bursa at each male pore in segment 15, which is capable of eversion through the pore to form a circular, flattened structure presumed to be a clasper. Into each bursa opens a coiled tubular prostate gland which is enveloped in a thin muscular sheath. In the three preceding segments there are similar and clearly homologous "prostate" glands but each opens into the follicle of a bundle of genital setae. The "prostates" in *Kynotus* would thus appear to be the equivalents of the genital seta glands of *Microchaetus* (Fig. 1. 9B), though probably acquired by parallelism in related forms rather than as homologues derived from primitive glands of a common ancestor. In those species of *Glossoscolex* which possess copulatory sacs, these may be sufficiently well developed to be reminiscent of the atria of the Alluroidinae or of the Tubificidae. As an example, those of *Glossoscolex catharinensis* (*v.* Michaelsen, 1918) are stoutly "bean-shaped" in the long axis of the worm and extend through three segments (16-18), opening to the exterior through an indistinct stalk in segment 17. They have a distinct, long but narrow lumen and a very thick, muscular wall. From the base of each copulatory sac a strong muscular band extends forwards to the region of the gizzard. The vasa deferentia open into the base of the copulatory sac at its anterior aspect. No mention is made of a glandular component in the sac and in a description of *G. lojanus*, Michaelsen specifically states that the spherical copulatory sac, which occupies segments 18 and 19, is entirely muscular, lacking glands even in the vicinity of the small pyriform lumen. These muscular sacs of *Glossoscolex* and *Kynotus* appear to the writer to be quite distinct from the terminal bursae of *Criodrilus* although Michaelsen sought to homologize the two types and considered them to indicate descent of *Criodrilus* from *Kynotus*. The bursae of *Criodrilus lacuum* are primarily glandular organs and Benham (1887) went so far as to describe them as prostates. He noted that the corresponding sperm duct passes to the dorsal surface of the gland, dips down through its mass, and opens to the exterior at the male pore, which is situated on a prominent rounded papilla which seemed to be merely the outer half of the "prostate". The gland itself consisted of cells similar to those forming the epidermis of the clitellum and continuous with them; the overlying muscular layers of the body wall were noted to be thin and the gland appeared to be formed merely by a hemispherical thickening of the epidermis. The external appearance of this glandular bursa is shown in Fig. 15. 11A. Oerley (1887) considered the function of the bursa to be secretion of the cocoon but it is more probable that it secretes the large spermatophores (Fig. 15. 11F) of this species. McKey-Fender and McNab (1953) have justifiably drawn attention to the similarity of this gland to the "prostate" described by Moore (1895) for *Bimastos palustris*. "Chitogenous glands" in the latter were shown by Moore to produce the spermatophores, and he, too, ascribed a similar function to the bursae of *Criodrilus*. A close phylogenetic relationship between the Lumbricidae and the Glossoscolecidae is especially well supported by the morphology of *Bimastos palustris* (see p. 189).

The "copulation glands" (Nagase and Nomura, 1937) of *Biwadrilus* are glandular prostate-like organs associated with the genital setae (Fig. 15. 12A, B) and are similar in internal structure to the genital seta glands of *Microchaetus*. They probably serve to cement the genital setae in the tissues of the concopulant as do smaller genital seta glands in *Lumbricus*. Nagase and Nomura (1937) have applied the term "prostate gland" to diffuse masses of gland cells (Fig. 15. 12B) grouped around the ectal extremity of the vasa deferentia. These glandular masses are scarcely more developed than similar diffuse glands observed by the writer on the internal surfaces of the male porophores in *Drilocrius* and *Glyphidrilocrius* and probably should be referred to as diffuse prostatic (?) gland-cells rather than as distinct prostate glands. Discrete, compact prostate glands have been described by Horst (1889) for *Glyphidrilus weberi* but their presence, at least in Javanese material, is not confirmed in the present study. *Callidrilus* thus appears unique in possessing compact glands which may be considered true prostate glands though not homologous with the prostates of other megadriles. Those of *Callidrilus ugandaensis* are a pair of dorso-ventrally depressed, lobed but compact, spindle-shaped glands restricted to segment 17. Each extends far dorsally into its segment and enters the body wall immediately lateral to the corresponding seta *b* without the intervention of a distinct externally differentiated duct. It consists of masses of glandular cells permeated by ramifying ducts of circular cross-section. Each duct is lined by a low, apparently cuboidal, unciliated epithelium. Around each duct are grouped greatly enlarged cells, each with a distinct nucleus and numerous eosinophil secretory granules. The attenuated proximal ends of these cells appear to arise individually from the duct. The glandular elements are thus similar to those in *Alluroides* but, whereas the glands of the Alluroidinae include layers of muscle fibres, these are absent from the prostates of *Callidrilus*. The entire gland is invested in a very low peritoneum. Within the body wall, the single ectal duct of each prostate dilates slightly but does not form a definite terminal chamber. The corresponding vas deferens opens to the exterior through a capacious chamber very shortly behind, but quite separately from the prostate aperture, in the same segment. The terminal chamber of the vas deferens, though large, is confined to the circular muscle of the body wall and has no separate musculature. Its walls are composed of a tall columnar epithelium which is ciliated dorsally, where the vas deferens enters it (Jamieson, 1968c).

The posterior prostate-like glands of *Sparganophilus tamesis* (Fig. 1. 5D-F; 15. 13, 15. 14) and all of these glands in *S. smithi* (Fig. 1. 5G) resemble those of *Callidrilus* in having glandular cells grouped around an epithelial layer which lines an intraglandular duct. *Sparganophilus* differs, however, in the unbranched condition of the duct, and, in *S. tamesis*, by differentiation of a terminal, externally recognizable duct. The posterior glands of *S. tamesis* (=*S. eiseni*) lie in some of segments 15 to 17 and 22 to 27. Hague (1923), who made a detailed examination of their structure, uses the term "accessory reproductive gland" for the anterior and posterior glands, considering that the term "prostate" should be reserved for glands connected with the sperm duct. She has shown that a posterior gland (Fig. 1. 5E, F) consists of a tubular and somewhat convoluted glandular part and a duct which opens to the exterior close beside and normally through the follicle of seta *b*. (In specimens examined by the writer the pore is always slightly lateral of the seta *b* and distinct from its follicle.) The duct consists of an epithelial layer which is continuous with the epidermis, an investing muscular layer, and outside this a prominent peritoneal layer. The glandular part consists of epithelial cells with attenuated processes which extend towards or to the lumen. Because of the regularity of cell walls and of the position of the nuclei around the lumen, there appears to be an epithelial lining surrounded by a glandular layer.

The anterior accessory reproductive or prostate-like glands occupy some of segments 3 to 10. Each is a spherical or somewhat elongated mass from which a duct opens into the follicle of the corresponding seta *a*. The structure of the glandular region (Fig. 1. 5D) accords with that of the posterior glands with the exception that a muscular layer, continuous with the circular muscle of the body wall, intervenes between the epithelium lining the lumen and the glandular layer. The proximal ends of the glandular cells are attenuated, can be traced into the muscular layer, and probably extend into the epithelial layer. Hague provides evidence that the extensive series of prostate-like glands in *S. tamesis* has the function of contributing to the copulatory slime-tube. This is of interest in view of previous suggestions that certain earthworms may previously have had prostate glands in many more segments than they now occupy (e.g. Jamieson, 1963). It is conceivable that tubular prostate glands, such as are seen in the Ocnerodrilinae, may have developed, phylogenetically, by specialization of slime-tube glands in the vicinity of the male pores to produce a secretion facilitating sperm ejaculation or for maintenance of sperm in the spermathecae of the concopulant. The function of prostate glands, it may be noted, is still imperfectly understood.

Spermathecae. Spermathecae are absent in *Criodrilus* and *Biwadrilus* and in some Glossoscolecini. In the Sparganophilinae they are pre-testicular, are paired or multiple in each segment occupied, and project freely into the coelom; diverticula are absent. In most Glossoscolecini they are pre-testicular, are usually restricted to one pair per segment, and commonly are adiverticulate and freely protuberant into the coelom but exceptions to all these conditions occur. In *Rhinodrilus fafner* the spermathecal pores are paired in intersegmental furrows 6/7-14/15 but in no Glossoscolecini are the spermathecal pores solely post-testicular. In *Martiodrilus crassus* the spermathecae are multiple at intersegments 5/6-8/9 or 9/10 and are completely sunk within the body wall. A pair of pyriform diverticula occurs on the ectal region of the ampulla, and opens into the proximal region of the duct in each of the two spermathecae of *Glossoscolex schutti* (*v.* Michaelsen, 1918); each diverticulum houses spermatozoa. In *Hormogaster* the spermathecae lie in, and frequently extend behind, the testis segments and are sessile on the parietes. In *Microchaetus* and *Tritogenia*, and in the Almini, the spermathecae are post-testicular (the sole supposedly diagnostic character of the former Microchaetinae or Microchaetidae) though in *Glyphidrilus stuhlmanni* the spermathecal pores extend forwards through the testis segments. In all Microchaetini and Almini they are wholly or largely concealed in the body wall, and lack diverticula. They are commonly subspherical sacs with short, usually solely intraparietal ducts, but in *Microchaetus* are often serpentine and in *Glyphidrilocrius ehrhardti* they may be three lobed or may have irregularly developed diverticula. Khalaf El Duweini (1950b) gives a detailed description of the spermathecae of *Alma nilotica*. In *Alma* a most remarkable backward transposition of the spermathecae occurs and a very considerable variation exists in the number of intersegments occupied both inter- and intra-specifically. For instance, the numbers of transverse spermathecal rows in *A. emini* are 10-33, in *A. nilotica*, 22-27, and in *A. stuhlmanni*, 7-27. The smallest recorded number of transverse rows is 7 (*stuhlmanni*) and the greatest 37 (*eubranchiata*) or, including rudimentary spermathecae, 137 (*eubranchiata cattharactae*). The most anterior intersegment occupied in *Alma* is 18/19 (*millsoni*) and the most posterior 253/254 (*eubranchiata*). The spermathecae are paired or multiple in each segment occupied in *Microchaetus*, *Kynotus* and *Hormogaster* and are multiple in the Almini.

REFERENCES

(other than Glossoscolecidae, for which see p. 71)

AIYER, K. S. P. 1925. Notes on the aquatic Oligochaeta of Travancore. I. *Ann. Mag. nat. Hist.*, (9) **16**, 31.

ALTMAN, L. C. 1936. Oligochaeta of Washington. *Univ. Wash. Publs Biol.*, **4**, 1.

ATHESTON, L. 1899. The epidermis of *Tubifex rivulorum* Lamarck, with especial reference to its nervous structures. *Anat. Anz.*, **16**, 497.

AVEL, M. 1959. Classe des Annélides Oligochètes. Grassé P. P. *Traité de Zoologie*, **5** (1), 224.

BAHL, K. N. 1946. Studies on the structure, development and physiology of the nephridia of Oligochaeta. Pt. VIII: biochemical estimation of the nutritive and excretory substances in the blood and coelomic fluid of the earthworm and their bearing on the role of the two fluids in metabolism. *Q. Jl microsc. Sci.*, **87**, 357.

— 1947. Excretion in the Oligochaeta. *Biol. Rev.*, **22**, 109.

BEDDARD, F. E. 1890. On the anatomy, histology and affinities of *Phreoryctes*. *Trans. R. Soc. Edinb.*, **35**, 629.

— 1891. Anatomical description of two new genera of aquatic oligochaeta. *Trans. R. Soc. Edinb.*, **36** (Pt. 2, no. 11), 273.

— 1892. A new Branchiate oligochaete. *Q. Jl microsc. Sci.*, **30**, 325.

— 1894a. Preliminary notice of South-American Tubificidae collected by Dr. Michaelsen, including the description of a branchiate form. *Ann. Mag. nat. Hist.*, **13**, 205.

— 1894b. A contribution to our knowledge of the Oligochaeta of tropical Eastern Africa. *Q. Jl microsc. Sci.*, **36**, 201.

— 1895. *A monograph of the order Oligochaeta*. Oxford.

— 1896. Naiden, Tubificiden und Terricolen. I. Limicole Oligochäten. *Ergebnisse Hamburger Magalhaensischen Sammelreise.* **36** (6), 1. Hamburg.

— 1901. On a freshwater Annelid of the genus *Bothrioneuron* obtained during the "Skeat Expedition" to the Malay Peninsula. *Proc. zool. Soc. Lond.*, **1901**, 81.

— 1906. Zoological results of the Third Tanganyika Expedition, conducted by Dr. W. A. Cunnington, 1904-1905. Report on the Oligochaeta. *Proc. zool. Soc. Lond.*, **1906**, 206.

— 1908. A note on the occurrence of a species of *Phreatothrix* (Vejd.) in England, and on some points in its structure. *Proc. zool. Soc. Lond.*, **1908**, 365.

BENHAM, W. B. 1903. On some new species of aquatic Oligochaeta from New Zealand. *Proc. zool. Soc. Lond.*, **1902**, 202.

— 1904. On some new species of the genus *Phreodrilus*. *Q. Jl microsc. Sci.*, **48**, 271.

— 1905. On a new species of *Haplotaxis*, with some remarks on the genital ducts in the Oligochaeta. *Q. Jl microsc. Sci.*, **48**, 299.

— 1907. On the Oligochaeta from the Blue Lake, Mount Kosciusko. *Rec. Aust. Mus.*, **6**, 251.

— 1909. Report on Oligochaeta of the Subantarctic Islands of New Zealand. In: Chilton, C. *The Subantarctic Islands of New Zealand.* **1**, 251. Wellington.

— 1915. Oligochaeta from the Kermadec Islands. *Trans. Proc. N.Z. Inst.*, **47**, 174.

BERGH, R. S., AND DITLEVSEN, A. 1899. Om et hidtil ukendt Bygningsforhold i Epidermis hos Oligochaeta limicola. *Overs. K. danske Vidensk. Selsk. Forh.*, **1899** (4), 323.

BOLDT, W. 1928. Mitteilung über Oligochaeten der Familie Tubificidae. *Zool. Anz.*, **75** (7-10), 10.

BOVERI-BONER, Y. 1920. Beiträge zur vergleichenden Anatomie der Nephridien niederer Oligochäten. Inaug. Diss. Zürich. Jena. *Vjachr. naturf. Ges. Zürich*, **65**, 506.

BRINKHURST, R. O. 1962. A redescription of *Peloscolex variegatus* Leidy (Oligochaeta, Tubificidae) with a consideration of the diagnosis of the genus *Peloscolex*. *Int. Revue ges. Hydrobiol.*, **47**, 301.

BRINKHURST, R. O. 1963. Taxonomical studies on the Tubificidae (Annelida, Oligochaeta). *Int. Revue ges. Hydrobiol.*, (suppl.) **2**, 7.
— 1964a. Observations on the biology of the lake-dwelling Tubificidae. *Arch. Hydrobiol.*, **60**, 835.
— 1964b. A taxonomic revision of the Alluroididae (Oligochaeta). *Proc. zool. Soc. Lond.*, **142**, 527.
— 1965a. A taxonomic revision of the Phreodrilidae (Oligochaeta). *J. Zool. Lond.*, **147**, 363.
— 1965b Studies on the North American aquatic Oligochaeta II: Tubificidae. *Proc. Acad. nat. Sci. Philad.*, **117**, 117.
— 1966. A taxonomic revision of the family Haplotaxidae. *J. Zool. Lond.*, **150**, 29.
BRODE, H. S. 1898. A contribution to the morphology of *Dero vaga. J. Morph.*, **14**, 141.
CEKANOVSKAYA, O. V. 1962. The aquatic Oligochaeta fauna of the USSR. *Opred. Faune SSSR*, **78**, 1.
CERNOSVITOV, L. 1936. Oligochaeten aus Südamerika. Systematische Stellung der *Pristina flagellum* Leidy. *Zool. Anz.*, **113**, 75.
— 1939. Oligochaeta. In: The Percy Sladen Trust Expedition to Lake Titicaca in 1937. *Trans. Linn. Soc. Lond. Ser. III*, **1**, 81.
CHEN, Y. 1940. Taxonomy and faunal relations of the limnetic Oligochaeta of China. *Contr. biol. Lab. Sci. Soc. China (Zool.)*, **14**, 1.
CLAPARÈDE, E. R. 1862. Recherches anatomiques sur les oligochètes. *Mém. Soc. Phys. Hist. nat. Genève*, **16**, 217.
COOK, D. G. 1967a. Studies on the Lumbriculidae (Oligochaeta). Ph.D. Thesis. University of Liverpool.
— 1967b. Studies on the Lumbriculidae (Oligochaeta) in Britain. *J. zool. Lond.*, **153**, 353.
— 1968. The genera of the family Lumbriculidae Vejdovsky and the genus *Dorydrilus* Piguet (Annelida, Oligochaeta). *J. zool. Lond.*, **156**, 273.
— 1969. The Tubificidae (Annelida, Oligochaeta) of Cape Cod Bay, with a taxonomic revision of the genera *Phallodrilus* Pierantoni, 1902, *Limnodriloides* Pierantoni, 1903, and *Spiridion* Knollner, 1935. *Biol. Bull*, **136**, 9.
CORDERO, E. H. 1948. 1948. Zur Kenntnis der Gattung *Opisthocysta* Cern. (Archiolgochaeta). *Comun. zool. Mus. Hist. nat. Montev.*, **2** (50), 1.
DAHL, I. O. 1960. The Oligochaete fauna of 3 Danish brackish water areas. *Meddr. Danm. Fisk.-vg. Havunders*, **2**, 1.
DE BOCK, M. 1901. Observations anatomiques et histologiques sur les Oligochètes specialment sur leur système musculaire. *Revue suisse Zool.*, **9**, 1.
DEHORNE, A. 1923. Filaments végétatifs et appareil secrétant dans les néphridies de *Stylaria lacustris. C. r. Séanc. Soc. Biol.*, **89**, 173.
— 1925. Aspects du Chondriome de *Stylaria lacustris* Linn. *Cellule*, **36**, 359.
DEHORNE, L., 1916. Les Naidimorphes et leur reproduction asexuél. *Archs Zool. exp. gén.*, **56**, 25.
DIXON, G. C. 1915. Tubifex. *L.M.B.C. Memoirs*, **23**. London. *Proc. Trans. Lpool biol. Soc.*, **29**, 303.
EISEN, G. 1879. Preliminary notes on genera and species of Tubificidae. *Bih. K. svenska Vetensk Akad. Handl.*, **16**, 1.
— 1881. Eclipidrilidae and their anatomy. A new limicolide Oligochaeta. *Nova Acta. R. Soc. Scient. Upsal.*, **11**, 1.
— 1886. Oligochaetological researches. XX. *Rep. U.S. Commnr. Fish*, **1886**, 879.
— 1888. On the anatomy of *Sutroa rostrata*, a new annelid of the family Lumbriculina. *Mem. Calif. Acad. Sci.*, (2) **1**, 1.
— 1895. Pacific Coast Oligochaeta I. *Mem. Calif. Acad. Sci.*, (2) **1**, 122.
FORBES, S. A. 1890. On an American Earthworm of the Family Phreoryctidae. *Bull. Ill. St. Lab. nat. Hist.*, **3**, 107.
GODDARD, E. J. 1909a. Contribution to a further knowledge of Australian Oligochaeta. I. Descriptions of two species of a new genus of Phreodrilidae. *Proc. Linn. Soc. N.S.W.*, (1908) **33**, 767.
— 1909b. Contribution to a further knowledge of Australian Oligochaeta. II. Description of a Tasmanian Phreodrilid. *Proc. Linn. Soc. N.S.W.*, (1908) **33**, 845.

GODDARD, E. J., AND MALAN, D. E. 1913a. Contributions to a knowledge of South African Oligochaeta. I. On a Phreodrilid from Stellenbosch Mountain. *Trans. R. Soc. S. Afr.*, **3**, 231.

— — 1913b. Contributions to a knowledge of South African Oligochaeta. II. Description of a new species of *Phreodrilus*. *Trans. R. Soc. S. Afr.*, **3**, 242.

GOODRICH, E. S. 1895. On the structure of *Vermiculus pilosus*. *Q. Jl microsc. Sci.*, **37**, 253.

HAFFNER, K. 1927. Untersuchungen über die Morphologie und Physiologie des Blutgefäss-systems von *Lumbriculus variegatus* Müll. *Z. wiss. Zool.*, **130**, 1.

HESSE, R. 1894. Zur vergleichenden Anatomie der Oligochäten. *Z. wiss. Zool.*, **58**, 394.

— 1902. Untersuchungen über die Organe der Lichtempfindung bei niederen Thieren. VIII. Weitere Thatsachen und Allgemeines. *Z. wiss. Zool.*, **72**, 565.

HRABĚ, S. 1927. *Rhynchelmis komárekei*, eine neue Lumbriculiden-Art aus Macedonien. *Zool. Anz.*, **71**, 170.

— 1931. Die Oligochaeten aus den Seen Ochrida und Prespa. *Zool. Jb. (Syst.)*, **61**, 1.

— 1935. Über *Moraviodrilus pygmaeus* n.gen., n.sp., *Rhyacodrilus falciformis* Br., *Ilyodrilus bavaricus* Oschm. und *Bothrioneurum vejdovskyanum* St. *Spisy vydáv. přír. Fab. Masaryk. Univ.*, No **209**, 1.

— 1938. *Trichodrilus moravicus* und *claparedei*, neue Lumbriculiden. *Zool. Anz.*, **15**, 73.

— — 1939a. O vývohi samčiko vývodnéo aparátu u néktorých nitěnek a žižalik. *Sb. přír Klubu Třebiči.*, **3**, 56.

— 1939b. Vodni Oligochaeta z Vysokých Tater. (Oligochètes aquatiques des Hautes Tatras.) *Vest. csl. zool. Spol.*, **6-7**, 209.

— 1958. A new species of Oligochaeta from the Southwest of France. *Notes biospéol.*, **13**, 171.

— 1960. Oligochaeta limicola from the collection of Dr. S. Husmann. *Spisy přír. Fak. Univ. Brne.*, No. **415**, 245.

— 1961. Dva nové druky rodu *Rhychelmis* ze Slovenska. *Spisy přír. Fak. Univ. Brne.*, No. **421**, 129.

ISOSSIMOFF, V. V. 1926. Zur Anatomie des Nervensystems der Lumbriculiden. *Zool. Jb. (Anat.)*, **48**, 365.

— 1948. The Lumbriculidae of Lake Baikal. Synopsis of Doctors Thesis, Kazan.

— 1962. The Oligochaetes of the family Lumbriculidae of Lake Baikal. *Trans. Limn. Inst., Acad. Sci., U.S.S.R., Siberian Section*, **1** (21)(1), 3.

JACKSON, A. 1931. The Oligochaeta of South-Western Australia. *J. Proc. R. Soc. West. Aust.*, **17**, 71.

JAMIESON, B. G. M. 1968a. A taxonometric investigation of the Alluroididae (Oligochaeta). *J zool. Lond.*, **155**, 55.

— 1968b. Macquaridrilus: a new genus of Tubificidae (Oligochaeta) from Macquarie Island. *Pap. Dep. Zool. Univ. Qd.*, **3** (5), 55.

KEYL, F. 1913. Beiträge zur Kenntnis von *Branchiura sowerbyi* Beddard. *Z. wiss Zool.*, **107**, 199.

KNOLLNER, F. H. 1935. Ökologische und systematische Untersuchungen über litorale und marine Oligochäten der Kieler Bucht. *Zool. Jb. (Syst.)*, **66**, 425.

LIEBERMANN, A. 1932. Studien über die Topographie und den Bewegungsmechanismus des ventralen Borstenfollikels von *Stylaria lacustris* L. *Zool. Jb. (Physiol.)*, **50**, 151.

LIEBMANN, E. 1931. Weitere Untersuchungen über das Chloragogen. *Zool. Jb. (Anat.)*, **54**, 417.

— The coelomocytes of Lumbricidae. *J. Morph.*, **71**, 221.

MARCUS, E. 1943. Sôbre Naididae do Brazil. *Bolm. Fac. Filos. Ciênc. Univ. S Paulo (Zool. 7)*, **32**, 3.

— 1944. Sôbre Oligochaeta limnicos do Brazil. *Bolm. Fac. Filos. Ciênc. Univ. S Paulo (Zool. 8)*, **43**, 5.

MEHRA, H. R. 1922. Two new Indian species of the little-known genus *Aulodrilus* Bretscher of the aquatic Oligochaeta belonging to the family Tubificidae. *Proc. zool. Soc. Lond.*, **1922**, 943.

— 1924. The genital organs of *Stylaria lacustris*. *Q. Jl microsc. Sci.*, **68**, 147.

MEHRA, H. R. 1925. The atrium and the prostate gland in the Microrili. Q. Jl. microsc. Sci. **69**, 399.

MEYER, F., 1916. Untersuchungen über den Bau und die Entwicklung des Blutgefässytems bei *Tubifex tubifex* (Müll.) *Vjschr. naturf. Ges. Zürich*, **60**, 592.

MICHAELSEN, W. 1901. Oligochaeten der Zoologischen Museen zu St. Petersburg und Kiev. *Izv. imp. Akad. Nauk.*, (5) **15**, 145.

— 1903a. Eine neue Haplotaxiden-Art und andere Oligochäten aus dem Telezkischen See im Nördlichen Altai. *Verh. naturw. Ver. Hamb.*, **10**, 1.

— 1903b. Die Oligochäten der Deutschen Tiefsee-Expedition nebst Erörterung der Terricolenfauna oceanischer Inseln, insbesondere der Inseln des subantarktischen Meeres. *Wiss. Ergebn. dt. Tiefsee-Exped. "Valdivia"*, **3**, 131.

— 1905. Die Oligochaeten des Baikal-Sees. *Wiss. Ergebnisse Zool. Exped. Baikal*—See Prof. Alexis Korotneff J. 1900-1902. **1**, 1 Kiew & Berlin.

— 1907. Oligochaeta. *Fauna Südwest—Aust.*,1, 124.

— 1925. Ein Süsswasser-Hohlenoligochät aus Bulgarien. *Mitt. naturh. Mus. Hamb.*, **41**, 85.

— 1926. *Agrodrilus vermivorus* aus dem Baikal-See, ein Mittelglied zwischen typischen Oligochäten und Hirudeen. *Mitt. naturh. Mus. Hamb.*, **42**, 1.

MILLOTT, N. 1943. The visceral nervous system of the earthworm. I. Nerves controlling the tone of the alimentary canal. *Proc. R. Soc. B*, **131**, 271.

MOORE, J. P. 1905. Some marine Oligochaeta of New England. *Proc. Acad. nat. Sci. Philad.*, **57**, 373.

MRAZEK, A. 1900. Die Samentaschen von *Rhynchelmis*. *Sber. K. Böhm. Ges. Wiss.*, **35**, 1.

— 1926. Zur Morphologie der Nephridien der Annulaten. *Vest. csl. zool. Spol. 1926*, **2**, 1-8.

NAIDU, K. V. 1962. Studies on the Freshwater Oligochaeta of South India. I. Aeolosomatidae and Naididae. *J. Bombay nat. Hist. Soc.*, **59** (2), 520.

— 1965. Studies on the Freshwater Oligochaeta of South India. II. Tubificidae. *Hydrobiologia*, **26**, 463.

NOMURA, E. 1913. On two species of aquatic Oligochaeta. *Limnodrilus gotoi* Hatai and *Limnodrilus willeyi*, n.sp. *J. Coll. Sci. imp. Univ. Tokyo*, **35** (4), 1.

— 1915 On the aquatic oligochaete *Monopylephorus limosus* (Hatai). *J. Coll. Sci. imp. Univ. Tokyo*, **35** (9), 1.

— 1926. On the aquatic oligochaete, *Tubifex hattai*, n.sp. *Sci. Rep. Tohoku Univ.*, (4) **1**, 193.

— 1929. On *Limnodrilus motomurai*, nov. sp., an aquatic oligochaete. *Annotnes zool. Jap.*, **12**, 131.

POINTNER, H. 1911. Beiträge zur Kenntnis der Oligochaetenfauna der Gewässer von Graz. *Z. wiss. Zool.*, **98**, 626.

ROSA, D., 1898. I pretesi rapporti genetici tra i linfociti ed il cloragogeno. *Atti Accad. Sci. Torino*, **33**, 612.

SAJOVIK, G. 1907. Anatomie, Histologie und Ersatz der Borstenorgane bei *Lumbricus*. *Arb. zool. Inst. Univ. Wien*, **17**, 1.

SMITH, F. 1900a. Notes on species of North American Oligochaeta. III. List of species found in Illinois, and descriptions of Illinois Tubificidae. *Bull. Ill. St. Lab. nat. Hist.*, **5** (10), 441.

— 1900b. Notes on species of North American Oligochaeta. IV. On a new lumbriculid genus from Florida, with additional notes on the nephridial and circulatory systems of *Mesoporodrilus asymmetricus* Smith. *Bull. Ill. St. Lab. nat. Hist.*, **5**, 459.

— 1918. A new North American oligochaete of the genus Haplotaxis. *Bull. Ill. St. Lab. nat. Hist.*, **13**, 43.

SOUTHERN, R. 1909. Contributions towards a monograph of the British and Irish Oligochaeta. *Proc. R. Ir. Acad.*, **27** (B, 8), 119.

SPERBER, C. 1948. A taxonomical study of the Naididae. *Zool. Bidr. Upps.*, **28**, 1.

STEPHENSON, J. 1907a. Description of an Oligochaete worm, allied to *Chaetogaster*. *Rec. Indian Mus.*, **1**, 133.

— 1907b. Description of two freshwater Oligochaete worms from the Punjab. *Rec. Indian Mus.*, **1**, 233.

STEPHENSON, J. 1912. On a new species of *Branchiodrilus* and certain other aquatic Oligochaeta, with remarks on cephalisation in the Naididae. *Rec. Indian Mus.*, **7**, 219.

— 1915. On a rule of proportion observed in the setae of certain Naididae. *Trans. R. Soc. Edinb.*, **50**, 783.

— 1917. Aquatic Oligochaeta from Japan and China. *Mem. Asiat. Soc. Beng.*, **6**, 83.

— 1922a. On the septal and pharyngeal glands of the Microdrili (Oligochaeta). *Trans. R. Soc. Edinb.*, **53**, 241.

— 1922b. On some Scottish Oligochaeta, with a note on encystments in a common freshwater oligochaete, *Lumbriculus variegatus* (Mull.). *Trans. R. Soc. Edinb.*, **53**, 277.

— 1926. The sexual organs of the freshwater Oligochaete *Naidium breviseta* (A. G. Bourne). *Ann. Mag. nat. Hist.*, (9) **18**, 290.

— 1930. *The Oligochaeta.* Oxford. Clarendon Press.

STOLTE, H. A. 1933. Oligochaeta. In *Bronns Klassen und Ordnungen des Tierreichs.* **4** (3)(3)(1), 1.

STOUT, J. D. 1958. Aquatic Oligochaetes occurring in forest litter. II. *Trans. R. Soc. N.Z.*, **85**, 289.

TIMM, R. 1883. Beobachtungen an *Phreoryctes Menkeanus*, Hoffm., und *Nais*. Ein Beitrag zur Kenntniss der Fauna Unterfrankens. *Arb. Zool. Inst. Würzburg*, **6**, 109.

d'UDEKEM, J. 1853. Histoire naturelle du *Tubifex* des Ruisseaux. *Mém. Acad. r. Sci. Lett. Belg.*, **26**, 3.

VEJDOVSKY, F. 1884. *System und Morphologie der Oligochaeten.* Prague.

YAMAGUCHI, H. 1936. Studies on the aquatic Oligochaeta of Japan. I. Lumbriculids from Hokkaido. *J. Fac. Sci. Hokkaido Univ.*, **5**, 73.

— 1953. Studies on the aquatic Oligochaeta of Japan. VI. A systematic report with some remarks on the classification and phylogeny of the Oligochaeta. *J. Fac. Sci. Hokkaido Univ.*, **11**, 277.

YAMAMOTO, G., AND OKADO, K. 1940. *Tubifex* (*Peloscolex*) *nomurai*, sp. nov. *Sci. Rep. Tohoku Univ.*, (4) **15**, 427.

REFERENCES FOR GLOSSOSCOLECIDAE

(See Chapter 15 for references not listed here)

AVEL, M. 1959. In Grassé, P. *Traité de Zoologie*, **5**, Fasc. 1. Masson.

BAHL, K. N. 1947. Excretion in the Oligochaeta. *Biol. Rev.*, **22**, 109.

BEADLE, L. C. 1957. Respiration of the African swampworm *Alma emini* Mich. *J. exp. Biol.*, **34**, 1.

BEDDARD, F. E., AND FEDARB, S. M. 1899. Notes upon two earthworms, *Perichaeta biserialis* and *Trichochaeta hesperideum*. *Proc. zool. Soc. Lond.*, **1899**, 803.

BENHAM, W. B. 1896. On *Kynotus cingulatus*, a new species of earthworm from Imerina in Madagascar. *Q. Jl microsc. Sci.*, **38**.

CARTER, G. S., AND BEADLE, L. C. 1932. The fauna of swamps of the Paraguayan Chaco in relation to its environment. III. Respiratory adaptations in the Oligochaeta. *J. Linn. Soc. (Zool.)*, **37**, 379.

CERFONTAINE, P. 1890. Recherches sur le système cutané et sur le systéme musculaire du Lombric terrestre *Archs. Biol., Paris*, **10**, 327.

COGNETTI, L. 1911. Ricerche sulla distruzione fisiologica dei prodotti sessuali maschili. *Memorie Accad. Sci. Torino*, (2) **61**, 293.

DE RIBAUCOURT, E. 1901. Étude sur l'anatomie comparée des Lombricides. *Bull. scient. Fr. Belg.*, **35**, 211.

GATES, G. E. 1958. On a hologynous species of the earthworm genus Diplocardia, with comments on oligochaeta hologyny and consecutive hermaphroditism. *Am. Mus. Novit.*, **1886**, 1-9.

GRESSON, R. 1927. On the structure of the branchiae of the gilled oligochaete *Alma nilotica*. *Ann. Mag. nat. Hist.*, 9, **19**, 348.

GROVE, A. J., AND COWLEY, L. F. 1927. The relation of the glandular elements of the clitellum of the brandling worm (*Eisenia foetida*, Sav.) to the secretion of the cocoon. *Q. Jl microsc. Sci.*, **71**, 31.

IZOARD, F. 1952. Développement et structure de la musculature longitudinale des Lombriciens. Diplome d'études sup., Bordeaux.

— 1958. La structure de la musculature longitudinale des Lombriciens. *C. r. Acad. Sci., Paris*, **246**, 1598.

JAMIESON, B. G. M. 1968. A new species of *Glyphidrilus* (Microchaetidae: Oligochaeta) from East Africa. *J. nat. Hist.*, **2**, 387.

JANDA, V. 1926. Die Veränderungen des Geschlechtscharakters und die Neubildung des Geschlechtsapparates von *Criodilus lacuum* Hoffm. unter künstlichen Bedingungen. *Biol. Zbl.*, **46**, 200.

OMODEO, P. 1956b. Sistematica e distribuzione geografica degli Hormogastrinae (Oligocheti). *Arch. Bot. Biogeog. Ital.*, **32** (ser. 4), 1, Fasc. 4, 159.

PICKFORD, G. E. 1940. An account of the anatomy of a giant earthworm from Ecuador. *Turtox News*, **18**, 7, 6 pages.

PICADO, C. 1913. Les Broméliacées epiphytes considérées comme milieu biologique. *Bull. scient. Fr. Belg.*, **47**, 215.

PITZORNO, M. 1899. Sull' apparato circulatorio dell *Hormogaster redii* Rosa. *Monitore zool. ital.*, **10** suppl. 47.

POOL, G. 1937. *Eiseniella tetraedra* (Sav.) ein Beitrag zur vergleichenden Anatomie und Systematik der Lumbriciden. *Acta. Zool.*, **18**, 1.

RHODE, E. 1885. Die Muskulatur der Chaetopoden. *Zool. Anz.* **8**, 135.

ROSA, P. 1889. Sulla struttura dello *Hormogaster redii* del Dott. Daniele Rosa. *Memorie Accad. Sci. Torino*, (2) **39**, 49.

— 1891. Die exotischen Terricolen des K.K. naturh. Hofmuseums. *Annln naturh. Mus. Wien*, **6**, 379.

STOLTE, H. A. 1933. In *Bronns Klassen und Ordnungen des Tierreichs*. **4**, Abt. 3. B. 3. L. 2: 161.

WOODWARD, M. F. 1892. Description of an abnormal earthworm possessing seven pairs of ovaries. *Proc. zool. Soc. Lond.*, **1892**, 184.

— 1893. Further observations on variations in the genitalia of British earthworms. *Proc. zool. Soc. Lond.*, **1893**, 319.

2

EMBRYOLOGY

D. T. ANDERSON
School of Biological Sciences
The University of Sydney, Australia

INTRODUCTION

The developmental morphology of the aquatic oligochaetes has been neglected for more than thirty years, a victim of the decline in descriptive embryology that followed the rise of experimental embryology in the 1930s. Problems unresolved in 1938 remain problems today, in spite of greatly improved knowledge of the systematics, ecology and range of variation of aquatic oligochaetes. That any comment of present significance can be made on oligochaete embryology is, indeed, a testimony to the care and accuracy with which the few devotees of the subject in the late 1900s and the first three decades of this century pursued and presented their work. Kowalevsky and Vejdovsky, great names in the post-Darwinian era of comparative embryology, laid the foundations of knowledge of the embryonic development of aquatic oligochaetes. Davydov, Svetlov, Meyer and, above all, Penners, built on these foundations during the technologically more competent years that followed the First World War, to establish facts that are still the best available to us. Since these facts were all determined before the introduction into common usage of any of the histological techniques now regarded as competent, we must remain a little sceptical of new discussion of oligochaete embryos until further extension and revision of the subject become available.

At the same time, as knowledge advances and concepts change, old facts

73

frequently take on new meanings. Recent progress in polychaete embryology, against a background of improved understanding of the generalities of embryonic development, has reshaped our thinking on polychaete embryos (Anderson, 1966a) in a way which sets the classical data of oligochaete embryology in a new mould (Anderson, 1966b). The vagaries of cell lineage, which so confused the comparison of embryos of different species among the oligochaetes, even for Penners (1922, 1924, 1929), can now be eschewed in favour of a simple interpretation of the formation and fates of presumptive areas in development, from which the variations manifested in different families emerge as modifications associated with differing modes of embryonic nutrition.

As is well known, all aquatic oligochaetes deposit their eggs in cocoons and their development proceeds directly to hatching as a juvenile with numerous segments. Comparative considerations dictate that we interpret as the primitive condition one in which the eggs are relatively large and yolky, in which development is sustained by the yolk, and in which cleavage, gastrulation and organogeny are adapted in a particular way to the presence of voluminous yolk reserves. This type of development is manifested by the tubificids and lumbriculids and has been studied embryologically in *Tubifex*, *Rhynchelmis* and, to a lesser extent, *Peloscolex* and *Limnodrilus*.* The embryo manifests no precociously differentiated structures having a temporary role vis-à-vis the surrounding albumen, and the role of the latter in embryonic nutrition is not clear.

Also, in several families of aquatic oligochaetes, the egg has less yolk and the embryo is adapted to feeding on the albuminous contents of the cocoon. Extremely little is known about embryonic development in these families, other than a few investigations among the naidids and a little on branchiobdellids and aeolosomatids, but the available facts give some indication of modes of secondary modification and point to many fascinating problems awaiting investigation.

CLEAVAGE

THE FIRST FIVE CLEAVAGE DIVISIONS IN TUBIFEX AND RHYNCHELMIS The spiral pattern of cleavage is retained in *Tubifex* and *Rhynchelmis*, but is modified in ways which relate to the magnitude of the large yolky egg and its development in a protective cocoon. Both in *Tubifex* (Penners, 1922, 1924) and in *Rhynchelmis* (Vejdovsky, 1886, 1888-92; Bergh, 1890; Penners, 1922; Svetlov, 1923), the first two cleavage divisions are at right angles along the animal-vegetal axis of the egg, the next four perpendicular to this axis and, except for a precocious onset of bilateral cleavage in the D-quadrant, successively dexiotropic, laeotropic, dexiotropic and laeotropic in the classical spiral cleavage sequence. In association with the large size of the egg, 300-500 μ in *Tubifex*, 1000 μ in

---

* Includes *L. newaensis*, now placed in *Tubifex*—R.O.B.

Rhynchelmis, the first cleavage division is very unequal, AB being much smaller than CD, and the second division maintains this inequality in the D-quadrant, D remaining large while C is only a little larger than A and B.

In *Tubifex,* at the four cell stage, D is dorsal and A, B and C are left-lateral, ventral and right-lateral respectively, relative to the antero-posterior axis of the embryo (Fig. 2. 1A). Later cleavages then lead to displacement of the B-quadrant cells to an anterior and the D-quadrant cells to a posterior position. In *Rhynchelmis,* B is already anterior, and D already posterior, at the four-cell stage (Fig. 2. 1B).

Associated with the large size of the D-cell, early cleavages after the four-cell stage manifest precocious division of the D-quadrant cells, and to a lesser extent of the C-quadrant cells, so that the cells of each quartette cut off at the third and subsequent divisions are formed in succession, the D-quadrant leading, the A- and B-quadrants lagging.

The third cleavage division is dexiotropic and very unequal, a first quartette of very small cells, 1a-1d, being cut off from stem cells 1A-1D. In *Tubifex,* 1a-1d lie bilaterally on either side at the anterior pole, 1d and 1a to the left, 1c and 1b to the right (Fig. 2. 1C). In *Rhynchelmis,* the cells are similarly bilateral, but dorsally placed (Fig. 2. 1D).

The laeotropic fourth division begins with the division of 1D into 2d and 2D. The cell 2d is much larger than any cell of the first quartette, though in *Tubifex* (Fig. 2. 1E) it is only about one quarter, and in *Rhynchelmis* (Fig. 2. 1F) less than a quarter, of the size of 2D. In *Tubifex,* 2d pushes forward antero-dorsally between 1d and 1c to lie in the dorsal midline, with the first quartette cells as an arc around its anterior face. In *Rhynchelmis,* 2d lies mid-dorsally behind the first quartette. 2a, 2b and 2c are now cut off laeo-tropically as small cells, so that 2a is added to the left, 2b to the front and 2c to the right of the micromeres bordering 2d.

At the same time, the fifth division begins in the D-quadrant. Division of 2D differs in the two embryos. In *Tubifex,* the resulting 3d cell is a small cell cut off dexiotropically and added to the left end of the micromere arc (Fig. 2. 1E). In *Rhynchelmis,* 3d is cut off in the dorsal midline as a cell directly behind and equal in size to 2d (Fig. 2. 1F). The latter cell, on the other hand, behaves similarly in both species, dividing unequally, with its smaller daughter pushing forwards into the micromere arc while its larger daughter remains mid-dorsal. This is repeated at least twice in subsequent division, a cell being added on each occasion to the micromere arc, leaving the large cell in the dorsal midline. The final designation of the large cell as $2d^{111}$ in *Tubifex* and $2d^{22}$ in *Rhynchelmis* does not belie the fundamental similarity of the process in the two species.

Meanwhile, the fifth cleavage divisions continue in the remaining quadrants, the stem cells 2A, 2B and 2C cutting off the small cells 3a, 3b and 3c to lie on the left, anteriorly and on the right respectively, at the outer margin of the micromere arc. In *Tubifex,* the cells 1a and 1c, on the right and left of this arc, also divide equally, increasing the number of micromeres. In

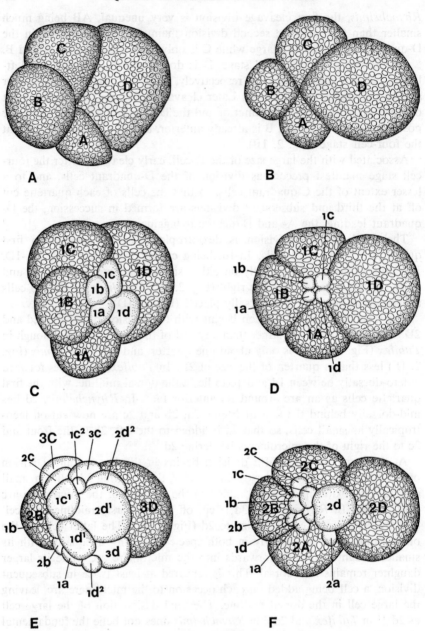

Fig. 2.1.

Tubifex. A. 4-cell stage, anterior view; C. 8-cell stage, antero-dorsal view; E. during formation of the third quartette, dorsal view;

Rhynchelmis. B. 4-cell stage, dorsal view; D. 8-cell stage, dorsal view; F. after formation of the second quartette, dorsal view.

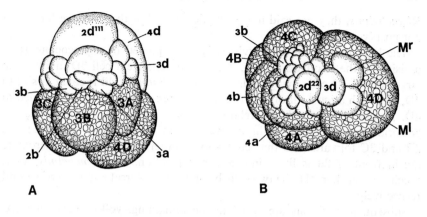

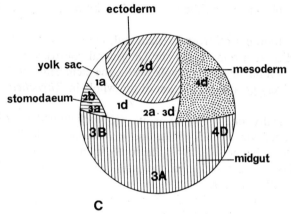

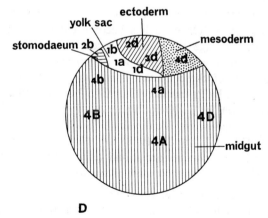

Fig. 2.2.

Tubifex. A. after formation of 4d, left anterior view; C. presumptive areas, lateral view;

Rhynchelmis. B. after formation of the mesoteloblasts, dorsal view; D. presumptive areas, lateral view.

Rhynchelmis, they are said not to divide again, but the evidence for this is not positive.

The later part of the fifth division is accompanied by a precocious sixth division in the D-quadrant (Fig. 2. 2A, 2B). The stem cell 3D divides into large 4d and 4D cells, lying in the midline. In *Tubifex*, 4d is equal in size to 4D, although almost all of the yolk in the parent 3D stem cell is confined to 4D. In association with the relatively large size of 4d, this cell fills the posterior midline of the embryo, and 4D moves ventrally, displacing 3A, 3B and 3C forwards. 3B thus comes to lie anteriorly, while the micromere arc in front of 2d is lifted into a dorsal position. In *Rhynchelmis*, 4d is much smaller than 4D, the two cells being postero-dorsal and postero-ventral respectively.

Subsequent divisions are equal in the remaining, yolky, stem cells 3A, 3B and 3C, simply increasing the number of yolky cells. 4D now divides in the same way.

Once 4d has been cut off from its stem cell, the embryos of *Tubifex* and *Rhynchelmis* display a remarkably similar general construction. Each has:

1. Ventrally, a number of large, yolky cells, the stem cells 3A-3C and 4D, forming a relatively greater proportion of the whole in *Rhynchelmis* than in *Tubifex*.

2. Postero-dorsally, one (*Tubifex*) or two (*Rhynchelmis*) large cells in the midline, segregated via 2d in *Tubifex* and via 2d and 3d in *Rhynchelmis*, surrounded anteriorly and laterally by an arc of micromeres cut off from the stem cells as first, second and third quartette cells, through a similar series of divisions in both embryos.

3. Posteriorly, in the dorsal midline, a large cell 4d, relatively greater in size in *Tubifex* than in *Rhynchelmis*.

THE PRESUMPTIVE AREAS OF THE BLASTULA IN TUBIFEX AND RHYNCHELMIS   Once 4d has become cut off from its stem cell early in the sixth cleavage division, it becomes possible to assign the blastomeres to presumptive areas of specific subsequent fate (Fig. 2. 2C, 2D). The designation of these areas is deducible from the facts of later development described by Penners (1922, 1924), Penners and Stablein (1930) and Meyer (1929, 1931), for *Tubifex*, Penners (1929) for the embryologically similar *Peloscolex benedeni* and Vejdovsky (1888-92), Bergh (1890), Penners (1922), Svetlov (1923) and Ivanov (1928) for *Rhynchelmis*. Differences between *Tubifex* and *Rhynchelmis* are found only in the constitution of the presumptive ectoderm. The presumptive areas are:

1. A dorsal area of presumptive ectoderm.
2. At the anterior edge of this area, in the dorsal midline, an area of presumptive stomodaeum.
3. A posterior area of presumptive mesoderm, incorporating presumptive germ cells.
4. A ventral area of presumptive midgut.

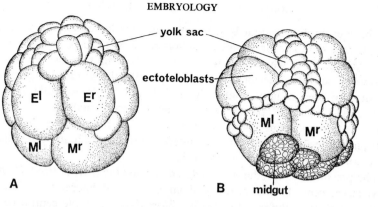

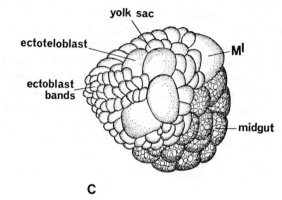

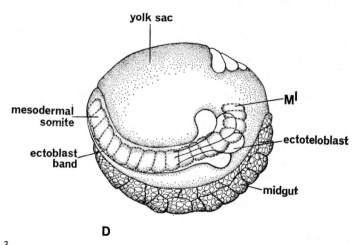

Fig. 2.3.

Tubifex. A. early ectoteloblast formation, postero-dorsal view; B. completion of ectoteloblast formation, posterior view; C. early germ band formation, lateral view; D. stage with 17 pairs of somites, lateral view.

The presumptive midgut is segregated into the large, ventral, yolky cells 3A-3C and 4D. The presumptive mesoderm is segregated posteriorly into the single cell, 4d, which, before it assumes its definitive mesodermal fate, throws off into the interior a pair of small cells, the primordial germ cells. 4d then divides equally bilaterally in M^1 and M^r, mesoteloblasts which subsequently bud off paired, ventro-lateral mesodermal bands. The presumptive stomodaeum comprises a group of micromeres at the anterior margin of the micromere arc, descended mainly from 2b and 3b, with possible contributions in *Tubifex* from 3a.

The presumptive ectoderm is made up of one (*Tubifex*) or two (*Rhynchelmis*) large, central cells surrounded anteriorly and laterally by an arc of micromeres. In *Tubifex*, and also in *Peloscolex*, the single central cell, the main product of 2d, divides equally bilaterally into a pair of cells, each of which gives rise by further division to a transverse row of four ectoteloblasts (Fig. 2. 3A, 3B). In *Rhynchelmis*, the anterior central cell, the main product of 2d, divides into a pair of cells which migrate laterally, while the posterior cell, 3d, divides into a pair of cells each of which divides again into three cells in a transverse row, with the large 2d cell at the end of it. The eventual product, although resulting from a slightly different series of divisions, is again a row of four ectoteloblasts on either side of the dorsal midline. Each ectoteloblast now begins to bud off a row of small cells forwards, external to the corresponding mesodermal band (Fig. 2. 3C).

While the ectoteloblasts are forming, the arc of micromeres multiplies and spreads:

1. Back mid-dorsally, as a strip separating the two ectoteloblast groups.
2. Up towards the dorsal midline from the ends of the arc, behind the ectoteloblast groups.

The back-growth mid-dorsally finally meets the upgrowth from the sides. When this stage is attained, the following subareas can be seen to have segregated within the presumptive ectoderm:

1. The ectoteloblasts, incorporating presumptive material of the prostomium and cerebral ganglia and of the major part of the ectoderm of the peristomium and trunk segments, and already beginning to bud off this material as ectodermal bands.
2. The presumptive temporary yolk sac ectoderm, mid-dorsally and laterally, later incorporated into the segmental epithelium.
3. The presumptive pygidial epithelium, incorporating presumptive proctodaeum, posterior to the ectoteloblasts. This sub-area, as noted above, has a paired bilateral origin and is also of temporary yolk sac function at this stage.

The adaptation of the majority of the ectoderm to a temporary yolk-sac function is associated with the large volume of presumptive midgut soon to be accommodated within the interior of the embryo.

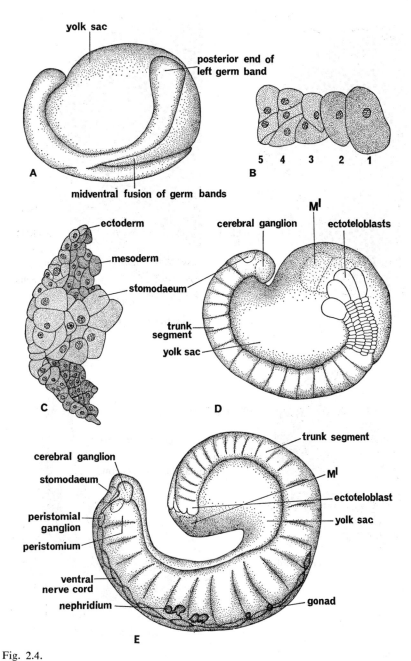

Fig. 2.4.
Tubifex. A. diagrammatic lateral view during mid-ventral fusion of germ bands;
B. successive early stages in somite formation; C. early ingrowth of stomodaeum,
frontal section; D. stage with 24 pairs of somites, lateral view; E. stage with 36 pairs
of somites, lateral view.

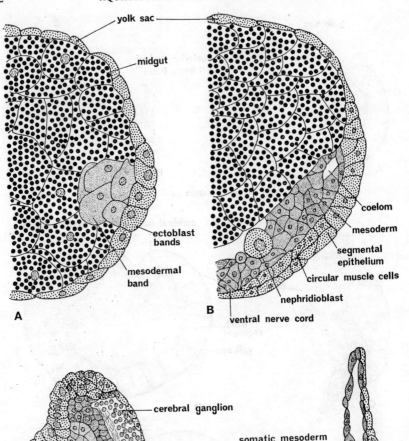

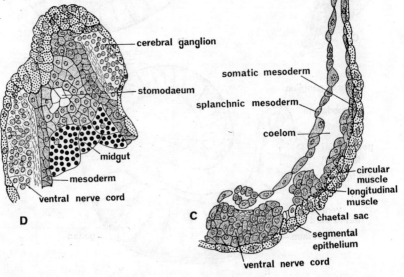

Fig. 2.5.
Tubifex. A. transverse section through middle region of a stage similar to Fig. 3D;
B. transverse section through middle region of a stage similar to Fig. 4D; C.
transverse section through middle region of a stage similar to Fig. 4E; D. sagittal
section through anterior end of a stage similar to Fig. 4D.

CLEAVAGE AND PRESUMPTIVE AREAS IN OTHER AQUATIC OLIGOCHAETES As already mentioned, the embryonic development of aquatic oligochaetes other than the tubificids and *Rhynchelmis* remains poorly known. The development of *Bdellodrilus* was studied in part by Tannreuther (1915), and on present indications differs little from that of tubificids. In the naidids and aeolosomatids, however, the eggs are less yolky and development, in so far as we know anything at all about it, is severely modified. So little is known of aeolosomatid development at present (Bunke, 1967) that it is scarcely possible to dwell on the modifications entailed, but the naidids have been slightly better documented in the works of Svetlov (1926) and Davydov (1942). Their fundamental modification is the formation, through cleavage, of a provisional cellular envelope within which the ectoteloblasts, M-cells and presumptive midgut become established and express their fate in the same general manner as in *Tubifex*, save only for the secondary modification of midgut development in association with the precocious ingestion of albumen. It seems highly probable that the external provisional envelope is a secondary modification of temporary yolk sac ectoderm, and that it has an absorptive function, rapidly supplemented by further, precociously formed, temporary organs subserving feeding on the ambient nutrient albumen of the cocoon.

Branchiobdellidae:

A consideration of cleavage in *Bdellodrilus* points the way to the basis of the developmental modifications expressed in naidids. Tannreuther (1915) made a detailed study of cleavage in *Bdellodrilus*, although his description of later development is more fragmentary and inconclusive. In part, the neglect to which Tannreuther's work has been subjected is a consequence of the somewhat obscure numerical system that he adopted for the designation of successive blastomeres. When his results are reinterpreted in accordance with the description of cleavage and presumptive areas given above for *Tubifex* and *Rhynchelmis*, the similarities and differences between *Bdellodrilus* and the larger-egged species become abundantly clear.

The egg of *Bdellodrilus*, laid in an individual small, stalked cocoon, is yolky, but smaller than that of *Tubifex*, being about $280 \times 240 \mu$ in diameter. As we shall see, the reduction in size is primarily a reduction in the proportion of midgut-forming material in the egg. Early cleavage retains exactly the pattern expressed in *Tubifex* and *Rhynchelmis*, with the quadrants being posterior (D), anterior (B), left (A) and right (C) at the four cell stage, as in *Rhynchelmis*, and the first quartette of micromeres 1a-1d being cut off dorsally (Fig. 2. 6A, 6B). 1D remains much the largest cell at the eight-cell stage.

With the onset of the fourth cleavage, the first aberration is observed (Fig. 2. 6C). As 1D divides at the posterior end, the resulting postero-dorsal 2d is considerably larger than its postero-ventral sister, 2D, in contrast to *Tubifex* and *Rhynchelmis*, where 2d is the smaller cell. The stem cell 2D is pushed forward into a left ventral position during this division, displacing 1A forwards and upwards to join 1B at the anterior end as a bilateral pair.

Disproportionately large though it is, 2d retains the fate previously described for it in *Tubifex*, as the mother cell of paired rows of four ectoteloblasts. The remainder of the fourth cleavage is also typical, as 1C, 1B and 1A in succession add micromeres 2c, 2b and 2a to the dorsal micromere cap. This cap is further supplemented by three cells added to its posterior

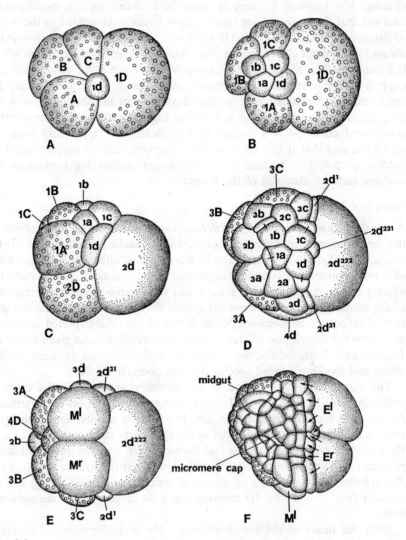

Fig. 2.6.

Bdellodrilus. A. 5-cell stage, dorsal view; B. 8-cell stage, dorsal view; C. 9-cell stage, lateral view; D. after formation of 4d, dorsal view; E. during division of 4d into M^l and M^r, ventral view; F. early ectoteloblast formation, dorsal view.

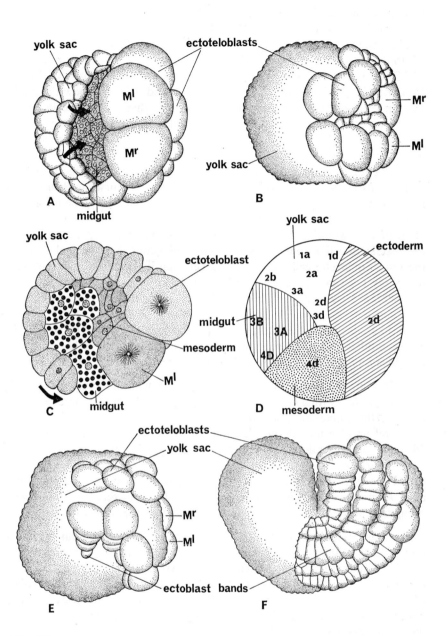

Fig. 2.7.
Bdellodrilus. A. gastrulation, ventral view; B. after ectoteloblast formation, dorsal view; C. sagittal section of embryo at stage shown in Fig. 7A; D. presumptive areas, lateral view; E. early budding of ectoteloblasts, dorsal view; F. stage with well developed germ bands, lateral view.

margin through divisions of 2d, whose final designation as the ectoteloblast mother cell thus becomes 2d^{222} (Fig. 2. 6D).

The fifth cleavage begins, as usual, with division of 2D to add a micromere, 3d, to the left of the micromere cap, followed by successive similar divisions of 2C, 2B and 2A to produce marginal micromeres 3c, 3b and 3a. In this respect, *Bdellodrilus* is identical with *Tubifex* and does not show the modification seen in *Rhynchelmis*, where 3d is larger than its sister cells of the third quartette and plays the role of a second ectoteloblast mother cell. As part of the fifth cleavage, the first quartette cells also divide, further increasing the number of cells in the micromere cap.

Except for the disproportionate size of 2d, therefore, cleavage up to the point of formation of the third quartette follows a course in *Bdellodrilus* identical with that of *Tubifex*. The stem-cells 3A-3D are proportionately much smaller, however, a reflection of the reduced volume of the egg. The onset of the sixth cleavage further emphasizes this feature. As usual, 3D is the first stem-cell to enter this division, segregating into 4d and 4D, but the fourth quartette cell 4d is much larger than its stem cell and continues to occupy the ventral surface (Fig. 2. 6E). 4D is squeezed forward between and behind 3A and 3B as a small yolky cell almost wholly displaced into the interior.

4d, like 2d, retains the fate previously seen in *Tubifex* and *Rhynchelmis*, as the mother cell of paired mesodermal teloblasts. Once it has been cut off, presumptive areas (Fig. 2. 7D) can be specified as:

1. Presumptive ectoderm, identical in arrangement and composition with that of *Tubifex*, with a dorsal area of micromeres in front of a large posterior cell, 2d, the ectoteloblast mother cell.

2. Presumptive mesoderm, identical with that of *Tubifex* as a single cell 4d, but more ventrally disposed and occupying a larger proportion of the ventral surface.

3. Presumptive midgut, comprising the yolky stem cells 3A-3C and 4D as in *Tubifex*, but displaced to the anterior end as four relatively small cells occupying proportionately much less of the embryo than in *Tubifex*.

Thus, as pointed out above, the smaller egg of *Bdellodrilus* can be interpreted as secondarily reduced, the reduction being in the volume of presumptive midgut material, basically, in the volume of yolk. The ultimate consequence of this is relatively precocious hatching with a small number of segments as compared with *Tubifex*. Intervening corollaries concern gastrulation and the role of the "temporary yolk sac ectoderm" in development, and will be further discussed below.

Naididae:

With the Naididae, we enter into a discussion of extreme secondary modification in the embryonic development of aquatic oligochaetes. Only Svetlov (1926) and Davydov (1942) have investigated these modifications, finding a

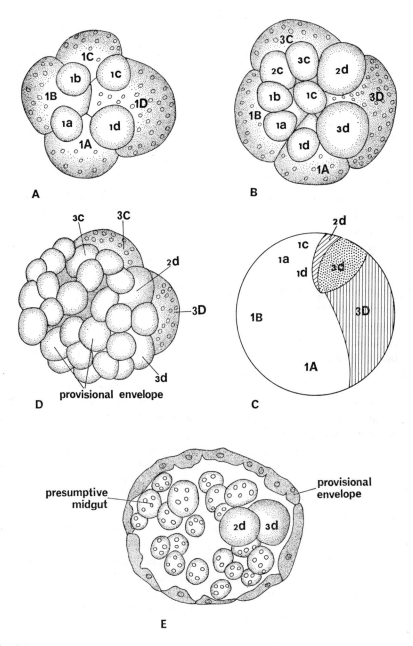

Fig. 2.8.
Stylaria. A. 8-cell stage, dorsal view; B. 12-cell stage, dorsal view; C. presumptive areas, lateral view; D. early stage in overgrowth by provisional envelope, dorsal view; E. early cell divisions within completed provisional envelope, dorsal view.

similar general pattern, but some differences in detail, in *Stylaria* and *Chaetogaster*.

The species of *Stylaria* studied by Davydov lays an egg 350-400 μ in diameter, thus little smaller than that of *Tubifex*, but relatively poor in yolk. A single egg is enclosed, with albumen, in a small cocoon. Cleavage follows the pattern first described by Svetlov (1926) for *Chaetogaster diaphanus*. The first two cleavage divisions are typical, yielding a large posterior D-cell and smaller, equal A, B and C cells in front of it. A typical third cleavage then follows, with formation of a dorsal quartette of small micromeres 1a-1d (Fig. 2. 8A).

At this stage, however, cleavage departs from the spiral regularity observed in *Tubifex*. The cells of the first quartette, together with 1A and 1B, temporarily cease to divide. The stem cells of the C and D quadrants, on the other hand, cut off two second quartette cells 2c and 2d, and two third quartette cells, 3c and 3d, dorsally behind the first quartette (Fig. 2. 8B). Of these cells, 2d and 3d are larger than the remaining micromeres.

The cleavage blastomeres now diverge along two paths of development. The stem cells 1A and 1B, together with the first quartette cells 1a-1d, begin to divide irregularly, with tangential spindles. The resulting cells spread over the surface of the remaining blastomeres as a uniform external layer which soon covers the entire surface of the embryo as a provisional envelope (Fig. 2. 8D). The cells of the provisional envelope make no structural contribution to the subsequent development of the embryo. It seems clear that the envelope has a nutrient function relative to the surrounding albumen.

The remaining blastomeres, 2c, 2d, 3c, 3C and 3D, develop within the envelope into definitive embryonic structures. As further division of these cells takes place, the division products tend at first to float freely as isolated cells (Fig. 2. 8E) and the structural integrity of the embryo is re-established only at a later stage. Davydov was able, however, to trace in some detail the contributions made by each of the six cells to the structure of the embryo. 3d remains temporarily undivided. 2d, lying adjacent, divides into a small plate of relatively large cells. The remaining cells, 2c, 3c, 3C and 3D, undergo repeated divisions, establishing a more or less spherical mass of small, somewhat yolky cells, in which 3d is embedded at the surface, with the products of 2d just above it. 3d now divides equally into paired M-cells. At the same time, the superficial cells of the cell mass become arranged as a distinct outer layer, the provisional epithelium, around a central mass, the presumptive midgut. Later events, to be described below, show that the descendants of 2d develop into the definitive ectoderm of the embryo, gradually replacing the provisional epithelium.

The relationship of embryonic development in *Stylaria* to the less modified development seen in *Tubifex*, *Rhynchelmis* and *Bdellodrilus* is at first sight rather tenuous, but can easily be resolved by a consideration of presumptive areas. The presumptive areas of *Stylaria*, as subsequent events described by Davydov show, are (Fig. 2. 8C):

1. Presumptive ectoderm, segregated into 2d.
2. Presumptive mesoderm, segregated into 3d.
3. Presumptive midgut, segregated into 2c, 3c, 3C and 3D, that is, into 1C and 3D.
4. Presumptive provisional envelope, segregated into 1A, 1B and the first quartette cells 1a-1d.

Before discussing how this pattern of presumptive areas might be functionally derived from the type of primitive pattern seen in *Tubifex*, it is instructive to examine the results obtained by Davydov for *Chaetogaster*. Even though, in the species studied by Davydov, the egg is relatively large, 400-500 μ in diameter, and contains a considerable amount of yolk, development is even more aberrant than in *Stylaria*. The first three cleavage divisions proceed in the same manner, yielding an eight-cell stage with a dorsal quartette of micromeres, 1a-1d, a large posterior stem-cell 1D and smaller, equal stem-cells 1A-1C. Now, however, all cells except 1d and 1D proceed into formation of a provisional envelope. Within the latter, 1d remains temporarily undivided, while 1D cuts off 2d and 3d. The 3d cell now also remains undivided while 2d and 3D enter into numerous divisions which result in an undifferentiated mass of yolky cells with 1d and 3d adjacent to one another at the surface. When the mass is well formed, 1d divides into a plate of superficial cells above 3d, while 3d divides bilaterally into paired M-cells, and the generalized cell mass resolves itself into an outer layer of provisional epithelium around a central mass of presumptive midgut. As in *Stylaria*, the provisional epithelium is later replaced by definitive ectoderm, formed by the descendants of 1d.

Thus, the presumptive areas in *Chaetogaster* are similar to those of *Stylaria*, but are segregated through a slightly different sequence of divisions, as:

1. Presumptive ectoderm, segregated into 1d.
2. Presumptive mesoderm, segregated into 3d.
3. Presumptive midgut, segregated into 2d and 3D.
4. Presumptive provisional envelope, segregated into 1a-1c and 1A-1C, namely, into A, B and C at the four-cell stage.

Variations in the manner in which fundamentally similar patterns of presumptive areas are segregated during cleavage are a universal feature of spiral cleavage embryos (Anderson, 1966a, 1966b) and do not need to be taken into account in comparing the presumptive areas of related species. The only important difference between *Stylaria* and *Chaetogaster* is in the greater proportion of cleavage blastomeres in *Chaetogaster* contributing to the presumptive provisional envelope, functionally associated with a larger egg. Otherwise, the development of the two embryos is virtually identical. We can now examine the question whether naidid embryonic development is a functional derivative of a more primitive *Tubifex*-like development, through a comparison of the formation and fates of presumptive areas.

Presumptive ectoderm in *Tubifex* comprises ectoteloblastic ectoderm 2d and a micromere cap of temporary yolk sac ectoderm in front of 2d. The temporary yolk sac ectoderm is later incorporated into the definitive ectoderm. In *Stylaria*, the definitive ectoderm is again segregated into 2d (1d in *Chaetogaster*) and has the same fate as in *Tubifex*, except that teloblast formation is omitted in favour of more generalized spreading from the posterior end, and that all the definitive ectoderm arises from this source. The latter specialization is associated with a more specialized role for the micromere cap, namely, formation of the provisional envelope. In this process, not only the first quartette, but also the entire A and B quadrants in *Stylaria*, and the entire A, B and C quadrants in *Chaetogaster*, have become involved, save only for the changed fate of 1d in *Chaetogaster* as presumptive ectoderm, more typically a feature of 2d. Thus, it is reasonable to regard the provisional envelope in naidids as a specialization of the temporary yolk-sac ectoderm of the more primitive *Tubifex* pattern of development, subserving a temporary nutritional role in association with secondary reduction of yolk. With this specalization, the cells of the component no longer contribute to the definitive ectoderm of the embryo and, associated with enlargement of the component, the presumptive midgut is reduced.

A functional, though not a phylogenetic, intermediate is exemplified in *Bdellodrilus*, with a reduced presumptive midgut but a less modified development of other components. The micromere cap of *Bdellodrilus* is established as temporary yolk sac ectoderm identical with that of *Tubifex*. It then spreads rapidly (Fig. 2. 7A) to cover most of the surface of the embryo (see below, Gastrulation) and finally becomes incorporated into the definitive ectoderm (see below, Organogeny). The initial rapid spread is a precocious exaggeration of the spreading which gradually takes place during gastrulation in the large yolky egg of *Tubifex*, resulting in rapid establishment of a surface epithelium which seems likely to have a nutritive function relative to the albumen. From this condition it is but a short step to the mode of development expressed in naidids, with their precocious, provisional envelope and restriction of definitive ectoderm to 2d.

Presumptive mesoderm, as in general among annelids, preserves a conservative mode of development, whatever the peculiarities of other components. This conservatism is probably a reflection of the central role played by the mesoderm in the causal development of the segments in annelids. The fact that the presumptive mesoderm is segregated into 3d in naidids, in contrast to 4d in *Tubifex*, *Rhynchelmis* and *Bdellodrilus*, is only a minor variation in an otherwise uniform segregation of a single posterior, mesoteloblast mother cell which then divides into paired M-cells. The same precocious segregation of mesoderm into 3d is seen in Hirudinea (Anderson, 1966b).

Presumptive midgut in *Tubifex* comprises four large, yolk-filled cells 3A, 3B, 3C and 4D which become internal to the ectoderm and mesoderm and undergo numerous divisions to form a yolky midgut mass. There is some indication of the segregation of small cells at the surface of the mass, acting as vitellophages, while the remainder of the cells gradually become grouped around a space which develops in the mass, and give rise to the midgut epithelium. *Rhynchelmis*, with a more voluminous midgut rudiment, exhibits the same features in a more exaggerated way (see below). In naidids, associated with reduced yolk, the presumptive midgut shows secondary specialization derivable from this basic pattern. The rudiment is reduced to 1C and 3D in *Stylaria* and to 2d and 3D in *Chaetogaster*, with associated specialization of the A and B, and in *Chaetogaster* also C, cells as provisional envelope cells. Numerous equal divisions then proceed as usual. The resulting cell mass lies within the provisional envelope. Now, precociously, before the presumptive ectoderm and mesoderm begin to spread over the surface of the midgut mass, the latter segregates a provisional epithelium at the surface, recalling the surface layer of vitellophages in *Tubifex*. It can reasonably be assumed that this layer has a nutritive role. Within it, the central cells hollow out as an epithelial sac which develops an antero-ventral opening through the provisional epithelium, the embryonic mouth. The hollow midgut sac thus forms and opens to the exterior when the ectoderm and mesoderm of the embryo are still simple posterior rudiments, and there seems little doubt that the midgut, too, plays a role in the precocious embryonic feeding on albumen. In the present context, the important point is that development of the midgut in naidids is interpretable as a direct modification of the tubificid pattern of midgut development associated with a transition from utilization of intracellular yolk to utilization of extracellular albumen.

Presumptive stomodaeum development in naidids cannot be usefully discussed, since it is too little understood at the present time. In sum, however, the formation and fates of presumptive areas in naidid embryos is a secondary modification of that of tubificids, functionally adapted to a transition from entolecithal development to development partially at the expense of albumen, and naidid embryos are not so aberrant among oligochaetes as they previously appeared to be.

GASTRULATION

GASTRULATION IN TUBIFEX AND RHYNCHELMIS Commensurate with the similarity of their presumptive areas, gastrulation takes an identical course in *Tubifex* and *Rhynchelmis*, and also in the closely related *Peloscolex*. Accounts of gastrulation have been given for *Tubifex* by Penners (1924) and Meyer (1929), for *Peloscolex* by Penners (1929) and for *Rhynchelmis* by Kovalesky (1871), Vejdovsky (1888-92) and Ivanov (1928).

As might be expected, the large, yolky presumptive midgut rudiment becomes internal as a result of overgrowth by other cells (Fig. 2. 3D). Cell divisions continue while the rudiment is still exposed at the surface, producing a solid mass of polygonal, yolky cells. Presumptive ectoderm, as it continues to develop, spreads down on either side of the mass, eventually meeting in the midline ventrally to enclose the midgut rudiment (Fig. 2. 4A, 4D). Complete enclosure is not attained until most of the trunk segments of the embryo have been delineated.

The mesoteloblasts begin their teloblastic activity while still at the surface (Fig. 2. 3C), so that budding overlaps the gastrulation movement of these cells into the interior. The latter comes about partly by an inward sinking of the M-cells on either side of the midgut posteriorly, but mainly by overgrowth by the posterior edge of the presumptive ectoderm, which spreads back over the M-cells as far as the edge of the exposed presumptive midgut (Fig. 2. 3D).

The presumptive stomodaeum, formed as a flat plate of cells at the surface in front of the presumptive midgut, grows into the interior by cell divisions as a solid mass (Fig. 2. 3D, 4C), formation of a lumen being much delayed.

The overgrowth and enclosure role of the presumptive ectoderm in gastrulation in *Tubifex* and *Rhynchelmis* is attained mainly through cell division and attenuation of the temporary yolk sac component of this area. The temporary assumption of this form by the majority of the presumptive ectoderm is, in fact, an adaptation to rapid completion of an ectoderm surface layer around the large mass of yolky midgut. The ectoteloblasts, in contrast, proceed directly into processes underlying adult organogeny, while being moved from a dorsal-lateral to a lateral position as the temporary yolk sac ectoderm expands. As already indicated, the teloblastic activity of the ectoteloblasts begins as soon as the cells are formed, four parallel rows of cuboidal cells being budded on either side, which push forward towards the anterior end (Fig. 2. 3C). The cell rows remain superficial, cleaving a path through the temporary yolk sac ectoderm, which at the same time pushes the rows ventrally as it spreads (Fig. 2. 3D). The ectoblast rows of each side eventually meet in the midline, first anteriorly in front of the stomodaeal rudiment, then mid-ventrally behind the stomodaeum and progressively posteriorly, completing enclosure of the midgut (Fig. 2. 4A, 4D). Most of the temporary yolk sac ectoderm now lies dorsally between the two ectoblast bands, but a narrow strip also persists along the ventral edge of each band and, at the completion of gastrulation by mid-ventral ectodermal fusion, comes to occupy the ventral midline as far forwards as the stomodaeal rudiment. Posteriorly, this mid-ventral strip of temporary yolk sac ectoderm remains contiguous with the pygidial temporary yolk sac ectoderm.

As the ectoblast bands grow in length, the mesodermal bands budded from the M-cells grow beneath them and are similarly carried, first to a lateral, then to a ventro-lateral position. The development of these bands as paired segmental somites proceeds in antero-posterior succession and precedes corresponding segment delineation of the overlying ectoblast bands. Later in development, the temporary yolk sac ectoderm is incorporated segmentally into the segmental epithelium, save for the most posterior part, which transforms into the pygidial epithelium. A short in-tucking of the latter at the anus gives rise to the proctodaeum.

GASTRULATION IN OTHER AQUATIC OLIGOCHAETES Associated with the secondary reduction of yolk and its attendant specializations of development, gastrulation processes are modified in both branchiobdellid and naidid oligochaetes. *Bdellodrilus*, with presumptive areas little different from those of *Tubifex* save in a relative reduction in size of the presumptive midgut, provides a convenient intermediate. As described by Tannreuther (1915), the major gastrulation process in *Bdellodrilus* is rapid overgrowth of the anterior end of the embryo by the temporary yolk sac ectoderm (Fig. 2. 7A, 7C). As a result, the presumptive midgut becomes covered, and therefore internal, when teloblastic budding of the germ bands is just beginning. The presumptive mesoderm, in the form of paired postero-ventral M-cells, is still superficial when budding begins, and becomes internal in the usual way as a result of gradual overgrowth by the presumptive ectoderm at the posterior end (Fig. 2. 7C). Associated with the small volume of presumptive midgut, the ectoteloblasts lie laterally to ventro-laterally at the posterior end and the ectoblast bands grow forward in their final position, mainly after gastrulation by ectodermal overgrowth has been completed. Formation of the ectoteloblasts from the 2d cell proceeds as in *Tubifex*, first by equal bilateral division into two cells (Fig. 2. 6F), each of which then divides again, and the resulting median cell twice again, to give a transverse row of four on each side (Fig. 2. 7B). After the first division of 2d, each of the resulting cells adds three micromeres to the posterior edge of the micromere cap (Fig. 2. 6D), preliminary to rapid proliferation of this cap in anterior overgrowth, and also cuts off a pair of postero-ventral micromeres whose proliferation plays a part in overgrowth of the presumptive mesoderm. In general, the ectodermal overgrowth through which gastrulation is attained in *Bdellodrilus* is rapid and precocious, an indication of the new role of the temporary yolk sac ectoderm as an external absorptive epithelium, and the presumptive midgut, presumptive mesoderm and ectoteloblasts no longer manifest gastrulation movements.

In the Naididae, the modification is carried to an extreme in which gastrulation is wholly eliminated. The overgrowth by temporary yolk sac ectoderm is even more precocious and rapid (Fig. 2. 8D, 8E), but can no longer be regarded as a part of gastrulation, because the resulting provisional envelope does not contribute structurally to the embryonic ectoderm. Presumptive midgut, mesoderm and ectoderm attain their definitive locations directly as a result of cleavage, and proceed directly into the organogenetic processes of germ band formation and midgut formation. Growth of the germ band ectoderm is simplified, with elimination of ectoteloblasts and direct spreading of the ectoderm from the posterior end over the exposed midgut rudiment. In one sense this can be regarded as a secondary gastrulation by overgrowth, but it is perhaps best regarded as the initial phase of organogeny.

Little can be said of the gastrulation activities of the stomodaeum and proctodaeum in either branchiobdellids or naidids, since only meagre information is available. In *Bdellodrilus* the stomodaeum arises as a small invagination of ectoderm at the anterior end, and the proctodaeum at the posterior end, both late in development. The naidid stomodaeum is formed earlier, by an ectodermal invagination at the site of the embryonic mouth, after the definitive ectoderm has spread forwards to this level.

ORGANOGENY

Once the presumptive areas of the blastula have become re-orientated through gastrulation, each proceeds into its own pattern of organogenetic activity. Overlap between the final phases of gastrulation movement of some rudiments and the initial phases of organogeny of others almost always occurs, and is especially noticeable when extreme lecithotrophy is combined with retention of total cleavage, as in *Tubifex* and *Rhynchelmis*. Secondary complications are also introduced into organogeny in species exhibiting precocious function of temporary embryonic organs, as in naidids, where development deviates temporarily for nutritional reasons from the direct path to the adult configuration of organs.

FORMATION OF TRUNK SEGMENTS It has already been described above how, with the onset of gastrulation, the ectoteloblasts and mesoteloblasts begin to proliferate ectodermal and mesodermal bands. The ectoblast bands are superficial and the mesodermal bands lie beneath them. In *Tubifex* (Fig. 2. 3C, 3D), the bands are lateral when first proliferated (Penners, 1924; Meyer, 1929), in *Rhynchelmis*, dorso-lateral (Kovalevsky, 1871; Vejdovsky, 1888-92; Ivanov, 1928), but become ventro-lateral as gastrulation overgrowth of the ectoderm proceeds. The ectoblast bands in these species remain superficial, growing forward on either side of the stomodaeal rudiment to meet in the anterior midline. In the smaller embryo of *Bdellodrilus*, the bands grow forward ventro-laterally (Fig. 2. 7E, 7F), and the ectoblast components plunge beneath the precociously expanded, temporary yolk sac ectoderm, which continues to spread back until it covers the entire embryo. Naidids (Davydov, 1942) lack definitive ectoteloblasts and display a more generalized forward growth of ectoderm from the posterior end.

When the ectoderm and mesoderm of the full complement of segments formed before hatching is budded off, the teloblasts become indistinguishable in the ectoderm and mesoderm of the posterior end.

From the paired mesodermal bands develop the paired somites of the trunk segments. In *Tubifex* (Penners, 1924; Meyer, 1929), *Rhynchelmis* and *Limnodrilus* (Ivanov, 1928) each somite originates as a single large cell cut off from the M-cell of its side (Fig. 2. 3D). The somite cell first undergoes a series of equal divisions, forming a group of cells (Fig. 2. 4B) which then becomes hollow by internal splitting. The walls of the somite develop as somatic, splanchnic and septal mesoderm respectively (Fig. 2. 5A, 5B, 5C). Similar separation of somite blocks which become hollow in antero-posterior sequence is seen in naidids (Davydov, 1942). According to Tannreuther (1915), however, the first sign of somite delineation in the small embryo of *Bdellodrilus* is, as in earthworms (Wilson, 1889; Bergh, 1890; Vejdovsky, 1888-92; Staff, 1910; Svetlov, 1928), the formation of paired coelomic cavities in already well developed mesodermal bands.

Ivanov (1928) attempted to demonstrate that the anterior segments of the trunk of *Rhynchelmis* each contain a pair of rudimentary larval somites developed from ectomesoderm, in addition to the normal pair of somites developed from the mesodermal bands, but his evidence for this cannot now be seriously entertained. It is probable that reinvestigation would reveal some peculiarities in mesoderm at the anterior end of the trunk, associated with specializations of the gut, blood system and musculature in the region, but it is unlikely that these segments are fundamentally different from the more posterior segments in the heteronomous sense implied by Ivanov.

As the ectoblast bands of *Tubifex*, *Rhynchelmis* and *Bdellodrilus* elongate, the cells of each of the four rows making up the band on either side undergo repeated divisions. The products of the most ventral row on each side remain distinct, but those of the three more lateral rows merge into a single band of small cells (Vejdovsky, 1889-92; Tannreuther, 1915; Penners, 1924; Meyer, 1929). Segment delineation in the ectoblast bands of *Tubifex* (Fig. 2. 4D, 4E) and *Rhynchelmis* occurs by the formation of transverse, intersegmental grooves, following formation of the underlying somites. In *Tubifex*, for example, when 24 pairs of somites are present, only 16 segments are delineated externally (Penners, 1929). The evocation of ectodermal segmentation by mesodermal somites has been well established experimentally for oligochaetes for several years (Watterson, 1955). External segmentation in the specialized embryos of branchiobdellids and naidids is delayed, and is not clearly understood.

FURTHER DEVELOPMENT OF THE TRUNK SEGMENTS The respective roles played by the somites, segmental ectoblast and temporary yolk sac ectoderm in the further development of the trunk segments of aquatic oligochaetes remain controversial, due mainly to technical inadequacies in past studies and to lack of critical evidence.

The somatic musculature:

The paired somites of each segment enlarge and spread upwards around the gut and downwards towards the ventral midline (Fig. 2. 5C), their coelomic cavities progressively expanding (*Tubifex*, Penners, 1924; *Rhynchelmis*, Kovalevsky, 1871; Vejdovsky, 1888-92). The anterior and posterior walls of each somite contribute to intersegmental septa, and their upper and lower edges to dorsal and ventral mesenteries. The somatic wall shows differentiation of outer myoblasts from inner peritoneal epithelium, and the muscle fibrils of the longitudinal muscles of the body wall differentiate within the myoblasts, first ventro-laterally, then dorso-laterally. When the chaetal sacs develop, some of the somatic myoblasts cluster around them and differentiate as chaetal sac musculature.

The origin and early development of the circular muscles of the body wall is unresolved. These muscles develop from cells lying laterally between cells definitely assignable to the somatic wall of the somite and cells definitely descended from the ectoteloblasts (Fig. 2. 5B, 5C). Several authors (Vejdovsky, 1888-92); Penners, 1922, 1924; Meyer, 1929) have interpreted the cells as derivatives of the ectoblast bands, but their evidence, when examined closely, is inconclusive. Earlier interpretations of a mesodermal origin of the circular muscle cells from the outer part of the somite are even more poorly supported, and a final solution awaits fresh evidence. If a final judgement is passed in favour of an ectoblastic origin of circular muscles in oligochaetes, the presumptive area maps of the oligochaete blastula presented above will require modification.

The splanchnic musculature:

The splanchnic walls of the somites become applied to the outer surfaces of the stomodaeum and midgut as splanchnic mesoderm. The musculature of the midgut wall presumably develops from these cells, but its development has not yet been described. The pharyngeal musculature develops in a complex manner from the walls of the anterior somites (Wilson, 1889; Ivanov, 1928), but the details of its formation are not clearly understood.

The blood vascular system:

The development of blood vessels in oligochaetes was reviewed by Hanson (1949). *Tubifex* is the only aquatic oligochaete in which the formation of blood vessels has been given any attention (Penners, 1924; Meyer, 1929). The blood vascular space occupies the site of the former blastocoel. The ventral longitudinal vessel forms through separation of the apposed walls of the ventral mesentery. The dorsal vessel develops precociously, as a paired, lateral, dorsal vessel between the upper edges of the somites and the lateral surfaces of the yolky midgut. Later, as the edges of the somites come together in the dorsal midline, the two half-vessels combine into a single, dorsal longitudinal vessel in the resulting mesentery. Commissural vessels are formed by separation of the apposed walls of the intersegmental septa.

The surface epithelium, chaetal sacs and ventral ganglia:

The ectoblast bands lying external to the somites of each segment, while they do not constitute the entire ectoderm of the segment (temporary yolk sac ectoderm is later incorporated, see below), contain the rudiments of all segmental structures of ectodermal derivation other than the surface epithelium, namely the chaetal sacs and ventral ganglia. Each component of segmental ectoderm (Fig. 2. 5A) comprises a ventro-lateral row of cells budded off by the most ventral of the ectoteloblasts, and a more lateral area of cells originating from the remaining ectoteloblasts.

The ventro-lateral cells are the neuroblasts of the ventral ganglia of the segments. These cells come to lie on either side of the midline, beneath the surface, and bud off ganglion cells as two separate strands which later sink inwards and fuse in the midline before neuropile formation begins (Fig. 2. 5B, 5C). In *Tubifex*, it is not clear whether the superficial cells covering the developing ganglia are cut off from the neuroblasts (Penners, 1924) or derived by spreading of the ventral band of temporary yolk sac ectoderm (Meyer, 1929). The point has not been examined for other aquatic oligochaetes.

The lateral ectoblast cells above the neuroblasts are those from which the circular muscle cells have been said to arise, and they have often been referred to as myoblasts (Penners, 1924; Meyer, 1929). Omitting this controversial point, these cells, once segmentally demarcated, develop in *Tubifex* as lateral segmental epithelium, in continuity below with the epithelial cells covering the developing ganglia and above with the temporary yolk sac cells (Penners, 1924). Groups of the lateral epithelial cells multiply to produce in-growths into the interior, the chaetal sacs (Vejdovsky, 1888-92; Penners, 1924; Meyer, 1929, for aquatic species; also Kovalevsky, 1871; Vejdovsky, 1886; Wilson, 1889; Bergh, 1890; Bourne, 1894; Staff, 1910; Svetlov, 1928; Vanderbroek, 1936, for earthworms).

FURTHER DEVELOPMENT OF THE TEMPORARY YOLK SAC ECTODERM The temporary yolk sac ectoderm of *Tubifex* and *Rhynchelmis*, which becomes spread broadly over the dorsal and lateral surfaces of the embryo during gastrulation, is gradually incorporated into the segmental surface epithelium (Figs. 2. 3D, 4D, 4E) as the embryo elongates and becomes tubular (Vejdovsky, 1888-92; Penners, 1924). In *Bdellodrilus*, where this component forms a complete external layer outside the ectoblast bands (Fig. 2. 7F), it is said by Tannreuther (1915) to differentiate as the entire surface epithelium, but Tannreuther did not follow the later development of the ectoblast bands in sufficient detail to uphold such a claim, and a reinvestigation of the question is required. The naidid temporary yolk sac ectoderm, highly specialized as an external provisional envelope, remains external to the embryo throughout development (Davydov, 1942). As the embryo grows, the cells of the envelope become attenuated and later begin to regress, with vacuolation of the cytoplasm and loss of nuclei. When the fully-developed embryo hatches, the remains of the provisional envelope are left within the cocoon.

DEVELOPMENT OF THE NEPHRIDIA No point in oligochaete embryology has been more controversial than the origin and development of nephridia. The early workers on this problem, basing doubtful conclusions on unsatisfactory evidence, bitterly argued their interpretations in a series of papers, mainly on earthworms (Hatschek, 1878; Vejdovsky, 1884, 1886, 1887, 1888-92, 1900; Bergh, 1888, 1890, 1899; Wilson, 1887, 1889; Staff, 1910; Tannreuther, 1915; Bahl, 1922). The majority opinion, favouring a unitary origin of each nephridium from a single large nephridioblast cell, was finally conclusively vindicated in studies on aquatic oligochaetes by Penners (1924; *Tubifex*), Ivanov (1928; *Rhynchelmis, Limnodrilus*) and especially by Meyer, (1929; *Tubifex*), and was confirmed for earthworms by Goodrich (1932) and Vanderbroek (1932, 1934). The description of nephridial development in *Tubifex* given by Meyer and confirmed for *Eisenia* by Vanderbroek, makes it clear (Goodrich, 1946) that nephridial development begins with such a cell, embedded in the septal wall (Fig. 2. 5B). A row of small cells budded off from the nephridioblast develops into the nephridial duct growing out towards the body surface, while the nephridioblast itself finally divides into a group of cells which form the nephrostomal funnel. Since the nephridioblast first becomes distinguishable in the septal wall, it is generally accorded a mesodermal origin, but the possibility that it is first segregated wthin the segmental ectoderm and subsequently migrates into the septum cannot be disregarded (Goodrich, 1946).

During gastrulation, most oligochaete embryos develop a pair of provisional protonephridia on either side at the anterior end, between the surface ectoderm and the gut. Each protonephridium comprises a solenocyte, a ciliated, usually coiled, intracellular duct and an opening to the exterior ventrolaterally or laterally (*Tubifex*, Ivanov, 1928; *Rhynchelmis*, Vejdovsky, 1886, 1887, 1888-92; also well described for earthworms (Vejdovsky, 1887, 1888-92; Lehmann, 1887; Bergh, 1888; Wilson, 1889; Hoffman, 1899; Ivanov, 1928; Svetlov, 1928)). The protonephridia persist until the anterior segmental nephridia have become functional, and are then resorbed. The origin and development of the provisional protonephridia is still obscure, but the work of Ivanov (1928) and Svetlov (1928) on earthworms indicates that they arise from laterally placed cleavage micromeres which migrate into the interior before the mesodermal bands begin to form.

DEVELOPMENT OF THE GONODUCTS The gonoducts of oligochaetes develop independently of the nephridia, which are generally absent from the genital segments. In all cases which have been adequately investigated, the gonoducts can be identified as coelomoducts (*Tubifex*, Gatenby, 1916; Ivanov, 1928; Meyer, 1929; also earthworms, Vejdovsky, 1884; Bergh, 1886; Lehmann, 1887). Each gonoduct originates as a thickening of the coelomic epithelium opposite a gonad. The thickening develops into a funnel, and an outgrowth from the base of the thickening forms the duct. At the end of each male duct, a small ectodermal invagination establishes the opening to the exterior. Receptacula seminales develop as simple invaginations of the ventro-lateral surface epithelium (Vejdovsky, 1884; Bergh, 1886, Gatenby, 1916; Mehra, 1924).

DEVELOPMENT OF THE GONADS   While the gonoducts arise late in development in oligochaetes, the primordial germ cells are set aside at a very early stage, according to the evidence of tubificid development (Ivanov, 1828; Penners, 1929; Meyer, 1929, 1931; Penners and Stablein, 1930). A pair of small cells cut off, according to Penners and Stablein, from 4d, is closely associated with the presumptive mesoderm at the posterior end of the blastula. This pair proliferates further cells which spread forwards through the mass of yolky midgut cells as gastrulation proceeds and eventually clump and settle in the vicinity of the genital segments, embedded in the walls of the somites. Most of the cells in the clumps, together with some which remain scattered among the yolky, midgut cells, appear to degenerate. Two pairs of cells persist in each genital segment and proliferate to form testes and ovaries respectively, projecting into the segmental coelom and covered by a thin peritoneal epithelium.

FURTHER DEVELOPMENT OF THE GUT   In aquatic oligochaetes with large, yolky eggs (*Tubifex, Peloscolex, Limnodrilus, Rhynchelmis*), the stomodaeum develops in two parts, giving rise in succession to the lining epithelia of the pharynx and buccal cavity (Penners, 1924, 1929; Ivanov, 1928). The stomodaeal presumptive area first buds off a mass of cells which pushes inwards against the yolky midgut rudiment (Fig. 2. 4C, 5D) and subsequently differentiate as pharyngeal epithelium, after growing back through the first few trunk segments and gaining a lumen. Meanwhile, the stomodaeal cells remaining at the surface thicken and invaginate as a plate, the lining epithelium of the buccal cavity. Continuity between the buccal and pharyngeal lumina is established late, by a breakthrough at the base of the buccal invagination.

In branchiobdellids (Tannreuther, 1915) and naidids (Davydov, 1942) the development of the stomodaeum is inadequately understood and requires further study.

The development of the midgut, in contrast, is well analysed for several species, and shows a variety of specializations. In *Tubifex*, the large mass of yolky midgut cells filling the interior of the embryo (Fig. 2. 5A) develops more or less directly into midgut epithelium as the yolk is utilized. A central split appears, around which the cells become arranged as an epithelium (Penners, 1924, 1934). Some evidence exists, however, of temporary vitello-phage activity associated with the yolk cells. Small cells cut off in the outer part of the yolk mass probably have this function, and are later resorbed rather than incorporated into the midgut epithelium.

Vitellophage specialization is more marked in *Rhynchelmis* (Vejdovsky, 1888-92; Schmidt, 1922; Ivanov, 1928). The yolky midgut cells fuse to form a syncitium, retaining at the surface a number of yolk-free cells with large nuclei, which act as vitellophages and are later resorbed. A number of distinct cells also persist within the syncitial mass and become arranged around a small space formed near the anterior end of the mass. As the yolk mass shrinks, the epithelial vesicle extends to form the lining epithelium of the midgut.

Anteriorly, in yolky oligochaete embryos (*Tubifex, Peloscolex, Limnodrilus, Rhynchelmis*; Penners, 1924, 1929, 1934; Ivanov, 1928; Meyer, 1929), the yolk disappears early from a group of midgut cells immediately behind the pharyngeal rudiment. These cells develop into the lining epithelium of the oesophagus. Posteriorly, the elongating midgut rudiment makes contact with the pygidial ectoderm, and late in development the anus breaks through at this point.

Bdellodrilus, with a smaller midgut rudiment, displays a simpler mode of midgut development (Tannreuther, 1915). Divisions of the midgut cells continue after gastrulation and the resulting cells spread backwards as the embryo elongates. A central lumen then develops within the midgut mass. The peripheral cells of the mass become arranged as an epithelium which differentiates as the midgut epithelium, while the more central cells around the lumen gradually degenerate and are resorbed.

The Naididae retain traces of the mode of midgut development seen in *Tubifex* and *Rhynchelmis*, but expressed after the precocious development of the midgut rudiment as a functional embryonic gut. The precise fate of the provisional epithelium formed around the embryonic midgut sac is not clear. The epithelial wall of the sac, however, fuses in later development into a syncitial mass, in which nuclei gradually emerge at the surface and form the focus for differentiation of a superficial layer of cells, the epithelium of the definitive midgut. Some nuclei remain within the syncitial mass enclosed by this epithelium, acting apparently as vitellophage nuclei while the enclosed mass is finally resorbed.

DEVELOPMENT OF THE HEAD

The prostomium:

The prostomium in oligochaetes has a bilateral origin from the anterior ends of the ectoblast bands (Figs. 2. 3D, 4D, 4E). Each half contributes a cerebral ganglion from its ventral, neuroblast component (*Tubifex, Rhynchelmis, Bdellodrilus*; Tannreuther, 1915; Penners, 1924; Ivanov, 1928). The developing ganglia separate from the overlying epithelium (Fig. 2. 5D), fuse and move back above the pharynx. The epithelium remaining at the surface forms the externally visible part of the prostomium. Beneath it lies a mesenchyme, in front of the first pair of somites, derived from the anteriormost cells of the mesodermal bands (Penners, 1924; Ivanov, 1928; Meyer, 1929). The contribution of this pre-segmental mesoderm to the structure of the head has not been elucidated, but the general indication of present evidence is that the oligochaete prostomium is a pre-segmental structure, like that of polychaetes (Anderson, 1966a).

The peristomium:

The development of the peristomium in oligochaetes has not been investigated in detail, but the evidence provided by Penners (1924) and Ivanov (1928) indicates that the peristomium is the first definite segment of the body, being established as a pair of somites and overlying segmental ectoderm just behind the prostomial rudiments, with the neuroblast components of the segmental ectoderm giving rise to a single pair of ventral ganglia (Figs. 2. 3D, 4D, 4E, 5D). The rudiments of the segment, budded off by posterior teloblasts, move forwards along either side to lie behind the prostomium and stomodaeum, but do not migrate in front of the stomodaeum.

The developing ganglia of the peristomial segment are in continuity, from the beginning, with the developing cerebral ganglia, since both arise from the ventro-lateral neuroblast rows of the ectoblast bands. The parts of these rows intervening between the two pairs of ganglia give rise directly to the circum-pharyngeal commissures. It must be emphasized, however, that a thorough investigation of the embryonic development of the oligochaete head is still wanting, and that the relationship between the embryonic peristomial segment and the fully-formed adult peristomium is still obscure.

DEVELOPMENT OF THE PYGIDIUM Pygidial development in oligochaetes involves little of significance (Meyer, 1929), save for the fact that the pygidial presumptive ectoderm first takes part in formation of the temporary yolk sac ectoderm between and behind the posterior ends of the ectoblast bands. The pygidial ectoderm retains this temporary form until the ectoteloblasts have become indistinguishable among the residual cells of the growth zone persisting behind the last trunk segment formed before hatching. The pygidial cells then transform from a flattened to a cuboidal form, making up the epithelium of the definitive pygidium. It is within this structure that the anus later appears as the aperture of a short proctodaeal invagination.

CONCLUSION

The embryonic development of aquatic oligochaetes is a modification of the spiral cleavage pattern of development exhibited in polychaetes, associated basically with an increase in size and yolk content of the egg and with development within a protective albuminous cocoon. Secondarily, in branchiobdellids, naidids and aeolosomatids, and also in earthworms, yolk is reduced and the embryo is adapted in various ways to feeding on the ambient albumen of the cocoon. These adaptations are functional modifications of the paths of development in the larger-egged species.

REFERENCES

ANDERSON, D. T. 1966a. The comparative embryology of the Polychaeta. *Acta Zool. Stockh.*, **47**, 1-42.

— 1966b. The comparative early embryology of the Oligochaeta, Hirudinea and Onychophora. *Proc. Linn. Soc. N.S.W.*, **91**, 10-43

BAHL, K. N. 1922. On the development of the entonephric type of nephridial system found in earthworms of the genus *Pheretima*. *Q. Jl microsc. Sci.* **66**, 49-103.

BERGH, R. S. 1886. Untersuchungen über den Bau und die Entwicklung der Geschlechtsorgane der Regenwürmer. *Z. wiss. Zool.*, **44**, 303-32.

— 1888. Zur Bildungsgeschichte der Excretionsorgane bei *Criodrilus*. *Arb. zool. zootom. Inst. Würzburg*, **8**, 223.

— 1890. Neue Beiträge zur Embryologie der Anneliden. I. Zur Entwicklung und Differenzierung des Keimstreifens von *Lumbricus*. *Z. wiss. Zool.*, **50**, 469-526.

— 1899. Nochmals über die Entwicklung der Segmentalorgane. *Z. wiss. Zool.*, **66**, 435-449.

BOURNE, A. G. 1894. On certain points in the development and anatomy of some earthworms. *Q. Jl microsc. Sci.*, **36**, 11-34.

BUNKE, D. 1967. Zur Morphologie und Systematik der Aeolosomatidae Beddard 1895 und Potamodrilidae nov. fam. (Oligochaeta). *Zool. Jb. Syst.*, **94**, 187-368.

DAVYDOV, C. 1942. Étude sur l'embryologie des Naididae indochinois. *Arch. Zool. exp. gén., Notes et Revue*, **81**, 173-94.

GATENBY, J. B., 1916. The development of the sperm duct, oviduct and spermatheca in *Tubifex rivulorum*. *Q. Jl microsc. Sci.*, **61**, 317-36.

GOODRICH, E. S. 1932. On the nephridiostome of *Lumbricus* *Q. Jl microsc. Sci.*, **75**, 165-79.

— 1945. The study of nephridia and genital ducts since 1895. *Q. Jl microsc. Sci.*, **86**, 113-301.

HANSON, J. 1949. The histology of the blood system in oligochaetes and polychaetes. *Biol. Rev.*, **24**, 127-73.

HATSCHEK, B. 1878. Studien über die Entwicklungsgeschichte der Anneliden. *Arb. zool. Inst. Univ. Wien*, **3**, 277-404.

HOFFMANN, R. W. 1899. Beiträge zur Entwicklungsgeschichte der Oligochäten. *Z. wiss. Zool.*, **66**, 335-57.

IVANOV, P. P. 1928. Die Entwicklung der Larvalsegmente bei den Anneliden. *Z. Morph. Ökol. Tiere.*, **10**, 62-161.

KOVALEVSKY, A. 1871. Embryologische Studien an Würmer und Arthropoden. *Zap. imp. Akad. Nauk.*, **16**, No. 12: 70 pp.

LEHMANN, O. 1887. Beiträge zur Frage von der Homologie der Segmentalorgane und Ausführgänge der Geschlechtsprodukte bei den Oligochaeten. *Jena Z. Naturw.*, **21**, 322-60.

MEHRA, H. R. 1924. The genital organs of *Stylaria lacustris*, with an account of their development. *Q. Jl microsc. Sci.*, **68**, 147-86.

MEYER, A. 1929. Die Entwicklung der Nephridien und Gonoblasten bei *Tubifex rivulorum* Lam. nebst Bemerkungen zum natürlichen System der Oligochäten. *Z. wiss. Zool.*, **133**, 517-62.

— 1931. Cytologische Studien über die Gonoblasten in der Entwicklung von *Tubifex*. *Z. Morph. Ökol. Tiere.*, **22**, 269-86.

PENNERS, A. 1922. Die Furchung von *Tubifex rivulorum* Lam. *Zool. Jb. Anat.*, **43**, 323-68.

— 1924. Die Entwicklung des Keimstriefs und die Organbildung bei *Tubifex rivulorum* Lam. *Zool. Jb. Anat.*, **45**, 251-308.

— 1929. Entwicklungsgeschichte Untersuchungen an marinen Oligochäten. I. Furchung, Keimstreif, Vorderdarm und Urkeimzellen von *Peloscolex benedeni* Udekem. *Z. wiss. Zool.*, **134**, 307-44.

— 1934. Die Ontogenese der entode malen Darmepithele bei limicolen Oligochäten. *Z. wiss. Zool.*, **145**, 497-507.

—, and STABLEIN, A. 1930. Über die Urkeimzellen bei Tubificiden (*Tubifex rivulorum* Lam. und *Limnodrilus udekemianus* Claparède). *Z. wiss. Zool.*, **137**, 606-26.

SCHMIDT, G. 1922. Zur Frage über die Entwicklung des Entoderms bei der *Rhynchelmis limosella* Hoffm. *Russk. zool. Zh.*, 3, 74-93.

STAFF, F. 1910. Organogenetische Untersuchungen über *Criodrilus lacuum*. *Arb. zool. Inst. Univ. Wien*, 18, 227-56.

SVETLOV, P. 1923. Sur la segmentation de l'oeuf de *Rhynchelmis limosella* Hoffmstr. *Izv. biol. nauchno-issled. Inst. biol. Sta. Perm. gosud. Univ.*, 1, 141-52.

— 1926 Über die Embryonalentwicklung bei den Naididen *Izy. biol. nauchno-issled. Inst. biol. Sta. Perm. gosud. Univ.*, 4, 359-72

— 1928. Untersuchungen über die Entwicklungsgeschichte der Regenwürmer. *Trudy osob. zool. Lab. sebastop. biol. Sta.*, 13, 95-329.

TANNREUTHER, G. W. 1915. The early embryology of *Bdellodrilus philadelphicus*. *J. Morph.*, 26, 143-216.

VANDERBROEK, G. 1932. Origine et dévelopement des saccules mesodermiques et des néphridies chez un oligochète tubicole: *Allolobophora foetida* Sav. *C. r. Ass. f. Advanc. Sci.*, 56, 292-96.

— 1934. Organogénèse du système néphridien chez les oligochètes et plus specialement chez *Eisenia foetida* Sav. *Recl. Inst. zool. Torley-Rousseau*, 5, 1-72.

— 1936. Organogénèse des follicules setigères chez *Eisenia foetida* Sav. *Mem. Mus. r. Hist. nat. Belg.*, 3, 559-68

VEJDOVSKY, F. 1884. *System und Morphologie der Oligochaeten.* Prague.

— 1886. Die Embryonalentwicklung von *Rhynchelmis*. *Sber. K. Bohm. Ges. Wiss.*, 2, 227-39.

— 1887. Das larval und definitiv Excretions system. *Zool. Anz.*, 10, 681-85.

— 1888-92. *Entwicklungsgeschichte-Untersuchungen.* Prague.

— 1892. Zur Entwicklungsgeschichte des Nephridial Apparates von *Megascolides australis*. *Arch. mikrosk. Anat. EntwMech.*, 40, 552-62.

— 1900. Noch ein Wort über die Entwicklung der Nephridien. *Z. wiss Zool.*, 67, 247-54.

WATTERSON, R. L. 1955. Selected Invertebrates. In Willier, B. H., Weiss, P. A., and Hamburger, V. (Eds.) *Analysis of development.* W. B. Saunders, Co., Philadelphia.

WILSON, E. B. 1887. The germ bands of *Lumbricus*. *J. Morph.*, 1, 183-92.

— 1899. The embryology of the earthworm. *J. Morph.*, 3, 387-462.

APPENDIX

Specific names as employed by the authors of embryological papers mentioned in the text.

Tubifex rivulorum Lam. (=*T. tubifex*)	Penners, 1922, 1924
	Meyer, 1929
Rhynchelmis limosella Hoffm.	Schmidt, 1922
	Svetlov, 1928
Peloscolex benedeni Udekem	Penners, 1929
Limnodrilus hoffmeisteri, L. newaënsis	
(=*Tubifex newaenesis*)	Ivanov, 1928
Bdellodrilus philadephicus	Tannreuther, 1915
Stylaria sp.	Davydov, 1942
Chaetogaster diaphanus Gr.	Svetlov, 1926
Chaetogaster sp.	Davydov, 1942

3

DISTRIBUTION AND ECOLOGY

1. MICRODRILES

R. O. Brinkhurst

The majority of publications in which the aquatic oligochaetes have been identified have been systematic studies from which a reasonable amount of information concerning geographical distribution can be derived. Ecologists have had little opportunity to explore the more detailed local distribution of even the commonest species or to establish the basic life histories, predator—prey relationships or other fundamental attributes of oligochaete population in rivers and lakes owing to the scattered nature of this systematic literature. In recent years some regional guides to the identification of oligochaetes have been produced, and there have been several systematic reviews of families.

Much of this activity has been stimulated by the growing interest in biological methods of detecting and assessing water pollution, as the development of very large tubificid populations below sources of organic pollution has been recognized for many years.

The following account recognizes that the present state of knowledge is inadequate in respect to the distribution of worms in certain regions, particularly Australia and New Zealand, and that the study of their ecology is in its infancy. Indeed, the major purpose of this book is to provide a basis for future ecological studies by summarizing the systematics and making possible the creation of more regional keys.

The Aelosomatidae and Opistocystidae are omitted as studies by Bunke (1967), E. du B. Marcus (1944), and Marcus (1944) have indicated that many species have yet to be described. There have been no detailed studies of these groups in North America, Africa or Australasia, and these small worms tend not to be included in material from surveys distributed for identification.

GEOGRAPHICAL DISTRIBUTION

THE WORLD

Despite the fact that the fauna of the southern hemisphere is poorly known, there does seem to be a genuine concentration of genera and species in the northern hemisphere. The whole family Lumbriculidae is represented in Africa, Australia and New Zealand only by the peregrine species *Lumbriculus variegatus*, and there are none reported from S. America. In contrast, the Phreodrilidae (Fig. 3. 1) occur only in the southern hemisphere with one poorly known representative reported from Ceylon. This southern circumpolar distribution might be interpreted in terms of Wegener's theory of Continental Drift. The other family with a notable concentration of species in Australia and New Zealand, the Haplotaxidae (Fig. 3. 2), is also well represented in Eurasia. The only western hemisphere records of this family to date, however, are of immature specimens, some of which may be referable to the only widely distributed species, *H. gordioides*.

Most genera of the Tubificidae and Naididae are cosmopolitan (Tables 3. 1, 3. 2) and several species have now been reported from every continent. Amongst these are the naidids: *Nais communis, N. variabilis, N. elinguis, Slavina appendiculata, Dero digitata, Aulophorus furcatus, Pristina aequiseta, P. longiseta* and the tubificids: *Limnodrilus hoffmeisteri, L. udekemianus, Tubifex tubifex* and *Aulodrilus pluriseta.*

Despite this, there still seems to be a greater diversity of species in the Northern hemisphere as, even within those genera listed as cosmopolitan, many species are restricted to a Palaearctic, Nearctic or even Holarctic distribution. Most of the genera with a more tropical and/or southern distribution are naidids (the three subgenera of *Dero*, plus *Allonais, Stephensoniana* and *Branchiodrilus*) and it is perhaps worthy of note that two of them are characterized by the possession of gills. One phreodrilid (*P. branchiatus*) and one tubificid that may have originated in tropical if not southern localities (*Branchiura sowerbyi*) also have gills, which are otherwise absent in the small aquatic oligochaetes.

THE CONTINENTS

While most naidid species are cosmopolitan, or are recorded from more than one continent, several of them occur only in S. America. There, more than twenty distinct species have been described, whereas only nine, at most, are limited to Europe, and even fewer are thought to be restricted to Africa or N. America. In the Tubificidae the monospecific genera, *Siolidrilus, Jolydrilus* and *Parandrilus* are South American, the genera *Antipodrilus* and *Macquaridrilus* (also monotypic) are limited to Australia. The strongest representation of genera in tropical and southern localities is restricted to

Table 3. 1. Distribution of Naididae by genus

NAIDIDAE

	Europe	Asia	Africa	N. Am.	S. Am.	Australasia
Chaetogaster	x	x	x	x	x	
Amphichaeta	x	x(B)		x		
Paranais	x	x	x	x	x (Titicaca)	
Waspa		x			x	
Homochaeta	x	x	?		x	
Ophidonais	x	x		x	x	
Specaria	x	x		x		
Uncinais	x	x(B)		x		
Nais	x	x	x	x	x	x
Slavina	x	x	x	x	x	x
Vejdovskyella	x	x	x	x		
Arcteonais	x	x		x		
Ripistes	x	x(B)				
Piguetiella	x	x				
Stylaria	x	x	x	x		
Neonais		x(B)				
Haemonais	x	x	?	x	x	
Branchiodrilus		x	x			x
Allodero	x	x	x	x		
Dero	x	x	x	x	x	x
Aulophorus	x	x	x	x	x	x
Allonais		x	x	(x)	x	x
Stephensonia		x	x			
Pristina	x	x	x	x	x	x

(B) = Baikal.

Table 3. 2. Distribution of Tubificidae by genus

TUBIFICIDAE

	Europe	Asia	Africa	N. Am.	S. Am.	Australasia
Tubifex	x	x	x	x	x	x
Psammoryctides	x	x		x		
Isochaeta	x	x		x		
Limnodrilus	x	x	x	x	x	x
Peloscolex	x	x		x	(x)	
Potamothrix	x	x	?	x		(x)
Ilyodrilus	x		?	x		
Antipodrilus						x
Paranadrilus					x	
Epirodrilus	x		?		x (Titicaca)	
Rhyacodrilus	x	x	x	x		x
Monopylephorus	x	x	x	x	x	x
Bothrioneurum	x	x	x	x	x	
Branchiura	x	x	x	x	x	x
Aulodrilus	x	x	x	x	x	x
Siolidrilus					x	
Telmatodrilus	x	x		x		x (Tasmania)
Clitellio						
Phallodrilus	x			x		
Spiridion	x					
Smithsonidrilus				x		
Thalassodrilus	x					
Jolydrilus					x	
Limnodriloides	x		x	x		
Maquaridrilus						x

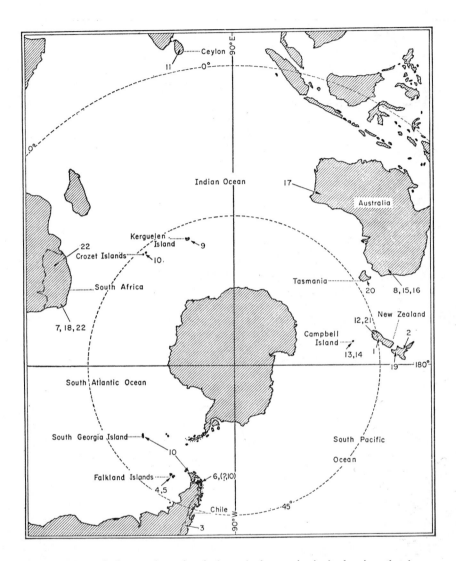

Fig. 3.1. Map of the southern hemisphere (polar projection) showing the known distribution of *Phreodrilus* species.

1=*subterraneus*. 2=*beddardi*. 3=*branchiatus*. 4=*niger*. 5 to 7=*niger* (as *albus, pellucidus, africanus*). 8=*notabilis*. 9=*kerguelenensis*. 10=*crozetensis*. 11=*zeylanicus*. 12=*lacustris*. 13=*campbellianus*. 14=*litoralis*. 15=*goddardi*. 16=*fusiformis*. 17=*novus*. 18=*africanus*. 19=*Schizodrilus* spp. 20=*Tasmaniaedrilus*. 21=*P. mauianus*. 22=*niger*, new material. The new species *mauienensis* is known from New Zealand, North and South Islands.

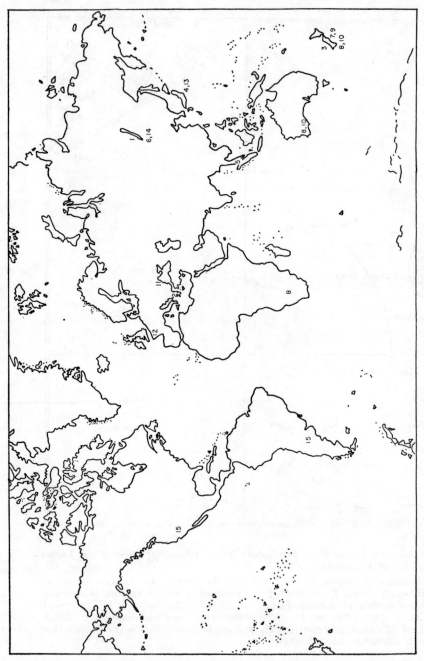

Fig. 3.2. The geographical distribution of *Haplotaxis* species: *H. gordioides gordioides*—at least holarctic, not shown. 1. *H. gordioides ascaridoides* (L. Baikal); 2. *H. gordioides dubius* (L. Ochrid); 3. *H. heterogyne*; 4. *H. gastrochaetus*; 5. *H. vermivorus*; 6. *H. ignatovi*; 7. *H. violaceus*; 8. *H. africanus*; 9. *H. smithii*; 10. *H. hologynus*; 11. *H. bureschi*; 12. *H. leruthi*; 13. *H. glandularis*; 14. *H. den-*

Table 3.3. Check list of Aquatic Oligochaetes from Australia, New Zealand, Tasmania, Auckland, Campbell, Stewart Islands, Macquarie Island—excluding Enchytraeidae and "megadriles".

NAIDIDAE
Nais
  elinguis
Allonais
  inaequalis
  pectinata
  ?paraguayensis
Dero
  nivea
  digitata
  dorsalis
Slavina
  appendiculata
Branchiodrilus
  hortensis
Aulophorus
  flabelliger
  furcatus

Pristina
  aequiseta
  longiseta
  proboscidae
  idrensis

TUBIFICIDAE
Tubifex
  tubifex
Limnodrilus
  hoffmeisteri
  claparedeianus
  udekemianus
Aulodrilus
  pigueti
  pluriseta
  Rhyacodrilus
  coccineus
  simplex
Monopylephorus
  rubroniveus
  irroratus
?Clitellio
  abjornseni sp. inq.

Macquaridrilus
  bennettae
Potamothrix
  bavaricus
Antipodrilus
  davidis
  timmsi
Branchiura
  sowerbyi
Telmatodrilus
  multiprostatus
  pectinatus

AEOLOSOMATIDAE
Aeolosoma
  niveum
  hemprichi

LUMBRICULIDAE
Lumbriculus
  variegatus

PHREODRILIDAE
Phreodrilus
  subterraneus
  beddardi
  mauienensis

HAPLOTAXIDAE
Haplotaxis
  africanus
  hologynus
  heterogyne
  violaceus
  smithii

  lacustris
  notabilis
  goddardi ⎫ on *Astacopsis*
  fusiformis ⎭
  novus
  campbellianus
  litoralis
  nothofagi
  major
?Tasmaniaedrilus
  ?tasmaniaensis sp. ing.

Aulodrilus and *Bothrioneurum*, but again these are also well represented in the northern hemisphere.

In contrast, the Lumbriculidae are more limited in distribution, with over 35 species known only from Asia, nearly 40 in Europe and about 10 in N. America. The fauna of each continent is distinct, with the three following exceptions. In addition to *Lumbriculus variegatus* previously mentioned, two species are represented in two continents. *Stylodrilus heringianus* is known from western Asia and from Europe and its presence in the St. Lawrence drainage area of N. America may be due to introduction via shipping, although a recent record from Manitoba is difficult to include in such a theory. The American species *Eclipidrilus lacustris* has now been reported in Britain.

The Australasian fauna is particularly poorly known, but thanks to responses to an appeal for collections, a preliminary list of species can be given here which includes over 50 names (Table 3. 3). Most of the species found to date are either cosmopolitan or endemic to the region, and only a few species such as *Allonais inaequalis* and *A. pectinata* have intermediate distributions which include Africa, Asia or S. America.

DISCONTINUOUS DISTRIBUTIONS AND ENDEMICS

One or two genera seem to have distributions worthy of special mention. The tubificid genus *Telmatodrilus* has a discontinuous distribution, being known from the Sierra Nevada mountains of California, Onega Lake, and Kamchatka, U.S.S.R., and Lake Pedder, in south-west Tasmania, where two species were discovered (Fig. 3. 3). This genus is distinguishable from the rest of the Tubificidae by several fundamental characters, including the form of the prostate glands, which resemble those of some Lumbriculid genera. The genus *Ilyodrilus*, the taxonomic limits of which have only recently been clarified, is primarily limited to N. America west of the Rocky Mountains, with one species which is probably cosmopolitan, *I. templetoni*. *Epirodrilus* has been reported from Lake Titicaca, Greece, Czechoslovakia and (probably) South Africa, with a different species in each locality (Fig. 3. 4). *Isochaeta* species are known from a series of limited localities in Europe, N. America and Asia, but otherwise such strongly localized distributions are rare, except in groundwater species (many *Haplotaxis* species, *Trichodrilus*) and species restricted to known centres of endemicity. Of these, the most significant are, of course, Lake Baikal, the Balkans Lake—series (including L. Ochrid, L. Prespa, L. Dojran and L. Skadar) together with Lake Titicaca and Lake Tahoe (Fig. 3. 5) although the supposed endemics in the last two may prove to be more widely distributed when the areas adjacent to them are studied in more detail. The whole subject of endemic speciation has to be regarded with some caution, however, as many systematists readily fall into the trap of deliberately looking for minute differences between specimens from localities known to contain endemic species and specimens from other, less

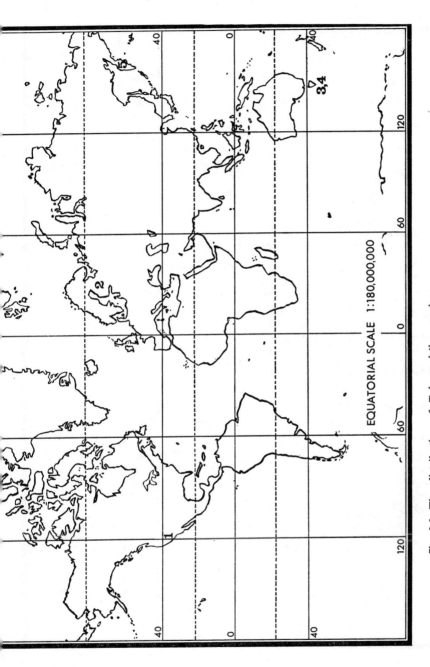

Fig. 3.3. The distribution of *Telmatodrilus* species.
1. *T. vejdovsky*; 2. *T. onegensis*; 3. *T. pectinatus*; 4. *T. multiprostatus*; 5. *T. ringulatus*.

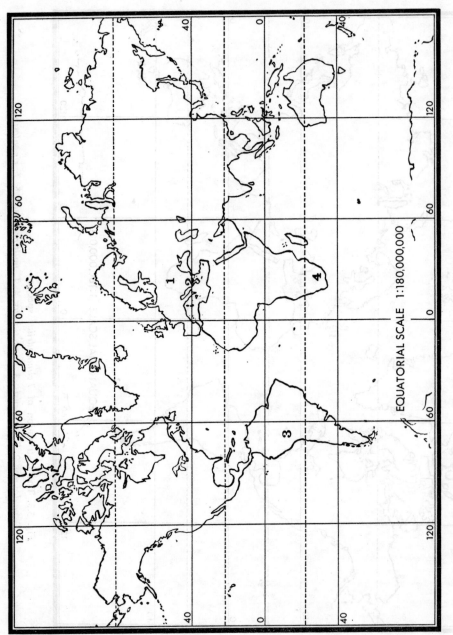

EQUATORIAL SCALE 1:180,000,000

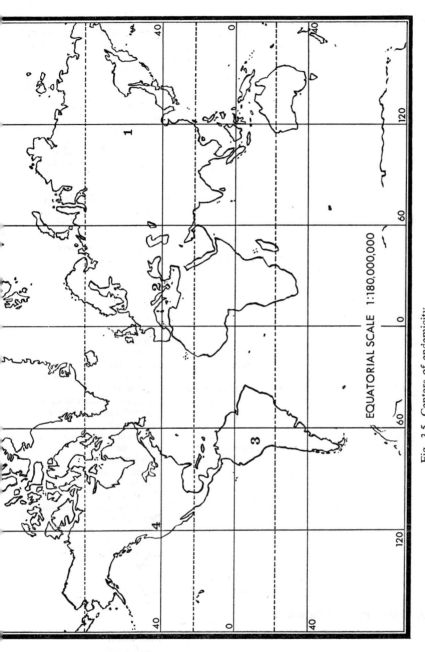

Fig. 3.5. Centers of endemicity.
1. Baikal; 2. Balkans Lakes; 3. Titicaca; 4. Tahoe.

EQUATORIAL SCALE 1:180,000,000

Table 3. 4. Lists of species and subspecies regarded as endemic to certain localities.

Lake		*Species*
1. Lake Baikal	Dorydrilidae (?):	*Lycodrilus* sp.
	Lumbriculidae:	*Lamprodrilus*—18 endemic, two with wider distribution (includes former genera *Teleuscolex* and *Agriodrilus*)
		Stylodrilus (=*Bythonomus*).
		crassus, asiaticus, opisthoannulatus
		Rhynchelmis
		brachycephala, olchonensis
		Styloscolex
		asymmetricus, baikalensis, chorioidalis, kolmakovi, swarczewski
	Tubificidae:	*Peloscolex*
		werestschagini, inflatus
		Rhyacodrilus (?)
		korotneffi, multispinus
		Isochaeta
		baicalensis, arenaria
	Naididae:	*Amphichaeta*
		magna
		Uncinais
		minor
		Nais
		baicalensis
		?tygrina
		Neonais
		elegans
	Haplotaxidae:	*Haplotaxis*
		gordioides ascaridoides
2. Ochrid and adjacent Balkans	Lumbriculidae:	*Lamprodrilus*
		pygmaeus intermedius
		pygmaeus ochridanus
		Stylodrilus
		?leucocephalus sp. ing.
	Tubificidae:	*Isochaeta*
		dojranensis
		Psammoryctides
		ochridanus
		Potamothrix
		prespaensis, ochridanus
		Peloscolex
		stankovici, tenuis tenuis
		?cernosvitovi
		Rhyacodrilus
		punctatus
	Naididae:	None
	Haplotaxidae:	*Haplotaxis*
		gordioides dubius
3. Titicaca and adjacent waters	Tubificidae:	
		Epirodrilus
		antipodum
		Isochaeta
		?lacustris (=*baicalensis*)
		Limnodrilus
		neotropicus

Lake		Species
	Naididae:	*Pristina*
		peruviana
		Paranais
		salina, macrochaeta
		Slavina
		isochaeta
		Nais
		andina
4. Tahoe	Lumbriculidae:	*Kincaidiana*
		freidris
	Tubificidae:	*Peloscolex*
		beetoni
		Isochaeta
		nevadana
		Psammoryctides (?)
		minutus
		Rhyacodrilus
		brevidentatus
		(*Ilyodrilus frantzi* now known from several lakes in western N. America.)

exciting localities. Whilst geographical considerations may be included in attempts to define the limits of taxonomic groups, discussion of zoogeographical distribution should flow from systematic decisions based on all the available evidence and should not be relied upon as lending too much support to those decisions. Again, the distinction between the survival of ancient forms and local speciation in ancient lakes is one which may be readily confused. The lists of endemic species in Table 3. 4 is shorter than some authors might expect, but several species thought to be endemic to certain water bodies have been found elsewhere (*I. frantzi* from Lake Tahoe has been found in L. Washington and possibly another western American lake for example) or have been identified as belonging to species with a wider distribution.

REGIONAL PATTERNS

Very few other regional centres can be identified. The Ponto-Caspian-Aralian fauna may contain a number of distinct forms but many of the species involved (in *Potamothrix* and *Psammoryctides*, for example) are not adequately separable from other European forms. Within N. America, several groups of tubificid species can be recognized (Table 3. 5). The western species are often the same as, or related to, species known to occur in Asia, whereas the eastern species are often the same as or related to European species. The pan-American group consists of species almost entirely restricted to the continent. One exception (*Limnodrilus cervix*) has almost certainly been introduced into Europe (Kennedy, 1965) and another (*P. gabriellae*) is also known from S. America. There is also a north-south axis, with species such as *Rhyacodrilus montana* and *Aulodrilus americanus* restricted to the northern states and Canada so far as has been established.

Table 3. 5. The Tubificidae of N. America, excluding those endemic to Lake Tahoe (see Table 3.4).

Genus	Widely distributed or Cosmopolitan	Eastern	Western	Pan-American
Tubifex	tubifex	newaensis		
		ignotus		
		kessleri		
		americanus		
		pseudogaster		
		longipenis		
Psammoryctides		curvisetosus	californianus	
Limnodrilus	hoffmeisteri	maumeensis	silvani	cervix
	udekemianus	angustipenis		
	claparedeianus			
	profundicola			
Isochaeta		hamata		
Peloscolex		ferox	oregonensis	variegatus
		aculeatus		gabriellae
		benedeni		apectinatus
		intermedius		nerthoides
		carolinensis		multisetosus
		freyi		
		superiorensis		
		dukei		
Potamothrix		hammoniensis		
		bavaricus		
		moldaviensis		
		vejdovskyi		
Ilyodrilus		templetoni	perrierii	
			fragilis	
			frantzi	montana
Rhyacodrilus		coccineus		sodalis
Branchiura	sowerbyi			
Monopylephorus	rubroniveus	lacteus		
	irroratus			
	parvus			
Bothrioneurum	vejdovskyanum			
Aulodrilus	limnobius	americanus		
	pigueti			
	pluriseta			
Telmatodrilus				vejdovskyi
Clitellio		arenarius		
Limnodriloides		arenicolus		
		medioporus		
Phallodrilus		coeleprostatus		
		obscurus		
Spiridion		?insigne		
Smithsonidrilus		marinus		
Adelodrilus		anissoetosus		

PEREGRINE SPECIES

A consideration of the species found in the St. Lawrence Great Lakes raises the possibility that several of them have been introduced to these relatively recent lakes which have proved susceptible to successful invasion by several groups of organisms. Many of the species which occur quite commonly in these lakes are little-known outside them in N. America, but are well-known European species. Amongst these are the tubificids *Peloscolex ferox, Potamothrix moldaviensis, P. vejdovskyi* (otherwise known only from the R. Danube) and possibly even *P. bavaricus* and *P. hammoniensis* (Brinkhurst, et al., 1968). The last two have been found in various parts of N. America, Africa and Australia, usually as scarce semi-mature specimens that cannot be identified with certainty. As these two, and especially the latter, are such common members of the bottom communities of reasonably productive lakes and rivers in Europe, their scarcity in a few isolated localities in other continents suggests that they have been introduced. Their discontinuous distribution and scarcity in each locality outside of Europe may indicate that they have either not had time to become adapted or that they fail to adapt well to new environments. There is some evidence to suggest that *P. bavaricus*, at least, may reproduce asexually (if the discovery of specimens in which the reproductive organs are shifted forward of their usual position is any indication, as seems possible—see p. 492), and this may, in part, account for its ability to survive in places to which it has been introduced. A similar ability to reproduce by fission may account for the successful spread of *Lumbriculus variegatus* noted above, and it seems possible that *Branchiura sowerbyi* may also be able to reproduce vegetatively (to judge by the way in which it breaks up under stress).

The story of the spread of *B. sowerbyi* from S. E. Asia into European botanic gardens along with tropical aquatic lilies has been accepted for many years, but the discovery of the species in a wide variety of localities in Europe, Africa, Australia, and America (where it is now known from Canada as well as from states from the eastern, western and southern limits of the U.S.A. and from Brazil) has made this story less firmly established (Aston, 1968). The species is often larger in artificially warmed waters in temperate regions than in normal lakes and streams, but this may also be used as evidence to support the view that it was merely overlooked in the past in many temperate localities. After the first report of the species in the U.S.A. from Ohio in 1931, it was reported from as far away as California as early as 1950 and by 1968 it was known from practically every state. This seems to suggest that the species was present but unreported, as the speed at which it would have spread would be phenomenal if Ohio were the initial introduction site.

The distribution of several species known from N. America and Europe calls for particular mention. The first, *Potamothrix vejdovskyi* is known from the R. Danube, Lake Geneva (Lac Leman) in Switzerland, and from Moldavia, close to the Black Sea. It is not an abundant species, neither is the more widely distributed *P. moldaviensis*, and yet both form a significant part of the fauna of various regions within the St. Lawrence-Great Lakes. *Tubifex newaensis* is only rarely found in the Great Lakes and in Central Europe, but it is inexplicably absent from Britain and Western Europe, suggesting that it is not just a scarce holarctic form. However, specimens from the Great Lakes must be studied in greater detail before the identification can be completely justified. The distribution of *L. cervix* in Britain strongly suggests that it was introduced via Liverpool and London, and that it competes successfully only in the man-made canals through which it has spread (Kennedy, 1965). Two other predominantly American genera are represented in Britain and, indeed, mainland Europe, in places which scarcely support the view that they were introduced in recent times. The lumbriculid *Eclipidrilus lacustris* is known from the St. Lawrence-Great Lakes and from Bala Lake, Merioneth, a lake used to supply water to Liverpool via the R. Dee and a system of aqueducts. There is, therefore, a remote chance that specimens introduced into Liverpool docks could travel to Bala Lake, but it is much less reasonable to suppose that the introduction took place via the earlier Elizabethan harbour at Neston in the R. Dee. Similar problems exist in relation to the distribution of *Sparganophilus tamesis* (q.v.).

HABITAT

There are few instances in which habitat preferences for aquatic oligochaetes can be established with reference to the physico-chemical parameters of the environment or to specific plant associations or any other readily recognizable ecological criteria. One distinction that can be made is between the marine, the salt tolerant (largely brackish-water dwelling) and the freshwater species within the Tubificidae.

SALT-WATER SPECIES

In this category may be classed several species of *Peloscolex, Tubifex,* and the genera *Clitellio, Limnodriloides, Thalassodrilus, Adelodrilus, Phallodrilus,* and *Spiridion,* but some of the latter have been found in groundwater in inland localities (Hrabě, 1960). Some of these species have been found in samples taken on a transect line between Cape Cod and Bermuda (Sanders *et al.*, 1965) at depths of as much as 5000 m., and many are abundant on parts of the continental shelf, as in Cape Cod Bay and San Francisco Bay (Brinkhurst and Simmons, 1968). This may surprise many marine biologists, as there is a tendency to regard all marine oligochaetes as *Clitellio arenarius*.

ESTUARIES

Amongst the brackish water forms may be considered several species of

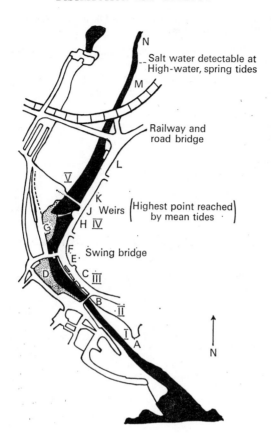

Fig. 3.6. Map of the Silverburn estuary, Castletown, Isle of Man. (The two weirs are shown as white lines on either side of J. A to N—sites of collections listed in Table 3.6.; stippled areas=mud banks).

Monopylephorus and others such as *Tubifex costatus, Peloscolex benedeni,* and several *Isochaeta* species. The transition between salt- and fresh-water has been studied in detail in two estuaries in Britain (Tables 3. 6, 3. 7 Fig. 3. 6, and Brinkhurst and Kennedy, 1962), and also in San Francisco Bay (Brinkhurst and Simmons, 1968).

The abrupt transition between the fresh- and salt-water species observed in the R. Stour and the Silverburn is related to human interference with the environment. The creation of weirs and locks to control flooding, to prevent the encroachment of salt-water over low-lying farmland, and to permit navigation has transformed most estuaries so that the transition zone between salt- and fresh-water has been restricted to a narrow zone below the lowest weir. The effect of this and other human "tidying" of the landscape has been to reduce the number of sites at which certain brackish-

Table 3. 6. Analysis of samples of about 50 worms from 13 stations in the Silverburn estuary. Isle of Man.

stations

Fresh-water species: Figures as percentages—Stations in Fig. 3·6

	A	B	C	D	E	F	G	H	J	K	L	M	N
Tubifex tubifex	—	—	—	—	—	—	—	—	2	—	42	41	28
T. nerthus	—	—	—	—	—	—	—	—	—	—	8	4	4
Limnodrilus hoffmeisteri	—	—	—	—	—	—	—	—	—	87	17	17	23
L. udekemianus	—	—	—	—	—	—	—	—	—	13	8	7	11
Stylodrilus heringianus	—	—	—	—	—	—	—	—	—	—	—	6	6
Enchytraeidae	—	—	—	—	—	—	—	—	—	—	17	25	28

Salt-water species:

	A	B	C	D	E	F	G	H	J	K	L	M	N
Tubifex costatus	19	33	22	29	59	29	45	41	32	—	—	—	—
T. pseudogaster	42	9	40	45	31	44	49	36	30	—	—	—	—
Clitellio arenarius	37	58	38	12	2	27	2	23	36	—	—	—	—
Peloscolex benedeni	2	—	—	14	8	—	4	—	—	—	8	—	—

Table 3. 7. Analysis of samples of about 25 to 35 worms from 16 stations in the Stour estuary, Essex, Great Britain.

stations

Fresh-water species:

	A	B	C	D	E	F	G	H	J	K	L	M	N	O	P	Q
Tubifex tubifex	—	—	—	—	—	—	—	—	3	30	—	—	16	3	—	—
T. ignota	—	—	—	—	—	—	—	—	—	—	—	—	—	3	—	—
Limnodrilus hoffmeisteri	—	—	—	—	—	84	10	82	85	60	28	88	80	88	92	46
L. udekemianus	—	—	—	—	—	—	—	—	6	—	—	—	—	3	8	—
Psammoryctides barbatus	—	—	—	—	—	—	—	—	6	64	—	4	—	—	—	5
Potamothrix hammoniensis	—	—	—	—	—	—	—	15	6	—	8	3	—	—	—	44
Aulodrilus pluriseta	—	—	—	—	—	—	—	3	—	—	—	9	—	—	—	5
Bothrioneurum vejdovskyanum	—	—	—	—	—	—	—	—	—	4	—	—	—	—	—	—
Peloscolex velutinus	—	—	—	—	—	—	—	—	—	—	—	—	3	—	—	—

Salt-water species:

	A	B	C	D	E	F	G	H	J	K	L	M	N	O	P	Q
Tubifex costatus	86	100	100	100	80	16	90	—	—	—	—	—	—	—	—	—
T. pseudogaster	14	—	—	—	—	—	—	—	—	—	—	—	—	—	—	—
Monopylephorus rubroniveus	—	—	—	—	20	—	—	—	—	—	—	—	—	—	—	—

water species can be found, as evidenced by the difficulty experienced in locating *Monopylephorus* species on Cape Cod, whereas they were apparently quite abundant at the turn of the century (Moore, 1905) .

In the study of San Francisco Bay, the predominantly salt-water areas with relatively constant salinity were inhabited by three species, *Peloscolex gabriellae*, *P. nerthoides*, and *P. apectinatus*, whereas the area subjected to the greatest fluctuation in salinity contained very few oligochaetes (unfortunately not available for identification). The innermost part of the Bay affected by the influx of fresh-water from the Sacramento and San Joaquin rivers is inhabited by *Limnodrilus hoffmeisteri* and *Ilyodrilus frantzi* and small numbers of *Peloscolex gabriellae*, *P. nerthoides* and *Paranais frici*, a naidid that has been found in both fresh- and brackish-water localities.

FRESH-WATER SPECIES

Whilst these and other qualitative studies of coastal sites and estuaries (Bulow, 1955, 1957; Lasserre, 1966, 1967, for example) serve to identify marine, and brackish- and fresh-water species, the variation in the proportional representation of species within each of these three associations is caused by factors as yet little understood. Some of this variation is undoubtedly due to pollution, as will be indicated later, but attempts to define habitat preferences between the many fresh-water species that inhabit unpolluted sites have, until very recently, met with limited success. At the superficial level, there is an obvious tendency for some naidids and lumbriculids to occur in stony streams whereas tubificids are more often found in the softer sediments of rivers and lakes. However, the fauna of small ponds is frequently limited to the ubiquitous *Tubifex tubifex* and *Limnodrilus hoffmeisteri*, possibly because of limitations in the ability of some species to invade such sites (the distribution mechanisms involved being unknown, however).

SUBTERRANEAN SPECIES

Species listed in publications on subterranean localities, ground waters, springs and wells are frequently those found in many other sorts of localities (Cernosvitov, 1939; Botea, 1963, for example) but sometimes species that appear to be primarily limited to such sites are described. Some of these, like *Haplotaxis gordioides*, are widely distributed, but many are known from a very restricted area. Many of them belong to families that may be thought of as ancient, or at least descended from ancient forms (*Haplotaxidae*, *Lumbriculidae*). Relatively few tubificids or phreodrilids have been described from groundwater localities (*Rhyacodrilus balmensis*, *R. subterraneus*, *R. lindberghi*, ?*Tubifex flabellisetosus* sp. dub., *Phreodrilus subterraneus*, with species such as *Isochaeta israelis* and *Epirodrilus michaelseni* from springs or wells) but several fall into the interesting group found in both brackish- and fresh-water springs (*Tubifex nerthus*, *Phallodrilus monospermathecus*, *Phallodrilus aequaedulcis*, *Spiridion insigne*) as noted by Hrabe (1960).

TERRESTRIAL SPECIES

Few species are found in localities approaching terrestrial conditions, but *Telmatodrilus vejdovskyi* was originally described from organically enriched wet places in alpine meadows in the Sierra Nevada mountains of California. Stout (1951) reported the discovery of *Aelosoma kashyapi* and *A. niveum* in two soil samples taken from the floor of the main crater on Raoul Island in the Kermadec group off New Zealand. The sites were thought to have been occupied by a lake at one time, and they lie close to streams draining into an existing lake, so that the discovery is perhaps not so extraordinary as one might at first suppose. Three aeolosomatids and some naidids and phreodrilids were reported by the same author (Stout, 1956, 1958) from forest litter cultures. The identity of the species concerned is, unfortunately, rather doubtful, but the record is sufficient to alert the attention of biologists to the existence of a few somewhat amphibious species such as *Pristina amphibiotica, P. idrensis* and *Aelosoma hemprichi.*

RIVERS

In the R. Thames at Reading, England, fourteen species were found. Whilst some species were found at most stations (*L. hoffmeisteri, Potamothrix moldaviensis*) several were present at only half the stations or less. Despite this, there was no obvious correlation between the distribution of the various species and the nature of the environment, as determined by visual inspection of the sediment (which varied very considerably) or by identification of the dominant emergent vegetation (Table 3. 8). The one exception to this is the clear relationship between the abundance of *B. sowerbyi* and the warm-water effluent from a power station (see below). The proportional representation of each *Limnodrilus* species was shown to vary through wide limits in surveys at the same localities in 1961-62 carried out by Kennedy (1965) but the overall relative abundance of worms from all stations sampled in 1959, 1961 and 1962 remained quite constant. Most of the species found in these surveys were listed by Hrabe (1941) in a comparative account of the species recorded from the Danube, Elbe, Moldau, Eger, Volga, Kama, Dneiper and Oka rivers (Table 3. 9), to which I have added the lists from the Susaa in Denmark (Berg, 1948) and the upper Danube (Korn, 1963). The lists for smaller British rivers are shorter than those for the larger European rivers cited, and the list from the R. Weaver is quite representative. Lists for the R. Stour and the Silverburn have been presented already. Wachs (1967) summarized a good deal of the literature on the occurrence of oligochaetes in running water in reporting on the fauna of the Fulda and Isar rivers, and the same author studied the fauna of the Werra, and also the Weser, which is formed by the union of the Werra and Fulda rivers (Wachs, 1963, 1964). These lists have also been added to Table 3. 9.

Many of these studies reveal quantitative as well as qualitative differences between the various stations with respect to the tubificid species present (and often species of other families as well—but there is so little comparative information that these are not included here). They agree in demonstrating an overall lack of correlation between the data obtained from chemical analyses of the overlying water and the kinds and proportions of the species present, and most conclude that the nature of the sediment is the most significant factor in determining the distribution of species. The dominance of Naididae amongst stones and plants, the occurrence of species like *Stylodrilus heringianus* and *Rhyacodrilus coccineus* in sandy sediments, and the frequency of *Potamothrix* and *Aulodrilus* species in

Tran-sect	Yards from south or north bank	Substratum	Vegetation	Depth in feet	1	2	3	4	5	6	7	8	9	10	11	12
1	S 1	Gravel	—	1	1	—	—	—	6	—	1	3	—	—	—	—
	5	Stones	—	2	3	10	—	—	3	—	—	—	—	—	—	—
	10	"	—	7	21	6	1	—	10	—	2	—	—	—	—	—
	20	"	—	12	2	—	—	—	—	—	—	C	—	—	—	—
	40	"	—	12·5	—	C	—	—	—	—	—	C	—	—	—	—
	55	Black mud and twigs	—	4	—	C	—	—	—	—	—	—	—	—	—	—
2	S 1	Mud	Acorus	1·5	C	—	—	—	C	—	1	C	1	—	—	1
	2	"	"	2	—	8	—	—	1	1	1	—	—	1	—	1
	5	Black, soft mud	Nuphar	3	2	20	—	—	8	—	2	2	—	—	—	—
	Middle	Shell gravel	—	10·5	—	C	2	—	2	—	1	2	—	—	—	—
	N 5	Black mud over stiff clay	—	4	2	2	1	—	6	—	—	6	—	—	—	—
	3	"	—	2	—	2	—	—	—	—	—	—	—	—	—	—
	1	"	—	1	2	1	—	—	6	—	—	—	—	—	—	1
3	S 1	Mud	Acorus, Glyceria	2	1	6	—	—	—	1	—	—	—	—	—	2
	S 4	Black mud	edge of Acorus	2·5	1	C	—	—	C	1	1	—	—	—	—	—
	N 1	(as N Tr. 2)	—	6	2	C	—	—	C	—	1	2	—	—	—	—
5	S 1	Oily black mud	Carex, Acorus	1	5	C	—	4	5	—	—	—	—	—	C	—
	3	"	Nuphar	2	5	C	4	1	5	—	—	—	—	—	2	2
	5	"	"	4	—	C	—	—	2	—	—	—	—	—	—	—
	8	"	"	6	—	1	—	—	C	—	—	4	—	1	—	—
	20	Stones	"	8	—	12	4	—	5	—	—	2	—	—	—	—
	N 1	(as N 2)	—	4	7	12	1	—	5	—	—	2	—	1	1	1
	Total (C=25)				79	248	13	5	123	1	9	96	1	2	28	6
	Percentage composition				12·9	40·6	2·1	0·8	20·1	0·2	1·5	15·7	0·2	0·3	4·6	1·0
	" in 1962 (C. Kenedy)				21·8	34·5	2·6	0·2	28·7	0·6	0·4	5·0	0·2	0·2	2·2	3·0

*1. Limnodrilus claparedeianus L. cervix
2. L. hoffmeisteri
3. Potamothrix hammoniensis
4. P. bavaricus
5. P. moldaviensis
6. Tubifex ignotus
7. T. tubifex
8. Psammoryctides barbatus
9. Psammoryctides albicola
10. Rhyacodrilus coccineus
11. Branchiura sowerbyi
12. Limnodrilus udekemianus

Table 3. 9. The Tubificidae recorded from some European rivers. (The absence of a record for a species may only mean that it has been overlooked.)

	UPPER DANUBE	DANUBE	ELBE	VLTAVA	EGER	VOLGA	KAMA	DNIEPER	OKA	FULDA	WERRA	WESER	ISAR	WEAVER	THAMES
Tubifex															
tubifex	x	x	x	x	x	x	x		x	x	x	x	x	x	x
newaensis						x	x	x	x						
ignotus	x	x	x	x		x			x	x					x
nerthus										x					
Psammoryctides															
barbatus	x	x	x	x		x	x		x	x	x	x	x	x	x
albicola		x	x						x	x	x				x
moravicus		x		x		x				x					
Potamothrix															
hammoniensis		x	x			x	x	x	x	x		x	x	x	x
bavaricus		x			x						x			x	x
vejdovskyi		x													
moldaviensis		x				x		x	x	x		x		x	
isochaetus		x													
Isochaeta															
michaelseni		x				x	x	x							
Ilyodrilus															
templetoni			x					x	x						
Limnodrilus															
hoffmeisteri	x	x	x	x	x	x	x	x	x	x	x	x	x	x	x
claparedeianus		x	x			x	x	x	x	x	x	x	x		x
udekemianus	x	x	x	x	x	x	x	x	x	x		x	x	x	x
profundicola		x						x					x		
Peloscolex															
ferox		x	x		x	x	x	?	x	x					
Aulodrilus															
pluriseta	x							x		x			x		
limnobius	x							x		x					
Rhyacodrilus															
coccineus	x	x	x	x	x	x			x	x			x		x
falciformis										x					
Bothrioneurum															
vejdovskyanum			x	x											
Branchiura															
sowerbyi		x		x											x

mud suggests that more detailed analyses of sediments may reveal patterns of distribution. Most of these studies, however, demonstrate neither the variation in the proportional representation of species in successive samples from a single locality, nor the wide variety of sediments that most of the commoner species can inhabit. The ability of many species to co-exist in the same sediment suggests that micro-habitats may not be recognizable by analyses of the physical nature of the habitat.

The variation in the abundance of three tubificid species in samples from Ditton Brook, a small tributary of the R. Mersey near Liverpool, England, was studied by Brinkhurst and Kennedy (1965). This stream is polluted by farm wastes and the coal-washing plant of a small mine. The sediments consist of an intimate mixture of fine silt and coal dust. They probably vary less than those found in an unaltered habitat, although this was not established by analysis. The proportions in which the three species were present varied widely from sample to sample, but consistent differences between stations with respect to the relative abundance of the species were established by averaging the data from at least six samples from each. These studies serve to reinforce the conclusion that the worms show a marked tendency to clump together in the sediment, as can be demonstrated quite readily by a visual inspection of any laboratory culture or polluted stream. During the period of investigation, which extended with varying degrees of intensity of effort from 1959 to 1964, *Limnodrilus* species, primarily *L. udekemianus*, increased in abundance throughout the system. All three species co-existed at all four stations studied throughout the six years, but the proportions present varied markedly from place to place and from time to time. The temporary invasion of the habitat by *Chironomus* larvae seemed to affect the tubificid population adversely, virtually rendering the patches of stream bed occupied by the larvae uninhabitable to the worms. The worms rapidly recolonized these patches once the midges emerged. The diversity of the fauna in this stream was severely reduced by the degree of pollution suffered, but equally significant local variations were established in the Thames, where the invertebrate fauna was much more diverse and several species of fish were present. Because of this sort of variation great care has to be exercised in interpreting estimates of the abundance of tubificids found in river surveys. No satisfactory description of the micro-habitats of river-dwelling tubificids has been published to date.

LAKES

The list of species present in European lakes does not differ markedly from that derived from studies of rivers in the same region (Table 3. 10). The number of species appears to increase with the size of the lake, probably because of the greater diversity of microhabitats present in the larger lakes. Many small lakes, and those in mountainous or sub-alpine regions, contain *T. tubifex*, *P. ferox*, *Ps. barbatus* and few other species (Brinkhurst, 1964a).

Table 3. 10. The Tubificidae recorded from some European Lakes. (The absence of a record for a species may only mean that it has been overlooked.)

	ANNECY	MAGGIORE	MERGOZZO	THUNNERSEE	VATTERN	WIGRYSEE	ZURICH	BODENSEE	GENFERSEE (le Léman)	NEUCHATEL	LUGANO	ESROM	DISNAI	BALA	WINDERMERE
Tubifex															
tubifex	X	X	X	X	X	X	X	X	X	X	X	X	X		X
newaensis															
ignotus	X	X					X	X		X	X	X	X		
Psammoryctides															
barbatus	X	X	X	X	X	X	X		X	X	X	X	X		
albicola		X					X					X			
Potamothrix															
hammoniensis	X	X	X	X		X		X	X	X	X	X	X		
bavaricus									X				X		
vejdovskyi									X						
moldaviensis								X							
heuscheri		X					X	X	X						
Ilyodrilus															
templetoni		X							X						X
Limnodrilus															
hoffmeisteri	X	X		X	X	X	X	X		X	X	X	X	X	X
claparedeianus		X				?	X	X	X	X		?			
udekemianus		X				X	X	X	X	X		X			
profundicola			X					X	X	X					
Peloscolex															
ferox	X	X	X	X		X	X	X	X	X	X			X	X
velutinus	X			X				X	X	X	X	X			
Aulodrilus															
pluriseta									X					X	X
limnobius									X						
Rhyacodrilus															
coccineus		X						X							
falciformis	X				X			?	X						
ekmani					X			X							
Bothrioneurum															
vejdovskyanum	X	X							?				X		
Branchiura															
sowerbyi		X	X				X		X						

In lakes studied in some detail, certain species may appear to be distributed in a characteristic way in relation to depth, but there are frequent exceptions to each supposed distribution. *Potamothrix hammoniensis* is often found from the littoral to profundal, in Ochrid, Dojran, Prespa (Sapkarev, 1963, 1964, 1965) and other European lakes (Brinkhurst, 1964a), and this species together with *T. tubifex* was reckoned to be ecologically tolerant by Rzoska (1936). The former species is often found in productive lakes, replacing *P. ferox*, which is common in oligotrophic or mesotrophic lakes. The ecologically tolerant, widely distributed *T. tubifex* is not always numerically abundant, except in the most productive or polluted lakes, and also the least productive lakes where few competing species are present. It is less abundant in moderately productive lakes in which a number of species occur. *Psammoryctides barbatus* is often present at most, if not all depths (Ekman, 1915; Lastockin, 1927; Valle, 1929; Berg, 1938) but it is limited to the zone from 0-100 m in Genfer See (maximum depth 309 m). Similarly, *L. udekemianus* is frequently restricted to littoral situations, but it was reported in sub-littoral and profundal zones by Valle (1927). *Peloscolex velutinus* and *Limnodrilus profundicola* (=*helveticus*) are often found in the profundal zones of lakes, but may be limited to shallow water on occasion (L. Dojran-Sapkarev, 1963).

In Rostherne Mere, a small lake in England, the bottom fauna is absent from the sediments of the deepest part of the lake, presumably because of the large quantity of bird faeces that create a severe oxygen demand by the sediments (Brinkhurst and Walsh, 1967). In this lake, all of the common benthic organisms may be found in samples taken from 1 m to 23 or 25 m, the zone from 25 m to the deepest point (30 m) being uninhabited. Different species do reach maximal abundance at different depths, however (Fig. 3. 7), and it is easy to see that a cursory sampling programme might suggest spurious depth-limitations.

A few species seem to occupy recognizeable sediments, *Rhyacodrilus coccineus* and the lumbriculid *Stylodrilus heringianus* being mostly restricted to sandy sediments whilst *Aulodrilus pluriseta* seem to occur most often in mud rich in plant fragments. Most of the studies on which correlations between the distribution and abundance of oligochaetes and the nature of the sediments are discussed have been based on a superficial examination of those sediments. Where careful analyses have been made, few very clear correlations between the variations in particle size or total organic matter present and the distribution and abundance of worms have been demonstrated (Della Croce, 1955; Brinkhurst, 1965a, 1967).

Sediment-preference tests have been carried out by Wachs (1967) and Zahner (1967). Most of the experiments were performed on *T. tubifex*, although a species with a more limited distribution might have been a more reasonable choice of experimental animal. In work with Bodensee sediments, Zahner showed that *T. tubifex* preferred the sandy sediments from 4 m. to the muddy ones from 40 m. despite the fact that, in the field,

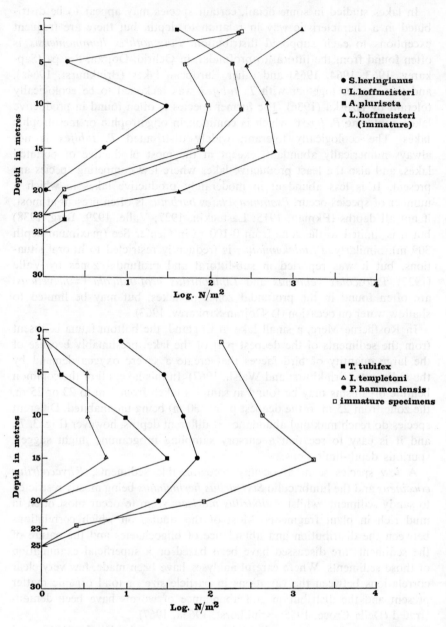

Fig. 3.7. The distribution of various oligochaete species in relation to depth in Rostherne Mere, England.

it was limited to the lower sediment. In the study by Wachs the worms were shown to move into the sediment with the highest nutritional potential in terms of organic carbon and nitrogen regardless of the texture of the sediment. Neither study considered the bacterial populations of the sediment or the possibility of species interractions, and only some of the possible sediment-type combinations were tested by Wachs, who used 25 different substrate types.

ST LAWRENCE—GREAT LAKES

Some evidence of changes in the relative abundance of species with differing degrees of eutrophy of lakes has recently been demonstrated in the St Lawrence—Great Lakes (Brinkhurst *et al.*, 1968). Thirty-one tubificid species have been recorded, including the cosmopolitan, pan-American and eastern fresh-water species listed above (Table 3. 5). *Stylodrilus heringianus* is the only common lumbriculid. with *Eclipidrilus lacustris* and *Lumbriculus variegatus* apparently restricted to the littoral. Some species are thought to have been introduced as they are often restricted to the lakes, and are absent from the other large Canadian lakes (Table 3. 11).

Several species are entirely or mainly restricted to the unproductive upper lakes (L. Superior, L. Michigan except the southern end, L. Huron

Table 3. 11. The occurrence of tubificids in some Canadian Lakes. (Numbers indicate number of samples containing each species.)

	NIPIGON	GREAT SLAVE	ATHABASKA	WOLLASTON	CREE	PATRICIA DISTRICT LAKES
Limnodrilus hoffmeisteri	15	6	21	8	16	16
L. udekemianus	1	0	1	0	0	1
L. claparedeianus	2	2	8	0	0	6
L. profundicola	18	2	2	0	0	0
Tubifex tubifex	0	0	0	0	0	1
T. kessleri americanus	1	2	14	0	1	0
Ilyodrilus templetoni	14*	0	1	3	6	0
Rhyacodrilus montana	19	12	54	43	38	18
R. coccineus	0	2	0	0	0	0
Aulodrilus americanus	9	0	5	2	4	10
Number of samples examined	133	40	83	65	80	58

* including 8 in a lake on an island within Nipigon.

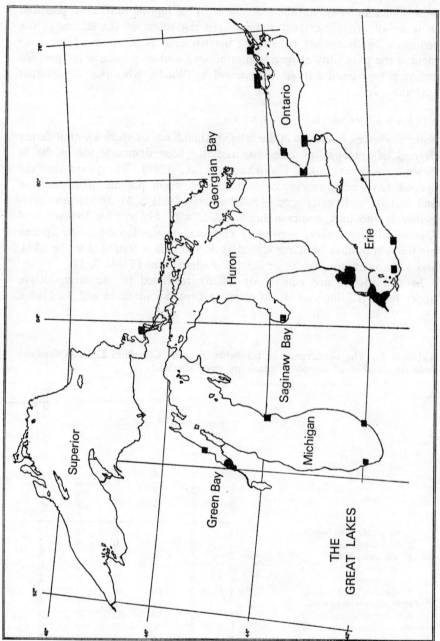

Fig. 3.8. Areas of the St. Lawrence Great Lakes inhabited by species such as *Limnodrilus cervix* and *Peloscolex multisetosus*

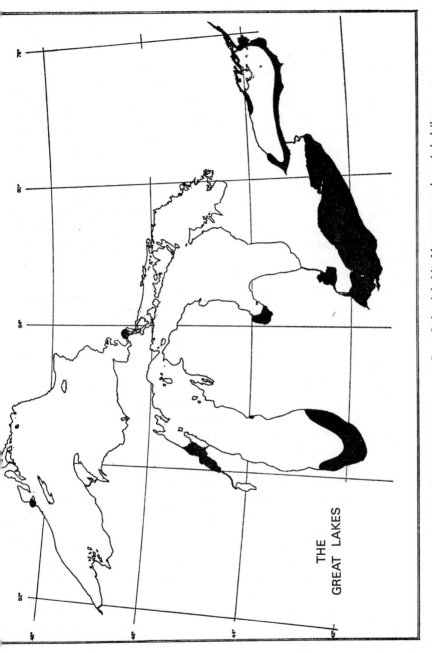

Fig. 3.9. Areas of the St. Lawrence Great Lakes inhabited by genera such as *Aulodrilus*, *Potamothrix* and species like *Peloscolex ferox*.

and Georgian Bay), including *Rhyacodrilus* species, *Tubifex kessleri americanus* and *Peloscolex variegatus*. Two other species, *T. tubifex* and *S. heringianus*, occur with these, but also inhabit the less productive parts of the lower lakes. More productive areas such as southern Lake Michigan, central Lake Erie, parts of eastern Lake Erie, the littoral of Lake Ontario and many bays, such as the outer parts of Saginaw Bay (Lake Huron) and the Bay of Quinte (Lake Ontario), are colonized by *Aulodrilus* and *Potamothrix* species, together with *Peloscolex ferox*. Western Lake Erie, Hamilton Bay, Green Bay, the inner parts of Saginaw Bay and Bay of Quinte contain *Limnodrilus* species in abundance and species such as *L. cervix*, *L. maumeensis*, and *Peloscolex multisetosus* are largely restricted to such places (Figs. 3. 8, 3. 9).

In the short period following the revision of the systematics of the N. American aquatic oligochaetes several accounts of the fauna of the Great Lakes have been published. Hiltunen (1967, 1969) reported on the fauna of parts of Lake Michigan and western Lake Erie, Johnson and Matheson (1968) studied the fauna of Hamilton Bay and the adjacent parts of Lake Ontario, and Veal and Osmond (1969) surveyed the inshore fauna of Lake Ontario and Lake Erie. Study collections from Georgian Bay, Lake Erie and Lake Ontario provided by the Great Lakes Institute, University of Toronto, formed the major basis of a report on the oligochaetes of the Great Lakes (Brinkhurst *et al.*, 1968) which also included data on samples from L. Michigan, L. Huron, the Bay of Quinte, Saginaw Bay and a few samples from L. Superior, most provided by collaboration with Dr A. M. Beeton.

In western L. Erie, the abundance of oligochaetes in the mouths of the three principal rivers was noted by Stillman Wright in 1930, and the extension of the zones dominated by worms was plotted some 30 years later (Carr and Hiltunen, 1965). The specific composition of the worm populations occupying these zones was demonstrated by Hiltunen (1969), and Veal and Osmond (1969). *Limnodrilus* species dominate the assemblage in the river mouths, and *Potamothrix* and *Aulodrilus* species are common in the rest of the basin. *Branchiura sowerbyi* is also present here. The differences between the western basin and the shallow-water areas of the central and eastern basins of L. Erie can be illustrated by reference to data presented by Veal and Osmond (Table 3. 12). This shows that, whilst the relative abundance of species differs significantly, most can be found at each location, which explains the apparent lack of correlation between lake type and the qualitative species lists available for European lakes (Brinkhurst, 1964). The same result was obtained in a detailed comparison of the fauna of Saginaw Bay based on four samples at each station, and the list of species discovered in an analysis of 60 samples from one of those stations (Brinkhurst, 1967). In that study, only four species were recorded from an analysis of four samples, but a further eleven species were found to be present when 60 samples were analysed. These species were present

**Table 3. 12. The inshore fauna of Lake Erie (from Veal and Osmond, 1969).
Percentage of sampling locations at which each species was found.**

		Western basin	Central basin	Eastern basin
Limnodrilus				
	hoffmeisteri	98	64	44
	cervix	76	19	4
	maumeensis	33	2	0
Tubifex				
	tubifex	7	38	24
Peloscolex				
	ferox	5	64	8
Potamothrix				
	moldaviensis	5	43	32
Stylodrilus				
	heringianus	0	38	24

at an average of much less than one per sample, however, and the four common species were found in almost the same average number per sample in the two analyses.

Several of the changes in abundance of species will be discussed later under the heading of pollution biology, but the problem of coexistence of a number of species at one locality remains; apparent changes from lake to lake or even from sampling point to sampling point seem to be revealed as changes in relative abundance alone if enough samples are analysed.

POLLUTION BIOLOGY

The abundance of tubificids in organically polluted water may be readily explained. In the first instance, competition for food and living space is reduced once the majority of the benthos have been killed by protracted exposure to water lacking oxygen. As long as some oxygen is available from time to time, and the poisonous products of anaerobic breakdown of organic matter and metabolic wastes do not accumulate, then the rich food supply permits spectacular growth in the tubificid population. If deoxygenation becomes too severe or toxic products are allowed to accumulate, then the worms will also be killed, as in septic streams. When a mass of tubificids, together with some of the substratum, is removed from a stream-bed red with worms and kept in a shallow pan with a little water, the worms rapidly die, which may indicate the need for flowing water. The poor representation of tubificids in small ponds may also reflect this, the duration of total deoxygenation and high temperature acting together being the most likely limiting factors other than means of dispersal.

In rivers less grossly polluted with organic matter, the standing crop of tubificids may be low, but the productivity high. This stage is reached in the benthic community so characteristic of many rivers—that comprising leeches (particularly *Erpobdella*), *Asellus*, tubificids, chironomids, and *Cladophora*. The unusual abundance of predatory leeches is probably the key to the relatively small number of tubificids present at any one time.

The absence of worms, or even a marked reduction in the normal complement, may well indicate the presence of traces of poisonous substances, such as the heavy metals. Some of the literature on the intolerance of oligochaetes to such poisons was reviewed elsewhere (Hynes, 1960) and has been confirmed in recent studies (including Learner and Edwards, 1963).

The simple phenomena described above show how little can be achieved by simply recording "oligochaetes" in a survey. When the species are identified, the range of observations regarding the "health" of the environment may be extended to include the less drastic disturbances that are frequently the cause of disagreement between parties concerned about the condition of a water body.

As indicated above, certain species are usually found in salt or brackish localities, or in estuaries where the salinity fluctuates, and the presence of these species in inland sites may suggest that they are enriched by salt either naturally or as a result of human activity. The best documented instance of this is the R. Werra, investigated by Wachs (1963), and the discovery of specimens of the brackish-water species *Paranais litoralis* in Saginaw Bay, L. Huron, fairly close to the salt-enriched Saginaw River (Brinkhurst, 1967).

When collections of worms from organically enriched sites are identified, it soon becomes apparent that some species are more tolerant to the situation than others. During a study of the reclamation of part of the R. Derwent, England (Brinkhurst, 1965) it was noticed that *T. tubifex* became less abundant as conditions improved but that other species came back into the community (*Potamothrix* species, *Rhyacodrilus coccineus*, *Psammoryctides barbatus*, *Aulodrilus pluriseta* and others—Table 3. 13).

In western L. Erie it was established that some species tolerated polluted inflows quite readily, that others seemed only to occur in the open lake, and that a third category of species seemed to be distributed along the shore line with no reference to polluted inflows (Hiltunen, 1969). The affect of the polluted Saginaw River is less noticeable in Saginaw Bay, L. Huron, in which the contribution of water from the open lake and the relatively small size of the inflow combine to produce a situation far less drastic than in western L. Erie (Brinkhurst, 1967). The three different species-associations recorded in the St Lawrence—Great Lakes seemingly reflect the degree of pollution or organic enrichment (eutrophication) of various parts of the system, as discussed above.

The presence of *Branchiura sowerbyi* in a warm water effluent in the R. Thames at Reading, England (Mann, 1965) and the frequency with which it is reported along with *L. hoffmeisteri* in organically enriched localities in tropical regions suggest that it may be a good "indicator" of thermal pollution in temperate zones. Its supposed introduction via tropical houses

Table 3.13. The Tubificidae recorded from the R. Derwent in 1958 and in 1959-1962

After improvement (7 surveys)				Before improvement (2 surveys)				
4	6	7	13	4	6	7	13	Stations
*	*	*	*	*	*	*	*	T. tubifex
—	*	—	—	—	—	—	—	T. ignota
*	*	*	*	*	*	*	*	L. hoffmeisteri
—	*	—	*	—	—	—	—	L. udekemianus
—	*	*	*	—	—	—	—	L. claparedeianus
—	—	*	—	—	—	—	—	I. templetoni
*	*	*	*	—	—	—	—	P. hammoniensis
*	—	—	—	—	—	—	—	P. moldaviensis
*	*	*	—	—	—	—	—	R. coccineus
—	*	*	*	—	—	—	—	B. vejdovskyanum
*	*	*	*	—	—	—	—	P. barbatus
—	—	—	*	—	—	—	—	A. pluriseta
6	9	8	8	2	2	2	2	Number of species
5	7	5	5	2	2	2	2	Maximum in any one survey

in botanic gardens and the like may be suspect now the species has been found in all parts of the world. It would appear to be commoner in Asia than in Europe or N. America, and it does seem to grow much larger in the heated effluent at Reading than in any cold-water sites examined by me. Its use as an indicator of warm-water conditions may, therefore, be limited, and this should be investigated further. The size of the worms and the presence of the sexually mature specimens may prove to indicate more than the mere presence of the species.

An undue abundance of worms in relation to the arthropods in a com-

munity, together with a wide diversity of species of worms, may be taken
to indicate pollution by insecticides by indiscriminate aerial spraying, wash-
ing water from hot-houses, the release of sheep-dip to a stream (Hynes,
1961), and similar activities. This result is borne out by a laboratory study
of the resistance of worms to insecticides (Whitten and Goodnight, 1966),
in which DDT, for instance, was found to be non-toxic to worms even at
concentrations exceeding 100 mg/1.

Further studies of polluted and unpolluted sites, and of reclamations,
need to be made, but it seems likely that the presence of certain species in
the community and the absence of others may well suggest certain minimum
conditions that are to be found in that situation in the same way as has
already been suggested many times in studies on other families of invertebrates.

The special value of tubificids lies in the detection of pollution in low-
land, muddy streams and rivers and on lake-beds where silt-dwelling
communities are normally present.

The mere presence or absence of species reveals a great deal about the
condition of the benthos, particularly when suspect areas are compared to
known natural localities, previous records from the same locality, or records
from closely similar situations at a nearby locality, but the data is very
greatly increased in value if a roughly quantitative assessment of the fauna
can be made, even if by a selective technique.

All sampling techniques are selective, but so long as the method employed
is standardized then the results should be comparable. Absolute quantitative samples, reflecting the quantitative relationship between each species
and a known area of the habitat, are very difficult, if not impossible, to
obtain, but they are only required for detailed research programmes on
productivity and need not confuse the issue here, or deter the investigator
from making biological surveys.

The detailed study of the quantitative relationships between four species
of tubificid in a stream near Liverpool indicated constant differences
between four adjacent sites so far as *T. tubifex* and two *Limnodrilus*
species (of which *L. hoffmeisteri* was by far the most abundant) were concerned. The fourth species (*L. udekemianus*) grew more abundant with
time, but was always most abundant at one station and became progressively
less abundant upstream (Brinkhurst and Kennedy, 1965). This data was
collected during a long and tedious research programme, but so far as
pollution assessment is concerned, a few small samples would have shown
that the common species were *T. tubifex* and *L. hoffmeisteri* and that
tubificids usually made up 100% of the fauna. Under these circumstances
careful "research quality" quantitative samples would not have been
necessary.

Unless a long series of accurate samples is collected, a statistical evaluation of variation between samples taken at one time and place cannot be
made; there is no need to quote the actual numbers obtained from less

rigorous sampling programmes. Plotting the frequency of organisms on an appropriate logarithmic scale is an effective way of recording data in a simplified way which prevents attention being focussed on small variations in numbers which may not be statistically signficant. Code letters may be apportioned to these logarithmic scales in order of magnitude, i.e. $P=1$, $F=5$, $C=25$, $A=125$, etc (where P means present, F means few, C means common and A means abundant). In lake studies, the abundance of species may be usefully plotted as contours based on such scales. Such plots show quite clear zones in some instances, which correlate well with known data on the distribution of other benthic species and with physico-chemical aspects of the environment. For instance, it is quite clear that *L. hoffmeisteri* is largely responsible for the heavy concentration of tubificids in the river mouths in western L. Erie, and that the plot of the percentage abundance of *L. hoffmeisteri* as compared to other oligochaetes reflects the known distribution of river-water leaving Saginaw Bay during prevailing weather conditions (Fig. 3. 10).

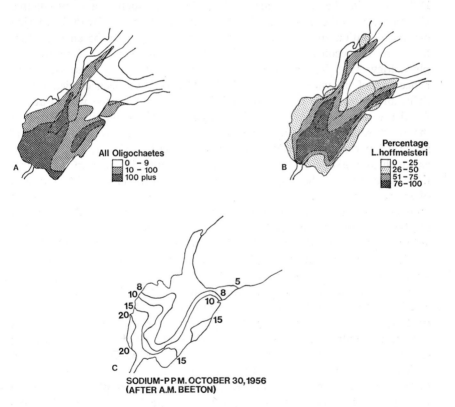

Fig. 3.10. The distribution of oligochaetes in general, (A) *L. hoffmeisteri* in particular (B) in relation to the flow of river water as identified by its sodium content, (C) Saginaw Bay, L. Huron.

The use of the percentage occurrence of *L. hoffmeisteri* may prove to be of general application. This is the commonest, most widely distributed, tubificid and its presence in any locality should not occasion surprise. There may be anything from ten to twenty other tubificids in an unpolluted locality, but some of these will be relatively scarce. The same scarce species may well be amongst the first to disappear where the environment is disturbed, so that their presence alone is a useful guide, but some of the more tolerant species (such as *Potamothrix hammoniensis*) may be present in significant numbers, even in waters mildly polluted with organic matter. Hence the numbers of the individual species in the samples will be revealing, and the proportion of all others to *L. hoffmeisteri* may be a useful index.

RESPIRATORY PHYSIOLOGY IN RELATION TO ECOLOGY

Differences in the species associations in unproductive and productive situations, and the reduction in the number of species with increasing organic pollution suggest that respiratory physiology may be an important factor in the ecology of tubificids. Most of the studies of respiratory physiology have ignored the specific identity of the worms used, and hence are of no value to the ecologist. Others have been carried out in light, and some have used apparatus involving shaking the worms about. As the worms are sensitive to both light and vibration, it is clear that the most representative results can be obtained only by using an inert substrate, performing the experiments in the dark, and reducing vibration as much as possible by using a probe for determining oxygen concentrations. In a study of the respiratory adaptability of the larvae of two species of mayfly, Eriksen (1963) demonstrated that these insects were respiratory regulators when provided with substrates consisting of glass tubes simulating burrows. Earlier studies had suggested that the larvae were respiratory adjustors, the contradiction being due entirely to the faulty methodology employed. In a recent study of the respiratory physiology of *Branchiura sowerbyi*, the worms were provided with glass tubes, and oxygen was determined by means of the Winkler method, not by the Warburg method which involves agitating the worms. The results of this study (Aston, 1966) confirm the general finding that tubificids are respiratory regulators at oxygen concentration down to a critical level, which may be variously expressed as 1·5-3·0% oxygen in a gas mixture or 7·5-17% saturation with oxygen (Palmer, 1968; Berg *et al.*, 1962). In *B. sowerbyi* the majority of the respiratory uptake (about 90%) apparently takes place through the tail region, which is not surprising in view of the fact that the tail is normally extended into the water whilst the rest of the worm remains buried in the mud during respiration. The tail plus gills in *B. sowerbyi* makes up only 18% of the total body surface however. Respiratory exchange is enhanced by the presence of haemoglobin and by undulating movements of the tail. In rare instances the body wall is extended into gill filaments with or without cilia (*B. sowerbyi, Phreodrilus branchiatus* and several naidid species). In earlier reports (Alsterberg, 1922; Kawaguti, 1932; Fox and Taylor, 1955, Berg *et al.*, 1962) increased respiratory movements of the tail with decreasing oxygen concentration were noted, but Aston (1966) found this to be a transient phenomenon in *B. sowerbyi*. When first exposed to deoxygenated water, worms of this species undulated the tail vigorously, but after a short time-interval this activity was modified, the worms undulating less often during periods of activity and resting more often. Periods of inactivity increased from five minutes in every 60 at high oxygen saturations to 45 minutes in every 60 at 1% saturation with oxygen at 20°C. The work of Kawaguti (1932) showed that the haemoglobin of *B. sowerbyi* could be loaded at oxygen tension down to 16% saturation, but that it was inefficient below that concentration. Attempts to block the action of the haemoglobin with carbon monoxide were made by Dausend (1931) and Koenen (1951), the latter demonstrating that changes in oxygen utilization demonstrated by Dausend were due to massive doses of carbon monoxide interfering with respiratory enzymes as with more restricted doses sufficient to affect only the haemoglobin, no differences in oxygen utilization could be established. The ability of tubificids to withstand periods of oxygen deprivation, including complete anaerobic conditions, for up to four weeks has been demonstrated under laboratory conditions (Alsterberg, 1922; Dausend, 1931; Aston, 1966) and might be anticipated from the observations on organically polluted rivers and eutrophic lakes. Indeed the ecological data suggests that the worms are more tolerant of oxygen shortage than the laboratory data would support, even though the species used in laboratories are invariably the same ones that survive the worst conditions in the field (because they may be obtained in very considerable numbers with the minimum of effort). According to Aston (1966) defaecation in *B. sowerbyi* ceases when the oxygen saturation is below 22% at 20°C, from which he deduced that feeding activity also ceases when oxygen concentration falls close to the critical level. There seems to be a paradox here, in that worms seem to survive and even to breed in summer in eutrophic lakes in which the bottom sediments become devoid of oxygen for several weeks, at least. On the other hand, Jonasson and Kristiansen (1967) showed that the larvae of the midge *Chironomus*, which also have haemoglobin and live in lake sediments, cease to grow during the period of oxygen depletion in late summer in Esrom Lake. One lake in which the deepest sediments remained deoxygenated throughout the year maintained a wide diversity of tubificids and other benthic invertebrates except in the oxygen-deficient region of the bottom, where even the tubificids failed to survive the prevailing conditions. Periods of severe organic pollution combined with rising temperatures may drastically reduce the number of individuals of even the hardiest tubificid species, but the popu-

lation usually recovers swiftly. Tubificids cannot, therefore, respire anaero-
bically for any length of time, but it seems possible that they can be
facultative anaerobes. The lack of tubificids in small ponds liable to periods
of oxygen depletion may give a clue to the oversight in most of the
laboratory studies carried out to date. The circulation of water over sediments containing the worms may play a role in maintaining the population,
presumably by removing the metabolic wastes which must accumulate in
the small aquaria and closed bottles normally used in the laboratory. The
study of Berg *et al.* (1962) includes information on *Psammoryctides barbatus*, which consumes more oxygen than *T. tubifex* and *P. hammoniensis*,
but most work has been restricted to studies on *T. tubifex* and *L.
hoffmeisteri*.

From the work described above it is immediately apparent that nothing
has been done on the respiratory tolerance of species known to occur only
in unpolluted and unproductive habitats, nor have any physiological
mechanisms been described which would account for the ability of a few
species like *L. hoffmeisteri* and *B. sowerbyi* to withstand a wider range of
environmental conditions than all other fresh-water tubificids.

SYMPATRIC SPECIES

Studies of the geographical distribution, habitat selection, and effect of
pollution on oligochaete populations have revealed that at least several
species are usually found in each habitat. Several aspects of the ecology
of oligochaetes may throw light on the mechanism by which species coexist
in the sediment. Life history studies must be examined to check on the
possibility of temporal succession of species. Detailed analyses of cores
are required in order to establish the presence or absence of differences in
vertical distribution in the sediment. Finally, differences in nutritional
requirements might be reflected in selective feeding or differences in the
ability to digest and absorb the organic material or microflora of the
sediment where ingestion is unselective.

LIFE HISTORY

The lack of discrete age classes has tended to discourage studies of the
life history of oligochaetes, but the lack of systematic studies is probably
also a contributing factor.

The earliest records of this aspect of tubificid ecology are limited to
notes on the presence of mature specimens in field populations, but no
conclusions can be drawn from these scattered notes, except that mature
specimens are often present all through the year. Recent studies have
recognized the presence of cocoons, newly hatched worms, immature worms,
mature worms, and breeding individuals (containing spermatophores or
sperm in the spermathecae derived from copulation). Such studies have
suggested that breeding in tubificids and some lumbriculids may be restricted

to brief periods for certain species (Brinkhurst, 1964, 1966; Cook, 1969), and in some of these species (*P. ferox, A. pluriseta*) fully mature individuals are only present just before breeding commences. Others, such as *Limnodrilus hoffmeisteri* and *Tubifex tubifex* may be found in mature condition and even breeding at all times in some localities, but even these species often have periods of intensive breeding. These periods may not be the same from one place to another, however. Kennedy (1966a) showed that the life history of *L. hoffmeisteri* varied from site to site and even from year to year at one site. In some instances it seemed that most of the worms became mature and reproduced within twelve months, but at other localities, worms might take as long as two years to mature. Seasonal peaks of breeding activity were reported in some places, but continuous production of cocoons in other places. Poddubnaya (1959a) established that this species bred once a year in Rybinsk reservoir but the arguments relating to the age of the worms at maturity is not convincing. Her data is unusual, in that mature specimens were absent in the winter months, whereas mature (if not breeding) specimens are usually to be found at all times in a wide variety of localities. In this reservoir, *L. hoffmeisteri* breeds at the same time as *Tubifex newaensis*, cocoons appearing from late May or early June into July. The claim that small *T. newaensis* could be distinguished from *L. hoffmeisteri* does not seem to be supported by the data, as the post-reproductive increase in numbers of worms was observed in the second size-class recognized, but not in the first. In the absence of data on these immature specimens it is difficult to accept the claim that the life-cycle goes to completion in a single year in the reservoir. Aston (1968) reported a one-year breeding cycle for *B. sowerbyi* both in the field (where the smaller worms may have been lost as samples were taken with a dip-net) and in the laboratory. *Limnodrilus udekemianus* appears to take two years to mature and breed (Kennedy, 1966b). Breeding takes place in late winter to early spring, but the timing and duration of reproductive activity varies from site to site. Wachs (1967) recorded peaks of breeding activity for *T. tubifex* and *Limnodrilus* spp. in March to June in the Fulda River, but these began well before the temperature rose to 12° or 15°C. Various authors (Poddubnaya, 1959; Grigelis, 1961; Timm, 1962) have claimed that, in lakes in northern Europe, breeding is initiated after the water temperature reaches this level, and Kennedy (1966) noted that low temperatures could interrupt breeding which could, however, occur at temperatures lower than 12°C. in autumn and spring in productive habitats where food was presumably abundant.

In those species, such as *Tubifex costatus, Aulodrilus pluriseta, Peloscolex ferox* (Brinkhurst, 1964 a, b), *Stylodrilus heringianus*, and *Eclipidrilus lacustris* (Cook, 1969) that have brief, well-defined breeding periods, the available evidence suggests that the worms take two years to mature and breed. Most of these studies suggest that the majority of adult, breeding, worms die after sexual reproduction although, in some species, worms are known to be capable of surviving as immature worms and breeding a second time, having developed a new set of reproductive organs (Brinkhurst and Kennedy, 1965).

In essence, all of these studies suggest that breeding behaviour varies within wide limits in most species, and that there is no evidence of a clearcut separation of species by virtue of alternating periods of abundance. Most survey data supports this contention, as species characteristic for any particular habitat can be found at all seasons of the year, often with mature specimens making up a proportion of the population at all times.

Asexual reproduction is more frequent than sexual reproduction in the Naididae and Aeolosomatidae, but has been held to be impossible in many other families. The occurrence of asexual reproduction in the Lumbriculidae is well established, especially in *Lumbriculus variegatus* and *Lamprodilus mrazeki* (Hrabe, 1937; Cook, 1969). In these species, the distribution and number of reproductive organs varies widely from specimen to specimen, as it does in the Aeolosomatidae (Bunke, 1967). The position of the reproductive organs in the Naididae is characteristically anterior to the position in which they were found in most tubificids, although the sequential arrangement and form of the parts is remarkably similar. Intermediates are known, and the record of reproductive organs of a tubificid in segments apparently anterior to those in which they are normally found may relate to the frequency of asexual division in the species. The correlation between the occurrence of fragmentation and asexual reproduction was noted by Christensen (1964). Asexual reproduction has been reported in *Aulodrilus* species (Hrabe, 1937) and in *Tubifex newaensis* (Poddubnaya, 1959). The ease with which *B. sowerbyi* breaks up when handled may suggest that this species often reproduces by fragmentation, as its apparent ability to invade man-made habitats also indicates. Whilst asexual reproduction may be quite widespread in the Oligochaeta (Christensen, 1960), it is unlikely to be a significant means of reproduction for many tubificids other than *Aulodrilus* species. Future studies will undoubtedly indicate that there are many peculiarities in the reproductive biology of oligochaetes. Gavrilov (1931, 1935, 1959) has made a detailed study of auto-fecundation in oligochaetes, demonstrating self-fertilization both with and without copulation. Polyploidy has been established in the Enchytraeidae almost as the rule rather than the exception, and parthenogenesis has been reported for the tubificids and enchytraeids as well as in earthworms (Christensen, 1961). In earthworms, the majority of polyploids are parthenogenetic, but of the 41 polyploid cytotypes in the enchytraeids, 27 are cross breeding. However, of these, all but four show abnormal bivalent formation during meiosis. The parthenogenetic triploid form of *Lumbricillus lineatus* does not produce mature spermatozoa, but the spermathecae of specimens in natural populations contain sperm which prove to be derived by copulation with diploid individuals. These sperm are necessary for complete development of triploid eggs, but they do not contribute nuclear material to them (Christensen, 1960). Sub-amphimictic reproduction occurs in one form of *Enchytraeus lacteus* (Christensen and Jensen, 1964). In this form somatic nuclei contain 170 chromosomes, eight being larger than the rest. The large chromosomes form bivalents at gametogenesis but the small ones remain as univalents. The female pronucleus contains only four large nuclei but all 162 small ones, and the small ones divide equationally at the second oocyte division. The male pronucleus contains four large chromosomes, the process of spermatogenesis having produced two small nuclei with four large chromosomes each (which then proceed into the second meiotic division) and one large nucleus containing the small chromosomes that fails to develop into sperm. At fertilization the somatic nuclear complement is restored by the addition of four large nuclei from the sperm to four large nuclei plus 162 small ones in the egg.

VERTICAL DISTRIBUTION

A few studies have been made on the distribution of species in relation to the depth to which they penetrate the sediment. All species require access to the mud–water interface for respiration, but they may penetrate to various depths in the sediment, mainly to feed. *Potamothrix hammoniensis* is recorded as present at depths of up to 26 cm in Prespa Lake, but it does not penetrate so deeply in Ochrid or Dojran (Sapkarev, 1959). Records of the depth distribution of a series of species from Ochrid show very little evidence of a division of the sediments into spheres of influence.

A similar study in Ditton Brook (Brinkhurst and Kennedy, 1965) demonstrated that *Tubifex* and *Limnodrilus* could be found throughout cores 11 cm deep. Forty cores from Toronto Harbour were analysed in four groups of ten cores each. In this locality, pollution has limited the oligochaete fauna to three species, *T. tubifex*, *L. hoffmeisteri* and *Peloscolex multisetosus*. All three species are most abundant in the first 2 cm of each core. The maximum depth penetration recorded was somewhere between 14-20 cm for one or two specimens of *L. hoffmeisteri*. Very few worms of any species were found below 10 cm.

NUTRITION

The food of the predatory *Chaetogaster* species has been investigated by Green (1954) and Poddubnaya (1965), both of whom found a wide variety of prey organisms, with Chydoridae predominating according to the most recent account. Many Naididae may be herbivorous or may graze the "aufwuchs" that develops on aquatic vegetation. *Stylaria* was said to feed on algae, diatoms and plant fragments by Yoshizawa (1928), and the tubificid *Psammoryctides albicola* was full of green plant fragments in the R. Thames at Reading. The lumbriculid *Lamprodrilus vermivorus* is predatory.

Most oligochaetes probably derive the bulk of their nutritional requirements from micro-organisms grazed off plants or more commonly, ingested along with allochthonous and autochthonous organic matter in sediments and soils. The role of tubificids and lumbriculids in circulating and irrigating the superficial layers of sediments has been discussed many times (Alsterberg, 1924; Solowiev, 1924; Raverra, 1955) but most investigation into their activity has been concerned with the changes in organic carbon, organic nitrogen and total nitrogen induced by the worms, the rate of transport of sediments and the depth of the mud in relation to the abundance of worms. A linear correlation between the abundance of worms and the depth of the oxidized layer of sediment has recently been established by Schumacker (1963), and between the organic carbon content and the abundance of worms by Wachs (1967), but which is cause and which is effect is difficult to assess although Schumacker believes that the worms are responsible for extending the oxidized layer by their feeding activity.

Poddubnaya (1961) found that the proportion of organic matter in faeces was higher than in mud prior to ingestion (if the translation available to me is accurate in this respect), but this may be because the worms were ingesting only the richer part of the sediments analysed before ingestion. Wachs (1967) reported a reduction of 21% in organic carbon and 34% in total nitrogen in experiments with *T. tubifex* at 12°C. Ivlev (1939) calculated the percentage assimilation by worms to be just about half of the nutritive value of the sediment in terms of calorific value, and calculated the coefficient of energy consumption at 31·6%. The net growth efficiency has been calculated by Welch (personal communication) at 65% using data from Ivlev and respiration data from Berg *et al.* (1962). None of these studies have taken into account the nature of the organic matter involved or the microflora associated with it. The variations in the rate of defaecation observed by several authors (reviewed by Raverra, 1955) have been shown to be caused by differences in carbon and nitrogen content of the sediment (Poddubnaya, 1961) and temperature (Wachs, 1967). Some attempts to follow feeding habits using tracers have indicated the ability of worms to mix sediments, and Poddubnaya and Sorokin (1961) suggested slight differences between *L. hoffmeisteri* adults and juveniles and *Tubifex newaensis* in regard to the depth at which feeding is concentrated. If the worms always feed in the "head-down" position with the tails in the mud–water interface (in the position adopted for respiration) then the depth at which they feed will be restricted by their body-length, and the sediments would not be effectively mixed below this depth. Whilst the majority of worms may be found in the uppermost (0-6 cm) layer of sediment in any sample, as described above, a significant number may be found down to at least 20 cm. Poddubnaya and Sorokin (1961) believe that these deep migrations are not related to feeding, and this observation was repeated by Sorokin (1966).

The role of micro-organisms in the diet of tubificids was suggested by Brinkhurst (1967) and a correlation between the abundance of oligochaetes and bacteria (*E. coli* counts) has been established by Brinkhurst and

Simmons (1968). The role of microflora as food of worms has recently been investigated by Brinkhurst and Chua (1969). The assemblage of species in the organically polluted sediments of Toronto Harbour were studied for three reasons, the locality was close to the laboratory, the worms exceedingly abundant, and the three species involved (*T. tubifex, L. hoffmeisteri, P. multisetosus*) can be distinguished readily by use of a lower-power stereomicroscope. The first stage of the study was to establish the identity of the aerobic, heterotrophic bacteria that would grow on plates using standard culture media. The medium used after a variety of trials was nutrient agar, cultured at 25°C. The bacteria were identified by morphological and biochemical tests and eight species were identified both in the laboratory and by an independent analysis by the Ontario Water Resources Commission bacteriologist. Most of the bacteria in the mud may be found in the guts of worms killed in the field* but only one species is not digested by the worms. The fact that a different species of bacterium survives digestion in each species of worm strongly suggests differences in the ability of the worms to utilize the nutritional resources of the mud.

Table 3.14 Bacteria present in mud, in samples of worms killed in the field (after some digestion), and in worms starved for a week (complete digestion). Samples from Toronto Harbour on four dates in May-June 1967. Numbers represent number of plates showing positive growth.
(From Brinkhurst and Chua, 1969)

Bacteria	No. mud samples	Samples from worms killed in the field			Samples from faeces and gut after 1 week		
		T.t[a]	*L.h*[a]	*P.m*[a]	*T.t*	*L.h*	*P.m*
Flavobacterium sp.	20	8	5	16	0	0	0
Pseudomonas fluorescens	10	0	2	5	0	0	0
Bacillus mycoides	19	2	1	1	0	0	0
Aeromonas sp.	20	0	0	14	0	0	20
Micrococcus sp.	16	10	12	10	0	15	0
Pseudomonas sp.	20	15	16	20	15	0	0
Bacillus cereus	20	2	4	10	0	0	0
No. of species present	7	5	6	7	1	1	1
Expected[b]	140	140	140	140	140	140	140
Observed[b]	125	37	40	76	15	15	20

[a]*T.t*=*Tubifex tubifex*, *L.h*=*Limnodrilus hoffmeisteri*, *P.m.*=*Peloscolex multisetosus*.
[b]Expected based on 5 plates per run, 4 runs, 7 species recognized equals 140 possible positive records. Observed equals number of plates on which positive growth was observed.

* These data are also summarized in Table 3. 14, where it is more immediately apparent that both *Micrococcus* and *Pseudomonas* are ingested by all three worm species but survive only in *L. hoffmeisteri* and *T. tubifex* respectively. The *Aeromonas* surviving in *Peloscolex* may be selectively ingested, as it was not found in the gut of either of the other two species, the alternative explanation being that it is digested very rapidly by *T. tubifex* and *L. hoffmeisteri*.
See also Wavre and Brinkhurst (1971).

This was further emphasized by the discovery that *P. multisetosus* could absorb amino acid directly from solution without the mediation of the gut microflora (Fig. 3. 11, 12). This study indicated that more success in identifying micro-habitats should follow studies relating worms to the specific components of the sediments.

2. GEOGRAPHICAL DISTRIBUTION OF THE ALLUROIDIDAE

B. G. M. Jamieson

The Alluroididae is a little-known group consisting of two subfamilies. The Alluroidinae include two genera in the Ethiopian Region, *Alluroides*, with four species, and the monotypic *Standeria*, together with a monotypic neotropical genus, *Brinkhurstia*. All Alluroidinae are aquatic or occur in swampy ground. The second subfamily, the Syngenodrilinae, is known from a single specimen (of unknown habitat) from tropical E. Africa, the type of *Syngenodrilus lamuensis*. Although records for the Alluroidinae are few, it is probable that members of the subfamily are both widespread and abundant in limnic habitats in S. America and in Africa and that they have frequently been overlooked by collectors. Their external appearance would cause them to be disregarded by workers on the better known aquatic microdrile families while workers on megadriles might easily dismiss them as juvenile earthworms. The writer was able to collect numerous specimens of *Alluroides pordagei* in a few hours collecting near Nairobi, Kenya, though when first encountered, they were taken for juveniles of associated Eudrilids because of their small size and the absence, to the naked eye, of a recognizable clitellum. *A. pordagei* has now been recorded from Mombasa to Mt. Kenya (Beddard, 1894; Michaelsen, 1914a, b; and new record, p. 000), from the Zambesi River (Michaelsen, 1913) and from several localities in the Republic of the Congo (Michaelsen, 1935, 1936); *A. brinkhursti* has a subspecies on Mt. Elgon, Uganda, and one in the highlands of Ethiopia; while A. *ruwenzoriensis* is known only from Mt. Ruwenzori (Brinkhurst, 1964) and *A. tanganyikae* from Lake Tanganyika (Beddard, 1906).

The known range of the genus *Alluroides* was greatly extended on the discovery by Cernosvitov (1936) of a species in Argentina, which was subsequently placed (Jamieson, 1968) in the monotypic genus *Brinkhurstia*. Cernosvitov invoked the theory of continental drift to explain division of the Alluroidids between Africa and S. America but it is at least as feasible that they reached these regions from the Holarctic. Similarities with the Japanese glossoscolecid *Biwadrilus* (*q.v.*) add support to a northerly origin. Discovery of *Standeria transvaalensis* as an endemic species in Lake Chrissie in the Transvaal (Jamieson, 1968) further extended the known range of the family.

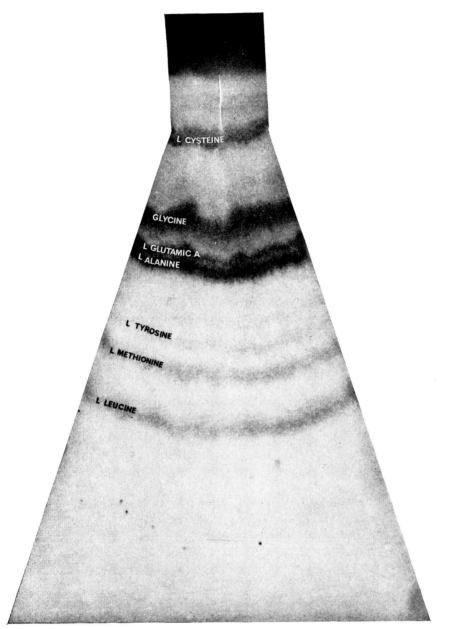

Fig. 3.11. Amino acids identified by paper chromatography from samples of sediment from Toronto Harbour.

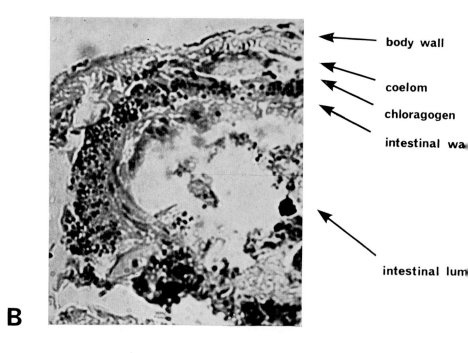

body wall

coelom

chloragogen

intestinal wa[l]

intestinal lum[en]

B

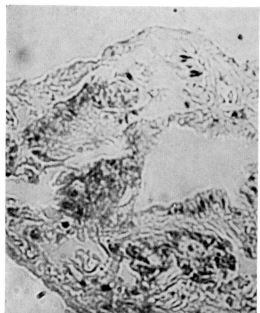

A

Fig. 3.12. Uptake of amino acid (glycine—2-C^{14}) by *P. multisetosus* autoradiographs. (A) control, (B) positive, showing accumulation of C^{14} in chloragogen.

3. GEOGRAPHICAL DISTRIBUTION OF THE GLOSSOSCOLECIDAE

B. G. M. Jamieson

A discussion of the geographical distribution of the Glossoscolecidae necessitates anticipation of new taxonomic groupings recognized in the present work. It will therefore be appropriate to preface the discussion with the following synopsis of the classification and distribution of the family (numbers of species in parentheses).

FAMILY GLOSSOSCOLECIDAE

Holarctic; Neotropical; Palaeotropical (Ethiopian and Oriental Regions); and Malagasian.

SUBFAMILY GLOSSOSCOLECINAE

Neotropical Region: Central and S. America, the Bermudas and the W. Indies including Barbados. Ethiopian Region: South Africa.

TRIBE GLOSSOSCOLECINI *Neotropical:* as for the subfamily but excluding S. Africa. Some species peregrine in warmer latitudes. 21 genera: All terrestrial—about 160 species.

TRIBE GLOSSOSCOLECINI Neotropical: as for the subfamily but (22), *Tritogenia* (5) (both genera terrestrial).

SUBFAMILY KYNOTINAE

Malagasian Region: Madagascar. *Kynotus* (13; Terrestrial).

SUBFAMILY HORMOGASTRINAE

Palaearctic (Mediterranean region): Italian Peninsula; Tuscan Archipelago; Sicily; Corsica; Sardinia; Tunisia; Algeria; S. E. Spain; France (E. Pyrenees). *Hormogaster* (2; Terrestrial).

SUBFAMILY ALMINAE

Ethiopian Region: tropical Africa and the Nile Valley. Neotropical: Central and S. America. Oriental Region: India; Burma; China (Hainan); Malaysia and Indonesia. Palaearctic: Europe; Syria and Palestine. (All aquatic or facultively moist-terrestrial.)

TRIBE ALMINI Ethiopian Region: Tropical Africa and the Nile Valley. Neotropical: Central and S. America south as far as the Juramento-Salado River. 5 genera: *Callidrilus* (2; Eastern tropical Africa); *Glyphidrilus* (15; E. Africa; India; Ceylon; Burma; Hainan; Malaya; Sumatra; Java; Borneo; Celebes); *Alma* (13; Tropical Africa and the Nile as far as Cairo); *Drilocrius* (5; Neotropical: Costa Rica to Argentina); *Glyphidrilocrius* (1; Brazil).

TRIBE CRIODRILINI   Palaearctic to as far east as Syria and Palestine (one record from N. America, probably transported). *Criodrilus* (1, and 1 dubious species).

SUBFAMILY BIWADRILINAE

Palaearctic: Japan. *Biwadrilus* (1).

SUBFAMILY SPARGANOPHILINAE

Holarctic. Two species indigenous in the Nearctic (N. and Central America, from Guatemala and Mexico to Ontario) of which one occurs also in England and France, probably as an introduction. *Sparganophilus*.

The Glossoscolecidae are absent, as native species, from only one of Wallace's faunal regions, the Australian, with the single exception of a species of *Glyphidrilus* which has crossed Wallace's line and has reached Celebes (Fig. 3.13). This absence, together with other evidence which will be presented below, indicates that although a large proportion of the species occur south of the Equator (in S. America and Africa), the family has had a northern origin, as appears to be true of most of the microdrile families. Absence from the Chilean-Patagonian sub-region of S. America appears to support the latter view and probably cannot be ascribed merely to the adverse climate of the subregion which at present appears to limit southward migration of south American glossoscolecids. If northern relatively cold-tolerant groups such as the Sparganophilinae, Criodrilini and Hormogastrinae, or even the circum-equatorial Almini, had previously occurred in the subregion it seems unlikely that extermination by climatic deterioration would have been complete. The area is capable of supporting earthworms of other families and there is no evidence of competitive exclusion. The northerly origin of the Glossoscolecidae is endorsed by the occurrence of four of the five subfamilies in the Holarctic of which three are endemic. The Neotropical region, in contrast, has only two subfamilies, the Glossoscolecinae and the Alminae, both of which are shared with, and form the total complement of, the Ethiopian region. The Malagasian region (Madagascar) has a single, endemic subfamily, the Kynotinae; while the Oriental region shares its single genus, *Glyphidrilus* (Alminae) with the Ethiopian region.

Though the distribution of the Glossoscolecidae agrees with that of the microdriles in being predominantly northerly at the subfamilial level, an interesting contrast with the microdriles exists at generic and specific levels for though the subfamilies are more numerous in the Palaearctic than elsewhere, the overwhelming majority of genera (23) and species (over 160) occur in the Neotropical. There are only ten genera with a total of 81 species, in all other regions. These are four endemic genera, with a total

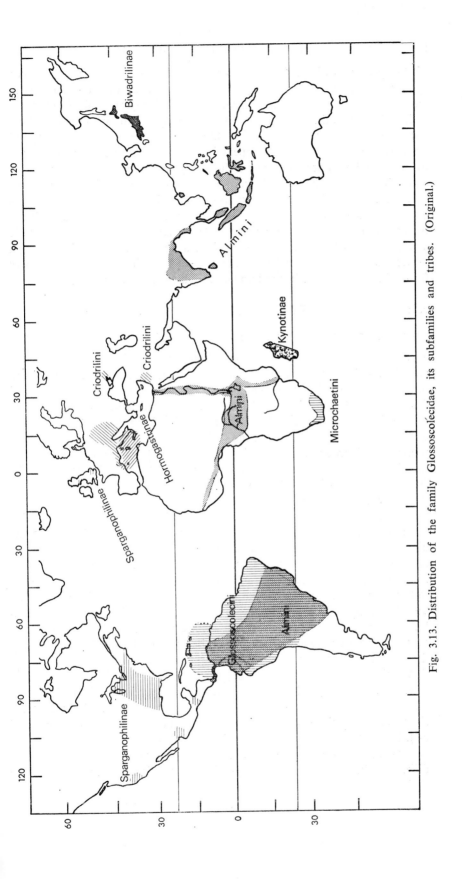

Fig. 3.13. Distribution of the family Glossoscolecidae, its subfamilies and tribes. (Original.)

of 46 species in the Ethiopian region; four genera, with six or seven species, in the Palaearctic; the genus *Sparganophilus*, with two species, in the Nearctic; *Kynotus*, with 13 species, in Madagascar; and *Glyphidrilus* with 14 species in the Oriental region and one in the Ethiopian region. There thus appears to have been a great proliferation of species in the Neotropical whereas the Holarctic fauna appears to be of a relict nature, as judged by the small number of species, their apparently primitive morphology, and a diversity which merits the recognition of four subfamilies. Despite this relict-nature, individual Holarctic species may be highly successful. *Sparganophilus* is a significant element of the limnic fauna of N. America, being by far the most common limnic megadrile in that continent, and has climatic and geographical ranges equalled by few, if any other megadriles. Though rather local, *Criodrilus*, in Europe, and *Biwadrilus*, in Japan, are successful forms, being abundant in the habitats in which they occur.

REGIONAL DISTRIBUTION

THE NEOTROPICAL REGION

The glossoscolecid fauna of the Neotropical region (Fig. 3. 14) consist of an endemic division of the Glossoscolecinae, the tribe Glossoscolecini, and representatives of the tribe Almini, of the subfamily Alminae. A detailed discussion of the distribution of the Glossoscolecini is beyond the scope of this work, as they are terrestrial, but something further may be said of the absence of glossoscolecids from the Chilean and Patagonian subregion. The present southern limit of the Glossoscolecini and, so far as is known, of the neotropical Almini, corresponds approximately with the course of the Juramento-Salado River (Fig. 3. 14). The lower limit on the west coast is thus very much more northerly (at approximately the Tropic of Capricorn) than that on the east coast which lies at about latitude 35° S, in the vicinity of the mouth of La Plata. It is tempting to relate this southerly limit to the distribution of rainfall but it coincides only very approximately with the 250-500 mm band of isohyets. Correspondence with the 10°C July (Winter) isotherm is, on the other hand, remarkably close. In favour of temperature rather than precipitation as a limiting factor in distribution is the fact that the Glossoscolecidae have not exploited the wet corridor into the Chilean-Patagonian subregion formed by the Andes, presumably because it is also cold. Temperature might, nevertheless, be working indirectly, for instance through vegetational or edaphic factors but correspondence of the distribution with vegetational regions is not close.

The distribution of the Glossoscolecini corresponds closely with that of other tropical S. American groups. Thus many tropical fish follow the same river system southward into N. Argentina but only a few characins and catfishes occur further south. South of this, on the southern tip of S. America, there are no fish of strictly fresh-water groups but a few very different fish, of salt-tolerant groups, do occur in fresh-water south even to

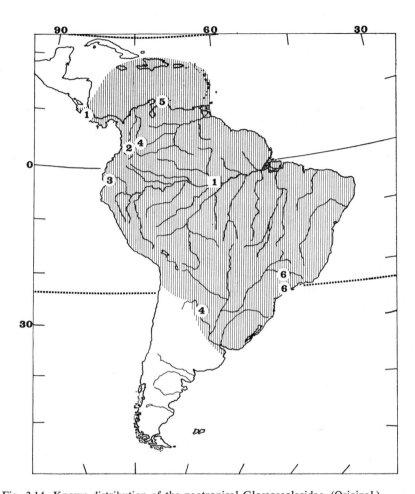

Fig. 3.14. Known distribution of the neotropical Glossoscolecidae. (Original.)
Crosshatched—Tribe Glossoscolecini (Glossoscolecinae).
Circles—*Drilocrius* (*Alminae*). 1. *D. alfari*; 2. *D. breymanni*; 3. *D. buchwaldi*; 4. *D. burgeri* (Argentinian location only approximate); 5. *D. hummelincki*; 6. *D. iheringi*.
White square—1. *Glyphidrilocrius ehrhardti*.

Tierra del Fuego. Most of these fish belong to a few families which do
not occur in the tropics but are represented in southern Australia and New
Zealand as well as southern S. America (Darlington, 1965). This pattern is
exactly that seen in the S. American oligochaetes. The glossoscolecids are
replaced in the southern tip of the continent by Acanthodrilinae of which
Microscolex and *Eodrilus* have a circum-antarctic distribution, and occur in
S. America, S. Africa and Australia, *Microscolex*, at least, being euryhaline.
Possibly the Chilean-Patagonian subregion has never been colonizable
from the north by relatively thermophilic species. As Darlington (1965) has

said, the far-southern area has probably always been isolated from the wet tropics by a barrier of relatively arid country and has been cool since the late Tertiary, glacial in the Pleistocene, and is still cold temperate or, in the mountains, glacial. The elevation of the Andes, in the Tertiary, still did not provide a suitable corridor for southerly penetration of the Glossoscolecidae.

To the north, the Glossoscolecidae extend into the Antillean subregion and the circum-mundane *Pontoscolex corethrurus* enters Mexico, but there are no endemic genera and very few endemic species in these two subregions (*Andiodrilus biolleyi* Cognetti and *A. orosiensis* Michaelsen in Costa Rica and *Diachaeta barbadensis* Beddard and *D. thomasi* Benham, in the W. Indies).

The Alminae are represented in S. America by only two genera, *Drilocrius* and *Glyphidrilocrius* (Fig. 3. 14). *Drilocrius* has a single species in Central America, the endemic Costa Rican, *D. alfari,* and the remaining species have a distribution in S. America which, though imperfectly known, appears to coincide with that of the Glossoscolecini. One species is found on the island of Aruba. *Glyphidrilocrius* is known from a single, endemic species, in Brazil.

THE ETHIOPIAN REGION

The glossoscolecids of the Ethiopian region are the Alminae (Tribe Almini) of the tropics and the Nile Valley, and the Glossoscolecinae (Tribe Microchaetini) of S. Africa (Fig. 3. 15). Of the three African genera of Almini, *Alma* and *Callidrilus* are endemic and only *Glyphidrilus* occurs elsewhere, having an oriental distribution. *Alma* is dominant, with 13 species spread throughout tropical Africa and along the Nile as far as its delta. (This Nile corridor must be regarded as an incursion of the Ethiopian region into the Palaearctic.) The genus is almost ubiquitous in limnic habitats enriched with organic matter and is only locally rivalled by some species of the ocnerodrile *Pygmaeodrilus* and of *Stuhlmannia* (Eudrilidae). *Callidrilus*, with only two species, occurs through a relatively narrow strip from Kenya to Mozambique with its western limit in Malawi. Collection of oligochaetes in Africa has been sufficiently intensive and widespread to suggest that patterns of distribution of *Alma* which can at present be discerned are real though much detail remains to be filled in, absence from the E. African and much of the W. African coastal plains being especially questionable. The distribution of *Alma* and the other aquatic Glossoscolecidae dispels the frequently sustained illusion that limnic species are of no value in zoogeographical studies. Within tropical Africa there is a remarkable degree of local endemism in *Alma* species. The most frequently recorded species, *Alma emini*, has a wide but nevertheless distinctly circumscribed distribution. Like the ocnerodrile *Pygmaeodrilus affinis* (v. Jamieson, 1957) it is widespread in the drainage basin of lakes Albert and Victoria but, unlike the latter species, it has not been found further north in the Nile than the

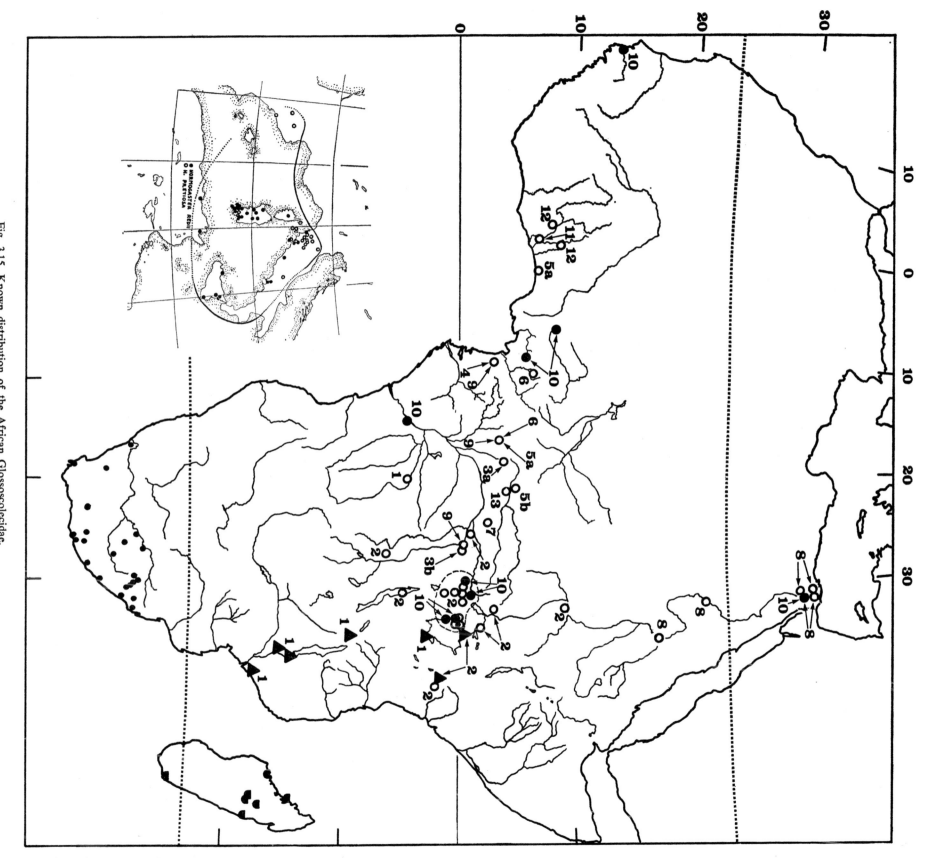

Fig. 3.15. Known distribution of the African Glossoscolecidae.
Large map—Ethiopian and Malagasian Glossoscolecidae. (Original.)
Large circles—*Alma* 1. *A. basongonis*; 2. *A. emini*; 3a *A. eubranchiata* eubranchiata;
4. *A. kamerunensis*; 5a. *A. millsoni millsoni*; 5b. *A. millsoni zebangui*; 6. *A. multi-
setosa*; 7. *A. nasuta*; 8. *A. nilotica*; 9. *A. pooliana*; 10. (black circles) *A. stuhlmanni*;
11. *A. tazelaari*; 12. *A. togoensis*; 13. *A. ubangiana*.
Triangles—*Callidrilus*. 1. *C. scrobifer*; 2. *C. ugandaensis*.
Small black circles (South African only)—*Microchaetus* and *Tritogenia* (Microchaetini).
Black half circles (Madagascar only)—*Kynotus*.
Inset map—Distribution of *Hormogaster* (from Omodeo, 1954).

southern Sudan, despite exhaustive collecting near Cairo which yielded
P. affinis and two other species of *Alma*. Like *P. affinis* it has entered the
Athi River but it has also entered the headwaters of the Congo. In explanation of the failure of *A. emini* to spread further along the course of the
Congo and Nile rivers may be offered the fact that its present limits of
distribution correspond with those of an area which throughout the year
maintains temperatures which for tropical Africa are relatively low (between
approximately 15 and 25°C) and which, perhaps more importantly, are
constant, within narrow limits. The higher prevailing temperatures of the
lower Nile in summer and of the Congo basin probably prevent survival of
worms or cocoons carried downstream into these regions. Zonation of
precipitation can be ruled out as a controlling factor in distribution as the
worm is aquatic. Edaphic and vegetational diversity over its range also
exclude these as primary factors in controlling its dispersal. In contrast with
A. emini, A. nilotica occurs in the Nile (at least from Khartoum northwards)
but has not entered the cooler uplands. The most widely distributed species
of *Alma*, *A. stuhlmanni*, is clearly tolerant of a relatively wide range of
temperatures as it occurs in the following river and lake systems: the
Gambia, the Niger, the Congo, Lake Victoria and the Nile (at Cairo).
Khalaf and Ghabbour (1965) have shown that it has a temperature pre-
ference of 24-26°C but that, when wet, it can survive 38°C. The remaining
Alma species are limited to single river systems, with the exception of
A. millsoni (Niger and Congo river systems). The great majority of species
of *Alma* occur in the Congo and its tributaries and no species is known
south of these. The Zambezi, in which considerable collections have
been made, has not yielded *Alma* whereas *Callidrilus* is characteristic of
this river and its tributary Lake Nyasa, has an extension to Lake Victoria
and Nairobi (both localities uncertain) but has never been reported, despite
much sampling, from the R. Congo.

The remaining Ethiopian glossoscolecids, the S. African Microchaetini
(Fig. 3. 15), are terrestrial and their geographical distribution need not be
discussed here. It may be noted, however, that unlike their S. American
relatives, the Glossoscolecini, they here co-exist with Acanthodrilinae and
that they are more cold-tolerant.

THE MALAGASIAN REGION

The 12 or 13 species of *Kynotus* are the only glossoscolecids reported from
this region (Madagascar; Fig. 3. 15). It appears that the 250 mile wide
Straits of Mozambique prevents interchange with the glossoscolecids of
mainland Africa. It must be pointed out, however, that only *Callidrilus* has
been reported from the E. African coastal plains.

THE ORIENTAL REGION

Apart from the circum-mundane *Pontoscolex corethrurus*, the only Glossos-

colecidae in the Oriental region belong to the aquatic genus *Glyphidrilus*
(Almini) (Fig. 3. 16). This genus has a disjunct distribution, with only a
single, morphologically aberrant species in E. Africa and the remaining
14 species spread throughout the Oriental region. The phylogeny of the
genus is discussed in Chapter 4 and it will suffice here to call attention to
the distribution of the oriental species. The evidence points to the genus
having entered the Indian subregion before the Pleistocene, but probably not
earlier than the Eocene regression of the Tethys Sea, as it is represented by
a morphologically primitive species in Ceylon, which island has been
separate from India since the Pleistocene. *G. annandalei*, in southern India
appears to be a closely related but more modified species. Absence of the
genus from the Deccan seems real as collecting in the area has yielded
numerous species of other families, and this absence is perhaps ascribable
to the formation, by extreme vulcanism, of the Deccan traps which continued into the Eocene, and which totally annihilated the fauna of this vast
region (de Beaufort, 1951). North of this region are three species, *gangeticus*
and the more modified *spelaeotes* in the Gangetic Basin to as far north as

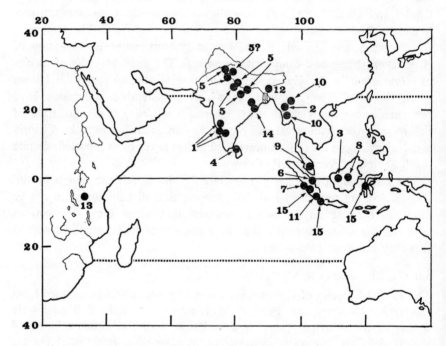

Fig. 3. 16. Known distribution of *Glyphidrilus*. (Original.)
1. *G. annandalei*; 2. *G. birmanicus*; 3. *G. buttikoferi*; 4. *G. ceylonensis*; 5. *G. gangeticus*;
6. *G. horsti*; 7. *G. jacobsoni*; 8. *G. kukenthali*; 9. *G. malayanus*; 10. *G. papillatus*;
11. *G. quadrangulus*; 12. *G. spelaeotes*; 13. *G. stuhlmanni*; 14. *G. tuberosus*; 15.
G. weberi.

Nepal which appear closely related to each other and to the southern Indian and Selanese species, and *G. tuberosus*, from the region of Orissa, which agrees closely with other species of the genus in its gross anatomy but does not appear to be a close relative of any of them. The Burmese species, *G. birmanicus* and *G. papillatus*, show close relationship with *gangeticus*. Records of *Glyphidrilus* from the remainder of the Oriental region are limited to Hainan Island (material doubtfully identified as *papillatus*), to Malaya (*G. malayanus*), and to Indonesia. Our knowledge of the Indonesian forms is unsatisfactory but they appear to be derivable from the *ceylonensis* plan. *G. weberi* has been reported from Sumatra, Java, Flores and Celebes but it appears (p. 767) that Horst (1889) confused two or more species. The record of this species for Celebes is the only record of a Glossoscolecid (other than *Pontoscolex*) east of Wallace's line but collecting in New Guinea and northern Australia has been inadequate. Other species of *Glyphidrilus* recorded from Indonesia are *G. jacobsoni* (Sumatra), the interesting hologynous *G. kukenthali* (Borneo), and *G. buttikoferi* (Borneo), all of which are redescribed in the present work, and two little-known species *G. horsti* and *G. quadrangulus* (both Sumatran).

THE HOLARCTIC

The distribution of the Palaearctic genera (Fig. 3. 13) *Criodrilus* (Alminae), *Biwadrilus* (Bowadrilinae) and *Hormogaster* (Hormogastrinae) (Fig. 3. 15, inset) are discussed on pp. 147 and 148. The genus *Sparganophilus* has previously been considered to have an endemic species in England and France, *S. tamesis*, but this is shown in the present work to be conspecific with and not even subspecifically distinct from *S. eiseni* of the eastern United States south to Guatemala over which its name has chronological priority. Lack of subspecific distinction of European populations suggests recent introduction from N. America as do other data (p. 000). A north American focus of endemicity is indicated by the occurrence of an endemic species, *S. smithi*, with two subspecies, in the Californian subregion which is characterized by five endemic species of the megascolecid earthworm genus *Plutellus*. The known distribution within America of *Sparganophilus tamesis* (=*eiseni*) almost fills the Alleghany subregion which is noted for the high endemicity of its fauna, having, according to Omodeo (1963), some 20 endemic species of *Diplocardia* (Megascolecidae) and eight endemic Lumbricidae. Origin of the Sparganophilinae or of their progenitors may, nevertheless, have been palaearctic (see p. 148).

DISPERSAL ROUTES OF THE GLOSSOSCOLECIDAE

Michaelsen (1918, 1928) explained the present distribution of the Glossoscolecidae (s. lat.) in the following terms (names recognized in the present work are substituted for those employed by Michaelsen): Ancestral S.

American Glossoscolecini, with calciferous glands, an oesophageal gizzard, muscular copulatory sacs and pretesticular spermathecae gave rise to all the existing Glossoscolecidae. The Sparganophilinae arose by secondary adaptation to an aquatic life, losing the gizzard and calciferous glands, developing a zygolobous prostomium, retaining the pretesticular spermathecae and taking up their present holarctic distribution. The early Glossoscolecini traversed Africa from S. America and gave rise, with loss of calciferous glands, to *Kynotus* in Madagascar. *Microchaetus* and *Tritogenia*, in S. Africa, arose from the Glossoscolecini by movement of the gizzard into segment 7, retention of the calciferous glands, loss of the copulatory sacs and translocation of the spermathecae to a post-testicular position, and from them arose *Glyphidrilus* which spread from Africa into the Oriental region; *Callidrilus* (shown in the present work to be closely similar to *Glyphidrilus*) arose from *Kynotus* (from his map, before the latter genus entered Madagascar) by extension of the gizzard and loss of the copulatory sacs. From *Kynotus* arose, in Africa, a hypothetical genus, *Archidrilocrius*, which retained an oesophageal gizzard, and the copulatory sacs, while developing an intestinal gizzard. A great diversification of *Archidrilocrius* and attendant extensive migrations occurred. A branch with reduced oesophageal gizzard returned to S. America as *Drilocrius*, while complete loss of the oesophageal gizzard gave *Alma*, in Africa. A further branch retained the copulatory sacs, but lost the oesophageal gizzard and spermathecae, developed grooved genital setae, and became the palaearctic genus *Criodrilus*, distributed from western Europe to Japan. The Lumbricidae developed from this palaearctic stock. The Hormogastrinae, with three oesophageal gizzards, no copulatory sacs and spermathecae in the testis segments, represent a persistence in the Mediterranean region of the early Criodrilin tribe, which was like them characterized by grooved genital setae.

Although the classification of the Glossoscolecidae (s. lat.) by Omodeo (1956b) differed significantly from that of Michaelsen (1928), he agreed with Michaelsen's view that the centre of dispersion of the Microchaetidae (*Hormogaster, Callidrilus, Glyphidrilus, Kynotus* and *Tritogenia*) was central and S. Africa. He considered palaeogeographic evidence to support migration of *Kynotus* to Madagascar in the Triassic but attributed much later dates due to migration of the other microchaetids. Migration of *Glyphidrilus* to India was placed in a transgressive period in the early Tertiary and colonization of Malaysia and Indonesia in successive periods. His reasons for placing the migration of *Glyphidrilus* as late as the Tertiary were the congeneric status and close similarity of its species despite their distribution over a vast area. With regard to *Hormogaster*, Omodeo points out that it has a relatively restricted but much fragmented, discontinuous distribution (see facing p. 152 and Fig. 3.15) in the Mediterranean region and suggests that it arose from a primitive michrochaetid form dispersed, before the Nummulitic [Palaeogene], throughout E. and N. Africa and that it owes its present isolated and fragmented distribution to the Nummulitic marine transgression. In this connection it may be noted that Corsica and Sardinia, which harbour *Hormogaster*, were never flooded by the sea (de Beaufort, 1951). Omodeo suggests that the progenitor of *Hormogaster* and *Glyphidrilus* may have occupied the northern part of Africa and have been eliminated by the transgression and raises the interesting point that collecting in the · Haggar and Tibesti massifs may reveal relics of this ancient fauna. It appears to Omodeo that *Sparganophilus* may be identifiable with Michaelsen's *Archidrilocrius* and be, therefore, representative of the ancestor of *Alma* and *Drilocrius* on the one hand, and *Criodrilus* and the Lumbricidae on the other. He points out that its distribution straddling the region between the Palaearctic and the Neotropical regions particularly suits it to the role of ancestor to these Neotropical, Ethiopian and Palaearctic genera. He reminds us, as evidence of the basic position of *Spargonophilus*, that it has been confused with *Haplotaxis* (=*Pelodrilus*), from which he considers it to differ little, by Tétry and considers that *Criodrilus bathybates* could equally well be placed in *Sparganophilus* or in the genus *Alluroides*.

An especially close morphological proximity of *Sparganophilus* to the Haplotaxidae is not recognized in the present work but Omodeo's implication of a Palaearctic origin of *Alma* and *Drilocrius* and a N.E. African origin of *Glyphidrilus* is of particular interest in relation to an hypothesis of the origin of the Glossoscolecidae developed in the present study. It would be specious to attempt a detailed reconstruction of the palaeogeography and phylogeny of the Glossoscolecidae but we may at least question an austral origin of the family. Three genera (all aquatic) seem to represent, and to be relics of, early diversification of the Glossoscolecinae and all are Holarctic. They are *Criodrilus* (W. Palaearctic), *Sparganophilus* (Nearctic and W. Palaearctic) and *Biwadrilus* (E. Palaearctic, Japan). All differ from a presumed haplotaxid ancestor in the multi-layered clitellum, more posterior male pores, small-yolked eggs and metagynous condition but, despite specializations, must be considered relatively primitive (see Chapter 4). The fact that the Lumbricidae, which are generally regarded as close relatives of the Glossoscolecidae, are Palaearctic with only a small endemic incursion into eastern N. America, adds support to an hypothesis of a Palaearctic origin of the Glossoscolecidae.

Although the distribution of the Lumbricidae, Biwadrilinae and Criodrilini indicates a Palaearctic centre of dispersal of the Glossoscolecidae the presence of endemic species of *Sparganophilus* in N. America and their absence (see p. 148) from the Palaearctic indicate that the Sparganophilinae or their progenitors spread from the Palaearctic to N. America not later than the early or mid-Tertiary and possibly much earlier. Certain affinities between the Biwadrilinae and the Sparganophilinae (p. 187) suggest origin of the latter family or its progenitors in the eastern Holarctic and favour colonization of N. America by a Bering Sea land bridge rather than by the N. Atlantic land bridge postulated by Omodeo. Supposed specific distinction of *S. eiseni* and *S. tamesis*, here disproved, on the two sides of the Atlantic formerly favoured the Atlantic bridge. The Pacific Coast distribution of the wholly endemic *Sparganophilus smithi* adds weight to an hypothesis of a westerly entry into N. America.

An early Glossoscolecid stock dispersed throughout the Holarctic would be better situated to give rise to the present day Glossoscolecid faunas of the Holarctic, Neotropical, Ethiopian and Oriental regions than the ancestor in S. America envisaged by Michaelsen. *Hormogaster* is probably a Tyrrhenian descendant of a stock which gave rise to the Glossoscolecinae. It is difficult to evaluate the significance of the greater simplicity of the Microchaetini (S. Africa) over the Glossoscolecini (S. America) but, it would appear (Chapter 4) that the S. American Glossoscolecini arose from a form resembling the Microchaetini but with the gizzard in 6 and pre-testicular spermathecae. The stock of which this proto-glossoscolecin was a part was presumably variable or labile with respect to the position of the gizzard and of the spermathecae. Restriction of the Microchaetini to S. Africa raises the problem of why they are absent from tropical Africa if their ancestors traversed Africa from the north.

The available evidence does not permit an answer to this question but the possibility deserves mention that a proto-microchaetin entered S. America from the Nearctic and invaded S. Africa (by the Triassic (?) connection of the advocates of Continental Drift?) only to be replaced in the Neotropical by its descendants, the Glossoscolecini. Movement could equally well have been from Africa to S. America with extinction of ancestral Microchaetini north of the Kalahari. The Kynotinae may have entered Madagascar at the same or an earlier period but their origins are obscure (p. 185). If they traversed Africa, either from S. America or from the north, we must accept that they became extinct in Africa as was indeed the case in some Malagasian vertebrate groups with fossil histories. Michaelsen would have regarded *Callidrilus* and its relatives as modern descendants of early African Kynotinae but evidence already alluded to of the primitiveness of *Criodrilus* and its relationship to the Almini, including *Callidrilus*, seems to preclude such an origin.

It seems likely that the Alminae did enter Africa from the north, very possibly from a Palaearctic focus near the present distribution of the Criodrilini. *Callidrilus* has been described on p. 187 as virtually a primitive *Glyphidrilus* but the only known African species of *Glyphidrilus* is in many ways the most highly evolved species in the genus and the species with the most basic morphology occurs in Ceylon with a close relative in southern India. It is highly probable that a *Callidrilus*-like form in the Tyrrhenian region or its vicinity gave rise to *Callidrilus* and to an early *Glyphidrilus* which entered Africa and the Indian subregion. The primitive nature of *G. ceylonensis* and the proposed eastward direction of movement suggest that Omodeo may have been correct in putting the invasion of Indonesia by *Glyphidrilus* at a later date than the (early Tertiary?) invasion of India but the hologynous condition of G. *kukenthali* (Borneo) perhaps suggests an early dispersal over the whole Oriental region.

The fact that the Alminae appear to be absent from Madagascar, if they have not existed there previously, might be taken to indicate that their invasion of Africa postdated invasion by the ancestors of the Kynotinae and occurred after the pre-Eocene separation of Madagascar from Africa.

Precursors of *Alma* and *Drilocrius* may have originated in the Holarctic and moved thence to the present sites of *Drilocrius* in the Neotropical and of *Alma* in tropical Africa. The present distribution of *Drilocrius* with maximum species diversity in the north of S. America and a species, and a related genus, in Central America, is not unfavourable to entry from the Nearctic. Absence of Almini in N. America and in the Palaearctic, other than the warm, moist, Nile valley, is probably due to the lower prevailing temperatures which would presumably have extinguished any northerly survivors of their migration.

ECOLOGY

Nothing of significance is known of the ecology of the Alluroididae beyond the little which can be gleaned from collectors' notes and this is true of the Glossoscolecidae with the exception of *Alma emini* and *A. stuhlmanni*. *A. emini* is especially common in papyrus swamps, in E. Africa, and lives in a substratum of decomposing vegetable matter which is strongly reducing and probably lacks free oxygen. Very often only the posterior end of the worm projects from the mud into the air or surface water. The dorsal surface of this rear extremity can be infolded to form a tube in which bubbles of air are frequently trapped and taken down when the worm retreats below the surface. The haemoglobin has a very low unloading tension, remaining oxygen-saturated at a partial pressure of 2 mm Hg oxygen in the absence of carbon dioxide. The animal is thus able to make full use of the oxygen which is available in the oxygen-deficient substrate in which it lives. Correlated with this need to be able to oxygenate at low oxygen pressures, the Bohr effect is very small; only at a tension of 200 mm Hg of carbon dioxide is the oxyhaemoglobin dissociation curve displaced to the right (i.e. is the unloading tension increased). The worms can live for up to 48 hours in water completely saturated with carbon dioxide. Laboratory experiments show that this species can survive under completely anaerobic conditions (e.g. an atmosphere of pure nitrogen) but behaviour in mud-water culture suggests that in normal conditions, oxygen is utilized (Beadle, 1957).

Khalaf El Duweini and Ghabbour (1963, 1964, 1965) have investigated some aspects of the ecology (edaphic factors) and physiology (temperature preferences) of a species of *Alma* in irrigation ditches in the vicinity of the Nile near Cairo and this species has been identified by Jamieson and Ghabbour (1969) as *A. stuhlmanni*. Wasawo and Visser (1959) comment on the formation of tussock-mounds by *Alma emini, A. stuhlmanni,* and an unidentified species of *Glyphidrilus* in swamps in northern Uganda. Carter and Beadle (1932) discuss respiratory adaptations of an unidentified species of *Drilocrius.*

REFERENCES

ALSTERBERG, G. 1922. Die respiratorischen Mechanismen der Tubificiden. *Lunds. Univ. Arsskrift,* **18,** 1.
— 1924. Die Nahrungszirkulation einiger Binnenseetypen. *Arch. Hydrobiol.,* **15,** 291.
ASTON, R. J. 1966. Temperature relations, respiration and burrowing in *Branchiura sowerbyi* Beddard (Tubificidae Oligochaeta). Unpublished Ph.D. Thesis. Univ. of Reading.
— 1968. The effect of temperature on the life cycle, growth and fecundity of *Branchiura sowerbyi* (Oligochaeta: Tubificidae). *J. Zool., Lond.,* **154,** 29.
BEDDARD, F. E. 1894. A contribution to our knowledge of the Oligochaeta of tropical eastern Africa. *Q. Jl. microsc. Sci. (N.S.),* **36,** 201.
— 1906. Zoological results of the Third Tanganyika Expedition, conducted by Dr. W. A. Cunnington, 1904-1905. Report on the Oligochaeta. *Proc. zool. Soc. Lond.,* **1906,** 206.
BERG, K. 1938. Studies on the bottom animals of Esrom Lake. *K. danske Vidensk. Selsk. Sbr.,* **8,** 1.
— 1948. Biological studies on the River Susaa. *Folia Limnol. scand.,* **4,** 1.
JONASSON, P. M., AND OCKELMAN, K. W. 1962. The respiration of some animals from the profundal zone of a lake. *Hydrobiologia,* **19,** 1.
BOTEA, F. 1963. Contributions to the study of Oligochaeta found in the phreatic waters. *Revue Biol. Buc.,* **8,** 335.
BRINKHURST, R. O. 1964a. Observations on the biology of lake dwelling Tubificidae. *Arch. Hydrobiol.,* **60,** 385.
— 1963b. Observations on the biology of the marine oligochaete worm *Tubifex costatus* (Clap.). *J. mar. biol. Ass. U.K.,* **44,** 11.
— 1964. A taxonomic revision of the Alluroididae (Oligochaeta). *Proc. zool. Soc. Lond.,* **142,** 527.
— 1965a. The biology of the Tubificidae in relation to pollution. *Proc. 3rd Seminar on Water Quality Criteria Cincinnati,* **1962,** 57.
— 1965b. Observations on the recovery of a British river from gross organic pollution. *Hydrobiologia,* **25,** 9.
— 1966. The tubificidae (Oligochaeta) of polluted water. *Verh. int. Verein. theor. angew. Limnol.,* **16,** 854.
— 1967. The distribution of aquatic Oligochaeta in Saginaw Bay, Lake Huron. *Limnol. Oceanogr.,* **12,** 137.
BRINKHURST, R. O., AND CHAU, K. E. 1969. A preliminary investigation of some potential nutritional resources by three sympatric tubificid oligochaetes. *J. Fish. Res. Bd. Can.,* **26** (10), 2659.
BRINKHURST, R. O., HAMILTON, A. L., AND HERRINGTON, H.B. 1968. Components of the bottom fauna of the St. Lawrence Great Lakes. *Univ. Toronto Gt. Lakes Inst. P.R.,* **33,** 1.
BRINKHURST, R. O. AND KENNEDY, C. R. 1962. Some aquatic Oligochaeta from the Isle of Man. *Arch. Hydrobiol.,* **58,** 367.
BRINKHURST, R. O., AND SIMMONS, M. L. 1968. The aquatic Oligochaeta of San Francisco Bay. *Calif. Fish and Game,* **54,** 180.
BRINKHURST, R. O., AND WALSH, B. 1967. Rostherne Mere, England—A further instance of guanotrophy. *J. Fish. Res. Bd. Can.,* **24,** 1299.

BULOW, T. 1955. Oligochaeten aus den Endgebieten der Schlei. *Kieler Meeresforsch.*, **11**, 253.

— 1957. Systematisch-autokologische Studien an eulitoralen Oligochaeten der Kimbrischen Halbinsel. *Kieler Meeresforsch.*, **13**, 69.

BUNKE, D. 1967. Zur Morphologie und Systematik der Aeolosomatidae Beddard, 1895 und Potamodrilidae nov. fam. (Oligochaeta). *Zool. Jb. Systematik.* **94**, 187.

CARR, J. F., AND HILTUNEN, J. K. 1965. Changes in the bottom fauna of western Lake Erie from 1930 to 1961. *Limnol. Oceanogr.*, **10**, 551.

CARTER, G. S. AND BEADLE, L. C. 1931. The fauna of the Paraguayan Chaco in relation to its environment. *J. Linn. Soc.*, **37**, 379.

CERNOSVITOV, L. 1936. Notes sur la distribution mondiale de quelques Oligochètes. *Mém. Soc. zool. tchecosl.*, **3**, 16.

— 1939. Catalogues des Oligochètes hypoges. *Bull. Mus. r. Hist. nat. Belg.*, **15**, 1.

CHRISTENSEN, B. 1960. A comparative cytological investigation of an amphimictic diploid and parthenogenetic triploid form of *Lumbricillus lineatus* O.F.M. (Oligochaeta, Enchytraeidae). *Chromosoma*, **II**, 365.

— 1961. Studies on the cyto-taxonomy and reproduction in the Enchytraeidae. *Hereditas*, **47**, 387.

— 1964. Regeneration of a new anterior end in *Enchytraeus bigeminus* (Enchytraeidae, Oligochaeta). *Vidensk. Meddr dansk naturh. Foren.*, **127**, 259.

CHRISTENSEN, B., AND JENSEN, J. 1964. Sub-amphimictic reproduction in a polyploid cytotype of *Enchytraeus lacteus* Nielsen and Christensen (Oligochaeta Enchytraeidae). *Hereditas*, **52**, 106.

COOK, D. G. 1967. Studies on the Lumbriculidae (Oligochaeta). Ph.D. Thesis, University of Liverpool.

DARLINGTON, P. J. 1965. *Biogeography of the southern end of the world.* Harvard University Press.

DAUSEND, K. 1931. Über die Atmung der Tubificiden. *Z. vergl. Physiol.*, **14**, 557.

de BEAUFORT, L. F. 1951. *The zoogeography of land and inland waters.* Sidgwick and Jackson. London.

DELLA CROCE, N. 1955. The conditions of sedimentation and their relationships with Oligochaeta populations of Lake Maggiore. *Memorie Ist. ital. Idrobiol. Suppl.*, **8**, 39.

ERIKSEN, C. H. 1963. The relation of oxygen consumption to substrate particle size in two burrowing mayflies. *J. exp. Biol.*, **40**, 447.

FOX, H. M., AND TAYLOR, A. E. R. 1955. The tolerance of oxygen by aquatic invertebrates. *Proc. R. Soc. B.*, **143**, 214.

GAVRILOV, K. 1931. Selbstbefruchtung bei *Limnodrilus*. *Biol. Zbl.*, **51**, 199.

— 1935. Contributions a l'étude de l'autofecondation chez les Oligochètes. *Acta zool., Stockh.*, **16**, 21.

— 1959. La sexualidad y oligoquetos. *Acta zool. S. America*, **1**, 145.

GREEN, G. 1954. A note on the food of *Chaetogaster diaphanus* Gruit. *Ann. Mag. nat. Hist.*, **1**, 842.

GRIGELIS, A. I. 1901. The oligochaete fauna and the dynamics of incidence and biomass of *Ilyodrilus hammoniensis* and *Psammoryctes barbatus* in Lake Disnai. *Liet. TSR Mokslu Akad. Darb. Serija B*, **3**, 145.

HILTUNEN, J. 1967. Some oligochaetes from Lake Michigan. *Trans. Am. microsc. Soc.*, **86**, 433.

— 1969. Distribution of oligochaetes in Western Lake Erie. *Limnol. oceanogr.* **14**, 260.

HORST, R. 1889. Over een nieuwe soort order de Lumbricinen door Prof. Max Weber uit Nederl. Indie medegebracht. *Tijdschr. Ned. dierk. Vereen.*, **2**, 77.

HRABE, S. 1937. Zur Kenntnis des *Lamprodrilus mrazeki* Hr., *Aulodrilus pluriseta* Pig. und *Aulodrilus pigueti* Kow. *Sb. Klubu prir. Brne*, **19**, 1.

— 1941. Zur Kenntnis der Oligochaeten aus der Donau. *Acta Soc. Sci. nat. moravosiles*, **13**, 1.

— 1960. Oligochaeta limicola from the collection of Dr Husmann. *Spisy prir. Fak. Univ. Brne*, **415**, 245.

HYNES, H. B. N. 1960. *The biology of polluted waters.* Liverpool.

— 1961. The effect of sheep-dip containing the insecticide BHC on the fauna of a small stream including *Simulium* and its predators. *Ann. trop. Med. Parasit.,* **55**, 192.

IVLEV, V. S. 1939. Transformation of energy by aquatic animals. Coefficient of energy consumption by *Tubifex tubifex* (Oligochaeta). *Int. Revue ges. Hydrobiol. Hydrogr.,* **38**, 449.

JAMIESON, B. G. M. 1957. Some species of *Pygmaeodrilus* (Oligochaeta). from East Africa. *Ann. Mag. nat. Hist.,* (12) **10**, 449.

— 1968. A taxonometric investigation of the Alluroididae (Oligochaeta). *J. Zool., Lond.,* **155**, 55.

JOHNSON, M. G., AND MATHESON, D. H. 1968. Macroinvertebrate communities of the sediments of Hamilton Bay and adjacent Lake Ontario. *Limnol. Oceanogr.,* **13**, 99.

KAWAGUTI, S. 1936. On the respiration of *Branchiura sowerbyi. Mem. Fac. Sci. Taihoku imp. Univ.,* **14**, 91.

KENNEDY, C. R. 1965. The distribution and habitat of *Limnodrilus* Claparède (Oligochaeta, Tubificidae). *Oikos,* **16**, 26.

— 1966a. The life history of *Limnodrilus hoffmeisteri* Claparède (Oligochaeta, Tubificidae) and its adaptive significance. *Oikos,* **17**, 159.

KHALAF EL DUWEINI, A., and GHABBOUR, S. I. 1963. A study of the specific distribution of megadrile oligochaetes in Egypt and its dependence on soil properties. *Bull. zool. Soc. Egypt,* **18**, 21.

— 1964. Effect of pH and of electrolytes on earthworms. *Bull. zool. Soc. Egypt,* **19**, 89.

— 1965. Temperature relations of three Egyptian Oligochaete species. *Oikos,* **16**, 9.

KORN, H. 1963. Studien zur Ökologie der Oligochaeten in der oberen Donau unter Berücksichtigung der Abwassereinwirkungen. *Arch. Hydrobiol. (suppl.),* **28**, 131.

LASSERRE, P. 1966. Oligochètes marins des côtes de France. I. Bassin d'Arcachon: Systematique. *Cah. Biol. mar.,* **7**, 295.

— 1967. Oligochètes marins des côtes de France. II. Roscoff, Penpoull, Etangs saumatres de Concarneau: Systematique, Ecologie. *Cah. Biol. mar.,* **8**, 273.

LASTOCKIN, D. A. 1927. Beiträge zu den Oligochaeten Russlands III, IZV. *Ivan.-Voznesensk. politekh. Inst.,* **10**, 65.

LEARNER, M. A., AND EDWARDS, R. W. 1963. The toxicity of some substances to *Nais* (Oligochaeta). *Proc. Soc. Wat. Treat. Exam.,* **12**, 161.

MANN, K. H. 1965. Heated effluents and their effects on the invertebrate fauna of rivers. *Proc. Soc. Wat. Treat. Exam.,* **14**, 45.

MARCUS, E. du B. R. 1944. Notes on fresh-water Oligochaeta from Brazil. *Comun. zool. Mus. Hist. nat. Montev.,* **1** (20), 1.

MARCUS, E. 1944. Sôbre Oligochaeta Limnicos do Brazil. *Bolm. Fac. Filos. Ciênc. Univ. S. Paulo,* **43**, 5.

MCKEY FENDER, D. 1953. The aquatic earthworm *Criodrilus lacuum* Hoffmeister in North America (Oligochaeta, Glossoscolecidae). *Wasmann J. Biol.,* **11**, 373.

MICHAELSEN, W. 1913. Oligochaeten vom tropischen und südlich-subtropischen Africa. *Zoologica, Stuttg.,* **25** (68), 1.

— 1914a. Oligochaeten vom tropischen Afrika. *Mitt. naturh. Mus. Hamb.,* **31** (2), 81.

— 1914b. Oligochaeta. In: *Beiträge z. Kenntnis der Land und Süsswasserfauna D.-Südwestafrikas. Lief* **1**, 137. Hamburg.

— 1918. Die Lumbricidae. *Zool. Jb.,* **41**, 1.

— 1928. Clitellata. *Hamb. Zool.,* **2** (2) (8), 1.

— 1935. Oligochaeten von Belgisch-Kongo. *Rev. Zool. Bot. afr.,* **27** (1), 33.

— 1936. Oligochäten von Belgisch-Kongo. II. *Rev. Zool. Bot. afr.,* **28**, 213.

MOORE, J. P. 1905. Some marine Oligochaeta of New England. *Proc. Acad. nat. Sci. Philad.,* **1905**, 373.

OMODEO, P. 1954. Problem faunistici riguardenti gli Oligocheti terricoli della Sardegna. *Atti Soc. Tosc. Sci. nat.,* **61**, 1.

— 1963. Distribution of the terricolous oligochaetes on the two shores of the Atlantic. In *North Atlantic Biota and their History.* Pergamon Press Oxford.

OMODEO, P. 1965. Sistematica e distribuzione geografica degli Hormogastrinae (Oligocheti). *Arch. Bot. Biogeog. Ital.*, **32** (ser. 4), vol. 1, Fasc. IV, 159.

PALMER, M. F. 1968. Aspects of the respiratory physiology of *Tubifex tubifex* in relation to its ecology. *J. zool., Lond.*, **154**, 463.

PODDUBNAYA, T. L. 1959a. Concerning the dynamics of the Tubificidae populations (Oligochaeta, Tubificidae) in Rybinsk Reservoir. *Trudy Inst. Biol. Vodokhran.*, **2**, 102.

— 1959b. Autotomy and regeneration of Tubificidae. *Byull. Inst. biol. Vodokhr.*, **5**, 15.

— 1961. Data on the nutrition of the prevalent species of tubificids on the Rybinsk Reservoir. *Trudy Inst. Biol. Vodokhran.*, **5**.

— 1965. Feeding of *Chaetogaster diaphanus* Gruit. (Naididae, Oligochaeta) in the Rybinsk Reservoir. *Trudy Inst. Biol. Vodokhran.*, **9**, 178.

PODDUBNAYA, T. L., AND SOROKIN, V. S. 1961. The thickness of the nutrient layer in connection with the movements of the Tubificidae in the ground. *Byull. Inst. biol. Vodokhr.*, **10**, 14.

RAVERRA, O. 1955. Amount of mud displaced by some freshwater Oligochaeta in relation to the depth. *Memorie Ist. ital. Idrobiol. suppl.*, **8**, 247.

RZOSKA, J. 1936. Über die Ökologie der Bodenfauna im Seenlitoral. *Arch. Hydrobiol. Rybact.*, **10**, 76.

SANDERS, H. L., HESSLER, R. R., AND HAMPSON, G. R. 1965. An introduction to the study of deep sea benthic faunal assemblages along the Gay Head—Bermuda transect. *Deep Sea Res.*, **12**, 845.

SAPKAREV, J., 1959. *Ilyodrilus hammoniensis* Mich. (Oligochaeta) in the large lakes of Macedonia (Ochrid, Prespa and Dojran). *Rev. Trav. Sta. Hydrobiol. Ohria.*, 7 1459, 1.

— 1963. Quantitative Zusammensetzung der Oligochaeten fauna des Prespa-Sees. *God. Zborn. Skopje*, **14**, 31.

— 1964. Quantitative Untersuchungen über die Oligochaeten des Dojran-Sees. *Izd. Zav. Ribarst. N.R. Maked.*, **3**, 1.

— 1965. Die Oligochaeten fauna des Ohrida-Sees. *God. zborn. Skopje Biol.*, **15**, 5.

SCHUMACHER, A. 1963. Quantitative Aspekte der Beziehung zwischen Stärke der Tubificidenbesiedlung und Schichtdicke der Oxydationszone in den Süsswatten der Unterelbe. *Arch. Fischereiwiss.*, **14**, 48.

SOLOWIEV, M. M. 1924. Über die Rolle der *Tubifex tubifex* in der Schlammerzeugung. *Int. Revue ges. Hydrobiol. Hydrogr.*, **12**, 90.

SOROKIN, J. I. 1966. Carbon 14 method in the study of nutrition of aquatic animals. *Int. Revue ges. Hydrobiol. Hydrogr.*, **51**, 209.

STEPHENSON, J. 1930. *The Oligochaeta*. Oxford. Clarendon Press.

STOUT, J. D. 1951. The occurrence of aquatic oligochaetes in soil. *Trans. R. Soc. N.Z.*, **80**, 97.

— 1956. Aquatic oligochaetes occurring in forest litter—I. *Trans. R. Soc. N.Z.*, **84**, 97.

— 1958. Aquatic oligochaetes occurring in forest litter—II. *Trans. R. Soc. N.Z.*, **85**, 289.

TIMM, T. 1962. Über die Fauna, Ökologie, und Verbreitung der Süsswasser Oligochaeten der Estnischen SSR. Tartu Riikliku Ulikooli Toimetised. *Zoologica-Alaseid Toid.*, **120**, 67.

VALLE, K. J. 1927. Ökologische-Limnologische Untersuchungen über die Boden- und Tiefenfauna in einigen Seen nördlich vom Ladoga. *Acta zool. fenn.*, **2**, 1.

VEAL, D. M., and OSMOND, D. S. 1969. Bottom fauna of the Western Basin and near-shore Canadian waters of Lake Erie. *Proc. 11th. Conf. on Great Lakes Res. Univ. Michigan*, Great Lakes Res. Div. Pub. **1968**, 151.

WACHS, B. 1963. Zur Kenntnis der Oligochaeten der Werra. *Arch. Hydrobiol.*, **59**, 508.

— 1967. Die Oligochaeten Fauna der Fliessgewässer unter besonderer Berücksichtigung der Beziehungen zwischen der Tubificiden Besiedlung und dem Substrat. *Arch. Hydrobiol.*, **63**, 310.

WASAWO, D. P. S., AND VISSER, S. A. 1959. Swamp worms and tussock mounds in the swamps of Teso Uganda. *East Afr. agric. J.*, **25**, 86.

WHITTEN, B. K., AND GOODNIGHT, C. J. 1966. Toxicity of some common insecticides to tubificids. *J. Wat. Pollut. Control Fed.*, **1966**, 227.

YOSHIZAWA, H. 1928. On the aquatic oligochaete *Stylaria lacustris* L. *Sci. Rep. Tuhoku Univ. Ser. 4. Biol.*, **3**, 587.

ZAHNER, R. 1967. Experimente zur Analyse biologischer, chemischer und physikalischer Vorgänge in der Wasser-Sediment-Grenzschicht stehender und langsam strömender Gewässer. I. Beschreibung der Versuchsanlage mit vorfläufigen Ergebnissen über das Verhalten der Tubificiden in Wahlversuchen. *Int. Revue ges Hydrobiol. Hydrogr.*, **52**, 627.

4

PHYLOGENY AND CLASSIFICATION

Part. 1.   R. O. Brinkhurst

Logically, perhaps, the material dealt with here should form the concluding chapter of the book. The reasons for presenting it here are that it enables a clear separation of the more general material from the detailed systematic reviews, and that it may be used as structure upon which a detailed study of those reviews may be based. Even within the chapter the arrangement may be thought to be a reversal of the logical process of creating a classification on anatomical evidence and then deducing the possible phylogenetic paths that might have been involved in the evolution of the modern Oligochaeta. Again, this has been done with a view to clarifying the classification and allowing the reader to benefit by the experience of those who have already trodden this difficult path. By presenting the broad outlines of a possible evolutionary past and illustrating the way in which the various oligochaete taxa represent the end products of this process we are not attempting simply to construct a classification to fit the proposed phylogeny, but rather to present what we feel to be the logical test of our classification as a means to signposting its major dichotomies. Despite the lack of fossil evidence, we still feel that the attempt to construct a phylogeny is a useful exercise in that it has considerable predictive value and should serve to focus attention on certain key areas requiring detailed investigation in order to substantiate some of our bolder statements. Whilst a truly artificial classification might be erected in order to deal with the mechanical problem of putting a name to any given organism, such a classification would not seem to have the same predictive value, nor does it lead to the same intellectual fun as the attempt to create a natural classification even where we must confess our total ignorance of the fossil history of the group, as we must here.

Another major barrier to the creation of a classification of the Oligochaeta is the lack of studies of the development of some of the major organ systems. A consideration of the homologies of the so-called atria with prostates of the Tubificidae and allied families with the prostates of many of the large terrestrial worms cannot be undertaken here because of our total ignorance of the development of these organs in all but a small number of species. The homologies of parts of the male duct described in different genera within the Tubificidae is a good example of a problem which could be resolved on the basis of such studies.

Despite this, it does seem posssible, within broad limits, to discuss the evolution of the Oligochaeta to the family level, and beyond that in some instances. This discussion will encompass the classification down to and including families, the major emphasis on the erection of subfamilies being placed at the beginning of the relevant systematic review. As the recognition of the family Glossoscolecidae entails a consideration of several groupings formerly diagnosed as families a more detailed treatment of this particular problem belongs here.

In order to present a clear picture of the proposed classification, little reference will be made to those that preceded it, valuable guides though these may have been. We recognize that, without these to act as alternative models, our problems would have been much greater but again, in deference to clarity and brevity, we merely ask that our classification be recognized as one derived from its antecedents, but with some unique features.

In the past few years, there has been a dramatic re-awakening in functional anatomy which has led to a critical reappraisal of some of the basic tenets of classification. The recent analysis of the function of the coelom and the septa of annelids by Clark (1964) for example has made possible a reconsideration of one of the basic pieces of dogma of annelid systematics—that the oligochaetes arose from the polychaetes. This belief seems to be founded on the concept of the marine origin of invertebrate phyla followed by the adoption of special modifications which allowed the invasion of land. Fresh waters have clearly been invaded from the sea via estuaries and from the surrounding land. The different invasion routes have usually left their mark, as in the Mollusca, but the origin of the fresh-water Oligochaeta is not, perhaps, immediately apparent.

The most profound difference between the polychaetes and the oligochaetes is usually held to be the development of copulation and the hermaphrodite condition of the latter. These tendencies are also widespread among the Platyhelminthes, another phylum that has undergone considerable evolution in fresh water and as parasites. Indeed, in the absence of any convincing evidence for the adaptive value of the separation of the sexes, one wonders why so much emphasis has been placed on this aspect of Annelid biology. According to the view of the evolution of first a coelomate and then a segmented coelomate advanced by Clark (1964), the earliest annelid could have had an anatomy resembling that of an oligochaete with a reproductive apparatus of a polychaete. The worm would have had a relatively simple body-wall with almost complete layers of longitudinal and circular muscle. The septa would have been complete except for provision for the passage of the gut and other organs through sphincters. The setation would have been rudimentary. The blood system would have included a long series of simple commissural vessels. There would have been no specialized reproductive ducts, only coelomoducts, and no evidence of restriction of the gametocytes to discrete gonads. We may, perhaps, admit to a marine ancestry for the group, as the presence of a ciliated larva in the polychaetes probably represents the retention of early mode of dispersal in the reproductive cycle. The suppression of larval stages in fresh-water invertebrates is quite predictable, as is their loss in terrestrial forms.

We may envisage the diversity of a group of simple annelids of this indeterminate form leading to two possible evolutionary paths. The first leads to progressive specialization towards a burrowing and surface crawling existence, the second towards a more fully aquatic mode of life involving the evolution of swimming by undulating the body. In the sea, increasing mobility led to the elaboration of the body wall musculature, the development of parapodia, the origin of complex setae of a variety of shapes (including hair setae) and some elaboration of the reproductive organs. The huge success of these organisms probably led to the gradual suppression of the ancestral types once worms recognizable as polychaetes exploited the sedimentary environment, producing such specialists as the Capitellidae. The other line of annelids that invaded the interstitial environment and the groundwater beneath the land masses produced both the specialized burrowing forms that became adapted to relatively dry soils and a second line of truly aquatic species. From each of these lines the paths of evolution continued to cross the boundary between aquatic and terrestrial habitats. Even the major distinction between the marine and the fresh-water or terrestrial realms has broken down, with the penetration of fresh-water environments by several small, highly specialized polychaetes like *Manyunkia* and of marine habitats by many oligochaetes. The successful invasion of the sea by oligochaetes, especially the Tubificidae, was emphasized by the recent discovery of specimens on the ocean floor at a depth of 2,000 m off Bermuda.

With this pattern of evolutionary paths to consider, it is possible to examine the existing families of oligochaetes in order to identify traces of their ancestry. In so doing, one point requires considerable emphasis. In the search for "primitive" forms, one should not be surprised to find features thought to be primitive in species also presenting various specialized or advanced characteristics. We would be fortunate to find species alive today in which all of the primitive characteristics were to be found simultaneously.

To start with, it is necessary to set out the anatomical features which might be worthy of special consideration, and those that may be regarded with suspicion. Beginning with the reproductive system, as so much emphasis has been placed on this in early classification, the first factor to consider is the wide variation in the position of the gonad-bearing segments. This has sometimes been used as evidence for the existence of a multigonadal ancestral condition, presumably because of a belief in the fixity of the segmental sequence. A study of regeneration and the results of asexual reproduction suggest that whatever determines the exact position in which gonads are eventually developed may be altered in existing species. Further-more, if the exact location of the gonads is ignored but the relative position of parts is considered, it is immediately clear that all save a few oligochaetes can be derived from a single type with two pairs of testes and two pairs of ovaries in four succeeding segments. The exceptions all prove to be species in which reproduction by architomy (splitting apart followed by regeneration) occurs, or species that reproduce parthenogenetically. These factors seem to upset the mechanism that determines the location of and the number of pairs of gonads, as can be shown by examining the wide variation in gonads in species such as *Lumbriculus variegatus* and *Lamprod-rilus mrazeki*. This basic octogonadal plan is illustrated in Fig. 4. 1, together with the patterns represented by existing families. (One exception to this derivation is the Aeolosomatidae but these may not be related to the other oligochaetes for a variety of reasons.) One further piece of evidence here is the development of four pairs of gonads in early stages of development of several species in which some are later resorbed.

The number and position of the spermathecae is variable, even within some species. The location of the spermathecal pores on the segment also varies, the variation in position along the long axis apparently being at the generic level but variation in position in the dorso-ventral plane being reported within species (*Potamothrix hammoniensis* for example). The oligochaetes seem to be remarkably adaptable in regard to the spermathecae, their position presumably being dependent on the position adopted during copulation. This being the case, it is not possible to read too much signifi-cance into the position of these organs.

The position of the male pores in relation to the testes with which their ducts are associated is a key character in discussing the relationships of the families. The acquisition of sperm storage organs, nutritive glands and copulatory devices in the form of penes of various types is also a fruitful source of information. The position of the clitellum in relation to the pores, and the size and amount of yolk in the eggs has some significance as well.

The setae show a tendency to become either paired in each bundle or indefinite in number and complex in form. The blood vascular system would provide more useful evidence if its form in the aquatic families were better understood.

Among the larger oligochaetes other anatomical features have been studied

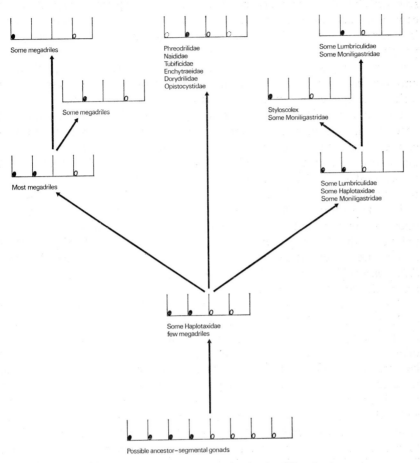

Fig. 4.1. The arrangement of testes and ovaries in the Oligochaeta.
Solid circles—testes; open circles—ovaries. Note that all known arrangements are
derivable from a form with two pairs of testes, two pairs of ovaries. The derivation
of multigonadial lumbriculids is explained in the test.
The positions of the names of modern families indicate the arrangement of gonads
only, no evolutionary significance is implied.

which extends the range of characters that can be utilized, but few
anatomical studies of the small aquatic worms have been made.

The two modern oligochaete families in which most traces of a supposedly
ancestral form can be found are the Haplotaxidae, and the Lumbriculidae,
many of which inhabit ground waters in various parts of the world. Most
of the Haplotaxidae are Australasian but the family is cosmopolitan. Most
species have two setae per bundle, and these may be simple or bifid or
sometimes both. In some species the setae are only slightly sigmoid, and in
one they are sickle shaped. There may be only a single seta in each

so-called bundle. There are no complex glands, sperm-storing atria, or copulatory devices in most members of the family. In *H. gordioides* there is a simple gizzard in IV-V, traces of one in other species, but a dorsal pharyngeal pad and associated pharyngeal glands in others. The eggs are large and yolky as in the majority of aquatic families, and the clitellum is only one cell thick and is found in the region of the male pores. A long series of simple commissural vessels is found in *H. gordioides.*

The Lumbriculidae are less obviously derived from our hypothetical ancestral annelid in so far as most morphological characteristics are concerned. The blood vascular system, for example, is rather specialized in many species in which blind-ending lateral vessels are attached to the dorsal vessel posteriorly, presumably as an aid to respiratory exchange. The presence of multigonadal forms has created much interest in this family, as has the form of the male ducts. The Lumbriculidae have all of the male ducts opening in testis-bearing segments, a situation that is practically unique to this family. There are well-developed atria with prostates and all but three species (*Lumbriculus variegatus, L. multiatriatus* and *Lamprodrilus mrazeki*) have lost the posterior pair of ovaries, neither of which are particularly "ancestral" characters.

A consideration of the origin of the male ducts provides convincing evidence that the ancestral lumbriculids diverged from an ancestral annelid independently to the direct ancestor of the haplotaxids, from which all the other oligochaete families (other than the Moniligastridae) can ultimately be derived. The first consideration regards the primitive condition of the male duct. The commonest condition is that the funnels associated with each testis-bearing segment lead to male ducts that penetrate the septum on which the funnels lie and open in a segment behind that bearing the related testes. Only in the lumbriculids do we find vasa deferentia running from funnels to atria and pores within the same segment as the associated testes. In many lumbriculids there are four vasa deferentia draining two segments entering two atria (in one instance one atrium) in the posterior testes-bearing segment. This is clearly an example of the condensation of reproductive organs seen throughout the oligochaetes, but how did it arise? There are two possibilities. Either the anterior vasa deferentia have abandoned the former male pores in the anterior testis segment, or the posterior vasa deferentia no longer open via the ovarian segment but have shifted forwards. If the latter situation represents the truth, the form of the male ducts of all other families is normal, that of the lumbriculids is aberrant. Alternatively, if the first condition is shown to be more acceptable, then the male ducts in lumbriculids show signs of a more primitive arrangement than that found elsewhere.

Three points convince us that the lumbriculid male duct arrangement, at its simplest, reveals something of the original plan found in annelids as a whole. First, these ducts are coelomoducts, and as such might be expected to open in the same segment as the funnel. Secondly, all oligochaetes show an increasing trend towards loss of the anterior elements of the male system, the posterior elements of the female system in a progressive concentration of the reproductive system. The male ducts show a strong tendency to open further and further back, especially in the megadrile forms. Some megadriles, including the Alluroididae, have retained only the anterior testes and the posterior ovaries. As a final piece of confirmation for this position, the genus *Rhynchelmis* includes species with rudimentary atria in the anterior testes-bearing segment and two pairs of vasa deferentia associated with the remaining atria. The third line of argument is that, whilst the haplotaxid condition may now be derivable from the same ancestral form as the lumbriculids by an extension of the same principle of rearward movement of the male ducts (affecting both pairs instead of just the anterior pair), it is very difficult to see how (or why) the lumbriculid situation arose from an ancestor with male ducts of the form found in all other families.

The presumed evolutionary sequence is illustrated in Fig. 4. 2, in which the names of existing genera indicate present-day representatives of each condition illustrated with no implication that one modern genus is derived from another. The sequence starts with an ancestral form with four pairs of gonads and coelomoducts all of the same form, all of which open into their own segment (possibly in the intersegmental furrow). The ancestral lumbriculid has developed atria and prostates. The female ducts have become simplified, but otherwise there has been little change. The loss of the posterior ovary and funnel is the next phase in concentration, followed by the rearward shift of the anterior vasa deferentia. Rudimentary atria in *Rhynchelmis* species provide living proof of the reality of this phase, leading to the commonest condition found in the family. A further continuation of the process of reduction of the anteriormost male elements follows, with reduction and loss of the testes and then the vasa deferentia and funnels. There are again living forms showing obvious evidence that this reduction happened, especially in *Rhynchelmis, Eclipidrilus* and *Lumbriculus variegatus*. This final reduction produces forms such as *Hrabea, Kincaidiana* and other *Eclipidrilus* species, but it is immediately obvious that the same condition could be reached directly from the ancestral form. The multiple atria of *Lumbriculus multiatriatus* are almost certainly secondary, produced by a tendency to asexual reproduction (and/or parthenogenesis), but the retention of both pairs of ovaries should be noted. The atria are derivable from the *L. variegatus* condition in which evidence of the earlier existence of the second pair of vasa deferentia is available. The alternative derivation of such a condition directly from the ancestral form is employed in order to explain the multigonadal state of some species of *Lamprodrilus*. Here, the asexually-reproducing *L. mrazeki* seems to be basically hologynous and there is no trace of an anterior pair of vasa deferentia entering the posterior atria. Most *Lamprodrilus* species have one or two pairs of male organs, but *L. satyriscus* has three or four male segments. Again, the female system

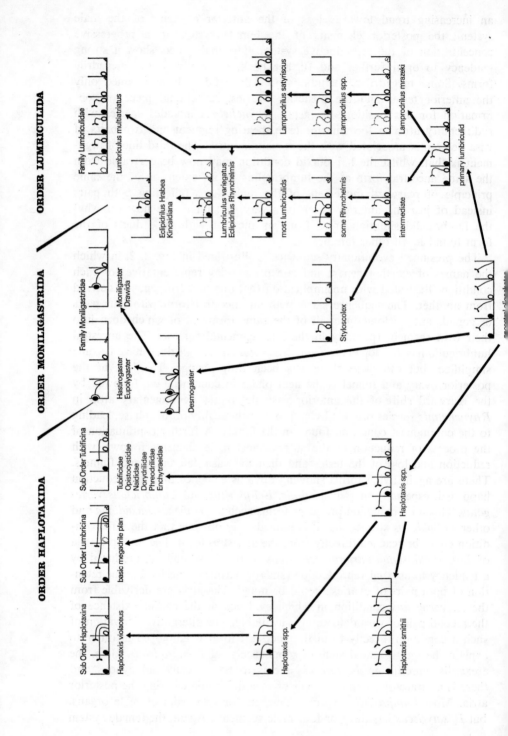

has been reduced to a single set of ovaries and funnels, atria are well developed, and a tendency towards asexual division is known in the genus, so that this multi-gonadal form is difficult to accept as a direct descendant of a multigonadal ancestor. The origin of *Styloscolex* is seen as a distinct development from the ancestral condition in which the posterior testes were lost, in contrast to the more normal contraction of the gonad row (Fig. 10. 2). It seems to have had little success as an innovation.

The situation in the Moniligastridae was clarified by a discussion of their phylogeny by Gates (1962). It is now believed that the testes, situated in the unique testis sacs, actually lie within the septa instead of growing back from them, and that the testes are still to be associated with the segment behind the testis sacs. The only space within the testis sacs arises from the sperm funnels, which are found closely associated with the testes. The male ducts traverse the segment to which the testes are attributed and open in or close to (presumably meaning just anterior to) the intersegmental furrow behind the testis segment in question and that immediately behind. The female ducts may also open in the segment behind that bearing the ovaries.

If the precise segment occupied by any particular gonad again be ignored it is apparent that the anterior pair of ovaries of the octogonadal basic set always remains, but the posterior one is lost in all modern members of the family. In this respect the family is distinct from the majority of megadrile oligochaetes. One or other pair of testes may be lost, producing two derived conditions, both represented by modern genera, as illustrated in Fig. 4. 2.

The unusual form and position of the testes and male funnels, together with the position of the male ducts in relation to their testes is sufficient evidence to merit recognition of a distinct taxon, derivable only from a common ancestor with both the Lumbriculidae, and the ancestral haplotaxid from which the rest of the Oligochaeta may be derived.

The haplotaxid condition is reached when both pairs of vasa deferentia migrate rearwards through a septum, and not just the anterior pair. In *Haplotaxis smithii* the female openings have also moved rearwards, but in most oligochaetes they open in the intersegmental furrow to the rear of each ovarian segment. In some other haplotaxids the posterior ovary is lost, and in *H. violaceus* the anterior pair of vasa deferentia open close to the posterior pair in the ovarian segment.

The majority of the aquatic worms are to be found in the Tubificidae and Naididae and related families (Dorydrilidae, Phreodrilidae, Opistocystidae, plus Enchytraeidae) in which the anterior testes and posterior ovaries and their associated ducts have been lost. The remaining gonads are then arranged in the same way as in the lumbriculid genera *Eclipidrilus*, *Kincaidiana* and *Hrabea*, but the male ducts open in the ovarian segment. The family Dorydrilidae was created for the genus *Dorydrilus* and the dubious *Lycodrilus* in which the male pores are in the ovarian segment in a worm in which superficial characteristics suggest an affinity with the

Lumbriculidae. This argument is detailed under the definition of the family.

Within this assemblage there are various evolutionary trends which can be suggested. The Enchytraeidae seem to have diverged at a stage when atria and prostates were rather ill-defined structures, or else they have independently evolved structures which carry out some of the functions of those organs. A trend away from primary aquatic forms (as represented by the supposedly primitive *Propappus* in which the setae were bifid and indefinite in number in each bundle) towards a more terrestrial mode of existence led to a simplification of the setae, which are usually simple-pointed and often straight rather than sigmoid. In the other families of this assemblage (Tubificidae, Naididae, Phreodrilidae, Opistocystidae), setal diversity developed which is matched only by the polychaetes. Hair setae are found in the dorsal bundles of representatives of each family, and the other dorsal setae may be bifid, pectinate, or even palmate. The vasa deferentia open into atria, which usually bear prostates derived from the lining cells of those atria. In the Phreodrilidae these prostate cells remain within the atria, perhaps echoing the early phase of development of this organ. All of these families have a greater or lesser development of the pharyngeal roof as an eversible pad, long held to be characteristic primarily of the Enchytraeidae. They also retain several supposedly primitive characteristics, such as the thin clitellum only one cell thick and large yolky eggs.

Differences between the families involve the segmental position of the reproductive organs, the relative position of the spermathecae (before or behind the gonads or even in the same segment as the ovaries at least) and setal differences (e.g. the special setal arrangement of the Phreodrilidae). The Naididae are perhaps the oligochaete family most highly adapted to an aquatic life out of the sediment. This development may have been responsible for the original development of hair-setae, which are difficult to understand as an adaptation to life in a burrow. The whole group may, therefore, have gone through a phase of evolution towards a swimming mode of life some time after the separation of the enchytraeids, but most have returned to a life spent entirely within the sediments. The naidids have added eyes, small size and asexual reproduction to their repertoire. The tubificids show signs of abandoning the hair setae and pectinate setae in many instances, they lack eyes and make much less use of asexual reproduction.

The link between the Tubificidae and the Naididae is very strong. About the only real distinctions are the anterior position of the reproductive organs in the latter, which is coupled with the occurrence of paratomy as a mode of asexual reproduction. The reproductive organs and their sequential arrangement are otherwise very similar, and one can point out the existence of tubificids in which the reproductive organs are apparently shifted forwards as a result of asexual reproduction (see *Potamothrix, Aulodrilus*). One or two small tubificids were originally described as naidids, which indicates that size is often a useful guide to the familial position of worms of this type. Many species have hair setae in the dorsal bundles in both families, and these may be accompanied by bifid or palmate or pectinate setae, whilst the ventral setae are almost always bifid. The prostate cells often coat the atria in the naidids, as they do in the tubificid genera such as *Rhyacodrilus*, but they are sometimes found on what appear to be the vasa deferentia, a location which might be equated to the situation in *Monopylephorus* and *Bothrioneurum*, but the nomenclature of parts in the last two is far from certain. The derivation of the Naididae via *Pristina* from tubificids like *Aulodrilus* ignores the differences in the atria and prostates, but points out the small changes that would have to occur in order to derive an early naidid from a tubificid, or vice versa.

Both of these families are cosmopolitan in distribution, but the tubificids are best represented in the north temperate lands. The tubificid genus *Telmatodrilus* has a discontinuous distribution, often taken to indicate antiquity, and has several rather unusual features. There are a series of small prostates on the atria, a feature otherwise restricted to certain lumbriculids, and the setae are thick and sometimes indistinctly bifid. Whereas the posterior setae are often bifid even in species where they are pectinate in anterior bundles in many tubificids, in one *Telmatodrilus* species only the posterior setae are pectinate. The body wall is thick in these worms, and the Californian species is apparently almost terrestrial in habit. The distinctive characters of this rather odd genus do not serve to link the tubificids to any other family, however, but only raise more questions than they solve.

The Phreodrilidae, confined to the southern tips of the continental land masses and southern islands (plus Ceylon), were originally classified as tubificids. The atria lack prostates or rather, the prostate cells line the atria and have not penetrated through the muscle layers as in other families. The dorsal setal bundles contain hair setae and minute needles that scarcely emerge from the setal sacs. The ventral setal bundles are commonly limited to one bifid and one simple-pointed seta each. Such an arrangement of the ventral setae is only approached by a few *Peloscolex* species in the Tubificidae and *Haplotaxis glandularis* (in which the same arrangement is also found in the dorsal bundles). The spermathecae lie posterior to the rest of the reproductive organs, and the ampullae are often on the end of long ducts that seem to be occluded. In separating the families the last character has to be used with caution, because the Lumbriculidae includes species with spermathecae behind the gonad-bearing segments as well as species with them in front. Although the single pairs of testes and ovaries are in succeeding segments, as in the tubificids, very few of the other characters suggest that the families should be merged. The similarities may be thought of more as a result of convergence than recent kinship, although both families may be ultimately descended from the same basic stem.

The Opistocystidae are a small family, too poorly known to be placed with any conviction in any scheme of classification. The two posterior appendages and single median process are unique to the family. The spermathecae lie behind the other reproductive organs, but the atria bear prostate cells, unlike the phreodrilids. The family seems to be as much a parallel of the Naididae as the phreodrilids are of the tubificids. The known distribution includes N. and S. America and N. Africa.

The Aeolosomatidae remain an enigma. Many of the anatomical features such as the divided ventral nerve cord set in the body wall are clearly primitive, but in many ways they are highly modified, miniaturized fresh-water specialists. If the origin of the oligochaetes, if not the annelids, be argued as outlined above, then the consideration of *Aeolosoma*, and especially *Rheomorpha*, as primitive forms, becomes untenable. If the ancestor of the annelids were a small ciliated coelomate lacking segmentation, the aeolosomatids might be considered as primitive. They have a ciliated prostomium which is used in locomotion. Most, other than *Rheomorpha*, have setae, but these are primarily hair setae which are found in ventral as well as dorsal bundles. The septa are wanting. The gametocytes arise throughout the coelom, but become limited to rudimentary testes lying in front of and behind the single unilateral ovary in most species. In most species the gametes escape through nephridia, but in *Potamodrilus* there are complex male ducts. As they are hermaphrodite and have relatively few setae they have usually been allied with the oligochaetes, but Sedgwick (1898) placed them in the Archiannelida, a rather unsatisfactory mixed-bag of little-known worms, some of which seem most likely to be simplified polychaetes.

The following reasons may be advanced for separating these worms from the other oligochaetes. While the gonad rudiments in oligochaetes may be derived from cells that migrate to a few loci, and hence the aeolosomatid situation might be regarded as primitive, no oligochaete family has testes in front of and behind the ovary. The so-called clitellum in this family is ventral in position, not dorso-lateral or annular. The "clitellum" of *Potamodrilus* resembles a copulatory gland similar to those found in addition to the clitellum of other genera.

If the nephridia are taken over for reproductive purposes in the aeolosomatids, it is difficult to see why the nephridia degenerate in the segments which subsequently develop gonoducts in other families. The ciliated prostomium enables these worms to swim only because they are so small. The loss of septa could well follow the adoption of a swimming habit, the development of cilia for this purpose being much more likely than their retention from an earlier ancestor. Cilia are present on gills on some oligochaetes and no-one would consider the possession of gills by *Dero*, *Branchiodrilus*, *Branchiura* and one *Phreodrilus* to represent survival of an ancient anatomical attribute. The ciliated postomial pit of *Bothrioneurum* is similarly considered to be a specialization. It would seem more reasonable to suppose that cilia are produced as organelles by most cells where selection favours their production. They must have been "re-evolved" many times over in most phyla. Furthermore, it is difficult to think of a selective value for the development of septa in a small ciliated coelomate, but the need for septa in a coelomate burrowing by deforming the body shape using the coelomic fluid and a muscular body wall is understandable. In other words, the lack of septa and the cilial mode of progression seem more related to adaptation to the aquatic existence of a worm small enough to exploit food resources beneath the notice of the naidids.

Because of the unique gonad series, and setal arrangement, the aeolosomatids cannot be thought of as the end-product of the tubificid to naidid trend. Rather, they seem to represent a unique branch of one of the earliest lines of annelid development. One clue seems to have been overlooked in most discussions to date. The ventral pharyngeal bulb, variously developed in the aeolosomatids, is similar to that found in *Protodrilus* and *Dinophilus* and quite unlike the pharyngeal plate formed as a thickening of the dorsal wall of the pharynx in most aquatic oligochaetes. A similar ventral pharyngeal bulb is present even in the peculiar genus *Stygocapitella* and the even more startling *Parergodrilus*. Bunke (1967) suggests that the similarity of the pharyngeal bulbs is superficial, but the matter needs careful investigation. The latter is found in humid forest litter, where aeolosomatids have also been reported. Whilst many of these archiannelids are dioeceus, some traces of hermaphroditism have been seen in *Parergodrilus*.

Without a detailed knowledge of the achiannnelids, it is impossible to re-classify them and include the Aeolosomatidae in any re-grouping of these obscure little worms, but enough is known about the latter to suggest that they are in no way related to the other oligochaetes and that their relationships seem more closely tied up with at least some of the genera presently relegated to this pigeon-hole.

Part 2.   B. G. Jamieson

The remaining oligochaete families (Alluroididae, Glossoscolecidae, Lumbricidae, Megascolecidae and Eudrilidae demonstrate their relationship with the Haplotaxidae by a remarkable fixity of the gonad-sequence, having not only two testis-segments and two ovarian segments in sequence but also the segmental enumeration of these (segments X-XIII) seen in the Haplotaxidae. Deletion of the anterior or posterior testes occurs in some species of all of the families, however, and the posterior testes are absent throughout one subfamily (the Alluroidinae).

Two characteristics indicate that these families may be encompassed in a single taxonomic group and distinguish them from the Haplotaxidae, Lumbriculidae and Moniligastridae. First, when a pair of ovaries is lost it is always the anterior pair and, secondly, the male pores are always more than one segment behind their corresponding testes and funnels (the opisthoporous condition). In only very few species is the trend towards loss of the anterior ovaries incomplete (p. 60) and in all the families the male pores have moved to a location two or more segments behind their haplotaxid location, the most anterior location being the thirteenth segment (Alluroididae and *Biwadrilus*). Primitiveness of the latter two taxa is indicated by persistence in them of lateral lines which are characteristic of microdriles but are known in no other megradriles.

Other trends are evident in the group and enhance its unity while distinguishing it from other groups, discussed earlier. Thus, when there are two pairs of vasa deferentia, the anterior and posterior ducts are almost always united, giving a single pair of male pores. Exceptions to this are limited to a single genus in each of three families (p. 46). A further trend in the group is loss of the large-yolked eggs seen in the Haplotaxidae and Moniligastridae, such eggs persisting in only the Alluroididae. Correlated with the production of relatively little-yolked eggs in the remaining families is a proliferation of the cells of the clitellum so that this becomes a multi-layered structure presumably permitting an increased secretion into the cocoon of nutrients for the developing young, in compensation for reduction of the amount of these substances in the eggs. A negative trend has been a failure to develop polymorphism of the somatic (as opposed to genital) setae, to multiply these in each bundle and to elongate them, in contrast with the Tubificidae and their relatives, and this failure is clearly associated with burrowing in soil or mud. A tendency to increase in number in the setae in a row, giving ultimately the perichaetine condition, is seen independently in the Megascolecidae and Glossoscolecidae, however. Development of bifid setae in some terrestrial glossoscolecids remains an enigma.

Whereas some of the families under discussion show indications of close interrelationship, the Alluroididae form a distinct entity set apart from these remaining familes by the retention of large-yolked eggs and, in the Alluroidinae, of lateral lines. They appear to be representatives of an early and not especially successful essay in the "opisthopore" condition by the presumed haplotaxid stock (Fig. 4. 2) and are separable from the Haplotaxidae almost solely by absence of the anterior ovaries and the development of atria, or prostates, at or near the male pores. In conjunction with the possession of lateral lines, the marked secondary annulation confers on the Alluroidinae remarkable similarity to the Haplotaxidae even in the external facies. The obviously primitive organization of the Alluroidinae and their endemism, at generic and specific levels, in both Africa and S. America, may indicate that they originated, and spread to their present distributions, before the Mesozoic separation of Africa and America postulated by advocates of continental drift. Alternatively they may be southerly survivors of a now extinct Holarctic stock. The present widely discontinuous distribution of the Haplotaxidae, which includes preglacial Holarctic lakes and the Neotropical and Ethiopian regions, permits either conclusion.

Phylogenetic affinities of the Syngenodrilinae (E. Africa) are very obscure. In a neo-Adansonian phenetic classification (Jamieson, 1968), their closest affinities were shown to lie with the Alluroidinae, though they had previously been placed in the Moniligastridae. We may envisage a haplotaxid-like ancestor, shared with the Alluroidinae, with large-yolked eggs and a single-layered clitellum, but differing from the Haplotaxidae in having the male pores in segment XIII. An independent origin of the Syngenodrilinae from the haplotaxid ancestors is, nevertheless, conceivable. They retain some primitive (i.e. haplotaxid) characters which are not retained in the Alluroidinae: elongated sperm sacs and holandry (testes in segments X and XI) but are highly modified in the location of the oesophageal gizzards (absent from most haplotaxids) in segments VIII and IX. The development of prostate-like glands (in XI-XIII) is a departure from haplotaxid organization seen also in *Sparganophilus* but unknown in the Alluroidinae unless the atria of *Brinkhurstia americanus* be considered homologous.

Trends in the Eudrilidae whether they are derivable directly or indirectly from haplotaxids, are extension of the male ducts to segment XVII, development of a communication between the spermatheca and the oviducal apparatus, so that internal fertilization often occurs and subsequent enclosure of the ovaries in the system thus produced. Development of these interrelationships of the female organs has apparently occurred a number of times (Jamieson, 1967, 1969) and before or after the commencement of a backward movement of the spermathecal pore, which usually is unpaired, and in some genera has come to lie behind the male pores. In *Hyperiodrilus* this backward movement has progressed unequally in different lineages, the pore occurring in any of segments IX to XIII though invariable in segmental location in each species. Evidence will be presented in a later volume that meronephry, so common in the Megascolecidae, has arisen in some eudrilids, though independently. A further departure from haplotaxid organization has been the development of "euprostate" glands, muscular and glandular terminal dilitations of the vasa deferentia strongly reminiscent of the atria of, for instance, *Alluroides pordagei*.

In the Megascolecidae the male pores have moved at least as far posteriorly as the anterior border of segment XVII and are usually located in segment XVIII. They open in conjunction with, or in the vicinity of, two to several pairs of tubular prostate glands or a pair (rarely two pairs) of glands with a branched system of ducts, known as racemose prostates. Recent authors (Gates, 1959; Sims, 1966) have conferred familial distinction on possessors of a racemose prostate, placing them in a restricted family Megascolecidae, but such distinction appears unwarranted despite unconfirmed embryological evidence that tubular prostates are ectodermal invaginations while racemose prostates are said to be mesodermal proliferations. Despite denial by several authors of the existence of morphological intermediates between the two types of prostates, those of *Perionyx egmonti* Lee, 1952 (a species which must be excluded from the genus as redefined by Gates, 1960) appear from Lee's description to be transitional between tubular and racemose prostates. Were it not for the compact form of its prostates, *P. egmonti* could be included in *Diporochaeta*. Significantly, an Indian *Perionyx* showed its closest affinities to lie with *Diporochaeta* in a computer analysis (Sims, 1966). Additional evidence against subdivision of the Megascolecidae (s. lat.) into distinct families is provided by *Travoscolides*. The prostates of this genus are stated by Gates (1940) to be tubular and yet its excretory system shows an organization, including advanced enteronephry, which precludes separation from genera with racemose prostates.

Similarly, division of the non-ocnerodrile Megascolecidae with tubular prostates into the families Megascolecidae, if holonephric, or Octochaetidae if meronephric, is debatable. These questions will be fully discussed in a second volume.

Other trends in the Megascolecidae (s. lat.) in addition to backward movement of the male pores are: frequent development of the perichaetine condition; the development of dorsal pores as has occurred independently in the Lumbricidae (presumably also in response to an environment more truly terrestrial than that of most Eudrilidae); a tendency to backward movement of the gizzard from the location in the fifth segment seen in *Haplotaxis gordioides* and, often, replication of this organ; extension of the oesophagus; development in many species, by splitting of the nephroblast rudiments, of meronephry, the meronephridia coming to open in many species into the gut; and reduction of the number of pairs of prostate glands to one, from the several pairs seen in the family in some Ocnerodrilinae.

The Megascolecidae and Eudrilidae were considered to be sufficiently closely related by Michaelsen (1903, 1928) to constitute a discrete assemblage (Familienreihe) within a suborder Opisthopora (other co-ordinate groups in the suborder were the Phreoryctina (Haplotaxidae, Alluroidae and Moniligastridae) and the Lumbricina (Glossoscolecidae and Lumbricidae)). Relationship of the Ocnerodrilinae, within the Megascolecidae and the Eudrilidae was considered especially close. Grounds for a proposed descent of the Eudrilidae from the Ocnerodrilinae were the presence in both groups of ventral calciferous glands and supposed homology of the euprostate glands of the Eudrilidae and terminal muscular bursae into which the vasa deferentia open in the ocnerodrile *Nannodrilus*. Relationship of the two groups was considered to be indicated with certainty by restriction of the male pores to the region of the seventeenth to nineteenth segments and from the location of the clitellum (in front of and often including these pores) and to be further supported by location of spermathecae near their megascolecid location, in the eudrilid *Hyperiodrilus*. Detailed consideration of the evolution of the Eudrilidae and of the Megascolecidae is beyond the scope of this volume but it must be mentioned that derivation of the Eudrilidae from the Ocnerodrilinae would necessitate accepting that they have lost the tubular prostate glands of the Ocnerodrilinae and of many other Megascolecidae and that the "ocnerodrile" ancestor possessed a long series of hearts. The first condition is not difficult to accept as the ocnerodrile *Malabaria* and related genera appear still to be in the process of reduction from several prostates to a single pair and in *Ilyogenia* there are species (e.g. *I. tepicensis*) with no prostates but with muscular "atria", not unlike euprostates, at the ectal ends of the sperm ducts. On the other hand the vascular system of the Eudrilidae presents serious objections to origin of the family from a form which would have been recognizable as an ocnerodrile as the only feature unequivocally distinguishing the Ocnerodrilinae from the remaining Megascolecidae is the restriction of hearts to segments IX-XI while *Hyperiodrilus* shows what is almost certainly a primitive condition of the vascular system (c.f. Chapter 1) with hearts in the sixth to eleventh segments (it differs further from the ocnerodrilinae in a derived character, the presence of a subneural blood vessel). Furthermore the malabarine ocnerodriles furnish evidence that the ancestral Ocnerodriles may have lacked extramural calciferous glands and suggesting either that ventral glands have developed independently in the Eudrilidae and Ocnerodrilinae or, if ventral calciferous glands are an indication of relationship between eudrilids and their acanthodrile possessors, requiring additional hypotheses. One such hypothesis would require separate origin of the Eudrilidae and Ocnerodrilinae with extramural calciferous glands, on the one hand, and the malabarine ocnerodriles, on the other, from an ancestor which was a megascolecid but, by virtue of its vascular system, was not an ocnerodrile. Though the Eudrilidae may have arisen from populations immediately ancestral to forms which we would now place in the Ocnerodrilinae we must thus consider the family has had at most an acanthodrilid rather than a specifically ocnerodriline origin. In view of the similarity of eudrilid euprostates and microdrile atria it seems worth considering that the tubular prostates of the Megascolecidae may have replaced euprostates rather than the reverse process postulated by Michaelsen and accepted by Stephenson (1930) or alternatively that the two families are descended from an ancestor which possessed both types of organ.

In conclusion, origin of the Eudrilidae and Megascolecidae independently from the presumed haplotaxid ancestors cannot be ruled out with certainty but phenetically the two families appear to form a single assemblage when contrasted with the Glossoscolecidae and the Lumbricidae. They differ from the latter families in location of the male pores at or behind the posterior border of the clitellum whereas the former condition is rare and post-clitellar male pores are found in only one genus in those families. The total absence of ventral calciferous glands from the other families is also striking as is the rarity of development of any form of prostate gland.

Turning to the Glossoscolecidae and Lumbricidae, the anatomy of the modern Haplotaxidae fulfils the requirements for derivation of these families from an ancestral haplotaxid though it presents one difficulty: derivation of the condition seen in many glossoscolecids and lumbricids in which the spermathecae are not, or not merely, pretesticular as in the Haplotaxidae but occur in or behind the testis-segments.

Michaelsen (1918) regarded the Glossoscolecinae (the Glossoscolecini of the present work) as ancestral to all other glossoscolecids precisely because they have the pretesticular spermathecae of the presumed haplotaxid ancestor. Location of spermathecae in the testis segments in *Hormogaster* and behind them in the Microchaetinae (s. Michaelsen) were looked upon as secondary conditions. The occurrence of either, or both, conditions in the Lumbricidae, was explained in terms of a totally new development of spermathecae in the Lumbricidae subsequent to derivation of the group from an ancestral *Criodrilus* which lacked spermathecae. The question of which spermathecal condition is primitive is probably insoluble and may be misconceived. It seems worth considering that the populations from which the Glossoscolecidae (s. lat) arose were variable or labile in this respect as are the Lumbriculidae and, indeed, the Glossoscolecini to this day. It may be assumed, nevertheless, that they would have been identifiable as haplotaxids. Modern haplotaxids appear to have stabilized with a pretesticular series of spermathecae but, from the examples of *Rhinodrilus fafner* and *Martiodrilus* (*Thamnodrilus*) *crassus* (p. 65), it is evident that even the S. American glossoscolecines, which have been defined by Stephenson by pretesticular spermathecae, are still labile in the position of the spermathecae.

The following phylogenetic scheme (see also Fig. 4. 4) for derivation of present day glossoscolecids from such a haplotaxid stock is here very tentatively advanced as being consistent with the classification adopted on morphological grounds (p. 190). For convenience in referring to Fig. 4. 3 the various inferred evolutionary stages are indicated, in text and figure, by bracketed capital letters. These letters are not arranged in hierarchical sequence.

In some populations of the haplotaxid stock the male pores may have moved one segment posteriorly into XIII, the segment which they occupy in the Alluroididae and Biwadrilinae. Retention of microdrile features and the modifications already alluded to gave the Alluroididae. The male ducts became embedded in the body wall in some populations (A) perhaps as an initial response to greater muscularization of the body wall, and the clitellum became multilayered (and extensive), probably as an adaptation providing sufficient food reserves in the cocoons to compensate for replacement of the large-yolked eggs of the Haplotaxidae with moderate-yolked megadrile eggs, these two changes giving the first Glossoscolecidae. As in modern haplotaxids, Sparganophilinae, and Alminae, anterior nephridia presumably were not developed in the adult. Such a morphology, with some specialization of the male reproductive apparatus, but with retention of the lateral lines of the Haplotaxidae and Lumbriculidae, is seen in *Biwadrilus*. In *Biwadrilus* extrinsic spermatophores were developed to an extent which rendered spermathecae obsolete. At some stage, the anterior ovaries were lost. The hologynous condition is known only in *Enantiodrilus* (Glossoscolecini) where it seems an abnormality and in one species of *Glyphidrilus*

(Almini) but lability with regard to retention of hologyny or adoption of metagyny may have persisted throughout glossoscolecid evolution.

The Sparganophilinae which, like *Biwadrilus*, have two pairs of latero-parietals, may have originated from other early glossoscolecids at a similar level of development. In *Sparganophilus* the tendency to backward movement of the male pores, so evident in the Glossoscolecidae as a whole, is well developed and the brain has moved back to segment III, but evidence of its primitive organs is still seen in the exceptionally short, unspecialized oesophagus, the intestinal origin being in IX (further forward than in modern haplotaxids or alluroidids), absence of a subneural vessel, and retention of precardiac commissural vessels. The lumbricid type of ovaries (p. 61) in the Sparganophilinae may indicate that the Lumbricidae arose from related populations of the early haplotaxid stock with multi-layered clitellum though with superficial sperm ducts, but may be simply an example of parallelism (see p. 188). Lateral lines were lost in the early spargano-philines or their predecessors.

The aquatic glossoscolecids which did not differentiate as spargano-philines or biwadrilines (or, no doubt, other lines now extinct) formed the evolutionary pool from which two major extant groups of the Glossosco-lecidae developed: the Alminae, which retained the aquatic mode of life, and the Glossoscolecinae, which became terrestrial.

This basal stock of the two subfamilies (B) may have had male pores in XV (the most anterior location in both extant groups), presumably retained the small dorsal intersetal/circumferential ratio, and may have had gizzard-like muscular thickening of the oesophagus, possibly in the region of IV-V as in *Haplotaxis gordioides*. It was, it appears, labile with regard to the location of the spermathecae. Nephropores were probably in *ab*.

Those populations (C) of this hypothetical stock which gave rise to the Alminae retained oesophageal gizzard (s) and probably had the anterior muscularization of the intestine seen in many present-day representatives. They also must have possessed a typhlosole and a median subneural vessel and have lost the (ventral?) pair of latero-parietals. Spermathecae were predominantly post-testicular. A quadrangular cross section developed, possibly allowing the maintenance of respiratory channels along the periphery of the body in submerged burrows shaped by the circular cross section of the anterior region of the body. The significance of the development of a dorsal or dorso-terminal anus is not apparent; especially as in *Alma nilotica* and *A. eubranchiata* it opens into the dorsal, gill-fringed respiratory groove.

Criodrilus represents one of the two lines of descent from these early Alminae of level C (see also p. 188). In it the brain has retained a primitive position, in segments I and II, and the male pores remain in segment XV but copulatory bursae of a glandular and muscular nature have been developed which must be responsible for production of its complex sperma-tophores. Spermathecae have been lost, perhaps, as suggested for *Biwadrilus*, because of the efficacy of the spermatophores. As spermatophores in *Alma* are deposited in the vicinity of the spermathecal pores even when the latter have moved to the posterior end of the body it may, perhaps, be inferred that the spermathecae of the ancestral Criodrilini (D) were in the vicinity of segments XIV-XVI on which the spermatophores are usually found. This location would add support to the taxonomic association in the present work with the Almini.

The other line of descent, the present day Almini (E) differentiated from level C by development of a supra-oesophageal vessel. The spermathecae have become exclusively post-testicular in all extant species excepting *Glyphidrilus stuhlmanni*. Most species of *Drilocrius* retained slight gizzard-like muscularization of the oesophagus and of the anterior region of the intestine while the male pores maintained a primitive location in XV or moved into XVI. In *Callidrilus* the male pores have moved into XVII and prostate glands have developed; thickening of the oesophageal musculature, though strong, is variable in extent, while intestinal thickening has been lost. Intrageneric variation in *Glyphidrilus* suggests the following trends: development of lamellar copulatory alae; backward migration of the male pores; increase in the longitudinal extent of the spermathecal pores; increasing development of the supra-oesophageal vessel until it takes over, in front of the hearts, the function of a now truncated dorsal vessel which terminates with the first hearts; loss of one or more pairs of anterior hearts; backward extension of the sinusoidal region of the oesophagus so that the intestinal origin moves from XV to XVIII; posterior movement of the oesophageal gizzard, which is well developed, from VII to VIII; loss of muscular thickening of the intestine; and a tendency to reduction and loss of the subneural vessel in front of its junction with the pair of latero-parietal vessels.

Alma is obviously very highly evolved as is shown by the development of unique setose claspers and, in some species, migration of the spermathecal pores and the clitellum to the posterior region of the body. The absence of a median subneural may therefore be regarded as secondary and it would appear that the paired neuro-parietals which are continuous anteriorly with the latero-parietals must represent a secondary pairing of a formerly un-paired, median subneural vessel. Intestinal muscular thickening has been retained but oesophageal gizzards are totally absent. The intestine commences in XVI to as far back as XXVIII. The dorsal vessel is probably always replaced in front of the hearts by the supra-oesophageal, and the anterior pair of hearts, in VII, is often lost. The genus is the only one in the megadriles to have developed gills (in two species); their development would appear to attest to a long aquatic history.

Glyphidrilocrius (p. 50) has taken elongation of the oesophagus to an extreme, intestinal origin being in segment XXXVIII in the only known species. Trends seen in other Almini are also apparent. Thus the hearts of VII have been lost and, as in some *Alma* and *Glyphidrilus* species, an additional pair has been added, in XIII; a median subneural vessel is absent as in *Alma* and the oesophageal gizzard has been lost while powerful thickening of the anterior intestinal musculature has been retained.

The Hormogastrinae, Kynotinae and Glossoscolecinae (Glossoscolecini and Microchaetini) appear to represent a terrestrial line of descent (F) from the ancestral populations of Glossoscolecidae (level B) postulated above. They presumably had an oesophageal gizzard, but had no typhlosole and probably no intestinal gizzards; and it seems likely that male pores occupied segment XV and that the position of the spermathecae relative to the testes was labile. They presumably retained the circular cross section and terminal anus of the Haplotaxidae and level B, and, from the evidence of *Hormogaster*, may have at first retained their small dorsal intersetal : circumferential ratio. Nephridiopores may have been shortly above *b* lines as in *Hormogaster*. An early departure from previous organization would seem to have been, from the evidence of *Hormogaster*, *Microchaetus*, *Kynotus* and many Glossoscolecini, the development of the nephridial caecum, possibly as an adaptation for water conservation.

Evolution from an ancestral form with nephridial caeca (F) to give the present day terrestrial Glossoscolecidae (*Kynotus*, *Hormogaster*, *Microchaetus*, *Tritogenia* and the Glossoscolecini) could have taken a number of paths and interpretation of which routes were followed requires subjective decisions as to whether the typhlosole found in all but *Kynotus* was evolved more than once, whether absence of subneural and supra-oesophageal vessels, in the Microchaetini alone, is a primitive or a derived condition and whether the nephridial sphincters (and even caeca) of *Microchaetus* and say, *Pontoscolex* have been derived from a common ancestor or are examples of parallelism, in the sense of Simpson (1961), that is the acquisition of similar structures independently but by virtue of genetic relationship. Phenetically the unity of these terrestrial taxa appears irrefutable and, as shown on p. 187, *Microchaetus* and *Tritogenia* are especially close morphologically, and, it is here considered, phylogenetically to the Glossoscolecini. It is here assumed that the absence of a typhlosole from *Kynotus*, in view of the primitive nature of much of the fauna of Madagascar, is a primitive feature and, adopting a principle of least change, it is also taken that a typhlosole was present in the presumed common ancestor of *Hormogaster*, the Microchaetini and the Glossoscolecini.

Hormogaster may then be derived from early typhlosolate populations (G) by development of subneural and a supra-oesophageal vessels, strong development of the oesophageal muscularization to give three gizzards, and stabilization of the location of the spermathecae in and behind the testis segments.

The Glossoscolecini appear to represent a further lineage (H), from this typhlosolate ancestor which developed nephridial sphincters that migrated to the vicinity of the dorsal setal couples, and other features of the nephridia characteristic of the Glossoscolecinae. In the glossoscolecine lineage, the dorsal intersetal : circumferential ratio must have increased and the sinusoidal region of the oesophagus have become so specialized as to produce a pair of extramural calciferous glands. One branch of this lineage, represented by the Microchaetini, has never developed (or, less probably, has lost) the subneural and supra-oesophageal vessel, has an oesophageal gizzard in VII and post-testicular multiple spermathecae. A second branch, constituting the Glossoscolecini, has developed the subneural and supra-oesophageal vessel, has an oesophageal gizzard in VI and is still labile with regard to the position of the spermathecae, which are pre-testicular but may or may not extend into or behind the testis segments. It rarely has the multiple spermathecae of the Microchaetini. Whether those Glossoscolecini which lack nephridial caeca and sphincters have lost these or are descendents of populations of level B which never developed them is likely to remain unanswerable. Affinities within the Glossoscolecini require much additional investigation.

Kynotus is shown in Fig. 4. 3 as having originated from atyphlosolar populations, below level (G), which developed subneural and supra-oesophageal vessels but did not develop the typhlosole of the ancestors of *Hormogaster* and the Glossoscolecinae nor the nephridial sphincters of the Glossoscolecinae.

The phylogeny postulated necessitates acceptance that subneural and suboesophageal vessels have been independently acquired in the Kynotinae, Hormogastrinae and Glossoscolecini. If, however, absence of these vessels in the Microchaetini were secondary, it would, alternatively, be possible to envisage a common ancestor of the Hormogastrinae and Glossoscolecinae possessing these vessels. If origin of the typlosole were independent in the Glossoscolecinae and Hormogastrinae, the latter ancestor could be placed at the root of all living terrestrial Glossoscolecidae. No firm opinion can be given with respect to these alternatives. The main thesis of the present work is the distinctness of the Alminae and an especially close affinity between the Microchaetini and Glossoscolecini.

CRIODRILUS AND THE ORIGINS OF THE LUMBRICIDAE

The Lumbricidae show so many features with the Glossoscolecidae which are not characteristic of other families that a close relationship between the two seems indisputable. Some glossoscolecid features of the Lumbricidae are as follows: the clitellum is often extensive and may occupy as many as 32 segments; the body may be, at least posteriorly, quadrangular in cross section (*Eiseniella tetraedra*; *Bimastos palustris*); longitudinal tubercula pubertatis are commonly present; the male pores are normally on segment XV (as in several Alminae, in the Kynotinae and Hormogastrinae) and, as in the Glossoscolecidae with the exception of *Opisthodrilus*, are never post-clitellar; the spermathecal pores vary in position from 5/6 to 19/20, and are often multiple, though usually in 9/10 and 10/11; the spermathecae are adiverticulate; the musculature at the anterior end of the intestine is thickened (as in most Alminae and, possibly, in the Hormogastrinae); some

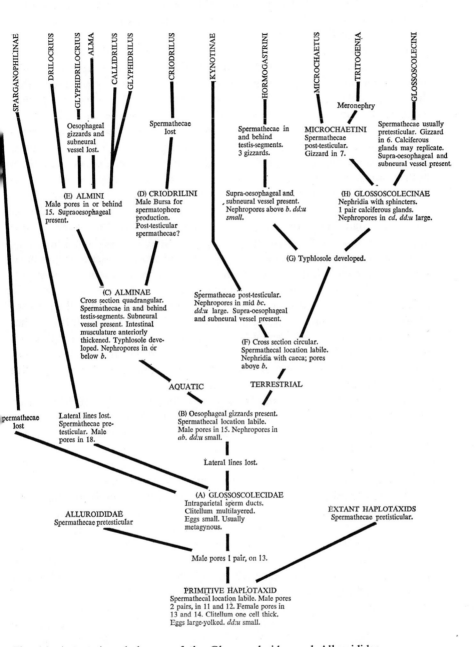

Fig. 4.3. A tentative phylogeny of the Glossoscolecidae and Alluroididae.

embedding of the vasa deferentia in the body wall may occur (see Chapter 1); and grooved copulatory setae are present (as in *Criodrilus* and as has been reported in *Hormogaster*).

Two genera of the Glossoscolecidae qualify for consideration as being especially closely related to the Lumbricidae. These are *Sparganophilus* and *Criodrilus*. Omodeo (1956a) considered it not improbable that the Lumbricidae had originated from a form related to *Sparganophilus*, while at the same time pointing out that *Criodrilus* is the only oligochaete known to have a haploid chromosome number of 11 other than the lumbricids *Eisenia foetida* and *E. spelaea*. In a further paper (1956 b) he recognized the close similarity of *Criodrilus* and the Lumbricidae, and derived them both from a common ancestor which was in turn derived from a form represented by *Sparganophilus*. It has been shown in Chapter 2 that *Sparganophilus*, alone in the Glossoscolecidae, resembles the Lumbricidae in having a single "tail" of oocytes at the free border of the ovary. It also approaches the Lumbricidae in having intraclitellar tubercula pubertatis, a saddle-shaped clitellum, and no oesophageal gizzard. Evidence is presented on p. 183 for regarding *Sparganophilus* as a relatively little-modified representative of the ancestral Glossoscolecidae and, accepting a Glossoscolecid origin of the Lumbricidae, it seems probable that the lumbricid-like features of *Sparganophilus* are an example of parallelism in related forms rather than of convergence of unrelated forms. An especially close relationship with the Lumbricidae is debatable, however, and the hypothesis of close relationships of the Lumbricidae with the Alminae and notably with *Criodrilus* deserves careful examination.

Stephenson (1930) has amply reviewed the arguments of Michaelsen (1918) for derivation of the Lumbricidae from *Criodrilus*. Points of similarity between *Criodrilus* and the Lumbricidae noted by Michaelsen and augmented by the writer are the anterior thickening of the musculature of the intestine; the copulatory sacs at the ends of the sperm ducts (seen in *Bimastos palustris*); location of the male pores, on conspicuous porophores, in segment XV; the presence (though not in all Lumbricidae) of spermatophores; the grooved copulatory setae (retaining, though, in *Criodrilus* traces of the usual transverse striation of glossoscolecid setae); and the absence of a supra-oesophageal blood vessel. Stephenson was not convinced of an origin of the Lumbricidae from the Glossoscolecidae, citing the usual location of the spermathecal pores in 9/10 and 10/11 in the Lumbricidae as an obstacle to such a derivation. He was clearly also doubtful both of the Glossoscolecid nature of *Criodrilus* and of its primitive position relative to the Lumbricidae. He preferred to consider *Criodrilus* as "descended in fact from *Allolobophora*", and as owing its peculiarities – loss of calciferous glands and gizzard, loss of anterior nephridia, and, possibly, loss of spermathecae – to adoption of an aquatic habitat.

In the present study the glossoscolecid nature of *Criodrilus* is accepted (p. 183) and location of the brain in segments I and II, together with the presence of slight thickening of the oesophageal musculature, are considered to preclude origin from the Lumbricidae. Close morphological affinities of Lumbricidae with the Glossoscolecidae have already been endorsed and the "lumbricid" features of *Criodrilus* alluded to are considered strongly to suggest origin of the Lumbricidae and of *Criodrilus* from an ancestor resembling the latter. *Bimastos palustris*, with its spermatophore-secreting copulatory sacs, its quadrangular cross section, absence of spermathecae, marginally intra-parietal sperm ducts, and aquatic habitat is here considered to owe these features to genetic relationship with *Criodrilus*.

Michaelsen, having derived the Lumbricidae from *Criodrilus*, explained the "reappearance" of spermathecae in the Lumbricidae an "atavism". Stephenson on the other hand, considered that if, contrary to his opinion, the Lumbricidae were derived from an ancestral *Criodrilus* it must have possessed spermathecae. It is here suggested that absence of spermathecae in *Criodrilus* and *Bimastos*, and their very variable position throughout the Lumbricidae (including "microchaetine" and "glossoscolecine" segmental distributions) is evidence of a lability with regard to presence and position of the spermathecae in common ancestral populations of the Criodrilini and Lumbricidae. The square cross section of the lumbricid *Eiseniella tetraedra* may also be interpreted as evidence of relationship with the Criodrilini though the concept of parallelism (Simpson, 1961) does not require it to have been a direct retention from such ancestors. Against regarding the square cross section as a character acquired convergently as an adaptation to an aquatic existence must be mentioned the fact that aquatic megradriles of other families, e.g. *Eukerria* and *Pygmaeodrilus*, among the ocnerodriles, and *Pontodrilus*, in the megascolecids, have a circular cross section. Although the writer favours the above hypothesis, one further derivation may perhaps be hazarded.

Tendency to an albeit parthenogenetic reversion to a haplotaxid location of the male pores in *Eiseniella*, while not contraindicating affinity with the Glossoscolecidae, could be interpreted as evidence of origin of the Lumbricidae (and the Criodrilini?) from a very early Glossoscolecid at or below level A in Fig. 4. 3. Such an interpretation would more easily explain the adumbration of lumbricid features in the very simple Sparganophilinae. There is little in the anatomy of *Eiseniella* which otherwise suggest that it is especially primitive, however.

Part 3. Conclusion

The following classification of the Oligochaeta is advanced with some hesitation, as it is difficult to assess the relative value of ranks without considerable experience of their application in other phyla. This scheme does serve to emphasize possible lines of descent rather than depending entirely on the degree of similarity between groups, and so it does attempt to reflect a natural classification. It is fully recognized that, in the absence of any fossil record in this ancient group, any classification should be viewed as a stimulus to further systematic study and not as a final statement of fact.

    Class Clitellata
      Subclass Oligochaeta
        Order Lumbriculida
          Family Lumbriculidae
        Order Moniligastrida
          Family Moniligastridae
        Order Haplotaxida
          Suborder Haplotaxina
            Family Haplotaxidae
          Suborder Tubificina
            Superfamily Enchytraeoidea
              Family Enchytraeidae
            Superfamily Tubificoidea
              Family Tubificidae
              Family Naididae
              Family Phreodrilidae
              Family Opistocystidae
              Family Dorydrilidae
          Suborder Lumbricina
            Superfamily Alluroidoidea
              Family Alluroididae
            Superfamily Lumbricoidea
              Family Glossoscolecidae
              Family Lumbricidae
            Superfamily Megascolecoidae
              Family Megascolecidae
              Family Eudrilidae

The relationships between these families are illustrated in Fig. 4. 4.

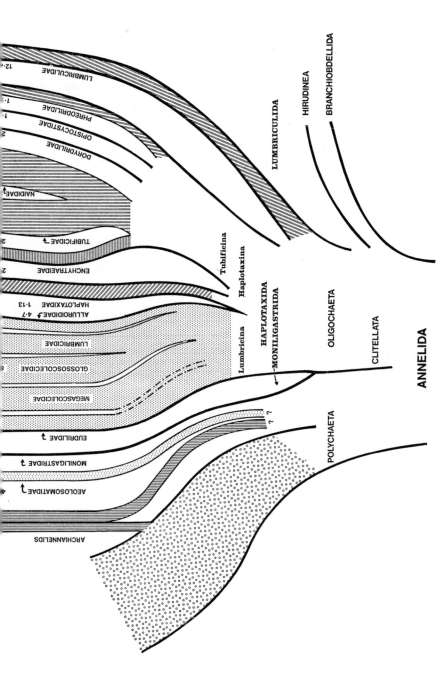

Fig. 4.4. Suggested relationships between the families of oligochaetes and their relationships to other major annelid groups. Some of the major evolutionary trends are indicated. For further clarification see text. Numbers represent total number of genera and species (aquatic genera and species in the Glossoscolecidae).

REFERENCES

CLARK, R. B. 1964. *Dynamics in metozoan evolution. The origin of the coelom and segments.* Oxford.

GATES, G. E. 1940. Indian earthworms. VIII-XI. *Rec. Indian Mus..* **42**, 115.

— 1959. On a taxonomic puzzle and the classification of the earthworms. *Bull. Mus. comp. Zool Harv.*, **121**, 229.

— 1960. On Burmese earthworms of the family Megascolecidae. *Bull. Mus. comp. Zool Harv.*, **123** (6), 203.

— 1962. On some Burmese earthworms of the moniligastrid genus. *Drawida. Bull Mus. comp. Zool. Harv.*, **127**, 297.

JAMIESON, B. G. M. 1967. A taxonomic review of the African megadrile genus *Stuhlmannia* (Eudrilidae, Oligochaeta). *J. Zool., Lond.*, **152**, 79.

— 1968. A taxonometric investigation of the Alluroididae (Oligochaeta). *J. Zool., Lond.*, **155**, 55.

— 1969. A new Egyptian species of *Chuniodriius* (Eudrilidae, Oligochaeta) with observations on internal fertilization and parallelism with the genus *Stuhlmannia*. *J. nat. Hist.*, **3**, 41.

LEE, K. E., 1952.. Studies on the earthworm fauna of New Zealand. *Trans. R. Soc. N.Z.*, **80**, 23.

MICHAELSEN, W. 1903. *Die geographische Verbreitung der Oligochaeten.* Berlin.

— 1918. Die Lumbricidae. *Zool. Jb.*, **41**, 1.

— 1928. Clitellata und Gürtelwürmer. Dritte Klasse der Vermes Polymera (Annelida). Kukenthal-Krumbach. *Handb. Zool.*, **2** (8), 1.

OMODEO, P. 1956a. Contributo alla revisione dei Lumbricidae. *Archo zool. ital.*, **41**, 129.

SEDGEWICK, A. 1898. *A student's textbook of zoology.* Vol. 1, London.

SIMPSON, G. G. 1961. *Principles of animal taxonomy.* Columbia University Press.

SIMS, R. W. 1966. The classification of the megascolecoid earthworms: an investigation of oligochaete systematics by computer techniques. *Proc. Linn. Soc. Lond.*, **177**, 125.

STEPHENSON, J. 1930. *The Oligochaeta.* Oxford. Clarendon Press.

PART TWO **SYSTEMATICS**

DEFINITIONS AND KEYS TO FAMILIES

CLASS CLITELLATA

Annelida without cephalic appendages, with or without oral sucker, (or posterior sucker in leeches) with or without eyes. Parapodia absent, setae relatively few or absent, hair setae in dorsal bundles present or absent, no jointed setae. Septa usually present, or coelom reduced to sinuses. Ventral nerve cords not sub-epidermal but within muscle layers at least, ganglia fused. Hermaphrodite reproductive system, usually concentrated in a few segments with complex gonoducts, glandular clitellum saddle-shaped and incomplete ventrally, or annular. Fertilized ova enclosed in cocoons. Development direct.

SUB-CLASS OLIGOCHAETA

Clitellata mostly with setae, with testes anterior to ovaries, usually a maximum of two pairs of each in four successive segments, male system occasionally replicated in several segments. Suckers absent (present in *Aspidodrilus*). Coelom capacious. Number of segments variable within specific limits.

1. ORDER LUMBRICULIDA

Setae four pairs per segment, simple or bifid. AT LEAST ONE PAIR OF MALE FUNNELS IN SAME SEGMENT AS THE MALE PORE OR PORES WITH WHICH THEY ARE ASSOCIATED. Clitellum one cell thick in region of male and female pores. Eggs large, yolky. Gut mostly simple with eversible pharynx.

A single family: Lumbriculidae

2. ORDER MONILIGASTRIDA

Setae four pairs per segment, simple. TESTES AND FUNNELS IN PAIRED DORSAL TESTIS SACS SUSPENDED IN A SEPTUM (or *Desmogaster*, in two successive septa); much elongated vasa deferentia opening, through capsular prostates, at male pores on or close to the intersegment immediately behind the testis sac containing the associated funnels and testes, i.e. opening in 10/11 (or in 11/12 and 12/13). Ovaries one pair, never more than two intersegments behind that bearing the posterior or only testis sac, i.e. in XI or XII, where testis sacs are at 9/10, or in XIII where testis sacs are at 10/11 or at 10/11 and 11/12. Clitellum one cell thick, including male and female pores. Eggs large, yolky. Gizzards oesophageal but posterior to the ovarian segment. Subneural vessel present. Spermathecae one or two pairs, anterior to the male pores.

A single family: Moniligastridae.

3. ORDER HAPLOTAXIDA

MALE FUNNELS ALWAYS AT LEAST ONE SEGMENT ANTERIOR TO THAT BEARING THE MALE PORE OR PORES WITH WHICH THEY ARE ASSOCIATED.

A. SUB-ORDER HAPLOTAXINA

Setae single or paired, mostly simple or bifid, rarely pectinate or keeled. Two pairs of testes, one or two pairs of ovaries (anterior pair of ovaries always present, the second pair sometimes lost), male pores open in the segment immediately behind that bearing associated funnels and testes, rarely both pairs of male pores opening separately in segment behind that bearing second pair of funnels and testes. No atria or penes, vasa deferentia opening via simple pores. Clitellum one cell thick, in region of genital pores. Eggs large, yolky. Gut simple, with an eversible dorsal pharyngeal plate with associated glands, or a muscular gizzard. Dorsal and ventral vessels linked by long commissurals.

A single family: Haplotaxidae

B. SUB-ORDER TUBIFICINA

Setae an indeterminate number per bundle, rarely single or paired (absent in *Achaeta*). Mostly bifid, pectinate or palmate, with or without hair setae in dorsal bundles, usually bifid in ventral bundles, or at least one bifid and one simple pointed; rarely all paired and simple pointed in all bundles. Testes and ovaries paired in successive segments. Male pore or pores in segments immediately behind that bearing associated testes and funnels, usually associated with glandular and muscular terminal apparatuses, usually atria; penes often present. Clitellum one cell thick, in region of genital pores. Eggs large, yolky. Gut simple, usually with an eversible dorsal pharyngeal plate and associated glands.

a. SUPER-FAMILY ENCHYTRAEOIDEA

Dorsal and ventral setae usually the same, sigmoid or straight, simple-pointed (bifid in *Propappus*, absent in *Achaeta*). Spermathecae anterior to the segments bearing reproductive organs, separated by a gap of five segments. Atria mostly absent, glands and penial bulbs present or absent.

One family: Enchytraeidae

b. SUPER-FAMILY TUBIFICOIDEA

Dorsal and ventral setae frequently dissimilar, dorsal bundles often with hair setae, ventrals without. Spermathecae at least with pores in segments immediately in front of or behind those bearing male pores, rarely in the same segment. Atria mostly present, penes present or absent.

Five families:

KEY TO FAMILIES OF THE TUBIFICOIDEA

1.	Setae paired, simple-pointed (rarely bifid, rarely single in *Lycodrilus* species endemic to L. Baikal)	Dorydrilidae
—	Setae usually numerous in each bundle, frequently bifid (or even pectinate in dorsal bundles), hair setae in dorsal bundles of many species, dorsal and ventral setae frequently dissimilar	2
2.	Spermathecae in front of segment bearing male pore	3
—	Spermathecae behind segment bearing male pore	4
3.	Asexual reproduction forming chains of individuals. Usually less than 2 cm long. Pectinate dorsal setae rarely present. Male pores between V and VIII.	Naididae
—	Asexual reproduction by fission or absent. Usually more than 2 cm long (marine species often smaller). Pectinate dorsal setae frequently present. Male pores normally on XI	Tubificidae
4.	Posterior end with two lateral and one median process. Dorsal bundles with hair setae only. Ventral setae more than two per bundle	Opistocystidae
—	No postero-lateral processes. Dorsal bundles with hair setae and short setae that scarcely project from the setal sacs. Ventral setae two per bundle, bifid or simple pointed, frequently one of each in each bundle	Phreodrilidae

C. SUB-ORDER LUMBRICINA

Setae eight per segment, or multiplied, but not more than two functional in a bundle; simple or (some Glossoscolecini) sporadically bifid. Testes two pairs, in X and XI, or one pair, in one of these; testis sacs, if present, never intraseptal. Male pores one pair (rarely two pairs), at least two segments behind that (or those) bearing associated testes and funnels (though in XI or XII in some parthenogenetic morphs of *Eiseniella*). Atria (prostates) present or absent. Ovaries one pair, in XIII (rarely two pairs in XII and XIII). Clitellum one or more cells thick, including the genital pores or not. Eggs large and yolky or small relative to body diameter. Gut with or without gizzards. Subneural vessel present or absent. Spermathecae anterior or posterior to the male pores.

a. SUPER-FAMILY ALLUROIDOIDEA

Setae four pairs per segment, simple crotchets. Testes one pair, in X or (*Syngenodrilus*) two pairs, in X and XI. Ovaries one pair, in XIII. Male pores one pair, on XIII. Vasa deferentia not embedded in the body wall, discharging through muscular, glandular atria, or into terminal chambers in common with these, or in the vicinity of prostate-like glands. Clitellum one cell thick; including the male and female pores. Eggs large and yolky. Gut with or without oesophageal gizzards. Subneural vessel absent. Spermathecae pretesticular; adiverticulate.

A single family: Alluroididae.

b. SUPER-FAMILY LUMBRICOIDEA

Setae eight (only in *Periscolex* numerous) per segment. Holonephric (exceptionally, *Tritogenia*, meronephric). Testes two pairs, in X or XI, or one pair, in one of these. Ovaries one pair, in XIII (only in *Glyphidrilus* two pairs, in XII and XIII). Male pores primarily in XIII or XIV or, secondarily, but frequently, further posteriorly. Vasa deferentia deeply embedded in the body wall or, if not, always with male pores on XV; seldom discharging through, or associated with, copulatory sacs or prostate-like glands. Clitellum more than one cell thick. Eggs small. Spermathecal pores very variable in number and location, pre- or post-testicular; if post-testicular more than one pair; spermathecae often intramural and frequently multiple in a segment; almost always lacking diverticula; not uncommonly absent.

Two families:

1. Male pores on XV (only in *Eiseniella* sometimes in front of this). Clitellum beginning in or behind XX. Ovaries with a single string of oocytes at the free edge. With an intestinal gizzard but never with an oesophageal gizzard.   Lumbricidae

—   Male pores if on XV intraclitellar or if

(*Kynotus*) in front of the clitellum, an oesophageal gizzard present. Ovaries with several
rows of oocytes at the free edge or if
(*Sparganophilus*) with a single string, gizzards
totally absent. Glossoscolecidae

5

FAMILY LUMBRICULIDAE

D. G. COOK

Type genus: Lumbriculus GRUBE

Prostomium prolobic or zygolobic, sometimes elongate, produced into proboscis. Setae four bundles per segment, beginning in II, two dorso-lateral, two ventro-lateral; two setae per bundle, single pointed or bifid, sigmoid with distinct node; setae rarely totally or partially absent. Without hair setae or modified genital setae. Male and spermathecal pores in region of VII to XV. Clitellum in region of genital pores, one cell thick.

Blind posterior lateral blood vessels often present. Testes variable in number and position. Atria paired or unpaired, always in a testis-bearing segment, associated with one or two pairs testes. Commonly one pair of atria in segment bearing posterior testes, two vasa deferentia and male funnels on each atrium, one funnel in anterior testes segment, the other on posterior septum of segment containing atrium. Anterior testes, funnels and vasa deferentia often absent, leaving testes, atrium, posterior funnels and vasa deferentia within a single segment; this arrangement sometimes serially repeated. Posterior vasa deferentia may penetrate posterior septum before returning to atrial segment. Sperm sacs present, extend for many segments posterior to testes. Ovaries one or two pairs, situated one or rarely two segments behind most posterior testes-bearing segment. Spermathecae variable in number and position.

KEY TO GENERA OF THE FAMILY LUMRICULIDAE

1.   With genital organs developed   3

— Without genital organs developed   2

2. Setae bifid. With dark green pigment in an-
    terior body wall. One pair blind, branched,
    lateral blood vessels present in anterior part
    of each segment, from about X to XV   *Lumbriculus* (in part)

— Immature, usually without this combination of
    characters   *unidentifiable material*

3. Spermathecae absent   4
— Spermathecae present   5

4. Atria paired, pear-shaped. Setae bifid   *Bichaeta*
— Atrium unpaired, ovoid to elongate, with long,
    often coiled duct. Setae single-pointed   *Tatriella* (in part)

5. All, or some, of the spermathecae open in
    segments anterior to the atrial segments   6

— All spermathecae situated in the same segment
    as, or in segments posterior to, the atrial
    segments   11

6. A single median, one pair, or rarely two pairs
    spermathecae open two segments in front of
    atrial segment, or on this and on the pre-atrial
    segment; i.e. on VIII or VIII and IX. Atria
    very narrow, elongate, with thin muscular
    walls, having appearance of vasa deferentia at
    least in part   7

— Spermathecae situated in the pre-atrial seg-
    ment, never two segments in front of atria.
    Atria often pear-shaped, but if elongate, with
    thick muscular walls   8

7. Atrium single, median, with ovoid to elongate,
    well-defined ampulla distally   *Tatriella* (in part)

— Atria paired, without well-defined ampullae
    distally   *Rhynchelmis*

8. Atria paired or single, elongate, with thick
    muscular walls, wholly or partially composed
    of muscle fibres arranged in opposing direc-
    tions in alternate layers, giving cross-hatched
    appearance in surface view   *Eclipidrilus*

— If atria elongate, musculature not of this form   9

9.   Atria open in VII or VIII. Ovaries situated
     two segments behind atrial segment   *Styloscolex*

—   Atria open in X. Ovaries situated in post-atrial
     segment   10

10. One pair testes, male funnels and vasa
     deferentia   *Hrabea*

—   Two pairs testes, male funnels and vasa
     deferentia   *Stylodrilus*

11. One pair spermathecae, situated in same seg-
     ment as atria. No spermathecae in post-atrial
     segments   *Guestphalinus*

—   Spermathecae present in segments behind atria   12

12. Spermathecae present in segment immediately
     posterior to atrial segment   13

—   Spermathecae usually absent in segment im-
     mediately posterior to atrial segment; begin
     two segments posterior to last atrial segment.
     If spermathecae present in first post-atrial
     segment, pores dorso-lateral in position   14

13. Two pairs testes. Atria open on IX or X.
     Prostomium rounded   *Trichodrilus*

—   One pair testes. Atria open on VIII or IX; if
     opening on IX, prostomium elongate, pro-
     duced into proboscis   *Kincaidiana*

14. Setae bifid. Spermathecal pores situated dorsal
     to ventral setae   *Lumbriculus* (in part)

—   Setae single pointed; sometimes wholly or
     partially absent. Spermathecal pores situated
     behind ventral setae   *Lamprodrilus*

GENUS **Lumbriculus** GRUBE, 1844
Type species: Lumbricus variegatus MÜLLER

Prostomium conical. Setae bifid with upper tooth reduced. One to four pairs
male pores between VII to XIII. Male pores surrounded by series of concentric
ridges in body wall. Number of spermathecal pores variable but usually two to
five, beginning two segments behind last atrial segment. Spermathecal pores
dorso-lateral to ventro-lateral, but always dorsal to line of ventral setae.

One pair blind, branched, lateral blood vessels present in anterior part of each
posterior segment, beginning between X to XV. One to four pairs testes, in region

of VII to XIII. One or two pairs testes associated with each pair atria. Atria cylindrical to saciform with inner layer of circular muscles and an outer layer of muscles arranged longitudinally to atrial axis. Penes dual; internally a protrusible penis composed of a pad of elongate cells, and externally a protrusible porophore, composed of concentric ridges in body wall which, when turgid, produce conical papilla. Prostate cells in discrete masses. One or two pairs ovaries, beginning in segment behind last pair of testes. Number of spermathecae variable, but usually two to five pairs and usually beginning two segments behind last atrial segment.

Holarctic. Introduced—S. Africa, Australia, New Zealand.

This, the type genus of the family, is perhaps the most problematical. Hrabe (1931, 1936a) claimed it to be monospecific, but Yamaguchi (1953) recognized five species.

It appears that, for any given population, the position of the genital elements varies and may be expressed by a series of overlapping normal curves. The mode for the atrial position lies between segments VIII to XII. Each different mode has been given a specific epithet by various authors (see Table under *L. variegatus*).

As four of these entities are morphologically identical, except for the variation in the pattern of the genital elements, and because the genital elements are so variable in position, only two species are here recognized; *L. variegatus* with four subspecies, and *L. multiatriatus.-*

Lumbriculus

1.	Two to four pairs male pores. Spermathecal pores situated just dorsally of line of ventral setae	*L. multiatriatus*
—	Usually one pair male pores. Spermathecal pores situated between lateral line and dorsal setae	*L. variegatus*

Lumbriculus multiatriatus Yamaguchi 1937

Lumbriculus multiatriatus YAMAGUCHI, 1937 a: 2, Fig. 1, 2, Pl. I.

Lumbriculus multiatriatus Yamaguchi. YAMAGUCHI, 1953: 301.

$l = 51$-70 mm, $w = 1.5$-1.8 mm, $s = 150$-190. Dark green pigment in anterior body wall present. Three or four pairs male pores, rarely two pairs, usually situated on X to XII or XIII. Four or five pairs spermathecal pores, beginning two segments behind last male pore, situated just dorsal to line of ventral setae.

Pharyngeal glands extend to VI. Three or four pairs testes, usually in X to XII or XIII. One pair testes associated with each pair atria. Atria elongate, extending posteriad in sperm sacs. Two pairs ovaries beginning in segment behind last testes-bearing segment. Spermathecae usually four or five pairs, usually beginning in second ovarian segment, with more or less spherical ampullae and discrete ducts.

Japan.

Yamaguchi (1937a) gave the following data on the variation of the position of the male genitalia.

Number male pores	Number individuals	Position most anterior pair male pores	Number individuals
1	1	seg. VII	1
2	4	seg. VIII	3
3	17	seg. IX	3
4	13	seg. X	27
		seg. XI	1

Lumbriculus variegatus (Müller, 1774)
Fig. 5. 1B, Fig. 5. 3A-C

Lumbricus variegatus MÜLLER, 1774: 26.

(?) *Nais variegata* SCHWEIGGER, 1820: 590.

(non) *Lumbricus variegatus* Müller. HOFFMEISTER, 1842: 1.

Lumbriculus variegatus (Müller). GRUBE, 1844: 207, Pl. VII, fig. 2a-d. LEIDY, 1850: 49. RATZEL, 1869: 585, Pl. XLII, fig. 10, 14, 19. EISEN, 1879: 9, Pl. VII, fig. 13a-e. VEJDOVSKY, 1884: 51, Pl. XII, fig. 16-32; 1895: 80. BEDDARD, 1895: 214; 1920: 230. SMITH, 1895: 293; 1905: 45; 1919: 39. HESSE, R., 1894: 355, Pl. XXII; 1902: 620. MICHAELSEN, 1900: 58; 1901a: 147; 1909: 55, Fig. 101; 1928: 42, Fig. 47. WENIG, 1902; 1. MRAZEK, 1906: 381, Fig. A-V; 1913 a: 1, Fig. 3-9. HESSE, E., 1909: 42. SOUTHERN, 1909: 121. FRIEND, 1914: 86; 1918: 124. SMITH and WELCH, 1919: 3a. BOVERI-BONER, 1920: 506. WELCH, 1921: 273. STEPHENSON, 1922: 291; 1930: 788. ISOSSIMOV, 1926: 365; 1962: 83. HAFFNER, 1928: 1, Fig. 1-39, Pl. IV. HRABE, 1929 a: 12; 1929 b: 200; 1931: 47; 1936 a: 11; 1937 a: 5; 1939 a: 210; 1941: 77; 1942: 26; 1952: 1; 1962: 278. SVETLOV, 1936: 92. YAMAGUCHI, 1936: 75; 1937 a: 1; 1953: 303. CERNOSVITOV, 1941 a: 214; 1941 b: 281. KENK, 1949: 37. CEKANOVSKAYA, 1962: 330, Fig. 206. BRINKHURST, 1962: 323; 1963 a: 44; 1963 b: 140. COOK, in BRINKHURST AND COOK, 1966: 11, Fig. 2d, 3a-c. COOK, 1967: 357, Fig. 4; 1968: 279.

Lumbriculus limosus LEIDY, 1850: 49, Fig. 16.

Lumbricus teres DALYELL, 1853: 140, Pl. XVII, Fig. 10-12.

(non) *Lumbriculus variegatus* (Müller). CLAPARÈDE, 1862: 255, Pl. III, IV.

Saenuris variegata JOHNSTON, 1865: 53.

Lumbricus lacustris VERRILL, *in* SMITH and VERRILL, 1871: 449. (in part)

Lumbriculus intermedius CZERNIAVSKY, 1880: 340.

Lumbriculus limosus Leidy. VEJDOVSKY, 1884: 51. SMITH, 1895: 295. MICHAELSEN, 1900: 65.

Thinodrilus inconstans SMITH, 1895: 292.

Thinodrilus inconstans Smith. SMITH, 1900a: 442; 1900b: 471. MOORE, 1906: 169.

Trichodrilus inconstans (Smith). MICHAELSEN, 1900: 59; 1901: 149.

Lumbricus lacustris (Verrill). MICHAELSEN, 1900: 65. (in part)

Lumbriculus inconstans (Smith). SMITH, 1905: 45; 1919: 33. GALLOWAY, 1911: 311. SMITH and GREEN, 1916: 83. SMITH and WELCH, 1919: 3. STEPHENSON, 1930: 789. ALTMAN, 1936: 7. YAMAGUCHI, 1936: 78; 1937 a: 1. CAUSEY, 1953: 423.

Lumbriculus japonicus YAMAGUCHI, 1936: 73, Fig. 1-3, 4c, 5, 7-9, Pl. IV, Fig. 1-4.

Lumbriculus japonicus Yamaguchi. HRABE, 1936 a: 12. YAMAGUCHI, 1937 a: 1; 1953: 301.

Lumbriculus mukoensis YAMAGUCHI, 1953: 302.

Lumbriculus variegatus typica COOK, *in* BRINKHURST and COOK, 1966: 11.

Lumbriculus variegatus inconstans (Smith) COOK, *in* BRINKHURST and COOK, 1966: 12.

Lumbriculus sachalinicus SOKOLSKAYA, 1967: 40, Fig. 1-3

l=up to 100 mm, w=1·5 mm, s=170. Dark green pigment in anterior body wall present. Setae about 210 μ long; ratio pt—node/total=0·34. Usually one pair male pores, situated between VII to XII. Usually four pairs spermathecal pores beginning two segments, or sometimes one segment, posterior to atrial segment; open between lateral line and line of dorsal setae.

Atria associated with one or two pairs testes. Atria cylindrical, usually confined to one segment; about 650 μ long, 180 μ wide, with muscle layer 20 μ thick. One or two pairs ovaries, beginning in post-atrial segment. Spermathecae with spherical to ovoid ampullae about 370 μ long, and discrete ducts, 220 μ long.

Northern hemisphere, S. Africa, Australia, New Zealand.

Five subspecies of *L. variegatus* are recognized. The different distributions of the genital elements in these entities is not thought to be sufficient justification to recognize them as distinct species. This contention is supported by the fact that the distribution of the genitalia is variable within the groups, while the morphology of the genitalia is the same throughout.

L. variegatus variegatus (Müller, 1774) (=*L. variegatus typica* Cook, 1965). Atria usually paired in VIII, associated with one pair testes. Usually one pair ovaries. Northern hemisphere (Australia, New Zealand, S. Africa—introduced).

L. variegatus inconstans (Smith, 1895). Atria usually paired in X, associated with two pairs testes. Usually two pairs ovaries. North America.

L. variegatus japonicus Yamaguchi, 1936. Atria usually paired in XI, associated with two pairs testes. Usually one pair ovaries. Japan.

L. variegatus mukoensis Yamaguchi, 1953. Atria usually paired in XII, associated with two pairs testes. Usually one pair ovaries. Japan.

L. variegatus sachalinicus Sokolskaya, 1967. Atria usually paired in XI, associated with two pairs testes. Usually one pair ovaries. One pair spermathecae. Sakhalin Is., USSR (possibly a variant of *L. v. japonicus*).

Below is a table comparing the distribution of the male pores in the various subspecies of *L. variegatus*.

PERCENTAGE FREQUENCY OF INDIVIDUALS WITH MALE PORES
IN SEGMENTS

segments	*variegatus*	*inconstans*	*japonicus*	*mukoensis*
VII	17·0	—	—	—
VIII	80·0	—	—	—
IX	2·25	18·2	1·9	3·7
X	0·75	72·7	3·7	—
XI	—	9·1	94·4	3·7
XII	—	—	—	92·6
Number of individuals on which figures based	200	11	161	27

GENUS **Lamprodrilus** MICHAELSEN, 1901
Fig. 5. 1 D

Type species: Lamprodrilus wagneri MICHAELSEN

Mostly white, unpigmented worms. Prostomium rounded or rarely elongate, produced into proboscis. Intersegmental furrows distinct (Lumbricid facies) or indistinctly developed (Ascarid facies). Secondary annuli usually present. Setae single pointed, or, rarely, wholly or partially absent. One to four pairs male pores between VII to XI, behind line of ventral setae. Openings of accessory copulatory glands sometimes present in region of male pores or on segment posterior to them. One to five pairs spermathecal pores between IX to XVII, in line of ventral setae, always beginning two segments behind last pair of male pores; i.e. with one clear segment, the ovarian segment, between last male segment and first spermathecal segment.

Blind lateral blood vessels usually present. One to four pairs testes between VII to XI. One to four pairs atria, one to each testis, in same segment as their gonad. Atria tubular, usually terminated by well-developed penes which are often enclosed in a sac composed of glandular, or glandular and muscular cells, termed penial bulbi. Prostate cells form compact layer surrounding atria. Accessory copulatory glands sometimes present in form of glandular ridges or structures resembling penial bulbi. One pair ovaries situated immediately behind last testes-bearing segment. One to five pairs spermathecae, more usually one pair, beginning in post-ovarian segment.

Lake Baikal, Jugoslavia, Czechoslovakia, USSR.

Lamprodrilus here includes Michaelsen's genera *Teleuscolex* and *Agriodrilus*. Hrabe (1931) considered that there was no justification for separating *Teleuscolex* from *Lamprodrilus* when he discovered *L. pygmaeus intermedius* which has only one pair of atria in IX. Likewise the author considers that *Agriodrilus* should be merged with *Lamprodrilus*, as the genus still remains homogeneous and easily definable by the relative positions of the elements of the genitalia.

Even after Isossimoff's studies on *Lamprodrilus*, the natural variation within species is not well known, and it is possible that at least some of the entities here described, are synonymous.

Many infraspecific categories have been described in this genus and for the most part are considered in this account as sub-species, except in those cases where the characters used to differentiate the entities are known to be unreliable or highly variable.

Lamprodrilus

1.	With three or four pairs male pores	*L. satyriscus*
—	With one or two pairs male pores	2
2.	With one pair male pores	3
—	With two pairs male pores	8
3.	One pair male pores between VI to IX. Dorsal setae distinctly shorter than ventrals	*L. mrazeki* (in part)
—	One pair male pores on X or XI. Dorsal setae do not differ markedly from ventrals	4
4.	Pair of large accessory copulatory glands present behind atria	*L. grubei*

—	Accessory copulatory glands absent	5
5.	Blind lateral blood vessels absent	*L. pygmaeus* (in part)
—	Blind lateral blood vessels present	6
6.	Male pores on XI	*L. vermivorus*
—	Male pores on X	7
7.	Intersegmental furrows indistinct (Ascarid facies)	*L. baicalensis*
—	Intersegmental furrows distinct (Lumbricid facies)	*L. korotneffi*
8.	Setae wholly or partially absent	9
—	Normal complement of setae present	11
9.	Setae totally absent	*L. achaetus*
—	At least some setae present	10
10.	Dorsal and ventral setae present on II, all setae absent in other parts of body	*L. pygmaeus* (in part)
—	Dorsal setae absent. Ventral setae very long, hair-like, 1·0 mm long	*L. bythius*
11.	Accessory copulatory glands present	12
—	Accessory copulatory glands absent	13
12.	One or two simple, median ventral, glandular furrows, in front of ventral setae, present on XII or XI and XII	*L. ammophagus*
—	Copulatory glands paired, latero-ventral, behind male pores, or a median ventral furrow from X to XIII	*L. pygmaeus* (in part)
13.	Dorsal setae considerably shorter than ventrals; maximum length of dorsals is shorter than minimum lengths of ventrals	14
—	Dorsal setae not considerably shorter than ventrals; maximum length of dorsals overlaps minimum length of ventrals	15
14.	Ventral setae up to 650 μ long. Prostomium elongate, produced into short proboscis	*L. inflatus*
—	Ventral setae up to 230 μ long. Prostomium rounded, as long as it is broad at peristomium	*L. mrazeki* (in part)

15. Penial bulbi asymmetrical, very large; up to
    400 μ diameter. Longitudinal muscle over
    200 μ thick   *L. bulbosus*

— Penial bulbi absent or of medium size; up
    to 300 μ diameter. Longitudinal muscle up
    to 160 μ thick   16

16. Blind lateral blood vessels completely absent   17
— Blind lateral blood vessels developed to
    greater or lesser extent   18

17. Intersegmental furrows distinct   *L. isoporus*
— Intersegmental furrows indistinct   *L. pygmaeus* (in part)

18. Anterior pair of atria half the size of posterior
    pair   *L. tolli*
— Both pairs of atria of same size   19

19. Blind blood vessels in anterior part of body
    very well developed, with numerous branches
    which fill body cavity and reach into pros-
    tomium   *L. dybowskii*

— Blind blood vessels do not fill whole of body
    cavity and do not extend into prostomium   20

20. Spots of pigment present in I to III. Body wall
    epidermis very thin compared with longitu-
    dinal muscle; 0·5 μ thick, and 120 μ thick
    respectively   *L. stigmatias*

— Pigment not confined to first three segments,
    or absent. Body wall epidermis not very thin
    compared with longitudinal muscle; over 10 μ
    thick, up to 160 μ thick respectively   21

21. Pigment present in body wall, especially
    anteriorly   22
— Pigment completely absent   23

22. Worms up to 4·0 mm diameter. Setae almost
    straight but bent abruptly distally, nearly into
    right angle   *L. melanotus*

— Worms up to 1·66 mm diameter. Setae sig-
    moid, not abruptly bent distally   *L. nigrescens*

23. Intersegmental furrows indistinct (ascarid
    facies)   *L. wagneri*

—   Intersegmental furrows distinct (lumbricid
     facies)   24

24. Worms over 50 mm long, up to 4·0 mm
     diameter   25

—   Worms under 50 mm long, up to 1·5 mm
     diameter   26

25. Anterior pair atria larger than posterior pair.
     Body composed of 150 segments   *L. semenkewitschi*

—   Both pairs of atria of same size. Body com-
     posed of 87 segments.   *L. polytoreutus*

26. Blind lateral blood vessels simple, or with one
     branch   *L. pallidus*

—   Blind lateral blood vessels with numerous
     caecae   *L. michaelseni*

Lamprodrilus achaetus Isossimov, 1948

Lamprodrilus achaetus ISOSSIMOV, 1948: 67.

Lamprodrilus achaetus Isossimov. ISOSSIMOV, 1962: 27. CEKANOVSKAYA,
     1962: 339

$l=5·0$ mm, $w=$up to 0·3 mm, $s=30$-32. Prostomium zygolobic. Secondary segmentation absent. Body smooth. Intersegmental furrows indistinct. Setae
completely absent. Two pairs male pores, on X and XI. One pair spermathecal
pores on XIII.
     Cuticle + epidermis, 3·5-4 μ thick. Circular muscle layer thin, longitudinal
muscle layer 16-20 μ thick. Inside longitudinal layer mesenchym, in places up
to 10 μ thick, posterior lateral blood vessels absent. Two pairs testes present in
X and XI. Vasa deferentia join atria proximally. Atria elongate, bent in a right
angle. Penes well developed, in penial sac. Accessory copulatory glands absent.
Ovaries in XII. One pair spermathecae present in XIII, with very large ampullae
which extend to XV, and long, thick ducts.
     Lake Baikal (U.S.S.R.)

     L. achaetus is unique among the Lumbriculidae because of the complete
absence of setae.

Lamprodrilus ammophagus Michaelsen, 1905

Lamprodrilus ammophagus MICHAELSEN, 1905: 2.

Lamprodrilus ammophagus Michaelsen. ISOSSIMOV, 1948: 90; 1962: 4.
     CEKANOVSKAYA, 1962: 346, Fig. 214.

$l=22$ mm, $w=$up to 1·0 mm, $s=$up to 50. Prostomium prolobic, rounded, as
long as it is broad at peristomium. Intersegmental furrows indistinct anteriorly,

more distinct posteriorly. Setae sigmoid, distally blunted, 80-120 μ long, 4-7 μ thick. Setae closely paired anteriorly and posteriorly, but more widely spaced in middle segments. Two pairs male pores on X and XI. On XII, in front of ventral setae a single glandular furrow present (accessory copulatory gland). Similar gland sometimes present in same position, on XI. One pair spermathecal pores on XIII.

Body wall thin. Two pairs testes in X and XI. Two pairs elongate atria, in X and XI, which are narrowed proximally and terminated by hemispherical, glandular penial bulbi. One or two accessory copulatory glands present in form of glandular furrows, in XII or XI and XII. Ovaries in XII. One pair spermathecae in XIII, with large bulbous ampullae and very narrow, short ducts.

Lake Baikal (U.S.S.R.)

Lamprodrilus baicalensis (Grube, 1873)
Fig. 5. 3D-H

Euaxes baicalensis GRUBE, 1873: 67. (in part)

Rhynchelmis baicalensis (Grube). VAILLANT, 1889: 221.

Euaxes baikalensis Grube. MICHAELSEN, 1900: 65.

Teleuscolex baicalensis (Grube). MICHAELSEN, 1901a: 170; 1905, 51. ISOSSIMOV, 1948: 22; 1962: 10, Fig. 2-4. CEKANOVSKAYA, 1962: 350. Fig. 217.

$l=$up to 75 mm, $w=4.0$ mm, $s=120$. Prostomium rounded, cone-shaped Intersegmental furrows indistinct. Setae 250 μ long, 8 μ thick. One pair male pores on X. One pair spermathecal pores on XII.

Cuticle 0·8 μ thick, epidermis 20 μ thick, circular muscle 15 μ thick, longitudinal muscle layer 260 μ thick. Blind blood vessels simple, tubular, begin in XXIV. Chloragogen cells surround gut from VI. One pair testes in X. Vasa deferentia penetrate atria proximally, run inside their walls and drain into the lumen apically. Atria cylindrical, terminated by small retractible penes. Bases of atria surrounded by large glandular penial bulbi. Ovaries in XI. One pair spermathecae, in XII, with spherical ampullae and wide, funnel-like, indistinctly separated ducts.

Lake Baikal (U.S.S.R.)

Lamprodrilus bulbosus Isossimov, 1948.
Fig. 5. 3E

Lamprodrilus bulbosus ISOSSIMOV, 1948: 45.

Lamprodrilus bulbosus Isossimov. ISOSSIMOV, 1962: 16. CEKANOVSKAYA, 1962: 341, Fig. 212.

$l=30$ mm, $w=3.0$ mm, $s=$up to 75. Prostomium small, prolobic, a little shorter than it is broad at peristomium. Intersegmental furrows distinct. Secondary annuli absent. Setae only slightly sigmoid, 228 μ long, 6 μ thick on XVIII. Two pairs male pores on X and XI. One pair spermathecal pores on XIII.

Epidermis in postclitellar segments, up to 20 μ thick. Circular muscle layer in anterior segments up to 38 μ thick, longitudinal muscle 200-250 μ thick. Two pairs testes in X and XI. Vasa deferentia join atria distally. Atria large, tubular, terminated by large, asymmetrical penial bulbi which are enclosed within penial sac. Penial bulbi 350-400 μ in diameter. Ovaries in XII. One pair spermathecae opening in XIII, with large ampullae which extend to XIV, and indistinctly separated ducts.

Lake Baikal (U.S.S.R.)

Lamprodrilus bythius Michaelsen, 1905

Lamprodrilus bythius MICHAELSEN, 1905: 2.

Lamprodrilus bythius Michaelsen. MICHAELSEN, 1928: 75. ISOSSIMOV, 1948: 111; 1962: 4, Fig. 26, 27. CEKANOVSKAYA, 1962: 348, Fig. 216.

$l=69$ mm, $w=$up to 8·0 mm, $s=110$. Prostomium retractable, produced into proboscis. Intersegmental furrows indistinct, but secondary annuli numerous. Dorsal setae completely absent. Ventral setae very slightly curved, almost hair-like 1·0 mm long, 10 μ thick. Two pairs male pores on X and XI, on the summit of large papillae. One pair spermathecal pores on XIII.

Body wall relatively very thin. In middle of body, epidermis 40 μ thick, circular muscle layer thinner than epidermis, longitudinal muscle 50 μ thick. In several middle segments, one pair long, narrow, unbranched, blind lateral blood vessels present. Two pairs testes, in X and XI. Short wide vasa deferentia join atria proximally. Atria tubular. Ovaries in XII. One pair spermathecae, opening in XIII, with narrow discrete ducts, and more or less spherical ampullae.

Lake Baikal (U.S.S.R.)

Lamprodrilus dybowskii Michaelsen, 1905

Lamprodrilus dybowskii MICHAELSEN, 1905: 2.

Lamprodrilus dybowskii Michaelsen. HRABE, 1929c: 173. ISOSSIMOV, 1948: 87; 1962: 4. CEKANOVSKAYA, 1962: 345.

$l=35$ mm, $w=$up to 1·3 mm, $s=110$. Short, rounded, prolobic prostomium. From III, some anterior segments biannulate. Setae distinctly curved distally, 350 μ long, 7 μ thick. Two pairs male pores on X and XI. One pair spermathecal pores on XIII.

In XX, epithelium, 20 μ thick, circular muscle layer thin, longitudinal muscle 50 μ thick. Blind lateral blood vessels well developed with many branches which fill the whole coelom and reach to anterior segments. Two pairs testes in X and XI. Vasa deferentia narrow. Atria thick, tubular. Penes cone-shaped, enclosed in penial sac. Ovaries in XII. One pair spermathecae, opening in XIII.

Laike Baikal (U.S.S.R.)

Lamprodrilus grubei (Michaelsen, 1901)
Fig. 5. 3G

Teleuscolex grubei MICHAELSEN, 1901 a: 173, Fig. b (error in legend).

Teleuscolex grubei Michaelsen. MICHAELSEN, 1905: 2. ISOSSIMOV, 1948: 32; 1962: 4. CEKANOVSKAYA, 1962: 351, Fig. 219.

$l=35$ mm, $w=1\cdot5$ mm, $s=177$. Prostomium short, cap-like. Intersegmental furrows distinct. Setae sigmoid, with indistinct node, 250 μ long, 14 μ thick. One pair male pores on X. One pair accessory copulatory papillae, oval-shaped, on X, behind male pores. One pair spermathecal pores on XII.

Cuticle thin, body wall fairly thick. Simple, very long blind blood vessels present from XVIII, four pairs per segment. One pair testes in X. Narrow vasa deferentia join atria medially. Atria elongate, irregularly curved, narrowed proximally, terminated by small, papillate penes. Without penial bulbi. One pair large, glandular, accessory copulatory glands, present in X, opening behind male pores. Ovaries in XI. One pair spermathecae, in XII, with long sacciform ampullae and shorter, narrow ducts.

Lake Baikal (U.S.S.R.)

Lamprodrilus inflatus Michaelsen, 1905

Lamprodrilus inflatus MICHAELSEN, 1905: 2.

Lamprodrilus inflatus Michaelsen. HRABE, 1929 c: 173. MICHAELSEN AND
VERESCHAGIN, 1930: 220. ISOSSIMOV, 1948: 102; 1962: 4. CEKANOV-
SKAYA, 1962: 347, Fig. 215.

$l=$up to 30 mm, $w=3\cdot3$ mm, $s=101$. Prostomium prolobic, elongate, retractible. Intersegmental furrows distinct in anterior and posterior body. Segments from III biannulate; the wide anterior parts may have one or two shallow annuli. Dorsal setae shorter than ventrals. Ventral setae only slightly sigmoid, with indistinct node; behind XV, up to 650 μ long, 8 μ thick. Setal length rapidly decreases towards anterior end of body. Dorsal setae 200-350 μ long, 12 μ thick. Two pairs male pores on X and XI. One pair spermathecal pores on XIII, in deep folds in epithelium.

In middle of body, epithelium 9 μ thick, circular muscle layer twice as thick as longitudinal layer which is 24 μ thick. Blind lateral blood vessels cylindrical, with thick covering of chloragogen cells. Two pairs testes in X and XI. Atria cylindrical, slightly dilated distally. Penes cone-shaped, surrounded by more or less spherical, glandular penial bulbi. Accessory copulatory glands absent. Ovaries in XII. One pair spermathecae in XIII, with sacciform ampullae and discrete ducts.

Lake Baikal (U.S.S.R.)

Lamprodrilus isoporus Michaelsen, 1901
Fig. 5. 3F, I

Lamprodrilus isoporus MICHAELSEN, 1901 b: 3.

Lamprodrilus isoporus Michaelsen. MICHAELSON, 1902 a: 47; 1905: 2.
HRABE, 1929 c: 173. SVETLOV, 1936: 87. ISOSSIMOV, 1948: 89; 1962: 4.
CEKANOVSKAYA, 1962: 339, Fig. 210.

Lamprodrilus isoporus f. *variabilis* SVETLOV, 1936: 89, Fig. 1, 2.

Lamprodrilus isoporus f. *variabilis* Svetlov. HRABE, 1936 a: 13; 1962: 278,
Fig. 42, 43, 46-51, 55. ISOSSIMOV, 1962: 6.

l=up to 40 mm, w=2·0 mm, s=45-52. Prostomium helmet-shaped, longer than it is broad at peristomium. Intersegmental furrows distinct. Secondary annuli absent. Setae sigmoid, 100-180μ long, 5-8μ thick. Two pairs male pores on X and XI. One pair spermathecal pores on XIII.

In postclitellar segments, epithelium 6 μ thick, circular muscle layer very thin, longitudinal muscle 50 μ thick. Blind lateral blood vessels absent. Two pairs testes present in X and XI. Narrow vasa deferentia penetrate into the segment posterior to their corresponding atria. Vasa deferentia join atria proximally. Atria tubular, terminated by more or less spherical penes. Accessory copulatory glands absent. Ovaries in XII. One pair spermathecae, opening in XIII, with narrow discrete ducts and spherical ampullae, which are located in XIV.

Lake Baikal, Chudskoie, Ladoga, Onega (U.S.S.R.)

Two subspecies are recognized.

L. isoporus isoporus Michaelsen, 1901. l=up to 40 mm, w=2·0 mm. One pair spermathecae in XIII.

L. isoporus variabilis Svetlov, 1936. l=20 mm, w=1·0 mm. One pair spermathecae in XIII, or a pair in XIII and a single on in XIV.

Lamprodrilus korotneffi (Michaelsen, 1901)

Teleuscolex korotnewi MICHAELSEN, 1901 a : 165, Pl. II, Fig. 16, 17.

Teleuscolex korotnewi f. *gracilis* MICHAELSEN, 1901 a : 167.

Teleuscolex korotneffi Michaelsen. MICHAELSEN, 1905 : 2; 1928 : 42. MICHAELSEN AND VERESCHAGIN, 1930 : 220. ISOSSIMOV, 1948 : 15; 1962 : 4, Fig. 1. CEKANOVSKAYA, 1962 : 351, Fig. 218.

Teleuscolex korotneffi f. *typica* MICHAELSEN, 1905 : 2.

Teleuscolex korotneffi f. *gracilis* Michaelsen. MICHAELSEN, 1905 : 2. MICHAELSEN and VERESCHAGIN, 1930 : 220. ISOSSIMOV, 1962 : 4. CEKANOVSKAYA, 1962 : 351.

l=115 mm, w=up to 4·6 mm, s=164. Prostomium cone-shaped. Intersegmentary furrows distinct. Secondary annuli present. Anterior part of body with circular rings of black or grey pigment. Setae sigmoid, up to 400 μ long, 18 μ thick. One pair male pores on X. One pair spermathecal pores on XII. Accessory copulatory glands absent.

In anterior segments, cuticle 12 μ thick, epidermis up to 20 μ thick, circular muscle 28 μ thick, longitudinal muscle layer 40 μ thick. Two or three pairs sparsely branched blind blood vessels present in middle segments. One pair testes in X. Atria elongate, pear-shaped, terminated by small cone-shaped penes. Accessory copulatory glands absent. Ovaries in XI. One pair spermathecae in XII, simple pear-shaped.

Lake Baikal (U.S.S.R.)

Michaelsen (1905) distinguished two forms of this species, which differed from each other only in the extent of the pigmentation. As pigmentation is known to be a variable character in other genera, it is proposed that those entities remain at an infrasubspecific rank.

Lamprodrilus melanotus Isossimov, 1948
Fig. 5. 3H

Lamprodrilus melanotus ISOSSIMOV, 1948 : 75.

Lamprodrilus melanotus Isossimov. ISOSSIMOV, 1962 : 7. CEKANOVSKAYA, 1962 : 343, Fig. 213.

l=24-32 mm, w=up to 4·0 mm, s=75-179. Body pigmented. Prostomium prolobic, broad and very short. Intersegmental furrows distinct. From VI, segments biannulate. Clitellum on IX to XV. Setae nearly straight except their distal end which is abruptly bent, almost to a right-angle. Setae 245 μ long, 6·6 μ thick. Two pairs male pores on X and XI. One pair spermathecal pores on XIII.

In postclitellar segments epidermis 33-55 μ thick, circular muscle layer is of uneven thickness. Pigment cells located in body wall. Numerous cylindrical, blind blood vessels begin in XIX. Chloragogen cells begin in VII on the gut, but reach into the prostomium on the dorsal vessel. Two pairs testes in X and XI. Vasa deferentia tortuous, join atria apically. Two pairs thick-walled atria situated entirely in X and XI. Anterior pair atria a little larger than posterior pair. Penial bulbi indistinct. Penes broadly conical. Accessory copulatory glands absent. Ovaries in XII. One pair large spermathecae, in XIII, twice as long as the atria, with ovoid ampullae and discrete ducts.
Lake Baikal (U.S.S.R.)

Lamprodrilus michaelseni Hrabe, 1929

Lamprodrilus michaelseni HRABE, 1929 c : 163, Fig. 1c, 2, 3, Pl. VI.

Lamprodrilus michaelseni Hrabe. HRABE, 1930 : 2; 1958 : 341; 1962 : 317.
SVETLOV, 1936 : 88. ISOSSIMOV, 1962 : 6.

l=up to 18 mm, w=1·0 mm, s=65. Prostomium rounded, as long as it is broad at peristomium. Segments biannulate from IV. Inter-segmental furrows in anterior part of body distinct. Clitellum on $\frac{1}{2}$IX to $\frac{1}{2}$XIV. Setae sigmoid, 150 μ long, 7 μ thick in anterior segments. Two pairs male pores on X and XI. One pair spermathecal pores on XIII.

Epidermis 10 μ thick, circular muscle 3 μ thick, longitudinal muscle layer 30 μ thick. In II to XV, one pair dorso-ventral blood vessels present, which from XX bear numerous blind caecae. In posterior part of body, two pairs lateral vessels present. Two pairs testes in X and XI. Vasa deferentia join atria proximally. Two pairs cylindrical atria, entirely situated in X and XI, with diameter of lumen 40 μ, become narrow proximally and are terminated by small conical penes. Without penial sac. Accessory copulatory glands absent. Ovaries in XII. One pair spermathecae, opening in XIII, with long discrete ducts and ovoid ampullae which are situated in XIV.
Yugoslavia.

Lamprodrilus mrazeki Hrabe, 1928

Lamprodrilus mrazeki HRABE, 1928: 2. (Nomen nudum)

Lamprodrilus mrazeki Hrabe. HRABE, 1929 b: 197, Fig. 1-5, Pl. I; 1936 a: 13; 1937 a: 1, Fig. 1; 1937 b: 11; 1939 a: 223; 1939 c: 57, Fig. 1-5. SVETLOV, 1936: 88; ISOSSIMOV, 1962: 6. BRINKHURST, 1962: 324. COOK, 1967: 359.

l=25-35 mm, w=1·0 mm, s=60-70. Prostomium conical, rounded, as long as it is broad at peristomium. Intersegmental furrows not well developed. Many fine secondary annuli present on all body segments. Setae sigmoid, abruptly curved distally; dorsal setae shorter than ventrals. Ventral setae 190-230 μ long; dorsal setae 120-160 μ long. One or two pairs male pores which are variable in position, but more usually on VII or VII and VIII. One to three pairs spermathecal pores, more usually one pair, present two segments behind the last atrial segment.

Cuticle 1 μ thick, epidermis 10-15 μ thick, circular muscle 1 μ thick, longitudinal muscle layer 30-50 μ thick. From VI to VIII, two pairs lateral blood vessels, covered in chloragogen cells and bearing blind caecae, present. Anterior pair of blood vessels in each segment larger than posterior pair. One or two pair testes present in segments containing atria. Vasa deferentia penetrate into the segments posterior to their corresponding atria before returning to join atria distally. Atria tubular, curved within their segments or partly in their succeeding segments. Penes conical, contained in shallow penial sac. One or two pairs ovaries, beginning in segment after last atrial segment. One to three pairs spermathecae with ovoid ampullae and discrete ducts, beginning in the segment after the first ovarian segment.

Czechoslovakia.

Hrabe (1929 b; 1937 a) studied the variation in the genitalia of *L. mrazeki* and found that the number and position of the genital elements were very inconstant, and concluded that this probably resulted from the fact that the species regularly undergoes asexual division.

L. mrazeki has the capacity to form cysts, composed of cuticular material, inside which asexual division may occur. Hrabe (1929) speculates that the cysts reported by Stephenson (1922) as being from *Lumbriculus variegatus*, and by Mrazek (1913 b) as being from a *Claparedeilla* (now *Stylodrilus*) species, are attributable to *L. mrazeki*. However, there is some evidence that encystment can occur in *Lumbriculus*, in response to dry conditions (see Ecological Section, and Cook, 1967).

Lamprodrilus nigrescens Michaelsen, 1903

Lamprodrilus nigrescens MICHAELSEN, 1903 a: 61. (Nomen nudum)

Lamprodrilus nigrescens Michaelsen. MICHAELSEN, 1905: 2. MICHAELSEN AND VERESCHAGIN, 1930: 220. HRABE, 1929 c: 173 ISOSSIMOV, 1948: 71; 1962: 4.

l=16-30 mm, w=up to 1·66 mm, s=64-83. Prostomium prolobic, shorter than it is broad at peristomium. Body orange-yellow colour, more intense in anterior part. Black pigment present, decreasing in quantity posteriorly. Intersegmental furrows distinct. Secondary annuli present from V. Clitellum on X to XIV. Setae sigmoid, 280 μ long, 5 μ thick. Two pairs male pores on X and XI. One pair spermathecal pores on XIII.

In XXX, epidermis up to 24 μ thick, circular muscle thin, longitudinal muscle layer 30 μ thick. Two to six simple tubular, blind blood vessels present. Two pairs testes in X and XI. Narrow vasa deferentia join atria proximally, run inside atrial wall and drain into lumen apically. Atria tubular, 291 μ long, 101 μ diameter, terminated by long narrow ducts which end in glandular bulbi. Penes cone-shaped, enclosed in well-developed penial sac. Accessory copulatory glands absent. Ovaries in XII. One pair spermathecae, opening in XIII, with egg-shaped ampullae and narrow strongly curved ducts, nearly as long as the ampullae.

Lake Baikal (U.S.S.R.)

Lamprodrilus pallidus Michaelsen, 1905

Lamprodrilus pallidus MICHAELSEN, 1905: 2.

Lamprodrilus pallidus Michaelsen. ISOSSIMOV, 1948: 98; 1962: 4. CEKANOV-SKAYA, 1962: 345.

l=16-30 mm, w=up to 1·66 mm, s=64-83. Prostomium prolobic, shorter than broad at peristomium. Intersegmental furrows distinct. Segments biannulate from V. Clitellum on X to XIII. Setae distinctly curved distally, 200 μ long, 5 μ thick. Two pairs male pores on X and XI. One pair spermathecal pores in deep folds in body wall, on XIII.

In XXV, epidermis 16 μ thick, circular muscle thin, longitudinal muscle layer 32 μ thick. Blind lateral blood vessels only slightly developed. Two pairs testes in X and XI. Narrow vasa deferentia join atria apically. Two pairs atria, in X and XI, which are narrow proximally and ensheathed in glandular bulbi. Penes conical, well developed, enclosed in penial sac. Prostate cell layer 90-110 μ thick. Ovaries in XII. One pair spermathecae, opening in XIII, with ovoid ampullae and narrow discrete ducts, as long as ampullae.

Lake Baikal (U.S.S.R.)

Lamprodrilus polytoreutus Michaelsen, 1901

Lamprodrilus polytoreutus MICHAELSEN, 1901 a: 163.

Lamprodrilus polytoreutus Michaelsen. MICHAELSEN, 1905: 2. HRABE, 1929 c: 173; ISOSSIMOV, 1948: 86; 1962: 4. CEKANOVSKAYA, 1962: 344.

l=58 mm, w=up to 4·0 mm, s=87. Prostomium pointed or elongated into short proboscis, as long at it is broad at peristomium. Intersegmental furrows very well developed. Segments biannulate. Setae slightly sigmoid, 360 μ long, 16 μ thick. Two pairs male pores on X and XI. One pair spermathecal pores on XIII.

In XXV, epidermis 50 μ thick, circular muscle 24 μ thick, longitudinal muscle layer 160 μ thick. From XX, blind, tubular lateral blood vessels are present, which bear caecae which increase in number and length in more posterior segments. Also branched, blind lateral blood vessels which originate from intestinal plexus are present. Two pairs testes present in X and XI. Two pairs tubular atria, in X and XI, become narrow proximally. Penial bulbi absent,

except in immature specimens. Ovaries in XII. One pair simple, pear-shaped spermathecae in XIII. (Specimens not completely mature.)
Lake Baikal (U.S.S.R.)

Lamprodrilus pygmaeus Michaelsen, 1901
Fig. 5. 4A, B

Lamprodrilus pygmaeus MICHAELSEN, 1901b: 68.

Lamprodrilus pygmaeus Michaelsen. MICHAELSEN, 1902 a: 46; 1905: 2, HRABE, 1929 c: 173; 1931: 2; 1958: 341. ISOSSIMOV, 1962: 23.

Lamprodrilus pygmaes f. *typica* MICHAELSEN, 1905: 2.

Lamprodrilus pygmaeus f. *glandulosa* MICHAELSEN, 1905: 2.

Lamprodrilus pygmaeus f. *ochridana* HRABE, 1931: 2.

Lamprodrilus pygmaeus f. *intermedia* HRABE, 1931: 2.

Lamprodrilus pygmaeus f. *glandulosa* Michaelsen. HRABE, 1931: 45; 1962: 314. ISOSSIMOV, 1948: 59; 1962: 4.

Lamprodrilus pygmaeus f. *ochridana* Hrabe. SVETLOV, 1936: 88. HRABE, 1958: 341; 1962: 314. ISOSSIMOV, 1962: 6.

Lamprodrilus pygmaeus f. *intermedia* Hrabe. SVETLOV, 1936: 88. HRABE, 1958: 341; 1962: 314, Fig. 44, 45.

Lamprodrilus pygmaeus f. *ochridanus* Hrabe. HRABE, 1962: 289.

Lamprodrilus pygmaeus f. *intermedius* Hrabe: HRABE, 1962: 289.

Lamprodrilus pygmaeus f. *typica* Michaelsen. HRABE, 1962: 314. ISOSSIMOV, 1962: 24, Fig. 13.

Lamprodrilus pygmaeus pygmaeus HRABE, 1962: 331, Fig. 52. CEKANOVSKAYA, 1962: 337, Fig. 209a.

Lamprodrilus pygmaeus f. *sulcata* ISOSSIMOV, 1948: 62.

Lamprodrilus pygmaeus f. *intermedia* ISOSSIMOV, 1948: 63.

Lamprodrilus pygmaeus f. *oligosetosa* ISOSSIMOV, 1962: 27, (=*intermedia* Is.)

Lamprodrilus pygmaeus glandulosus Michaelsen. CEKANOVSKAYA, 1962: 338, Fig. 209b.

Lamprodrilus pygmaeus sulcatus Isossimov. CEKANOVSKAYA, 1962: 338, Fig. 209b.

l=9-20 mm, w=0·5-0·85 mm, s=38-60. Prostomium rounded, as long as it is broad at peristomium. Intersegmental furrows indistinct. Secondary annulations absent. Clitellum on X to XIII. Setae sigmoid, 60-140 μ long, 2-5 μ thick, or absent except on II. One or two pairs male pores on X or X and XI. Accessory copulatory openings, in form of pits or furrows, present or absent. One pair spermathecal pores on XII or XIII.

Epidermis 3-4 μ thick, longitudinal muscle 30-80 μ thick. Blind lateral blood vessels absent. One or two pairs testes in X or X and XI. Male funnels large. Vasa deferentia narrow. Atria tubular, narrowed proximally and terminated by cone-shaped penes which are enclosed in glandular penial bulbi. Accessory copulatory organs in the form of penial bulbi or glandular strips, or absent. Ovaries in XI or XII. One pair spermathecae, opening in XII or XIII, with large sacciform ampullae and narrow discrete ducts.

Lakes Baikal and Ochrid (U.S.S.R., Jugoslavia).

Six subspecies of *L. pygmaeus* are recognized provisionally.

L. pygmaeus pygmaeus Michaelsen, 1905. l=9-13 mm, w= 0·5-0·6 mm. Setae 60 μ long, 2-3 μ thick, all of more or less equal size. Two pairs male pores on X and XI. Without accessory copulatory glands.

L. pygmaeus glandulosus Michaelsen, 1905. l=18-20 mm, w=0·85 mm. Setae 100-140 μ long, 5 μ thick, but shorter in middle and posterior segments. Two pairs male pores on X and XI. One or two pairs accessory copulatory glands present of similar structure to penial bulbi, behind male pores.

L. pygmaeus ochridanus Hrabe, 1931. l=13-20 mm, w=0·5-0·75 mm. Ventral setae longer than dorsals on corresponding segments, and setae in middle segments smaller than those in anterior and posterior segments. Two pairs male pores on X and XI. Without accessory copulatory glands.

L. pygmaeus intermedius Hrabe, 1931. One pair male pores on X. Without accessory copulatory glands. One pair spermathecal pores on XII.

L. pygmaeus sulcatus Isossimov, 1948. l=9-13 mm, w=0·5-0·6 mm. Setae 150-160 μ long, 5 μ thick. Two pairs male pores on X and XI. Mid-ventral furrow present from X to XIII; on each side of this, a glandular strip is developed.

L. pygmaeus oligosetosus Isossimov, 1962. Setae absent except on II. Two pairs male pores on X and XI without accessory copulatory glands.

Lamprodrilus satyriscus Michaelsen, 1901
Fig. 5. 1C, Fig. 5. 4D

Lamprodrilus satyriscus MICHAELSEN, 1901a: 151, Fig. a.

Lamprodrilus satyriscus f. *typica* MICHAELSEN, 1901a: 153.

Lamprodrilus satyriscus f. *decatheca* MICHAELSEN, 1901a: 153.

Lamprodrilus satyriscus f. *ditheca* MICHAELSEN, 1901a: 153.

Lamprodrilus satyriscus Michaelsen. MICHAELSEN, 1905: 28; 1928: 46.
ISOSSIMOV, 1948: 117, Fig. 28; 1962: 3. YAMAGUCHI, 1953: 302.
CEKANOVSKAYA, 1962: 334, Fig. 205b, 207.

Lamprodrilus satyriscus f. *typica* Michaelsen. MICHAELSEN, 1905: 2, Fig. 6. YAMAGUCHI, 1936: 85. CEKANOVSKAYA, 1962: 335.

Lamprodrilus satyriscus f. *decatheca* Michaelsen. MICHAELSEN, 1905: 2. CEKANOVSKAYA, 1962: 335.

Lamprodrilus satyriscus f. *ditheca* Michaelsen. MICHAELSEN, 1905: 2. CEKANOVSKAYA, 1962: 335.

Lamprodrilus satyriscus f. *tetratheca* MICHAELSEN, 1905: 2.

Lamprodrilus (Metalamprodrilus) satyriscus ISOSSIMOV, 1948: 118.

Lamprodrilus (Metalamprodrilus) decathecus ISOSSIMOV, 1948: 120.

Lamprodrilus (Metalamprodrilus) dithecus ISOSSIMOV, 1948: 121.

Lamprodrilus (Metalamprodrilus) tetrathecus ISOSSIMOV, 1948: 121.

Lamprodrilus satyriscus f. *tetratheca* Michaelsen. CEKANOVSKAYA, 1962: 335.

l=40-115 mm, w=up to 3·0 mm, s=100-166. Prostomium prolobic, short. Intersegmental furrows not well developed. Setae sigmoid, 250 μ long, 12 μ thick. Three or four pairs male pores: the most posterior pair situated on XI. One to five pairs spermathecal pores, the most anterior pair being situated on XIII. Three or four pairs accessory copulatory papillae present in anterior part of segments immediately posterior to male pore segments.

Cuticle thin, epidermis thick, circular muscle thin, longitudinal muscle layer very thick. Two or three pairs tubular, unbranched, blind blood vessels in middle and posterior segments. Three or four testes present in VIII to XI or IX to XI. Narrow vasa deferentia join atria medially. Atria tubular, narrowed proximally and terminated by penial bulbi, composed of glandular and muscular cells; bulbi 165 μ diameter, 129 μ long. Penes pointed, cone-shaped, 63 μ long, 48 μ diameter, contained in penial sac. Behind each penial bulbus, an accessory copulatory gland developed, enclosed in copulatory sac, situated in anterior part of each segment following each atrial segment. Ovaries in XII. One to five pairs bulbous spermathecae, the most anterior of which situated in XIII.

Lake Baikal (U.S.S.R.)

Four subspecies of *L. satyriscus* are recognized.

L. satyriscus satyriscus Michaelsen, 1901. Four pairs male pores on VIII to XI. Four pairs spermathecal pores on XIII to XVI.

L. satyriscus dithecus Michaelsen, 1901. Three pairs male pores on IX to XI. One pair spermathecal pores on XIII.

L. satyriscus tetrathecus Michaelsen, 1905. Three pairs male pores on IX to XI. Two pairs spermathecal pores on XIII and XIV.

L. satyriscus decathecus Michaelsen, 1901. Three pairs male pores on IX to XI. Five pairs spermathecal pores on XIII to XVII.

Lamprodrilus semenkewitschi Michaelsen, 1901.

Lamprodrilus semenkewitschi MICHAELSEN, 1901b: 69.

Lamprodrilus semenkewitschi Michaelsen. MICHAELSEN, 1902a: 47; 1905: 2. HRABE, 1929 c: 173. MICHAELSEN AND VERESCHAGIN, 1930: 220. ISOSSIMOV, 1948: 84; 1962: 4. CEKANOVSKAYA, 1962: 344.

$l=55$ mm, $w=$up to 3.0 mm, $s=150$. Prostomium broad and very short. Intersegmental furrows well developed. Secondary annuli absent. Clitellum on X to XII, not very well developed. Setae thin. Two pairs male pores on X and XI, the anterior ones behind ventral setae, the posterior pair situated lateral to outer ventral setae. One pair spermathecal pores on XIII.

In XX, epidermis 100 μ thick, circular muscle very thin, longitudinal muscle layer 105 μ thick. From XIX, about six blind lateral blood vessels with numerous branches present. Two pairs testes in X and XI. Atria narrow, tubular, in X and XI, without penial bulbi or distinct penial sacs. Anterior pair atria little larger than posterior pair. Ovaries in XII. One pair spermathecae, opening in XIII, with large ovoid ampullae and shorter discrete ducts.

Lake Baikal (U.S.S.R.)

Lamprodrilus stigmatias Michaelsen, 1901

Lamprodrilus stigmatias MICHAELSEN, 1901a: 151.

Lamprodrilus stigmatias Michaelsen. MICHAELSEN, 1905: 2. MICHAELSEN AND VERESCHAGIN, 1930: 220. HRABE, 1929 c: 173. ISOSSIMOV, 1948: 93; 1962: 4, Fig. 22. CEKANOVSKAYA, 1962: 347.

$l=28-32$ mm, $w=$up to 1.0 mm, $s=86$. Prostomium prolobic, as long as it is broad at peristomium. Between I to III, spot of pigment present. Intersegmental furrows nearly indistinct. From II, segments biannulate. Setae slightly sigmoid, sharply pointed, 130 μ long, 5 μ thick. Two pairs male pores on X and XI. One pair spermathecal pores on XIII.

In XXV, cuticle 6 μ thick, epidermis 0.5 μ thick, circular muscle 1 μ thick, longitudinal muscle layer 120 μ thick. From XXII two pairs blind lateral blood vessels developed. Chloragogen cells surround gut from VIII, but extend to prostomium on dorsal vessel. Two pairs testes in X and XI. Atria tubular, two pairs, in X and XI. Accessory copulatory glands absent. Ovaries in XII. One pair spermathecae, opening in XIII,, small, pear-shaped.

Lake Baikal (U.S.S.R.)

Lamprodrilus tolli Michaelsen, 1901
Fig. 5. 4C

Lamprodrilus tolli MICHAELSEN, 1901a: 151, Fig. c.

Lamprodrilus tolli Michaelsen. MICHAELSEN, 1905: 28. HRABE, 1929 c: 173; 1962: 319. ISOSSIMOV, 1962: 7. CEKANOVSKAYA, 1962: 340, Fig. 211. COOK, 1968: 285.

$l=17-30$ mm, $w=$up to 1.2 mm, $s=60$. Prostomium zygolobic, large, rounded.

Intersegmental furrows not very distinct. Anterior segments biannulate. Setae sigmoid, distally bent nearly into right angle, 140 μ long, 7 μ thick. Two pairs male pores on X and XI. One pair spermathecal pores on XIII.

From XII, short, thick, sparsely branched, blind blood vessels present. Gut covered by layer of very large chloragogen cells. Two pairs testes, in X and XI. Vasa deferentia narrow, do not penetrate into the segment posterior to their corresponding atria. Vasa deferentia join atria proximally. Atria tubular, two pairs, confined to X and XI, terminated by muscular, bulbous penes. Anterior pair of atria about half the length and diameter of the posterior pair. Accessory copulatory glands absent. Ovaries in XII. One pair spermathecae, opening in XIII, with pear-shaped ampullae and small discrete ducts.

River Iana (Northern Siberia), Liakhova Island (North Arctic Ocean) (U.S.S.R.)

Lamprodrilus vermivorus (Michaelsen, 1905) Nov. comb.

Agriodrilus vermivorus MICHAELSEN, 1905: 2.

Agriodrilus vermivorus Michaelsen. MICHAELSEN, 1928: 27, Fig. 93, 94. ISOSSIMOV, 1948: 183, CEKANOVSKAYA, 1962: 353, Fig. 220, 221.

$l=80$ mm, $w=2.25$ mm, $s=$up to 78. Prostomium small, rounded. Intersegmental furrows in first eleven anterior segments indistinct; distinct posteriorly. Clitellum indistinct, on XI to $\frac{1}{2}$XV. Setae slightly sigmoid, 350 μ long, 20 μ thick. One pair male pores on XI. One pair spermathecal pores, in deep furrows on XIII.

Cuticle, epidermis, circular and longitudinal muscle all rather thick, especially anteriorly. In first eleven segments, layer of diagonal muscles developed. Body cavity in anterior segments very much reduced, filled with mesenchym cells, derived from peritoneal epithelium. Anterior part of gut triangular in cross section. Dorso-ventral blood vessels in anterior segments very tortuous. Middle and posterior segments with blind, branched lateral blood vessels, and dorso-intestinal vessels with numerous long, unbranched caecae. One pair testes in XI. Male funnels very large. Vasa deferentia join atria proximally. Atria large, tubular, extending into XII, terminating in large, retractable muscular penes. Ovaries in XII. One pair spermathecae in XIII, with discrete ducts.

Lake Baikal (U.S.S.R).

This species is remarkable for its predatory habits. It is said to live on small lumbriculids.

Lamprodrilus wagneri Michaelsen, 1901
Fig. 5. 4E

Euaxes baicalensis GRUBE, 1873: 67. (partim).

Rhynchelmis baicalensis (Grube). VAILLANT, 1889: 221. (partim).

Lamprodrilus wagneri MICHAELSEN, 1901a: 151, Fig. d. (Error in legend).

Lamprodrilus wagneri Michaelsen. MICHAELSEN, 1905: 2, Fig. 7. HRABE, 1929c: 172; 1962: 314. ISOSSIMOV, 1948: 44; 1962: 4, Fig. 5, 6. CEKANOVSKAYA, 1962: 335, Fig. 208. COOK, 1968: 280.

Lamprodrilus wagneri f. *longus* ISOSSIMOV, 1948 : 44.

Lamprodrilus wagneri f. *longus* Isossimov, ISOSSIMOV, 1962 : 16.

l=30-42 mm, w=2·0-2·5 mm, s=up to 86. Prostomium prolobic, large. Intersegmental furrows often not well developed. From III, very fine secondary annuli present. Clitellum indistinct. Setae slightly sigmoid, with indistinct node, 155 μ long, 5 μ thick. Two pairs male pores on summit of ovoid papillae, on X and XI. One pair spermathecal pores on XIII.

Cuticle 9 μ thick, epidermis 12 μ thick, circular muscle 7 μ thick, longitudinal muscle layer 120 μ-140 μ thick. Up to 8 pairs tubular, simple or branched, blind blood vessels present from XXVIII. A second, accessory dorsal blood vessel present in middle of body. Two pairs testes in X and XI. Vasa deferentia, diameter 20 μ, join atria medially to sub-apically. Atria elongate, tubular, in X and XI, but partly in their succeeding segment. Penial bulbi, formed from glandular cells and muscle fibres, only slightly developed, 140 μ diameter. Ovaries ribbon-like, in XII, but may extend backwards in ovisacs to XVII. One pair spermathecae, opening in XIII, with large sacciform ampullae and long discrete ducts which reach to XIV.

Lake Baikal (U.S.S.R.)

As the length of Lumbriculidae is known to vary within wide limits for many species, Isossimov's form of *L. wagneri* f. *longus*, is here regarded as an infrasubspecific entity.

GENUS **Trichodrilus** CLAPARÈDE, 1862

Fig. 5. 2A

Type species: Trichodrilus allobrogum CLAPARÈDE

Usually small worms with rounded prostomium. Body wall unpigmented and bearing secondary annuli. Setae thin, sigmoid, usually single pointed. One pair male pores behind ventral setae on IX or X. One or two pairs spermathecal pores behind ventral setae on X or XI or XI and XII.

Posterior lateral blood vessels present or absent. Two pairs testes in VIII and IX or IX and X. Two pairs vasa deferentia open into one pair of spherical to tubular atria which are present in IX or X. One pair ovaries in segment behind the last testes bearing segment. One or two pairs spermathecae present, opening in ovarian segment or in ovarian and post-ovarian segments.

Europe, ? U.S.A.

As defined above and summarized as Lumbriculidae with two pairs of testes, one pair of atria and one or two pairs of spermathecae located behind the atrial segment, *Trichodrilus* is clearly separated from other members of the family. Species of this genus are typically medium to very small, unpigmented worms, with very slender, sigmoid setae. The genera *Dorydrilus* (Dorydrilidae) and *Bichaeta* however, are superficially similar to this, thus making identification of immature individuals, even to generic level on this basis, rather uncertain.

It should be pointed out that many of the specific characters used in this genus are somewhat dubious, but as no proof of their inadequacy is yet available most of the entities described have had to be retained. Characters of uncertain validity are as follows: (*i*) the extent of the pharyngeal glands, (*ii*) the penetration into the ovarian segment of the posterior vasa deferentia, (*iii*) the dimensions of the atria. Characters based on the male genitalia, however, are probably the most reliable but caution should be exercised in the use of the dimensions given in the following descriptions as they are subject to wide variation and depend greatly on the sexual maturity and activity of the individual. The extent of the pharyngeal glands is also dependent on the size and possibly the state of sexual maturity of the individual. In *Eclipidrilus lacustris* those cells which constitute the pharyngeal glands become more concentrated in anterior segments as the individual grows.

The anterior blood system in at least two species of this genus is peculiar. In *T. moravicus* and *T. hrabei* the anterior commissures branch off the dorsal vessel, penetrate into the succeeding segment and there join the ventral vessel, whereas in other genera the commissures join the dorsal and ventral vessels in the same segment.

Trichodrilus is divisible into four major categories thus:

1. Species possessing two pairs of spermathecae and with posterior lateral blood vessels.

> *T. allobrogum**
> *T. icenorum**
> *T. macroporophorus**
> *T. intermedius**
> *T. cantabrigiensis**
> *T. leruthi*

2. Species with two pairs of spermathecae but without posterior lateral blood vessels.

> *T. claparedei**
> *T. hrabei**
> *T. medius**
> *T. moravicus**
> *T. tenuis**
> *T. tacensis*

3. Species with one pair of spermathecae and with posterior lateral blood vessels.

T. pragensis
T. ptujensis

4. Species with one pair of spermathecae and without posterior lateral blood vessels.

T. cernosvitovi
T. sketi
T. stammeri
T. strandi
T. tatrensis

Within the first two divisions, two major species groups emerge (those with * above), the members of which are very closely related and can only be separated by difficult, and often dubious characters. It would be premature however, to synonomize the entities within the groups, or to erect superspecies, subgenera, or any other taxa to attempt to reflect this state of affairs, as the arrangement is further complicated. For example *T. pragensis* has two pairs of spermathecae in immature individuals, but the posterior pair are degenerate in mature specimens. Beddard, in his description of *T. cantabrigiensis*, stated that only one pair of spermathecae were present, but when the original material was redescribed (Cook, 1967) the species was found to possess, quite unmistakably, two pairs. It is certain that Beddard was not dealing with a mixture of species and it must be assumed that he was unfortunate enough to describe a peculiar individual. Thus even the number of spermathecae is not an absolutely reliable character and therefore throws some doubt on the validity of the major species groups.

Trichodrilus

1.	2 pairs spermathecae present in XI and XII	2
—	1 pair spermathecae present in X or XI	13
2.	Posterior lateral blood vessels present	3
—	Posterior lateral blood vessels absent	8
3.	Atria tubular; length/width=2·4	4
—	Atria pear-shaped; length/width up to 1·7	5
4.	Vasa deferentia join atria apically. Atrial muscle only 10 μ thick. Penes simple, protrusible	*T. intermedius*
—	Vasa deferentia join atria subapically. Atrial muscle 23 μ thick. Penes large, internal; swollen bulb at proximal end of atria	*T. leruthi*

5.   Atrial muscle 20 μ thick. Atrium, length/
     width=1·7   *T. macroporophorus*

—   Atrial muscle up to 10 μ thick. Atrium, length/
     width up to 1·5   6

6.   Posterior pair vasa deferentia penetrate into
     XI. Penes small, protrusible   7

—   Posterior pair vasa deferentia do not penetrate
     into XI. Penes large, internal, with porophore   *T. cantabrigiensis*

7.   Pharyngeal glands extend to VIII   *T. allobrogum*

—   Pharyngeal glands extend to VII   *T. icenorum*

8.   Atria tubular; length/width=3·2. Posterior
     pair vasa deferentia do not penetrate into
     XI   *T. tacensis*

—   Atria spherical to pear-shaped; length/
     width=1·2 to 1·9. Posterior pair vasa deferentia
     penetrate into XI   9

9.   Atrial muscle 23 μ thick   *T. claparedei*

—   Atrial muscle up to 10 μ thick   10

10. Prostomium with dorsal, transverse furrow   *T. medius*

—   Prostomium not of this form   11

11. Pharyngeal glands extend to VI. Atrial muscle
     only 2 μ thick   *T. tenuis*

—   Pharyngeal glands extend to VII or VIII.
     Atrial muscle 5-10 μ thick   12

12. Pharyngeal glands extend to VII. Atrial muscle
     10 μ thick. Atria laterally compressed, ovoid,
     with long axis in horizontal plane   *T. hrabei*

—   Pharyngeal glands extend to VIII. Atrial
     muscle 5 μ thick. Atria regular, pear-shaped   *T. moravicus*

13. Spermathecae in X. Setae bifid   *T. strandi*

—   Spermathecae in XI. Setae single pointed   14

14. Atria very long, tubular, extending into XI;
     length=350 μ. length/width=10. Vasa defer-
     entia joins atrium subapically   *T. cernosvitovi*

—   Atria pear-shaped to elongate: if elongate,
     maximum length/width=4, never 350 μ long,

and vasa deferentia join atria medially or
apically · 15

15. Atrial muscle 20-25 μ thick · · · · · · · · · · · · · · · 16
— Atrial muscle 5-10 μ thick · · · · · · · · · · · · · · · 17

16. Vasa deferentia join medially. Pharyngeal
glands extend to VII · · · · · · · · · · · · · · · · · · · *T. stammeri*
— Vasa deferentia join atria apically. Pharyngeal
glands extend to VIII · · · · · · · · · · · · · · · · · · · *T. ptujensis*

17. Atria elongate; length/width=3. Posterior
pair vasa deferentia do not penetrate into XI · · · *T. tatrensis*
— Atria pear-shaped; length/width up to 2.
Posterior pair vasa deferentia do penetrate into
XI · 18

18. Posterior lateral blood vessels present.
Pharyngeal glands extend into VIII · · · · · · · · · · *T. pragensis*
— Posterior lateral blood vessels absent.
Pharyngeal glands extend into VII · · · · · · · · · · · *T. sketi*

Trichodrilus allobrogum Claparède, 1862
Fig. 5. 11A

Trichodrilus allobrogum CLAPARÈDE, 1862: 267, Pl. III, fig. 6-9 and 15,
Pl. IV, fig. 3.

Trichodrilus allobrogum Claparède. VEJDOVSKY, 1884: 51. MICHAELSEN,
1900: 59; 1909: 54. SMITH, 1905: 45. KINDRED, 1918: 49. HRABE, 1937 b:
2; 1954 a: 185, Fig. 6, 7; 1960: 271. CEKANOVSKAYA, 1962: 355. BRINK-
HURST AND COOK, 1966: 11, Fig. 2c. COOK, 1967: 363; 1968: 280.

$l=20$-25 mm, $w=0.3$-0.5 mm, $s=70$. Prostomium longer than it is broad at
peristomium. Setae single pointed, 114 μ long; ratio lengths, pt-node/
total=0.36-0.40. Male pores paired on X. Two pairs spermathecal pores, on
XI and XII.

Pharyngeal glands extend to VIII. Three to six pairs unbranched, posterior
lateral blood vessels present. Testes in IX and X. Posterior pair vas deferentia
penetrate into XI. Pear-shaped atria paired in X, ending in indistinct, protrusible, ejaculatory duct. Atria 120 μ long, 110 μ wide, with muscle layer
5-10 μ thick. Ovaries paired in XI. Two pairs spermathecae in XI and XII,
with ampullae 90 μ long, 75 μ wide, with discrete ducts 95 μ long, 25 μ wide.

Czechoslovakia, France, Denmark, Germany, Italy, Switzerland, U.S.A. (?)

Trichodrilus cantabrigiensis (Beddard, 1908)
Fig. 5. 11B

Phreatothrix cantabrigiensis BEDDARD, 1908: 365, Fig. 76, 77.

Trichodrilus cantabrigiensis Beddard. SOUTHERN, 1909: 127. HRABE, 1937b: 2. BRINKHURST, 1962: 324. COOK, 1967: 362, Fig. 76, e.

$s = 76$. Prostomium rounded, slightly longer than broad at peristomium. Secondary annulus present on each segment from III. Setae single pointed, 95 μ long; Ratio of lengths, pt-node/total$=0.39$. Male pores paired on X. Two pairs spermathecal pores, on XI and XII.

Pharyngeal glands extend to VIII. Three to five pairs unbranched, posterior lateral blood vessels present. Testes in IX and X. Posterior pair vasa deferentia do not penetrate into XI. Vasa deferentia join atrium laterally. Atria pearshaped in X, 100 μ long, 80 μ wide, with muscle layer 2-5 μ thick. Protrusible penes distinct. Ovaries in XI. Two pairs spermathecae in XI and XII, with spherical to elongate ampullae, 120 μ-320 μ long, and discrete ducts 90 μ long.
Great Britain.

This entity was regarded as a *species dubia* until Beddard's original material was redescribed (Cook, 1967). This study showed that Beddard failed to see the posterior lateral blood vessels, and the second pair of spermathecae in XII.

Trichodrilus cernosvitovi Hrabe, 1937
Fig. 5.11E

Trichodrilus cernosvitovi HRABE, 1937b: 1, Fig. 10-12.

Trichodrilus cernosvitovi Hrabe. HRABE, 1960: 270.

$w = 0.17$-0.34 mm. Prostomium rounded, as long as peristomium. Clitellum weakly developed on X to XIII. Setae single pointed; dorsals shorter than ventrals. In VI, dorsal setae 68 μ long, ventral setae 88 μ long. Male pores paired on X. One pair spermathecal pores on XI.

In V, epidermis+circular muscle layer 3 μ thick; longitudinal muscle layer 4.4 μ thick. Pharyngeal glands extend to VII. Chloragogen cells begin in VII. Posterior lateral blood vessels absent, at least from XIII to XXXIII. Testes in IX and X. Atria very long, tubular, becoming narrower proximally. Atria penetrate into XI, where the vasa deferentia join them subapically. Atria approximately 350 μ long, 35 μ wide, with muscle layer 7μ thick. Penes small, hemispherical. Ovaries in XI. One pair spermathecae in XI, with large ampullae and very short, discrete ducts.
Belgium.

The description of this species is based on two fragments, only one of which was mature!

Trichodrilus claparedei Hrabe, 1937
Fig. 5. 11F

Trichodrilus claparedei HRABE, 1937b: 3, (Nomen nudum).

Trichodrilus claparedei Hrabe. HRABE, 1938: 73, Fig. 9, 10; 1960: 270.

$l=30$ mm, $w=0.6-0.8$ mm. Prostomium short. Clitellum developed on X to XIII. Setae single pointed; dorsals shorter than ventrals. Dorsals, 126-160 μ long; ventrals, 138-190 μ long. Male pores paired on X. Two pairs spermathecal pores, on XI and XII.

In VIII, cuticle, 1 μ thick; epidermis, 14-23 μ thick; circular muscle, 4·7 μ thick; longitudinal muscle, 23-47 μ thick; Pharyngeal glands extend to VIII. Chloragogen cells begin in VII. Posterior lateral blood vessels absent. Testes in IX and X. Posterior pair vasa deferentia penetrate into XI. Vasa deferentia join atria subapically. Atria spherical in X, 260 μ long, with muscle layer 23-40 μ thick. Prostate cells large, pear-shaped. Penes protrusible. Ovaries in XI. Two pairs spermathecae in XI and XII. with long ampullae and short ducts.

France.

Trichodrilus hrabei Cook, 1967
Fig. 5. 11C, D

Trichodrilus hrabei COOK, 1967: 364 Fig. 7c, d.

$l=8-15$ mm, $w=0.28-0.43$ mm, $s=55$. Prostomium rounded, longer than it is broad at peristomium. Secondary annulus present on each segment from III. Clitellum weakly developed on IX to XII. Setae single pointed, 75-100 μ long; Ratio of lengths, pt-node/total$=0.4$. Male pores paired on X. Two pairs spermathecal pores, on XI and XII.

Epidermis + circular muscle layer, 5-6 μ thick; longitudinal muscle layer 12-17 μ thick. Pharyngeal glands extend to VII. Chloragogen cells begin in VI or VII. Posterior lateral blood vessels absent. Testes in IX and X. Posterior pair vasa deferentia penetrate into XI. Vasa deferentia join atria subapically. Atria ovoid, laterally compressed, in X, 75-85 μ long, 85-150 μ wide, with muscle layer 7-13 μ thick. Prostate cells elongate, in large discrete masses. Penes small, protrusible. Ovaries in XI. Two pairs spermathecae in XI and XII, with large ovoid to elongate ampullae and discrete ducts, 50-60 μ long.

Great Britain.

Trichodrilus icenorum Beddard, 1920
Fig. 5, 11G

Trichodrilus icenorum BEDDARD, 1920: 227.

Trichodrilus lengersdorfi MICHAELSEN, 1933: 15, Fig. 3 and 4.

Trichodrilus lengersdorfi Michaelsen. HRABE, 1937b: 2; 1960: 246, Fig. 20, 21.

Trichodrilus icenorum Beddard. HRABE, 1937b: 2; 1938: 73; 1960: 270. BRINKHURST, 1962: 324. COOK, 1967: 363, Fig. 7a.

$l=25-40$ mm, $w=0.5-1.6$ mm, $s=110$. Prostomium short, rounded. Setae single pointed, up to 220 μ long. Male pores paired on X. Two pairs spermathecal pores, on XI and XII.

Pharyngeal glands extend to VII. Chloragogen cells begin in VI or VII. Posterior lateral blood vessels present which may bear small blind caecae. Testes in IX and X. Posterior pair vasa deferentia penetrate into XI. Vasa deferentia join atria laterally. Atria pear-shaped, in X, 120-235 μ long, 90-106 μ wide, with muscle layer 5-10 μ thick. Prostate cells large pear-shaped. Penes small, protrusible. Ovaries in XI. Two pairs spermathecae in XI and XII, with short discrete ducts and spherical ampullae.

Belgium, Germany, Great Britain.

A re-examination of Beddard's material (Cook, 1967) demonstrated that, contrary to the original description, posterior lateral blood vessels are present in this species, and that it therefore belongs to *T. allobrogum* group, and that *T. lengersdorfi* Michaelsen, 1933 is its synonym.

Trichodrilus intermedius (Fauvel, 1903)
Fig. 5. 12A

Trichodriloides intermedius FAUVEL, 1903: 221.

Trichodrilus intermedius (Fauvel). MICHAELSEN, 1933: 6. HRABE, 1937b: 1, Fig. 6-9; 1954a: 185. 1960: 260.

$l = 30\text{-}60$ mm, $w = 1.0$ mm. Prostomium short, rounded. Secondary annuli distinct, on each segment from III. Clitellum on X-XIV. Setae single pointed. Male pores paired on X. Two pairs spermathecal pores on XI and XII.

Cuticle 1 μ thick; epidermis 19-22 μ thick; circular muscle 5·4 μ thick; longitudinal muscle 23 μ thick. Pharyngeal glands extend to V. Chlorogogen cells begin in VI. Two to six pairs posterior lateral blood vessels present. Testes in IX and X. Posterior pair vasa deferentia penetrate into XI. Vasa deferentia join atria apically. Atria tubular, 188-255 μ long, 80 μ wide, with muscle layer 9 μ thick. Prostate cells cover atria in layer 70-95 μ thick. Penes small, protrusible. Ovaries in XI. Two pairs spermathecae, in XI and XII, with large, oval to pear-shaped ampullae, 630 μ long, 120 μ wide, and discrete ducts, 160 μ long, 60 μ diameter.

France.

Trichodrilus leruthi Hrabe, 1937
Fig. 5. 12B

Trichodrilus leruthi HRABE, 1937 b: 1, Fig. 2-5.

Trichodrilus leruthi Hrabe. HRABE, 1954 a: 185. 1960: 270.

$w = 0·67\text{-}0·85$ mm. Prostomium as long as width of peristomium. Secondary annuli not very distinct. Clitellum not developed. Setae single pointed; dorsals shorter than ventrals; ventral seta IV 198 μ long. Male pores paired on X. Two pairs spermathecal pores, on XI and XII.

In V to IX, epidermis + circular muscle. 12·6 μ thick; longitudinal muscle, 21 μ thick. Pharyngeal glands extend to V. Two to three pairs posterior lateral blood vessels present. (In one example, L to LX had two pairs; LX to LXXXI had three pairs.) Testes in IX and X. Posterior pair vasa deferentia penetrate into XI. Vasa deferentia join atria subapically. Atria tubular, 440 μ long, 180 μ wide, with muscle layer 23 μ thick. Prostate cells in small diffuse layer. Large thick penial bulb composed of lining cells (not muscle), present at proximal end of each atrium, 215 μ long, 90 μ wide. Ovaries in XI. Two pairs spermathecae present, in XI and XII, with pear-shaped ampullae, 400 μ long, 95 μ wide, and discrete ducts, 235 μ long.

Belgium.

The description is based on two fragments of a mature individual and one immature individual, but is remarkable in that the entity has large internal penes similar to those found in the Dorydrilidae.

Trichodrilus macroporophorus Hrabe, 1954
Fig. 5. 12C

Trichodrilus macroporophorus HRABE, 1954 a : 183, Fig. 1-4.

Trichodrilus macroporophorus Hrabe. HRABE, 1960 : 271.

$l=18\text{-}20$ mm, $w=0.4$ mm maximum, $s=70\text{-}80$. Prostomium longer than it is broad at peristomium. Anterior secondary annuli distinct. Clitellum not well developed. Setae single pointed, 130 μ long. Male pores paired on X on summit of large conical porophore. Two pairs spermathecal pores, on XI and XII.

In anterior segments, epidermis + circular muscle, 10-12 μ thick, longitudinal muscle, 15-30 μ thick. Pharyngeal glands extend to VII (sometimes $\frac{1}{2}$VIII). Chloragogen cells begin in VIII. Several pairs, unbranched posterior lateral blood vessels present. Testes in IX and X. Posterior pair vasa deferentia, penetrate into XI. Vasa deferentia join atria laterally. Atria pear-shaped, 125 μ long, 75 μ wide, with muscle layer, 20 μ thick. Penis small, protrusible, on large porophore. Ovaries in XI. Two pairs spermathecae, in XI and XII, with large ovoid ampullae.

Austria.

Trichodrilus medius Hrabe, 1960
Fig. 5. 12D

Trichodrilus medius HRABE, 1960 : 267, Fig. 22-25.

$l=18$ mm, $w=0.4\text{-}0.6$ mm, $s=80$. Prostomium longer than it is broad at peristomium, with dorsal, transverse furrow, dividing it into two parts; posterior part as long as peristomium. Secondary annuli present from II. Clitellum developed on IX to XII. Setae single pointed, 135 μ long. Male pores paired on X. Two pairs spermathecal pores, on XI and XII.

In X, epidermis 8-13 μ thick, longitudinal muscle up to 27 μ thick. Pharyngeal glands extend to VI. Chloragogen cells begin in VI. Posterior lateral blood vessels absent. Testes in IX and X. Posterior pair vasa deferentia penetrate into XI. Vasa deferentia join atria apically. Atria pear-shaped in X, 185 μ long, 100 μ wide, with muscle layer 9 μ thick. Prostate cells in thick diffuse layer. Penes small, internal. Ovaries in XI. Two pairs spermathecae, in XI and XII, with long ovoid ampullae and short discrete ducts.

Germany.

Trichodrilus moravicus Hrabe, 1937
Fig. 5. 12E

Trichodrilus moravicus HRABE, 1937 c : 2. (Nomen nudum).

Trichodrilus moravicus Hrabe, HRABE, 1937 b: 3. (Nomen nudum); 1938: 73, Fig. 1-8; 1960: 265.

l=10-30 mm, w=0·26-0·56 mm. Prostomium longer than it is broad at peristomium. Secondary annuli present from II. Clitellum developed on X to XII. Setae single pointed; dorsals shorter than ventrals; dorsals, 56-85 μ long, ventrals, 61-126 μ long; ratio of lengths, pt-node/total=0·43. Male pores paired on X. Two pairs spermathecal pores on XI and XII.

In VI, epidermis, 9·5 μ thick, longitudinal muscle 14-25 μ thick. Pharyngeal glands extend to VIII. Posterior lateral blood vessels absent. Testes in IX and X. Posterior pair vasa deferentia penetrate into XI. Vasa deferentia join atria apically. Atria pear-shaped, 117-188 μ long, 66-99 μ wide, with muscle layer 5 μ thick. Penes small, protrusible. Ovaries in XI. Two pairs spermathecae. in XI and XII, with long discrete ducts and ovoid ampullae.
Czechoslovakia.

Hrabe (1938) gives some interesting data on size variation in this species. He found that in some individuals, separated only slightly in space, the length and diameter of complete and fully mature specimens. were over twice that of other examples.

Trichodrilus pragensis Vejdovsky, 1875
Fig. 5. 12F

Trichodrilus pragensis VEJDOVSKY, 1875: 196.

Phreatothrix pragensis (Vejdovsky). VEJDOVSKY, 1876 b: 541, Pl. XXXIX; 1884: 54, Pl. XI, fig. 17-19. BEDDARD, 1908: 365.

Trichodrilus pragensis Vejdovsky. MICHAELSEN, 1900: 59; 1909: 54, Fig. 100. SMITH, 1905: 45. KINDRED, 1918: 49. HRABE, 1930: 6; 1937 b: 2; 1960: 259. CEKANOVSKAYA, 1962: 355. BRINKHURST, 1963 b: 145.

l=30-40 mm, w=0·6-0·7 mm, s=60-80. Prostomium 1·5 times longer than it is broad at peristomium. Setae single pointed. Male pores paired on X. One pair spermathecal pores on XI.

Pharyngeal glands extend to VIII. Four to six pairs branched, posterior lateral blood vessels present. Testes in IX and X. Posterior pair vasa deferentia penetrate into XI. Atria spherical, non-petiolate. Ovaries in XI. One pair spermathecae, in XI. In immature individuals, a second pair of spermathecae are present in XII, which at maturity degenerate.
Czechoslovakia, Italy.

Although *T. pragensis* was described over ninety years ago and has been mentioned many times in the literature, it has apparently seldom been found since then, and no reliable, modern description of this species can be traced.

Trichodrilus ptujensis Hrabe, 1963
Fig. 5. 13A

Trichodrilus ptujensis HRABE, 1963: 67, Fig. 4 and 5.

$l=18.5$ mm, $w=0.2$-0.4 mm, $s=76$. Prostomium conical, as long as it is broad at peristomium. Secondary annuli present from XII to XIV. Clitellum developed on X to XIII. Setae single pointed, 160-180 μ long. Male pores paired on X, situated a little dorsally to ventral setae. One pair spermathecal pores on XI, situated a little ventrally to lateral line.

Pharyngeal glands extend into VIII. Chloragogen cells begin in VIII. Four pairs unbranched, posterior lateral blood vessels present in XL to LV. Testes in IX and X. Posterior pair vasa deferentia do not penetrate into XI. Vasa deferentia join atria apically. Atria elongate. Atrial ampulla spherical, 166 μ external diameter, with muscle layer 26 μ thick. Proximal duct long, ending in small conical, protrusible penes. Ovaries in XI. One pair spermathecae in XI, with ovoid ampullae and discrete ducts, 40 μ diameter.

Yugoslavia.

The description is apparently based on three individuals, only one of which was mature.

Trichodrilus sketi Hrabe, 1963
Fig. 5. 13B

Trichodrilus sketi HRABE, 1963: 67, Fig. 1-3.

$l=32$ mm, $w=0.6$ mm, $s=96$. Prostomium shorter than it is broad at peristomium. Secondary annuli present on II to VIII. Clitellum developed on X to XII. Setae single pointed, 177-200 μ long. Male pores paired on X. One pair spermathecal pores on XI.

In X, epidermis 13 μ thick, longitudinal muscle 25 μ thick. Pharyngeal glands extend to VII. Posterior lateral blood vessels absent. Testes in IX and X. Posterior pair vasa deferentia penetrate into XI. Vasa deferentia join atria apically. Atria ovoid, 175 μ long, 95 μ wide, with muscle layer 5 μ thick. Atria non-petiolate; penes internal. Prostate cells in six discrete groups, all discharging apically into atrium. Ovaries in XI. One pair spermathecae, in XI, with very long ampullae and discrete ducts, 206 μ long, 43 μ diameter.

Yugoslavia.

Trichodrilus stammeri Hrabe, 1937
Fig. 5. 13C

Trichodrilus stammeri HRABE, 1937b: 3. (Nomen nudum).

Trichodrilus stammeri Hrabe, HRABE, 1938: 73 (Nomen nudum); 1942: Fig. 15 and 16 (No description); 1960: 270 (Keyed out); 1963: 68 (Mention only). BRINKHURST, 1963 b: 146.

$w=0.4$-0.7 mm. Prostomium shorter than it is broad at peristomium. Secondary annuli indistinct. Clitellum not developed. Setae single pointed, dorsals shorter than ventrals; inner ventrals, 112-127 μ long, inner dorsals, 82-119 μ long; ratio lengths, pt-node/total$=0.47$. Male pores paired on X. One pair spermathecal pores on XI.

Pharyngeal glands extend to VII. Chloragogen cells begin in VI. Posterior lateral blood vessels absent, at least up to LVI. Testes in IX and X. Vasa

deferentia join atria medially. Atria elongate, 264 μ long, 69 μ wide, with muscle layer 20 μ thick. Ovaries in XI. One pair spermathecae in XI, with large ampullae which extend into XII, and short discrete ducts, 38 μ in diameter.
Austria.

This entity was never formally described by Hrabe although he keys it out in his 1937b and 1960 papers, and a sketch of its atrium appears in a paper on the spring fauna of Czechoslovakia (1942). However, from a single specimen loaned to the author by Prof. Hrabe, the above description was compiled. It is thought that this specimen is the only available example of the entity and that a formal description was delayed until further material was obtained.

Trichodrilus strandi Hrabe, 1936
Fig. 5. 13D

Trichodrilus strandi HRABE, 1963 a: 11. (Nomen nudum).

Trichodrilus strandi Hrabe. HRABE, 1963 b: 404, Fig. 1-4; 1937 b: 2; 1938: 73; 1942: 28; 1960: 270. BRINKHURST, 1963 b: 146.

$w=0.4-0.76$ mm. Prostomium small, conical. Secondary annuli present from VII. Clitellum not developed. Setae bifid, with upper tooth reduced. Male pores paired on IX. One pair spermathecal pores on X.
Pharyngeal glands extend to VI. Posterior lateral blood vessels absent. Testes in VIII and IX. Posterior vasa deferentia penetrate into X. Vasa deferentia join atria subapically. Atria pear-shaped, with long, discrete proximal duct. Atrial ampullae, 60 μ long, 56 μ diameter, with muscle layer 8 μ thick; ducts, 69 μ long, 26 μ maximum diameter. Penes small, protrusible, on small conical poro- phore. Prostate cells small diffuse layer. Ovaries in X. One pair spermathecae in X, with ovoid ampullae and discrete ducts 145 μ long.
Austria, Italy.

This species is unique among *Trichodrilus* as its setae are bifid and as its genitalia are displaced one segment anterior of the normal position in the genus.

Trichodrilus tacensis Hrabe, 1963
Fig. 5. 13E

Trichodrilus tacensis HRABE, 1963: 67, Fig. 6.

$l=13-21$ mm, $w=0.44-0.48$ mm, $s=56-68$. Prostomium conical, as long as it is broad at peristomium. Secondary annuli present on III to IX. Clitellum deve- loped on X to XII. Setae single pointed. Male pores paired on X. Two pairs spermathecal pores, on XI and XII.
Pharyngeal glands extend to VII or VIII. Posterior lateral blood vessels absent. Testes in IX and X. Posterior pair vasa deferentia do not penetrate into XI. Vasa deferentia join atria apically. Atria tubular, 255 μ long, 80 μ wide, with muscle layer 7 μ thick. Penes conical porophores, 48 μ long, 19 μ wide at distal end. Ovaries in XI. Two pairs spermathecae in XI and XII, with ovoid to

spherical ampullae and short discrete ducts.
Yugoslavia.

Trichodrilus tatrensis Hrabe, 1937
Fig. 5. 13F

Trichodrilus tatrensis HRABE, 1937 b: 3, (Nomen nudum, but keyed out).

Trichodrilus tatrensis Hrabe. HRABE, 1938: 73 (Nomen nudum); 1939 a: 211, (Nomen nudum); 1939 b: 2, (Nomen nudum); 1942: 26, Fig. 11-14; 1960: 246, Fig. 27, 28; 1962: 289; 1963: 68.

$l=10$ mm, $w=0.25-0.3$ mm, $s=50-60$. Prostomium rounded, as long as it is broad at peristomium. Secondary annuli not clearly marked. Clitellum only slightly developed on X to XII. Setae single pointed, 100 μ long. Male pores paired on X. One pair spermathecal pores on XI.

In V to VIII, body wall only 8 μ thick. Pharyngeal glands extend to VI. Chloragogen cells begin in VII. Posterior lateral blood vessels absent. Testes in IX and X. Posterior pair vasa deferentia join atria medially. Atrial ampullae elongate, pear-shaped, with long proximal duct; ampullae, 150 μ long, 50 μ wide, with muscle layer 5 μ thick. Penes conical porophores, 23 μ long. Prostate cells in diffuse layer. Ovaries in XI. One pair spermathecae in XI, with large ovoid ampullae and short discrete ducts.
Czechoslovakia.

Trichodrilus tenuis Hrabe, 1960
Fig. 5. 13G

Trichodrilus tenuis HRABE, 1960: 271, Fig. 26.

$w=0.19-0.25$ mm. Prostomium conical. Secondary annuli present from III. Clitellum not developed. Setae single pointed, 60-80 μ long; ratio lengths, pt-node/total$=0.48$. Male pores paired on X. Two pairs spermathecal pores, on XI and XII.

In VI, epidermis+circular muscle, 5.5 μ thick; longitudinal muscle layer, 6 μ thick. Pharyngeal glands extend to VI. Chloragogen cells begin in VI. Posterior lateral blood vessels absent. Testes in IX and X. Posterior pair vasa deferentia penetrate into XI. Atria ovoid, 73 μ long, 58 μ wide, with muscle layer 2 μ thick. Penes protrusible. Prostate cells large, covering apical part of atria. Ovaries in XI. Two pairs spermathecae in XI and XII, with long ovoid ampullae and short discrete ducts.
Germany, Austria.

SPECIES INQUIRENDA

Trichodrilus spelaeus Moszynski, 1936

Trichodrilus spelaeus MOSZYNSKI, 1936: 214.

Trichodrilus spelaeus Moszynski. HRABE, 1937 b: 2: 1938: 73; 1960: 270.

Setae single pointed. One atrial pore on IX and one on X. One pair spermathecal pores on X.

A single atrium present in IX and X. One pair spermathecae in X. Silesia.

This entity was described from a single individual. It seems probable that Mosynski was dealing with an abnormal specimen, thus *T. spelaeus* is here regarded as a species inquirenda.

GENUS **Kincaidiana** ALTMAN, 1936
Fig. 5. 1E

Type species: Kincaidiana hexatheca ALTMAN

Small to medium-sized worms. Setae single pointed or bifid in some part of body. One pair male pores on VIII or IX, behind line of ventral setae. One to three pairs spermathecal pores, situated behind line of ventral setae, on the post-atrial segment, or on atrial and two post-atrial segments.

One pair testes and male funnels in VIII or IX. One pair atria in testicular segment. Prostate cells arranged in discrete pear-shaped glands on atrial surface. One pair ovaries in IX or X. One to three pairs spermathecae situated in segments behind atrial segment, or in the atrial and two post-atrial segments.

North America.

Kincaidiana

1. Prostomium produced into long proboscis.
    Three pairs spermathecal pores   *K. hexatheca*

—   Prostomium short, rounded. One pair sperma-
    thecal pores   *K. freidris*

Kincaidiana freidris Cook, 1965
Fig. 5. 4J-K, Fig. 5. 5A-C

Kincaidiana freidris COOK, in BRINKHURST AND COOK, 1966: 10, Fig. 2b, 5c, 5e-h.

l=8-12 mm, w=0·5-0·6 mm, s=65. Prostomium small, rounded, shorter than it is broad at peristomium. Anterior segments biannulate. Setae single pointed, sigmoid, 90 μ long. One pair male pores on VIII. One pair spermathecal pores on IX.

Posterior lateral blood vessels absent. One pair testes in VIII. One pair male funnels and vasa deferentia. Vasa deferentia join atria medially, run inside muscle layer of atria and drain into lumen apically. Atria elongate, tubular, often twisted asymmetrically in body cavity, 820-910 μ long, 60-70 μ diameter, with muscle layer 5 μ thick. Penes large protrusible, 40 μ long. Ovaries in IX. One pair spermathecae in IX, with sacciform ampullae and long, often coiled, narrow, discrete ducts, 12 μ diameter, with a proximal swelling 35 μ wide.

North America.

Kincaidiana hexatheca Altman, 1936
Fig. 5. 4F-I, Fig. 5. 5D

Kincaidiana hexatheca ALTMAN, 1936: 63, Pl. VII, VIII, IX.

Kincaidiana hexatheca Altman. BRINKHURST AND COOK, 1966: 10, Fig. 2a, 5b, 5d, 5i. COOK, 1968: 281.

$l=45$-50 mm, $w=0.7$-1·3 mm, $s=200$. Prostomium elongate, produced into long, pseudo-segmented proboscis. Intersegmental furrows only distinct anteriorly. No secondary annulation, or annuli very indistinct. Setae bifid in few anterior segments, single-pointed posteriorly, 230-300 μ long, 10 μ thick. Male pores on IX. Three pairs spermathecal pores on IX, X and XI, the most anterior pair being in front of male pores but behind ventral setae.

Epidermis 22 μ thick, circular muscle 12 μ thick, longitudinal muscle layer 94 μ thick. In middle and posterior segments, one pair dorso-ventral blood vessels present in anterior part of each segment, and one pair blind vessels in posterior part. Pharyngeal glands extend into VIII. Chloragogen cells begin in VII. One pair testes in IX. One pair male funnels and vasa deferentia. Vasa deferentia join atria medially. Muscular coat of atria extend for short distance along vasa deferentia. Atria elongate pear-shaped, 600 μ long, 88-100 μ wide, with muscle layer 7·5 μ thick. Penes protrusible, on summits of porophores. Ovaries in X. Three pairs spermathecae in IX, X and XI, with very elongate pear-shaped ampullae, 650 μ long, and discrete ducts, 230 μ long, 65 μ diameter.

North America.

GENUS **Guestphalinus** MICHAELSEN, 1933
Fig. 5. 2E

Type species: **Dorydrilus wiardi** MICHAELSEN

Medium-sized worms. Prostomium elongate. Setae single pointed, sigmoid. One pair male pores on IX. One pair spermathecal pores on IX, in front of male pores, but behind, and in line with ventral setae.

Two pairs testes in VIII and IX. Two pairs male funnels and vasa deferentia. One pair atria in last testes-bearing segment. One pair ovaries in post-atrial segment. One pair spermathecae in atrial segment.

Germany.

Guestphalinus is a mono-specific genus, elevated to generic status by Hrabe (1936a) from a subgenus of *Dorydrilus*. The relationships of the genus are uncertain owing to the position of the spermathecae which place it as an intermediate between *Trichodrilus* and *Stylodrilus*.

Guestphalinus wiardi (Michaelsen, 1933)

Dorydrilus (*Guestphalinus*) *wiardi* MICHAELSEN, 1933: 7, Fig. 1, 2.

Dorydrilus wiardi Michaelsen. YAMAGUCHI, 1953: 312.

Guestphalinus wiardi (Michaelsen). HRABE, 1936 a: 10; 1960: 259. COOK, 1968: 281.

$l=$50-65 mm, $w=$1·1-1·6 mm maximum, 0·4-0·6 mm in posterior segments, $s=$115-145. Prostomium elongate, forming a proboscis. Secondary annuli present from IV. Clitellum not developed. Setae single pointed, 300 μ long, 8 μ thick. Male pores paired on IX. One pair spermathecal pores on IX, in front of male pores, but behind ventral setae.

In XXXV, epidermis, 10 μ thick, circular muscle, 3 μ thick, longitudinal muscle layer, 100 μ thick. Pharyngeal glands present on IV to VII. In middle and posterior segments, one pair large, blind-ending blood vessels present. In XX to XXXIII, dorsal vessel is doubled. Testes in VIII and IX. Posterior pair vasa deferentia do not penetrate into X. Vasa deferentia join atria medially. Atria elongate, pear-shaped, in IX, 300 μ long, 90 μ wide, with muscle layer 16 μ thick. Penes absent, male pores on small, conical porophore. Prostate cells large, pear-shaped. Copulatory gland (Michaelsen's "Pubertätsdrüse"), present at base of ventral setae of IX. Ovaries in X. Female funnel on septum X/XI, but opens in anterior part of XI. One pair spermathecae in IX, with long, pear-shaped ampullae and long, not clearly delimited ducts.

Germany.

GENUS **Bichaeta** BRETSCHER, 1900

Fig. 5. 2D

Type species: Bichaeta sanguinea BRETSCHER

Prostomium rounded. Setae slender, bifid, with upper tooth reduced. Male pores paired on X. Spermathecal pores absent.

Two pairs testes in IX and X. One pair atria in X, with thick muscular walls. One pair ovaries in XI. Spermathecae totally absent.

Switzerland, Italy.

Bichaeta is a peculiar monotypic genus which is characterized by the complete absence of spermathecae, and therefore its relationships with other genera are obscure. Its setae are slender and very similar in form to those of *Trichodrilus strandi*. The structure of the atria, however, is very similar to that of *Eclipidrilus*, as its muscle wall is composed of muscle fibres which are arranged in opposing directions, in alternate layers. Hrabe (1939 a) pointed out that *Bichaeta* is one more example of the tendency for the genital organs in the Lumbriculidae to become reduced and more concentrated; a phenomenon which has occurred in *Eclipidrilus*.

Bichaeta sanguinea Bretscher, 1900
Fig. 5. 13H, I

Bichaeta sanguinea BRETSCHER, 1900: 444, Pl. XXXIII.

(?) *Bythonomus lemani* Grube. MICHAELSEN, 1903b: 1.

Bichaeta sanguinea Bretscher. BRETSCHER, 1903: 1. PIGUET, 1905: 619. HRABE, 1935: 10; 1936a: 3, Fig. 8, 9; 1936b: 404; 1939a: 233. BEDDARD, 1920: 230. BRINKHURST AND KENNEDY, 1962: 186. BRINKHURST, 1963b: 140. COOK, 1968: 281.

Athecospermia minuta PIERANTONI, 1904: 1.

Athecospermia minuta Pierantoni. PIERANTONI, 1905: 227. STEPHENSON, 1930: 792.

Trichodrilus sanguineus (Bretscher). PIGUET, 1913: 111. PIGUET AND BRETSCHER, 1913: 155, Fig. 39A. BEDDARD, 1920: 235. MICHAELSEN, 1933: 6.

l=7-13 mm, w=0·3-0·6 mm, s=40-70. Prostomium rounded, as long as it is broad at peristomium. Secondary annuli present. Clitellum developed on IX to XII (XIII). Setae bifid, with upper tooth reduced. Male pores paired on X. Spermathecal pores absent.

Epidermis+circular muscle, 11 μ thick, longitudinal muscle layer, 12-20 μ thick. Pharyngeal glands present in III to VII. Chloragogen cells begin in VII. Posterior lateral blood vessels absent. Testes in IX and X. Posterior pair vasa deferentia penetrate into XI. Vasa deferentia join atria apically. Atria pear-shaped in X, 150-175 μ long, 85-110 μ wide, with muscle layer 35-47 μ thick; ampullae small, ovoid; proximal duct narrow, 70 μ long. Penes absent. Ovaries in XI. Spermathecae absent.

Switzerland, Italy.

GENUS **Stylodrilus** CLAPARÈDE, 1862

Type species: Stylodrilus heringianus CLAPARÈDE

Medium sized worms. Prostomium rounded. Setae single pointed, or bifid, or both. One pair male pores on X. One pair spermathecal pores on IX.

Posterior lateral blood vessels often present. Two pairs testes in IX and X. Two pairs male funnels and vasa deferentia present. Vasa deferentia join atria proximally to apically. One pair relatively small, spherical to tubular atria present in X. One pair ovaries in XI. One pair spermathecae in IX.

Holarctic.

Stylodrilus

1.	Setae all or mostly bifid	2
—	Setae all single-pointed	8
2.	One pair long non-retractible penes present on X, 300 μ long, reflexed over ventral surface of body	*S. heringianus*

— Penes, if present, not more than 50 μ long   3

3. Atria spherical, non-petiolate   *S. parvus*

— Atria pear-shaped to elongate, petiolate   4

4. Vasa deferentia join atria medially   5

— Vasa deferentia join atria subapically or apically   6

5. Circular muscle of body wall, 40-150 μ thick. Secondary annuli present from VI   *S. crassus*

— Circular muscle of body wall less than 10 μ thick. Secondary annuli present from VIII   *S. opisthoannulatus*

6. Retractable penes present, 30-40 μ long, enclosed in evaginable sac. Thick muscular bulb present at base of penes. Atria pear-shaped   *S. brachystylus*

— Penes not of this form. Atria more elongate   7

7. Penis papillae 45 μ long. Posterior pair vasa deferentia do not penetrate into XI   *S. cernosvitovi*

— Penes indistinct or absent. Posterior pair vasa deferentia penetrate into XI   *S. asiaticus*

8. Atria very elongate, extending into XII   *S. subcarpathicus*

— Atria do not extend beyond X   9

9. Ventral setae sickle-shaped, twice as long as dorsals. Atria pear-shaped, with narrow duct   *S. mirus*

— Dorsal and ventral setae of same size. Atria tubular   10

10. Two pairs posterior lateral blood vessels present   *S. lemani*

— Posterior lateral blood vessels absent   *S. sulci*

Stylodrilus asiaticus (Michaelsen, 1901)
Fig. 5. 5E, Fig. 5. 6G

Claparedeilla asiatica MICHAELSEN, 1901 a: 181.

Bythonomus asiaticus (Michaelsen). MICHAELSEN, 1905: 60. ISOSSIMOV, 1948: 126; 1962: 5, Fig. 30-32.

Stylodrilus asiaticus (Michaelsen) HRABE, 1929 a: 9; 1950: 283. CEKANOV-SKAYA, 1962: 360. BRINKHURST, 1965 a: 436, Fig. 2g.

l=42 mm, *w*=up to 1·6 mm, *s*=114. Prostomium conical, about as long as it is broad at peristomium. Secondary annuli present from V. Setae bifid with upper tooth reduced, 200 μ long.

Circular muscle layer 4 μ thick. Longitudinal muscle layer 80 μ thick. Posterior lateral blood vessels absent. Posterior pair vasa deferentia penetrate into XI. Vasa deferentia wide, join atria more or less apically. Atria elongated, pear-shaped, 400 μ long, 130 μ wide. Penes small, indistinct. Prostate cells compact layer covering atria. Spermathecae with very large ampullae and short thick-walled ducts.

Lake Baikal (U.S.S.R.)

Stylodrilus brachystylus Hrabe, 1928
Fig. 5. 5F, Fig. 5. 6H

Stylodrilus brachystylus HRABE, 1928: 2, (Nomen nudum)

Stylodrilus brachystylus Hrabe. HRABE, 1929 a: 10, Fig. 1; 1950: 282. BRINKHURST, 1965 a: 434, Fig. 1b, 2h.

$l=20$ mm, $w=0.5$ mm, $s=65\text{-}75$. Prostomium conical, rounded. Clitellum on $\frac{1}{2}$IX to $\frac{1}{2}$XII. Setae bifid with upper tooth reduced, except in ventral bundles of II to IV, where they are single pointed. Setae 70-120 μ long.

Epidermis 5-7μ thick, circular muscle 1 μ thick, longitudinal muscle 10-20 μ thick. One pair short, blind, pouch-like, lateral blood vessels in anterior part of segments in posterior third of body. Posterior pair vasa deferentia penetrate into XI. Vasa deferentia join atria sub-apically. Atria pear-shaped, 110 μ long, 70 μ wide, with muscle layer 12-18 μ thick. Atria terminated by muscular penial bulb with muscle layer 21 μ thick, and penes 30-40 μ long, enclosed in evaginable sac. Spermathecae with short thick ducts and large ampullae.

Czechoslovakia.

Brinkhurst (1965 a) mentioned the possibility that *S. parvus* and *S. brachystylus* may be immature or neotenic forms of *S. heringianus*. It has been shown, however (Cook, 1967) that *S. heringianus* does not pass through the *brachystylus* stage in its development phase and that the *parvus* stage is represented only by a mass of undifferentiated cells. Thus both entities are here regarded as valid species.

Stylodrilus lemani (Grube, 1879)
Fig. 5. 5J, Fig. 5. 6J

Bathynomus lemani GRUBE, 1879: 116

Bythonomus lemani Grube. GRUBE, 1880: 228. VEJDOVSKY, 1884: 51. BEDDARD, 1895: 208. MICHAELSEN, 1900: 65; 1909: 51; PIGUET AND BRETSCHER, 1913: 153. HRABE, 1929 a: 9; 1936 a: 13; 1939 a: 211; CERNOSVITOV, 1941: 214. DELLA CROCE, 1955: 48. RAVERA, 1955: 247. BRINKHURST, 1963 b: 140.

Lumbriculus integrisetosus CZERNIAVSKY, 1880: 340

Claparedilla meridionalis VEJDOVSKY, 1883: 226.

Claparedilla meridionalis Vejdovsky. VEJDOVSKY, 1884: 53, Pl. XI, Fig. 20. BEDDARD, 1895: 220.

Pseudolumbriculus claparedianus DIEFFENBACH, 1886: 81.

Claparedeilla integrisetosa (Cerniavsky). MICHAELSEN, 1900: 61; 1901 a: 181.

Stylodrilus lemani (Grube). BRINKHURST, 1965 a: 438, Fig. 1g, 2j.

$l=25$-40 mm, $w=0.5$-0.9 mm, $s=65$. Prostomium conical, longer than it is broad at peristomium. Setae single pointed, or sometimes very faintly bifid, 120 μ long, 3.5-5.0 μ thick.

Circular muscle 3-5 μ thick, longitudinal muscle 35-60 μ thick. Two pairs posterior lateral blood vessels present; anterior pair with few large diverticulae, posterior pair with many small caecae. Posterior pair vasa deferentia penetrate into XI. Vasa deferentia join atria medially. Atria elongate, 300-350 μ long, 65-80 μ wide, with muscle layer 12-18 μ thick. Penes small, protrusible. Prostate cells in discrete masses. Spermathecae with large barrel-shaped ducts, 250 μ long, 100 μ diameter at widest part, and elongate ampullae 300 μ long, 150 μ wide.

U.S.S.R., Czechoslovakia, Switzerland, Austria, Yugoslavia, Italy, France.

Stylodrilus cernosvitovi Hrabe, 1950
Fig. 5. 6A

Stylodrilus cernosvitovi HRABE, 1950: 253, Fig. 45-50.

Stylodrilus cernosvitovi Hrabe. CEKANOVSKAYA, 1962: 356, Fig. 223. BRINKHURST, 1965a: 436, Fig. 2a.

$l=15$-18 mm, $w=$ up to 0.7 mm, $s=90$. Prostomium short, conical. Setae bifid with upper tooth reduced.

Epidermis 9.6 μ thick, longitudinal muscle layer 20-25 μ thick. Posterior pair vasa deferentia do not penetrate into XI. Vasa deferentia join atria apically. Atria elongate, with small pear-shaped ampullae and thick muscular walls. Atria terminated by penes 45 μ long, but without distinct penial bulb.

Caspian Sea.

Stylodrilus crassus (Isossimov, 1948)

Bythonomus crassus ISOSSIMOV, 1948: 136, Fig. 36, 37.

Bythonomus crassus var. *crassior* ISOSSIMOV, 1948: 136.

Stylodrilus crassus (Isossimov). CEKANOVSKAYA, 1962: 356, Fig. 227. BRINKHURST, 1965a: 437.

Bythonomus crassus Isossimov. ISOSSIMOV, 1962: 48, Fig. 36, 37.

Bythonomus crassus var. *crassior* Isossimov. ISOSSIMOV, 1962: 48, Fig. 38.

$l=$ up to 23 mm, $w=1.83$ mm, $s=113$. Prostomium rounded, shorter than it is broad at peristomium. Secondary annuli present from VI. Setae bifid with upper tooth reduced, 286 μ long, 11·4 μ thick.

Epithelium 34μ thick, circular muscle layer 40-150 μ thick, longitudinal muscle layer 122 μ thick. Vasa deferentia join atria medially. Atria pear-shaped with discrete duct. Total length of atria about 400 μ. Penes small papillae. Prostate cells compact layer covering atria. Spermathecae large, with discrete ducts and ampullae twice as long as latter.

Lake Baikal (U.S.S.R.)

Provisionally two subspecies are recognized.

S. crassus crassus Isossimov, 1948.
In V, circular muscle layer of body wall is about 40 μ thick.

S. crassus crassior Isossimov, 1948.
In V, circular muscle layer of body wall is about 150 μ thick.

Stylodrilus heringianus Claparède, 1862
Fig. 5.5G, Fig. 5.6D, E. K

Stylodrilus heringianus CLAPARÈDE, 1862: 224, Pl. III, Fig. 11, Pl. IV, Fig. 2, 13-17.

(?) *Enchytraeus annellatus* KESSLER, 1868: 105, Pl. VI, Fig. 1a, 1b.

Stylodrilus gabretae VEJDOVSKY, 1883: 225.

Stylodrilus heringianus Claparède. VEJDOVSKY, 1884, 51. BEDDARD, 1895: 222. FRIEND, 1896: 143. MICHAELSEN, 1900: 62; 1909: 53, Fig. 98, 99; 1928: 4, Fig. 3, 57. SOUTHERN, 1909: 142. MRAZEK, 1926: 1. HRABE, 1929 a: 9; 1930: 6; 1937 c: 3; 1939 a: 210; 1941: 4; 1942: 26; 1950: 267; 1952: 1; 1960: 246; 1962: 278. CEKANOVSKAYA, 1962: 356, Fig. 222. BRINKHURST, 1962: 323; 1963a: 44; 1965a: 431, Fig. 1c, 2d, 2e. BRINKHURST AND COOK, 1966: 9, Fig. 2e, 3d, 3e. COOK, 1967: 359, Fig. 5 a-d. COOK, 1968: 282.

Stylodrilus gabretae Vejdovsky. VEJDOVSKY, 1884: 51, Pl. XI, Fig. 9-16. BEDDARD, 1895: 222. FRIEND, 1896: 143. MICHAELSEN, 1900: 63; 1909: 53; 1928, 23, Fig. 23. MARTIN, 1907: 21. SOUTHERN, 1909: 142.

Stylodrilus vejdovskyi BENHAM, 1891: 209, Pl. VII, Fig. 42-44.

Bythonomus lemani Grube. ZSCHOKKE, 1891: 120.

Stylodrilus vejdovskyi Benham. BEDDARD, 1895: 222. FRIEND, 1896: 143, Fig. 1-3. MICHAELSEN, 1900: 63; 1909: 53. SOUTHERN, 1909: 127.

Stylodrilus zschokkei BRETSCHER, 1905: 665.

Stylodrilus hallissyi SOUTHERN, 1909: 121, Pl. IX, Fig. 8a-g.

l=25-40 mm, w=0·7-1·0 mm, s=70-110. Prostomium rounded, shorter than it is broad at peristomium. Secondary annuli present from IV. Clitellum on IX to XII. Setae mostly bifid with upper tooth reduced, 85-160 μ long; ratio lengths, pt-node/total=0·34. Male pores on summit of non-retractable penes, 300 μ long, which project backwards along the ventral body wall.

Circular muscle 4-8 μ thick, longitudinal muscle 20-30 μ thick. Two pairs short, blind, pouch-like lateral blood vessels present in posterior third of body. Pharyngeal glands extend to VII or VIII. Chloragogen cells begin in VI. Posterior pair vasa deferentia penetrate into XI. Relatively large vasa deferentia (32 μ wide) penetrate atria sub-apically. Atria pear-shaped, 100 μ long, 68 μ wide, with muscle layer 7-10 μ thick. Thick rings of muscle, continuous with circular layer of body wall, surround base of atria. Penes very long, formed from deep fold in body wall. Prostate cells in discrete masses, draining into apical half of atria. Spermathecae with discrete ducts, 250 μ long, and pear-shaped ampullae.

Europe, W. Asia, N. America.

Stylodrilus mirus (Cekanovskaya, 1956)
Fig. 5. 6I

Bythonomus mirus CEKANOVSKAYA, 1956: 662, Fig. 2-4.

Bythonomus mirus Cekanovskaya. CEKANOVSKAYA, 1962: 363, Fig. 230-232.

Stylodrilus mirus (Cekanovskaya). BRINKHURST, 1965a: 436, Fig. 2i.

l=18-20 mm, w=0·5 mm, s=100-140. Setae single pointed. Ventral setae bent back at node and strongly hooked distally, somewhat sickle-shaped. Dorsal setae half the size of ventrals, not strongly hooked.

Circular muscle layer thin, longitudinal muscle, 15-24 μ thick. Anterior male funnels smaller than posterior pair. Posterior pair vasa deferentia penetrate into XI. Vasa deferentia join atria medially. Atria pear-shaped, petiolate. Penes absent. Spermathecae large with short discrete ducts.

U.S.S.R.

S. mirus is unique among *Stylodrilus* species, not only in the form of its setae, but also in that its anterior pair of vasa deferentia are smaller than its posterior pair. This is the only indication in this genus that the male genitalia are liable to reduction by the disappearance of the anterior testes and vasa deferentia, as has happened in *Eclipidrilus* and *Rhynchelmis,* and suggests that the Japanese genus *Hrabea* is derived from a *Stylodrilus*-like ancestor.

Stylodrilus opisthoannulatus (Isossimov, 1948)
Fig. 5. 6B

Bythonomus opisthoannulatus ISOSSIMOV, 1948: 133, Fig. 33-35.

Bythonomus opisthoannulatus Isossimov. CEKANOVSKAYA, 1956: 662. ISOSSIMOV, 1962: 48, Fig. 33-35.

Stylodrilus opisthoannulatus (Isossimov). CEKANOVSKAYA, 1962: 361, Fig. 226. BRINKHURST, 1965a: 436, Fig. 2b.

w=1·1 mm. Prostomium cap-like, as long as it is broad at peristomium.

Secondary annuli distinct from VIII. Setae bifid, 200 μ long, 8 μ thick.

Posterior pair vasa deferentia penetrate into XI. Vasa deferentia join atria medially. Atria elongate, oval-shaped. Penes absent. Prostate cells compact layer covering atria. Spermathecae with discrete tubular ducts 100 μ long, and large ampullae 350 μ long.

Lake Baikal (U.S.S.R.)

Stylodrilus parvus (Hrabe and Cernosvitov, 1927)
Fig. 5. 5E, Fig. 5. 6F

Anastylus parvus HRABE and CERNOSVITOV, 1927: 203, Fig. 1-4.

Stylodrilus parvus (Hrabe and Cernosvitov). HRABE, 1929a: 11; 1930: 6; 1931: 2; 1942: 26; 1950: 254; 1962: 278. SAPKAREV, 1961: 9. BRINK-HURST, 1963a: 45; 1965a: 435, Fig. 1a, 2f.

Bythonomus parvus (Hrabe and Cernosvitov). STEPHENSON, 1930: 790.

(?) *Stylodrilus leucocephalus* HRABE, 1931: 6, Fig. 5. 5F.

l=10-15 mm, w=up to 0·4 mm, s=54-68. Prostomium rounded, one and a half times longer than it is broad at peristomium. Secondary annuli present from V. Clitellum on ½IX to ½XII. Setae bifid with upper tooth reduced.

Posterior lateral blood vessels absent. Posterior pair vasa deferentia penetrates into XI. Vasa deferentia join atria apically. Atria small, spherical to ovoid. Penes absent. Spermathecae large with short discrete ducts.

Czechoslovakia, U.S.S.R., Jugoslavia.

See remarks under *S. brachystylus* and *S. leucocephalus*.

Stylodrilus subcarpathicus (Hrabe, 1929)
Fig. 5. 5H, Fig. 5. 6C

Bythonomus subcarpathicus HRABE, 1929a: 17, Fig. 5, 6.

Bythonomus subcarpathicus Hrabe. HRABE, 1930: 6; 1939c: Fig. 5b. CEKANOVSKAYA, 1962: 364, Fig. 229.

Stylodrilus subcarpathicus (Hrabe). BRINKHURST, 1965 a: 438, Fig. 1d, 2c.

l=20-30 mm, w=up to 0·9 mm, s=75-90. Prostomium rounded. Secondary annuli present from V. Clitellum on IX to ½XV. Setae single pointed; dorsal setae shorter than ventrals.

In XX, epidermis 8-13 μ thick, circular muscle 2 μ thick, longitudinal muscle 20-50 μ thick. Two pairs lateral blood vessels present in middle and posterior segments, bearing 0 or 1 short blind diverticulum. Posterior pair vasa deferentia penetrate into XI. Vasa deferentia join atria sub-medially to proximally. Atria very elongate, extending into XII. Spermathecae large, extending into XI, with long discrete ducts and elongate pear-shaped ampullae.

U.S.S.R.

S. subcarpathicus has very elongate atria with comparatively thick muscular walls, and large elongate spermathecae which bear a strong resemblance to the organs of *Eclipidrilus* species. However, no description is given of the arrangement of the atrial musculature, which is the major generic distinction between *Stylodrilus* and *Eclipidrilus*. Thus, until more information is available on this species, it must be remembered that *S. subcarpathicus* may be an *Eclipidrilus*.

Stylodrilus sulci (Hrabe, 1932)
Fig. 5. 5E, Fig. 5. 6J

Bythonomus sulci HRABE, 1932: 1.

Bythonomus sulci Hrabe. ISOSSIMOV, 1962: 115.

Stylodrilus sulci (Hrabe). BRINKHURST, 1965a: 438, Fig. 1a, 2j.

l=12-22 mm, w=0·6 mm, s=60. Prostomium conical, rounded. Clitellum on $\frac{1}{2}$IX to $\frac{1}{2}$XIV. Setae single pointed.

Posterior lateral blood vessels absent. Vasa deferentia join atria medially. Atria tubular.

Yugoslavia.

SPECIES INQUIRENDAE

Stylodrilus aurantiacus (Pierantoni, 1904)
Fig. 5. 5K, 1, 2, 3, 4, Fig. 5. 6J

Aurantina aurantiaca PIERANTONI, 1904: 2.

Aurantina aurantiaca Pierantoni. PIERANTONI, 1905: 232, Pl. XIV, fig. 5-11.

Bythonomus aurantiacus (Pierantoni). HRABE, 1929 a: 19. BRINKHURST, 1963b: 146.

Stylodrilus aurantiacus (Pierantoni) BRINKHURST, 1965a: 438, Fig. 1f, 2j.

l=15-20 mm, s=60-70. Setae single pointed.
Two pairs dorso-ventral blood vessels in post-clitellar segments, except few terminal segments where there is one pair only. In middle segments, lateral blood vessels are simple, in more posterior segments they bear small blind caecae. Vasa deferentia join atria medially. Atria tubular.

Italy.

Stylodrilus lankesteri (Vejdovsky, 1877)
Fig. 5. 5I, Fig. 5. 6J

Lumbriculus variegatus (Müller). CLAPARÈDE, 1862: 255, Pl. III, fig. 1-5, 14, Pl. IV, fig. 4. (Partim).

Lumbriculus lankesteri VEJDOVSKY, 1877: 464.

Claparedilla lankesteri (Vejdovsky). VEJDOVSKY, 1844: 54, Pl. XII, fig. 10-15. BEDDARD, 1895: 221.

Claparedeilla lankesteri (Vejdovsky). MICHAELSEN, 1900: 61; 1901a: 181.

(?) *Aurantina aurantiaca* PIERANTONI, 1904: 2.

Bythonomus lankesteri (Vejdovsky). MICHAELSEN, 1909: 52, Fig. 95-97. HRABE, 1929a: 15.

(?) *Euaxes obtusirostris* Menge. MRAZEK, 1927: 1.

Stylodrilus lankesteri (Vejdovsky). BRINKHURST, 1965a: 439, fig. 1e, 2j.

l=40 mm. Prostomium conical, twice as long as it is broad at peristomium. Setae single pointed.
  Two pairs posterior lateral blood vessels present bearing small blind caecae. Vasa deferentia join atria medially. Atria tubular.
  Czechoslovakia.

  S. lankesteri, with its probable synonym *S. aurantiacus*, may be synono-mus with *S. lemani* as their only difference is in the posterior blood system. The described differences in the posterior lateral blood vessels, between *S. lankesteri* and *S. lemani*, are probably valid as specific criteria, but the reliability of the descriptions are questionable, especially in the case of *S. lemani*. The only material of *S. lemani* available to the author was not well preserved, rendering it impossible to study the blood system and there-fore *S. lankesteri* and *S. aurantiacus* are retained, for the present, as species inquirendae.

Stylodrilus leucocephalus Hrabe, 1931

Stylodrilus leucocephalus HRABE, 1931: 2.

Stylodrilus leucocephalus Hrabe, HRABE, 1950: 283; 1958: 341. SAPKAREV, 1961: 9. BRINKHURST, 1965a: 435, Fig. 1b.

l=up to 36 mm, w=0·65 mm, s=136. Secondary annuli present from IV. Setae bifid except for ventral setae of II and III, which are single pointed.
  One pair short blind, pouch-like, posterior lateral blood vessels present. Atria small, spherical to ovoid.
  Lake Ochrid (U.S.S.R.)

  Brinkhurst (1965a) regards *S. leucocephalus* as a probable synonym of *S. parvus*, and indeed the only described differences between the two entities are size and the presence or absence of the short, rudimentary posterior lateral blood vessels. Even the latter distinction is suspect as *S. heringianus* has been variously described as having 0 to 2 pairs of such blood vessels.

GENUS **Eclipidrilus** EISEN, 1881
Fig. 5. 2C

Type species: Eclipidrilus frigidus EISEN

Medium sized worms. Prostomium rounded or produced into a proboscis. Segments biannulate, but may be indistinctly so. Setae single pointed. One pair male pores or a single median male pore on X. A single median, or two median, or a pair of spermathecal pores on IX, or very rarely, two median on VIII and IX.

Two pairs posterior lateral blood vessels present. One or two pairs testes in X or IX and X. Vasa deferentia drain into atrial lumen apically, but run inside muscular coat proximal to this. Atria tubular, with thick muscular coat, wholly or partially composed of many layers of muscle fibres arranged in spirals around the atria. Prostate cells in compact layer surrounding atria. One pair ovaries in XI. A single median, two median, or a pair of spermathecae present in IX.

U.S.A., Britain, U.S.S.R. ?

Eclipidrilus is regarded as being distinct from *Stylodrilus*, although the distribution of the genital elements is similar, because of the characteristic arrangement of the atrial musculature in at least some part of the male apparatus. The muscle fibres are arranged in opposing directions in alternate layers, diagonal to the long axis of the atrium, and this imparts the characteristic cross-hatched appearance of the atrium in surface view. Outside these opposing layers there is a thin covering of fibres arranged longitudinally to the atrial axis.

Eclipidrilus also differs from *Stylodrilus* by the fact that the male and spermathecal elements are very large and the former usually extend for many segments posteriad, whereas in *Stylodrilus* the genital elements are usually small compared with the body size. There is in this genus, however, a tendency for at least the male elements to become reduced to single structures.

Eclipidrilus

1. Atrium single, unpaired, opening mid-ventrally   2
—   Atria paired, opening ventro-laterally   3

2. Prostomium produced into proboscis. A pair
    of spermathecae present, opening mid-ven-
    trally. Atrium with constriction about half
    way along its length   *E. asymmetricus*
—   Prostomium rounded. An unpaired sperma-
    theca present, opening mid-ventrally. Atrium
    without constriction   *E. lacustris*

3. Prostomium rounded. Atria with constriction.
    Penes small   *E. frigidus*

— Prostomium produced into proboscis. Atria without constriction. Penes very large, retractable   4

4. Atrial muscle very thick, becoming suddenly thin distally; ratio, diameter lumen/external diameter=0·3. Muscle layer 60-100 μ thick   *E. palustris*

— Atrial muscle thin, not markedly thinner distally; ratio, diameter lumen/external diameter=0·6-0·9. Muscle layer, 6-20 μ thick   *E. daneus*

Eclipidrilus asymmetricus (Smith, 1896)
Fig. 5. 7C

Mesoporodrilus asymmetricus SMITH, 1896: 402, Pl. XXXVI, fig. 7-10, Pl. XXXVII, fig. 11, 12.

Mesoporodrilus asymmetricus Smith. SMITH, 1900b: 462; 1919: 33. MICHAELSEN, 1900: 61. YAMAGUCHI, 1937a: 8.

Eclipidrilus asymmetricus (Smith). MICHAELSEN, 1901a: 149. GALLOWAY, 1911: 31. HRABE, 1929a: 11; 1936a: 11, Fig. 8; 1939: 233. BRINKHURST AND COOK, 1966: 3, Fig. 1c.

$l=30$ mm, $w=0·5$ mm, $s=65$. Prostomium elongate, forming a proboscis. Clitellum well developed on IX to XIII. Setae single pointed. Male pore single, median, on X. One pair spermathecal pores on IX both in mid-ventral line, one behind the other.

XVIII to XXIII possess one pair lateral blood vessels in anterior part of segments; XXIV to terminal possess two pairs lateral blood vessels, bearing blind caecae. Testes one pair in X. A single (?) male funnel in X. Atrium long, cylindrical, extending to XIV, divided into two parts by constriction in XI or XII. Distal part of atrium with wide lumen, proximal part with thick lining cells. Atrial muscle layer thick. Penes protrusible. Ovaries in XI. One pair spermathecae in IX, both opening medially.

U.S.A.

The description of this species was apparently based on two individuals, one of which was in bad condition and probably immature. As far as the author is aware, the species has not been seen since 1896.

Eclipidrilus daneus Cook, 1966
Fig. 5. 7E, F, Fig. 5. 8E, F
Eclipidrilus daneus COOK. *in* BRINKHURST AND COOK, 1966: 3, Fig. 1d, 3g, 4e, 4f.

$l=40$ mm, $w=1·0$ mm. Prostomium elongate, forming a proboscis. Secondary annuli present in anterior segments. Setae single pointed, 200-250 μ long. Male pores paired on X. One pair spermathecal pores on IX.

Two pairs posterior lateral blood vessels, bearing blind caecae, present.

Testes one pair in X. Two pairs male funnels and vasa deferentia, anterior pair very small, rudimentary. Vasa deferentia penetrate atrial muscle layer proximally, run along inside wall of atria, and discharge into lumen distally. Atria elongate, cylindrical, usually curved or twisted, 1·5-2·0 mm long, 120-220 μ diameter, with muscle layer 6-20 μ thick; ratio, diameter lumen/external diameter=0·6-0·9. Penes very long, retractable, contained within deep fold in body wall, including longitudinal muscle layer. Penis retractor muscle present distally, joining body wall dorsally. Ovaries in XI. One pair spermathecae in IX, with long discrete ducts and elongate ampullae which are divided into two parts; thin walled proximal part and thick-walled distal part. Spermathecae usually coiled in IX.
U.S.A.

E. daneus is very closely related to *E. palustris* and differs from it only in the proportions of its atria. It must be pointed out that *E. daneus* is a morphospecies in the sense of Cain (1954) and it is possible that further work may invalidate this entity or reduce it to an infraspecific rank.

Eclipidrilus frigidus Eisen, 1881
Fig. 5. 7A

Eclipidrilus frigidus EISEN, 1881: 1, Pl. I and II.

Eclipidrilus frigidus Eisen. VEJDOVSKY, 1884: 51. BEDDARD, 1891: 284; 1895: 208. EISEN, 1895: 84, Pl. XLIII-XLV. SMITH, 1896: 403; 1900b: 462; 1919: 37. MICHAELSEN, 1900: 60; 1901a: 149. HESSE, 1909: 43. GALLOWAY, 1911: 311. SMITH AND DICKEY, 1918: 207. BRINKHURST AND COOK, 1966: 3, Fig. 1a. COOK, 1968: 282.

l=20-30 mm, w=1·0-1·5 mm. Prostomium rounded, very short. Setae single pointed, 140-180 μ long; ratio lengths, pt-node/total=0·3. Male pores paired on X. One pair spermathecal pores on IX.

Two pairs posterior lateral blood vessels present. Two pairs testes in IX and X. Both pairs male funnels small. Vasa deferentia penetrate atrial muscle layer proximally but open into atrial lumen apically. Atria elongate, cylindrical, divided into two parts by looped constriction in XIII or XIV. Atria 1·8-2·3 mm long, 170-200 μ diameter, with muscle layer 15-20 μ thick. Distal part of atria with wide lumen, and spiral muscle fibres, proximal part with thick lining cells which more or less occlude lumen, and without spiral muscle fibres. Penes small, eversible. Ovaries in XI. One pair spermathecae in IX, with ovoid ampullae, 450 μ long, 270 μ wide, with discrete, thick-walled ducts, 500 μ long.
U.S.A.

Eclipidrilus lacustris (Verrill, 1871)
Fig. 5. 7B, Fig. 5. 8A-D

Lumbricus lacustris VERRILL. *in* SMITH AND VERRILL, 1871: 449.

Lumbricus lacustris Verrill. MICHAELSEN, 1900: 65. SMITH, 1919: 33.

Mesoporodrilus lacustris (Verrill). SMITH, 1919: 33. LASTOCKIN, 1937: 235. YAMAGUCHI, 1937a: 8.

Eclipidrilus lacustris (Verrill). HRABE, 1936a: 12, Fig. 8; 1939a: 234. LASTOCKIN, 1937: 233. BRINKHURST, 1965a: 441. BRINKHURST AND COOK, 1966: 6, Fig. 1b, 4a-d. COOK, 1967: 361, Fig. 6. 1968: 285.

l=35-50 mm, *w*=0·8-1·4. Prostomium truncated cone shaped; as long as it is broad at peristomium. Anterior segments weakly biannulate. Setae single pointed, 180 μ long; ratio lengths, pt-node/total=0·34. Male pore single, median, on X. One spermathecal pore, median ventral, on IX.

Pharyngeal glands present in IV to VI. Chloragogen cells begin in VI. XVI to XXIII possess one pair large lateral blood vessels in anterior part of segments, XXIV to terminal segment, possess two pairs lateral blood vessels, bearing small blind caecae. Two pairs testes in IX and X. Two pairs vasa deferentia penetrate atrial muscle layer proximally, run within this layer and penetrate lumen apically. Atrium large, cylindrical, median, opening in X, 1·5-2·5 mm long, 200-300 μ wide, with muscle layer 40-60 μ thick. Whole atrium covered in muscle fibres arranged in opposing spirals. Penis small, protrusible. Prostate cells diffuse layer covering atrium. Ovaries in XI. Single median spermatheca in IX, (occasionally one in VIII also), with pear-shaped ampulla 0·7-1·0 mm long, and discrete duct, 500-600 μ long, 75-90 μ diameter.

U.S.A., British Isles, Western U.S.S.R. (?).

The record of *E. lacustris* from the Crimea is dubious, but is included here for the sake of completeness.

Eclipidrilus palustris (Smith, 1900)
Fig. 5. 7D, E

Premnodrilus palustris SMITH, 1900b: 459, Pl. XLI, fig. 1.

Premnodrilus palustris Smith. MICHAELSEN, 1901a: 149. SMITH, 1919: 38.

Eclipidrilus palustris (Smith). GALLOWAY, 1911: 310. HRABE, 1929a: 11; 1936a: 12; 1939a: 233. BRINKHURST AND COOK, 1966: 3, Fig. 1d.

l=50 mm, *w*=1·0 mm. Prostomium elongate, forming a proboscis. Anterior segments distinctly biannulate. Setae single pointed, 150-300 μ long. Male pores paired on X. One pair spermathecal pores on IX.

Two pairs posterior lateral blood vessels, bearing small blind caecae, present. One pair testes in X. Two pairs male funnels and vasa deferentia, anterior pairs very small, rudimentary. Vasa deferentia penetrate atrial muscle proximally, run within this layer and penetrate lumen apically. Atria large, cylindrical, extending for a number of segments posteriad, 2·4-3·0 mm long, 270-300 μ diameter, with muscle layer 60-100 μ thick; ratio, diameter lumen/external diameter=0·3. Distally atrial muscle becomes thin. Penes very long, retractable, contained within deep fold in body wall, including longitudinal muscle layer. Penis retractor muscle present distally, joining body wall dorsally, and conspicuous muscular pad surrounding it proximally. Ovaries in XI. One pair spermathecae in IX, with long discrete ducts and elongate ampullae which are divided into two parts; thin walled proximal part and thick-walled distal part. Spermathecae may be coiled within IX, or straight, extending for many segments posteriad.

U.S.A.

GENUS **Hrabea** YAMAGUCHI, 1936
Fig. 5. 1F

Type species: Hrabea ogumai YAMAGUCHI

Rather large worms. Prostomium rounded. Setae single pointed, sigmoid. One pair male pores on X, behind line of ventral setae.

One pair testes in X. One pair male funnels and vasa deferentia. One pair atria in testes segment. One pair ovaries in XI. One pair spermathecae in pre-atrial segment.

Japan.

Hrabea is probably closely related to *Stylodrilus* and differs from it in possessing only one pair of testes. In *Stylodrilus mirus* the anterior pair of male funnels are smaller than the posterior pair, which indicates that the change from the bitesticular to the monotesticular condition can occur in *Stylodrilus*. Thus, if evidence could be found that *Hrabea* was derived from a bitesticular ancestor, then the justificaton for its generic status would cease to exist.

Hrabea ogumai Yamaguchi, 1936

Hrabea ogumai YAMAGUCHI, 1936: 73, Fig. 14, 15, Pl. V, fig. 4-7.

Hrabea ogumai Yamaguchi. YAMAGUCHI, 1953: 303.

l=70-80 mm, w=1·5-2·0 mm, s=146-183. Prostomium rounded. Secondary annuli present. Setae single pointed; 300-350 μ long; ratio lengths, pt-node/total =0·33. Male pores paired on X. One pair spermathecal pores on IX.

Posterior lateral blood vessels absent. Testes in X. One pair very large male funnels present on septum X/XI. Vasa deferentia do not penetrate into XI. Vasa deferentia join atria apically. Atria pear-shaped; 460 μ long, 150 μ wide, with muscle layer 7-11 μ thick. Penes absent. Prostate cells in discrete, pear-shaped glands, opening into atria in its apical area. Ovaries in XI. One pair spermathecae, in IX, with discrete ducts, 120-370 μ long, and spherical ampullae which fills whole of IX.

Japan.

GENUS **Rhynchelmis** HOFFMEISTER, 1843
Fig. 5. 2F

Type species: Rhynchelmis limonosella HOFFMEISTER

Medium to large sized worms. Prostomium rounded or prolonged into proboscis. Secondary annuli present on at least some segments. Setae single pointed or bifid. One pair male pores on X. One or two pairs spermathecal pores on VIII or VIII and IX, or a single median pore on VIII. A single or paired, rudimentary male pore may be present on IX or X.

Longitudinal muscle of body wall divided into eight bands and usually curling inwards at their margins. Posterior lateral blood vessels usually present. One or two pairs testes in X or IX and X. One or two pairs vasa deferentia and male funnels; the anterior pair may or may not be associated with a pair of testes. Vasa deferentia drain into atria apically, but may run inside muscle coat of atria proximal to this. Atria, long, narrow tubular structures, usually extending for many segments posteriad. Prostate glands discrete, pear-shaped cell masses, covering atria. A single or paired rudimentary atrium often present anterior to functional atria, without connexion with vasa deferentia. Ovaries paired in XI. A single unpaired, or, one or two pairs spermathecae present in VIII or VIII and IX Ampullae of spermathecae usually communicate with gut cavity.

Holarctic.

Rhynchelmis

1.	Single median spermatheca, bearing digitiform diverticulae, present in VIII. Penes large, retractable, enclosed in penial sac	*R. rostrata*
—	One or two pairs spermathecae present in VIII or VIII and IX. Penes small	2
2.	Two pairs spermathecae present	3
—	One pair spermathecae present	4
3.	Single rudimentary atrium present, opening mid-ventrally on IX. Two pairs testes	*R. tetratheca*
—	Rudimentary atrium absent. One pair testes	*R. granuensis*
4.	Single or paired rudimentary atria present in IX, opening on IX or X	5
—	Rudimentary atrium absent	8
5.	Paired rudimentary atria present in IX. Anterior male funnels absent. Anterior vasa deferentia end blindly on septum IX/X or in X	*R. brachycephala*
—	Single rudimentary atrium present. Anterior male funnels present. Anterior vasa deferentia do not end blindly	6
6.	Setae bifid with upper tooth reduced. One pair testes	*R. komareki*
—	Setae single pointed. Two pairs testes	7

7. In first 33-41 anterior segments, one pair dorso-ventral blood vessels present, without caecae, in posterior segments, from about XXXIV, one pair dorso-intestinal vessels present bearing one to three caecae　　*R. vagensis*

— From IX, one pair dorso-ventral vessels with many caecae. Posteriorly only dorso-intestinal vessels　　*R. limosella*

8. Prostomium short, rounded. Atria short, extending to XI　　*R. olchonensis*

— Prostomium long, produced into proboscis. Atria long, extending through at least three segments　　9

9. Spermathecal ampullae do not communicate with gut cavity　　*R. orientalis*

— Spermathecal ampullae do communicate with gut cavity　　10

10. One pair male funnels and vasa deferentia. Longitudinal muscle bands of body wall not curled inwards at their margins　　*R. elrodi*

— Two pairs male funnels and vasa deferentia. Longitudinal muscle bands curl inwards at their margins　　*R. vejdovskyi*

Rhynchelmis brachycephala Michaelsen, 1901
Fig. 5. 9C

Rhynchelmis brachycephala MICHAELSEN, 1901a : 176, Fig. e, Pl. II, fig. 18, 19.

Rhynchelmis brachycephala f. *typica* MICHAELSEN, 1905 : 2, Fig. 9.

Rhynchelmis brachycephala f. *bythia* MICHAELSEN, 1905 : 2, 62; 1920 : 130, Fig. 1.

Rhynchelmis brachycephala Michaelsen. MICHAELSEN, 1905 : 2; 1928 : 48, Fig. 54. SMITH AND DICKEY, 1918 : 207, Pl. XVI. HRABE, 1931 : 6; 1936a : 12, Fig. 8. ISOSSIMOV, 1948 : 167, Fig. 45-57; 1962 : 4, Fig. 52, 57. CEKANOVSKAYA, 1962 : 371, Fig. 238.

Rhynchelmis brachycephala f. *bythia* Michaelsen. SMITH AND DICKEY, 1918 : 207. HRABE AND CERNOSVITOV, 1926 : 267. ISOSSIMOV, 1948 : 167; 1962 : 4. HRABE, 1961 : 137.

Rhynchelmis brachycephala f. *typica* Michaelsen. HRABE AND CERNOSVITOV, 1926: 267. MICHAELSEN AND VERESCHAGIN, 1930: 220. ISOSSIMOV, 1948: 167; 1962: 4, Fig. 45-47.

Rhynchelmis bythia Michaelsen. MICHAELSEN, 1920: 130, Fig. 1; 1928: 48, Fig. 54.

Rhynchelmis brachycephala var. *tentaculata* ISOSSIMOV, 1948: 167; 1962. 66, Fig. 48-51.

Rhynchelmis brachycephala brachycephala HRABE, 1961: 137.

l=up to 185 mm, w=4·5 mm, s=300. Prostomium short, rounded, or produced into proboscis. Clitellum on segments XI to $\frac{1}{2}$XIX. Setae single pointed, 400 μ long, 10 μ thick. One pair rudimentary male pores on IX. One pair spermathecal pores on VIII.

Epidermis+circular muscle layer, 30 μ thick; longitudinal muscle layer 150 μ thick, divided into eight bands which curl inwards at their margins and may form a spiral. Posterior lateral blood vessels present. One pair testes in X. One pair male funnels in X. Two pairs vasa deferentia, anterior pair ending blindly. Atria extend to XVII or further. One pair rudimentary tubular atria present in IX, opening to exterior on glandular protruberances. One pair spermathecae present in VIII, with discrete ducts and ovoid ampullae bearing two or three diverticulae, one of which opens into the gut cavity.

Lake Baikal (U.S.S.R.).

Michaelsen (1905) distinguished two forms of *Rhynchelmis brachycephala*, f. *typica* and f. *bythia*. He separated these on the basis of size differences, the extent of curling in the longitudinal muscle layer and the length of the functionless, anterior pair of vasa deferentia. Isossimov (1948), however, showed that there was continuous variation in these characters between the two forms. In the same paper he described *R. brachycephala* f. *tentaculata* in which the prostomium is produced to form a proboscis. It appears that f. *tentaculata* exists only in northern Baikal, while the typical form is found in the other regions of the lake and that the two forms are rarely found together. Thus two subspecies are here recognized.

Rhynchelmis brachycephala brachycephala Michaelsen, 1901.
Without proboscis.

Rhynchelmis brachycephala tentaculata Isossimov, 1948.
With proboscis

Rhynchelmis elrodi Smith and Dickey, 1918
Fig. 5. 9D

Rhynchelmis elrodi SMITH AND DICKEY, 1918: 208.

Rhynchelmis elrodi Smith and Dickey. MICHAELSEN, 1920: 130, Fig. 1; 1928: 48, Fig. 54, 93. HRABE, 1929a: 11. ALTMAN, 1936: 62. YAMAGUCHI, 1936: 85; 1937a: 8. BRINKHURST AND COOK, 1966: 7, Fig. 1e, 4h-i.

Rhynchelmis glandula ALTMAN, 1936: 59, Pl. VIII, fig. 60-64, Pl. IX, fig. 68-73.

Rhynchelmoides elrodi (Smith and Dickey). HRABE, 1936 a: 11, Fig. 8.

Rhynchelmoides elrodi (Smith and Dickey). HRABE, 1939 a: 233; 1961: 136.

Rhynchelmoides glandula (Altman). HRABE, 1939 a: 233.

$l=$50-100 mm, $w=$1·0-1·3 mm, $s=$175. Prostomium produced into proboscis. Setae single pointed, 250 μ long. Without rudimentary male pores. One pair spermathecal pores on VIII.

Longitudinal muscle bands not curled at their margins. Two pairs posterior lateral blood vessels present from XII; anterior pair with numerous caecae, posterior pair with few caecae. One pair testes in X. One pair male funnels and vasa deferentia, in X. Vasa deferentia run inside atrial muscle layer to apical end. Atria 2·0-2·5 mm long, 100-200 μ diameter, with thin muscle layer. Penes small, protrusible. No rudimentary atria. One pair spermathecae in VIII, with ovoid ampullae which communicate with the gut cavity.

U.S.A.

Hrabe (1936a) withdrew *R. elrodi* from the genus *Rhynchelmis* and erected *Rhynchelmoides* to accommodate it. He justified this by making the generic criterion of *Rhynchelmis* the possession of eight longitudinal muscle bands, some of which curled inwards at their margins. In view of all the other characters which *R. elrodi* has in common with the rest of the *Rhynchelmis* species, the genus *Rhynchelmoides* is rejected here.

Rhynchelmis granuensis Hrabe, 1954

Rhynchelmis granuensis HRABE, 1954b: 310. (Nomen nudum).

Rhynchelmis granuensis Hrabe. HRABE, 1961: 129, Fig. 1, 2.

Rhynchelmis granuensis f. *granuensis* HRABE, 1961: 129.

Rhynchelmis granuensis f. *onegensis* HRABE, 1961: 137.

Rhynchelmis granuensis onegensis Hrabe. HRABE, 1962: 278, Fig. 57; 1963: 74.

Rhynchelmis granuensis granuensis Hrabe. HRABE, 1963: 67.

$l=$up to 50 mm, $w=$1·3 mm, $s=$120-160. Prostomium produced into long proboscis. Clitellum on IX to XVII (XVIII). Setae single pointed or bifid. No rudimentary male pores on IX. Two pairs spermathecal pores on VIII and IX.

In anterior segments, epidermis, 4-17 μ thick, circular muscle, 3 μ thick, longitudinal muscle, 30-40 μ thick. Longitudinal muscle bands curled inwards at their margins. From I to between XXV-XXXIII, one pair dorso-ventral blood vessels present in posterior part of each segment. From between XXV-XXXIII to terminal segment, the vessels become dorso-intestinal, and bear short caecae from LX to XC. One pair testes in X. Two pairs male funnels, anterior one smaller than posterior. Two pairs equally wide (59 μ) vasa deferentia. Atria, 65 μ diameter, extend to XVI. Rudimentary atria anterior to functional ones absent. Two pairs spermathecae, in VIII and IX, with long narrow ducts and ampullae which communicate with gut cavity by a short duct.

Czechoslovakia. L. Onega (Finno-Karrelia).

Two subspecies are recognized.

R. granuensis granuensis Hrabe, 1961.
With single pointed setae.
Czechoslovakia.

R. granuensis onegensis Hrabe, 1961.
With bifid setae.
Lake Onega (Finno-Karrelia).

Rhynchelmis komareki Hrabe, 1927

Rhynchelmis komareki HRABE, 1927: 170, Fig. 1-5.

Rhynchelmis komareki Hrabe. HRABE, 1931: 2; 1958: 341; 1961: 131; 1962: 335. ISOSSIMOV, 1962: 67.

Rhynchelmis komareki f. *typica* HRABE, 1931: 2.

Rhynchelmis komareki f. *brevirostra* HRABE, 1931: 2.

Rhynchelmis komareki f. *komareki* HRABE, 1958: 341.

Rhynchelmis komareki f. *brevirostris* Hrabe, HRABE, 1958: 341.

Rhynchelmis komareki f. *typica* Hrabe. SAPKAREV, 1961: 9. ISOSSIMOV, 1962: 67.

Rhynchelmis komareki f. *brachycephala* HRABE, 1961: 137. (error for *brevirostra*).

Rhynchelmis komareki f. *komareki* Hrabe. HRABE, 1961: 137.

Rhynchelmis komareki f. *brevirostra* Hrabe. ISOSSIMOV, 1962: 67.

l=up to 32 mm, w=1·5 mm, s=100-150. Prostomium rounded or produced into proboscis. Clitellum on VII to XVII. Setae bifid with upper tooth reduced, 150-210 μ long. A single, more-or-less median rudimentary male pore present on IX. One pair spermathecal pores on VIII.

Epidermis 9 μ thick, circular muscle 3 μ thick, longitudinal muscle 75 μ thick. Longitudinal muscle bands curled inwards at their margins. Posterior lateral blood vessels present. One pair testes in X. Two pairs male funnels and vasa deferentia present, anterior pair functionless, smaller than posterior. Atria extend through four or five segments. Penes protrusible. One unpaired rudimentary atrium, covered with prostate cells, present in IX. One pair spermathecae, in VIII, with short narrow ducts and large sacciform ampullae which communicate with gut cavity by short diverticulae.

Lake Onega (Finno-Karrelia). Macedonia.

Two subspecies of *R. komareki* are recognized.

R. komareki komareki Hrabe, 1927.
With prostomium produced into proboscis. (=f. *typica*.) Shallow water.

R. komareki brevirostra Hrabe, 1931.
Prostomium rounded. (=f. *brevirostris* and *brachycephala*.) Deep water.

Rhynchelmis limosella Hoffmeister, 1843
Fig. 5. 9A

Rhynchelmis limosella HOFFMEISTER, 1843: 192, Pl. IX, fig. 8.

Euaxes filirostris GRUBE, 1844: 204, Pl. VII, fig. 1.

Euaxes filirostris Grube. MENGE, 1845: 24, Pl. III, fig. 14-17.

Rhynchelmis limosella Hoffmeister. VEJDOVSKY, 1876a: 332, Pl. XXI-XXIV; 1884: 51, Pl. XII, fig. 33, 34, Pl. XIII, fig. 1, 2, Pl. XVI, fig. 1-6. EISEN, 1888: 2; 1892: 324. BEDDARD, 1892: 197. MRAZEK, 1900: 1. MICHAELSEN, 1900: 63; 1909: 51, Fig. 93, 94; 1920: 130, Fig. 1; 1928: 26, 48, Fig. 28, 54. SMITH AND DICKEY, 1918: 207. HRABE AND CERNOSVITOV, 1925: 2. HRABE, 1927: 170; 1938: 77; 1941: 4; 1942: 23; 1960: 246; 1961: 129, Fig. 4; 1962: 278. PERCIVAL AND WHITEHEAD, 1930: 286. GAVRILOV, 1935: 26. ALTMAN, 1936: 60. ISOSSIMOV, 1962: 6. CEKANOVSKAYA, 1962: 367. COOK, 1967: 368; 1968: 283.

Rhynchelmis limosella f. *limosella* HRABE, 1961: 137.

Rhynchelmis limosella f. *occidentalis* HRABE, 1961: 137.

l=80-140 mm, w=2·0-3·0 mm, s=160-200. Prostomium produced into proboscis. Clitellum on VIII to XVI. Setae single pointed. Single, unpaired, rudimentary male pore on IX or X. One pair spermathecal pores on VIII.

Epidermis and circular muscle layer thin. Longitudinal muscle bands curled inwards at their margins. From IX, one pair lateral blood vessels present which communicate with ventral vessel and with gut plexus, and which have many blind caecae. In posterior segments blood vessels become dorso-intestinal only. Two pairs testes in IX and X. Atria extend through many segments. One, unpaired, rudimentary atrium present in IX, opening in IX or X. One pair spermathecae, in VIII, with pear-shaped ampullae which communicate with gut cavity.

Germany, Czechoslovakia, U.S.S.R., Belgium, Italy, Poland, British Isles (?).

Hrabe (1961) keyed out a new form of this species, *R. limosella* f. *occidentalis* but no details were given. Thus provisionally, two subspecies are recognized.

R. limosella limosella Hoffmeister, 1843.
With rudimentary male pores opening on IX.

R. limosella occidentalis Hrabe, 1961.
With rudimentary male pore opening on X.

Rhynchelmis olchonensis Burov, 1932

Rhynchelmis olchonensis BUROV *in* BUROV AND KOZHOV, 1932: 82.

Rhynchelmis olchonensis Burov. HRABE, 1936a: 12, Fig. 8; 1961: 137. ISOSSIMOV, 1962: 4. CEKANOVSKAYA, 1962: 371.

l=over 30 mm, w=0·46-0·62 mm. Prostomium rounded as long as, or shorter than it is wide at peristomium. Clitellum on X to XII. Setae single pointed, 120μ long, 4·5 μ thick. Without rudimentary male pores. One pair spermathecal pores on VIII.

Two pairs testes in IX and X. Two pairs male funnels and vasa deferentia; latter 18-20 μ diameter. Atria 70 μ diameter, extend to XI. Penes small, retractable. No rudimentary atrium. One pair spermathecae in VIII, with discrete ducts, 35 μ wide, and long ampullae which have no connexion with gut cavity.
Lake Baikal (U.S.S.R.).

Rhynchelmis orientalis Yamaguchi, 1936

Rhynchelmis orientalis YAMAGUCHI, 1936: 73, Fig. 11-13, Pl. IV, fig. 5, 6, Pl. V, fig. 2, 3.

Rhynchelmis orientalis Yamaguchi. HRABE, 1936 a: 12, Fig. 8; 1961: 137. YAMAGUCHI, 1953: 303. ISOSSIMOV, 1962: 115.

l=up to 110 mm, w=2·0 mm, s=195-245. Prostomium produced into proboscis. Clitellum on VIII to XVI. Setae single pointed. No rudimentary male pores. One pair spermathecal pores on VIII.

In X, epidermis is 18-23 μ thick, circular muscle, 6-13 μ thick, longitudinal muscle, 50-75μ thick. Longitudinal muscle bands not curled inwards at their margins. From XII, in posterior part of each segment, one pair lateral blood vessels with few caecae present. From VIII in anterior part of each segment, a pair of lateral blood vessels with many caecae present. One pair testes present in X. Two pairs male funnels, the anterior pair small, functionless. Anterior pair vasa deferentia absent or at least disappear in mature specimens. Vasa deferentia run inside muscle coat of atria before draining into their lumen. Atria extending through 10-13 segments, 50-120 μ diameter, with muscle layer 2·5-4 μ thick. Without rudimentary atria. One pair spermathecae in VIII, with ducts 90 μ diameter, and long ampullae which do not connect with the gut cavity.
Japan.

This species is very closely related to *R. elrodi* and differs from it only by the possession of the rudimentary male funnels and the absence of a connexion between the spermathecae and the gut.

Rhynchelmis rostrata (Eisen, 1888)
Fig. 5. 8G-I, Fig. 5. 9F

Sutroa rostrata EISEN, 1888: 1, Pl. I, II.

Sutroa alpestris EISEN, 1892: 321, Pl. XXIV-XXVI.

Sutroa rostrata Eisen. EISEN, 1892: 322. BEDDARD, 1892: 195, Pl. I; 1895: 208. MICHAELSEN, 1900: 64. GALLOWAY, 1911: 311.

Sutroa alpestris Eisen. BEDDARD. 1895: 208. SMITH, 1900b: 463. MICHAELSEN, 1900: 65. GALLOWAY, 1911: 312. HRABE, 1936a: Fig. 8.

Rhynchelmis rostrata (Eisen). BRINKHURST AND COOK, 1966: 3, Fig. 1f, 4g, 5a.

$l=50$-70 mm, $w=1.0$-2.0 mm. Prostomium produced into proboscis. Setae single pointed, 250 μ-320 μ long, 8·8 μ thick. Without rudimentary male pores. A single median spermathecal pore on VIII.

In anterior segments, epidermis about 30 μ thick, circular muscle 12 μ thick, longitudinal muscle 150 μ thick. Longitudinal muscle bands not curled inwards at their margins. Two pairs lateral blood vessels present in post-clitellar segments, bearing blind caecae. One pair testes in X. Two pairs male funnels and vasa deferentia, anterior pair small, rudimentary. Atria very long, often coiled 50 μ diameter, with muscle layer 6-10 μ thick. Penes very large, retractable, enclosed in large penial sac. A single, unpaired spermatheca present in VIII, with pear-shaped ampulla which communicates with the gut cavity, and with a number of digitiform diverticulae joining it at junction with its short thick duct.
    U.S.A.

Rhynchelmis tetratheca Michaelsen, 1920
Fig. 5. 9B

Rhynchelmis tetratheca MICHAELSEN, 1920: 130, Fig. 1.

Rhynchelmis tetratheca Michaelsen. HRABE, 1927: 172; 1950: 260; 1961: 130. MICHAELSEN, 1928: 48, Fig. 54. ISOSSIMOV, 1962: 105. CEKANOVSKAYA, 1962: 367, Fig. 233, 234.

$l=25$-40 mm, $w=1.0$-1.5 mm, $s=120$. Prostomium produced into long proboscis. Clitellum on VIII to XIII. Setae bifid with upper tooth very much reduced. Single median rudimentary male pore on IX. Two pairs spermathecal pores, on VIII and IX.

Two pairs testes in IX and X. Two pairs male funnels and vasa deferentia, anterior pair smaller than posterior. Atria long, thick walled. A single rudimentary atrium present in IX, with narrow duct opening under ventral nerve cord. Two pairs spermathecae in VIII and IX, with ampullae which communicate with gut cavity.
    Germany, U.S.R.R.

Rhynchelmis vagensis Hrabe, 1954

Rhynchelmis vagensis HRABE, 1945b: 310. (Nomen nudum).

Rhynchelmis vagensis Hrabe, HRABE, 1961: 129, Fig. 3, 4.

l=40-70 mm, w=1·4 mm, s=165. Prostomium produced into proboscis. Setae single pointed. A single rudimentary male pore, more or less median on IX. One pair spermathecal pores on VIII.

Longitudinal muscle bands curl inwards at their margins. From between XXXIII to XXXV, to terminal segment, one pair dorso-intestinal blood vessels present which bear one to three blind caecae. From I to between XXXII to XLI, one pair dorso-ventral blood vessels, without caecae, present. Two pair testes in IX and X. Atria extending into XV. A single rudimentary atrium in IX, opens more or less under ventral nerve cord. One pair spermathecae in VIII, with ampullae which communicate with gut cavity.

Czechoslovakia.

R. vagensis is very closely related to *R. limosella* and differs from it only in the arrangement of the posterior lateral blood vessels and in size.

Rhynchelmis vejdovskyi Hrabe and Cernosvitov, 1925
Fig. 5. 9E.

Rhynchelmis vejdovkyi HRABE AND CERNOSVITOV, 1925: 1, Fig. a, b, Pl. I, fig. 1-5. *and* 1926: 265, Fig. 1.

Rhynchelmis vejdovskyi Hrabe and Cernosvitov. HRABE, 1927: 172; 1936a: 12, Fig. 8; 1961: 128. YAMAGUCHI, 1936: 88. ISOSSIMOV, 1962: 73.

l=50-60 mm, s=90-120. Prostomium produced into proboscis. Clitellum on VIII to XIV. Setae single pointed, 150-210 μ long, 8 μ thick. Without rudimentary male pores. One pair spermathecal pores on VIII.

Longitudinal muscle bands curled inwards at their margins. Two pairs dorso-intestinal blood vessels, without blind caecae, present in post-clitellar segments. One pair testes in X. Two pairs male funnels and vasa deferentia present, anterior pair functionless. Atria extend through about three segments. Without a rudimentary atrium. One pair spermathecae, in VIII, whose pear-shaped ampullae communicate with gut cavity.

U.S.S.R.

GENUS **Tatriella** HRABE, 1936
Fig. 5. 2G

Type species: Tatriella slovenica HRABE

Setae single pointed, sigmoid. Male pore single, median, on X. Spermathecal pore single, median, on VIII, or absent.

Two pairs testes in IX and X. Two pairs vasa deferentia join single atrium near distal end. Atrium unpaired, in X, or often extending a few segments posteriad. One pair ovaries in XI. Spermathecae, if present, single, median.

Czechoslovakia.

The appearance of the proximal part of the atrium and the position of the spermatheca indicate that this genus is closely related to *Rhynchelmis*, but the extreme reduction of the genitalia is here regarded as sufficient justification for its generic status. It may be argued that *Eclipidrilus* contains species with paired, or with unpaired atria and spermathecae, but here evidence of their relationship, in the form of the unique histology of the atrial musculature, is unequivocable. It is suggested, however, that as well as the reduction of the genitalia, the form of the atrium is sufficiently distinct from that of *Rhynchelmis* to preclude the inclusion of *Tatriella* in *Rhynchelmis*, unless species with characters intermediate between these two are discovered.

Tatriella slovenica Hrabe, 1936
Fig. 5. 10A-D

Tatriella slovenica HRABE, 1936 a : 12, Fig. 8.

Tatriella slovenica Hrabe. YAMAGUCHI, 1937 a : 8. HRABE, 1939 a : 211, Fig. 9-14.

$l=15$-28 mm, $w=0.65$ mm, $s=65$-85. Prostomium rounded, shorter than it is broad at peristomium. III to VII, biannulate. Setae single pointed, sigmoid, 100 μ long. Male pore single, mid-ventral, on X. Spermathecal pore single, mid-ventral, on VIII, or absent.

In VI, epidermis 9 μ thick, circular muscle 28 μ thick, longitudinal muscle layer 22·4 μ thick. Chlorogogen cells begin in VI. Blind posterior lateral blood vessels absent. Two pairs testes in IX and X. Two pairs male funnels and narrow vasa deferentia. Posterior pair vasa deferentia do not penetrate into XI. Two pairs vasa deferentia join single, narrow, atrial duct, near junction with wide lumen. Atrium unpaired, in X, composed of long narrow duct and large, ovoid to elongate ampulla. Atrium coiled within its segment or extends posteriad in sperm sac. Penis small, conical. Prostate cells elongate, covering distal half of atrial duct. One pair ovaries in XI. Spermatheca usually absent. About 8% of individuals have a single median spermatheca, in VIII, with pear-shaped to ovoid ampullae 387 μ long, 146 μ wide, and ducts 140 μ long, 21·5 μ diameter.

Czechoslovakia.

GENUS **Styloscolex** MICHAELSEN, 1901
Fig. 5. 1G

Type species: Styloscolex baicalensis MICHAELSEN

Medium sized worms, without pigment. Prostomium rounded. Setae single pointed, sigmoid.

One pair testes in VII or VIII. One pair male funnels and vasa deferentia. Atria elongate, may extend posteriad for about three segments. Penes distinct, retractable, contained within penial sac, and often with thickened cuticular sheath. One pair ovaries in IX or X, always with a clear segment between testes and ovaries. One or two pairs spermathecae present in VI or VII, or VII and XI.

L. Baikal, Japan.

Styloscolex is remarkable among the Lumbriculidae as it is the only genus which has an intervening segment between the testicular and ovarian segments. Yamaguchi (1953) speculated that this condition is derived from the loss of a posterior pair of testes and male ducts from a multiatrial ancestor.

The atria in *Styloscolex* are usually thick muscular, elongate structures, superficially resembling those of *Eclipidrilus* species. The muscle fibres, however, are all arranged longitudinally to the atrial axis.

Styloscolex

1.	One pair spermathecae present in VI or VII	2
—	Two pairs spermathecae present in VII and XI	*S. tetrathecus*
2.	Spermathecae in VI, atria in VII. Anterior part of body considerably swollen	*S. swarczewski*
—	Spermathecae in VII, atria opening in VIII. Anterior part of body not swollen	3
3.	Atria confined to VIII. Two pairs rudimentary female funnels present in XI and XII	*S. kolmakowi*
—	Atria extend beyond VIII. Without rudimentary female funnels	4
4.	Penes slender cone-shaped	5
—	Penes tubular	6
5.	Secondary annuli present from III. Spermathecal ampullae always asymmetrically disposed; one directed forwards, the other backwards	*S. asymmetricus*
—	Secondary annuli present from IV. Spermathecal ampullae not always disposed asymmetrically	*S. japonicus*
6.	Prostate cells in compact layer	*S. chorioidalis*
—	Prostate cells in discrete masses	*S. baikalensis*

Styloscolex asymmetricus Isossimov, 1948
Fig. 5. 10E-H

Styloscolex asymmetricus ISOSSIMOV, 1948: 151, Fig. 39-42.

Styloscolex asymmetricus Isossimov, ISOSSIMOV, 1962: Fig. 39-42. CEKA-NOVSKAYA, 1962: 377, Fig. 240.

l=18-20 mm, w=up to 0·6 mm, s=85-90. Prostomium rounded, longer than it is broad at peristomium. Secondary annuli present from III. Clitellum on VII to XII. Dorsal and ventral setae of same shape, 115-135 μ long. Male pores on VIII. One pair spermathecal pores on VII.

Circular muscle thin, longitudinal layer thickened. Testes in VIII. Vasa deferentia join atria proximally, run inside muscle layer, and drain into lumen apically. Atria extend into XI. Penes long, funnel-like, with thickened cuticular sheaths, enclosed in a penial sac. Prostate cells form compact layer covering atria. One pair ovaries and female funnels in X. One pair spermathecae with very large ampullae and narrow discrete ducts, opening in VII. Ampullae asymmetrically situated, one directed forwards, the other backwards.

Lake Baikal (U.S.S.R.).

It is possible that this entity is a synonym of *S. japonicus* Yamaguchi, as the only difference between the two is the extent of the biannulations and the fact that in *S. asymmetricus* one spermathecal ampulla is always directed posteriorly and the other anteriorly. This asymmetrical disposition of the spermathecae has however, been observed by the author in some specimens among the type material of *S. japonicus*. Further studies on the Lake Baikal entities are necessary before a definite decision is made.

Styloscolex baikalensis Michaelsen, 1901

Styloscolex baikalensis MICHAELSEN, 1901b: 4.

Styloscolex baikalensis Michaelsen. MICHAELSEN, 1902a: 49; 1905: 57. MICHAELSEN AND VERESCHAGIN, 1930: 220. BUROV, 1931: 80. YAMAGUCHI, 1937b: 169; 1953: 311. ISOSSIMOV, 1948: 145; 1962: 4. CEKANOVSKAYA, 1962: 375. COOK, 1968: 283.

l=20-40 mm, w=0·56-1·0 mm, s=90-105. Prostomium rounded, about as long as it is broad at peristomium. Secondary annuli present on V to XI. Clitellum on ½VII to ½XII. Setae 160 μ long, 6 μ thick. Male pores on VIII. One pair spermathecal pores on VII.

Body wall thin. Posterior lateral blood vessels absent. Testes in VIII. Narrow vasa deferentia joint atria proximally. Atria elongate; distally with thin muscular wall, without prostate cells. Prostate cells in discrete masses, confined to middle portion of atria. Penes cylindrical, narrow, retractable, 280 μ long, 20 μ diameter, contained within penial sac. One pair ovaries and female funnels in X. One pair spermathecae opening in VII, with short, narrow ducts and long sacciform ampullae.

Lake Baikal (U.S.S.R.).

Styloscolex chorioidalis Isossimov, 1948
Fig. 5. 10.I, J.

Styloscolex chorioidalis ISOSSIMOV, 1948: 159, Fig. 43, 44.

Styloscolex chorioidalis Isossimov. ISOSSIMOV, 1962: 62, Fig. 43, 44.
CEKANOVSKAYA, 1962: 376, Fig. 239.

$l=25$ mm, $w=0.9$ mm, $s=92$. Prostomium rounded, as long as it is broad at peristomium. Secondary annuli present from III. Clitellum on VII to XII. Male pores on VIII. One pair spermathecal pores on VII.

Testes in VIII. Vasa deferentia join atria proximally. Atria extend to XI. Prostate cells cover atria in compact layer. Penes narrow, tubular, within penial sac. One pair ovaries and female funnels in X. One pair spermathecae, opening in VII, with discrete ducts and elongate ampullae extending to IX.

Lake Baikal (U.S.S.R.).

S. chorioidalis is probably a synonym of *S. baikalensis*. Isossimov (1948) stated that he did not find the latter species in Lake Baikal, and considered it a dubious entity. In the same work, however, he described *S. choryoidalis* which differs from *S. baikalensis* only in a few minor details which are probably more apparent than real. The condition and extent of the prostate cells, and the extent of secondary annulation, for example, probably depend on the sexual state and the state and type of fixation of the material studied.

Styloscolex japonicus Yamaguchi, 1937

Styloscolex japonicus. YAMAGUCHI, 1937 b: 167, Fig. 1, 2, Pl. X.

Styloscolex japonicus Yamaguchi. YAMAGUCHI, 1953: 304.

$l=20-30$ mm, $w=0.75-1.0$ mm, $s=52-72$. Prostomium rounded, conical, longer than it is broad at peristomium. Secondary annuli present from IV. Clitellum on $\frac{1}{2}$VII to XI. Male pores on VIII. One pair spermathecal pores on VII.

Chloragogen cells begin in VI. Pharyngeal glands extend to VI. In middle and posterior segments, two pairs pouch-like, bi- or tri-lobed diverticulae of dorsal vessel present. Testes in VIII. Short thick vasa deferentia join atria proximally. Atria tubular, extending to X, 85-100 μ diameter, with muscle layer 12 μ thick. Penes elongate, cone-shaped, 120-140 μ long, 12 μ in diameter in its middle region, 30 μ diameter proximally. One pair ovaries and female funnels in X. One pair spermathecae, filling VII, with large ampullae which may be asymmetrically situated, and discrete ducts 43 μ diameter.

Japan.

Styloscolex kolmakowi Burov, 1931

Styloscolex kolmakowi BUROV, 1931: 84.

Styloscolex kolmakowi Burov. YAMAGUCHI, 1937b: 169; 1953: 311.
ISOSSIMOV, 1948: 149; 1962: 4. CEKANOVSKAYA, 1962: 374.

l=50 mm, w=up to 3·0 mm, s=105. Prostomium rounded, as long as it is broad at peristomium. IV to VII, biannullate, following segments multiannullate. Clitellum on VII to XV. Setae 236 μ long, 15 μ thick. Male pores on VIII. One pair spermathecal pores on VII.

Cuticle 2 μ thick, epidermis 37 μ thick, circular muscle 30 μ thick, and longitudinal muscle layer 190 μ thick. Testes in VIII. Vasa deferentia join atria subapically. Atria sacciform confined to VIII, 200 μ diamater, covered with layer of prostate cells. Penes retractable. One pair ovaries in X. One pair female funnels in X, 200 μ long. Two pairs functionless, rudimentary, female funnels present in XI and XII, 150 and 90 μ long respectively, which may or may not open to the exterior. One pair spermathecae in VII with discrete ducts and sacciform ampullae.

Lake Baikal (U.S.S.R.).

Styloscolex swarczewski Burov, 1931

Styloscolex swarczewski BUROV, 1931 : 82.

Styloscolex swarczewski Burov. YAMAGUCHI, 1937b: 169; 1953: 311. ISOSSIMOV, 1948: 147; 1962: 4. CEKANOVSKAYA, 1962: 376.

l=50 mm, w=up to 4·0 mm. Anterior part of body considerably swollen. s=81. Prostomium conical, as long at it is broad at peristomium. Secondary annuli present from II. Clitellum on VI to XI. Setae 300 μ long, 18·5 μ thick. Male pores on VII. One pair spermathecal pores on VI.

In XIX, epidermis 75 μ thick, muscle 76 μ thick, longitudinal muscle layer 190 μ thick. Testes in VII. Vasa deferentia join atria distally. Atria cylindrical to sacciform, do not extend beyond VII. Penes narrow, stylet-shaped, 110 μ diameter at base. One pair ovaries and female funnels in IX. One pair spermathecae in VI, with ducts 100 μ diameter (swollen to 200 μ in part) and sacciform ampullae 800 μ diameter, with small proximal diverticulae.

Lake Baikal (U.S.S.R.).

Styloscolex tetrathecus Burov, 1931

Styloscolex tetrathecus BUROV, 1931 : 80.

Styloscolex tetrathecus Burov, YAMAGUCHI, 1937b: 169; 1953: 311. ISOSSIMOV, 1948: 164; 1962: 4, 119. CEKANOVSKAYA, 1962: 374.

l=up to 25 mm, w=0·72 mm, s=70. Prostomium large, longer than it is broad at peristomium. Clitellum on VII to XI. Setae 140 μ long, 7·5 μ thick. Male pores on VIII. Two pairs spermathecal pores on VII and XI.

In XVIII, cuticle 1 μ thick, epidermis 4-5 μ thick, circular muscle 3·5 μ thick and longitudinal muscle layer 30 μ thick. Chloragogen cells begin in VI. Testes in VIII. Vasa deferentia 20 μ diameter. Atria elongate, muscular, extending to X posteriorly and into VII anteriorly where they change gradually into long narrow penes. Atria 75 μ diameter. Thick layer of prostate cells cover atria. One pair ovaries and female funnels in X. Two pairs spermathecae in VII and XI, with discrete ducts and sacciform ampullae.

Lake Baikal (U.S.S.R.), N.E. China.

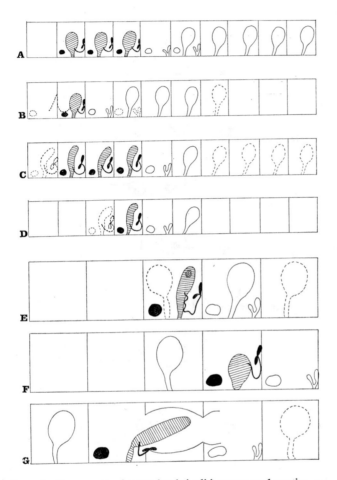

Fig. 5.1. Reproductive organs of some lumbriculid genera and species.
A. *Lumbriculus multiatriatus*; B. *L. variegatus*; C. *Lamprodrilus satyriscus*; D. *Lamprodrilus spp.*; E. *Kindcaidiana*; F. *Hrabea*; G. *Styloscolex.*
Black circles—testes, open circles—ovaries, atria cross hatched, broken outlines—present in some specimens or in some species.

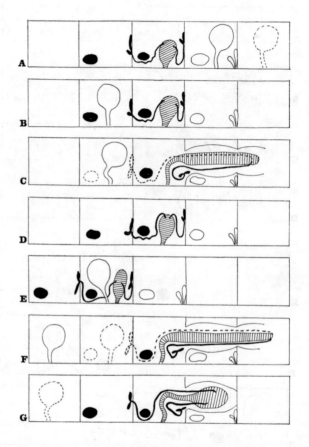

Fig. 5.2. Reproductive organs of some lumbriculid genera (cont'd.).
A. *Trichodrilus*; B. *Stylodrilus*; C. *Eclipidrilus*; D. *Bichaeta*; E. *Guestphalinus*; F. *Rhynchelmis*; G. *Tatriella*.

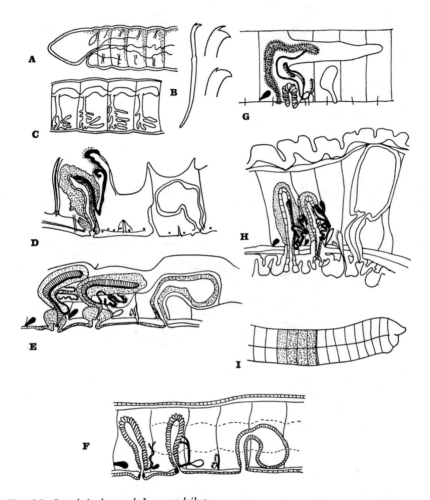

Fig. 5.3. *Lumbriculus* and *Lamprodrilus*.
L. variegatus. A—prostomium, B—setae, C—posterior lateral blood vessels.
Lamprodrilus spp. D to H—reproductive organs. D—*L. baicalensis,* E—*L. bulbosus,*
    F—*L. isoporus,* G—*L. grubei,* H—*L. melanotus.*
L. isoporus. I—anterior end, whole worm.

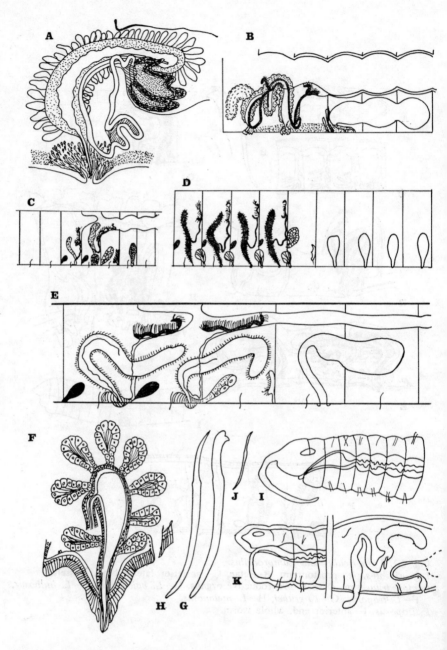

Fig. 5.4. *Lamprodrilus* and *Kincaidiana.*
Lamprodrilus spp. A to E—reproductive organs.
A—*L. pygmaeus pygmaeus*, B—*L. pygmaeus sulcatus*, C—*L. tolli*, D—*L. satyriscus*,
    E—*L. wagneri.*
K. hexatheca. F—atrium, G—anterior seta, H—posterior seta, I—anterior end.
K. freidris. J—seta, K—anterior end.

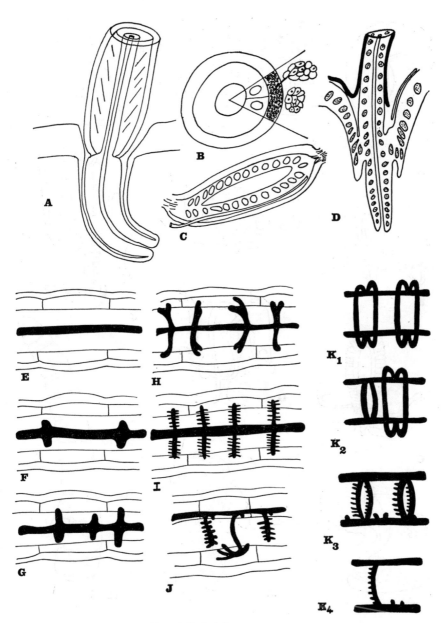

Fig. 5.5. *Kincaidiana* (cont'd.) and *Stylodrilus*.
K. freidris (cont'd.). A—penis, B—transverse section of atrium, C—longitudinal section of atrium.
K. hexatheca (cont'd.). D—penis.
Stylodrilus spp. E to K—posterior dorso-lateral blood vessels. E—*S. parvus, S. asiaticus* and *S. sulci*, F—*S. leucocephalus, S. brachystylus*, G—*S. heringianus*, H—*S. subcarpathicus*, I—*S. lankesteri*, J—*S. lemani*, K—1 to 4—*S. aurantiacus*, successively more posterior segments.

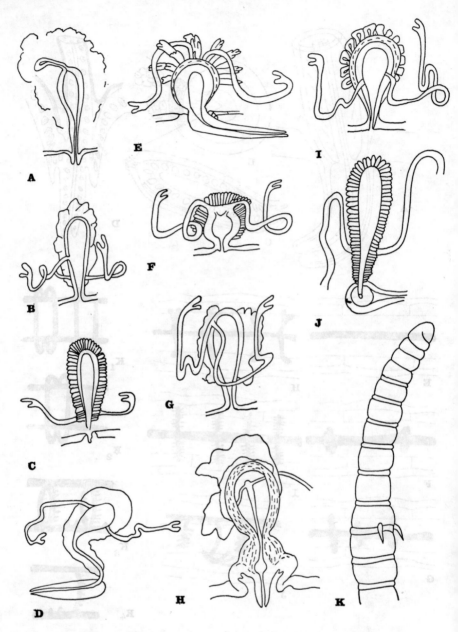

Fig. 5.6. *Stylodrilus* (cont'd.).
Stylodrilus spp. A to J—atria.
A—*S. cernosvitovi*, B—*S. opisthoannulatus*, C—*S. subcarpathicus*, D—*S. heringianus* acc. to Claparède, E—*S. heringianus* acc. to Hrabě, F—*S. parvus*, G—*S. asiaticus*, H—*S. brachystylus*, I—*S. mirus*, J—*S. lemani, S. sulci, S. lankesteri, S. aurantiacus. S. heringianus.* K—anterior end, showing penes.

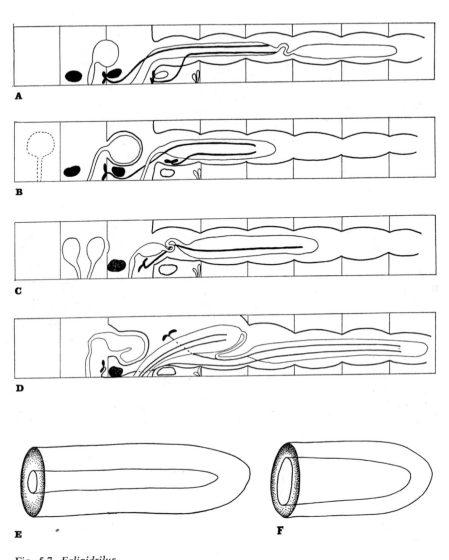

Fig. 5.7. *Eclipidrilus.*
Eclipidrilus spp. A to D—reproductive organs.
A—*E. frigidus,* B—*E. lacustris* (all median), C—*E. asymmetricus* (all median), D—
E. palustris and *E. daneus.*
E. palustris. E—optical section, distal end of atrium.
E. daneus. F—the same.

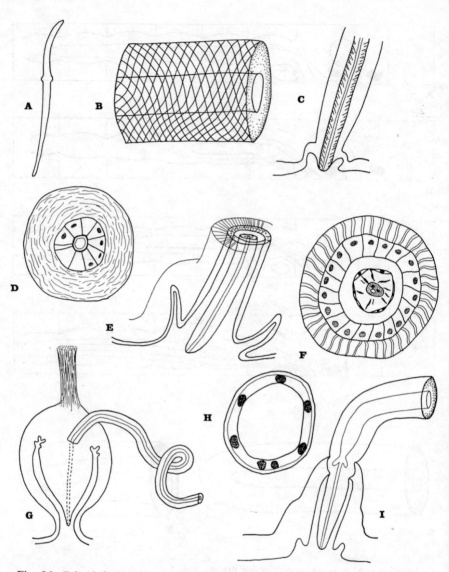

Fig. 5.8. *Eclipidrilus* (cont'd.) and *Rhynchelmis*.
E. lacustris. A—seta, B—spiral muscle of atrium, C—longitudinal section of penis,
  D—transverse section of penis.
E. daneus. E—longitudinal section of penis, F—transverse section of penis.
R. rostrata. G—penis and part of atrium, H—distal end of atrium, transverse section,
  I—longitudinal section of penis.

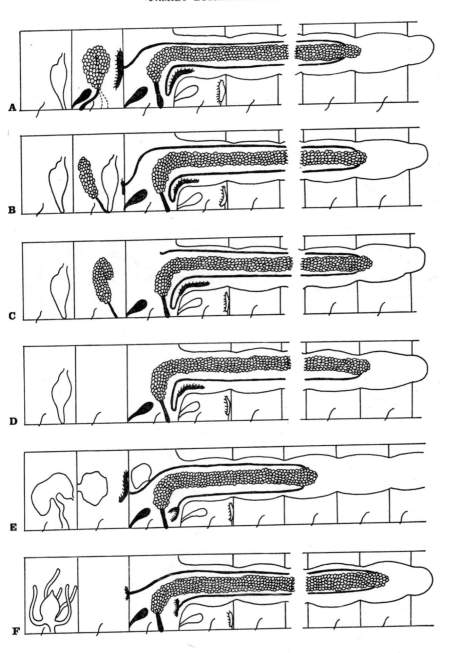

Fig. 5.9. *Rhynchelmis* (cont'd.).
Rhynchelmis spp. A to F—reproductive organs, A—*R. limosella*, B—*R. tetratheca*,
(? 2 pairs of testes, not shown in illustration), C—*R. brachycephala*, D—*R. elrodi*,
E—*R. vejdovskyi*, F—*R. rostrata*.

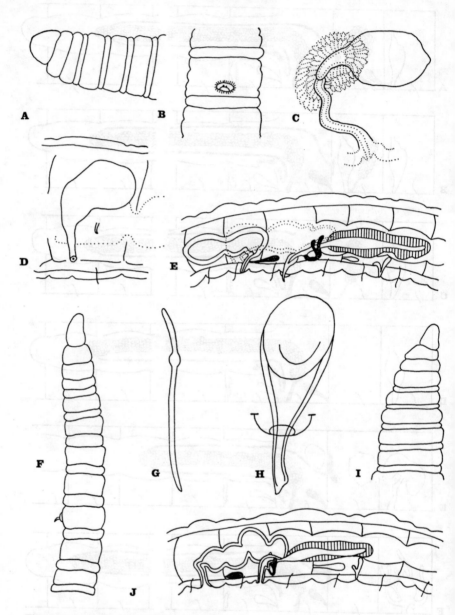

Fig. 5.10. *Tatriella* and *Styloscolex*.
T. slovenica. A—anterior end, B—male pore, C—atrium, D—spermatheca.
S. asymmetricus. E—reproductive organs, F—anterior end, G—seta, H—penis.
S. chorioidalis. I—anterior end, J—reproductive organs.

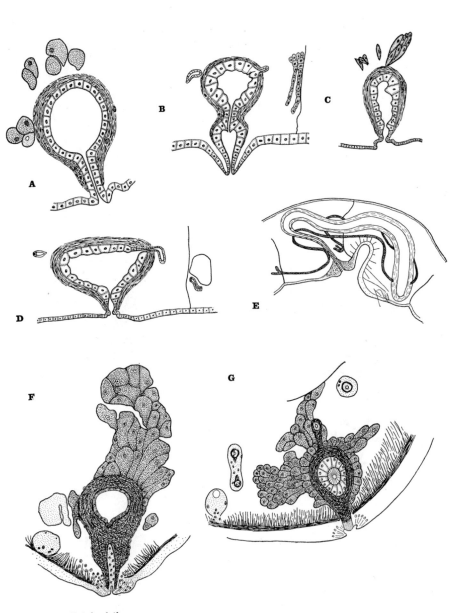

Fig. 5.11. *Trichodrilus.*
Trichodrilus spp. A to G—atria, A—*T. allobrogum*, B—*T. cantabrigiensis*, C, D—*T. hrabei*, transverse and longitudinal section, E—*T. cernosvitovi*, F—*T claparedei*, G—*T. icenorum.*

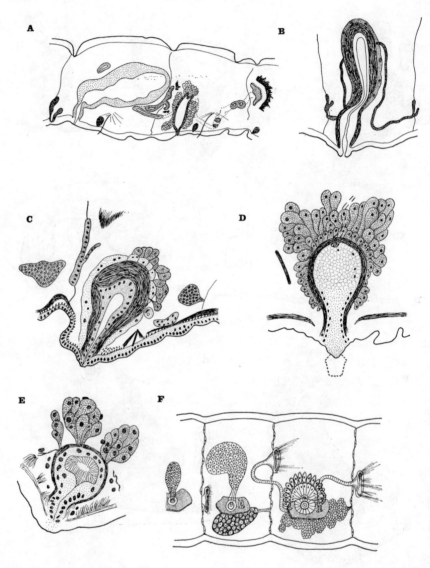

Fig. 5.12. *Trichodrilus* (cont'd.).
Trichodrilus spp. A to F—atria, A—*T. intermedius*, B—*T. leruthi*, C—*T. macroporo-phorus*, D—*T. medius*, E—*T. moravicus*, F—*T. pragensis*.

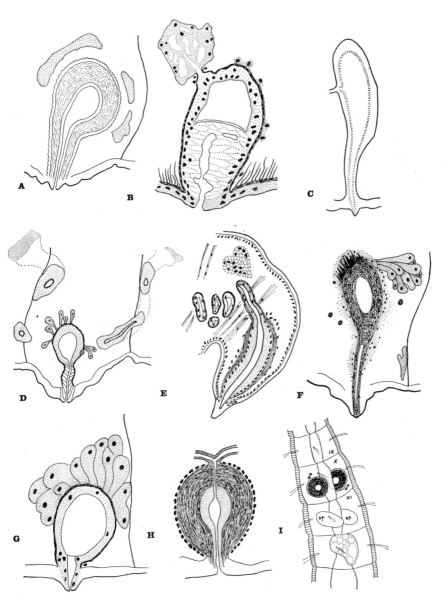

Fig. 5.13. *Trichodrilus* (cont'd.) and *Bichaeta.*
Trichodrilus spp. A to G—atria, A—*T. ptujensis*, B—*T. sketi*, C—*T. stammeri*, D—*T. strandi*, E—*T. tacensis*, F—*T. tatrensis*, G—*T. tenuis.*
B. sanguinea. H—atrium, I—reproductive organs (sp—spermathecae, ♂—atria).

REFERENCES

References enclosed in brackets were unavailable to the author.

ALTMAN, L. C. 1936. Oligochaeta of Washington. *Univ. Wash. Pubs Biol.*, **4**, 1.

BEDDARD, F. E 1891 Anatomical descriptions of two new aquatic Oligochaeta. *Trans. R. Soc. Edinb.*, **36**, 273.

— 1892. A contribution to the anatomy of *Sutroa*. *Trans. R. Soc. Edinb.*, **37**, 195.

— 1895. *A monograph of the order of Oligochaeta*. Oxford.

— 1908. A note on the occurrence of a species of *Phreatothrix* (Vejd.) in England, and on some points in its structure. *Proc. Zool. Soc. Lond.*, **1908**, 365.

— 1920. On the genus *Trichodrilus* and on a British species of the genus. *Ann Mag. nat. Hist.*, **6**, 227.

BENHAM, W. B. 1891. Notes on some aquatic Oligochaeta. *Q. Jl microsc. Sci.*, **33**, 187.

BOVERI-BONER, Y. 1920. Beiträge zur Vergleichenden Anatomie der Nephridien niederer Oligochäten. *Vjschr naturf. Ges Zurich*, **65**, 506.

BRETSCHER, K. 1900. Südschweizerische Oligochaeten. *Revue suisse Zool.*, **8**, 435.

— 1903. Beobachtungen über die Oligochaeten der Schweiz. *Revue suisse Zool.*, **11**, 1.

— 1905. Beobachtungen über die Oligochaeten der Schweiz. *Revue suisse Zool.*, **13**, 663.

BRINKHURST. R. O. 1960. Introductory studies on the British Tubificidae (Oligochaeta). *Arch. Hydrobiol.*, **56**, 395.

— 1962. A check list of British Oligochaeta. *Proc. Zool. Soc. Lond.*, **138**, 317.

— 1963a. A guide for the identification of British Aquatic Oligochaeta. *Scient. Publs Freshwat. biol. Ass.*, **22**, 1.

— 1963b. The aquatic Oligochaeta recorded from Lake Maggiore with notes on the species known from Italy. *Memorie Ist. ital. Idrobiol.*, **16**, 137.

— 1964. Observations on the biology of lake-dwelling Tubificidae. *Arch. Hydrobiol.*, **60**, 385.

— 1965a. A revision of the genera *Stylodrilus* and *Bythonomus* (Oligochaeta, Lumbriculidae). *Proc. Zool. Soc. Lond.*, **144**, 431.

— 1965b. A taxonomic revision of the Phreodrilidae (Oligochaeta). *J. Zool., Lond.*, **147**, 363.

— 1966. A contribution towards a revision of the aquatic Oligochaeta of Africa. *Zool. Afr.*, **2**, 131.

BRINKHURST, R. O., and COOK, D. G. 1966. Studies on North American aquatic Oligochaeta. III. Lumbriculidae and additional notes and records of other families. *Proc. Acad. nat. Sci. Philad.*, **118**, 1.

BRINKHURST, R. O., and KENNEDY, C. R., 1962. A report on a collection of aquatic Oligochaeta deposited at the University of Neuchâtel by Dr. E. Piguet. *Bull. Soc. Neuchâtel. Sci. nat.*, **85**, 183.

BUROW, V. S. 1931. The Oligochaetes of the Baikal area (II) three new species of *Styloscolex* from Lake Baikal. *Izv. biologo-geogr. nauchno-issled. Inst. Irkutsk*, **5**, 79 (in Russian).

BUROW, V. S., and KOZKOV, M. 1932. Concerning the distribution of bottom fauna in the Maloye More of Baikal. *Trudy vist.-sib. gos. Univ.*, **1**, 60 (in Russian with German summary).

CAUSEY, D. 1953. Microdrili in artificial Lakes in Northwest Arkansas. *Am. Midl. Nat.*, **50**, 420.

CEKANOVSKAYA, O. V. 1956. Contribution to the Oligochaet-fauna in the Enisei-basin. *Zool. Zh.*, **35**, 657 (in Russian with English summary).

— 1962. Aquatic Oligochaetous worms of the fauna of the U.S.S.R. *Opred Faune SSSR*, **78**, 3 (in Russian).

CERNOSVITOV, L. 1941a. Oligochaeta from various parts of the world. *Proc. Zool. Soc. Lond.*, **111**, 197.

— 1941b. Oligochaeta from Tibet. *Proc. Zool. Soc. Lond.*, **111**, 281.

CLAPARÈDE, E. 1862. Recherches sur les Oligochètes. *Mém. Soc. Phys. Hist. nat. Genève,* **16,** 217.

COOK, D. G. 1966. In Brinkhurst, R. O., and Cook, D. G. 1966.

— 1967. Studies on the Lumbriculidae (Oligochaeta) in Britain. *J. Zool., Lond.,* **153,** 353.

— 1968. The genera of the family Lumbriculidae and the genus *Dorydrilus* (Annelida, Oligochaeta) *J. Zool. Lond.,* **156,** 273.

CZERNIAVSKY, V. 1880. Materialia ad Zoographiam Ponticam Comparatuam. *Byull. mosk. Obshch. ispyt. Prir.,* **55,** 291.

DALYELL, J. G. 1853. *The powers of the creator displayed in the creation; or, observations on life admidst the various forms of the humbler tribes of animated nature; with practical comments and illustrations.* London.

DELLA CROCE, N. 1955. The conditions of sedimentation and their relations with oligochaete populations of Lake Maggiore. *Memorie Ist. ital. Idrobiol. (supl. 8),* 39.

DIEFFENBACH, O. 1886. Anatomische und Systematische Studien an Oligochaetae luniculae. *Ber. oberhess. Ges. Nat.-u. Heilk.,* **1885,** 65.

DITLEVSEN, A. 1904. Studien an Oligochaeten. *Z. wiss. Zool.,* **77,** 398.

(EISEN, G. 1879. On Oligochaeta collected during the Swedish expeditions to the Arctic regions in the years 1870, 1875 and 1876. *Kongl. Svenska Vetenakaps-Akad. Handl.,* (7) **15,** 1.)

— 1881. Eclipidrilidae and their anatomy. A new limicolide Oligochaeta. *Nova. Acta R. Soc. Scient. upsal.,* **11,** 1.

— 1888. On the anatomy of *Sutroa rostrata,* a new annelid of the family Lumbriculina. *Mem. Calif. Acad. Sci.,* (2) **1,** 1.

— 1892. Anatomical notes on *Sutroa alpestris,* a new Lumbriculide Oligochaete from Sierra Nevada. *Zoe,* **2,** 322.

— 1895. Pacific Coast Oligochaeta. *Mem. Calif. Acad. Sci.,* (2) **1,** 1.

FAUVEL, P. 1903. Un nouvel Oligochète de puits (*Trichodriloides intermedius* n.g. n. spc.) *C. r. Ass. fr. Avanc. Sci.,* **32,** 221.

FRIEND, Rev. H. 1896. New and little-known Oligochaeta. *Naturalist, Hull,* **1896,** 141.

— 1914. Some East Sussex Oligochaeta. *Zoologist,* **18,** 81.

— 1918. Oligochaete worms of Lancashire and Cheshire. *Lancs. Chesh. Nat.,* **11,** 121.

GALLOWAY, T. W. 1911. The common fresh water Oligochaeta of the United States. *Trans. Am. microsc. Soc.,* **30,** 285.

GAVRILOV, C. 1935. Contributions a l'étude de l'autofécondation chez lez Oligochètes. *Acta Zool. Stockh.,* **16,** 21.

GRUBE, E. 1844. Über den *Lumbricus variegatus* Müller's und ihm verwandte Anneliden. *Arch. Naturgesch.,* **10,** 198.

— 1873. Über einige bisher unbekannte Bewohner des Baikalsees. *Iber schles. Ges. vaterl. Kult.,* **50,** 66.

— 1879. Untersuchungen über die physikalische Beschaffenheit und die Flora und Fauna der Schweizerseen. *Jber. schles. Ges. vaterl. Kult.,* **56,** 115.

— 1880. Neue Ermittlungen uber die Organisation von *Bythonomus lemani. Jber schles. Ges. vaterl. Kult.,* **57,** 228.

HAFFNER, K. 1928. Untersuchungen über die Morphologie und Physiologie des Blutgefäss—systems von *Lumbriculus variegatus.* Müll. *Z. wiss. Zool.* **130,** 1.

HESSE, E. 1909. Contribution a l'étude des Monocystidees des Oligochètes. *Archs Zool. exp. gén.,* (5) **3,** 27.

HESSE, R. 1894. Die Geschlechtsorgane von *Lumbriculus variegatus.* Gr. *Z. wiss. Zool.,* **58,** 355.

— 1902. Zur Kenntnis der Geschlechtsorgane von *Lumbriculus variegatus. Zool. Anz.,* **25,** 620.

(HOFFMEISTER, W. 1842. De vermibus quibusdam Genus Lumbricorum pertinentibus (Dissertatio inaugur. 4). Berolini.)

— 1843. Beiträge zur Kenntniss deutscher Landanneliden. *Arch. Naturgesch.,* **9,** 183.

HRABE, S. 1927. *Rhynchelmis Komáreki,* eine neue Lumbriculiden Art aus Macedonien. *Zool. Anz.,* **71.** 170.

— 1928. Přispevek k poznáni moravských tubificid a lumbriculid. *Biol. Listy,* **14,** 1.

HRABE, S. 1929a. Zwei neue Lumbriculiden-Arten, sowie einige Bemerkungen zur Systematik eineger bereits bekannter. *Zool. Anz.*, **84**, 9.

— 1929b. Lamprodrilus mrázeki, eine neue Lumbriculiden-Art (Oligo.) aus Böhmen. *Zool. Jb.* (Syst.), **57**, 197.

— 1929c. Lamprodrilus Michaelseni, eine neue Lumbriculiden-Art aus Mazedonien. *Arch. Hydrobiol.*, **20**, 163.

— 1930. Příspevek k poznáni oligochaet z jezera Janiny, jeho okoli a z ostrova Korfu. *Mem. Soc. r. Sci. Boheme*, **2**, 1 (in Czech).

— 1931. Die Oligochaeten aus den Seen Ochrida und Prespa. *Zool Jb.* (*Syst.*), **61**, 1.

— 1932. Bythonomus sulci, n. sp. nový jeskynní sporostetinatý cerv. *Příroda, Brno*, .25, 1 (in Czech with French summary).

— 1935. Über *Moraviodrilus pygmaeus* n. gen., n. sp., *Rhyacodrilus falciformis* Br., *Ilyodrilus bavaricus* Oschm. und *Bothrioneurum vejdovskyanum* St. *Spisy vydáv. přír. Fak. Masaryk Univ. No.* **209**, 1.

— 1936a. Über Dorydrilus (*Piguetia*) *mirabilis* n. sub. gen., n. sp. aus einem Sodbrunnen in der Umgebung von Basel sowie über Dorydrilus (*Dorydrilus*) *michaelseni* Pig. und *Bichaeta sanguinea* Bret. *Spisy vydáv, přír. Fak. Masaryk. Univ. No.* 227, 3.

— 1936b. *Trichodrilus strandi* n. sp. ein neuer Vertreter der Höhlen-Lumbriculiden. *Festschrift j. Embrik Strand Riga*, **1**, 404.

— 1937a. Příspevek k poznáni žižalice *Lamprodrilus mrázeki* a nitenek rodu *Aulodrilus. Sb. Klubu přír. Brne*, **19**, 1 (in Czech with German summary).

— 1937b. Contribution a l'étude du genre *Trichodrilus* (Oligo. Lumbriculidae) et description de deux espèces nouvelles. *Bull. Mus. r. Hist. nat. Belg.*, (No. 32) **13**, 1.

— 1937c. Příspevek k poznáni zvířeny Ktálickéko Snežniku. *Sb. Klubu přír Brne*, **20**, 1 (in Czech).

— 1938. *Trichodrilus moravicus* und *claparedei*, neue Lumbriculiden. *Zool. Anz.*, **15**, 73.

— 1939a. Vodni Oligochaeta z Vysokých Tater. *Věst. čsl. zool. Spol.*, **6-7**, 209 (in Czech with French resume).

— 1939b. Bentinkà zvířena tatranských jezer. *Sb. Klubu přír. Brne*, **22**, 1 (in Czech).

— 1939c. O vyvoji samicho vyvodneho aparatu u nekterych nitenck a zizalic. *Sb. přír Klubu Třebici*, **3**, 56 (in Czech with German summary).

— 1941. K. poznáíí dunajských Oligochaet. *Acto Soc. Sci. nat. moravo-siles.*, **13**, 1 (in Czech).

— 1942. Poznanky a zvirene be studní a pramenu no Slovensku. *Sb. Klubu přír. Brne*, **24**, 23 (in Czech with German summary).

— 1950. Oligochaeta Kaspického jezera. *Acta Acad. Sci. nat. moravo-siles.*, **22**, 251 (in Czech with English Summary).

— 1952. Oligochaeta. 2 Limnicola. *Zoology Iceland*, (20b) **2**, 1.

— 1954a. *Trichodrilus macroporophorus* n. sp. eine neue Lumbriculiden-Art aus Österreich. *Zool. Anz.*, **153**, 185.

— 1954b. Klíc zvířeny CSR. 1. Pragu. (in Czech.)

— 1958. Die Oligochaeten aus den Seen Dojran und Skadar. *Spisy vydáv. přír. Fak. Masaryk. Univ. No.* **397**, 337.

— 1960. Oligochaeta limicola from the collection of Dr. S. Husmann. *Spisy přír. Fak. Univ. Brne. No.* **415**, 245.

— 1961. Dva nové druhy rodu *Rhynchelmis* ze Slovenska. *Spisy přír. Fak. Univ. Brne. No.* **421**, 129 (in Czech with English summary).

— 1962 Oligochaeta limicola from Onega Lake collected by Mr. B. M. Alexandrov. *Spisy přír. Fak. Univ.. Brne, No.* **435**, 277.

— 1963. Oligochaeta limicola from Slovenija. *Biol. Vestnik*, **11**, 67.

HRABE, S., and CERNOSVITOV, L. 1925. Un nouveau représentant europeén du genre *Rhynchelmis* Hoffm. *Rhynchelmis Vejdovskyi. Mém Soc. r. Sci. Bohème*. **2**, 1.

— 1926. Über eine neue Lumbriculiden-Art (*Rhynchelmis Vejdovskyi* n. sp.) *Zool. Anz.*, **65**, 265.

— 1927. Über eine neue Lumbriculiden-Gattung *Anastylus parvus* n.g. n. sp. aus Karpathorussland. *Zool. Anz.* 71: 203.

ISOSSIMOV, V. V. 1926. Zur Anatomie des Nervensystems der Lumbriculiden. *Zool. Jb.* (*Anat.*), **48**, 365.

— 1948. The Lumbriculidae of Lake Baikal. Synopsis of Doctors Thesis, Kazan. (In Russian.)

— 1962. The Oligochaetes of the family Lumbriculidae of Lake Baikal. *Trans. limn. Inst., Acad. Sci. U.S.S.R. Siberian Sect.*, **1** (21) (1), 3. (In Russian.)

JANDA, V. 1924. Die Regeneration der Geschlechtsorgane bei *Rhynchelmis limosella* Hoffm. *Zool. Anz.*, **59**, 257.

JOHNSTON, G. 1865. *A catalogue of British non-parasitical worms in the collection of the British Museum*, London.

KENK, R. 1949. The animal life of temporary and permanent ponds in Southern Michigan. *Misc. Publs Mus. Zool. Univ. Mich.*, **71**, 5.

(KESSLER, K. 1868. (The fauna of Lake Onega. Oligochaeta.) *Trudy I-go. S. R. E. St. Petersburg*, **1868**, 102.)

KINDRED, J. E. 1918. A representative of the genus *Trichodrilus* from Illinois. *Bull. Ill. St. Lab. nat. Hist.*, **13**, 49.

LASTOCKIN, D. 1937. New spp. of Oligochaeta limicola in the European part of the U.S.S.R. *Dokl. Akad. Nauk. SSSR*, **17**, 233.

LEIDY, J. 1850. Descriptions of some American Annelida abranchia. *J. Acad. nat. Sci. Philad.*, (2) **2**, 43.

MARTIN, C. H. 1907. Notes on some Oligochaeta found on the Scottish Loch Survey. *Proc. R. Soc. Edinb.*, **28**, 21.

MENGE, A. 1845. Zur Rothwürmer-Gattung *Euaxes*. *Arch. Naturgesch.*, **11**, 24.

MICHAELSEN, W. 1900. Oligochaeta. *Tierriech.*, **10**, 1.

— 1901a. Oligochaeten der Zoologischen Museen zu St. Petersburg und Kiev. *Izv. imp. Akad. Nauk*, (5) **15**, 145.

— 1901b. Fauna Oligochaet'Bajkala. 50 jähriges Jubiläum der ostsibirischen Abteilung der Kaiserl. Russischen Geographischen Gesellschaft, Jubiläums-Festschrift, red. von A. V. Korotneff, Kiev.

— 1902a. Die Oligochaeten Fauna des Baikal-Sees. *Verh. naturw. Ver. Hamb.*, (3) **9**, 43.

— 1902b. Die Fauna des Baikal-Sees. *Verh. naturw. Ver. Hamb.*, (3) **9**, 17.

— 1903a. *Die geographische Verbreitung der Oligochaeta*, Berlin.

— 1903b. Neue Oligochaeten und neue Fundorte altbekannter. *Mitt. naturh. Mus. Hamb.*, **19**, 1.

— 1905. Die Oligochaeten des Baikal-Sees. Wissenshaftliche Ergebnisse einer Zoologischen Expedition nach dem Baikal-See unter Leitung des Profs. Alexis Korotneff in den Jahren 1900-1902. Kiev and Berlin.

— 1909. Oligochaeta: in BRAUER, Die Süsswasserfauna Deutschlands, **13**, 1.

— 1920. Zur Stammesgeschichte und Systematik der Oligochaten insbesondere der Lumbriculidae. *Arch. naturg.* 1920 (A) **8**, 130.

— 1928. Vermes Polymers (8) *in Handb. Zool.* (2) (2) **2**, 1.

— 1933. Über Höhlen Oligochäten. *Mitt. Höhl.-u. Karstforsch.* **1933**, 1.

MICHAELSEN, W. and VERESCHAGIN, G. 1930. Oligochaeten aus dem Selenga-Gebiete des Baikalsees. *Trudy Kom. Izuch. Ozera Baikala* **3**, 213.

MOORE, J. P. 1906. Hirudinea and Oligochaeta collected in the Great Lakes region. *Bull. Bur. Fish., Wash.* **25**, 153.

MOSZYNSKI, A. 1936. Ein neuer Vertreter der Gattung *Trichodrilus* Clap. (*Trichodrilus spelaeus* n. sp.) aus dem Stollen in Neu-Klessengrund. *Beitr. Biol. Glatzer Schneeberges*, **2**, 214.

MRAZEK, A. 1900. Die Samentaschen von *Rhynchelmis*. *Sber. K. Böhm Ges. Wiss.*, **35**, 1.

— 1906. Die Geschlechtsverhältnisse und die Geschlechtsorgane von *Lumbriculus variegatus*. Gr. *Zool. Jb.* (*Anat.*) 23, 382.

— 1913a. Beiträge zur Naturgeschichte von *Lumbriculus variegatus*. *Sber K. böhm. Ges. Wiss.*, **14**, 1.

— 1913b. Enzystierung bei einem Süsswasseroligochaeten. *Biol. Zbl.*, **33**, 658.

— 1926. K. biologii rodu *Stylodrilus*. *Vest ceske Akad. Ved. Umeni*, **5**, 1.

— 1927. *Euaxes obtusirostris* Menge 1845. *Mem. Soc. r. Sci. Boheme*, **1**, 1.

MÜLLER, O. F. 1774 *Vermium terrestrium et fluviatilium, Hevniae et Lipsiae 1773-74*.

PERCIVAL, E., and WHITEHEAD, H. 1930. Biological survey of the River Wharfe. *J. Ecol.*, **18**, 285.

PIERANTONI, U. 1904. Sopra alcuni Oligocheti raccolti nel fiume Sarno. *Annuar. R. Mus. zool. R. Univ. Napoli No.* **26**, 1.

— 1905. Oligocheti del fiume Sarno. *Archo zool. ital.*, **2**, 227.

PIGUET, E. 1905. Le *Bythonomus lemani* de Grube. *Revue suisse Zool.*, **13**, 617.

— 1913. Notes sur les Oligochaetes. *Revue suisse Zool.*, **21**, 111.

PIGUET, E., AND BRETSCHER, K. 1913. Oligochètes. *In Catalogue des Invertébrés de la Suisse*, **7**, 1.

RATZEL, F. 1869. Beiträge zur anatomischen und systematischen Kenntnis der Oligochaeten. *Z. wiss. Zool.*, **18**, 563.

RAVERA, O. 1955. Amount of mud displaced by some fresh water Oligochaeta in relation to depth. *Memorie Ist. ital. Idrobiol. Supl.*, **8**, 247.

SAPKAREV, J. 1961. Oligochaetenansidlung der Gehängägewasser des Ohrider Talkessels. (Abstract in *Bull. Sci.*, **6**, 9.)

(SCHWEIGGER, 1820. In *Handb. Naturh.*, **1820**, 590.)

SMITH, F. 1895. Notes on species of North American Oligochaeta. *Bull. Ill. St. Lab. Nat. Hist.*, **4**, 292.

— 1896. Notes on species of North American Oligochaeta II. *Bull. Ill. St. Lab. Nat. Hist.*, **4**, 396.

— 1900a. Notes on species of North American Oligochaeta III. List of species found in Illinois and descriptions of Illinois Tubificidae. *Bull. Ill. St. Lab. Nat. Hist.*, **5**, 441.

— 1900b. Notes on species of North American Oligochaeta IV. On a new lumbriculid genus from Florida, with additional notes on the nephridial and circulatory systems of *Mesoporodrilus asymmetricus* Smith. *Bull. Ill. St. Lab. Nat. Hist.*, **5**, 459.

— 1905. Notes on species of North American Oligochaeta V. The systematic relationships of *Lumbriculus (Thinodrilus) inconstans* (Smith). *Bull. Ill. St. Lab. Nat. Hist.*, **7**, 45.

— 1919. Lake Superior lumbriculids, including Verrill's *Lumbricus lacustris*. *Proc. biol. Soc. Wash.*, **32**, 33.

SMITH, F., and DICKEY, L. B. 1918. A new species of *Rhynchelmis* in North America. *Trans. Am. microsc. Soc.*, **37**, 207.

SMITH, F., and GREEN, B. R. 1916. The Porifera, Oligochaeta and certain other groups of invertebrates in the vicinity of Douglas Lake, Michigan. *Rep. Mich. Acad. Sci.*, **17**, 81.

SMITH, F., and WELCH, P. S. 1919. Oligochaeta collected by the Canadian Arctic Expedition 1913-18. *Rep. Can. arc. Exped.*, **9**, 1.

SMITH, S. I., and VERRILL, A. E. 1871. Notice of the invertebrata dredged in Lake Superior in 1871 by the U.S. Lake Survey, under the direction of Gen. C. B. Comstock, S. E. Smith, naturalist. *Amer. J. Sci. Arts*, **3**, 448.

SOKOLSKAYA, N. L. 1967. New species of the genus *Lumbriculus* Grube (Lumbriculidae, Oligochaeta) from South Sarkhalin Reservoirs. *Byull. mosk Obshch Ispyt Přír*, **72**, 40.

SOUTHERN, R. 1909. Contributions towards a monograph of the British and Irish Oligochaeta. *Proc. R. Ir. Acad.*, **27**, 119.

STEPHENSON, J. 1922. On some Scottish Oligochaeta, with a note on encystment in a common freshwater oligochaete, *Lumbriculus variegatus* (Müll). *Trans. R. Soc. Edinb.*, **53**, 277.

— 1930. The Oligochaeta. Oxford.

SVETLOV, P. von 1936. *Lamprodrilus isoporus* Mich. (Lumbriculidae) aus dem Lodoga und dem Onegasee. *Zool. Anz.*, **113**, 87.

VAILLANT, L. 1889. *Histoire naturelle des annelés marins et d'eau douce.* Vol. 3. Lombriciniens, Hirudiniens, Bdellomorphes, Térétulariens et Planariens, Paris.

VEJDOVSKY, F. 1875. Beiträge zur Oligochaetenfauna Böhmens. *Sber. K. böhm Ges. Wiss.*, **1875**, 191.

— 1876a. Anatomische Studien an *Rhynchelmis limosella*, Hoffm. (*Euaxes filirostris*, Grub.). *Z. wiss. Zool.*, **32**, 332.

VEJDOVSKY, F. 1876b. Über *Phreatothrix* eine neue Gattung der Limicolen (ein Beitrag zur Brunnenfauna von Prag.). *Z wiss Zool.*, **27**, 541.

— 1877. O. faune vod studicných. Casopis ceskeho Muses.)

— 1883. Revisio Oligochaetorum Bohemiae. *Sber. K. böhm. Ges. Wiss.*, **1883** (1884), 215.

— 1884. *System und Morphologie der Oligochaeten*. Prag.

— 1895. Zur Kenntnis des Geschlechtsapparates von *Lumbriculus variegatus*. *Z. wiss Zool.*, **59**, 80.

WAGNER, F. von. 1900. Beiträge zur Kenntnis der Reparationsprozesse bei *Lumbriculus variegatus*. *Zool. Jb.* (*Anat.*), **13**, 603.

— 1906. Zur Oekologie des *Tubifex* und *Lumbriculus*. *Zool. Jb.* (*Syst.*), **23**, 295.

WELCH, P. S. 1921. Oligochaeta collected in Greenland by the Croker Lane Expedition. *Bull. Am. Mus. nat. Hist.*, **44**, 269.

WENIG, J. 1902. Beiträge zur Kenntnis der Geschlechtsorgane von *Lumbriculus variegatus*. *Sber. K. böhm. Ges. Wiss.*, **14**, 1.

YAMAGUCHI, H. 1936. Studies on the aquatic Oligochaeta of Japan. I. Lumbriculids from Hokkaido. *J. Fac. Sci. Hokkaido Univ.*, **5**, 73.

— 1937a. Studies on the aquatic Oligochaeta of Japan. III. A description of *Lumbriculus multiatriatus* n. sp. with remarks on the distribution of genital organs in the Lumbriculidae. *J. Fac. Sci. Hokkaido Univ.*, **6**, 1.

1937b. Studies on the aquatic Oligochaeta of Japan. IV. *Styloscolex japonicus* n. sp. *Annotnes zool. Jap.*, **16**, 167.

— 1953. Studies on the aquatic Oligochaeta of Japan. VI. A systematic report, with some remarks on the classification and phylogeny of the Oligochaeta. *J. Fac. Sci. Hokkaido Univ.*, **11**, 277.

ZSCHOKKE, F. 1891. Weiterer Beitrag zur Kenntnis der Fauna von Gebirgsseen, *Zool. Anz.*, **14**, 119.

6

FAMILY HAPLOTAXIDAE

Type genus: Haplotaxis HOFFMEISTER

Setae single or closely paired, S-shaped or distally hooked, dorsals sometimes smaller than ventrals, sometimes absent posteriorly, or even completely absent. Genital setae in some species. Body wall with cuticle thick in at least some species, longitudinal muscle thick. Some anterior septa may be thickened. Meganephridia from a few pregonadal segments to the distal end except for the gonadal segments, pores near ventral setae. Gut simple with an eversible pharynx or a ring of muscles around the pharynx forming a gizzard. Septal glands in those species with eversible pharynx. Dorsal and ventral blood vessels connected by a single pair of long commissural vessels often in each segment. Clitellum of a single layer of cells, mostly in the region of the genital pores.

Testes in X and XI (or IX and X), ovaries in XII orXII and XIII (or XI and XII). Spermathecae one to four pairs, anterior to the gonads, simple, without diverticulae. Spermathecal pores frequently lateral or in the line of the dorsal setae. Male pores two pairs, ventro-lateral or lateral, very small and frequently invisible externally, on segments XI and XII (or on X and XI) or both on XII. Female pores in 12/13, 12/13 and 13/14 (or 11/12 and 12/13). Male efferent ducts very short or short and slightly coiled, without atria or prostate glands, but distal end lined with cuticle.

Cosmopolitan.

The family was recently revised by Brinkhurst (1966).

GENUS **Haplotaxis** HOFFMEISTER, 1843
Fig. 6.1

Type species: Lumbricus gordioides HARTMAN
Defined and distributed as the family.

Haplotaxis and *Pelodrilus* were both created as monospecific genera, apparently differing widely in character. The third species to be described, *H. smithii*, was placed in *Haplotaxis* as the male pores were on separate segments, i.e. not all located on a single segment as in the solitary species of *Pelodrilus*. The position of this species was never subsequently questioned, despite the fact that, with the addition of several species between 1900 and 1910, it should clearly have been placed in *Pelodrilus* rather than *Haplotaxis* as formerly defined, as it has paired setae and no gizzard. The description of new species gradually widened the original generic limits, particularly with respect to the arrangement and number of the reproductive organs. Despite this, no critical examination of the generic limits was carried out prior to the general revision of the Oligochaeta by Stephenson (1930). By then, considerable doubt about the justification for retaining the two genera had been expressed by Michaelsen (1907) and Benham (1909), and both Stephenson (1930) and, subsequently, Hrabe (1958) echoed this doubt. The sole criterion upon which the recognition of *Pelodrilus* was based was the absence of the gizzard, held to be characteristic of *Haplotaxis*. The description of *H. smithii* contained no reference to a gizzard, but re-examination of the type specimens has now established that the pharynx is of the eversible type and that septal glands are present.

The type material of *H. vermivorus* in the Zoological Museum, Amsterdam, consists of some fragments, which clearly constitute the residue of the original specimen, the anterior end of which has been sectioned. The sections have not been traced to date and so I have not been able to illustrate the form of the pharynx in this species. However, the description clearly refers to the pharynx as being eversible, consisting of a thickened dorsolateral pad as in the Enchytraeidae but this was attributed to post-mortem changes by Michaelsen (1932).

The significance of these facts lies in their relationship to the combination of the other characters of these species. The prostomium of *H. smithii* bears a furrow, all four pairs of gonads are present (as in *Haplotaxis*) but the setae are paired, and all of a size. In *H. vermivorus* the setae are absent dorsally, sickle-shaped and single ventrally, and the prostomium is grooved, but there is only a single pair of ovaries. These two species are, therefore, intermediate between *Haplotaxis* and *Pelodrilus*, and the last justification for the recognition of *Pelodrilus* has been removed.

A third genus, *Heterochaetella*, was established by Yamaguchi (1953) by reference to the number of pairs of spermathecae, the absence of a gizzard and the presence of bifid setae. The existing genera were accepted without comment, and as the present findings indicate that the retention of *Pelodrilus* can no longer be justified, *Heterochaetella* can only be separated from

Haplotaxis by the single setal character. It is accordingly merged in *Haplotaxis*.

The most recent genus to be established was defined by Cekanovskaya (1959) under the name *Adenodrilus*. I have found no mention of *Heterochaetella* in this account, and the other two genera were accepted without detailed examination. The separation of *Adenodrilus* was largely based on the presence of bifid or pectinate setae and of large glands filling segments X to XIII. The setal character is of much less significance now that *Heterochaetella* is merged in *Haplotaxis*, and the glands differ only in size (so far as can be determined) from those of other *Haplotaxis* species. I therefore united all of the species within the single genus *Haplotaxis* (Brinkhurst, 1966).

Confusion arose over the use of the names *Haplotaxis* and *Phreoryctes* in the past, because the name *Haplotaxis* had been introduced into botanical literature. However, Michaelsen (1900) re-introduced the name *Haplotaxis* and his action was subsequently validated by the International Code of Zoological Nomenclature (Article 2).

Haplotaxis

1.	Some setae bifid or pectinate	2
—	All setae simple-pointed, often hooked	3
2.	Setae of each bundle one simple-pointed, one bifid	*H. glandularis*
—	All setae bifid or pectinate except perhaps some anterior ventral setae	*H. denticulatus*
3.	Setae all of similar size, paired, at least anteriorly, and present in dorsal and ventral bundles	4
—	Setae dissimilar, with large single ventral setae, dorsal setae small and frequently reduced in number or absent	11
4.	Setae paired anteriorly, single in mid and posterior bundles	*H. ignatovi*
—	Setae paired in all segments	5
5.	Setae ornamented or keeled distally	6
—	Setae unornamented	7
6.	Setae ornamented distally	*H. bureschi*
—	Setae keeled distally	*H. leruthi*

| 7. | Two pairs of ovaries | 8 |
| | One pair of ovaries | 10 |

| 8. | Two or three pairs of spermathecae, no copulatory glands | H. smithii |
| | One pair of spermathecae, copulatory glands in XI to XIII | 9 |

| 9. | Spermathecae elongate, in VII | H. hologynus hologynus |
| | Spermathecae rounded, in VIII | H. hologynus bipapillatus |

| 10. | Both pairs of male pores in XII close together | H. violaceus |
| | Male pores in XI and XII | H. africanus |

| 11. | One pair of ovaries | 12 |
| | Two pairs of ovaries | H. gordioides 14 |

| 12. | Gizzard absent | H. vermivorus |
| | Gizzard present, if vestigial | 13 |

| 13. | Dorsal setae present, smaller than ventrals, all sickle-shaped | H. heterogyne |
| | Dorsal setae absent, ventrals fairly straight, enlarging from II to XX, whereupon they rapidly decrease in size | H. gastrochaetus |

| 14. | Four pairs of spermathecae | H. gordioides ascaridoides |
| | Three pairs of spermathecae | 15 |

| 15. | Genital setae present | H. gordioides dubius |
| | Genital setae absent | H. gordioides gordioides |

Haplotaxis gordioides (Hartmann, 1821)
Fig. 6. 2A-C

Lumbricus gordioides HARTMANN, 1821: 45.

Tubifex uncinarius DUGÈS, 1837: 33, Pl. 1, fig. 28-30.*

Haplotaxis menkeana HOFFMEISTER, 1843: 193, Pl. IX, fig. 7.*

Phreoryctes menkeanus (Hoffmeister). HOFFMEISTER, 1845: 40*; LEYDIG, 1865: 249, Pl. XVI-XVII, figs. 9-11, 12a, 14-17, Pl. XVIII, figs. 18-22, 24-27.* VEJDOVSKY, 1884: 50. BEDDARD, 1890: 629; 1895: 189. FORBES, 1890b: 107.

Georyctes menkei SCHLOTTHAUBER, 1860: 122.*

Georyctes lichtensteini SCHLOTTHAUBER, 1860: 122.*

Nemodrilus filiformis CLAPARÈDE, 1862: 275, Pl. III, fig. 16.

Phreorcytes heydeni NOLL, 1873: 131.*

Phreoryctes heydeni Noll. NOLL, 1874: 260, Pl. 7, figs. 1-4.*

Lumbricogordius hartmanni HEYDEN (? Mss. in Noll).*

Phreoryctes filiformis (Claparède). VEJDOVSKY, 1875: 198*; VEJDOVSKY, 1884: 49, Pl. XII, figs. 3-9. BEDDARD, 1890: 630; 1895: 189.

Clitellio uncinarius VAILANT, 1889.*

Phreorcytes emissarius FORBES, 1890a: 477.

Phreorcytes emissarius Forbes. FORBES, 1890b: 108, Fig. 1-15. BEDDARD, 1895: 190.

Phreoryctes endeka GIARD, 1894: 310.*

?*Dichaeta curvisetosa* FRIEND, 1896: 110.

Phreoryctes endeka var *pachyderma* FERRONNIÈRE, 1899: 233, Pl. XIX, figs. 1-2.

Phreoryctes gordioides (Hartmann). MICHAELSEN, 1898.*

Phreoryctes gordioides (Hartmann). MICHAELSEN, 1899: 105; 1929: 328.

Haplotaxis gordioides (Hartmann) MICHAELSEN, 1900: 108; 1901: 201; 1903 a; 1903 b: 64; 1905 a: 63; 1925 b: 272. HESSE, 1923: 138. STEPHENSON, 1930: 804. ALTMAN, 1936: 72(?). YAMAGUCHI, 1937: 68, Pl. IV, fig. 1; 1953: 304. CEKANOVSKAYA, 1962: 322, Fig. 203. BRINK-HURST, 1966: 34.

?*Haplotaxis intermedia* PIERANTONI, 1904: 2.

Haplotaxis ascaridoides MICHAELSEN, 1905a: 63.

?*Haplotaxis intermedia* Pierantoni. PIERANTONI, 1905: 239, Fig. 15-22. HRABE, 1931: 51.

Haplotaxis emissarius (Forbes). SMITH, 1918: 43. STEPHENSON, 1930: 804. HRABE, 1931: 56.

Haplotaxis forbesi SMITH, 1918: 44.

Haplotaxis ascaridoides Michaelsen. MICHAELSEN, 1925 b: 274. STEPHENSON, 1930: 804. HRABE, 1931: 58. CEKANOVSKAYA, 1962: 323.

Haplotaxis forbesi Smith. STEPHENSON, 1930: 804.

* Seen in Michaelsen, 1899.

Phreoryctes dubius HRABE, 1931 : 51, Fig. 8, 9a-d.

?*Haplotaxis curvisetosa* (Friend). CERVOSVITOV, 1942 : 274.

$l=180$-400 mm, $w=0.3$-2.0 mm. Prostomium long, with a transverse groove. Segment I short, anterior segments biannulate. Cuticle with unicellular glands. Setae mostly single, four per segment (with partially developed replacement setae ventrally). Dorsal setae small, varying in extent (absent behind XI or as far as LXXX). Setal formula aa=al, al $\geq$ 1d, aa=$\frac{1}{2}$dd. Ventral setae large, more or less sickle-shaped. ? Genital setae. Clitellum XI to XXVIII. Spermathecal pores 6/7 to 8/9, lateral. Male pores in front of ventral setae of XI and XII, minute. Female pores in 12/13 and 13/14 in line of ventral setae. Fore-gut with a ring of circular muscle forming a simple gizzard in IV to V. No septal glands. Dorsal and ventral blood vessel linked by a pair of long commissural vessels in each segment. Nephridia IX to X, XV onwards, opening in front of ventral setae. Timm's glands in XI to XVI or every segment, mid-ventral beneath the nerve cord. Setal glands in XII to XIV, copulatory glands in XI to XIV. Testes and male funnels paired in X and XI. Males efferent ducts make several turns. Sperm sacs on 9/10 in IX, 10/11 in XI, 11/12 to XVIII. Ovaries and female funnels paired in XII, XIII. Egg-sac present, visible behind XVIII. Spermathecae in VII to IX (? in VI), ducts short, not clearly separable from ampullae.

Holarctic.

The supposed gonads of *H. forbesi* were re-investigated from the type material by Brinkhurst (1966), who showed that the organs concerned (in XV, XVI and even XVII) were probably glands associated with normal nephridia. The same features have since been observed in another immature N. American specimen which suggests that the original observation was not a freak.

Consideration of the differences between the type species and both *H. dubius* and *H. ascaridoides* led to the separation of three subspecies.

Haplotaxis gordioides gordioides (Hartmann, 1821). Three pairs of spermathecae. No genital setae, but setal glands on XII to XIV.

Holarctic.

Haplotaxis gordioides dubius (Hrabe, 1931). Three pairs of spermathecae. Genital setae and glands on XII to XIV.

Lake Ochrid.

Haplotaxis gordioides ascaridoides Michaelsen, 1905. Four pairs of spermathecae. No genital setae but setal glands on XII to XIV.

Lake Baikal.

Haplotaxis heterogyne Benham, 1903
Fig. 6. 2E

Haplotaxis heterogyne BENHAM, 1903 : 233, Fig. 23.

Haplotaxis heterogyne Benham. BENHAM, 1905 : 299, Fig. 1-46. STEPHENSON, 1930 : 804. BRINKHURST, 1966 : 37.

$l=20$ mm, $w=0.3$ mm, $s=60$. Prostomium very long, but not divided by a furrow. Setae four per segment, ventrals two to three times larger than the dorsals except in most anterior segments. Dorsal setae on all segments of the body. Clitellum XI to $\frac{1}{2}$XIV, complete, ring-like. Spermathecal pores anterior on VIII and IX lateral. Male pores not seen. Female pores XIII lateral to ventral setae. Gizzard in IV. No septal glands. Dorsal and ventral vessel linked by a pair of long, looping commissural vessels in each segment.

Nephridia in X, and from XIV onwards, pores near ventral setae. Timm's glands absent? Setal glands absent? "Copulatory glands" of XI to XIII paired, opening laterally near ventral setae in XII, but below the nerve cord in XI and XIII. Testes and male funnels paired in X and XI, anterior to the ventral setae. Sperm sacs on 10/11 reaches XII, and on 11/12 reaches the anterior end of XIII. Ovaries and female funnels paired in XII. Egg sac median on 12/13 reaches XIV. Spermathecae paired in VIII and IX.

Lake Wakatipu, S. Island, New Zealand.

Two worms at a depth of 550 feet.

The species badly needs re-investigation as several features of the anatomy are inadequately described in the literature.

Haplotaxis gastrochaetus (Yamaguchi, 1953)
Fig. 6. 2D

Haplotaxis gastrochaetus YAMAGUCHI, 1953: 304-6; Figs. 13, 14. Plate 7, fig. 8.

Haplotaxis gastrochaetus Yamaguchi. BRINKHURST, 1966: 38.

$l=50\text{-}72$ mm, $w=0.6$ mm, $s=93\text{-}119$. No transverse furrow on prostomium, round, cone-like, longer than base is broad. Intersegmental furrows not strongly marked, no secondary annulation. No dorsal setae. Ventral setae single, very slightly curved, get larger from II to XX and then rapidly smaller in size, buried in body wall, no nodulus. Cuticle 10 μ thick, large unicellular epidermal glands from II to about XX. Nephridial pores in front of setae. Spermathecal pores lateral in 7/8, 8/9. ? Male pores and female pores. Gizzard vestigial (?). Long transverse commissural vessels from II to XXII at least. Timm's glands, copulatory glands and setal glands ? absent. Testes in X and XI, ovaries in XII. Spermathecae in VIII and IX.

Subterranean water in wells, Sapporo, Japan.

One partially mature and three immature specimens.

According to Yamaguchi, this species was placed in *Haplotaxis* because of the presence of single implanted setae of characteristic shape, the absence of dorsal setae, the distribution of the gonads and spermathecae and the presence of the "vestigial" gizzard. No description of the gizzard is given, so that the use of the term "vestigial" in the original description cannot be explained further. The large setae lacking the nodulus are certainly characteristic, but dorsal setae have been found wanting in other species. The lack of glands and of references to the genital pores and efferent ducts is undoubtedly due to the shortage of mature specimens.

Haplotaxis vermivorus (Michaelsen, 1932)

Phreoryctes vermivorus MICHAELSEN, 1932: 1.

Haplotaxis vermivorus (Michaelsen). BRINKHURST, 1966: 38.

$l=140$ mm, $w=0.8-0.9$ mm, $s=300$. Cylindrical anteriorly, but dorso-ventrally flattened in mid and hind region (0.6×0.9 mm), posteriorly awl-shaped. Prostomium prolobic, dome-shaped, and divided by a groove. Segment I short, all segments simple, no secondary annulation. No dorsal setae. Ventral setae single, hooked or sickle-shaped, 0.45 mm $\times 26$ μ. No external sign of genital pores. Body wall thick, circular muscle thin. No thickened septa. Fore-gut with a dorsal pad, slight thickening laterally. Dorsal and ventral blood vessels linked by a long pair of commissurals in each segment. First nephridium in XVI, gonads in X and XI (? testes) and XII (? ovaries). Timm's glands, copulatory glands and setal glands not observed. Spermathecae not developed.

Poeloe Berhala (N. coast Sumatra, Malacca Straits).

Michaelsen stated that the ventral setae resemble those of *H. dubius* as illustrated by Hrabe (1931, Fig. 8(g)). The fore-gut was said to bear a dorsal pad like that of the Enchytraeidae, which Michaelsen suggested might be due to post-mortem changes. He concluded that the gizzard was laterodorsal in extent.

This obscure and badly described worm appears to forge a link between the former genera *Haplotaxis* and *Pelodrilus* in so far as the structure of pharynx, the prostomium, and the setae are concerned.

Haplotaxis ignatovi (Michaelsen, 1903)

Pelodrilus ignatovi MICHAELSEN, 1903 a: 3.

Pelodrilus ignatovi Michaelsen. STEPHENSON, 1930: 805. MALEVICH, 1949: 123. HRABE, 1958: 177. CEKANOVSKAYA, 1962: 321.

Haplotaxis ignatovi (Michaelsen). BRINKHURST, 1966: 39.

$l=30-35$ mm, $w=1.2$ mm maximum, $s=68$. Prostomium round, shorter than broad, zyglobic. Segments from VI biannulate, first ring short. Cuticle thick. Setae closely paired anteriorly, single in mid and posterior bundles. aa$=2/3$ bc$=$dd. Setae all thin, S-shaped and blunt-ended, nodulus in the distal $1/4$, all of equal size (360 $\mu \times 18$ μ in XX). Clitellum scarcely developed, but $11/12$ furrow indistinct. Spermathecal pores in setal line, laterally in $7/8$ and $8/9$. Male pores just in front of the ventral setae of XI and XII, female pores on the setal line in $12/13$. Thin-walled pharynx, no gizzard. Septal glands in III to VI. An irregular heart on the dorsal vessel. Copulatory glands in XI numerous, of irregular shape. Testes and male funnels paired in X and XI, sperm sacs on $10/11$ and $11/12$. Ovaries and female funnels paired in XII, a median ovi-sac on $12/13$. Spermathecae two pairs on VIII and IX, with pear-shaped ampullae, narrow ducts half the length of the ampullae.

Teletsk Lake, Altai Mountains, Central Asia.

The species approaches *H. violaceus*, *H. monticola*, *H. vermivorus*, *H.*

heterogyne and *H. gastrochaetus* in the absence of the posterior ovaries, but the number and shape of the setae serve to distinguish it from them.

Haplotaxis violaceus (Beddard, 1891)
Fig. 6. 2F

Pelodrilus sp. BEDDARD, 1891 a : 91.

Pelodrilus violaceus BEDDARD, 1891 b : 292, Plate II, figs. 17, 20 to 29, Plate III, figs. 35, 36, 38 to 41.

Pelodrilus violaceus Beddard. BEDDARD, 1895 : 192. MICHAELSEN, 1900 : 107. STEPHENSON, 1930 : 805. HRABE, 1958 : 177.

Haplotaxis violaceus (Beddard). BRINKHURST, 1966 : 39.

$l=26$-52 mm, slender. Prostomium short and blunt. Setae closely paired, simple-pointed, ? without nodulus. Clitellum XI to XIII, dorsal only. Two mid-ventral glands, one behind the other on X. Nephridial pores in front of the ventral setae. Spermathecal pores above the line b close to 7/8 on VIII. Male pores minute, two pairs on XII above the line b. Female pores 12/13 above the line b. Septa present from 4/5, those of 5/6 to 8/9 or 9/10 thickened and muscular. Fore-gut with an eversible pharynx, septal glands in V to VII. Nephridia from VII except XI and XII. Testes and male funnels paired in X and XI. Vasa deferentia long and coiled, those originating in the funnels on 10/11 traversing XI to enter XII anteriorly, close to the body wall, those from 11/12 coiled in XII and opening within that segment, close to the opening of the others, but independent of them. Sperm sacs in XI to XII, ovaries in XII, fused beneath nerve cord; oviducts short, ova large and yolky. Ovi-sacs thin-walled, in XIII. Spermathecae in VIII, long and bent upon themselves, ducts not distinct.

Ashburton, New Zealand, in rich wet soil near a swamp.

The type material was discovered in the collections of the British Museum (Natural History) and many features of the anatomy confirmed from both the original sections and new sections from type material stored in spirit (Brinkhurst, 1966).

Haplotaxis africanus (Michaelsen, 1905)
Fig. 6. 2I

Pelodrilus africanus MICHAELSEN, 1905 a : 19.

Pelodrilus darlingensis MICHAELSEN, 1907 : 134.

Pelodrilus africanus Michaelsen. MICHAELSEN, 1905 b : 34; 1907 : 136; 1913 : 480, Fig. A. STEPHENSON, 1930 : 805. HRABE, 1931 : 177; 1958 : 177.

Pelodrilus monticola MICHAELSEN, 1908 : 33.

Pelodrilus tuberculatus BENHAM, 1909 : 263, Pl. X, fig. 12-14.

Pelodrilus darlingensis Michaelsen. MICHAELSEN, 1913: 484. JACKSON, 1931: 81. HRABE, 1958: 177.

Pelodrilus tuberculatus Benham. MICHAELSEN, 1924: 217. BENHAM, 1950: 28. HRABE, 1958: 177.

Pelodrilus monticola Michaelsen. HRABE, 1958: 177.

Haplotaxis africanus (Michaelsen). BRINKHURST, 1966: 40.

(non) *Pelodrilus africanus* GODDARD, 1924: 12 (nomen nudum).

l=25-70 mm, w=1-1·45 mm, s=70-128. Prostomium zyglobic, short and broad. Segments simple. Setae simple-pointed, closely paired. Clitellum ½XI to XIII or XIV. Mid-ventral glands variable, on VI to X or less, or absent. Spermathecal pores in 7/8, lateral. Male pores on XI and XII in or a little above ventral setal line. Female pores 12/13 in line of ventral setae. Septa 5/6 and 9/10 thickened, those of other anterior segments less so. Fore-gut with eversible pharynx in II to IV. Septal glands from V to VII (or IX). Nephridia from V or VI, absent in ?X, XI to XII. Testes and male funnels paired in X and XI. Vasa deferentia narrow and coiled. Ovaries and female pores paired in XII. Spermathecae large, paired in VIII, ducts more or less indistinct.

Auckland Island, Adam Island, Stewart Island; Simonstown, S. Africa; Collie, South West Australia.

The differences between *darlingensis, monticola, tuberculatus* and *africanus* are trivial and depend on characters subject to variation within the species. The name *africanus* takes precedence over the other three by priority.

A single haplotaxid, collected by B. Jamieson in an acid stream in a *Melaleuca* swamp opposite Bribie Island, coastal Queensland (2.IX.67), seems to resemble *H. africanus*, but the position of the male pores cannot be determined from the sections, and there are two pairs of spermathecae (in VII and VIII) with the pores at the anterior edge of the segment. The sperm sac in XI extends from the septum 10/11, but that from 11/12 extends back to XVIII. Septal glands extend from V (or even IV) to IX. Testes are present in X and XI, ovaries in XII.

Haplotaxis smithii (Beddard, 1888)
Fig. 6. 2G, H

Phreoryctes smithii BEDDARD, 1888.

Phreoryctes smithii Beddard. BEDDARD, 1890: 629, Figs. 1 to 2; 1895: 190, FORBES, 1890 b: 107.

Haplotaxis smithii (Beddard). MICHAELSEN, 1900: 109. BRINKHURST, 1966: 41.

l=50-200 mm, w=1·25 mm. Prostomium with a transverse furrow. Setae closely paired on four bundles per segment, dorsal setae the shorter posteriorly. Setae

S-shaped, distal bend pronounced. Clitellum a complete girdle one cell thick from ½X to XIII. Spermathecal pores lateral on VII and VIII (sometimes on VI as well). Male pores minute, above ventral setal line of XI and in the setal line of XII. Female pores on setal line of 12/13 and 13/14. Body wall thick, circular muscle thin, lateral line present. Large gland cells present, pharynx eversible, no gizzard. Septal glands present (? from V to IX). Dorsal and ventral vessels joined by a pair of long commissural vessels in each segment. Nephridial pores in front of setae (? ventral), nephridia from XIV onwards. Timm's glands, setal glands and copulatory glands absent. Testes and male funnels paired in X and XI. Male efferent ducts very simple, short tubes in XI and XII. Ovaries and female funnels paired in XII and XIII, female ducts the same as male ducts. Sperm sacs in IX and XIV, egg-sacs in XIV to XVI. Spermathecae in VII and VIII (sometimes a third pair in VI), pyriform pouch with a ? glandular lining.
Ashburton, New Zealand, in marshy soil and a pool.

Several anatomical details were confirmed after examination of new material collected at Canterbury, New Zealand, in Lake Coleridge by Mr. M. Flain.

Haplotaxis hologynus (Michaelsen, 1907)
Fig. 6. 2J

Pelodrilus hologynus MICHAELSEN, 1907: 136.

Pelodrilus aucklandicus BENHAM, 1909: 265, Fig. 15.

Pelodrilus hologynus Michaelsen. BENHAM, 1909: 266. MICHAELSEN, 1924: 216. STEPHENSON, 1930: 805. JACKSON, 1931: 82. HRABE, 1958: 177.

Pelodrilus bipapillatus MICHAELSEN, 1924: 213, Fig. 2.

Pelodrilus aucklandicus Benham. MICHAELSEN, 1924: 216. HRABE, 1958: 177.

Pelodrilus bipapillatus Michaelsen. STEPHENSON, 1930: 805. HRABE, 1958: 177.

Haplotaxis hologynus (Michaelsen). BRINKHURST, 1966: 41.

l=48-55 mm, w=1·0-1·25 mm, s=100-140. Prostomium conical, about as long as broad, with a transverse furrow (? always). No secondary annulation. Setae two per bundle, all simple-pointed, lateral setal distance less than ventral median distance. Clitellum XII to ½XIV. Spermathecal pores lateral in 6/7 or 7/8. Male pores in line with the ventral setae, posterior on XI, anterior on XII. Female pores on 12/13 and 13/14. Pharynx eversible? Septal glands in V to VIII or IX. Septa 7/8 to 11/12 thickened. Nephridia in VIII to X, XV onwards. Copulatory glands pear-shaped in XI and XII around male pores, obscuring 11/12 furrow. Testes and male funnels paired in X and XI. Sperm sacs paired in IX, 11/12 to XIX plus. Vasa deferentia coiled. Ovaries and female funnels paired in XII and XIII. Ovaries in XIII, XIV. Spermathecae paired in VII, with long twisted ampullae partially filling VIII, ducts not clearly separable from ampullae, or in VIII and of other form.

S.W. Australia, Auckland Isles, Adams Island, Stewart Island.

Benham (1909) claimed that both male pores lay in segment XI, but examination of his own illustration as well as re-examination of the type specimens is sufficient to disprove this. Michaelsen (1924) showed that the pores lay on XI and XII externally, associated with the copulatory glands, and regarded *aucklandicus* as a synonym of *hologynus*. The only difference between *P. bipapillatus* and *hologynus* is the presence of oesophageal diverticulae in the former, and the difference in both position and shape of the spermathecae in the two entities. The difference may be regarded as sufficient for the recognition of subspecies.

Haplotaxis hologynus hologynus (Michaelsen, 1907). No oesophageal diverticulae. Spermathecae in VII, elongate with indistinct ducts.

Haplotaxis hologynus bipapillatus (Michaelsen, 1924). Oesophageal diverticulae in XIV to XIII. Spermathecae in VIII (? with distinct ducts).

Haplotaxis bureschi (Michaelsen, 1925)
Fig. 6. 2L

Pelodrilus bureschi MICHAELSEN, 1925a: 85.

Pelodrilus bureschi Michaelsen. MICHAELSEN, 1926: 57. STEPHENSON, 1930: 805. CERNOSVITOV, 1937: 77, Figs. 10 to 12. HRABE, 1958: 171, 177, Figs. 2, 8 to 9; 1963a: 75; 1963b: 77. BOTEAU and BOTOSANEANU, 1966: 223, Fig. 2.

Haplotaxis bureschi (Michaelsen). BRINKHURST, 1966: 42.

$l=50$-170 mm, $w=5$ mm, $s=205$ maximum. Prostomium prolobic. Anterior segments two-ringed, anterior ring the longer and bears setae. Body wall iridescent. Setae closely paired, sigmoid, hooked distally, simple-pointed but finely ornamented distally, aa>bc, dd very large. Clitellum complete on XI to XVII. Spermathecal pores in the line cd in 5/6, 6/7, 7/8, and sometimes also 8/9. Male pores between b and c on XI and XII, female pores anterior to ventral setae on XIII and XIV. No gizzard. Septal glands (poorly developed) from ? to VIII. Nephridial pores in setal line a. Very variable setal glands, one to six pairs or singly in some segments. Testes and male funnels paired in X and XI, vasa deferentia irregularly winding, enter body wall by ventral setae of XI and XII. Ovaries and female funnels paired in XII and XIII. Spermathecae three to four pairs on VI to VIII or IX.

Groundwater in Yugoslavia, Bulgaria, Roumania.

This species is readily distinguished from all others mainly in the thickness of the body and ornamented setae but is close to *H. leruthi* in which the setae are keeled.

Haplotaxis leruthi (Hrabe, 1958)
Fig. 6. 2K

Pelodrilus leruthi HRABE, 1958: 171, Figs. 1, 3 to 7, 10.

Haplotaxis leruthi (Hrabe). BRINKHURST, 1966: 42.

$l=100$ mm, $w=2.8$ mm, $s=180$. Prostomium prolobic, short and wide. All segments biannulate, the anterior ring about three times smaller than the posterior. Cuticle iridescent. Setae paired, sigmoid, bearing a keel in all segments, or only those from XX to XXV unless where worn off. In segment 100 aa:bc:dd $=1:0.8:6$. Clitellum indistinct in X to XVI. Spermathecal pores in 5/6 to 7/8 in the line cd. Male pores open laterally on XI and XII. Eversible pharynx present in II to III, no gizzard. Septal glands in IV to VIII. Long commissural vessels in a few anterior segments. Setal glands on IX to XIII. Testes and male funnels paired in X and XI, testes at first in a thin pouch. Vasa deferentia long, penetrating the body wall in front of the ventral setae of XI and XII, running between the muscle layers a little behind the mid-point of the line b-ç. Sperm sacs paired in IX, and a pair from each of 10/11, 11/12, 12/13 reach XIII. Ovaries and female funnels paired in XII and XIII, the posterior ovaries may be long, reaching XVII. Mature eggs large.

Caves in France.

The species closely resembles *H. bureschi*, but the setae are distinctively keeled.

Haplotaxis glandularis (Yamaguchi, 1953)
Fig. 6. 2O

Heterochaetella glandularis YAMAGUCHI, 1953: 306, Figs. 15 to 19, Plate VII, Figs. 9 to 10.

Haplotaxis glandularis (Yamaguchi). BRINKHURST, 1966: 43.

$s \geq 176$. (No complete specimens.) Prostomium short and round, subspherical. Segments biannulate except I to IV, posterior annulus divided into three secondary annuli in all except I to II, in which only two. Body wall thin and transparent, with gland cells which penetrate body wall. Setae eight per segment in four pairs. Each pair with one bifid and one simple seta, very closely paired, aa $=1/5$ to $1/7$ overall, dd $=2$ to $3 \times$ aa; aa usually more than bl; bl always greater than lc. ? Setal glands in each segment just above cd. Spermathecal pores in 5/6 to 8/9. Pharynx in II to III. Septal glands in IV to VII. Dorsal and ventral vessels linked by long commissural vessels in each segment, posteriorly as intestinal plexus from the dorsal vessel. Meganephridia in (VIII) IX, XV and thence posteriorly, pores on the line of the ventral setae. Testes and male funnels paired in X and XI, ovaries and female funnels in XII and XIII. Spermathecae in VI to IX.

Japan.

The combination of bifid and simple-pointed setae in each bundle serves to distinguish this from all other *Haplotaxis* species.

Haplotaxis denticulatus (Cekanovskaya, 1959)
Fig. 6. 2M, N

Adenodrilus denticulatus CEKANOVSKAYA, 1959: 1156, Figs. 2-5.

Adenodrilus denticulatus Cekanovskaya. CEKANOVSKAYA, 1962: 324, Fig. 204.

Haplotaxis denticulatus (Cekanovskaya). BRINKHURST, 1966: 43.

$l=30$ mm, $w=1.25$ mm, $s=80$-114. Prostomium conical. First two segments biannulate, rest triannulate. Dorsal and ventral setae paired, bifid, with upper tooth more or less shorter and thinner than the lower, with one or two intermediate teeth. Some anterior ventral setae simple-pointed. Spermathecal pores dorsal in VIII, sometimes also IV. Male pores invisible externally, female pores on setal line in 11/12 and 12/13. Pharynx wide on II to IV. No gizzard. Septal glands on IV to VI of some specimens. Large, tubular, copulatory glands open near ventral setae on X to XIII. Testes and male funnels paired on IX and X, ovaries and female funnels paired in XI and XII. Spermathecae small with short ducts, in VIII, sometimes also in IV. (Description based on the short account in Cekanovskaya, 1962.)
    Fergan Valley, C. Asia.

Cekanovskaya (1962) wrote that the male pores are invisible externally, and so their position on X and XI is conjectural unless they were identified in sections (as indicated by the illustrations in Cekanovskaya, 1962). Assuming that the male pores are sited in the usual place, the gonads are all shifted one segment forward of their position in other species, but the sequential arrangement is that of other haplotaxid species with two pairs of ovaries.

NOMINA DUBIA

Haplotaxis carnivorus Omodeo, 1958—nomen nudum.

Haplolumbriculus Omodeo, 1958—Ovaries in XI or XI and XII.

Haplolumbriculus insectivorus Omodeo, 1958—nomen nudum.

Pelodrilus carnivorus Omodeo, 1958—nomen nudum.

Pelodrilus falcifer Omodeo, 1958: 19, Fig. 4.
    The position of the gonads (testes XI, ovaries XIII) is peculiar.

Pelodrilus americanus Cernosvitov, 1939.

Pelodrilus kraepelini (Michaelsen): Hrabe, 1931; Cernosvitov, 1939.
    The last three were regarded as species inquirendae outside the Haplotaxidae by Brinkhurst (1966).

The specimens named *Pelodrilus cuenoti* Tétry, 1934, proved to belong to *Sparganophilus tamesis* (Sparganophilidae).

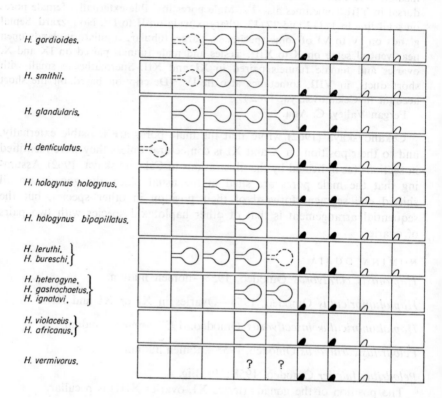

H. gordioides.

H. smithii.

H. glandularis.

H. denticulatus.

H. hologynus hologynus.

H. hologynus bipapillatus.

H. leruthi.
H. bureschi.

H. heterogyne.
H. gastrochaetus.
H. ignatovi.

H. violaceus.
H. africanus.

H. vermivorus.

Fig. 6.1. Relative positions of spermathecae, testes (solid circles) and ovaries (open circles) in *Haplotaxis* species. The first segment shown on the left is V in all instances bar *H. denticulatus*, where it is IV. All structures illustrated are paired.

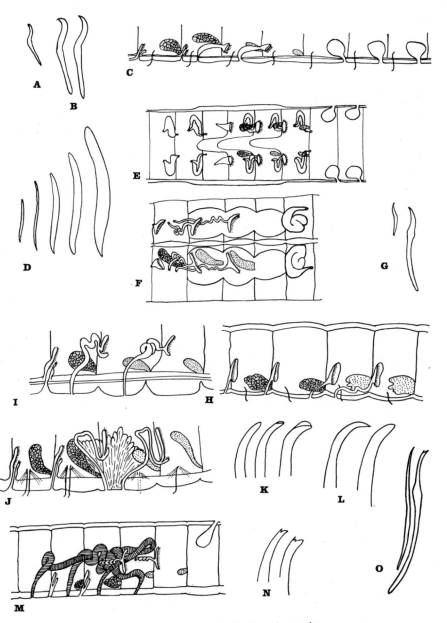

Fig. 6.2. Reproductive organs and setae of *Haplotaxis* species.
A,B—Ventral setae of II and VIII-X, *H. gordioides dubius*. C—reproductive organs,
H. gordioides. D—setae of *H. gastrochaetus* (from left to right, segments II, V, X,
XV, XIX). E—reproductive organs, *H. heterogyne*. Note nephridia in anterior
testes segment resembles male ducts. F—reproductive organs, *H. violaceus*. Note
rearward shift of anterior vasa deferentia. G—setae of *H. smithii*, small anterior
ventral seta, larger dorsal median seta. H—reproductive organs, *H. smithii*. Note
similarity of male and female ducts. I—reproductive organs, *H. africanus*. J—
reproductive organs, *H. hologynus*. K—setae of *H. leruthi*, showing progressive
development of keel on posterior setae (at right), ?due to wear. L—setae of *H.
bureschi*, worn (right) and unworn form (left). M—reproductive organs, *H.
denticulatus*, N—setae, *H. denticulatus*, O—setae, *H. glandularis*.

REFERENCES

ALTMAN, L. C. 1936. Oligochaeta of Washington. *Univ. Wash. Publs Biol.*, **4**, 1.

BEDDARD, F. E. 1888. On the reproductive organs of *Phreoryctes*. *Ann. Mag. nat. Hist.*, (6) **1**, 389.

— 1890. On the anatomy, histology and affinities of *Phreoryctes*. *Trans. R. Soc. Edinb.*, **35**, 629.

— 1891a. Abstract of some investigations into the structure of the Oligochaeta. *Ann. Mag. nat. Hist.*, (6) **7**, 88.

— 1891b. Anatomical description of two new genera of aquatic Oligochaeta. *Trans. R. Soc. Edinb.*, **36**, 629.

— 1895. *A monograph of the Order Oligochaeta*. Oxford, Clarendon Press.

BENHAM, W. B. 1890. An attempt to classify earthworms. *Q. Jl. microsc. Sci.*, **31**, 201.

— 1903. On some new species of aquatic Oligochaeta from New Zealand. *Proc. zool. Soc. Lond.*, **1903** (2), 202.

— 1905. On a new species of *Haplotaxis*, with some remarks on the genital ducts in the Oligochaeta. *Q. Jl. microsc. Sci.*, **48**, 299.

— 1909. Report on Oligochaeta of the subantarctic Islands of New Zealand. In Chilton, C. *The subantarctic Islands of New Zealand*. **1**, 251. Wellington.

— 1950. The Oligochaeta of the Auckland and Campbell Islands. *Cape Exped. Ser. Bull. No.* **10**, 27.

BOTEA, F., and BOTESANEANU. 1966. *Pelodrilus bureschi* Mich. 1924 dans les grottes du Banat. *Int. J. Speleol.*, **2**, 223.

BRIGGS, J. 1953. Behaviour and reproduction of Salmonid fishes in a coastal stream. *Fish Bull. Calif.*, **94**, 47.

BRINKHURST, R. O. 1966. A taxonomic revision of the family Haplotaxidae (Oligochaeta). *J. Zool. Lond.*, **150**, 29.

CEKANOVSKAYA, O. V. 1959. On the Oligochaeta of the water-bodies of Central Asia (Fergan Valley and River Murgab). *Zool. Zh.*, **38**, 1152 (in Russian).

— 1962. The aquatic Oligochaeta fauna of the USSR. *Opred. Faune SSSR No.*, **78**, 1 (in Russian).

CERNOSVITOV, L. 1937. Die Oligochaeten Fauna Bulgariens. *Izv. tsarsk. prirodonauch. Inst. Sof.*, **10**, 69.

— 1939. The Oligochaeta in the Percy Sladen Trust Expedition to Lake Titicaca in 1937. *Trans. Linn. Soc. Lond. (Zool.)*, **1**, 81.

— 1942. A revision of Friend's types and descriptions of British Oligochaeta. *Proc. zool. Soc. Lond.*, (B) **111**, 237.

— 1945. Oligochaeta from Windermere and the Lake District. *Proc. zool. Soc. Lond.*, **11c**, 523.

CLAPARÈDE, E. R. 1862. Recherches anatomique sur les Oligochaetes. *Mèm. Soc. Phys. Hist. nat. Genève*, **16**, 217.

CLAUS, C. 1880. *Grundzüge der Zoologie*, **1**, 482. 4th ed. Marburg: N. G. Elwert'sche Univ.

FERRONNIÈRE, G. 1899. Contribution a l'étude de la faune de la Loire-Inferieur. *Bull. Soc. Sci. nat. Quest Fr.*, **9**, 229.

FORBES, S. A. 1890a. Note on an American species of *Phreoryctes*. *Am. nat.*, **24**, 477.

— 1890b. On an American earthworm of the family Phreoryctidae. *Bull. Ill. St. Lab. nat. Hist.*, **3**, 107.

FRIEND, H. 1896. Notes on Essex earthworms. *Essex Nat.*, **9**, 110.

GODDARD, E. J. 1924. On *Pelodrilus africanus*, a new haplotaxid from South Africa. *Trans. R. Soc. S. Afr.*, **12**, xii.

HESSE, E. 1923. Sur l'habitat de *Haplotaxis gordioides*. *Bull. Soc. zool. Fr.*, **48**, 138.

HRABE, S. 1931. Die Oligochaeten aus den Seen Ochrida und Prespa. *Zool. Jb. (Syst.)*, **61**, 1.

— 1933. Zur Kenntnis der *Pelodrilus kraepelini* (Michaelsen). *Zool. Anz.*, **104**, 225.

— 1958. A new species of Oligochaeta from the Southwest of France. *Notes biospéol.*, **13**, 171.

HRABE, S. 1963a. Oligochaeta limicola from Slovenija. *Biol. Vest.*, **11**, 77.

JACKSON, A. 1931. The Oligochaeta of South-Western Australia. *J. Proc. R. Soc. West. Aust.*, **17**, 71.

MALEVITCH, I. 1949. The oligochaeta fauna of the Teletsk Lake. *Trudy zool. Inst., Leningr.*, **7**, 119 (in Russian).

MICHAELSEN, W. 1899. Beiträge zur Kenntnis der Oligochäten. *Zool. Jb. (Syst.)*, **12**, 105.

— 1900 Oligochaeta. *Tierreich*, **10**, 1.

— 1901. Oligochaeten der Zoologischen Museen zu St. Petersburg und Kiew. *Izv. imp. Akad. Nauk*, **15**, 137.

— 1903a. Eine neue Haplotaxiden-Art und andere Oligochäten aus dem Telezkischen See im Nördlichen Altai. *Verh. naturw. Ver. Hamb.*, **10**, 1.

— 1903b. Die geographische Verbreitung der Oligochaeten. Berlin, Friedländer und Sohn.

— 1905a. Die Oligochaeten des Baikal Sees. *Wiss. Ergebn. Exped. Baikal See, 1900-2*, **1**, 1.

— 1905b. Die Oligochaeten der Deutschen Südpolar Expedition 1901-3. *Dt. Südpol.-Exped.*, **9**, *Zool.* 1 (1), 1.

— 1907. Die Fauna Südwest-Australiens. *Ergebn. Hamb. Sudwest Austral. Forsch. (1905)* **1**, 124.

— 1908. Annelida A: Oligochaeten aus den Westlichen Kapland. *Denkschr. med.-naturw. Ges. Jena*, **13**, 30.

— 1913. Die Oligochaeten des Kaplands. *Zool. Jb. (Syst.)*, **34**, 473.

— 1914. Oligochaeten vom tropischen Africa. *Mitt. naturh. Mus. Hamb.*, **31**, 81.

— 1924. Oligochaeten von Neuseeland und dem Auckland-Campbell Inseln, nebst einigen anderen Pacifischen Formen. *Vidensk. Meddr. dansk naturh. Foren.*, **75**, 197.

— 1925a. Ein Süsswasser-Hohlenoligochät aus Bulgarien. *Mitt. Zool. St. Inst. Hamb.*, **41**, 85.

— 1925b. Zur Kenntnis einheimischer und ausländischer Oligochaeten. *Zool. Jb. (Syst.)*, **51**, 255.

— 1926. *Pelodrilus bureschi*, ein Süsswasser-Hohlenoligochät aus Bulgarien. *Trud. bulg. prir Druzh.*, **12**, 57.

— 1929. Oligochaeten der Kamtschatka-Expedition 1908-9. *Exheg. zool. Muz.*, **30**, 315.

— 1932. Ein neuer *Phreoryctes* von der Tropinsel Poeloe Berhala. *Miscnea zool. sumatr.*, **71**, 1.

OMODEO, M. 1958. La reserve Naturelle intégrale du Mont Nimba. IV. 1. Oligochetès. *Mém. Inst. fr. Afr. noire*, **53**, 9

PIERANTONI, U. 1904. Sopra alcuni Oligocheti raccolti nel fiume Sarno. *Annuar. R. Mus. zool. R. Univ. Napoli*, **1**, 1.

— 1905. Oligocheti del fiume Sarno. *Archo. zool. ital.*, **2**, 227.

SMITH, F. 1918. A new North American oligochaete of the genus *Haplotaxis*. *Bull. Ill. State Lab. nat. Hist.*, **13**, 43.

STEPHENSON, J. 1930. *The Oligochaeta*. Oxford, University Press.

TÉTRY. A. 1934. Description d'une espèce francaise du genre *Pelodrilus*. *C.r. hebd. Séanc. Acad. Sci., Paris*, **199**, 322.

— 1938. *Contribution à l'étude de la faune de l'Est de la France (Lorraine)*. Nancy.

VEJDOVSKY, F. 1884. *System und Morphologie der Oligochaeten*. Prague.

YAMAGUCHI, J. 1937. Study on the Aquatic Oligochaeta of Japan. II. Occurrence of the subterranean oligochaete *Haplotaxis gordioides* (Hartmann) in Japan. *Annotnes zool. jap.*, **16**, 68.

— 1953. Studies on the Aquatic Oligochaeta of Japan. VI. A systematic report, with some remarks on the classification and phylogeny of the Oligochaeta. *J. Fac. Sci. Hokkaido Univ., Zool.*, **11**, 277.

7

FAMILY NAIDIDAE

Type genus: Nais MÜLLER

Prostomium usually well developed, with or without proboscis. Eyes present or absent. An indefinite number of ventral setae per bundle, beginning in II; bifid, or simple pointed crotchets in a few species. Dorsal setae, beginning in II, III, IV, V, or VI, farther back or totally absent; consisting of an indefinite number of hair setae, accompanied by needle setae of various shapes, or of needle setae only. Clitellum in a few segments in the region of the gonads; testes and ovaries, one pair each, in IV-V, V-VI, or VII-VIII; male efferent apparatus paired; funnels in segment with testes, atria in segment with ovaries, usually opening apart. No penes, penial setae often present; often diffuse prostate gland cells on vasa deferentia or atria; spermathecae in segment bearing testes, usually an un-paired sperm-sac and ovisac formed. Asexual reproduction by budding or fragmentation. Ontogeny deviating, with a periblast forming round the embryo.

Cosmopolitan.

KEY TO THE GENERA OF THE FAMILY NAIDIDAE

1.	No dorsal setae present	*Chaetogaster*
—	Dorsal setae present	2
2.	Hair setae absent	3
—	Hair setae present	8

304

3.	Dorsal setae begin in II or III	4
—	Dorsal setae begin in V or VI	5
4.	Dorsal setae begin in II	*Homochaeta*
—	Dorsal setae begin in III	*Amphichaeta*
5.	Dorsal setae straight, thick, and single in each bundle	*Ophidonais*
—	Dorsal setae more than one per bundle, not thick and straight	6
6.	Dorsal setae from VI	*Uncinais*, and the doubtful genus *Neonais*
—	Dorsal setae from V	7
7.	Nephridia absent. Body wall naked	*Paranais*
—	Nephridia present, no funnel. Body wall papillate or encrusted	*Waspa*
8.	Dorsal setae beginning in II	9
—	Dorsal setae beginning behind II	10
9.	Body wall encrusted with foreign matter. Ventral setae with upper tooth becoming progressively longer posteriad. Testes in IV, ovaries in V. Four segments formed anteriorly	*Stephensoniana*
—	Body wall naked. Ventral setae with upper tooth becoming relatively shorter posteriad. Testes in VII, ovaries in VIII. Seven segments formed anteriorly	*Pristina*
10.	Dorsal setae shed anteriorly after budding, retained from about XVIII	*Haemonais*
—	Dorsal setae from IV, V or VI	11
11.	Gills present	12
—	Gills absent	13
12.	Gills ciliated, lamelliform or digitiform processes around anus	*Dero*
—	Gills ciliated, digitiform projections along all or most of the body	*Branchiodrilus*
13.	Anterior end forming a proboscis	14
—	No proboscis	16

14.   All dorsal setal bundles with up to 30 setae
each   *Arcteonais*

—   Most setal bundles with fewer setae, mostly
less than 10   15

15.   Dorsal setae of VI to VIII with giant hair
setae and up to 40 per bundle, the rest with
2-6 setae   *Ripistes*

—   Dorsal setae of all bundles less than 10 per
bundle   *Stylaria*

16.   Hair setae of VI elongate. Body wall encrusted
with foreign matter   *Slavina*

—   No elongate hair setae. Body wall usually
naked   17

17.   Hair setae thick, stiff and strongly serrated   *Vejdovskyella*

—   Hair setae non-serrate, non-thickened   18

18.   Needle setae more or less resemble ventral
setae   19

—   Needle setae distinctly different in form to
ventral setae   20

19.   Eyes present. Hair setae short, often absent.
Needles closely resemble ventral setae   *Piguetiella*

—   Eyes absent. Hair setae twice as long as
needles, needles slightly straighter than ventral
setae with shorter teeth   *Specaria*

20.   Eyes usually present. Anterior ventral setae
usually differ from median and posterior setae.
Stomach present. Prostate present   *Nais*

—   Eyes absent. Ventral setae all more or less
alike. Prostate and stomach absent   *Allonais*

SUBFAMILY **Chaetogastrinae** SPERBER, 1948

Type genus: Chaetogaster VON BAER

Prostomium often weakly developed. Segment III elongated. No eyes. Dorsal setae bifid crotchets, or absent. Pharynx in I-III, attached to the body-wall by numerous radial muscular strands; no dorsal diverticulum; no pharyngeal, oesophageal or septal glands; stomach of special structure. Chloragogen beginning in V. Commissural vessels one pair or absent. Nephridia closed. No

coelomocytes. Clitellum absent between male pores; testes and spermathecae in V, ovaries and atria in VI; spermathecal and atrial ducts well-defined; no prostate; five segments formed at anterior end, by budding. No swimming.
Cosmopolitan.

GENUS **Chaetogaster** VON BAER, 1827

Type species: Chaetogaster limnaei VON BAER

Prostomium weakly developed. No dorsal setae; ventral setae bifid or simple crotchets, absent in III-V. Septa incomplete. Oesophagus in IV; stomach well-defined. Intestinal plexus on the stomach conspicuous, with a ventral trunk; one pair of commissural vessels in IV; in some instances vascular system in the pharynx region atrophied. Connectives of the ventral nerve-cord not fused in the anterior segments; cerebral ganglia usually weak, without posterior lobes. Clitellum in ½ V-VI; vasa deferentia joining atria at the proximal end; penial setae present. Budding and fission continuing during earlier stages of maturity.
Cosmopolitan.

Chaetogaster

1.	Parasitic or commensal on snails. Setae numerous with long, strongly curved teeth	*C. limnaei*
—	Free living. Setae normally shaped, fewer	2
2.	Setae simple-pointed	*C. setosus*
—	Setae bifid	3
3.	Prostomium conspicuous	*C. diastrophus*
—	Prostomium inconspicuous	4
4.	Prostomium with median incision	*C cristallinus*
—	Prostomium without median incision	5
5.	Setae large (more than 145 μ long in II)	*C. diaphanus*
—	Setae small (less than 100 μ long in II)	*C. langi*

Chaetogaster diastrophus (Gruithuisen, 1828)
Fig. 7. 1A-1

(?) *Nais vermicularis* MÜLLER, 1773: 20 (? partim).

(?) *Nais vermicularis* Müller. GMELIN, 1788-1793: 3120 (? partim). MODEER, 1798: 118 (? partim). OKEN, 1815: 364 (? partim). LAMARCK, 1816: 223 (? partim). BLAINVILLE, 1825: 223 (? partim); 1828: 497 (? partim).

Nais diastropha GRUITHUISEN, 1828: 416, Pl. XXV, Figs. 7-9.

Copopteroma nais CORDA, 1837: 390 (vide MICHAELSEN, 1900: 21).

(?) *Chaetogaster vermicularis* (Müller) (partim). GRUBE, 1851: 105.

(?) *Chaetogaster Mülleri* D'UDEKEM, 1855: 554.

Chaetogaster LEYDIG, 1857: 344, Fig. 184.

(?) *Chaetogaster Mülleri* d'Udekem. D'UDEKEM, 1859: 24; 1861: 248, Figs. 2-3.

(?) *Chaetogaster gulosus* LEIDY, 1852a: 124.

Chaetogaster diastrophus (Gruithuisen). VEJDOVSKY, 1833: 221; 1884: 38, Pl. VI, Figs. 11-15. BEDDARD, 1895: 307. BRETSCHER, 1896: 511. MICHAELSEN, 1900: 21; 1909a: 10. SMITH, 1900: 443; 1918: 638. PIGUET, 1906: 200. SOUTHERN, 1909: 131. GALLOWAY, 1911: 306. PIGUET and BRETSCHER, 1913: 18. POINTNER, 1914b: 93. SCHUSTER, 1915: 8, 62. SVETLOV, 1924: 189. MALEVICH, 1927: 4. UDE, 1929: 22. KNOLLNER, 1935: 427. SZARSKI, 1936a: 103, Pl. IV, Figs. 1-3; 1936b: 393. CHEN, 1940: 31; 1944: 2. MARCUS, 1943: 11, Pl. I, Fig. 1. SPERBER, 1948: 59, Figs. 3c, 6, 7A, 7B, 7G, Pl. I, Fig. 1; 1950: 52, Fig. 3A, B, Pl. I, Fig. 1; 1960: 155. ERCOLINI, 1956: 5. TIMM, 1959: 24. BRINKHURST, 1962: 318; 1963a: 18, Fig. 2a; 1963b: 143; 1964: 201, Fig. 1A. CEKANOVSKAYA, 1962: 205, Fig. 117, 118. MOSZYNSKA, 1962: 8. NAIDU, 1962a: 133, Fig. 4.

Chaetogaster Mülleri d'Udekem. TIMM, 1883: 154.

Chaetogaster vermicularis (Müller). VAILLANT, 1890: 446.

(?) *Chaetogaster gulosus* Leidy. VAILLANT, 1890: 451.

Chaetogaster punjabensis STEPHENSON, 1907a: 133, Pl. V, figs. 1-11.

Chaetogaster punjabensis Stephenson. STEPHENSON, 1907b: 246; 1920: 196.

Chaetogaster palustris POINTNER, 1914b: 272, Pl. XXVIII, Figs. 2-3.

Chaetogaster palustris Pointner. SCHUSTER, 1915: 9. UDE, 1929: 22. SPERBER, 1943: 63. NAIDU, 1962: 132.

Chaetogaster annandalei STEPHENSON, 1917: 88.

Chaetogaster annandalei Stephenson. STEPHENSON, 1918: 9; 1923: 9.

Chaetogaster langi Bretscher. STEPHENSON, 1922b: 278; 1923: 50.

l=1-5 mm, s=10-16. Prostomium fairly well developed, pointed, with long sensory hairs. Four to eight setae in II, three to seven per bundle in the other segments. Transverse stomachal vessels 15-20. Cerebral ganglia with large hind lobes: a statocyst in the brain. Spermathecal ampulla elongate, vasa deferentia long, atrial ampulla fairly large.

Europe. India, China, N. and S. America, Afghanistan.

Chaetogaster palustris of Pointner (1914) was accepted by Sperber (1948) with some reservations, was not keyed out in a later paper (Sperber 1950— probably identical to *C. diastrophus*) and was regarded as a synonym of *C. diastrophus* by Cekanovskaya (1962).

Chaetogaster langi Bretscher, 1896
Fig. 7. 1F-H

Chaetogaster langi BRETSCHER, 1896: 512, Fig. 1.

Chaetogaster langi Bretscher. BRETSCHER, 1900: 18, Pl. I, Fig. 15. MICHAEL-SEN, 1900: 21; 1909 a: 10. MUNSTERHJELM, 1905: 10. PIGUET, 1906: 202. WALTON, 1906: 690. GALLOWAY, 1911: 306. PIGUET and BRETSCHER, 1913: 21. POINTNER, 1914 b: 93. SCHUSTER, 1915: 64. SVETLOV, 1924: 189. MALEVICH, 1927: 5. WOLF, 1928: 387, Figs. 2-3. UDE, 1929: 23. CERNOS-VITOV, 1930: 9; 1938 b: 268. SCIACCHITANO, 1938: 256. CHEN, 1940: 32; 1944: 2. MARCUS, 1943: 13, Pl. I, Fig. 2. SPERBER, 1948: 63, Fig. 7c, h, Pl. I, Fig. 2; ?1950: 52, Fig. 3c, Pl. I, fig. 2; 1958: 46; 1960: 155. ERCOLINI, 1956: 4. TIMM, 1959: 24, BRINKHURST, 1962: 318; 1963 a: 18, Fig. 2b; 1963 b: 143; 1964: 202, Fig. 1b. CEKANOVSKAYA, 1962: 208, Fig. 122. MOSZYNSKA, 1962: 8. NAIDU, 1962 a: 134, Fig. 5a- c.

Chaetogaster spongillae ANNANDALE, 1906: 188, Fig. 1a.

(?) *Chaetogaster* sp. ANNANDALE, 1906: 189, Fig. 1b.

(?) *Chaetogaster pellucidus* WALTON, 1906: 690, Fig. 4a-d.

Chaetogaster spongillae Annandale. STEPHENSON, 1907 b: 248; 1911, 205, Fig. 1; 1923: 52, Fig. 10. MARCUS, 1943: 16, Pl. I, fig. 4.

Chaetogaster Annandale. STEPHENSON, 1907 b: 248; 1923: 53.

(?) *Chaetogaster pellucidus* Walton. GALLOWAY, 1911: 306.

(?) *Chaetogaster parvus* POINTNER, 1914 a: 606, Pl. XVIII, Fig. 1.

(?) *Chaetogaster parvus* Pointner. POINTNER, 1914 b: 94. MARCUS, 1943: 15, Pl. I, fig. 3.

(non) *Chaetogaster langi* Bretscher. STEPHENSON, 1922 b: 278; 1923: 50.

$l=0.8-2$ mm, $s=8-21$. Prostomium inconspicuous. Setae in II 3-9, in the rest 3-6 per bundle. Transverse stomachal ducts 8-10. A statocyst in the brain. Spermathecae small, with round ampulla; vasa deferentia short, straight; atria small, round; penial setae 2 per bundle.

Europe, Asia, Africa, N. and S. America. Found also in brackish water (the Baltic).

Sperber (1948) regarded *C. parvus* as extremely doubtful, and later (Sperber, 1950) merely mentioned it in considering *C. langi* and did not key it out. Marcus (1943) rejected this and other synonymies, but Sperber (1948) pointed out the inadequacy of some of the characters used in establishing these species.

Chaetogaster diaphanus (Gruithuisen, 1828)
Fig. 7. 1I-K

Das madenähnliche Schlänglein RÖSEL VON ROSENHOF, 1755: 578, Pl. XCIII, figs. 1-7.

Nais vermicularis Müller. GMELIN, 1788-1793: 3120 (partim); MODEER, 1798: 118 (? partim). LAMARCK, 1816: 223 (partim). BLAINVILLE, 1825: 129 (partim). TABL. ENC. METH., 1827: 134, Pl. LII, figs. 1-7.

Nais diaphana GRUITHUISEN, 1828: 409, Pl. XXV, figs. 1-5.

Chaetogaster niveus EHRENBERG, 1828: 109, foot-note.

Lurco PRITCHARD, 1832: 78 ff., Pl. VIII, fig. 1.

Chaetogaster nivaeus Ehrenberg. GERVAIS, 1838: 15.

Nais diaphana et *perversa* Gruithuisen. GERVAIS, 1838: 15.

Blanonais vermicularis (Roesel). GERVAIS, 1838: 15.

Chaetogaster diaphanus (Gruithuisen). ØRSTED, 1842: 138, Pl. III, figs. 2, 15-17 (partim). LEIDY, 1864: 10, Pl. III, figs. 6-7. LANKESTER, 1869 b: 284, Pl. XV, figs. 5, 6, 8. TAUBER, 1873: 384, Pl. XIII, figs. 8-20; 1879: 26. SEMPER, 1877: 75, Pl. IV, figs. 1-8. VEJDOVSKY, 1883: 220; 1884: 37, Pl. V, Pl. VI, figs. 19-21. TIMM, 1883: 154. LEVINSEN, 1884: 216. VAILLANT, 1890: 449. BEDDARD, 1895: 306. BRETSCHER, 1896: 511. MICHAELSEN, 1900: 21; 1909 a: 10; 1926 a, 21; 1927 a: 7, Fig. 1. SMITH, 1900: 443; 1918: 638. DITLEVSEN, 1904: 461. MUNSTERHJELM, 1905: 10. PIGUET, 1906: 203; 1919 a: 789; 1928: 80. PIGUET AND BRETSCHER, 1913: 19-20. POINTNER, 1914 b: 94-97. SCHUSTER, 1915: 8, 62. EKMAN, 1915: 147. DEHORNE, 1916: 35, Figs. 1-34. STEPHENSON, 1922 b: 279. SVETLOV, 1924: 190; 1926 b: 360, Figs. 1-15. MALEVICH, 1927: 5. WOLF, 1928: 387. UDE, 1929: 24, Fig. 19. CHEN, 1940: 29, 1944: 2. SPERBER, 1948: 66, Figs. 5, 7d, I, Pl. 1, figs. 5, 6; 1950: 52, Fig. 3d, Pl. 1, fig. 5. ERCOLINI, 1956: 5. TIMM, 1959: 24. SOKOLSKAYA, 1961: 50; 1962: 129. BRINKHURST, 1962: 318; 1963 a: 17, Fig. 2c; 1963 b: 143; 1964: 202, Fig. 1c. CEKANOVSKAYA, 1962: 206, Fig. 119. MOSZYNSKA, 1962: 8. NAIDU, 1962 a: 132.

Nais lacustris DALYELL, 1853: 130, Pl. XVII, figs. 1-5.

Nais scotia JOHNSTON, 1865: 71.

(non) *Chaetogaster niveus* Ehrenberg. LANKESTER, 1869 a: 641, Pl. XLVIII, figs. 9-11.

Vetrovermis hyalinus IMHOF, 1888: 48.

Chaetogaster pellucidus STEPHENSON, 1907 b: 237, Figs. 2-6, Pl. IX, figs. 1-10.

Chaetogaster orientalis STEPHENSON, 1910 a: 68, Fig. 4, Pl. VIII, figs. 3-4.

Chaetogaster orientalis Stephenson. STEPHENSON, 1922 c: 109, Figs. 1-6; 1923: 51, Fig. 9; CERNOSVITOV, 1942 c: 281.

Chaetogaster diaphanus (Gruithuisen) var. *cyclops* HAYDEN, 1922: 168.

$l=2.5$-25 mm, $s=14$-15. Prostomium inconspicuous. Setae of II 6-13 per bundle, in the other segments 4-10. Stomachal plexus forming transverse ducts only anteriorly. A statocyst in the brain. Spermathecal ampulla large, roundish or pear-shaped, duct forcibly dilated before the opening; vasa deferentia short; atrial ampulla large, round. Penial setae 3-5 per bundle.

Europe, Asia, N. America. Found in brackish water also in the Baltic.

Chaetogaster crystallinus Vejdovsky, 1883. Fig. 7. 1 L-N.

Chaetogaster Mülleri d'Udekem. TAUBER, 1879: 76.

(?) *Chaetogaster niveus* Ehrenberg. LANKESTER, 1869 a: 641, Pl. XLVIII, figs. 9-11.

Chaetogaster crystallinus VEJDOVSKY, 1883: 220.

Chaetogaster crystallinus Vejdovsky. VEJDOVSKY, 1884: 37, Pl. VI, figs. 1-10. VAILLANT, 1890: 450. BEDDARD, 1895: 307. BRETSCHER, 1896: 511. SPERBER, 1948: 68, Fig. 7e, k, Pl. I., fig. 3. BRINKHURST, 1962: 318; 1963 a: 17, fig. 2d; 1963 b: 143; 1964: 202, fig. 1d. CEKANOVSKAYA, 1962: 207, fig. 120, 121. MOSZYNSKA, 1962: 9. NAIDU, 1962 a: 135, fig. 6a-h.

Chaetogaster crystallinus Vejdovsky. MICHAELSEN, 1900: 21; 1909 a: 10. MUNSTERHJELM, 1905: 10. PIGUET, 1906: 203. SOUTHERN, 1909: 132. PIGUET AND BRETSCHER, 1913: 21. POINTNER, 1914 b: 96. SCHUSTER, 1915: 9, 64. STEPHENSON, 1922 b: 278; 1932: 229. SVETLOV, 1924: 190. WOLF, 1928: 386. UDE, 1929: 23. TIMM, 1959: 24.

Chaetogaster sp. SVETLOV, 1924: 190.

(?) *Chaetogaster crystallus* (Gruith). CHEN, 1944: 2.

$l=2.5$-7 mm. Prostomium inconspicuous, with a median incision. Setae in II 4-13 per bundle, in the rest 4-6. Transverse stomachal ducts 20-22. A statocyst in the brain. Spermathecae with large ampulla and short duct; vasa deferentia long, narrow, straight; atrial ampulla round, duct short; penial setae 2 per bundle; a small ovisac formed in VII.

Europe, Asia, N. America, Africa (Abyssinia). Found also in brackish water (Finland).

The identity of *C. crystallus* of Chen (1944), attributed to Gruithuisen, cannot be determined as no description was given. I assume that this is simply an erroneus citation of *C. cristallinus*.

Chaetogaster setosus Svetlov, 1925
Fig. 7. 1Q, R

Chaetogaster setosus SVETLOV, 1925: 473, Pl. I, figs. 1-2.

Chaetogaster setosus Svetlov. LASTOCKIN, 1937: 233. SPERBER, 1948: 71; 1950: 53. CEKANOVSKAYA, 1962: 210, Fig. 124. FINEGENOVA, 1962: 221. NAIDU, 1962 a: 132.

$l=0.66$ mm, $s=13$. Prostomium reduced. Setae simple-pointed, curved distally; those of II 9-10 per bundle, with nodulus $\frac{1}{3}$ from the proximal end, in the following segments 5-8 per bundle, without a nodulus. Oesophagus well developed. Brain with a deep median incision in the hind border.
    Russia.

Chaetogaster limnaei von Baer, 1827
Fig. 7. 1O, P

Prostomium vestigial. Setae usually numerous, with strongly hooked teeth. Oesophagus very short. Commensal on pulmonates.

Sperber (1948) decided that the simplest way of dealing with the variation within the forms inhabiting snails as commensals or parasites (subspecies *vaghini*) was to split them up as a series of subspecies. It should be noted that the parasitic form *vaghini* is separated ecologically from *C. limnaei limnaei*, which may occupy the outer surface of a snail also parasitized internally by *vaghini* (Gruffydd, 1965). The separation of other subspecies of *C. limnaei* from *C. limnaei limnaei* is based on few reliable characters and would appear to be in need of reappraisal following a study of the types, if they exist.

Chaetogaster limnaei limnaei von Baer, 1827
Fig. 7. 1O, P

Chaetogaster Limnaei VON BAER, 1827: 611, Pl. XXX, figs. 23-24.

Chaetogaster furcatus EHRENBERG, 1828: 109, foot-note.

Nais vermicularis ? Müller. DUGÈS, 1837: 30, Pl. I, figs. 21-23.

Chaetogaster Linnaei Baer. GERVAIS, 1838: 15.

Chaetogaster furcatus Ehrenberg. GERVAIS, 1838: 15.

Mutzia heterodactyla Agassiz. VOGT, 1841: 36, Pl. II, figs. 13-15.

Gordius inquilinus Müller. GOULD, 1841: 213.

Chaetogaster diaphanus (Gruithuisen) (partim). ØRSTED, 1842: 138, Pl. III, figs. 2, 15-17.

Chaetogaster vermicularis Müller (partim). GRUBE, 1851: 105.

Choetogaster Linnaei Baer. D'UDEKEM, 1855: 554.

Choetogaster Limnaei Baer. D'UDEKEM, 1859: 24.

Chaetogaster Limnaei Baer. LANKESTER, 1869 a: 631, Pl. XLVIII, figs. 1-3, 12-13, Pl. XLIX, figs. 14-15, 17-27, 29-37; 1869 b: 272, Pl. XIV, Pl. XV, figs. 1-8; 1869 c: 102; 1871: 99. TAUBER, 1873: 399, Pl. XIII, figs. 1-7; 1879: 76. TIMM, 1883: 154. VEJDOVSKY, 1883: 220; 1884: 36, Pl. VI, figs. 16-18. LEVINSEN, 1884: 216. BEDDARD, 1895: 306. BRETSCHER, 1896: 511. MICHAELSEN, 1900: 22; 1903 b: 172; 1909 a: 11, Figs. 12-13; 1909 b: 131; 1926 a: 21; 1927, 8, Fig. 2. SMITH, 1900: 443; 1918: 638. WILLCOX, 1901: 905. LEVANDER, 1901: 12. MOORE, 1906: 169. PIGUET, 1906: 205. PIGUET AND BRETSCHER, 1913: 20. SCHUSTER, 1915: 7, 64. STEPHENSON, 1923: 50. SVETLOV, 1924: 190. WOLF, 1928: 388. KONDO, 1936: 383, Pl. XXIII, fig. 3. SCIACCHITANO, 1938: 257. CHEN, 1940: 28; 1944: 2. MARCUS, E., 1947: 2. BRINKHURST, 1962: 318; 1963 a: 17, Fig. 2e; 1964: 202, Fig. 1e. NAIDU, 1962 a: 132.

Chaetogaster Limnaeae Baer. VAILLANT, 1890: 447, Pl. XXII, figs. 24-25.

(?) *Chaetogaster limnaei* ? Baer. STEPHENSON, 1918, 9.

Chaetogaster limnaei limnaei Baer. SPERBER, 1948: 73, Fig. 3a, b, 7f, Pl. 1, fig. 4; 1950: 54, Fig. 3f, Pl. 1, fig. 4; 1960: 155. CEKANOVSKAYA, 1962: 208, Fig. 123. MOSZYNSKA, 1962: 9.

$l = 1.2$-5 mm, $s = 13$. Setae of II 5-20 per bundle, in the following segments 4-20. Intestinal plexus on the stomach forming an irregular network. No statocyst in the brain. Spermathecae with long duct; atria small, elongated; penial setae with a double hook.

Europe, Asia, N. America. Found also in brackish water (Finland).

Chaetogaster limnaei bengalensis Annandale, 1905

Chaetogaster bengalensis ANNANDALE, 1905: 117-120, Pl. III.

Chaetogaster bengalensis Annandale. STEPHENSON, 1918: 10-11; 1920: 195; 1923: 49-50. CHEN, 1940: 26-28, Fig. 3.
Chaetogaster limnaei bengalensis Annandale. SPERBER, 1948: 74. SOKOLSKAYA, 1961: 51, Fig. 1.

$l = 8$-10 mm. Setae 11-20 in all bundles. Vascular plexus on the stomach forming transverse and longitudinal ducts (?). A statocyst in the brain.
Asia.

Chaetogaster limnaei australis Davies, 1913

Chaetogaster australis DAVIES, 1913: 89, Pl. IX, figs. 1, 4, 6, 7a, 8-16.

Chaetogaster limnaei australis Davies. SPERBER, 1948: 75.

$l=0.88$-1.83 mm. Setae 8-11 per bundle, with long teeth. Nephridia with a large dilatation, and a diverticulum before the opening.
Australia. (In *Limnaea, Planorbis, Isidora.*)

Chaetogaster limnaei victoriensis Davies, 1913

Chaetogaster victoriensis DAVIES, 1913: Pl. IX, figs. 2-3, 5, 7b.

Chaetogaster limnaei victoriensis Davies. SPERBER, 1948: 75.

$l=6$-9 mm. Setae 5-9 per bundle, those of II longer than the rest, all with short, parallel teeth, and nodulus median or slightly distal.
Australia. (In *Isidora texturata.*)

Chaetogaster limnaei vaghini Gruffydd, 1965

Chaetogaster limnaei vaghini GRUFFYDD, 1965: 193, Fig. 2.

$l=0.7$-3.0 mm. Setae up to 8 per bundle, arranged in semi-circles. Genital setae on segment VI, 3 per bundle. No statocyst in brain. Found in kidneys of freshwater snails.
Great Britain.

Mature worms had 13-18 segments, with the clitellum from most of V, all of VI, part of VII. Ripe ova were seen in VI, possibly derived from a single ovary. Spermathecae occur in V, opening behind 4/5, and the atria lie in the anterior half of VI.

SPECIES INQUIRENDA

Chaetogaster krasnopolskiae Lastockin. LASTOCKIN, 1937: 233. SPERBER, 1948: 72; 1950: 54. CEKANOVSKAYA, 1962: 209. FINEGENOVA, 1962: 221. NAIDU, 1962 a: 132.
Russia.

GENUS **Amphichaeta** TAUBER, 1879

Type species: Amphichaeta leydigii TAUBER

Prostomium well developed. Dorsal setae from III on. Setae all bifid crotchets of similar form. Septa well developed. Oesophagus in IV and part of V, stomach in V-VI, intestine dilated in VII. No transverse commissural vessels. Cerebral ganglia fused anteriorly, with two long posterior lobes. Clitellum in V-VI; male funnels in VI opening into V; vasa deferentia straight, joining atria at lower part of anterior face; no penial setae; sperm-sac and ovisac formed. No budding during maturity.
Holarctic (?).

Amphichaeta

1. Setae up to IV with upper tooth shorter than lower — *A. magna*

— Setae up to IV with upper tooth as long as or longer than lower — 2

2. Posterior setae with upper tooth nearly three times longer than lower — *A. americana*

— Posterior setae with teeth about equally long — 3

3. Setae mostly four per bundle, reduced to two in ventral bundles of IV and V. Nephridia paired — *A. leydigii*

— Setae anteriorly four per bundle, posteriorly three. Nephridia single — *A. sannio*

Amphichaeta leydigii Tauber, 1879

Amphichaeta Leydigii TAUBER, 1879: 76.

Amphichaeta leydigii Tauber. VAILLANT, 1890: 443. BEDDARD, 1895: 304. MICHAELSEN, 1900: 20. SPERBER, 1948: 76; 1950: 54. CEKANOVSKAYA, 1962: 202.

Amphichaeta leydigi Tauber. KOWALEWSKI, 1910: 804; 1917 a: 39, Figs. 1-3; 1917 b: 77, Figs. 1-3. MICHAELSEN, 1926 b: 257 (partim); 1927 a: 8 (partim). WOLF, 1928: 385, Fig. 1. PIGUET, 1928: 79. UDE, 1929: 25 (partim).

Amphichaeta sp. GOLANSKI, 1911.

Amphichaeta sannio var. *multisetosa* LASTOCKIN, 1924 a: 14.

l (mature specimens)= up to 1·7 mm, (chains)=up to 4 mm, *s*=13-14. Setae of II four per bundle; in III dorsal setae usually 5, ventral ones 4 per bundle; in IV and V usually 2 per bundle; in the other segments most often 4, occasionally 1 to 3 per bundle. No vascular loop in the prostomium. Nephridia mostly paired, not connected with the ventral vessel, opening immediately in front of the ventral setae. Spermathecal ampulla tubular, duct indistinct; proximal part of the atrial ampulla bag-like, distal part tubular; vasa deferentia joining atria above atrial duct; atrial duct indistinct, with very wide lumen.

Europe. Fresh water.

Amphichaeta sannio Kallstenius, 1892
Fig. 7. 2A, B

Amphichaeta sannio KALLSTENIUS, 1892: 42, Figs. 1-5.

Amphichaeta sannio Kallstenius. BEDDARD, 1895: 304. MICHAELSEN, 1900: 20. KOWALEWSKI, 1917 a: 79; 1917 b: 47. SPERBER, 1948: 76, Fig. 71, m, Pl. 2, figs. 4, 5; 1950: 55, Figs. 4, 5. DAHL, 1960: 8. CEKANOVSKAYA, 1962: 203, Fig. 115, 116.

Amphichaeta leydigi Tauber. MICHAELSEN, 1926 b: 257 (partim); 1927 a: 8 (partim). UDE, 1929: 25 (partim). KNÖLLNER, 1935: 428.

(?) *Amphichaeta sannio* Kallstenius. LASTOCKIN, 1924 a: 14.

? *Amphichaeta asiatica* LIANG, 1958: 41, Pl. 1, fig. 1-6.

l (the first zooid of a chain)=1·5 mm. Ventral setae of II usually 4 per bundle; dorsal setae of III 4 per bundle; other setae usually 3 per bundle. Left branch of the ventral vessel making a loop into the prostomium. Nephridia frequently unpaired, closely connected with the ventral vessel, opening far in front of the ventral setae. Spermathecal ampulla globular or baglike, spermathecal duct distinct; vasa deferentia joining atria above the male duct; atria globular, atrial duct distinct, with narrow lumen.

Europe. In brackish water (mostly?), ? China.

A. asiatica differs from *A. sannio* in being larger (up to 3 mm) with setae longer and thicker and with a nodulus, and with two loops of the ventral vessel entering the prostomium. It was also fresh water in habit.

Amphichaeta americana Chen, 1944
Fig. 7. 2C-E

Amphichaeta americana CHEN, 1944: 2-3, Fig. 1.

Amphichaeta americana Chen. BRINKHURST, 1964: 203.

l=1·5 mm (3 mm in life). Prostomium somewhat elongate, ciliated anteriorly. No eyes. Setae all bifid crotchets, all alike, almost straight, very slightly curved toward distal end. Dorsal setae from III, all 3-4 per bundle. Ventral setae of segments II and III 2-4 per bundle with the upper tooth thinner than the lower, but only slightly longer. Setae of remaining bundles 3-5 per bundle with the upper tooth nearly three times longer than the lower. No coelomic corpuscles.

U.S.A. (1 locality).

Amphichaeta magna Sokolskaya, 1962
Fig. 7. 2F-I

Amphichaeta magna SOKOLSKAYA, 1962: 129, Fig. 1-2.

l=4-5 mm, *s*=14-19 (12 in a mature specimen). Prostomium elongate. No eyes. Dorsal setae from III, ventral setae from II, all bifid crotchets, up to IV, setae with upper tooth thinner and shorter than the lower, 4-7 per bundle. From IV onwards teeth of setae equally long, or the upper the longer, 3-4 per bundle.

Lake Baikal.

SUBFAMILY PARANAIDINAE SPERBER, 1948

Type genus: Paranais CZERNIAVSKY

No eyes. Dorsal setae from V onwards, all double-pronged crotchets. Pharynx with dorsal diverticulum; pharyngeal and septal glands present; stomach present. Nephridia absent or present, closed. Coelomocytes present. Testes and spermathecae in IV, ovaries and atria in V; no prostate gland cells. Four segments formed at anterior end, by budding.

Cosmopolitan.

GENUS: **Paranais** CZERNIAVSKY, 1880

Type species: Nais litoralis MÜLLER

Ventral setae all of one shape. One pair of septal glands, in IV. Dorsal vessel mid-dorsal; commissural vessels in I-IV forming a plexus; in some following segments a free loop. Clitellum present between male pores; penial setae present. No nephridia.

Cosmopolitan.

The limits of the genus were reduced by the re-erection of *Uncinais* Levinsen and *Homochaeta* Bretscher by Sperber (1948).

Paranais

1.	Segments II-III narrower than rest, with cutaneous glands. Ventral setae of 11, 8-9 per bundle	*P. macrochaeta*
—	Segments II-III as usual, no glands. Ventral setae fewer	2
2.	All setae with upper tooth longer than lower	*P. frici*
—	Some anterior setae with upper tooth longer than lower, the rest or even all setae with teeth about equal or the upper tooth shorter than the lower	3
3.	Ventral setae of 11 with teeth equally long, rest with shorter upper tooth than lower	*P. simplex*
—	Ventral setae of 11 with upper tooth longer than the lower, but progressively shorter until teeth equal posteriorly	4
4.	Commissural vessels in V-VII long, winding. Chloragogen very dense	*P. botniensis*
—	Commissural vessels in V-VII short, straight. Chloragogen less dense	*P. litoralis*

Paranais litoralis (Müller, 1784)

Nais litoralis MÜLLER, 1784: 120, Pl. LXXX, figs. 2-6.

Nais littoralis Müller. GMELIN, 1788-1793: 3122 (partim). LAMARCK, 1816: 223 (partim). TABL. ENC. METH., 1827: 134, Pl. LIV, figs. 4-10 (partim). BLAINVILLE, 1825: 129 (partim); 1828: 498 (partim). GRUBE, 1851: 104.

Blanonais littoralis (Müller). GERVAIS, 1838: 15 (partim).

Nais littoralis Ørsted (Müller ex parte). ØRSTED, 1842: 136.

Enchytraeus triventralopectinatus MINOR, 1863: 36.

Paranais littoralis (Ørsted). CZERNIAVSKY, 1880: 311. DE VOS, 1922: 276.

Uncinais littoralis (Ørsted). LEVINSEN, 1884: 218.

Paranais littoralis (Müller). BOURNE, 1891: 348, Pl. XXVI, fig. 2, Pl. XXVII, figs. 3-6. MOORE, 1905: 376.

Uncinais littoralis (Müller). BEDDARD, 1895: 295.

Paranais litoralis (Müller, Ørsted). MICHAELSEN, 1900: 18; 1909 a: 12, Fig. 16; 1927 a: 8, Fig. 3. WOLF, 1928: 388. UDE, 1929: 25, Fig. 20. KONDO, 1936: 384, Pl. XXIII, fig. 7.

Paranais litoralis (Müller). SMITH, 1918: 639. KNÖLLNER, 1935: 429, Figs. 1-4. HRABE, 1936: 1279. CERNOSVITOV, 1937 a: 71, Fig. 1. SPERBER, 1948: 83, Figs. 8a-d, Pl. 3. fig. 1; 1950: 55, Fig. 6a-c. TIMM, 1959: 24. DAHL, 1960: 8. BRINKHURST, 1962: 319; 1963 a: 18, Fig. 2h; 1963 b: 143; 1964: 204, Fig. 2a; 1966: 12. CEKANOVSKAYA, 1962: 199, Fig. 113. MOSZYNSKA, 1962: 9.

(?) *Paranais litoralis orientalis* (Müller). SOKOLSKAYA, 1964: 57, Fig. 1.

l=9-14 mm (living, mature), 2-3·5 mm (preserved), s=13-46. Dorsal setae slightly thinner than ventral. Ventral setae with nodulus distal; those of II 5-7 per bundle, slightly longer than the rest, with upper tooth longer than lower; in the remaining segments usually 2-3 per bundle, upper tooth posteriorly equally long as, or slightly longer than lower. Stomachal epithelium with intra-cellular canals. Simple commissural loops in V-VI, or V-VII. Spermathecal ampulla long, cylindrical, duct short; atrial ampulla large, cylindrical, extremely muscular, duct short, narrow; vasa deferentia extremely narrow, joining atria immediately above atrial duct; penial setae very stout, with a simple hook, or bifid.

Africa, Europe, Asia, N. America. Brackish or salt water.

Sokolskaya (1964) raises some doubts about the identity of Asian material after studying mature specimens, the characters of which differ in some details from the European forms. Mature N. American and Japanese forms have yet to be described.

Paranais frici Hrabe, 1941

Paranais frici HRABE, 1941: 20, Figs. 16-19.

Paranais frici Hrabe. SPERBER, 1948: 86, Fig. 8e-h; 1950: 56, Fig. 5e-h. CEKANOVSKAYA, 1962: 201, Fig. 114. TIMM, 1965: 38. BRINKHURST, 1966: 12.

$l=2.7$-5.7 mm (preserved), 9 mm (living), $s=32$. Ventral setae of II 2-4 per bundle, with nodulus median, and upper tooth more than twice as long as lower; in all other segments dorsally and ventrally 1-2 setae per bundle, or dorsally in V, 3 per bundle, with nodulus distal and upper tooth longer than lower. Stomach in VII, without intra-cellular canals. Simple, non-winding commissural vessels in V-VI. Penial setae shaped like ordinary ventral setae.
Africa, Europe, North America. In fresh and brackish water.

Paranais simplex Hrabe, 1936

Paranais litoralis (Müller). LASTOCKIN, 1922: 279.

Paranais simplex HRABE, 1936: 1268.

Paranais simplex Hrabe. HRABE, 1941: 22. SPERBER, 1948: 87; 1950: 56. CEKANOVSKAYA, 1962: 200.

l (mature)$=4$-4.5 mm, $s=28$-41. Ventral setae of II 5-6 per bundle, not longer than in neighbouring segments, with equally long teeth; in the remaining segments and dorsally 3-4 per bundle, with upper tooth shorter than lower; setae in anterior segments longer than the rest. No canals in the stomachal epithelium, no distinct stomachal dilatation. Simple commissural vessels in V-VII. Spermathecal ampulla large, ovoid, duct short, narrow; vasa deferentia short, straight, joining atria above atrial duct; atria ovoid or cylindrical, atrial duct short, narrow; penial setae shaped like ordinary setae, but fewer.
Russia. Found in fresh water.

Paranais botniensis Sperber, 1948

Paranais botniensis SPERBER, 1948: 87, Fig. 8i-m, Pl. 3, figs. 2-7.

Paranais botniensis Sperber. POPESCU-MARINESCU *et. al.* 1966: 163.

$l=6$-10 mm, $s=22$. Dorsal setae with distinct nodulus, 2-4 per bundle, slightly thinner than ventral, with teeth equally long. Stomachal dilatation in VIII, VII, without intracellular canals; intestine extremely wide; chloragogen excessively dense and thick. Commissural vessels in V-VI, or V-VII, simple but winding; pharyngeal plexus very dense and complicated. Clitellum in $\frac{1}{2}$ IV-VI; spermathecal ampulla small, globular, duct relatively long and strong, male funnels large, vasa deferentia very narrow and thin-walled, entering atria immediately above atrial duct; atrial ampulla oblong, muscular layer not extremely thick, duct long, well-defined; penial setae two per bundle, long, straight, double-pronged, with nodulus strongly distal.

Europe.

Paranais macrochaeta Cernosvitov, 1939
Fig. 7. 2J-M

Paranais macrochaeta CERNOSVITOV, 1939: 86, Fig. 15-19.

Paranais macrochaeta Cernosvitov. SPERBER, 1948: 90.

l (preserved)=3-4 mm, *s*=30-40. Dorsal setae 2-3 per bundle with teeth equally cutaneous glands, and with coelom traversed by numerous peritoneal strands. Dorsal setae 4-6 per bundle; ventral setae in II 8-9 per bundle, longer, straighter and thicker than the rest, displaced laterally; in III-IV 1-3 per bundle, in the remainder 4-5 per bundle; all setae with upper tooth much longer than lower. Stomach in VI (?).

South America (Titicaca basin). In fresh water. (2 specimens.)

SPECIES INQUIRENDAE

Paranais salina Cernosvitov, 1939
Fig. 7. 2N-Q

Paranais salina CERNOSVITOV, 1939: 84, Fig. 8-14.

Paranais salina Cernosvitov. SPERBER, 1948: 91.

l (preserved)=3-4 mm, *s*=30-40. Dorsal setae 2-3 per bundle with teeth equally long, ventral setae of II displaced laterally, 5 per bundle, with upper prong twice as long as lower, in III 3-4 per bundle, with upper tooth 1½ as long as lower, in the remainder 2-3 per bundle, with teeth equally long. Stomach in VIII (?); anus opening into a fossa with a rectangular incision in the dorsal wall. Septal glands present (?). Budding not observed (12 specimens).

South America (Titicaca basin). In oligohaline water.

Sperber (1948) noted that the location of this species in *Paranais* was uncertain owing to the inadequacy of the description.

Paranais palustris Udalicov, 1907
No description available.

Paranais papillosa Cernosvitov, laps. pro. *Pristina papillosa* Cernosvitov, Brinkhurst (1963),=*Peloscolex swirenkowi* (Tubificidae).

GENUS **Waspa** MARCUS, 1965

Type species: Waspa evelinae MARCUS

No eyes. Setae all bifid crotchets, starting in V dorsally, II ventrally. Glandular pharyngeal pouch in III, septal glands in IV. Nephridia closed, paired from V or IV on. Coelomocytes present. Spermathecae and testes in IV, atria and ovaries in V. No prostate. Four segments formed anteriorly in budding.

Brazil, China.

The genus differs from *Paranais* in possessing nephridia, but these lack funnels as in the Chaetogastrinae. It is impossible to distinguish the genus from *Neonais* Sokolskaya, 1962 because of the lack of any description of the reproductive organs and nephridia in the latter, but the nature of the body wall in all three species involved suggests a possible relationship, as well as the arrangement of the setae.

Body wall papillate, no swimming	*W. evelinae*
Body wall with foreign matter attached, swims	*W. mobilis*

Waspa evelinae Marcus, 1965
Fig. 7. 3A-C

Waspa evelinae MARCUS, 1965: 62, Fig. 1-10.

$l=5$ mm (single), up to 12 mm (chains), $s=40$-42 (mature worm). Prostomium broader than long, rounded. Body wall papillate. All setae bifid crotchets. Dorsal setae from V, shorter than ventrals, but otherwise the same in number and form. Ventral setae of II distinct, 3-6 per bundle, nodulus median or just above, upper tooth about twice as long as lower and thinner. In the rest, 2-4 (posteriorly 2-3), shorter than in II, teeth equal. Ventral bundle of V with 2 penial setae per bundle, thicker than the others, upper tooth longer and thinner than the lower. Pharynx in III, with glandular diverticulum. Septal glands in IV. Chloragogen from V. Stomach in VII or VIII. Coelomocytes 10-15 μ in diameter. Nephridia from V, paired, ventro-lateral, but lacking funnels. Clitellum from IV to VI. Short unpaired sperm-sac from 4/5, egg-sac from 5/6. Spermathecae in IV, spermathecal pores near ventral setae.
No swimming.
Brazil, brackish water.

Waspa mobilis (Liang, 1958)
Fig. 7. 3D-G

Paranais mobilis LIANG, 1958: 43, Pl. II, Fig. 20-25.

$l=4$·5-8 mm, $s=13$-16. Brownish, with foreign matter. Prostomium bluntly triangular. No eyes. Dorsal setae from V, 2 per bundle thinner than ventrals, nodulus median or proximal, upper tooth about twice as long as lower. Ventral setae of II 3-4, median nodulus, on III 2, shorter, nodulus more proximal, rest 2 per bundle, curved shaft upper tooth twice as long or longer than lower, 2 penial setae in each bundle of V. Few coelomocytes. Septal glands in IV. Stomach in VII. Dorsal vessel from VII on left. Nephridia from VI, no funnels. Sperm funnels in IV, vas deferens enters centre of atrium, atrium with large ampulla, sperm sac to VII or VIII. Spermathecae in IV, globular—ovoid. Sperm sacs to VIII or X.
Swims with spiral motion, fresh water.
Nanking, China.

This species is undoubtedly close to *W. evelinae* and may prove to be no more than a variant of it.

SUBFAMILY NAIDINAE LASTOCKIN, 1924

Type genus: Nais MÜLLER

Eyes present or absent. Pharynx with dorsal diverticulum; pharyngeal and oesophageal or septal glands present. Commissural vessels in I-V, more or less simple, or forming a plexus, additional free loops often present. Testes and spermathecae in IV or V, ovaries and atria in V or VI. Four of five segments formed anteriorly, by budding or fragmentation.

Cosmopolitan.

GENUS Homochaeta BRETSCHER, 1896

Type species: Homochaeta naidina BRETSCHER

No eyes. Dorsal setae from II, all double-pronged crotchets of similar shape as the ventral setae. Stomachal dilation probably present. Vascular commissures in anterior segments simple, or forming a network. Spermathecae and atria without ducts; no prostate gland cells; penial setae present.

Eurasia, ? Africa, ? S. America.

Homochaeta

1.	Prostomium pointed, elongate. Anterior dorsal setae with upper tooth longer than lower (or all simple pointed)	*H. naidina*
—	Prostomium short. No setae with upper tooth longer than lower.	2
2.	All setae with the upper tooth shorter than the lower	*H. setosa*
—	All setae with teeth equally long	*H. lactea*

Homochaeta naidina Bretscher, 1896

Homochaeta naidina BRETSCHER, 1896: 508.

Naidium naidina (Bretscher). BRETSCHER, 1899: 393.

Paranais naidina (Bretscher). MICHAELSEN, 1900: 18; 1909a: 12. PIGUET and BRETSCHER, 1913: 16. UDE, 1929: 26. KONDO, 1936: 384, Pl. XXIII, fig. 6.

(?) *Osaka shimasakii* KONDO, 1936: 383, Pl. XXIII, fig. 5.

(?) *Paranais heteroseta* KONDO, 1936: 384, Pl. XXIII, fig. 8.

(?) *Paranais shimasakii* (Kondo). CERNOSVITOV, 1939: 87.

(?) *Paranais japonica* CERNOSVITOV, 1939: 88.

Homochaeta naidina Bretscher. SPERBER, 1948: 92; 1950: 57. BRINKHURST, 1962: 319; 1963 a: 18; 1963 b: 143. CEKANOVSKAYA, 1962: 198. MOSZYNSKA, 1962: 9.

$l=8$ mm, $s=20$. Prostomium elongated, pointed. Anterior segments pigmented. Dorsal setae of II-V 5-6 per bundle, longer and slenderer than the rest, with longer upper tooth; in the other dorsal bundles, and in all ventral bundles setae 3-5 per bundle, with teeth equally long. Some or all setae simple-pointed. Stomach in VIII (?). Transverse vessels in I-V forming a capillary net-work.

Europe (Switzerland), Japan.

The setae of *O. shimasakii* are all simple-pointed, those of *P. heteroseta* (=*P. japonica*) are simple-pointed in posterior dorsal bundles.

Homochaeta setosa (Moszynski, 1933)

(?) *Paranais multispinus* MICHAELSEN, 1914: 150, Pl. IV, fig. 4.

Paranais setosa MOSZYNSKI, 1933: 141.

Paranais setosa Moszynski. POPESCU-MARINESCU *et al.*, 1966: 163.

Homochaeta setosa (Moszynski). SPERBER, 1948: 93, Fig. 27c; 1950: 57. MOSZYNSKA, 1962: 10. BRINKHURST, 1963b: 143.

$l=$up to 10 mm, $s=$up to 50. Prostomium short, obtuse. No pigment. All setae of the same shape, with lower tooth longer than upper; in II and III 4-8 setae per bundle, in the other segments up to 12. Stomach in VIII (?). Clitellum in V-$\frac{1}{2}$VIII; spermathecal ampulla pear-shaped, thin-walled, opening directly outwards; vasa deferentia short, straight, narrow; atria elongated, lower part narrow, curved forwards, distal part wide, muscular, opening directly outwards into an invagination; penial setae 5 per bundle, wholly resembling ordinary setae.

Europe, ? Africa.

SPECIES INCERTA SEDIS

Homochaeta lactea (Cernosvitov, 1937)

Paranais lacteus CERNOSVITOV, 1937b: 141.

Homochaeta lactea (Cernosvitov). SPERBER, 1948: 93.

$l=$ 5 mm, $s=30$. Prostomium short, rounded. Setae all with teeth equally long, in the anterior segments 6-8 per bundle, in the rest 4-5; in II-VI longer than the rest. Pharynx in II-III, stomach in VII (?). Dorsal vessel mid-dorsal, simple vascular loops in III-VI. Chloragogen from VII. First nephridia in IX. Coelomocytes present. No budding observed.

Swims with lateral movements.

S. America (Argentine).

Described from a single specimen, this species is only provisionally placed here (Sperber, 1948).

SPECIES INQUIRENDAE

Considered here because of the reconstruction of the genus *Paranais*, q.v.

Paranais tenuis Cernosvitov, 1937 (see *Aulodrilus*—Tubificidae).

Paranais elongata Pierantoni, 1909 (? tubificid).

Paranais (?) *chilensis* Michaelsen, 1926 d (=*P.* ? *filiformis* (Beddard) Michaelsen, 1926 b: (partim).
The above may properly be tubificids (Sperber, 1948).

Chaetogaster filiformis Schmarda=*Schmardaella filiformis* (Schmarda), Michaelsen, 1925=*Paranais* ? *filiformis* (Schmarda), (Beddard), Michaelsen 1926b: (partim).

Probably not a naidid (Sperber, 1948), or ? in *Allodero* (q.v.)

GENUS **Specaria** SPERBER, 1939

Type species: Nais josinae VEJDOVKSY

No eyes. Dorsal setae from VI, hairs and bifid needles; ventral setae all of one type. Pharyngeal and oesophageal glands present; a stomachal dilatation present. Vascular system in I-V forming a net-work. Coelomocytes present. Clitellum absent round male pores; spermathecae opening laterally; vasa deferentia with prostate gland cells, joining atria immediately above atrial duct; atria without prostate cells, opening into eversible pockets; penial setae present.
Holarctic.

Specaria josinae (Vejdovsky, 1883)
Fig. 7. 3H-J

Nais Josinae VEJDOVSKY, 1883: 218.

Nais Josinae Vejdovsky. VEJDOVSKY, 1884: 29, Pl. II, figs. 25-28, Pl. III, figs. 1-4. VAILLANT, 1890: 373. PIGUET, 1906: 229. PIGUET AND BRETSCHER, 1913: 30.

Nais josinae Vejdovsky. BEDDARD, 1895: 288. MICHAELSEN, 1900: 26; 1909 a: 16. FUCHS, 1907: 452. SCHUSTER, 1915: 40, 70, Fig. 25. STEPHENSON, 1925 b: 1310. UDE, 1929: 44.

Specaria josinae (Vejdovsky). SPERBER, 1939: 3, Fig. 2; 1948: 95, Figs. 9, 26d, Pl. IV, Pl. V, fig. 1; 1950: 58, Fig. 7. HRABE, 1941: 4, Figs. 13-15. TIMM, 1959: 25. SOKOLSKAYA, 1961: 52; 1962: 131. BRINKHURST, 1962: 319, 325; 1963 a: 25, Fig. 5a; 1963 b: 143; 1964: 205, Fig. 2b. CEKANOVSKAYA, 1962: 191, Fig. 107. MOSZYNSKA, 1962: 10.

$l=3$-10 mm, $s=14$-53. No pigment. Dorsal setal bundles consisting of 2-6 hairs and 2-6 needles, curved, bifid, with nodulus distal and teeth about equally long; ventral setae all of one type, with a faint nodulus, slightly distal, or almost median, and upper tooth longer than lower. Pharynx in II-III, oesophagus in IV-VI, stomach in VII-VIII. A pair of branching transverse vessels in VI; dorsal vessel ventrally to the left. Nephridial tubes packed closely round ventral vessel. Clitellum in V-$\frac{1}{2}$VIII; spermathecal ampulla, roundish, duct distinct; vasa deferentia with thick prostate, opening laterally into lower part of atria; atria thin-walled, small; atrial duct short; penial setae 1-3 per bundle, medially from pockets, with a simple hook distally.

No swimming.

Europe, Bear Island, N. America, Asia.

GEVAIS **Uncinais** LEVINSEN, 1884

Type species: Nais uncinata ØRSTED

Eyes present. Dorsal setae from VI on, double-pronged crotchets only; ventral setae all of one type. A stomachal dilatation present; pharyngeal and oesophageal gland cells present. Dorsal vessel to the left. Coelomocytes present. Clitellum absent round male pores; vasa deferentia with prostate gland cells, joining atria immediately above atrial duct; atria without prostate, duct poorly defined; penial setae present.

Holarctic (?).

Uncinais

1. All setae with upper tooth longer and thinner than lower   *U. uncinata*

— All setae with the upper tooth shorter or as long as the thicker lower tooth   *U. minor*

Uncinais uncinata (Ørsted, 1842)
Fig. 7. 4A

Nais uncinata ØRSTED, 1842: 136.

Nais uncinata Ørsted. GRUBE, 1851: 104. TAUBER, 1879: 75. TIMM, 1883: 154.

Paranais uncinata (Ørsted). CZERNIAVSKY, 1880: 311. MICHAELSEN, 1900: 19; 1903 b: 170; 1909 a: 12; 1923: 31; 1926 c: 3; 1927 a: 9. PIGUET, 1906: 194, Pl. IX, figs. 1-7; 1909: 173, Pl. III, fig. 1; 1919 a: 789. PIGUET AND BRETSCHER, 1913: 16. WOLF, 1928: 389, Fig. 4. MALEVICH, 1929: 43. UDE, 1929: 26.

Uncinais uncinata (Ørsted). LEVINSEN, 1884: 218. BEDDARD, 1895: 296.

BRETSCHER, 1899: 392. SPERBER, 1948, 98, Fig. 26g, Pl. V, fig. 2-4; 1950: 58, Fig. 8. CAIN, 1959: 193, Fig. 1. SOKOLSKAYA, 1961: 52, Fig. 2; 1962: 131. BRINKHURST, 1962: 319; 1963 a: 18, Fig. 2f; 1963 b: 143; 1964: 205, Fig. 2c.

?*Ophidonais uncinata* (Ørsted). VEJDOVSKY, 1884: 24.

Ophidonais uncinata (Ørsted). VAILLANT, 1890: 355.

$l=5$-18 mm, $s=31$-54. Dorsal setae 2-4 per bundle, slightly shorter than ventral with nodulus distal; ventral setae 2-7 per bundle, those of II longer than the rest, with nodulus proximal; behind VI nodulus distal; all setae with upper tooth longer, and lower tooth thicker. Stomach beginning in VIII. Transverse vessels in II-V forming a network; a network in the body wall in I-VIII; dorsal vessel to the left. Clitellum in V-VIII; spermathecal ampulla roundish or ovoid, duct fairly long, well-defined; vasa deferentia straight, with sparse prostate cells, joining atria immediately above atrial duct; atria very small, sometimes opening into a pit; penial setae with a simple hook; female funnels unusually large. Body tapering backwards.

N. America, Europe, Turkestan. Found in brackish water also (the Baltic).

Cain (1959) described a series of brown pigment bands on the dorsal side of the anterior end, which is held in a bent position protruding from the sediment in life.

Uncinais minor Sokolskaya, 1962
Fig. 7. 4B-E

Uncinais minor SOKOLSKAYA, 1962a: 131, Fig. 3.

$l=2.8$-4.4 mm (mature), chains 3.7-4.4 mm, $s=26$-44. Prostomium elongate, triangular, about as broad as long. Eyes present. Dorsal setae from VI (V in one specimen), bifid setae with upper tooth shorter and thinner than the lower, nodulus distal. Ventral setae 5-7 per bundle in II-V, those of II slightly longer than the rest, from VI, setae shorter and thicker, 2-6 per bundle; upper tooth thinner than the lower and as long or shorter than the same. Penial setae in VI large, straight, simple-pointed, 4 per bundle. Clitellum from setal line of V to end of VII or setal line of VIII. Male pores on VI in depressions near setal line. Vas deferens straight, surrounded by prostatic cells, opens to anterior ventral part of atrium. Atria spherical, thin walled. Spermathecae in V, ampullae elongate, thin walled, duct short, thick walled. Coelomocytes present.

Swims with spiral motion.
Lake Baikal.

GENUS **Ophidonais** GERVAIS, 1838

Type species: Nais serpentina MÜLLER

Eyes present. Body wall with scattered papillae. Dorsal setae from VI, all stout, straight needles; ventral setae, all of one form. Pharyngeal and oesophageal glands present; stomachal dilatation present. Vascular system with transverse commissural vessels in II-V, simple or anastomosing; dorsal vessel in the middle line. Coelomocytes present. Clitellum absent round male pores; vasa deferentia, covered with prostate glands, joining atria immediately above atrial duct; atria devoid of prostate gland cells, atrial duct poorly defined; penial setae present. Europe, Asia, N. America, S. America.

Ophidonais serpentina (Müller, 1773)
Fig. 7. 4F-H

Das Mercurschlängelein RÖSEL VON ROSENHOF, 1755: 568, Pl. XCII, figs. 1-13.

Die geschlängelte Naide MÜLLER, 1771: 84, Pl. IV, figs. 1-4.

Nais serpentina MÜLLER, 1773: 20.

Nais serpentina Müller. GMELIN, 1788-1793: 3121. MODEER, 1798: 121. OKEN, 1815: 364. LAMARCK, 1816: 223. BLAINVILLE, 1825: 129; 1828: 498. TABL. ENC. METH., 1827: 134, Pl. LIII, figs. 1-4. JOHNSTON, 1845: 443. GRUBE, 1851: 104. D'UDEKEM, 1855: 551; 1859: 21. LANKESTER, 1869 c: 102; 1871: 99. TAUBER, 1879: 74. TIMM, 1883: 154. LEVINSEN, 1884: 220. BEDDARD, 1895: 285. BRETSCHER, 1896: 507. SMITH, 1900: 443.

Ophidonais serpentina (Müller). GERVAIS, 1838: 19. VEJDOVSKY, 1883: 217; 1884: 27, Pl. III, figs. 14-16 STOLC, 1886 b: 503. VAILLANT, 1890: 354, Pl. XXIII, fig. 12. BOURNE, 1891: 345. MICHAELSEN, 1900: 22; 1903 b: 172, Figs. 1-2. PIGUET AND BRETSCHER, 1913: 22. SCHUSTER, 1915: 9, Figs. 1-4. SMITH, 1918: 639. WOLF, 1928: 388. UDE, 1929: 27, Figs. 21-24. CORDERO, 1931: 349. SCIACCHITANO, 1938: 257. SPERBER, 1948: 100, Fig. 10, 26f, Pl. V, fig. 5, Pl. VI, fig. 1; 1950: 59, Fig. 9, Pl. I, fig. 6; 1958: 46. ERCOLINI, 1956: 6. TIMM, 1959: 6. BRINKHURST, 1962: 319; 1963 a: 18, Fig. 2g; 1963 b: 143; 1964: 206, Fig. 2d. CEKANOVSKAYA, 1962: 195 Fig. 111. MOSZYNSKA, 1962: 11. SOKOLSKAYA, 1962: 133.

Serpentina quadristriata ØRSTED, 1842: 134.

Serpentina quadristriata Ørsted. JOHNSTON, 1865: 70.

Slavina serpentina (Müller). BOUSFIELD, 1886 a: 266, Pl. XXXIII, fig. 5.

Ophidonais reckei FLOERICKE, 1892: 469.

Nais reckei (Floericke). BEDDARD, 1895: 289.

Ophidonais reckei Floericke. MICHAELSEN, 1900: 23; 1909 a: 11. SOUTHERN, 1909: 132, Pl. VII, fig. 1.

Ophidonais serpentina (Müller) var. *meridionalis* PIGUET, 1906: 206, Pl. IX, figs. 8-17.

Ophidonais serpentina (Müller) f. *typica.* MICHAELSEN, 1909 a: 11. POINTNER, 1914 b: 98. SVETLOV, 1926 a: 250.

Ophidonais serpentina (Müller) var. *meridionalis* Piguet. MICHAELSEN, 1909 a: 11, Figs. 14-15. PIGUET AND BRETSCHER, 1913: 24. SVETLOV, 1926 a: 251. MALEVICH, 1927: 5; 1929: 44.

Ophidonais sp. KLEIBER, 1911: 22.

l =6-36 mm, *s*=35-97. Anterior end of body with 3-4 transverse pigment stripes. Body wall with sensory papillae; often a thin crust of foreign particles. Dorsal setae 1 per bundle, very stout, with blunt, double-pointed or simple distal end; ventral setae 2-6 per bundle, with longer upper prong, those of II longer than the rest, in II-V with nodulus proximal or median, in the following segments nodulus median or distal. Pharynx stretching into IV; stomachal dilatation in VIII-X or IX-XI. Transverse vessels in II-IV, or II-V, simple or anastomosing. Clitellum in V-VII or V-VIII; spermathecal ampulla thin-walled, elongated, sometimes entering the sperm sac, duct long, narrow; vasa deferentia long, winding, with thick prostate covering; atria small, ejaculatory duct short, indistinct; penial setae 2-3 per bundle, distal end broadened or forming a simple hook; filamentous spermatophores formed.

Europe, N. and S. America, Siberia, Turkey.

GENUS **Nais** MÜLLER, 1773

Type species: Nais barbata MÜLLER

Eyes normally present. Anterior segments usually pigmented. Dorsal setae beginning in VI, hairs and palmate, pectinate, double or simple-pronged needles. Ventral setae of II-V mostly well differentiated from those of the following segments: Pharynx in II-III; pharyngeal and oesophageal glands present; stomach beginning in VII (occasionally VIII or IX). Vascular system with simple or anastomosing transverse vessels in I-V. Coelomocytes present. Clitellum in ½ V-VII, absent between the male pores; spermathecae with distinct ducts; vasa deferentia with prostate glands, joining atria immediately above atrial ducts; atria without prostate; penial setae present, with a simple or double hook.

Cosmopolitan.

Nais

1. Needles spatulate or pectinate   2
— Needles bifid or simple-pointed   3

2. Needles spatulate   *N. schubarti* (i.s.)
— Needles pectinate   *N. africana* (i.s.)

3. Needles simple-pointed   4
— Needles bifid   9

4. Ventral setae of II-V with long thin teeth, the upper very long, the lower half as long or less and closely applied to the upper, at least basally, or vestigial — *N. behningi*

— Ventral setae with (relatively) short teeth at an angle to each other, the lower not vestigial — 5

5. Needle setae hair-like with a thin tip — 6

— Needle setae with a broader, blunter tip — 7

6. Ventral setae behind V shorter, with teeth equally long. Hair and needle setae each up to 5 per bundle — *N. barbata*

— Ventral setae behind V with upper tooth longer than lower. Hairs and needles each 1-3 per bundle — *N. pseudobtusa*

7. Needles thick, with tip commonly reflected. Hairs serrate — *N. baicalensis* (i.s.)

— Needles without reflected tip. Hairs smooth — 8

8. All ventral setae with upper tooth as thick as but twice as long as the lower — *N. alpina*

— Ventral setae of II-V with upper tooth about twice as long as lower, but thinner, posterior ventral setae with upper tooth about as long as lower — *N. simplex*

9. Ventral setae of at least some segments behind V with some thickened setae or a single giant seta, the lower tooth short or rudimentary — 10

— Ventral setae behind V not thickened or replaced by giant setae — 11

10. Thickened setae from VII, some setae single giant setae with reduced lower tooth — *N. bretscheri*

— Thickened setae from VI, giant setae absent — *N. pardalis*

11. Needle tooth long and approximately parallel — *N. elinguis*

— Needle teeth short, diverging — 12

12. Ventral setae of II-V twice as long as the rest — *N. raviensis* (i.s.)

— Ventral setae of II-V slightly longer than or as
long as the rest   13

13. Spermathecae open into chamber intucked
from body wall   N. borutzkii
— Spermathecae do not open into such a
chamber   14

14. Stomach widening gradually   N. communis
— Stomach widening abruptly   N. variabilis

Nais communis Piguet, 1906
Fig. 7. 5E-H

(?) Die zungenlose Naide MÜLLER, 1771 : 74, Pl. II (? partim).

(?) Nais elinguis MÜLLER, 1773 : 22 (? partim).

(?) Nais elinguis Müller (? partim). MÜLLER, 1776 : 219. GMELIN, 1788-1793 :
3121. MODEER, 1798 : 130. BLAINVILLE, 1825 : 131; 1828 : 498. TABL. ENC.
METH., 1827 : 134, Pl. LIII, figs. 9-11. ØRSTED, 1842 : 135. GRUBE, 1851 :
104. D'UDEKEM, 1855 : 551; 1859 : 20. TAUBER, 1879 : 73. CZERNIAVSKY,
1880 : 308. TIMM, 1883 : 140, Pl. XI, figs. 20, 22. LEVINSEN, 1884 : 219.
DIEFFENBACH, 1885 : 98. STOLC, 1886 : 502. VAILLANT, 1890 : 369, Pl.
XXIII, fig. 11. BOURNE, 1891 : 344. BENHAM, 1892 : 212, Pl. VII, figs. 38-41.
BEDDARD, 1895 : 284. FERRONNIÈRE, 1899 : 254. SMITH, 1900 : 443. MICHAEL-
SEN, 1900 : 25. MOORE, 1906 a : 166. MUNSTERHJELM, 1905 : 13. GALLOWAY,
1911 : 303.

(?) Opsonais elinguis (Müller). GERVAIS, 1838 : 17. (? partim).

Nais heterochaeta BENHAM, 1893 : 383, Pl. XXXIII, figs. 1-5.

Nais communis PIGUET, 1906 : 247, Pl. X, fig. 9, Pl. XI, figs. 14-17, 19,
Pl. XII, fig. 11.

(?) Nais parvula WALTON, 1906 : 697, Fig. 7.

Nais parviseta WALTON, 1906 : 699, fig. 9.

Nais communis Piguet. MICHAELSEN, 1909 : 18, Fig. 25. PIGUET, 1913 : 116;
1928 : 80. PIGUET AND BRETSCHER, 1913 : 37. SCHUSTER, 1915 : 41, Fig.
27. SMITH, 1918 : 639. STEPHENSON, 1922 a : 255 : 1932 : 231. SVETLOV,
1924 : 193. STOLTE, 1927 : 1, Figs. 1-20. UDE, 1929 : 46, Fig. 56. MICHAEL-
SEN AND BOLDT, 1932 : 591. REDECKE AND DE VOS, 1932 : 15. SZARSKI,
1936 b : 389, Pl. XVIII, figs. 1-5, 9. PREU, 1937 : 258, Figs. 1-13.
WESENBERG-LUND, 1938 : 7. SCIACCHITANO, 1938 : 259. CHEN, 1940 : 33,
Fig. 4; 1944 : 5 (?) MARCUS, 1943 : 21, Pl. I, fig. 5, Pl. II, figs. 6-8.
SPERBER, 1948 : 602, Pl. VII, fig. 1; 1950 : 60, Fig. 10, Pl. I, fig. 7; 1958 :

46; 1960: 156. MARCUS, E., 1949: 1. ERCOLINI, 1956: 7. SAPKAREV, 1957: 138. TIMM, 1959: 25; 1962: 191, Fig. 1a. BRINKHURST, 1962: 319; 1963 a: 28, Fig. 5h; 1963 b: 143; 1964: 206, Fig. 2f. CEKANOVSKAYA, 1962: 184, Fig. 100. MOSZYNSKA, 1962: 11. NAIDU, 1962 a: 140, Figs. a-f. BOTEA, 1963: 337.

Nais variabilis var. *punjabensis* STEPHENSON, 1909 a: 255, Figs. 1-3, Pl. XV, figs. 1-8, Pl. XVI, figs. 9-18.

Nais variabilis var. *punjabensis* Stephenson. STEPHENSON, 1910 a: 66, Pl. VIII, figs. 1-2.

Nais communis var. *punjabensis* (Stephenson). PIGUET, 1909: 198. STEPHENSON, 1910 b: 237, Pl. XI, figs. 2, 4; 1918: 12; 1920: 196; 1923: 55, Figs. 11, 12. MICHAELSEN and BOLDT, 1932: 591.

Nais communis var. *caeca* STEPHENSON, 1910 b: 238, Pl. XI, fig. 3.

Pterochaeta astronensis PIERANTONI, 1911: 4, Pl. IV, figs. 1-12.

(non) *Nais communis* Piguet. SOUTHERN, 1913: 4.

(?) *Nais communis* var. *acuta* POINTNER, 1914 b: 101.

Nais communis var. *caeca* Stephenson. STEPHENSON, 1918: 12; 1923: 57, Fig. 13; 1925c: 44.

(non) *Nais communis* Piguet. STEPHENSON, 1931b: 39.

(?) *Nais communis* Piguet f. *magenta* MARCUS, 1943: 23, Pl. II, fig. 8.

(?) *Nais inflata* LIANG, 1963: 560, Fig. 1a-d.

l (simple)=1·5-6·6 mm, (chains)=1·8-12 mm, s=12-32. Eyes present or absent. Usually brown pigment in I-V. Dorsal setae needles 1-2 per bundle, with nodulus often inconspicuous, $\frac{1}{5}$-$\frac{1}{3}$ from the distal end, and with teeth short, diverging, usually distinct; hairs 1-2 per bundle. Ventral setae 2-6 per bundle, those of II-V with median nodulus, hardly longer and slightly straighter and thinner than the rest—here nodulus distal. Stomach slowly widening. Commissural vessels of I-V with frequent anastomoses; dorsal vessel to the left. Nephridial funnel with oblique border, glandular part long. Spermathecal ampulla large, reaching back through VI, proximal part bag-like, distal part tubular, duct narrow; vasa deferentia thick, with prostate only on their posterior part; atrial ampulla roundish, duct about as long as ampulla, swollen beneath the opening of the vas deferens; penial setae 2-3 per bundle.

No swimming.

Cosmopolitan. Found also in brackish water (the Netherlands).

Naidu (1962 a) believed that specimens described by Piguet (1928) and Marcus (1943) under the names *N. communis* and *N. communis magenta* should be recognized as a separate species, as suggested by Sperber (1948). Chen (1940) suggested that this and *N. variabilis* are in fact a single species, apparently because the Chinese specimens differ somewhat from the European form of this species. The author has found this distinction to be amongst the most difficult to make when using old, preserved material.

Nais inflata of Liang (1963) may belong here, though the needles have short parallel teeth. The difference between anterior and posterior ventral setae is slight so far as length and breadth are concerned.

Nais variabilis Piguet, 1906
Fig. 7. 5T-L, Fig. 7. 8N-Q

(?) *Nais elinguis* Müller. (?partim) of several authors (see *N. communis*).

(?) *Nais rivulosa* LEIDY, 1850: 43, Pl. II, fig. 2.

(?) *Nais rivulosa* Leidy. MINOR, 1863: 36. VAILLANT, 1890: 370.

Nais elinguis Müller. VEJDOVSKY, 1883: 218; 1884: 28, Pl. II, figs. 16-22; Pl. III, figs. 5, 6. STOLC, 1886 b: 504.

(?) *Nais obtusa* (Gervais). MICHAELSEN, 1902: 139. MICHAELSEN AND VERESCHAGIN, 1930: 213.

Nais variabilis PIGUET, 1906: 253, Pl. X, figs. 10-18; Pl. XI, figs. 18, 20, 21, 23, Pl. XII, figs. 12-13.

Nais variabilis Piguet. PIGUET, 1909: 195, Pl. III, figs. 9-11; 1928: 81. MICHAELSEN, 1909 a: 19, Fig. 26; 1926 a: 21. PIGUET AND BRETSCHER, 1913: 38. SCHUSTER, 1915: 42, 74. STOLTE, 1921: 538. STEPHENSON, 1922 b: 280. SVETLOV, 1924: 193; 1926 a: 251. MALEVICH, 1927: 7; 1929: 46. WOLF, 1928: 395. UDE, 1929: 47, Fig. 57. REDEKE AND DE VOS, 1932: 16. KONDO, 1936: 384, Pl. XXIII, fig. 10. SZARSKIČ 1936 b: 396, Pl. XVIII, fig. 16. CERNOSVITOV, 1930: 10; 1937 b: 144; 1938 b: 269. HRABE, 1939: 221. SPERBER, 1948: 107, Pl. VI, fig. 3, Pl. VII, fig. 2; 1950: 60, Pl. I, fig. 8; 1958: 46; 1960: 156. ERCOLINI, 1956: 7. SOKOLSKAYA, 1961: 53; 1962: 133. BRINKHURST, 1962: 319; 1963 a: 28; 1963 b: 143; 1964: 207. CEKANOVSKAYA, 1962: 186, Fig. 102. MOSZYNSKA, 1962: 11. TIMM, 1962: 192, Fig. 1bc. NAIDU, 1962 a: 140. BOTEA, 1963: 337.

(?) *Nais communis* Piguet. SOUTHERN, 1913: 4.

(?) *Nais japonica* KONDO, 1936: 385, Pl. XXII, fig. 12.

(?) *Nais bekmani* SOKOLSKAYA, 1962 a: 133, Fig. 4; 1962 a: 662, Fig. 2.

(?) *Nais menoni* NAIDU, 1962 a: 142, Fig. 8a-f.

l=3-10 mm, *s*=18-38. Pigment in I-V present or absent. Dorsal setae, hairs and needles 1-2 per bundle, needles with conspicuous nodulus, at 1/6-1/3 from the distal end, teeth short but obvious. Ventral setae 2-7 per bundle, those of II-V with nodulus slightly proximal or median, longer, straighter and thinner than those of the following segments; the rest with distal nodulus. Stomachal widening abrupt. Dorsal vessel to the left of the middle line. Glandular part of nephridia short. Spermathecal ampulla ovoid, confined to V, duct strong, dilated; vasa deferentia completely surrounded by strong gland cells; atria pear-shaped, atrial duct usually short, narrow or swollen; penial setae 2-3 per bundle.

Swims with spiral movements.

Cosmopolitan. Found also in brackish water (the Netherlands).

The species is extremely variable in size, pigmentation, size of setae, and in the shape of the setae . . .' Sperber (1948). The separation of this from *N. communis* is firmly based on (*i*) the sudden dilatation of the stomach, (*ii*) longer, thicker hair setae, (*iii*) ability to swim—characters I would consider to be dubious to say the least.

Nais bekmani of Sokolskaya 1962 (a, b) seems to be indistinguishable from this species. The distinction between *bekmani* and other *Nais* species was not fully discussed in either of the descriptions, which refer solely to *N. barbata*. In the paper dealing with *N. bekmani* Sokolskaya claims that *N. obtusa* (Gervais) of Michaelsen (1902) and Michaelsen and Verschagin (1930) is identifiable as this same species *N. bekmani*, not as *N. barbata* (which includes most records of *N. obtusa*).

Naidu (1962) distinguished *N. menoni* from *N. communis* on the shape of the needle setae, the position of the stomach, the absence of a lateral contractile blood vessel, size, and the ability to swim. The teeth of the setae as illustrated do not seem to agree well with the written description cited above. The main characteristic separating it from *N. variabilis* (not discussed by Naidu, 1962) seems to be the lack of a difference in length between the ventral setae of II-V and the rest. In view of the variability of the setal length and shape in *N. variabilis*, the identity of *N. menoni* may be questioned, its identity with *N. variabilis* cannot be established because the stomach does not widen abruptly, as in all but newly-budded specimens of this widely variable species (Sperber, 1948). In the terminal portion (lettered J1-J3, KI-K2) of the key to the genus *Nais* in Naidu (1962 a) the distinction between *N. raviensis* and the three following species is inaccurate, and the distinction between *N. variabilis* and both *N. menoni* and *N. communis* depends on the supposed slight difference between the lengths of the anterior and posterior setae of *N. variabilis*, a character shown to vary by Sperber (1948). The name *menoni* is not available as it is used in *Pristina*, recorded as part of *Nais* by Udekem (1855).

Nais simplex Piguet, 1906
Fig. 7. 5M-O, 7. 6A

Nais variabilis var. *simplex* PIGUET, 1906 : 260, Pl. XI, figs. 22, 24; Pl. XII, figs. 1-3, 14.

Nais simplex Piguet. PIGUET, 1909 : 202, Pl. III, fig. 12. MICHAELSEN, 1909 a : 19, Fig. 27. PIGUET AND BRETSCHER, 1913 : 34. STEPHENSON, 1922 a : 255; 1922 b : 281. HAYDEN, 1922 : 167. MALEVICH, 1929 : 47. UDE, 1929 : 50, Fig. 61. SPERBER, 1948, 110, Fig. 11a-c, 26 e, Pl. VII, figs. 3-5;

1950: 61, Fig. 11a-c, Pl. II, fig. 1, 1958: 47; 1960: 156. BRINKHURST, 1962: 319; 1963 a: 26, Fig. 5e; 1964: 207, Figs. 2g. CEKANOVSKAYA, 1962: 182, Fig. 98. MOSZYNSKA, 1962: 11. NAIDU, 1962 a: 139. TIMM, 1962: 193, Fig. 2a. Botea, 1963: 337.

(?) *Nais tortuosa* WALTON, 1906: 698, Fig. 8.

(?) *Nais sp.* STEPHENSON, 1909 b: 108.

(?) *Nais andina* CERNOSVITOV, 1939: 89, Figs. 26-33.

(?) *Nais sp.* Stephenson. CERNOSVITOV, 1942 b: 282.

(?) *Nais andina* Cernosvitov. SPERBER, 1948: 112. NAIDU, 1962 a: 139.

$l = 4\text{-}8$ mm, $s = 18\text{-}37$. Pigmentation variable. Dorsal setae hairs and needles 1-2 per bundle, needles simple-pointed, with nodulus more than $\frac{1}{3}$ from the distal end. Ventral setae of II-V, 2-6 per bundle, longer, straighter and thinner than the rest, with nodulus proximal and upper tooth nearly twice the length of the lower one; those following 2-5 per bundle, stouter, with teeth about equally long. Stomachal dilatation sudden. Dorsal vessel to the left of the middle line. Nephridial funnels with a thin membrane round the border. Spermathecal ampulla small, roundish or ovoid, duct long, with a strong dilatation before the outer opening; vasa deferentia almost straight, with gland cells on their posterior half; atrial ampulla roundish, ejaculatory duct narrow; penial setae 2 per bundle.

Swims with spiral movements.

Europe, N. America, Tibet (?), Africa.

A description of *N. andina* is appended under *nomina dubia*.

Nais bretscheri Michaelsen, 1899
Fig. 7. 6B-H

Nais bretscheri MICHAELSEN, 1899: 121.

Nais bretscheri Michaelsen. BRETSCHER, 1899: 389, Fig. 1. PIGUET, 1906: 267, Pl. X, fig. 19, Pl. XII, figs. 6, 16; 1909: 205, Pl. III, fig. 13. PIGUET AND BRETSCHER, 1913: 29.

Nais bretscheri Michaelsen. MICHAELSEN, 1900: 26; 1909 3: 17, Fig. 22; 1923: 31. SVETLOV, 1924: 193. WOLF, 1928: 393. UDE, 1929: 43, Fig. 53. CHEN, 1940: 35, Fig. 5. CERNOSVITOV, 1930: 10. SPERBER, 1948: 120, Fig. 13a-m; 1950: 63, Fig. 14a-m, ERCOLINI, 1956: 9, Fig. 1. TIMM, 1959: 25; 1962: 196, Fig. 3b. SOKOLSKAYA, 1961: 55, Fig. 3. CEKANOVSKAYA, 1962: 189, Fig. 105. MOSZYNSKA, 1962: 12. NAIDU, 1962 a: 140. BRINKHURST, 1963 b. 143.

(? non) *Nais bretscheri* Michaelsen. MICHAELSEN AND BOLDT, 1932: 591. CHEN, 1940: 35, Fig. 5.

Nais iorensis PATARIDZE, 1957: 91, Fig. 1-3.

Nais iorensis Pataridze. CEKANOVSKAYA, 1962: 190, Fig. 106.

$l=3$-7 mm, $s=19$-34. Eyes present or absent. Anterior end usually heavily pigmented. Dorsal setae and needles 1-2 per bundle; needles double-pronged, with nodulus $\frac{1}{4}$ from the distal end; hairs short. Ventral setae of II-V, 2-7 per bundle, thin and straight, with upper tooth about double the length of lower; those of the following segments 1-6 per bundle, three types being distinguishable: (*i*) normal setae with distal nodulus of 3 sorts of different thicknesses, with upper tooth 1-3 times as long as lower, but thinner, (*ii*) thick setae, longer and thicker than the former, with upper tooth as thick as, and 2-3 times as long as lower; (*iii*) in some cases giant setae, very long and thick with upper tooth abruptly bent at the base, and reduced lower tooth; giant setae always alone, thick setae 1-2 per bundle, often together with 1-2 normal setae. Stomachal dilatation slow. Dorsal vessel to the left. Spermathecal duct not ciliated; vasa deferentia with prostate gland cells; atria globular, penial setae 2 per bundle.
Swimming never observed.
Europe, Turkestan.

Cekanovskaya (1962) indicated that *N. iorensis* was identifiable as *N. bretscheri*, the only difference being the apparent duplication of the lower tooth of the giant ventral setae and the abrupt transition between the oesophagus and the stomach, a character held to be of significance in the separation of *N. communis* and *N. variabilis* by Sperber (1948).

The whole of what is now recognized as *N. pardalis* may be, in reality, no more than variants of this species.

Nais pardalis Piguet, 1906
Fig. 7. 6I-L

Nais Bretscheri Michaelsen var. *pardalis* PIGUET, 1906: 270, Pl. X, fig. 20, Pl. XII, figs. 4, 5, 17.

Nais pardalis Piguet. PIGUET, 1909: 206, Pl. III, figs. 14-16. MICHAELSEN, 1909 a: 17, Fig. 23. PIGUET AND BRETSCHER 1913: 30. SCHUSTER, 1915: 40. SVETLOV, 1924: 193. PIGUET, 1928: 81. WOLF, 1928: 395. UDE, 1929: 48, Fig. 58. SZARSKI, 1936 b: 397. MARCUS, 1943: 31, Pl. III, figs. 15-16, Pl. IV, figs. 17-18. SPERBER, 1948: 124, Fig. 13n-p, Pl. VIII, figs. 5-6; 1950: 63, Fig. 14n-p; 1958: 47, Fig. 1-2. TIMM, 1959: 25; 1962: 197, Fig. 4a, b. SOKALSKAYA, 1961: 57, Fig. 4. CEKANOVSKAYA, 1962: 187, Fig. 103. MOSZYNSKA, 1962: 12. NAIDU, 1962 a: 140. BRINKHURST, 1963 a: 26, Fig. 6b; 1964: 208, Fig. 3a.

(?) *Nais bretscheri* Michaelsen. MICHAELSEN AND BOLDT, 1932: 591. CHEN, 1940: 35, Fig. 5.

Nais lastockini SOKOLSKAYA, 1958: 297, Fig. 2.

Nais lastockini Sokolskaya. CEKANOVSKAYA, 1962: 188, Fig. 104.

$l=2\cdot5$-7 mm, $s=19$-32. Brown pigment anteriorly. Dorsal setae needles 1-2 per bundle, with nodulus about $\frac{1}{3}$ from the tip, and two fine, equal teeth; hairs

1-2 per bundle. Ventral setae of II-V, 2-5 per bundle, with upper tooth 1 $\frac{1}{2}$-2 times as long as lower, and nodulus median or slightly proximal; from VI on 1-5 per bundle, with distal nodulus, of two sorts: (*i*) normal setae of three kinds, with teeth equal, or upper tooth slightly longer or shorter than lower; (*ii*) usually thick setae, with upper tooth 2-3 times as long as lower. Stomachal dilatation sudden, elongated cells projecting into the lumen. Dorsal vessel to the left. Spermathecal ampulla ovoid, duct well defined, with a distal swelling; vasa deferentia surrounded by gland cells in front of the atria; atrial ampulla pear-shaped, thick-walled, duct poorly defined, narrow, curving forwards; penial setae 3 per bundle, of the common type.

Swimming with spiral movements.

Europe, Asia, N. and S. America.

The description of enlarged setae in *N. pardalis* further suggests that the original decision by Piguet (1906) that this was no more than a form of *N. bretscheri* may have been correct. The situation closely resembles that of the gradual merging of the characters of *Vejdovskyella comata* and *V. intermedia*, and I would think it quite possible that the description of more variants will lead to the inclusion of *N. pardalis* in *N. bretscheri*.

Nais elinguis Müller, 1773
Fig. 7. 6M-P

(?) *Die zungenlose Naide* MÜLLER, 1771: 74, Pl. II (? partim).

(?) *Nais elinguis* MÜLLER, 1773: 22 (? partim).

(?) *Nais elinguis* Müller (? partim). Several authors (see *N. communis*). MICHAELSEN, 1905 b: 306.

(?) *Opsonais elinguis* (Müller). GERVAIS, 1838: 17 (? partim).

Nais elinguis Müller. ØRSTED, 1842: 135 (partim). MICHAELSEN, 1900: 25 (partim); 1903 b: 175, Pl. I, fig. 4; 1909 a: 18, Fig. 24; 1926 a: 21; 1927 a: 10, Fig. 5. PIGUET, 1906: 241, Pl. X, fig. 8, Pl. XI, figs. 8-13, Pl. XII, fig. 10. PIERANTONI, 1911: 2, Pl. IV, figs. 13-16. PIGUET AND BRETSCHER, 1913: 36. SCHUSTER, 1915: 40. SMITH, 1918: 639. STOLTE, 1921: 547. STEPHENSON, 1922 b: 279; 1923: 58. DE VOS, 1922: 276. MEHRA, 1924: 179, Fig. 5, Pl. IV, figs. 11, 13, Pl. V, fig. 23; 1925: 412, Pl. XXXV, fig. 3. WOLF, 1928: 394. UDE, 1929: 45, Figs. 54, 55. CERNOSVITOV, 1930: 10; 1937 a: 70; 1939: 91, Figs. 34-42. REDEKE AND DE VOS, 1932: 16. KNÖLLNER, 1935: 434, Fig. 1. KONDO, 1936: 384, Pl. XXIII, fig. 11. HRABE, 1936: 1270, Fig. 11; 1937 b: 4, Fig. 1. SZARSKI, 1936 b: 396. SCIACCHITANO, 1938: 258. CHEN, 1944: 5. SPERBER, 1948: 127, Fig. 14a-c, Pl. IX, figs. 1-3; 1950: 64, fig. 15; 1958: 47; 1960: 157. ERCOLINI, 1956: 10. TIMM, 1959: 24; 1962: 199, Fig. 4c. DAHL, 1960: 8. BRINKHURST, 1962: 319; 1963 a: 27, Fig. 5g; 1963 b: 143; 1964: 208, Fig. 3b. CEKANOVSKAYA, 1962: 185, Fig. 101. MOSZYNSKA, 1962: 13. NAIDU, 1962 a: 140. BOTEA, 1963: 337.

Nais aralensis LASTOCKIN, 1922: 279.

$l = 2 \cdot 2$-12 mm, $s = 15$-37. Eyes present or absent. Anterior end usually pigmented reddish brown. Dorsal setae hairs and needles 1-3 per bundle, needles with nodulus $\frac{1}{4}$-$\frac{1}{3}$ from the tip, and teeth long, upper tooth longer than lower. Ventral setae 2-5 per bundle, those of II-V hardly longer, slightly straighter and thinner than (or less than half as thick as) the rest, with nodulus $\frac{1}{3}$-$\frac{1}{2}$ from the distal end, and with upper tooth about twice as long as lower; from VI on with nodulus distal, and upper tooth about twice as long and $\frac{1}{2}$-1 as thick as lower. Stomachal dilatation slow. Dorsal vessel mid-dorsal. Spermathecal ampulla large, elongated, duct long, narrow; vasa deferentia long, curved, wholly surrounded by abundant gland cells, joining atria immediately above atrial duct; atrial ampulla globular, extremely strongly muscular, duct long, well defined, wide; penial setae 4-5 per bundle, with a simple distal hook.

Swims with lateral movements.

Cosmopolitan, frequently in brackish water, abundant in organically polluted situations.

Nais borutzkii Sokolskaya, 1964
Fig. 7. 5A-D

Nais borutzkii SOKOLSKAYA, 1964: 60, Fig. 2-4.

$l = 3 \cdot 1$-8-0 mm, $s = 26$-48. Prostomium rounded-triangular or skittle-shaped. Eyes present. Dorsal setae from VI, 0-2 hair and 1-2 bifid setae, hairs thin and short, bifid with upper tooth thinner and shorter than lower. Pharynx from II-IV, oesophagus from mid IV. Septal (?) glands in III, oesophageal (?) glands in IV. Stomach begins in VII; chloragogen behind 5/6. Coelomocytes present. Clitellum $\frac{1}{2}$ V-$\frac{1}{2}$ VIII. Male pores lateral to and behind penial setae. Spermathecal pores in line of ventral setae, in front of them. Vas deferens with prostate glands. Atria in VI, large and round, thick muscular walls. Spermathecal ampulla sharply separated from duct, which communicates with the exterior via an intuck of the body-wall forming a reservoir, into which the duct projects freely.

Kamchatka, USSR. Brackish water.

The reservoirs on the spermathecal pores serve to separate this species from others in the genus, otherwise it is close to *N. elinguis* according to Sokolskaya (1964).

Nais alpina Sperber, 1948
Fig. 7. 6Q-S, 7. 7A

Nais alpina SPERBER, 1948: 113, Figs. 11d, f, 12a, Pl. VII, figs. 6-7, Pl. VIII, figs. 1-3.

Nais alpina Sperber. SPERBER, 1950: 61, Fig. 11 d, f. BRINKHURST, 1962: 319, 325; 1963 a: 26, Fig. 5f. NAIDU, 1962 a: 139. BOTEA, 1963: 337.

l (preserved) $= 3 \cdot 6$-3·8 mm, $s = 19$-22. Whitish opaque, sometimes with brown pigment anteriorly. Dorsal setae needles 1-2 per bundle obtusely simple pointed, with nodules $\frac{1}{3}$-$\frac{1}{4}$ from the distal end; hairs 1-2 per bundle. Ventral setae of II-V, 4-7 per bundle, longer, straighter and thinner than the rest, with teeth equally thick, upper tooth twice the length of the lower, and nodulus proximal; in the following segments 3-7 per bundle, with nodulus distal, and upper tooth half as thick as lower. Stomachal dilatation strong and sudden. Dorsal vessel to the left. Spermathecal ampulla large, oblong, hanging backwards, duct very long, narrow; vasa deferentia long, surrounded by numerous gland cells; atrial ampulla pear-shaped, atrial duct short, indefinite; penial setae 3 per bundle, single-pointed.

No swimming observed.

Europe.

Nais barbata Müller, 1773
Fig. 7. 7F-I

(?) *Anguille blanchâtre* BONNET, 1779: 195.

Die bärtige Naide MÜLLER, 1771: 80, Pl. III, figs. 1-3.

Nais barbata MÜLLER, 1773: 23.

Nais barbata Müller. MÜLLER, 1776: 219. GMELIN, 1788-1793: 3122. MODEER 1798: 138. BLAINVILLE, 1825: 131; 1828: 497. ØRSTED, 1842: 135, Pl. III, fig. 13. GRUBE, 1851: 104. D'UDEKEM, 1855: 551; 1859: 20. SEMPER, 1877: 67, Pl. III, figs. 9-14 (partim). TAUBER, 1879: 74. VEJDOVSKY, 1883: 218; 1884: 29, Pl. II, fig. 24. TIMM, 1883: 135 (partim). LEVINSEN, 1884: 219 (partim). STOLC, 1886: 504. VAILLANT, 1890: 369, Pl. XXII, figs. 14-15. BOURNE, 1891: 344. BENHAM, 1892: 214. BEDDARD, 1895: 283. BRETSCHER, 1896: 507. FERRONNIÈRE, 1899: 251 (partim). LEVANDER, 1904: 200. SPERBER, 1948: 116, Pl. VIII, fig. 4; 1950: 62, Fig. 12a-c, Pl. II, fig. 2; 1960: 156. TIMM, 1959: 24; 1962: 194, Fig. 2b. HARMAN, 1961: 93. SOKOLSKAYA, 1961: 54; 1962: 136. BRINKHURST, 1962: 319; 1963 a: 25, Fig. 5c; 1963 b: 143; 1964: 207, Fig. 2e. CEKANOVSKAYA, 1962: 181, Fig. 97. MOSZYNSKA, 1962: 12. NAIDU, 1962 a: 139.

(non) *Nais barbata?* Müller. TABL. ENC. METH., 1827: 134, Pl. LIV, figs. 2-3.

(?) *Opsonais obtusa* GERVAIS, 1838: 17.

Uronais barbata (Müller). GERVAIS, 1838: 18.

(?) *Nais Greeffi* FLOERICKE, 1892: 469.

Nais obtusa (Gervais). MICHAELSEN, 1900: 25; 1903 b: 178; 1905 c: 5; 1909 a: 20, Fig 29. PIGUET, 1906: 234, Pl. X, figs. 2-4; Pl. XI, fig. 5; Pl. XII, fig. 8; 1909: 188, Pl. III, figs. 2-7. SOUTHERN, 1909; 133, Pl. VII, fig. 2. PIGUET AND BRETSCHER, 1913: 31. POINTNER, 1914 b: 99. SCHUSTER, 1915: 43. STEPHENSON, 1923: 60. SVETLOV, 1924: 195; MALEVICH, 1927: 7. WOLF, 1928: 394. UDE, 1929: 49, Fig. 59. SCIACCHITANO, 1938: 258. SAPKAREV, 1957: 138.

(?) non *Nais obtusa* (Gervais). MICHAELSEN, 1901: 139. MICHAELSEN AND VERESCHAGIN, 1930: 213.

$l=3.5$-6 mm, $s=25$-33. Yellowish brown pigment anteriorly. Dorsal setae needles 2-5 per bundle, single-pointed, with a long, sharp tip, and only slightly distal nodulus; hairs 1-5 per bundle, stiff. Ventral setae 2-5 per bundle, those of II-V much longer, thinner and straighter than the rest, with upper tooth longer than lower and slightly thinner, nodulus proximal; those in the following segments with teeth equally long, upper tooth much stouter, nodulus distal. Stomachal dilatation sudden. Dorsal vessel to the left. Spermathecal ampulla elongated, when filled stretching back through the whole of VI, duct long, well marked off, abruptly dilated; vasa deferentia long, sinuous, covered with gland cells; atrial ampulla roundish, when filled elongated backwards, walls muscular, thickened round the opening of the vas deferens; ejaculatory duct distinct, narrow; penial setae 2-3 on each side, with a simple hook.

Swims with spiral movements.

Europe, Asia, N. America, Afghanistan.

The use of the name *N. obtusa* is in accord with the opinion of Beddard (1895) and Michaelsen (1900), who doubted the identity of "Die bärtige Naide" of Müller (1771), but Sperber (1948) stated that the original illustration was quite consistent with later descriptions including those of *Nais obtusa* (Gerv.) of most subsequent authors. The identity of *Opsonais obtusa* Gervais is much less certain. The identity of *N. obtusa* of Michaelsen (1901) and Michaelsen and Vereschagin (1930) is doubtful. It was said to be attributable to the new species *N. bekmani* by Sokolskaya (1962), which I hold to be most probably identical to *N. variabilis* (q.v.).

Nais pseudobtusa Piguet, 1906
Fig. 7. 7B-E

Nais barbata Müller, FERRONNIÈRE, 1899: 251 (partim).

Nais obtusa (Gervais) var. *pseudobtusa* PIGUET, 1906: 238, Pl. X, figs. 5-7, Pl. XI, figs. 4, 6, 7, Pl. XII, fig. 9.

Nais pseudobtusa Piguet. PIGUET, 1909: 193, Pl. III, fig. 8. PIGUET AND BRETSCHER, 1913: 33. MALEVICH, 1927: 8. SPERBER, 1948: 118; 1950: 83, Fig. 12d-f; 1960: 157. HARMAN AND PLATT, 1961: 93: BRINKHURST, 1962: 319; 1963 a: 26, Fig. 5d; 1963: 143; 1964: 208, Fig. 2h. CEKANOVSKAYA, 1962: 180, Fig. 96. MOSZYNSKA, 1962: 12. NAIDU, 1962 a: 139. TIMM, 1962: 195, Fig. 3a. BOTEA, 1963: 337.

Nais pseudoobtusa Piguet. MICHAELSEN, 1909 a: 20, Fig. 28. POINTNER, 1914 b: 98. HAYDEN, 1922: 167. SVETLOV, 1924: 194. WOLF, 1928: 395. UDE, 1929: 50, Fig. 60. SZARSKI, 1936 b: 397, Pl. XVIII, fig. 10.

Nais pseudoobtusa (Piguet), var.? MICHAELSEN, 1914: 151.

(?) *Pristina variabilis* FRIEND, 1916: 25.

(?) *Nais incerta* CERNOSVITOV, 1942 b: 252.

Nais pseudoptusa (Piguet). SAPKAREV, 1957: 136.

Nais koshovi SOKOLSKAYA, 1962: 138, Fig. 6.

$l=1.7$-6 mm, $s=17$-28. Brown pigment often present. Dorsal setae hairs and needles 1-3 per bundle, needles with long, pointed tip, and nodulus 1/3 from the tip. Ventral setae II-V, 2-5 per bundle, longer straighter and slenderer than the rest, with upper tooth 1-$\frac{1}{2}$ as long as lower and with proximal nodulus, the rest 2-6 per bundle, with distal nodulus, stouter than the former, and with lower tooth thicker. Stomachal dilatation sudden. Dorsal vessel to the left. Spermathecal ampulla ovoid, duct long, dilated. Male funnels very thick, with a median, ciliated lobe; vasa deferentia relatively short, thick, and straight, covered with glands from funnel to atrium; atrial ampulla globular, wall not thickened, duct well defined, narrow; penial setae 2-3, often with a vestigial distal tooth.

Swims with spiral movements.

Europe, N. America, Africa, Afghanistan.

The distinctions between this species and *N. barbata* are that the ventral setae from VI backwards are thicker than the preceding ones with teeth about equally long in the latter, but are all thin with the upper tooth longer than the lower in *N. pseudobtusa*, and, in the needles, the nodulus is almost median in *N. barbata*, but one-third the way from the tip in *N. pseudobtusa* (Sperber, 1948).

Nais koshovi strongly resembles this species, the needle setae being said to be sabre-shaped in the definition, but illustrated as straight and described as such in the discussion. No attempt was made by Sokolskaya (1962 a) to distinguish between *N. koshovi* and either *N. barbata* or *N. pseudobtusa*, but the ventral setae strongly resemble those of the latter.

Nais behningi Michaelsen, 1923
Fig. 7. 7J-N

Nais behningi MICHAELSEN, 1923: 34.

Nais behningi Michaelsen. SVETLOV, 1924: 194. BEHNING, 1929: 250. HRABE, 1941: 19, Figs. 1-7. SPERBER, 1948: 119, Fig. 12b; 1950: 63, Fig. 13. CEKANOVSKAYA, 1962: 183, Fig. 99. NAIDU, 1962 a: 139.

Dorsal setae hairs and needles 1-2 per bundle, needles with thin, simple-pointed distal end. Stomachal dilatation sudden. Ventral setae of II-V, 6-10 per bundle, longer straighter, and half as thick as the rest, upper tooth enormously long, strongly curved, lower tooth reduced or vestigial; those following 2-5 per bundle, with nodulus slightly proximal, and upper tooth twice as long as, and thinner than lower, or thickened setae with lower tooth somewhat reduced in length, teeth about equally broad at base. Spermathecal ampulla small, globular, thin-walled, duct twice as long as ampulla, well-defined, narrow, with thick walls; vasa deferentia moderately long, strongly curved, but not winding, in their whole course surrounded by gland cells and joining atria immediately above

the atrial duct; atria roundish or pear-shaped, with a thin muscular layer, duct short, narrow; penial setae 2-4 per bundle, with a simple hook.
Swimming?
E. and Central Europe, N. America.

SPECIES INCERTA SEDIS

Nais raviensis Stephenson, 1914
Fig. 7. 8A-C

Nais raviensis STEPHENSON, 1914: 324, Figs. 1-2

Nais raviensis Stephenson. STEPHENSON, 1915 b: 785; 1923: 65. MICHAELSEN AND BOLDT, 1932: 595. SPERBER, 1948: 130.

(?) *Nais tenuidentis* WALTON, 1906: 700 Fig. 10.

(?) *Nais communis* Piguet. STEPHENSON, 1931 b: 39.

l=3 mm. No eyes. Dorsal setae from VI on, 1 short hair per bundle and 1-2 needles, double-pronged, with short, diverging prongs, and nodulus ⅔ from the distal end; ventral setae 3-4 per bundle, those of II-V twice as long as, and slightly thinner than the rest, with nodulus proximal, and upper tooth much longer than lower; those following with distal nodulus, and lower prong thicker and slightly longer than upper. Stomachal dilatation inconspicuous. At least one pair of simple transverse vessels, in VI. Brain deeply bifid behind, with straight anterior border. No coelomocytes.
India, Sumatra, Kenya, S. Africa.

Nais schubarti Marcus, 1944
Fig. 7. 8D-F

Nais schubarti MARCUS, 1944: 52; Pl. X, figs. 40-41.

Nais schubarti Marcus. SPERBER, 1948: 130, Fig. 14d.

l (preserved)=2·5 mm. Dorsal setal bundles with 2 hairs and 1 spatulate needle with distal nodulus; ventral setae of II-V 3-5 per bundle, twice as long as, and thinner than the rest, with nodulus proximal, and distal tooth longer than proximal; those following 3 per bundle, with distal nodulus and teeth about equally long. Stomachal dilatation in VIII. Anus terminal.
S. America (Brazil).

Nais africana Brinkhurst, 1966
Fig. 7. 8K-M

Nais africana BRINKHURST, 1966: 135, Fig. 1A-C

l=2-6 mm, *s*=*c*. 50. Eyes present or absent. Dorsal setae from VI, 1 or 2 hairs, 1 or 2 pectinate needles. Ventral setae 3-4 per bundle, those of II-V thin, with the upper tooth thinner than but as long as the lower, the rest thicker with the upper tooth a little shorter than the lower. Penial setae in VI. Coelomocytes present.
S. Africa.

The species was placed in *Nais* pending a description of live and mature specimens. The specimens without eyes may simply be a result of preservation.

Nais baicalensis Sokolskaya, 1962
Fig. 7. 8G-J

Nais baicalensis SOKOLSKAYA, 1962: 136, Fig. 5.

l=3-5 mm, s=15-27. Prostomium triangular but rounded tip. Eyes present. Cuticle with granules packed tightly. Dorsal setae from V or VI, hairs short and feathered, 1-3 per bundle, needles thick, simple-pointed with tip often bent back. Ventral setae with upper tooth at least twice as long as but a little thinner than the lower in all bundles, setae of II-V slender with proximal nodulus, 4-5 per bundle, from VI setae shorter and thicker, median nodulus, 3-5 per bundle. Pharyngeal diverticulum in III-IV, stomach widens abruptly in VII, chloragogen from 6/7.

Swimming with spiral motion.

Lake Baikal.

The species is distinguished from *N. alpina* and *N. simplex* by virtue of its size, thickness and form of needle setae (which are thicker than the feathered hair-setae) and the patterned cuticle (Sokolskaya, 1962 a). The nature of the cuticle and the feathering of the hair setae are very odd characteristics, reminiscent of some species of *Vejdovskyella* and the position of this species in the genus *Nais* seems doubtful, especially in view of the absence of a description of the male efferent ducts.

SPECIES INQUIRENDAE

Nais tygrina ISOSSIMOV, 1949.

Nais tygrina ISOSSIMOV, 1949: 9 (nomen nudum).

Nais sp. (? *tygrina* Isossimov). SOKOLSKAYA, 1962: 139, Fig. 7.

l=6 mm, s=18-20. Prostomium short. Eyes present. Pigmented brown anteriorly, in transverse rows dorsally. Cuticle patterned with granules tightly packed. Anterior segments annulate. Dorsal setae from VI, 1-2 hairs and 1-3 needles, thin simple-pointed tips. Ventral setae from 4 to 6 or 7 setae per bundle, all with the upper tooth longer than the lower with nodulus proximal, those of II straighter and longer than those of III-V, all these markedly shorter than those from VI on. Digestive glands present. Stomach widens in VII.

Lake Baikal.

Sokolskaya (1962 a) states that the original account was very brief, mentioning only blue stripes anteriorly and double the number of setae found in *N. barbata* (=*N. obtusa*). The specimens obtained by Sokolskaya were not attributed to this species for certain, owing to a lack of similarity to *N. barbata*, which the original material was supposed to resemble. Neither was it attributed to a new species, because of a description of the reproductive organs. This did not prevent the same author from describing *N. baicalensis*. Again, the nature of the cuticle perhaps suggests some relationship to *Vejdovskyella*.

Nais andina Cernosvitov, 1939

Nais andina CERNOSVITOV, 1939: 89; Figs. 26-33.

Nais andina Cernosvitov. SPERBER, 1948: 112. NAIDU, 1962 a: 139.

Pigment present. Ventral setae 3-4 per bundle, about equally long in all segments, distal tooth longer and thinner than proximal, more markedly so in II-V; needles 1-2 per bundle, bluntly single-pointed, with nodulus at $\frac{1}{3}$ of the setal length from the tip; hair setae 1-2 per bundle, about 1-2 times as long as the needles, frequently absent in some or all bundles. Stomachal dilatation sudden (?).
    S. America (Peru).

The species resembles *N. simplex* and may prove to be identical when the sexual organs are described.

NOMINA DUBIA

Probably Tubificidae:

Nais filiformis Blainville, 1825

Nais coecilia Mayer, 1859

Nais papillosa Kessler, 1868

Nais gigantea Kessler, 1868

Probably Enchytraeidae:
Nais albida Carter, 1858

Probably Polychaeta:
Nais quadricuspida Fabricius, 1780

Nais marina Fabricius, 1780

Nais aequisetina Dugès, 1837

Nais picta Dujardin, 1839

Nais Carolina Blanchard, (Gay, 1849)

Nais auricularia Bosc, 1802

Nais bipunctata Delle Chiaje, 1827

Nais coccinea Delle Chiaje, 1827

Nais de Horatiis Delle Chiaje, 1827

GENUS **Slavina** VEJDOVSKY, 1883

Type Species: Nais appendiculata D'UDEKEM

Body wall provided with rows of sensory papillae, and usually surrounded by adhering foreign matter. Eyes present or absent. Dorsal setae beginning in IV or VI, non-serrated hairs and fine, bifid or simple needles, slightly curved distally. Stomach present; pharyngeal and oesophageal glands present. Coelomocytes present. Clitellum absent between male pores; vasa deferentia without prostate, joining atria above atrial duct; atria with or without prostate; penial setae present.

Cosmopolitan.

Slavina

1.	Dorsal setae from IV	2
—	Dorsal setae from VI	3
2.	Needles bifid, hairs elongate in V	*S. sawayai*
—	Needles simple-pointed, no elongate hairs	*S. isochaeta*
3.	No elongate hair setae, 1 hair and 1 needle per bundle	*S. evelinae*
—	Elongate hair setae on VI, 1-2 or 3 hairs and needles	*S. appendiculata*

Slavina appendiculata d'Udekem, 1855
Fig. 7. 8R, S; 7. 9A-C

(?) *Nais escherosa* GRUITHUISEN, 1828: 409.

(?) *Nais gracilis* LEIDY, 1850: 43, Pl. II, fig. 1.

Nais appendiculata D'UDEKEM, 1855: Pl. I, fig. 3.

Nais appendiculata d'Udekem. D'UDEKEM, 1859: 21. VAILLANT, 1890, 371. BRETSCHER, 1896: 508.

Nais gracilis Leidy. D'UDEKEM, 1859: 22. VAILLANT, 1890: 372-373. BEDDARD, 1895: 287.

Slavina appendiculata (d'Udekem). VEJDOVSKY, 1883: 219; 1884: 30, Pl. III, figs. 17-26. STOLC, 1886 b: 503. BOUSFIELD, 1886 a: 264, Pl. XXXIII, figs. 1-4. BOURNE, 1891: 345. BEDDARD, 1895: 287. BRETSCHER, 1900: 14. MICHAELSEN, 1900: 32; 1903 b: 185; 1909 a: 13, Figs. 17-18; 1913: 207. PIGUET, 1906: 282, Pl. XII, fig. 20. GALLOWAY, 1911: 303. PIGUET and BRETSCHER, 1913: 47. SCHUSTER, 1915: 34. SMITH, 1918: 639. STEPHENSON, 1923: 82, Figs. 28-29. SVETLOV, 1924: 190. MALEVICH, 1927: 5. WOLF, 1928: 391. UDE, 1929: 42, Figs. 51-52. SCIACCHITANO, 1938: 258.

CHEN, 1940: 43, Fig. 10. MARCUS, 1944: 54, Pl. X, fig. 42, Pl. XI, figs. 43-44, 50. SPERBER, 1948: 133, Fig. 15a, b, 266; 1950: 65, Fig. 16. TIMM, 1959: 24. SOKOLSKAYA, 1961: 59; 1962: 140. BRINKHURST, 1962: 319; 1963 a: 23, Fig. 4d; 1964: 209, Fig. 3c. CEKANOVSKAYA, 1962: 169, Fig. 88. MOSZYNSKA, 1962: 13. HARMAN, 1965 a: 565.

Nais lurida TIMM, 1883: 153, Pl. XI, fig. 25.

(?) *Slavina gracilis* (Leidy). VEJDOVSKY, 1884: 30. MICHAELSEN, 1900: 33. MOORE, 1906: 167. GALLOWAY, 1911: 303.

Nais lurida Timm. DIEFFENBACH, 1885: 104. BRETSCHER, 1899: 390. BEDDARD, 1895: 287. BRETSCHER, 1900: 15.

Slavina lurida (Timm). BOUSFIELD, 1886 a: 268.

Slavina punjabensis STEPHENSON, 1909 a: 272, Pl. XIX, figs. 41-45, Pl. XX, figs. 50-52.

Slavina punjabensis Stephenson. STEPHENSON, 1915 a: 793, Pl. LXXX, figs. 4-5. MEHRA, 1925: 413, Pl. XXXV, fig. 1.

(?) *Slavina* sp. STEPHENSON, 1916: 301, Pl. XXX, fig. 1.

(?) *Slavina montana* STEPHENSON, 1923: 84, Fig. 30.

Slavina truncata HARMAN, 1965 a: 566, Figs. 1-2.

$l=2$-20 mm, $s=23$-46. Eyes present. Dorsal setal bundles from VI on, consisting of 1-2 stout hairs (in VI 1-3 per bundle, strongly elongated), and 1-2 straight needles with distal part effilated, and tip often slightly distended; ventral setae 2-5 per bundle, in II-V thinner, and in II slightly longer than the rest, all with proximal nodulus, an angular proximal bend, and upper tooth thinner and slightly longer than lower. Pharynx in II-IV; stomach in VII or VIII. Dorsal vessel situated mid-dorsally; commissural vessels in II-V forming a net-work. Clitellum in V-VII; spermathecal ampulla long, extending back into the sperm-sac, duct long, narrow; male funnels cup-shaped, vasa deferentia narrow, entering atria immediately above atrial duct; atria large, thick-walled, with scattered prostate gland cells, atrial duct short, narrow; penial setae 3 per bundle, with a simple, sharply curved hook.
No swimming.
Europe, N. and S. America, S. and E. Asia, Africa, New Zealand.

Slavina truncata differs very little from *S. appendiculata*, apparently only by the supposed absence of eyes according to the key in Harman (1965), despite the fact that the three specimens were examined preserved and one had what might have been eyes obscured by foreign matter. The posterior ventral setae are said to have the upper tooth longer and thinner than the lower, and the illustration shows this to be so but not to any marked extent. The illustration of an anterior ventral seta clearly indicates fore-shortening of the upper tooth as it is not figured directly from the side. I cannot see any distinction between them and the ventral setae of *S. appendiculata*. The measurements of all setae mentioned by Harman (1965) fall in the size-range quoted by Sperber (1948).

Slavina isochaeta Cernosvitov, 1939
Fig. 7. 9D-G

Slavina isochaeta CERNOSVITOV, 1939: 88, Figs. 20-25.

Slavina isochaeta Cernosvitov. SPERBER, 1948: 133.

l (preserved)=2·5-4 mm, *s*=up to 43. No eyes. Dorsal setal bundles from IV, containing 2-3 thick, stiff hairs; none especially elongated, and 2-3 needles, tapering towards the distal end; ventral setae 4-5 per bundle, in II-V longer and thinner than the rest, with upper tooth half as long again as lower; in the other segments upper tooth almost as long as lower. Stomach beginning in IX (?).
    S. America (Titicaca basin).

Slavina sawayai Marcus, 1944
Fig. 7. 9H, I

Slavina sawayai MARCUS, 1944: 59, Pl. XII, figs. 48-49.

Slavina sawayai Marcus. SPERBER, 1948: 133.

l=30 mm, *s*=180. No eyes. Dorsal setae from IV, 1-3 hairs per bundle, those of V much longer than the rest, and 1-3 needles, with finely bifid distal end; ventral setae 6 per bundle, all with median nodulus, and upper tooth longer and much finer than lower. Stomach beginning in VII.
    S. America (Brazil).

    The species was described from a single specimen.

Slavina evelinae (Marcus, 1942)
Fig. 7. 9J, K

Peloscolex evelinae MARCUS, 1942: 157, Pl. I.

Slavina evelinae (Marcus). MARCUS, 1944: 57, Pl. X, fig. 45, Pl. XI, figs. 46-47. MARCUS, E., 1947: 5. SPERBER, 1948. 136.

l=30 mm, *s*=150-180. No eyes. Dorsal setae from VI onwards, 1 hair, none elongated, and 1 fine needle with thickened distal end; ventral setae 1-4 per bundle, in II-V shorter and thinner than the rest in immature specimens, longer in mature ones, with teeth equally long, the lower being thicker; in the other segments lower tooth longer and much thicker than upper; on maturity all setae longer. Stomachal dilatation in X. Clitellum in V-VII; spermathecal ampulla long, entering the sperm sac, duct long, narrow; opening of male funnels oblique, dorsal wall extremely long, vasa deferentia very short, opening closely above atrial duct; atrial ampulla large, ovoid, strongly reflected backwards, with thick, glandular epithelium, without any prostate gland cells; atrial duct long, narrow, thin-walled, curved, the two ducts opening into a small common "ejaculatory chamber", eversible, with thick, muscular wall; one penial seta with a strongly curved distal hook present on each side, in front of the atrial ducts. Fragmentation.

S. America (Brazil).

The nature of the glandular part of the atrium is unusual, so much so that Sperber (1948) regards this species as only doubtfully placed in *Slavina*, but so closely related to it that the creation of a separate genus is not justified.

GENUS **Vejdovskyella** MICHAELSEN, 1903

Type species: Bohemilla comata VEJDOVSKY

Eyes present or absent. Dorsal setae from V or VI, stout, serrated hairs, and fine, straight, simple-pointed needles without nodulus, with effilated distal end or curved pectinate needles. Ventral setae of II-V of similar shape as the posterior. Stomach present; pharyngeal and oesophageal glands present. Dorsal vessel situated dorsally; vascular loops in several anterior segments. Coelomocytes present. Clitellum absent between male pores; atria with prostate gland cells; vasa deferentia without; penial setae present.

Europe, Asia, N. America, Africa.

The name *Bohemilla* of Vejdovsky was shown to be pre-occupied by Michaelsen (1903 b), who changed the genus name to *Vejdovskyella*. In a later decision made in ignorance of this change, Strand (1928) changed the name of the genus to *Bohemillula*.

Vejdovskyella

| 1. | Needles numerous, simple, hair-like | *V. comata* |
| — | Needles pectinate | *V. hellei* |

Vejdovskyella comata (Vejdovsky, 1883)
Fig. 7. 10A-E

Bohemilla comata VEJDOVSKY, 1883: 218.

Nais hamata TIMM, 1883: 152, Pl. XI, fig. 24.

Bohemilla comata Vejdovsky. VEJDOVSKY, 1884: 28, Pl. II, figs. 1-7.
VAILLANT, 1890: 376. BOURNE, 1891: 344. MICHAELSEN, 1900: 30.

Macrochaeta intermedia BRETSCHER, 1896: 509.

Macrochaetina intermedia (Bretscher). BRETSCHER, 1899: 392. MICHAELSEN,
1900: 31; 1909 a: 21, Fig. 30; 1926 c: 5. PIGUET, 1906: 279, Pl. XII,
fig. 19. PIGUET AND BRETSCHER, 1913: 46. EKMAN, 1915: 147. WOLF,
1928: 391.

Vejdovskyella comata (Vejdovsky). MICHAELSEN, 1903 b: 184; 1909 a: 22, Fig. 31. MAULE, 1906: 302. SOUTHERN, 1909: 134. KLEIBER, 1911: 23. PIGUET AND BRETSCHER, 1913: 45. SCHUSTER, 1915: 27, Figs. 15-21. HAYDEN, 1922: 167. PIGUET, 1928: 83. WOLF, 1928: 390, Fig. 5. UDE, 1929: 37, Figs. 44-46. CERNOSVITOV, 1942 a: 206. SPERBER, 1948: 137, Fig. 15c-f, 26a; 1950: 65, Fig. 17c-f, Pl. II, fig. 3. TIMM, 1959: 24. SOKOL-SKAYA, 1961: 59, Fig. 5. BRINKHURST, 1962: 319; 1963 a: 23, Fig. 4c; 1963 b: 143; 1964: 210, Fig. 3d. CEKANOVSKAYA, 1962: 165, Fig. 85. MOSZYNSKA, 1962: 13.

Vejdovskyella comata (Vejd.) var *scotica* STEPHENSON, 1922 b: 281, Fig. 1.

Vejdovskyella comata (Vejd.) var *scotica* Stephenson. PIGUET, 1928: 86.

Vejdovskyella intermedia (Bretscher). PIGUET, 1928: 86, Fig. 3. UDE, 1929: 38, Fig. 47. SPERBER, 1948: 140; 1950: 66, Fig. 18; 1958: 48, Figs. 3-4. CEKANOVSKAYA, 1962: 166, Fig. 86. BRINKHURST, 1966: 12, Fig. 6a.

Bohemillula comata (Vejd.). STRAND, 1928.

(?) *Vejdovskyella* sp? ALTMAN, 1936: 14.

Vejdovskyella macrochaeta LASTOCKIN, 1937: 233.

Vejdovskyella macrochaeta Lastockin. SPERBER, 1948: 141. CEKANOVSKAYA, 1962: 167, Fig. 87. FINEGENOVA, 1962: 220.

?*Vejdovskyella simplex* LIANG, 1958: 42, Pl. 1, figs. 7-13.

Vejdovskyella comata grandisetosa FINEGENOVA, 1962: 220, Fig. 1.

$l=1.3$-8 mm, $s=24$-34. Eyes present or absent. Body wall with scattered papillae and/or dense gland cells with adhering foreign matter. Dorsal setae from V or VI 4-9 serrate hairs, 1-12 needles, fine and hair-like. Ventral setae of II longer and thinner than the rest, with longer upper tooth and proximal nodulus as in other ventral setae. Ventral setae of IV fewer and smaller than others or missing. From VI on ventral setae of three types: (*i*) Normal ventral setae; (*ii*) thicker setae with long upper tooth; (*iii*) very thick setae with short upper tooth which may be subdivided, often 1 per bundle. Pharynx in II-IV, stomachal dilatation in VI-VII or VIII. Vascular commisures in IV-VI or more, anastomosing in II and III. Clitellum in V-½VIII. Spermathecal ampulla small, duct very long, vasa deferentia curved, entering atria at lower part of anterior face, ampulla spherical with short duct. Penial setae 1 per bundle, simple distal hook.
No swimming.
Europe, Asia, N. America, Africa.

The situation regarding *Vejdovskyella* species is thoroughly confused. In the account by Sperber (1948) the two species *V. intermedia* and *V. comata* were readily separated as follows:

	comata	*intermedia*
1.	Eyes present.	No eyes
2.	Dorsal setae from V.	Dorsal setae from VI.
3.	Ventral setae of IV sometimes missing, or just fewer and shorter.	In VI or others single ventral setae with large lower tooth, 2-3 thin upper teeth, setae very thick.

Sperber (1948) could not find a description of *V. macrochaeta* of Lastockin (1937). Later, Sperber (1958), described specimens of *V. intermedia* without thickened setae but with setae having elongate upper teeth. Finegenova (1962) published a description of *V. comata grandisetosa* which has eyes, ventral setae absent on IV, on V-VII, thick setae with a short, thin upper tooth and a large lower tooth, dorsal setae beginning in V. Simultaneously, Cekanovskaya (1962) published an account of the three previously existing species (*macrochaeta, intermedia, comata*), indicating, as had Finegenova (1962), that Hrabe had identified *macrochaeta* with *intermedia* without comment in a key to the Czech fauna published in 1954. Having translated the account in Cekanovskaya (1962), the situation is further confused, because *V. macrochaeta* has eyes, dorsal setae start in VI, and the enlarged setae have a single upper tooth and a large lower tooth, very much like those of *V. comata grandisetosa*. As we now have all combinations of the supposed diagnostic characters in a series of forms, the only course is to regard them as variants of a single species.

V. simplex of Liang (1958) is smaller than most *V. comata*, lacks papillae, has fewer, shorter dorsal setae but longer ventral setae lacking the "characteristic proximal bend".

The second species in the genus is distinguishable by virtue of the pectinate needles.

Vejdovskyella hellei sp. nov.
Fig. 7. 10F-H

$l=c$. 5 mm, $s=37$. Prostomium blunt. No eyes. Body covered with foreign matter. Dorsal setae from VI or VII, 1 short and 2 long serrate hairs, up to 3 needles, curved distally with teeth about equal in size, one to three small intermediate teeth. Anterior ventral setae longest in II, 4-6 per bundle, nodulus distal, with upper tooth much longer than lower but bent over, posteriorly upper tooth not so long, nodulus distal setae bent proximally. Coelomocytes present.

Alaska, Olsen Creek, Prince William Sound. In brackish water. Holotype USNM 35524.

This species is placed in *Vejdovskyella* pending a full description from more complete material, most of the specimens available being badly fragmented. The serrate hair setae and the covering of foreign matter seem to agree well with other species of the genus, but the needle setae are peculiar. Hair setae with lateral hairs are present in *N. baicalensis*, in which the cuticle has granules in it, but the needle setae of that species are distinctive, although the ventrals resemble those of *V. hellei*. The needles of *N. borutzkii* are similar to those of *V. hellei* but not pectinate, but pectinate setae have been observed in *N. africana*, neither of these species having serrate hair setae however.

GENUS **Arcteonais** PIGUET, 1928

Type species: Stylaria lomondi MARTIN

No pigment. Eyes present or absent. Prostomium forming a proboscis. Dorsal setae from VI, numerous, fine, smooth hairs, and fine, simple-pointed needles; ventral setae all of one shape. Stomachal dilatation present. Dorsal vessel situated mid-dorsally; vascular loops in II-V. Clitellum absent between male pores; vasa deferentia devoid of prostate; atria with prostate; penial setae present.

Europe, N. America, Asia.

Arcteonais lomondi (Martin, 1907)
Fig. 7. 10I-K

(non) *Caecaria brevirostris* FLOERICKE, 1892: 470.

Caecaria brevirostris Floericke, PLOTNIKOFF, 1901: 248, Fig. 2. MUNSTER-HJELM, 1905; 15, Pl. I, figs. 6-7.

Stylaria lomondi MARTIN, 1907: 25, Pl. II.

Stylaria brevirostris (Floericke). WOLF, 1928: 396, Figs. 7-9.

Arcteonais lomondi (Martin). PIGUET, 1928: 88, Fig. 4. UDE, 1929: Figs. 63-65. MALEVICH, 1929: 45. SPERBER, 1948: 142, Fig. 15g; 1050: 67, Figs. 17b, 19g. TIMM, 1959: 25. SOKOLSKAYA, 1961: 60. BRINKHURST, 1962: 319; 1963 a: 23, Fig. 4f; 1964: 210, Fig. 3e. CEKANOVSKAYA, 1962: 162, Fig. 82.

l=6-10 mm, *s*=42-50. Dorsal setal bundles containing 8-18 straight, slender hairs, and 9-12 very fine needles; ventral setae 3-7 per bundle, all very slightly curved, with nodulus median or proximal, and upper tooth longer and somewhat thinner than lower. Pharynx extending through IV; stomachal dilatation sudden, beginning in VIII. Clitellum in V-VII; spermathecal ampulla spherical, duct fairly short, well defined; vasa deferentia strongly curved, entering atria above atrial duct; atrial ampulla reflected backwards, covered with clusters of prostate gland cells, duct inconspicuous; penial setae 4 per bundle with a simple distal hook, arranged horizontally.

No swimming (?).

Europe, N. America, Asia.

GENUS **Ripistes** DUJARDIN, 1842

Type species: Stylaria parasita SCHMIDT

Eyes present. Pigment present. Prostomium forming a proboscis. Dorsal setae from VI, smooth hair setae, in VI-VIII strongly elongated, and straight, single-pointed needles without nodulus. Pharyngeal and oesophageal glands present; stomach present. Dorsal vessel situated dorsally; vascular commissures simple or branching. Clitellum absent between male pores; vasa deferentia without, atria with prostate; penial setae present.

Europe, Lake Baikal.

Ripistes parasita (Schmidt, 1847)

Fig. 7. 10L-N

Ripistes DUJARDIN, 1842 : 93.

Stylaria parasita SCHMIDT, 1847 : 320.

Nais parasita (Schmidt). GRUBE, 1851 : 104.

Pterostylarides parasita (Schmidt). CZERNIAVSKY, 1880 : 310. BOURNE, 1891 : 348.

Stylaria parasita Schmidt. VEJDOVSKY, 1883 : 219; 1884 : 31, Pl. II, figs. 8-12. VAILLANT, 1890 : 365.

Pterostylarides macrochaeta BOURNE, 1891 : 349, Pl. XXVI, fig. 1.

Ripistes parasitica (Schmidt). BEDDARD, 1895 : 293.

Ripistes macrochaeta (Bourne). BEDDARD, 1895 : 294. MICHAELSEN, 1900 : 32; 1909 a : 14. SCHUSTER, 1915 : 35, Figs. 22-24. LASTOCKIN, 1924 b : (vide MALEVICH, 1925). MALEVICH, 1925 : 66. CORI, 1928 : 67, Figs. 1-2.

Ripistes parasita (Schmidt). MICHAELSEN, 1900 : 31; 1905 c : 5; 1909 a : 14, Fig. 19. UDE, 1929 : 39, Fig. 48. MALEVICH, 1929 : 44. PESSON, 1938 : 39. SPERBER, 1948 : 143, Figs. 15h-k, 16, 26a, Pl. IX, figs. 4-7; 1950 : 68, Fig. 17h-k, Pl. II, fig. 4. SOKOLSKAYA, 1961 : 61. BRINKHURST, 1962 : 319; 1963 a : 23, Fig. 4g. CEKANOVSKAYA, 1962 : 164, Figs. 83, 84. MOSZYNSKA, 1962 : 13.

Ripistes rubra LASTOCKIN, 1918.

Ripistes rubra Lastockin. SVETLOV, 1924 : 191. UDE, 1929 : 40. LASTOCKIN, 1937 : 233.

$l=2\text{-}7.5$ mm, $s=23\text{-}30$ (in mature specimens 27-39). Dorsal setal bundles in VI-VIII consisting of 2-16 giant hair setae, 2-6 short hairs, and 10-18 needles; in the following segments 1-3 moderately long hairs, and 1-3 needles; ventral setae of IV-V absent, those of II and III 2-7 per bundle, finer than the rest, with proximal nodulus and longer upper tooth; in the remaining segments 3-8 per bundle, nodulus distal, and upper prong slightly shorter than lower. Stomachal dilatation sudden, beginning in VI or VII. Dorsal vessel mid-dorsal, or slightly to the left; vascular commissures in III-V, those of III branching. Clitellum in 1/2 V-1/2 VIII; spermathecal ampulla large, bag-like, duct fairly long, well defined; vasa deferentia opening above atrial duct; atrial ampulla reflected backwards, duct short, poorly defined; penial setae 2 per bundle, with a simple hook. Swims with sagittal movements. Constructs fixed, hyaline tubes.

Europe, L. Baikal.

GENUS **Stylaria** LAMARCK, 1816

Type species: Nereis lacustris LINNAEUS

Eyes normally present. Pigment present. Prostomium forming a proboscis. Dorsal setae from VI, hairs and straight, simple-pointed needles without nodulus; ventral setae all with proximal tooth weak, nodulus proximal, distal part of setae straight, proximal part angularly bent. Pharyngeal and oesophageal glands present; stomach present. Simple transverse vessels in II-V; dorsal vessel in the middle line. Coelomocytes present. Clitellum absent round male pores; vasa deferentia with or without prostate glands on their hindmost part; atria with prostate glands; penial setae present.

Europe, Asia, N. America, Africa.

Stylaria lacustris (Linnaeus, 1767)
Fig. 7. 11A-C

Mille-pieds á dard TREMBLEY, 1744: 80, Pl. VI, fig. 1. REAUMUR, 1748: 70.

Das Wasserschlänglein mit dem langen, zungenähnlichen Fühlhorn RÖSEL VON ROSENHOF, 1755: 483, Pl. XVIII, figs. 17-18, Pl. XIX, fig. 1.

Das Stachelschlängelein LEDERMÜLLER, 1763: 161, Pl. XXXII, f, g, h.

Nereis lacustris LINNAEUS, 1767: 1085 (partim).

Die gezüngelte Naide MÜLLER, 1771: 14, Pl. I.

Nais proboscidea MÜLLER, 1773: 21.

Nais proboscidea Müller. MÜLLER, 1776: 219. GMELIN, 1788-1793: 3121. MODEER, 1798: 124. OKEN, 1815: 364. GRUITHUISEN, 1823: 233. Pl. XXXV. BLAINVILLE, 1825: 131; 1828: 498. TABL. ENC. METH., 1827: 134, Pl. LIII, figs. 3-8. GRUBE, 1851: 104. DALYELL, 1853: 132, Pl. XVII, figs. 6-7. D'UDEKEM, 1855: 550; 1856: 53, Pl. III, figs. 17-21; 1859: 19. SEMPER, 1877: 73, Pl. III, figs. 1-3, 9-17. TIMM, 1883: 153. LEVINSEN, 1884: 218.

Stylaria paludosa LAMARCK, 1816: 223.

Stylaria proboscidea (Müller). EHRENBERG, 1828: 112. TAUBER, 1873: 405, Pl. XIV; 1879: 73. VEDOVSKY, 1883: 219. CHEN, 1940: 44, Fig. 11; 1944: 6.

Stylinais proboscidea (Müller). GERVAIS, 1838: 17.

Stylaria paludosa Lamark. ØRSTED, 1842: 133. LEIDY, 1852 c: 286. CZERNIAVSKY, 1880: 309.

Stylaria lacustris (Linnaeus). JOHNSTON, 1845: 443; 1865: 70. VEJDOVSKY,

1884: 30, Pl. III, fig. 27, Pl. IV. VAILLANT, 1890: 362. BOURNE, 1891: 346, Pl. XXVII, figs. 7-10. FERRONNIÈRE, 1899: 257. MICHAELSEN, 1900: 33 (partim); 1903 b: 186; 1905 c: 5; 1909 a: 14, Fig. 20; 1926 a: 22; 1927 a: 9, Fig. 4. TRYBOM, 1901: 25. DITLEVSEN, 1904: 459. MUNSTERHJELM, 1905: 14. MOORE, 1906: 167. PIGUET, 1906: 287; 1909: 209, Pl. III, fig. 19; 1919 a: 790; 1928: 87. WALTON, 1906: 693, Fig. 6 (partim). GALLOWAY, 1911: 303 (partim). PIGUET AND BRETSCHER, 1913: 48. SCHUSTER, 1915: 43, Figs. 28-32. DEHORNE, 1916: 75, Figs. 35-48. SMITH, 1918: 639. MEHRA, 1924: 147, Figs. 1-5, Pl. IV, V. WOLF, 1928: 295. UDE, 1929: 41, Figs. 49-50. CERNOSVITOV, 1930: 10. BERG, 1938: 49, Fig. 34. SCIACCHITANO, 1938: 258. SPERBER, 1948, 147, Fig. 15c, Pl. X, XI; 1950: 68, Fig. 171, Pl. II, fig. 5. SAPKAREV, 1957: 136. TIMM, 1959: 24. SOKOLSKAYA, 1961: 61; 1962: 141. BRINKHURST, 1962: 319; 1963 a: 23, Fig. 4h; 1963 b: 144; 1964: 210, Fig. 3f. CEKANOVSKAYA, 1962: 159, Figs. 78, 79. MOSZYNSKA, 1962: 14. NAIDU, 1962 b: 520.

Stylaria proboscidea Lamarck. LEYDIG, 1864: 10, Pl. IV, fig. 5.

Stylaria phyladelphiana CZERNIAVSKY, 1880: 309.

Stylaria scotica CZERNIAVSKY, 1880: 309.

Stylaria phyladelphiana Czerniavsky. VAILLANT, 1890: 364.

Nais lacustris (Linnaeus). BEDDARD, 1895: 284. BRETSCHER, 1896: 507.

Caecaria rara FLOERICKE, 1892: 470.

Caecaria silesiaca FLOERICKE, 1892: 470.

Caecaria brevirostris FLOERICKE, 1892:470.

(non) *Caecaria brevirostris* Floericke. PLOTNIKOFF, 1901: 244, Fig. 1. MUNSTERHJELM, 1905: 15, Pl. figs. 6-7.

$l=5.5$-18 mm. Eyes present. Proboscis projecting from a notch between two lateral lobes. Hairs 1-3 per bundle, finely serrated; needles 3-4 per bundle; ventral setae 4-7 per bundle, all alike. Stomachal dilatation sudden, beginning in VII or VIII. Clitellum in V-VII; spermathecal ampulla long, entering sperm sac, duct short, well defined; male funnels very large; vasa deferentia short, hind part covered with gland cells, opening on anterior face of atria; atrial ampulla large, with thick epithelium, covered with thick prostate glands; atrial duct inconspicuous; penial setae 2 per bundle, with a simple hook.

Swims with sagittal movements in the horizontal plane.

Europe, W. Asia, N. America. Found in brackish water also (the Baltic).

Stylaria fossularis Leidy almost certainly belongs here, as held by Michaelsen (1900). The sole difference of any note between the two appears to be the shape of the prostomium and proboscis.

SPECIES INQUIRENDA

Stylaria fossularis Leidy, 1852
Fig. 7. 11B

Stylaria fossularis LEIDY, 1852 c: 287.

Stylaria fossularis Leidy. LEIDY, 1852 d: 350. VAILLANT, 1890: 364. MOORE, 1906: 167, Fig. 3. SMITH, 1918: 639. HAYDEN, 1922: 167. CHEN, 1940: 44, Fig. 11; 1944: 6. SPERBER, 1948: 149; 1960: 157. SOKOLSKAYA, 1961: 62; 1964: 141. CEKANOVSKAYA, 1962: 160, Figs. 80, 81. NAIDU, 1962 b: 520, Fig. 9a-h. BRINKHURST, 1964: 211, Fig. 39.

Stylaria lacustris (L.). MICHAELSEN, 1900: 33 (partim). WALTON, 1906: 693, Fig. 6 (partim). STEPHENSON, 1909 a: 276, Pl. XIX, fig. 46; 1911: 209, Fig. 3; (?) 1920: 200; 1923: 85, Fig. 31 (partim). GALLOWAY, 1911: 303 (partim). AIYER, 1925: 31. YOSHIZAWA, 1928: 587, Fig. 1028. KONDO, 1936: 384, Pl. XXIII, fig. 9.

Stylaria kempi STEPHENSON, 1916: 303, Pl. XXX, fig. 2.

Stylaria kempi Stephenson. STEPHENSON, 1923: 86, Fig. 32.

l=up to 15 mm, s=18-36. Eyes normally present. Proboscis projecting from the tip of the pointed prostomium. Dorsal setae 2 hairs, and 1-3 short needles per bundle; ventral setae 5-14 per bundle. Stomachal dilatation beginning in VIII. Clitellum in V-VII; spermathecal ampulla elongated, not entering sperm sac, duct well defined; vasa deferentia without gland cells entering atria above ejaculatory duct; atria oval, covered with gland cells, duct inconspicuous; penial setae 2-3 per bundle, with a simple hook.

Swims with sagittal movements in the horizontal plane.

E. and S. Asia, N. America, Europe (Britain), Africa.

This species was merged with *S. lacustris* by Michaelsen (1900) but separated again by Chen (1940) on the basis of the distinction in the form of the prostomium at the point of origin of the proboscis. The discovery of this form in Britain in a locality in which *S. lacustris* was also found, further supports my view that the separation of these two species is invalid (Brinkhurst, 1964).

GENUS **Piguetiella** SPERBER, 1939

Type species: Nais Blanci PIGUET

Eyes present. Dorsal setae from VI on, hairs and double-pronged crotchets; ventral setae all of one type. Vascular system simple, with transverse vessels in I-V. Pharyngeal and oesophageal glands present; stomachal dilatation present. Coelomocytes present. Clitellum present round male pores; spermathecal setae present; vasa deferentia devoid of gland cells, entering atria on top; atria with prostate glands; no penial setae.

Europe, Asia.

Piguetiella blanci (Piguet, 1906)
Fig. 7. 11D-H

Nais Blanci PIGUET, 1906: 231, Pl. X, fig. 1, Pl. XI, figs. 1-3, Pl. XII, fig. 7.

Nais blanci Piguet. MICHAELSEN, 1909 a: 16, Fig. 21. UDE, 1929: 51, Fig. 62. MALEVICH, 1929: 46.

Nais Blanci Piguet. PIGUET AND BRETSCHER, 1913: 35.

Piguetiella blanci (Piguet). SPERBER, 1939: 1, Fig. 1; 1948: 151, Fig. 17, 27e, Pl. XII, XIII, XIV, Fig. 1; 1950: 69, Fig. 20. HRABE, 1941: 4, Figs. 8-12. TIMM, 1959: 25. BRINKHURST, 1962: 319; 1963 a: 25, Fig. 5b. CEKANOVSKAYA, 1962: 192, Fig. 108.

(?) *Piguetiella blanci amurensis* SOKOLSKAYA, 1958: 302, Fig. 3; 1961: 62, Fig. 6.

(?) *Piguetiella amurensis* Sokolskaya. CEKANOVSKAYA, 1962: 193, Fig. 109.

l=3-7 mm, s=24-42. No pigment. Dorsal bundles consisting of 0-3 hairs, and 2-6 bifid crotchets, entirely similar to the ventral setae, with nodulus slightly distal and upper tooth as long as, or slightly longer than lower, but considerably thinner; ventral setae 3-9 per bundle, all of one type, those of II slightly longer than the rest. Pharynx simple, or those of I joining those of II; dorsal vessel ventrally to the left. Clitellum in $\frac{1}{2}$V-VII; spermathecal ampulla, when filled, very long, entering sperm sac; duct distinct, with narrow lumen; spermathecal setae projecting through the openings of special pockets, 2 per bundle, single-pointed, or with a small proximal tooth, nodulus about median; male funnels large; vasa deferentia long; atria very large, oval, duct inconspicuous, opening on protrusion of clitellum.
No swimming.
Europe, Asia.

The difference between *P. blanci* and *P. amurensis* are (*i*) fewer dorsal setae in the latter, of slightly different form to the ventrals, (*ii*) more segments in the latter.

As *P. amurensis* was described from immature specimens, it may be best to reserve judgement on its identity. The dorsal setae showed a tendency to drop out of anterior segments with preservation, which may indicate some relationship to the subsequent genus.

GENUS **Haemonais** BRETSCHER, 1900

Type species: Haemonais waldvogeli BRETSCHER

No eyes. Dorsal setae originally from II or VI, hairs and stout, double-pronged crotchets; after separation on budding setae shed in a number of anterior segments in the posterior zooid; ventral setae in a number of anterior segments differing in shape from the posterior ones. No distinct stomachal dilatation. Chloragogen present in all segments. Vascular system complicated, with an anterior net-work and transverse loops in all other segments, giving rise to longitudinal and transverse vessels along the inner side of the body-wall; dorsal vessel situated to the left. Coelomocytes present. Clitellum absent between male pores; vasa deferentia entering atria at upper part of anterior wall; no prostate; penial setae present.

Europe, Asia, N. America, S. America, ? Africa.

Haemonais waldvogeli Bretscher, 1900
Fig. 7. 11M-P

Hämonais waldvogeli BRETSCHER, 1900: 16, Pl. I, figs. 11-14.

Haemonais waldvogeli Bretscher. PIGUET AND BRETSCHER, 1913: 54. LASTOCKIN, 1924 a: 16. MALEVICH, 1929: 48, Fig. 1. UDE, 1929: 51. SPERBER, 1948: 154, Fig. 18c, 27b; 1950: 70, Fig. 21. TIMM, 1959: 25. BRINKHURST, 1963 a: 19, Fig. 2j; 1964: 212, Fig. 4a; 1966: 12. CEKANOVSKAYA, 1962: 194, Fig. 110. NAIDU, 1962 b: 522, Fig. 10a-c.

Haemonais laurentii STEPHENSON, 1915 a: 769, Figs. 1-5, Pl. LXXIX.

(?) *Haemonais ciliata* HAYDEN, 1922: 169.

Haemonais laurentii Stephenson. MEHRA, 1920: 457. STEPHENSON, 1923: 79, Fig. 27. MARCUS, 1944: 63, Pl. XII, fig. 51-2. MARCUS E., 1947: 5, 1949: 2, Fig. 1-2.

(? non) *Haemonais laurentii* Stephenson. CHEN, 1940: 41, Fig. 9.

$l = 5$-20 mm, $s = 40$-60. Dorsal setal bundles originally beginning in VI (or II?), later shed in a number of anterior segments (usually those in front of XVIII-XX), containing as a rule 1 short hair, and 1 curved needle seta, with distal nodulus and long teeth, the upper tooth longer, or occasionally absent; ventral setae 2-4 per bundle, in anterior 15 or 17 segments slightly longer and thinner, with nodulus proximal, and upper tooth slightly longer than lower; in posterior segments setae more curved, with distal nodulus, and upper prong shorter and much thinner than lower, or occasionally absent. Pharynx in II-IV, with pharyngeal glands, oesophagus in V-XII, no discernible stomachal dilatation. Chloragogen present throughout the alimentary canal. Coelomocytes present. Clitellum in $\frac{1}{2}$VI-$\frac{1}{2}$VIII; spermathecal ampulla ovoid, with thick, irregular epithelium, some cells hanging into the lumen, duct short; male funnels cup-shaped; vasa deferentia short, wide, entering atria towards their upper surface; atrial ampulla small, ovoid, with thick, glandular epithelium, duct short; penial setae 1-3 per bundle, with distal nodulus, and distal end bifid, strongly curved.

No swimming.

Europe, Asia, N. and S. America, ? Africa.

Haemonais laurentii of Chen (1940) may well belong to a distinct species according to Sperber (1938).

GENUS **Neonais** SOKOLSKAYA, 1962

Type species: Neonais *elegans* SOKOLSKAYA

Eyes absent. Dorsal setae from VI, forked needles like ventral setae of VI on, no hairs. Ventral setae from II-V simple-pointed, from VI on bifid setae. Stomach enlarges. Dorsal vessel median, commissural vessels in II-VI anastomosing, in VII and VIII free, in VI intestinal vessel parallel to dorsal vessel ends blindly in 1. No coelomocytes.
    Lake Baikal.

    The definition of this genus is inadequate, so much so that it is impossible to decide on the true position of the two species of *Waspa*, which may well prove to be congeneric with *Neonais* once the reproductive organs and nephridia of the latter are described.

Neonais elegans Sokolskaya, 1962
Fig. 7-11I-L

Neonais elegans SOKOLSKAYA, 1962: 141, Fig. 8.

$l=6$ mm, maximum, $s=25$-40. No chains observed. Prostomium triangular, as long as the breadth at the base. Eyes absent. Cuticle from prostomium to VII strongly granular, intersegmental furrows obliterated, from VII granules less compact especially at the ends of each segment, producing annulate pigment bands. Dorsal setae from VI, nodulus distal, bifid with upper tooth shorter and thinner than the lower, 1-3 per bundle, shorter than ventral setae. Ventral setae of II-V long, slender, nearly straight with hooked simple-pointed tips, no nodulus. The setae decrease in length from II-V, those of II are thinner, of V thicker than those of III-IV. From VI setae sigmoid, bifid, with upper tooth shorter and thinner than lower, nodulus distal, shorter than those of II-V, up to 4 per bundle. Dorsal vessel median, lateral commissurals anastomose in II-VI, not in VII-VIII, intestinal vessel lateral from VI to I.
    Lake Baikal.

GENUS **Branchiodrilus** MICHAELSEN, 1900

Type species: Chaetobranchus semperi BOURNE

No eyes. Transverse pigment stripes on anterior segments. Branchial processes from (IV or) VI onwards on a number of segments, enclosing dorsal setae. Dorsal setae beginning in (IV or) VI, hairs, and except in a number of anterior segments, straight or curved, simple-pointed needles. No stomachal dilatation. Vascular system in I-V forming a plexus; in all other segments one pair of transverse loops, in the gill-bearing segments each giving off a branch into a gill, and other branches to the body-wall; dorsal vessel situated ventrally. Coelomocytes present. Clitellum absent between the male pores; vasa deferentia entering atria towards their upper surface; no prostate; atrial duct surrounded by gland cells; penial setae present. Budding incomplete; 5 segments normally formed at anterior end.
    Asia, Africa, Australasia.

Branchiodrilus

1.	Gills on most of the body	2
—	Gills restricted to anterior half of body	*B. semperi*
2.	Gills enclosing dorsal setae where present, gills differ little in length along body	*B. cleistochaeta*
—	One hair seta free of gills posteriorly, gills progressively reduced in length posteriorly	*B. hortensis*

Branchiodrilus semperi (Bourne, 1890)
Fig. 7.12A-D

Chaetobranchus semperi BOURNE, 1890: 83, Pl. XII.

Chaetobranchus semperi Bourne. BOURNE, 1891: 355. BEDDARD, 1895: 302.

Branchiodrilus semperi (Bourne). MICHAELSEN, 1900: 24. STEPHENSON, 1912, 228; 1923: 75. SPERBER, 1948: 156. CEKANOVSKAYA, 1962: 176. NAIDU, 1962 b: 526, Fig. 11a-f; 1965: 16.

Branchiodrilus menoni STEPHENSON, 1912: 219, Figs. 1-3, Pl. XI.

Branchiodrilus menoni Stephenson. STEPHENSON, 1921: 752; 1923: 76, Fig. 25; 1925 a: 882. AIYER, 1925: 31.

$l=8$-50 mm, $s=77$-200. Gills only on about the anterior half of the body. Dorsal setal bundles with 1-3 hairs and 1-3 needles, straight in anterior segments, curved distally in posterior segments; ventral setae 2-6 per bundle; in a varying number of anterior segments thinner and with longer teeth than the rest, with nodulus median or proximal, and upper tooth $1\frac{1}{2}$ times as long as lower; in posterior segments nodulus distal, and prongs approximately equally long. In the first 12-30 segments hair setae very fine, enclosed within the gills, in posterior segments thicker, projecting freely, or some setae enclosed in gills in the rest of the gill-bearing segments.

No swimming.

South Asia.

Branchiodrilus cleistochaeta Dahl, 1957
Fig. 7. 12E-G

Branchiodrilus cleistochaeta DAHL, 1957: 1155, Figs. 1, 2.

Branchiodrilus cleistochaeta Dahl. NAIDU, 1962 b: 525.

$l=9\cdot5$ mm, $s=c$. 100. Prostomium weakly developed. Gills from IV to LXXX, anteriorly $1\frac{1}{2}$ times body width, posteriorly shorter. Dorsal setae from IV, hairs 1 per bundle on IV-VI, 2 in the rest, one shorter than the other, all within gills where these present, 1 small simple-pointed or bifid needle per bundle. In posterior, non-gilled segments 1 hair and 1 needle. Ventral setae 3-4 per bundle, sigmoid, bifid, upper tooth longer or equal to but thinner than the lower, posteriorly upper tooth shorter.

Nyon, French Cameroons. 1 specimen (Zoological Museum, Copenhagen, Denmark).

The head end may not have been completely formed, hence the dorsal setae may in fact, begin in VI as in other species.

Branchiodrilus hortensis (Stephenson, 1910)

Lahoria hortensis STEPHENSON, 1910 a: 59, Figs. 1-3, Pl. VIII.

Branchiodrilus hortensis (Stephenson). STEPHENSON, 1912: 229; 1923: 77, Fig. 26. MEHRA, 1920: 463, Figs. 1 B, 3. CHEN, 1940: 65, Fig. 21. SPERBER, 1948: 157, Fig. 28a. SOKOLSKAYA, 1961: 65, Fig. 7. CEKANOVSKAYA, 1962: 177, Fig. 94, 95. NAIDU, 1962 b: 525.

Branchiodrilus hortensis (Stephenson) var. *japonicus* YAMAGUCHI, 1938; 530, Figs. 1-2, Pl. XIX.

Branchiodrilus hortensis bifidus LIANG, 1958: 47, Pl. IV, fig. 47-50.

$l =$ up to 50 mm, $s = 35$-120. Gills present on nearly all segments from VI onwards. Dorsal setae 2-5 per bundle, on anterior part of body enclosed within the gills, on posterior part usually one hair projecting freely; in these segments also needles, 1-2 per bundle, with straight tip; ventral setae all of one type with distal nodulus, the upper tooth being thinner and equal to or longer than the lower. Clitellum in V-VIII; spermathecal ampulla large, duct fairly long, narrow, well-defined; male funnels large, reaching back into the sperm-sac behind the atria; vasa deferentia long, strongly curved, entering atria near upper surface; atrial ampulla large, pear-shaped, duct narrow, well-defined, opening into a depression in the body wall; penial setae 2-3 per bundle, with a simple, distal hook.

No swimming.

East and South Asia, Australia, Africa.

In the original description of this species, the upper tooth of the ventral seta appears to be a little longer than the lower.

In the subspecies described by Liang (1958) there are 5-7 hair setae anteriorly, the needles appear only behind segment L and are single and bifid. There are up to 6 setae ventrally. Bifid needles were noted in material from Africa examined by the author.

GENUS **Dero** OKEN, 1815

Type species: Nais digitata MÜLLER

No eyes. Dorsal setae from IV, V, or VI onwards, hairs and double-pronged, pectinate or palmate needles; ventral setae of II-V usually of a shape different from those of the rest, with upper tooth longer than lower, and equally thick, compared with equally long or shorter, and thinner, in the latter. Pharynx in II-IV, with pharyngeal glands; often septal glands; stomachal dilatation slow, beginning in VIII, IX, or X, or inconspicuous. Anus opening into a ciliated branchial fossa, usually containing gills; angles of posterior border of the fossa in one subgenus projecting as palps. A pharyngeal vascular plexus; contractile transverse vessels in a number of segments from VI onwards; dorsal vessel situated ventrally to the left; blood reddish. Coelomocytes present or absent. Nephridia often invested with bladder-like peritoneal cells. Clitellum absent between male pores; vasa deferentia joining atria anteriorly or on top; usually no prostate gland cells; ejaculatory duct surrounded by gland cells; penial setae present or absent. Asexual reproduction by budding or fragmentation. Usually living in tubes of secreted mucus and foreign matter.

Cosmopolitan.

Sperber (1948) emphasized the variability of the branchial apparatus within species of *Dero*, which has led to very considerable confusion due to the description of these variants as species.

Dero

1.	Branchial fossa with palps	*Dero* (*Aulophorus*)
—	Branchial fossa absent, or present without palps	2
2.	Asexual reproduction by fragmentation. Some specimens parasitic, lacking branchial fossae. Ventral setae all of one shape	*Dero* (*Allodero*)
—	Asexual reproduction by budding. Non parasitic, with branchial fossae. Ventral setae of II-V mostly sharply differentiated from those of other segments	*Dero* (*Dero*)

The definitions of these three subdivisions overlap considerably, partly owing to the inadequate state of knowledge of *Allodero*. The distinction between *Aulophorus* and *Dero* is narrow, especially in view of the intermediate nature of the branchial fossa of *D. D. dorsalis*. The close parallel between many *Aulophorus* and *Dero* species in setal characteristics, together with the similarity of the reproductive organs and the tube-building habits, make it imperative that a detailed study of the systematics of this whole genus be made. As most of the species are tropical in distribution, material is not readily available at the moment. Cekanovskaya (1962) regarded *Dero* and *Aulophorus* as separate genera, as did Liang (1964) who also separated *Allodero* at the generic level.

SUB-GENUS *Allodero* SPERBER, 1948

SUB-GENERIC TYPE: *Nais malayana* STEPHENSON

Dorsal setae beginning in IV (or ? III), hairs and double-pronged needles or fewer or absent; ventral setae all of one shape. A branchial fossa present or absent, with or without gills. No palps. Parasitism occurring. Asexual reproduction by fragmentation.

Asia, Africa, N. America, S. America.

It would be premature at this time to attempt to key out this sub-genus, which is little more than an assemblage of parasitic and free living forms, one of which has been demonstrated to be capable of existing in both states and to be very different in form in the free-living state to those taken from frogs. Further study is needed to clarify the range of characters displayed by a single species in the transition from one habitat to the other.

Allodero malayana (Stephenson, 1931)

Nais malayana STEPHENSON, 1931 a : 266, Figs. 1-2.

Dero malayana (Stephenson). MICHAELSEN, 1933 : 334. CERNOSVITOV, 1938 b: 274.

Allodero malayana (Stephenson). SPERBER, 1948 : 160.

l (preserved)=3-5 mm, *s*=37-73. Dorsal setal bundles with 1-2 hairs, and 1-2 needles, bifid, with short teeth, and nodulus 1/4 from distal end; ventral setae 3-4 per bundle, all alike, with nodulus slightly distal, and upper tooth slightly longer than lower. Pharyngeal gland cells very long; "septal glands" present; anus dorsal, forming a ciliated fossa, without gills, but with a median ciliated projection from the bottom.

S.E. Asia (Malacca).

Allodero bauchiensis (Stephenson, 1930)

Nais bauchiensis STEPHENSON, 1930 : 367, Figs. 1-2.

Nais bauchiensis Stephenson. STEPHENSON, 1931 a : 283.

Dero bauchiensis (Stephenson). MICHAELSEN, 1933 : 334.

Allodero bauchiensis (Stephenson). SPERBER, 1948 : 160.

l=3·5-11 mm, *s*=26-93. Dorsal setal bundles containing 1-2 hairs, and 1-2 needles, bifid, with short, equally long teeth, and nodulus ⅓ from the distal end; ventral setae 4-7 per bundle, in some anterior segments with nodulus median, and teeth equally long; posteriorly nodulus distal, and upper tooth equal to, or shorter than lower; transition gradual. Alimentary canal degenerate; no septal glands; no branchial fossa. Dorsal vessel ventrally to the left; vessels on the inner surface of the body-wall in posterior segments.

Africa. Parasitic on eyes and in Harderian glands of frogs (genus *Phrynomerus*).

The free living form has not been described and it may prove to resemble *A. malayana* although that species has fewer ventral setae.

Allodero prosetosa Liang, 1964
Fig. 7.12H-J

Allodero prosetosa LIANG, 1964: 647, Pl. IV, fig. 1-6.

l=2-4 mm, s=18-30. Prostomium rounded-conical. No eyes. Pharyngeal and oesophageal glands present. Septal glands prominent in V or IV and V. Dorsal setae from III, 1-2 hairs 1-2 needles, bifid with equal teeth slightly diverging. Nodulus weak on distal $\frac{1}{4}$. Ventral setae 2-4 all similar or those on II-V or III-V shorter and slightly thinner than rest, upper tooth thinner but as long as lower anteriorly, shorter posteriorly. Stomach dilating in IX or IX and X, no intracellular canals. Dorsal vessel left-ventral. Coelomocytes present. Anus dorsal, opening slightly dilated, no branchial fossa or gills.

    Asexual reproduction? fragmentation, ? 5 new segments formed anteriorly. Sinkiang, China.

Allodero hylae Goodchild, 1951

Schmardaella hylae GOODCHILD, 1951: 205, Pl. 1, fig. 1-6.

l=4·2 mm, s=31-40. Prostomium rounded. Dorsal setae from VI-XII single thin bifid needles in each segment, upper tooth much shorter than the lower. Ventral setae from II, 3-5 bifid setae, with upper tooth as long as or longer than the thicker lower tooth. Paired dorsal ridges or papillae associated with dorsal setae. Coelomocytes present. Stomach (?) in VI or not distinguished from intestine, gut seemingly blocked at intervals. Anus dorsal, with rudimentary fossa.

    Florida, U.S.A. Parasite of frogs (*Hyla*).

Allodero lutzi (Michaelsen, 1926)

Schmardaella lutzi MICHAELSEN, 1926 d: 100, Fig. C.

Schmardaella lutzi Michaelsen. LUTZ, 1927: 485.

"Schmardaella" lutzi Michaelsen. STEPHENSON, 1931 a: 284.

Dero lutzi (Michaelsen). MICHAELSEN, 1933: 334.

Allodero lutzi (Michaelsen). SPERBER, 1948: 161.

l (preserved)=7-7·5 mm, s=34-60. Dorsal setae usually absent, present only in the free-living state, 1 hair and 1 double-pronged needle per bundle; ventral setae 3-4 per bundle, all alike, with equally long teeth. Alimentary canal in parasitic specimens degenerate, with enormous chloragogen. Anal end unsegmented, swollen, richly vascularized, with a respiratory fossa with 4 pairs of digitate gills, under non-parasitic conditions; in parasitic conditions neither fossa nor gills. Dorsal vessel alternatingly to the right and the left; commissural vessels in all segments.

    S. America; parasitic in ureters of frogs (*Hyla*).

Allodero bilongata (Chen, 1944)
Fig. 7. 12K-O

Naidium bilongatum CHEN, 1944: 5, fig. 3.

Naidium bilongatum Chen. BRINKHURST, 1964: 219, fig. 6e.

$l=6-8\cdot5$ mm (25 mm in life), $s=40-65$. Prostomium rounded, broad at base. Dorsal setae from III, 2 hairs and 2 bifid needles. Hair setae of III-IV very long. Ventral setae bifid, 6-10 per bundle anteriorly, falling to 4-6 post anteriorly with upper tooth longer than lower, progressively shorter posteriorly, always thinner than lower. Stomach in VII. Coelomocytes abundant.
Pennsylvania, U.S.A.

Liang (1964), a student of Professor Y. Chen, suggested that this inadequately described species properly belongs in *Allodero* together with the little-known species *Naidium palmeni* Munsterhjelm, 1905.

Schmardaella filiformis of Moore (1906) was thought to be attributable to *Allonais chelata* by Marcus (1947), and *Schmardaella filiformis* (Schmarda) of Michaelsen (1925), which lacks dorsal setae, may possibly be placed here. Sperber (1948) thought that it was not a naidid.

SUB-GENUS *Dero* OKEN, 1815

SUB-GENERIC TYPE: *Nais digitata* MÜLLER

Dorsal setae from IV, or VI; ventral setae of II-V as a rule sharply differentiated from those following. Stomachal dilatation present. No palps. No coelomocytes. Prostate gland cells sometimes present on atria; as a rule no penial setae. Budding present. Usually in fixed tubes.
Cosmopolitan.

Dero

1.	Dorsal setae from IV. Two divergent processes from posterolateral border of branchial fossa, gills, 5 pairs	*D. dorsalis*
—	Dorsal setae from VI. No divergent processes on branchial fossa, gills more or less than 5 pairs	2
2.	Seven or more pairs of gills	3
—	Four or less pairs of gills	5
3.	About forty pairs of gills	*D. botrytis*
—	Seven or eight pairs of gills	4
4.	Needle teeth fine and equal, eight pairs of gills	*D. evelinae*

—	Lower tooth of needles weak or vestigial, seven pairs of gills	*D. multibranchiata*
5.	Hairs plumose, lateral hairs on one side	6
—	Hairs smooth	7
6.	Needles pectinate	*D. pectinata*
—	Needles bifid	*D. plumosa*
7.	Needles palmate	*D. palmata*
—	Needles bifid or pectinate	8
8.	Needles pectinate	*D. asiatica*
—	Needles bifid	9
9.	Branchial fossa prolonged anteriorly or posteriorly	10
—	Branchial fossa not prolonged	12
10.	Branchial fossa prolonged anteriorly	*D. zeylanica*
—	Branchial fossa prolonged posteriorly	11
11.	Four pairs of gills. Ventral setae of II-V with upper tooth 1½ times as long as lower, posteriorly upper tooth thinner and slightly shorter than lower	*D. cooperi*
—	Three pairs of gills, or two pairs and one pair ciliated swellings. Ventral setae of II-V with upper tooth almost twice as long as lower, posteriorly teeth about equally long	*D. nivea*
12.	Four pairs of gills	13
—	Two or three pairs of gills	14
13.	Hairs and needles single	*D. digitata*
—	Hairs and needles paired in each bundle	*D. indica*
14.	Needles with upper tooth shorter and thinner than upper	*D. sawayi*
—	Needle teeth fine and equal	*D. obtusa*

Dero dorsalis Ferronnière, 1899
Fig. 7. 13A-C

(?) *Xantho decapoda* DUTROCHET, 1819: 155.

(?) *Nais decapoda* (Dutrochet). BLAINVILLE, 1825: 131.

(?) *Uronais decapoda* (Dutrochet). GERVAIS, 1838: 18.

(?) *Dero? decapoda* (Dutrochet). VAILLANT, 1890: 386.

Dero dorsale FERRONNIÈRE, 1899: 255.

Dero tubicola POINTNER, 1911: 274, Pl. XXVIII, figs. 4-5.

Dero tubicola Pointner. SCHUSTER, 1915: 18, Figs. 10-14. MALEVICH, 1929: 47. UDE, 1929: 36, Fig. 41.

Dero austrina STEPHENSON, 1925 a: 882. Pl. I, fig. 1.

Dero austrina Stephenson. AIYER, 1929: 34, Figs. 10-11. STEPHENSON, 1931 a: 269. MICHAELSEN, 1933: 334. CHEN, 1940: 57. Figs. 18-19.

Dero dorsalis Ferronnière. MICHAELSEN, 1933: 334. SPERBER, 1948: 162; 1950: 70, Fig. 22, 23a. CEKANOVSKAYA, 1962: 171, Fig. 90. NAIDU, 1962 b: 529, Fig. 12a-h. TIMM, 1962: 200, Fig. 5a.

l=10-30 mm, s=23-150. Dorsal setal bundles from IV on, 1 hair and 1 double-pronged needle, with upper tooth slightly longer than proximal; ventral setae of anterior segments with longer upper tooth, and nodulus slightly proximal or distal, posteriorly gradually changing, teeth becoming equally long, or upper tooth shorter, and nodulus slightly distal. Stomach beginning in IX, X, or XI. Septal glands present, in IV-V. Branchial fossa with two diverging processes from postero-lateral border, and normally 5 pairs of gills; 2 posterior, ventral foliate, 1 lateral foliate, and 2 anterior, dorsal, one of which is long, cylindrical. Dorsal vessel to the left; contractile loops in VI-XIV. Clitellum in V-VII; spermathecal ampulla small, pear-shaped, duct short, poorly defined; male funnels large, cup-shaped; vasa deferentia short, entering atria at upper part of anterior wall; atrial ampulla large, almost globular, thick-walled, duct fairly long, well-defined, opening into an invagination, and piercing an aggregation of gland cells, surrounded by muscles; no penial setae.

Tube-living.

Europe, S. and E. Asia.

Dero digitata (Müller, 1773)
Fig. 7. 13D-H

Die blinde Naide, Das Blumenthier MÜLLER, 1771: 90, Pl. V.

Nais digitata (*coeca*) MÜLLER, 1773: 22.

Nais digitata Müller. MÜLLER, 1776: 219. GMELIN, 1788-1793: 3121. MODEER, 1798: 133. BLAINVILLE, 1825: 131; 1828: 499. TABL. ENC. MÉTH., 1827: 134, Pl. LIII, figs. 12-18.

Dero digitata (coeca) (Müller). OKEN, 1815: 363.

Nais (*Dero*) *furcata, florifera*. OKEN, 1815: 363.

Uronais digitata (Müller). GERVAIS, 1938: 18.

Proto digitata Oken. ØRSTED, 1842: 133. JOHNSTON, 1845: 69. HOUGHTON, 1860: 393, Figs. 1-2.

Dero digitata (Müller). GRUBE, 1851: 105. (?) TAUBER, 1879: 75. TIMM, 1883: 254. STOLC, 1886 a: 310, Pl. I-II. STIEREN, 1892: 122. BRETSCHER, 1896: 510. MICHAELSEN, 1900: 28; 1905 b: 307; 1909 a: 24, Figs. 38-39. PIGUET and BRETSCHER, 1913: 42. UDE, 1929: 35, Figs. 38-39. SCIACCHI-TANO, 1938: 258. SPERBER, 1948: 165, Figs. 19a-e, 27a, Pl. XIV, fig. 2-5, Pl. XV-XVIII, figs. 1-3, 6; 1950: 71, Fig. 23b, Pl. III, fig. 1-2; 1958: 49. SOKOLSKAYA, 1961: 66. BRINKHURST, 1962: 319; 1963 a: 21, Fig. 4a; 1963 b: 144; 1964: 212, Fig. 4b; 1966: 13. CEKANOVSKAYA, 1962: 170, Fig. 89. MOSZYNSKA, 1962: 14. NAIDU, 1962 b: 531, Fig. 13a-h.

Dero limosa LEIDY, 1852 b: 226.

(non) *Dero digitata* (Müller). D'UDEKEM, 1855: 549; 1859: 19. LEVINSEN, 1884: 218. VAILLANT, 1890: 381, Pl. XXII, figs. 21, 22.

Dero limosa Leidy. MINOR, 1863: 38. LEIDY, 1880: 422, Figs. 1-2. BOUS-FIELD, 1886 b: 1098; 105, Pl. V, figs. 11-16. VAILLANT, 1890: 385. STIEREN, 1892: 122. BEDDARD, 1895: 298. MICHAELSEN, 1900: 28; 1903 b: 178; 1909 a: 23, Figs. 35-36. MOORE, 1905 a: 167. GALLOWAY, 1911: 304. PIERANTONI, 1911: 3. STEPHENSON, 1914: 330, Fig. 6 (?); 1915 b: 785; 1915 c: 789, Pl. LXXX, figs. 1, 3; 1923: 88 (?). SCHUSTER, 1915: 16, Fig. 7. SMITH, 1918: 640. MAYHEW, 1922: 159, Pl. XVI. MALEVICH, 1927: 8. WOLF, 1928: 389. UDE, 1929: 34, Figs. 38-39. AIYER, 1929: 33, Fig. 9. KONDO, 1936: 385, Pl. XXIV, fig. 13. CERNOSVITOV, 1938 b: 269, Figs. 1-2. BERG, 1938: 46, Figs. 27-32. SCIACCHITANO, 1938: 258. CHEN, 1940: 52, Fig. 15.

(?) *Dero philippinensis* SEMPER, 1877: 107.

(?) *Dero philippinensis* Semper. BOUSFIELD, 1886 b: 1098. VAILLANT, 1890: 386.

Dero acuta BOUSFIELD, 1886 b: 1098.

(?) *Dero Mülleri* BOUSFIELD, 1886 b: 1098.

Dero acuta Bousfield. BOUSFIELD, 1887: 105.

(?) *Dero mülleri* Bousfield. BOUSFIELD, 1887: 104, Pl. IV, figs. 9-10. BEDDARD, 1895: 298. STIEREN, 1892, 122. MICHAELSEN, 1900: 28.

(?) *Dero intermedius* CRAGIN, 1887: 32.

Dero (?) *obtusa* d'Udekem. MICHAELSEN, 1903 b: 181.

Dero incisa MICHAELSEN, 1903 b: 182, Fig. 3.

Dero incisa Michaelsen, 1909 a: 24, Fig. 37. SCHUSTER, 1915: 17, Figs.

8-9 (?). UDE, 1929: 35, Fig. 40. STEPHENSON, 1932: 234, Figs. 6, 7 (?).

Dero michaelseni SVETLOV, 1924: 195, Figs. 1-8.

(?) *Dero bonairiensis* MICHAELSEN, 1933: 336, Pl: I, figs. 3-6.

Dero kawamurai KONDO, 1936: 385, Pl. **XXIV**, fig. 14.

Dero tanimotoi KONDO, 1936: 386, Pl. **XXIV**, Fig. 15.

Dero quadribranchiata CERNOSVITOV, 1937 b: 145, Figs. 25-30. MARCUS, E. 1947: 6.

l=6-32 mm, s=20-105. Dorsal setae from VI on, 1 hair and 1 needle, double-pronged, with upper tooth 1-2 times as long as the lower; ventral setae in II-V, 3-6 per bundle, longer than the rest, with nodulus proximal, and upper tooth 1½-2 times as long as lower; those following, 2-5 per bundle, thicker and more curved, with distal nodulus, and upper tooth hardly longer than lower. 3 pairs of septal glands, in IV-VI; stomach in IX-X or X-I. Branchial fossa normally with 4 pairs of gills, 1 small, dorsal pair, and 3 ventral, foliate, sometimes one or more pairs lacking, or 1-3 pairs cleft, forming supernumerary gills. Contractile vascular loops in some or all of VI-XIII. Clitellum in V-VII; spermathecal ampulla globular, duct well-defined; vasa deferentia short, narrow, joining atria near the top; atrial ampulla when mature anteflected, duct well-defined; no penial setae.

Swims with spiral movements. In mucous tubes.

Cosmopolitan.

Naidu (1962 b) did not accept several of the synonymies claimed by Sperber (1948) in her very extensive review of the species. Instead, Naidu felt that several of these entities were more properly identifiable as *D. cooperi* (q.v.), seemingly based on the length of the setae, and that one (*D. quadribranchiata*) was probably a distinct species. *Dero indica* of Naidu may well belong here as well.

Dero indica Naidu, 1962
Fig. 7. 13I-M

Dero indica NAIDU, 1962 b: 533, Fig. 14a-g.

l=6·5-8·5 mm, s=36-70. Prostomium bluntly triangular. No eyes. Dorsal setae from VI, 2 hairs and 2 needles, posteriorly 1 of each, hairs shorter than width of body, bayonet-shaped; needles bifid, distal nodulus, upper tooth longer than lower. Ventral setae of II-V 4 per bundle longer, thinner, straighter than the rest, teeth equally thick, diverging, upper tooth 1½ times as long as the lower, nodulus median, rest 3-4 per bundle, nodulus distal, upper tooth thinner but little longer than lower. Pharynx in II-IV wide, oesophagus in V-VIII, stomach in IX-½X chloragogen from VI. Branchial fossa funnel-shaped, opening postero-dorsally, gills 4 pairs, foliate, 1 pair short ovoid and flat on supra-anal diverti-culum, 1 pair broad flat, on inner surface of lateral margins, 2 pairs long, flat, spindle shaped on floor of fossa. No coelomocytes. Septal glands in IV-V.

No tubes.
Southern India.

The number of dorsal setae is intermediate between those of *D. digitata* and *D. zeylanica*, but the species has more pronounced teeth on the needles than either of these. The difference in length of the upper and lower teeth of the needles, used to separate the species in the key in Naidu (1962) do not tally with the description of the figure of *D. indica* as both this and *D. digitata* seem to have longer upper teeth on the needles. I regard this as most likely to be identical to *D. digitata*, but must concede that Naidu had both species before him.

Dero zeylanica Stephenson, 1913
Fig. 7. 13N-P, Fig. 7. 14A

Dero zeylanica STEPHENSON, 1913 a : 252, Pl. I, figs. 1-4.

Dero zeylanica Stephenson. STEPHENSON, 1923 : 89, Fig. 33. AIYER, 1929 : 30, Fig. 8. SPERBER, 1948 : 178. NAIDU, 1962 b : 536, Fig. 15a-k.

l (simple)=up to 10 mm, (chains)=up to 14 mm, *s*=42-60. Dorsal setae from VI onwards, 1-4 hairs, and 1-4 needles, bifid, with very fine teeth, and nodulus slightly distal; ventral setae of II-V, 4-5 per bundle, longer than the rest, with nodulus median, and upper tooth twice as long as lower, the rest 4-6 per bundle, with nodulus distal, and upper tooth slightly longer than, and $\frac{1}{3}$-$\frac{1}{2}$ as thick as lower. No septal glands. Stomach in IX-X, X, or VIII-IX; but gland cell aggregations on the gut in IV and V, 4 pairs of foliate gills, 2 ventral, 1 lateral, and 1 dorsal; branchial fossa with a forward diverticulum above the anus. Dorsal vessel ventrally to the left; contractile loops in VI-IX. Brain widely indented in front, narrowly behind. Clitellum in $\frac{1}{2}$ V-$\frac{1}{2}$ VIII. Spermathecal ampulla long, club-shaped, often entering the sperm-sac, reaching through VI and VII, duct narrowing towards the opening; male funnels cup-shaped; vasa deferentia joining atria at their anterior face, slightly above the middle; atrial ampulla large, globular, thin-walled; atrial duct short, surrounded by peritoneal cells, opening into an invagination.

Swims with spiral movements. Usually not tube-dwelling.
Southern India, Ceylon.

Naidu (1962 b) described the needles as sickle-shaped with small teeth, the upper longer than the lower, the anterior ventral setae 4-6 per bundle, upper tooth slightly longer than the lower. In the key to species in that work *D. digitata* is separated from *D. cooperi*, *D. zeylanica* and a new species *D. indica* on the basis of the inequality of the teeth of the needles, the upper being stated to be no longer than the lower in *digitata* (from x1 to x2 the lower tooth according to Sperber, 1948) but its teeth "about equal" in the other three. It is clear that the distinction must be a fine one at best.

Sperber (1948) stated that *zeylanica* is similar to the Indian forms of *D. digitata*, but the species differ in the most usual number of hairs and needles (three each in *zeylanica*) and the presence of septal glands in *digitata*. There is also the fact that *zeylanica* does not seem to inhabit tubes.

Dero cooperi Stephenson, 1932
Fig. 7. 14B-E

Dero cooperi STEPHENSON, 1932: 231, Figs. 2-5.

Dero cooperi Stephenson. SPERBER, 1948: 179. NAIDU, 1962 b: 538, Fig. 16a-i.

? *Dero limosa* Leidy. STEPHENSON, 1914: 330, Fig. 6; 1923: 88.

? *Dero incisa* Michaelsen. SCHUSTER, 1915: 17, Figs. 8-9. STEPHENSON, 1932: 234, Figs. 6-7.

? *Dero bonairiensis* MICHAELSEN, 1933: 336, Pl. I, figs. 3-6.

♂ (preserved)=3.4-4·3 mm, *s*=33-46. Dorsal setae from VI onwards, 1 hair and 1 needle per bundle, needles bifid, with teeth short, equal; ventral setae of II-V, 3-5 per bundle, longer than the rest, with nodulus median, and upper tooth 1·5 times as long as lower; the rest 4 per bundle, with nodulus distal, and teeth much shorter, upper prong thinner and slightly shorter than lower. Branchial fossa with a spout-like posterior prolongation; anterior portion with 4 pairs of gills, the foremost dorsal, very small, the hindmost foliate, the second and third pair thicker; anterior pairs in the contracted condition hidden by the roof of the fossa; no diverticulum above anus.
Abyssinia.

Naidu (1962 b) regarded the names marked (?) above as being correctly assigned to *D. cooperi* on the basis of the lengths of the setae, whereas Sperber (1948) regarded all these as belonging to *D. digitata*. Again, according to Naidu, the posterior ventral setae have the upper teeth thinner and slightly longer or as long as the lower, not shorter as noted by Sperber (1948).

Dero obtusa d'Udekem, 1855
Fig. 7. 14F-I

(?) *Xantho hexapoda* DUTROCHET, 1819: 155.

Dero obtusa D'UDEKEM, 1855: 549, Pl. I, fig. 1.

Dero obtusa d'Udekem. D'UDEKEM, 1859, 18. PERRIER, 1872: 65, Pl. I. TAUBER, 1879, 75. VEJDOVSKY, 1883: 217; 1884: 27. LEVINSEN, 1884: 218. BOUSFIELD, 1886 b: 1098; 1887, 104, Pl. III, figs. 1-3. VAILLANT, 1890: 385, Pl. XXII, fig. 23. STIEREN, 1892: 122. BEDDARD, 1895: 300. MICHAELSEN, 1900: 28; 1909 a: 23, Fig. 34. GALLOWAY, 1911: 304. PIGUET and BRETSCHER, 1913: 43. SCHUSTER, 1915: 15, Figs. 5-6. SMITH, 1918: 640. WOLF, 1928: 390. UDE, 1929: 34, Fig. 35. CERNOSVITOV, 1942 a: 202, Figs. 21-31. MARCUS, 1943: 56, Pl. IX, fig. 41, Pl. X, fig. 42; 1944: 67, Pl. XIII, fig. 57. SPERBER, 1948: 180, Figs. 19f, 20a, Pl. XVIII, fig. 5; 1950: 71, Fig. 23c, Pl. III, fig. 3. ERCOLINI, 1956: 12, Fig. 2. BRINKHURST, 1962: 319; 1963 a: 21, Fig. 4b; 1963 b: 144; 213, fig. 4c. CEKANOVSKAYA, 1962: 172, Fig. 91. MOSZYNSKA, 1962: 14. NAIDU, 1962 b: 529.

Dero latissima BOUSFIELD, 1886 b: 1098.

Dero perrieri BOUSFIELD, 1886 b: 1098.

Dero perrieri Bousfield. BOUSFIELD, 1887: 104, Pl. IV, figs. 4-7. STIEREN,
    1892: 121. BEDDARD, 1889 b: 440, Figs. 1-3; 1895, 299. MICHAELSEN,
    1900: 27; 1909 a: 22, Figs. 32-33. PIGUET, 1906: 274, Pl. X, fig. 21,
    Pl. XII, fig. 18. PIGUET and BRETSCHER, 1913: 42. SMITH, 1918: 640.
    UDE, 1929: 37, Figs. 42-43. SCIACCHITANO, 1938: 258. CHEN, 1940: 57,
    Fig. 17.

Dero latissima Bousfield. BOUSFIELD, 1887: 104, Pl. IV, fig. 8. STIEREN,
    1892: 122. BEDDARD, 1895: 300. MICHAELSEN, 1900: 27.

Dero communis GOLANSKI, 1911 (vide SZARSKI, 1936 b).

Dero polycardia HAYDEN, 1922: 168.

Dero communis Golanski. SZARSKI, 1936 b: 388.

(? non) *Dero obtusa* d'Udekem. CHEN, 1940: 55, Fig. 16 A.

l=5-17 mm, s=21-35. Dorsal setae from VI on, usually 1 hair, and 1 needle,
finely bifid, with equal teeth, and with nodulus about $\frac{1}{3}$ from the tip; ventral
setae of II-V, 2-4 per bundle, longer and thinner than the rest, with upper
tooth twice as long as lower, nodulus proximal; from VI on, 3-6 per bundle, with
teeth about equally long, nodulus distal, stoutness and thickness of lower prong
gradually increasing, and length of upper prong gradually decreasing backwards. Stomach in IX-X or X. Septal glands in IV-VI. Branchial fossa normally
with 3 pairs of gills, 2 ventral digitate, and 1 lateral foliate, or 2 ventral, and
1 pair of dorsal ciliated swellings. Contractile loops in VI to VII, IX, or X.
Clitellum in V-VII; spermathecal ampulla roundish or elongated, duct well-
defined, swollen in the middle; vasa deferentia short, joining atria at anterior
side; atrial ampulla large, globular, duct well-defined, dilated at the middle,
surrounded by gland cells; no penial setae.
    Swimming. Tube-dwelling.
    Europe, Palestine, N. and S. America, China, Africa.

Dero nivea Aiyer, 1929
Fig. 7. 14J-M

Dero niveum AIYER, 1929: 40, Figs. 16-17.

Dero palestinica CERNOSVITOV, 1938 a: 541, Figs. 5-10.
(?) *Dero obtusa* d'Udekem. CHEN, 1940: 55, Fig. 16 A.

Dero nivea Aiyer. SPERBER, 1948: 184, Fig. 19g, Pl. XVIII, fig. 4; 1950:
    72, Fig. 23d, Pl. III, fig. 4; 1958: 49, fig. 5-7. BRINKHURST, 1964: 213,
    Fig. 4d. CEKANOVSKAYA, 1962: 173. NAIDU, 1962 b: 540, Fig. 17a-c;
    1965: 17.

l=2·5-10 mm, s=23-45. Dorsal setae from VI on, 1 hair, and 1 bifid needle
with equal teeth; ventral setae about 4 per bundle, in II-V longer and thinner

than the rest, with distal tooth almost twice as long as proximal, and nodulus proximal; in the remaining segments teeth about equally long, and nodulus distal. Stomach in VIII-IX; septal glands in IV-VI. Branchial fossa slightly prolonged backwards, with 3 pairs of short, stumpy gills, 2 ventral, 1 dorsal, or with 2 pairs of ventral gills, and 1 pair of dorsal, ciliated swellings. Dorsal vessel ventrally to left; commissural contractile vessels in VI-VIII.

Europe, Asia, America, Africa, Australia.

Dero sawayai Marcus, 1943
Fig. 7. 14N-Q

(?) *Dero heterobranchiata* MICHAELSEN, 1933: 332, Pl. I, fig. 2.

Dero sawayai MARCUS, 1943: 35, Pl. IV, fig. 19, Pl. V, fig. 20.

Dero sawayai Marcus. MARCUS, 1944: 50. SPERBER, 1948: 186. NAIDU, 1962 b: 541, Fig. 18a-g.

$l=$3-6 mm. Dorsal setae from VI on, 2 hairs, and 2 bifid needles, with lower tooth longer and thicker than upper; ventral setae of II-V, 4 per bundle, much longer thinner and straighter than the rest, with proximal nodulus, and upper tooth longer than lower; in the remainder 4 per bundle, with lower tooth longer and thicker than upper, and distal nodulus. Stomachal dilatation beginning in VIII. Branchial fossa with 2 pairs of finger-shaped gills, 1 ventral and 1 dorsal. Dorsal vessel ventrally on the left side; contractile loops in VI-VII.

Swims with brisk, wriggling movements. Constructs fixed mucous tubes with foreign matter.

S. America (Brazil), and India.

Dero multibranchiata Stieren, 1892
Fig. 7. 15A-D

Dero multibranchiata STIEREN, 1892: 107, figs. 1-4.

Dero multibranchiata Stieren. BEDDARD, 1895: 301. MICHAELSEN, 1900: 29. MARCUS, 1944: 65, Pl. XIII, figs. 55-56. MARCUS, E., 1947: 6. SPERBER, 1948: 186. NAIDU, 1962 b: 528.

l (preserved)$=$7-10 mm. Dorsal setae from VI onwards, 1 hair, and 1 needle, bifid, with lower tooth weak or vestigial, and nodulus ⅓ from the tip; ventral setae 4-6 per bundle, those of II-V longer than the rest, with nodulus proximal, and teeth long, upper one longer; posteriorly nodulus median or distal, upper tooth shorter. 7 pairs of digitiform gills.

Swims with serpentine movements. In mucous tubes.

Brazil, Trinidad.

Chen (1940) placed this species in *D. limosa*, mistakenly according to Marcus (1944) and Sperber (1948). Figs. 2 and 3 of Plate I, in Marcus, E. (1947) refer to *D. quadribranchiata* ($=D.$ *digitata*), not *D. multibranchiata* as indicated in the text.

Dero botrytis Marcus, 1943
Fig. 7. 15E-H

Dero botrytis MARCUS, 1943: 37, Pl. V, figs. 21-22.

Dero botrytis Marcus. SPERBER, 1948: 187. NAIDU, 1962 b: 528.

l=up to 30 mm. Dorsal setae from VI onwards, 1 hair and 1 finely bifid needle, with lower tooth slightly thicker than upper; ventral setae of II-V, 6-7 per bundle, longer than the rest, with nodulus proximal and upper tooth longer and thinner than lower; in the remaining segments 4-6 per bundle, with distal nodulus, and upper tooth shorter and thinner than lower. Contractile vascular loops in VII-XII. Branchial fossa disc-shaped, with about 40 pairs of finger-shaped gills. Clitellum in V-½VIII; spermathecal ampulla very large, extending forwards to III; no penial setae.
S. America (Brazil).

Dero pectinata Aiyer 1929
Fig. 7. 15I-M

Dero pectinata AIYER, 1929: 36, Figs. 12-13.

Dero pectinata Aiyer. MICHAELSEN, 1933: 335. SPERBER, 1948: 187. NAIDU, 1962 b: 528.

l=about 2 mm, *s*=19-25. Dorsal setae from VI on, 1 plumose hair, and 1 needle, with three equal teeth, and nodulus weak, about ⅙ from the tip; ventral setae of II-V, 4 per bundle, about twice as long as, and straighter and thinner than the rest, with longer upper tooth, and nodulus proximal; in the following segments 2-4 per bundle, with longer and thicker lower tooth, and nodulus distal. Septal glands in III-V; stomach in VIII. 2 pairs of small, knob-like gills from the ventral wall of the fossa. Dorsal vessel ventrally to the left; 2 pairs of contractile loops, in VI-VII; no penial setae.
Southern India, West-Indies (oligohaline), Australia.

Dero plumosa Naidu, 1962
Fig. 7. 16A-E

Dero plumosa NAIDU, 1962 b: 543, Fig. 19a-h.

l=1·2-1·5 mm, *s*=19-25. Prostomium bluntly triangular, longer than broad. Eyes absent. Dorsal setae from VI, 1 unilaterally feathered hair and 1 bifid needle with minute teeth. Ventral setae 4 per bundle anteriorly, in 11-V twice as long, less curved than rest, nodulus proximal, upper tooth 1½ times as long as lower but as thick, rest with distal nodulus, upper tooth shorter and thinner than lower. Pharynx in II-III, oesophagus in IV-VIII, stomach in VIII. Septal glands in IV-V. Branchial fossa convex and ciliated anteriorly, posteriorly pointed and non ciliate, 2 pairs digitiform gills, posterior pair long, projecting far beyond fossa in life.
Lives in gelatinous tubes covered in foreign matter. Swims by brisk, serpentine movements.
Southern India.

The species differs from *D. pectinata* in having bifid needles.

Dero palmata Aiyer, 1929
Fig. 7. 15N, O

Dero palmata AIYER, 1929: 39, Figs. 14-15.

Dero palmata Aiyer. SPERBER, 1948: 187, Fig. 18a. NAIDU, 1962 b: 528.

l=about 3 mm. Dorsal setae from VI on, 0 hair, and 1 palmate needle, with distal nodulus; ventral setae of II-V, 4 per bundle, nearly twice as long as the rest, with proximal nodulus, and upper tooth twice as long as lower; the rest 2-4 per bundle, with upper tooth shorter than lower, and nodulus distal. Stomach in VIII. Branchial fossa with 3 pairs of finger-shaped gills, 1 lateral, 2 ventral. 3 contractile loops, in VI-VIII.
Southern India.

Dero asiatica Cernosvitov, 1930
Fig. 7. 16F-J

Dero asiatica CERNOSVITOV, 1930: 10, Figs. 1-10.

Dero asiatica Cernosvitov. SPERBER, 1948: 188. CEKANOVSKAYA, 1962: 174, Fig. 92. NAIDU, 1962 b: 529.

l=about 4 mm, *s*=36. Dorsal setae from VI on, usually 1 hair and 1 needle, with strongly diverging prongs and 3-4 fine intermediate teeth; ventral setae in II-V straighter and thinner than the rest, with proximal or median nodulus, and longer upper tooth, those following about equally long, with upper tooth equally long as, or slightly shorter than lower, and nodulus median or distal, the shape changing gradually backwards. Stomach in X (?). Branchial fossa with median incision in the dorsal margin; 2 pairs of gills (?). Clitellum in V-VII. Spermathecal ampulla small, thin-walled, duct short, opening laterally; vasa deferentia short, wide; atrial ampulla small, globular, duct narrow, short, piercing an aggregation of gland cells surrounded by strong muscles; penial setae 4-5 per bundle, obtuse, single-pointed, without nodulus.
Turkestan.

Dero evelinae Marcus, 1943
Fig. 7. 16K-M

Dero evelinae MARCUS, 1943: 39, Pl. VI, Pl. VII, figs. 27-31, Pl. VIII, Pl. IX, figs. 34-38.

Dero evelinae Marcus. MARCUS, 1944: 64, Pl. XIII, figs. 53-54. SPERBER, 1948: 188. MARCUS, E., 1949: 2. NAIDU, 1962 b: 528.

l=8-18 mm. Dorsal setae from VI on, 1 hair seta, and 1 finely bifid needle per bundle; ventral setae of II-V, 3-4 per bundle, longer than the rest, with nodulus proximal and upper tooth longer than lower; in the remaining bundles 4-6 setae per segment, with nodulus distal, and lower tooth equally long as upper but thicker. Stomach beginning in IX. Branchial fossa with 8 pairs of finger-shaped gills, partly bilobed. Contractile vascular loops in VI-XVI. Clitellum in ½ V-½ VIII; spermathecal ampulla globular, duct long, well-defined, male funnels large, vasa deferentia entering atria at the top; atrial ampulla pear-

shaped, reflected backwards, covered with prostate gland cells; atrial duct long, surrounded by gland cells. No penial setae.
  Constructs fixed mucous tubes.
  S. America (Brazil).

The prostate gland covering the atria is unique in the genus.

SPECIES INQUIRENDA

Dero olearia FRIEND, 1912: 101.

The species is unidentifiable from the description.

SUB-GENUS *Aulophorus* SCHMARDA, 1861

SUB-GENERIC TYPE: *Nais furcata* MÜLLER

Dorsal setae present from IV, V, or VI onwards; ventral setae of II-V different or not different in shape from those of following segments. Stomachal dilatation present or absent. Posterior border of branchial fossa projecting into two palps. Coelomocytes present or absent. Spermathecae rarely absent; vasa deferentia joining atria at anterior side; no prostate glands; usually no penial setae. Budding or fragmentation. Usually living in portable tubes.
  Cosmopolitan.

Aulophorus

1.	Dorsal setae beginning in IV	*A. superterrenus*
—	Dorsal setae beginning in V or VI	2
2.	Dorsal setae beginning in V	3
—	Dorsal setae beginning in VI	7
3.	Needles bifid with intermediate tooth or teeth	4
—	Needles simply bifid	5
4.	Intermediate teeth fine, numerous. 4 pairs of gills	*A. pectinatus*
—	Intermediate tooth short, broad, 3 pairs of gills	*A. indicus*
5.	Needle teeth equally long	*A. borelli*
—	Needle teeth unequal	6
6.	Needles with upper tooth shorter than lower	*A. furcatus*
—	Needles with upper tooth straight, slightly longer than lower	*A. hymanae*
7.	Needles bifid	*A. gravelyi*
—	Needles pectinate or palmate	8

8.	Needles pectinate (bifid with intermediate teeth or web)	9
—	Needles palmate	10
9.	Needles with minute, short, blunt, intermediate teeth. Four pairs of gills	*A. beadlei*
—	Needles with concave webbing between teeth. Three pairs of gills	*A. caraibicus*
10.	Ventral setae of II-V more than twice the length of the rest, needles with a very wide palm	*A. flabelliger*
—	Ventral setae of II-V not more than twice the length of the rest, needles with relatively narrow palm	11
11.	Stomach in VIII, ventral anterior setae 7-14 per bundle	*A. vagus*
—	No stomach, ventral anterior setae up to 9 per bundle	12
12.	Atrium very long, reaching VIII	*A. tonkinensis*
—	Atrium short, globular	*A. carteri*

This key does not include the species inquirenda *A. schmardai* and its two "formae".

Aulophorus superterrenus Michaelsen, 1912

Aulophorus superterrenus MICHAELSEN, 1912: 112, Pl. III, figs. 5-6.

Aulophorus superterrenus Michaelsen. STEPHENSON, 1931 a: 270, Fig. 3. MICHAELSEN, 1933: 335. MARCUS, 1943: 92, Pl. XVII, XVIII, Pl. XIX, figs. 76-79, Pl. XXVI, figs. 107-108, Pl. XXVII, Pl. XXVIII, figs. 109-111, Pl. XXIX, fig. 114, Pl. XXX, fig. 121, Pl. XXXIII, figs. 126-127; 1944: 50. SPERBER, 1948: 190. NAIDU, 1962 c: 897.

l (preserved)=10-18 mm, (living, out-stretched)=up to 60 mm, *s*=68-128. Dorsal setae from IV onwards, 1-3 hairs and 1-3 bifid needles with thicker lower tooth; ventral setae 5-10 per bundle, in II-V slightly shorter than the rest, all of approximately the same shape, with distal nodulus, and lower prong thicker than upper, but equally long or shorter. 4 pairs of foliate gills, 2 ventral, 1 lateral, 1 dorsal; lateral pair occasionally lobed and fused. Palps fine, slightly diverging. Dorsal vessel on left side of the gut; contractile loops in VI-X; non-contractile, bifurcate loops in all segments from XI backwards. Clitellum in V-VII; spermathecal ampulla pear-shaped, duct moderately long; male funnels very large, entering atria at lower part of anterior face; atrial ampulla small, spherical, duct short, with gland cells; ducts of both sides opening into a common invagination of the body-wall; penial setae present, with a simple hook; one pair of pear-shaped glandular masses in IV, in front of the ventral setal bundles. Fragmentation.

No swimming. No tube-formation.

S. America, Malacca. In tree-holes and water-basins of epiphytes.

Aulophorus furcatus (Müller, 1773)
Fig. 7. 17A-D

Das geschmeidige Wasserschlänglein mit zwey Gabelspitzen RÖSEL VON ROSENHOF, 1755: 581, Pl. XCIII, figs. 8-16.

Die augenlose Naide MÜLLER, 1771: 95, foot-note.

Nais furcata MÜLLER, 1773: 23.

(non) *Nais (Dero) furcata, florifera* OKEN, 1815: 363.

Uronais furcata (Roesel). GERVAIS, 1838, 18.

Dero digitata (Müller). D'UDEKEM, 1855: 549; 1859: 18. VAILLANT, 1890: 381.

Dero palpigera GREBNITZKY, 1873: 268. (vide CZERNIAVSKY, 1880=D. *rodriguezii* Semper).

Dero Rodriguezii SEMPER, 1877: 106, Pl. IV, figs. 15-16.

Dero Rodriguezii Semper. CZERNIAVSKY, 1880: 312.

Dero furcata Oken. BOUSFIELD, 1887: 105, Pl. V, figs. 17-18. STIEREN, 1892: 119. BEDDARD, 1895: 299. MICHAELSEN, 1900: 29; 1903 b: 184. PIGUET, 1906: 278. GALLOWAY, 1911: 304.

Dero palpigera Grabnitzky. VAILLANT, 1890, 382.

Aulophorus furcatus (Oken). MICHAELSEN, 1905 b: 308; 1909 a: 25, Fig. 40. PIGUET and BRETSCHER, 1913: 45. STEPHENSON, 1914: 332; 1916: 306, Pl. XXX, fig. 3; 1923: 92, Fig. 34. SMITH, 1918: 639. UDE, 1929: 33, Fig. 34. AIYER, 1930: 43. MICHAELSEN and BOLDT, 1932: 597. CERNOSVITOV, 1937 b: 145; 1942 a: 201, Figs. 16-20. CHEN, 1940: 61. MARCUS, 1943: 87, Pl. XIV, fig. 61, Pl. XV, figs. 66-69, Pl. XVI; 1944: 50. SCIACCHITANO, 1938: 357. SPERBER, 1948: 191, Fig. 20b-d; 1950: 72; 1958: 49. SOKOLSKAYA, 1961: 67, Fig. 1. BRINKHURST, 1962: 320; 1963 a: 21, Fig. 4c; 1963 b: 144; 1964: 214, Fig. 5a. CEKANOVSKAYA, 1962: 175, Fig. 93. MOSZYNSKA, 1962: 15. NAIDU, 1962 c: 899, Fig. 20a-g.

(?) *Aulophorus palustris* MICHAELSEN, 1905 b: 308.

Dero sp. STEPHENSON, 1910 a: 71.

(?) *Aulophorus furcatus* var. *brevipalpus* GOLANSKI, 1911.

Aulophorus stephensoni MICHAELSEN, 1912: 116.

(?) *Aulophorus palustris* Michaelsen. STEPHENSON, 1913: 255, Pl. I, fig. 5; 1916: 306.

(?) *Aulophorus africanus* MICHAELSEN, 1914: 152, Pl. IV, figs. 1-3.

Dero roseola NICHOLLS, 1921: 90.

(?) *Aulophorus michaelseni* STEPHENSON, 1923: 93, Fig. 35.

(?) *Aulophorus michaelseni* Stephenson. AIYER, 1929: 43, Fig. 18.

l=6-20 mm, s=35-82. Dorsal setae from V onwards, 1 hair and 1 bifid needle with slightly longer lower tooth, and nodulus $\frac{1}{3}$ from the tip; ventral setae of II-IV, 2-5 per bundle, with long prongs, upper one longer than lower; from V onwards slightly shorter, with teeth subequal. Stomachal dilatation defined or not, beginning in VII or VIII. Branchial fossa with parallel palps, and 3 or 4 pairs of gills. Septal glands in IV-VI. 2-5 contractile loops from VI or VII to VII-X. Coelomocytes present or absent. Clitellum in V-VII; spermathecal ampulla large, ovoid or long, tubular, entering the spermsac, duct narrow; male funnels cup-shaped; vasa deferentia entering atria at anterior side; atria small, thin-walled, subspherical, duct short, surrounded by gland cells; no penial setae. Budding present.

Swimming with transverse horizontal movements. Constructs attached or portable tubes, with foreign matter.

Cosmopolitan.

Naidu (1962 c) re-validated *A. michaelseni* Stephenson (=*A. palustris* Mich. of Stephenson, 1913, 1916) on the basis of the form of the needles. A comparison of the illustrations of needle setae of both *A. furcatus* and *A. michaelseni* in Naidu (op. cit.) does little to clarify the issue, except that the upper tooth is slightly less shorter than the lower in *michaelseni* than in *furcatus*. In the key to species and in the description Naidu refers to the distal tooth (=upper) as longer than the proximal (=lower) in *michaelseni*, in contradiction to the illustration. The dimensions of the setae of *michaelseni* in Naidu's accounts lie within the range described for *furcatus*, and I fail to see the justification for the separation of these two.

Aulophorus hymanae Naidu, 1962
Fig. 7. 17E-H

Aulophorus hymanae NAIDU, 1962 c: 905, Fig. 22a-f.

l=8-10 mm, s=50-80. Prostomium longer than broad. No eyes. Dorsal setae from V, 1 hair and 1 needle, needle bifid with distal nodulus, upper tooth straighter and a little longer than the curved lower tooth. Ventral setae 4-5 per bundle, 2-3 posteriorly, those of II-IV longer, thinner, straighter than the rest, nodulus median; upper tooth longer and thinner than the lower (?), others, upper tooth thinner than but as long as the lower. Pharynx in II-V, oesophagus in VI, no stomachal dilatation. Branchial fossa funnel shaped with one pair of long non-vascular palps and three pairs of digitate gills all shorter than palps. No coelomocytes. Clitellum ½V-VII, sperm and egg sacs to X and XI, sperm funnels thick walled on 5/6, atria in VI, ampullae ovoid, spermathecal paired in V extended to VIII in sperm sac. No penial setae.

Swims with horizontal transverse undulations. Inhabiting mucous tubes. Southern India.

Again, the ventral setae of II as illustrated by Naidu (1962 c) have the upper tooth much thinner than the lower but at most equal to the lower in length or shorter, but they are described as longer than the lower in the text. The seta illustrated looks, in fact, something like a posterior seta, but Naidu does state that the setae vary.

Aulophorus borellii Michaelsen, 1900
Fig. 7. 17I-L

Dero sp. COGNETTI, 1900: 1, Fig. 1.

Dero borellii MICHAELSEN, 1900: 522.

Aulophorus borellii (Michaelsen). MICHAELSEN, 1927 b: 369. STEPHENSON, 1931 c: 305, Pl. XVII, fig. 4. MARCUS, E., 1947: 7, Pl. I, fig. 4-6. SPERBER, 1948: 194. NAIDU, 1962 c: 898. BRINKHURST, 1964: 215, Fig. 5b.

l (preserved)=up to 35 mm, s=up to 154. Dorsal setae from V on, 1 hair and 1 bifid needle or 2 of each, with teeth equally long, the lower being thicker; ventral setae of anterior segments 4-5 per bundle, with nodulus slightly distal, and upper tooth as long as but much thinner than lower; in posterior segments 2-3 per bundle, with nodulus more distal, teeth equally long, 4 pairs of ridge-like gills. Probably fragmentation.

S. America. ? N. America.

Marcus (1947) recorded considerable variation in the length of the hair setae in worms from different populations.

Aulophorus pectinatus Stephenson, 1931
Fig. 7. 17M-P

Aulophorus pectinatus STEPHENSON, 1931 c: 308, Pl. XVII, fig. 6. SPERBER, 1948: 194. MARCUS, E., 1949: 3, Fig. 3-6. NAIDU, 1962 c: 897.

l (preserved)=12-16 mm. s=108-113. Dorsal setae from V on, 1 hair and 1 bifid needle, or 2 of each with 2-4 intermediate teeth; ventral setae of anterior segments 3-4 per bundle, in posterior 2-5 per bundle, with nodulus more or less distal, and prongs about equally long, the lower being thicker. Gills 4 pairs. Palps diverging. Probably fragmentation.

S. America, ? Africa.

The African specimens have hair setae with lateral hairs and needles with single, intermediate tooth.

Aulophorus indicus Naidu, 1962
Fig. 7. 18A-C

Aulophorus indicus NAIDU, 1962 c: 909, Fig. 23a-d.

l=3-6 mm, *s*=47-60. Prostomium bluntly triangular, no eyes. Dorsal setae from V, 1 hair bayonet-shaped, 1 needle with two equal teeth and one intermediate tooth. Ventral setae of II-V, 3 per bundle, with proximal nodulus, upper tooth more than twice as long and thinner than the lower, in the rest 3-4 (1-2 posteriorly) with upper tooth thinner than thick lower tooth, decreasing from as long as lower to shorter than lower. Pharynx in II-V, oesophagus in VI-VII, stomach in VIII-IX. Branchial fossa wide, cup-like, short palps, three pairs of gills. Coelomocytes present. ? Fragmentation.

Southern India.

The needle setae resemble those of *A. beadlei* but the ventral setae of the latter were not fully described and it has four pairs of gills, and more ventral setae than *indicus*.

Aulophorus gravelyi Stephenson, 1925
Fig. 7. 18D

Aulophorus gravelyi STEPHENSON, 1925 c: 46, Pl. III, fig. 2.

Aulophorus gravelyi Stephenson. MICHAELSEN AND BOLDT, 1932: 597. SPERBER, 1948: 194. NAIDU, 1962 c: 898.

Aulophorus varians LIANG, 1958: 46, Pl. IV, figs. 39-42.

l (preserved)=5-8 mm, *s*=46-73. Dorsal setae from VI onwards, 2-3 hairs, and 2-3 finely bifid needles, with slightly longer upper prong, and slightly distal nodulus; ventral setae 3-4 per bundle, those of II-V longer and straighter than the rest, with longer upper tooth, and nodulus median, those of following segments with equally long teeth, and nodulus median or slightly distal. Branchial fossa with 4 pairs of gills, the first pair dorsal, small, the second and third cylindrical, the fourth broader. No stomachal dilatation. No coelomocytes. Probably fragmentation.

Southern India, Sumatra, China.

The species described by Liang (1958) has a fifth pair of gills which look more like duplicate palps, and has smaller, fewer setae.

Aulophorus beadlei Stephenson, 1931
Fig. 7. 18E

Aulophorus beadlei STEPHENSON, 1931 c: 306, Pl. XVII, fig. 5. SPERBER, 1948: 195. NAIDU, 1962 c: 898.

l (preserved)= 4 mm, *s*=26-44. Dorsal setae from VI on, 1 hair, and 1 bifid needle, with short teeth, with an irregularity between the latter, and an inconspicuous nodulus, ⅓ from the distal end; ventral setae of II-V, 5-6 per bundle, longer than the rest, with upper tooth longer than lower, and proximal nodulus; the rest 4-5 per bundle, with upper tooth slightly longer than lower, and nodulus distal. 4 pairs of foliate gills. No budding, probably fragmentation. In mucous tubes.

S. America.

Aulophorus caraibicus Michaelsen, 1933

(?) *Aulophorus discocephalus* SCHMARDA, 1861: 9, Pl. XVII, fig. 151.

(?) *Aulophorus discocephalus* Schmarda. VAILLANT, 1890: 388. MICHAELSEN, 1900: 35.

Aulophorus caraibicus MICHAELSEN, 1933: 338, Pl. I, figs. 7-9. SPERBER, 1948: 195. NAIDU, 1962 c: 898. BRINKHURST, 1964: 214.

l (preserved), $s=28$-40. Dorsal setae probably from VI on, 1 hair, and 1 bifid needle with short teeth, possessing an intermediate webbing with concave border; ventral setae of II-V, 3 per bundle, probably slightly longer than the rest, with longer and slightly thinner upper tooth, those following 2-3 per bundle, with teeth equally long and thick. Branchial fossa with moderately long palps, about parallel, and 3 pairs of gills, 2 ventral foliate, 1 dorsal smaller. Probably fragmentation.
W. Indies (in oligohaline water).

Aulophorus tonkinensis (Vejdovsky, 1894)
Fig. 7. 18F-I

(?) *Aulophorus oxycephalus* SCHMARDA, 1861: 9, Pl. XVII, fig. 152.

(?) *Dero? oxycephala* (Schmarda). VAILLANT, 1890: 387.

(?) *Dero sp.*, STUHLMANN, 1891: 925.

(?) *Dero stuhlmanni* STIEREN, 1892: 123.

Dero tonkinensis VEJDOVSKY, 1894: 244.

(?) *Dero stuhlmanni* Stieren. MICHAELSEN, 1900: 29.

Dero tonkinensis Vejdovsky. MICHAELSEN, 1900: 30; 1905 a: 353.

Aulophorus tonkinensis (Vejdovsky). MICHAELSEN, 1909 b: 132. STEPHENSON, 1911: 212; 1923: 91; 1931 b: 43; 1932: 236. AIYER, 1925: 35, Fig. 3. MICHAELSEN and BOLDT, 1932: 598. SPERBER, 1948: 196; 1958: 49, Fig. 8-9. NAIDU, 1962 c: 911, Fig. 24a-h; 1965: 18. BRINKHURST, 1964: 215.

(?) *Aulophorus stuhlmanni* (Stieren). MICHAELSEN, 1914: 155. CUNNINGTON, 1920: 574.

Aulophorus oxycephalus Schmarda. CHEN, 1940: 62, Fig. 20.

$l=2$-5 mm, $s=26$-45. Dorsal setae from VI onwards, 1 hair and 1-2 narrowly palmate needles with long prongs; which may appear ribbed, ventral setae of II-V, 3-9 per bundle, longer than the rest, with upper tooth longer than lower, and nodulus proximal; those following 3-7 per bundle, with upper tooth shorter and thinner than lower, and nodulus distal. No stomachal dilatation. Branchial fossa funnel-like, normally with 2 pairs of long, cylindrical gills, 1 dorsal and 1 ventral. Commissural vessels in VII-VIII. Coelomocytes present. Clitellum in V-VIII, absent or weak ventrally in V; no spermathecae; vasa deferentia probably very thin, entering atria at middle; atrial ampulla elongate, enormously long, when full, reaching back into VIII; atrial duct long or short, curved; no penial setae. Budding present. In tubes with or without foreign matter. Swims.

S. and E. Asia, Africa.

According to Naidu (1962 c) the spermathecae are absent. This is recognized intraspecific variation in the Tubificidae but is not often reported in the Naididae.

Aulophorus carteri Stephenson, 1931
Fig. 7. 18J-L

Aulophorus carteri STEPHENSON, 1931 c: 303, Pl. XVII, figs. 2-3.

Aulophorus carteri Stephenson. MARCUS, 1943: 60, Pl. X, figs. 43-46, Pl. XI, XII, XIII, Pl. XIV, figs. 60, 62-63, Pl. XV, figs. 64-65; 1944: 50. MARCUS, E., 1947: 6. SPERBER: 198. NAIDU, 1962 c: 898.

$l=$2-10 mm, $s=$21-24. Dorsal setae from VI onwards, 1-2 hair setae, and 1-2 palmate needles with long teeth and web without ribs; ventral setae of II-V, 5-7 per bundle, about twice as long as the rest, with upper tooth longer than lower, and nodulus proximal; in the remaining segments 3-6 per bundle, with upper tooth shorter and thinner than lower. Branchial fossa with slightly diverging palps, and 3 pairs of short gills, 1 ventral, 1 lateral, and 1 dorsal. Masses of gland cells in II-III; no definite septal glands. No stomach. Clitellum in V-½VIII; spermathecal ampulla globular, duct short, poorly defined; vasa deferentia long, curved, joining atria above atrial duct, atrial ampulla large, globular, duct short, surrounded by gland cells; no penial setae. Budding present.

Swims with horizontal movements. Constructs portable tubes of foreign matter (spores etc.).

S. America.

Aulophorus vagus Leidy, 1880
Fig. 7. 19A-D

Aulophorus vagus LEIDY, 1880: 423, Figs. 3-4.

Aulophorus vagus Leidy. REIGHARD, 1885: 88, Pl. I-III. CHEN, 1940: 65; 1944: 7. SPERBER, 1948: 198. NAIDU, 1962 c: 898. BRINKHURST, 1964: 215, Fig. 5c.

Dero vaga (Leidy). VAILLANT, 1890: 383. BEDDARD, 1895: 300. BRODE, 1898: 141, Pl. XIII-XV. GALLOWAY, 1899: 115, Pl. I-V; 1911: 304. MICHAELSEN, 1900: 29. WALTON, 1906: 692. SMITH, 1918: 639.

(non) *Dero vaga* (Leidy). STIEREN, 1892: 118.

$l=$5-10 mm, $s=$24-60. Dorsal setae from VI onwards, 1-3 hairs, and 1-3 palmate needles, with intermediate teeth; ventral setae of II-V, 7-14 per bundle, longer and straighter than the rest, with proximal nodulus, and upper tooth longer than lower; those following 4-7 per bundle with upper tooth much shorter and thinner than lower. A stomachal dilatation in VIII-X; no septal glands. Branchial fossa when expanded roundish, directed obliquely upwards, with strongly diverging palps, and 1 (2?) pair of small, ventral gills. Contractile loops in VIII-X. Coelomocytes present. Budding present. In portable tubes with foreign matter.

N. America.

The palmate needles figured by Brinkhurst (1964) were finely toothed distally. This may well be a variable characteristic in these species (see below).

Aulophorus flabelliger Stephenson, 1931
Fig. 7. 19E

Aulophorus flabelliger STEPHENSON, 1931 b: 44, Fig. 4. SPERBER, 1948: 199, Fig. 18B. NAIDU, 1962 c: 898.

Aulophorus heptobranchiatus LIANG, 1958: 47, Pl. IV, figs.43-46.

l (preserved)$=2.75$ mm. $s=27$. Dorsal setae from VI on, 1 hair, and 1 palmate needle with strongly diverging prongs, and broad, obliquely cut web; ventral setae of II-V, 5-7 per bundle, more than double the length of the rest, straighter, with nodulus proximal, and upper tooth longer than lower; those following with nodulus distal, upper prong $\frac{1}{3}$ as thick and $\frac{1}{2}$ as long as lower. A stomachal dilatation beginning in VIII or IX. Branchial fossa with 2 or 3 pairs of long, cylindrical gills, and long, parallel palps.

Budding present. In tubes with foreign matter.

Africa (Kenya), Australia, China.

Australian specimens seem only to have 2 long pairs of gills, the third pair being either much smaller and contracted, hence invisible on preserved material, or absent. The Chinese species has an extra median gill, the anterior ventral setae not much longer than the posterior setae and fewer in number. The web of the needles is also thinner, without the "cuts".

SPECIES INQUIRENDA

Aulophorus schmardai (Michaelsen, 1905)

(?) *Dero vaga* Leidy. STIEREN, 1892: 118.

Dero Schmardai MICHAELSEN, 1905 a: 350.

Aulophorus schmardai (Michaelsen). PIGUET, 1928: 81, Fig. 1. STEPHENSON, 1931: 295. SPERBER, 1948: 197. NAIDU, 1962 c: 898. BRINKHURST, 1964: 215.

l (simple, preserved)$=2$-2.6, (chains, preserved)$=2.8$ mm, $s=18$-21. Dorsal setae from VI onwards, 1 hair and 1 palmate needle with very long teeth, and web without ribs; ventral setae of II-V, 6-8 per bundle, twice as long as the rest, with upper tooth longer than lower; those following 3-6 per bundle, with upper tooth mostly shorter than lower. Septal glands in V-VII. Branchial fossa small, slit up dorsally, directed backwards, with fairly long, parallel palps, and at least 2 pairs of small gills, 1 dorsal, 1 ventral (possibly 2 ventral pairs).

Budding present. In tubes plastered with sand and plant matter.

S. America (Paraguay).

This species is close to *A. tonkinensis*, but has smaller gills and possibly one extra pair. Both Sperber (1948) and Stephenson (1931) doubted the separation of this from *tonkinensis* as more than a variant of the latter.

Aulophorus schmardai "forma" *huaronensis* (Piguet, 1928)

Aulophorus schmardai (Michaelsen) var. *huaronensis* PIGUET, 1928: 82, Fig. 2.

Aulophorus huaronensis (Piguet). SPERBER, 1948: 197. NAIDU, 1962 c: 898.

l (preserved)=2·5-3 mm. Dorsal setae from VI on, 1-2 hairs, and 1-2 palmate needles, with long teeth, well ribbed; ventral setae of II-V, 4-6 per bundle, twice as long as, and straighter than the rest, with upper tooth longer and thinner than lower; those following 2-4 per bundle, with upper tooth shorter and thinner than lower. No septal glands. Branchial fossa dilated, directed obliquely upwards, with moderately long, parallel palps, and 3 pairs of gills, 2 ventral, cylindrical, 1 small dorsal.

Budding present.

Sperber (1948) separated this species from *schmardai* because of the absence of septal glands, the ribbed web of the needles, the smaller number of ventral setae and differences in the branchial fossa. It is close to *A. carteri* but the latter has no ribs in the webs of the needles, a character which may, however, vary within a species, neither does it have true septal glands. The form of the atria separates *carteri* from *tonkinensis*, which it otherwise resembles.

If this form proves identical to *carteri* then the name *huaronensis* takes priority.

Aulophorus schmardai "forma" *costata* Marcus, 1944
Fig. 7. 19F-I

Aulophorus schmardai Mich. forma *costata* MARCUS, E., 1944: 3, Figs. 4-10.

l=4 mm, *s*=13-22. Prostomium triangular, no eyes. Dorsal setae from VI, 1 hair and 1 palmate needle with long teeth and web ribbed. Ventral setae of II-V, 4-6 per bundle, upper tooth longer and thinner than the lower, from VI on 3-5 per bundle about half as long, upper tooth thinner and a little shorter than lower. Branchial fossa with deep median notch, two pairs of gills, border of fossa as accessory gill on each side.

Brazil.

By describing this as a form of *schmardai* together with *huaronensis*, Marcus (1944) includes in this species forms with ribbed and unribbed palmate setae, with and without septal glands. The stomach of *costata* is not mentioned. As all of them resemble both *tonkinensis* and *carteri*, which can only be validly separated on the form of the atrium (assuming *carteri* material to have been fully mature!) the absence of a description of mature forms renders their identification impossible. Sperber (1948) thought that both *huaronensis* and *schmardai* were of uncertain identity, and did not see the above description.

GENUS **Allonais** SPERBER, 1948

Type species: Nais pectinata var inequalis STEPHENSON

No eyes. Ventral setae of II-V only slightly different from those of following segments; dorsal setae normally from V, VI, or VII, hairs and double-pronged or pectinate needles. No septal glands. No stomachal dilatation? Usually a plexus in II-V, and simple vessels in some following segments. Coelomocytes present. Vasa deferentia joining atria above atria duct; no prostate gland cells; penial setae present or absent. Fragmentation.

Asia, Africa, Australasia, S. America, N. America.

Allonais

1.	Needles simple pointed	*A. lairdi*
—	Needles pectinate or bifid, with at least a fine upper tooth	2
2.	Needle teeth horse-shoe shaped	*A. chelata*
—	Needle teeth not horse-shoe shaped	3
3.	Needle teeth equally long or upper longer than lower, 1-5 intermediate teeth	*A. pectinata*
—	Needle teeth unequal, upper teeth shorter than lower	4
4.	Needle teeth fairly narrow, more or less equal, diverging, with stomach	*A. gwaliorensis*
—	Needle teeth unequal, or needles pectinate. No stomach	5
5.	Needles with upper tooth rudimentary, or shorter than lower, or bifid	*A. paraguayensis*
—	Needles with upper tooth distinct and 1-4 intermediate teeth, or upper tooth appearing multiple	*A. inaequalis*

Allonais inaequalis (Stephenson, 1911).
Fig. 7. 20L

Nais pectinata Stephenson var. *inaequalis* STEPHENSON, 1911: 208, Fig. 2.

Nais pectinata var. *inaequalis* Stephenson. MEHRA, 1920: 458, Figs. 1-2.
STEPHENSON, 1923: 64, Fig. 20. AIYER, 1929: 21, Fig. 2e. MICHAELSEN
and BOLDT, 1932: 593.

Nais pectinata Stephenson. STEPHENSON, 1931 c: 302, Pl. XVII, fig. 1.

Nais pectinata var. *ranauana* Boldt. MICHAELSEN and BOLDT, 1932: 594,
Fig. 1.

Allonais inaequalis (Stephenson). SPERBER, 1948: 201, Fig. 21a-d.

$l=8$-18 mm, $s=40$-95. Dorsal setae 1-2 hairs, and 1-2 needles, pectinate, with
1-4 long intermediate teeth, (connected by a webbing?) lower tooth usually
longer than upper; ventral setae 4-8 per bundle, in II-V somewhat thinner and
straighter than the rest, with slightly proximal nodulus, and upper tooth slightly
longer than lower; those following with slightly distal nodulus, and teeth
equally long or the upper slightly shorter. Transverse vessels in II-V forming a
plexus; from VI on in a number of segments a pair of transverse vessels on
the anterior face of each septum; dorsal vessel ventrally to the left; clitellum in
$\frac{1}{2}$V-VIII, absent between male pores; spermathecal ampulla large, roundish,
duct thick; male funnels large, vasa deferentia short, thick, making a single
bend; atrial ampulla ovoid, duct well defined; penial setae 4-6 per bundle,
usually with a simple hook.

Southern Asia, Africa, S. America, Australia.

Allonais paraguayensis (Michaelsen, 1905)
Fig. 7. 20N, 7. 21A

Nais paraguayensis MICHAELSEN, 1905 a: 354.

Nais paraguayensis Michaelsen. MICHAELSEN, 1905 b: 306; 1909 b: 131.
STEPHENSON, 1909 a: 263, Pl. XVII, figs. 22-24; 1920: 197, Pl. IX, fig. 1;
1921: 750; 1923: 61, Figs. 15-16; 1931 c: 301. MICHAELSEN AND BOLDT,
1932: 592. HYMAN, 1938: 126. MARCUS, 1943: 23, Pl. II, figs. 11-12, Pl.
III, Figs. 9-10, 13.

(?) *Nais paraguayensis* var. *aequalis* STEPHENSON, 1920: 197, Pl. IX, fig. 2.

(?) *Nais paraguayensis* var. *barkudensis* STEPHENSON, 1921: 751, Pl.
XXVIII, fig. 1.

(?) *Nais paraguayensis* var. *aequalis* Stephenson. STEPHENSON, 1923: 62,
Fig. 17; 1924: 321; 1931 a: 265; 1932: 231. MARCUS, 1944: 51, Pl. IX,
fig. 38.

(?) *Nais paraguayensis* var. *barkudensis* Stephenson. STEPHENSON, 1923; 63,
Fig. 18. MICHAELSEN AND BOLDT, 1932: 593.

Nais paraguayensis f. *typica* Michaelsen. STEPHENSON, 1931 a: 265. CHEN, 1940: 36, Fig. 7. MARCUS, 1944: 50.

Dero (Aulophorus) paraguayensis (Michaelsen) var. *aequatorialis* CERNOSVITOV, 1938 b: 270, Fig. 3-10.

Allonais paraguayensis paraguayensis (Michaelsen). SPERBER, 1948: 203, Fig. 286. HARMAN AND PLATT, 1961: 93. NAIDU, 1962 c: 915. BRINKHURST, 1964: 216; 1966: 12.

Allonais paraguayensis aequatorialis (Cernosvitov). SPERBER, 1948: 204. NAIDU, 1962 c: 915.

l=4-60 mm, *s*=15-200. Dorsal setae beginning in V, VI or VII, 1-2 hairs per bundle, and 1-2 needles, simple, pointed or bifid, with long teeth, the lower being about twice as long as the upper, the upper bifid or single or strongly reduced; ventral setae 2-8 per bundle, all about equally long, or shorter in the anterior segments; in the latter slightly thinner, with nodulus median or distal, and upper tooth usually slightly longer; in the rest nodulus distal, and teeth equally long. Vascular system in anterior segments forming a plexus. Clitellum in V-VIII, spermathecal ampulla ovoid, short duct well defined. Atrial ampulla round, duct well defined, penial setae present 3-11 per bundle, simple and hooked. 5 or 6 segments at anterior end on regeneration.

Swims with transverse movements, rotating round its axis.

Asia, Africa, N. and S. America.

The variation within specimens seen by me from Africa makes it impossible to maintain the two sub-species as distinct. Sperber (pers. comm.) agreed that my specimens were intermediate in form. The species approaches *inaequalis* very closely.

Allonais chelata (Marcus, 1944)
Fig. 7. 20A-C

Nais paraguayensis forma *chelata* MARCUS, 1944: 51, Pl. X, fig. 39. MARCUS, E., 1947: 4.

?*Schmardaella filiformis* (Schmarda). MOORE, 1906: 168.

Allonais chelata (Marcus). SPERBER, 1948: 205. NAIDU, 1962 c: 914.

l (preserved)=up to 15 mm, *s*=50-115. Dorsal setae from VI on, 1-2 hairs, and 1-2 needles, bifid, with nodulus strongly distal, and prongs curved, about equally long; or dorsal setae missing in some or all bundles, ventral setae of II-V slightly longer than the rest, finer, with teeth equally thick and nodulus median, in the posterior segments upper tooth thinner, and nodulus distal. Anterior vascular plexus; in most other segments a transverse loop.

North (?) and South America.

Schmardaella filiformis of Moore (1906) was placed here by Marcus (1947), but *S. filiformis* of other accounts was stated to be a different species.

Allonais gwaliorensis (Stephenson, 1920)
Fig. 7. 20D-G

(?) *Nais fusca* CARTER, 1858: 20, Pl. II-IV, figs. 1-30.

(?) *Nais fusca* Carter. VAILLANT, 1890: 374. BOURNE, 1891: 344.

Nais gwaliorensis STEPHENSON, 1920: 198, Pl. IX, figs. 3-4.

Nais gwaliorensis Stephenson. STEPHENSON, 1923: 59, Fig. 14. MICHAELSEN AND BOLDT, 1932: 592. CHEN, 1940: 35, Fig. 6.

Allonais gwaliorensis (Stephenson). SPERBER, 1948: 205. NAIDU, 1962 c: 919, Fig. 27a-f; 1965: 20, Fig. Ie.

(?) *Allonais gwaliorensis* (Stephenson). SPERBER, 1958: 50, Fig. 10-12.

Allonais rayalaseemensis NAIDU, 1962 c: 917, Fig. 26a-f.

l (living)=4-12 mm, (preserved)=2·7 mm, *s*=25-59. Dorsal setal bundles from VI with 1-2 hairs, and 1-2 double-pronged needles, with tooth slightly longer or shorter than lower; ventral bundles of II-V containing 3-5 setae, with nodulus about median, and upper tooth slightly longer than lower; in the following segments 4-6 setae per bundle, with distal nodulus, and teeth equally long. Clitellum in V-VIII, weaker but not absent round male pores; spermathecal ampulla ovoid, duct thick; vasa deferentia slightly curved; atria round or ovoid, duct inconspicuous; penial setae present.

Asia, Madagascar, ? Africa.

Sperber (1958) described specimens from Madagascar that were rather small, and the upper tooth of the anterior ventral setae was much longer than the lower. There were some differences in the form of the needles, but in view of the variability of *A. paraguayensis* this may not be of too much significance. Naidu (1962 c) described his new species on the basis of the marked inequality of the needle teeth (although he did not, in fact, discuss the differences between *rayalaseemensis* and *gwaliorensis*). In his description of the latter he recorded the penial setae for the first time. He also described a stomach in both forms, in XI-XII in *rayalaseemensis* and (weak) in IX-X in *gwaliorensis*, and the former is larger with bigger setae.

Allonais pectinata (Stephenson, 1910)
Fig. 7. 20M

Nais pectinata STEPHENSON, 1910 b: 236, Pl. XI, fig. 1.

Nais pectinata Stephenson. STEPHENSON, 1920: 198; 1923: 63, Fig. 19. AIYER, 1930: 19, Fig. 2.

(non) *Nais pectinata* Stephenson. STEPHENSON, 1931 c: 302.

? *Nais pectinata* Stephenson. STEPHENSON, 1932: 229, Fig. 1.

Nais denticulata CHEN, 1940: 39, Fig. 8.

Allonais pectinata (Stephenson). SPERBER, 1948: 206. NAIDU, 1962 c: 915.

l=1·5-8 mm, *s*=15-65. Dorsal setae 1-2 hairs and 1-2 needles per bundle, needles pectinate, with 1-5 intermediate teeth, nodulus ⅓-⅕ from the distal end; ventral setae of II-V 3-5 per bundle, with upper tooth slightly longer than lower, and nodulus slightly proximal; those following 2-7 per bundle, thicker, with nodulus distal, and teeth equally long. Dorsal vessel to the left; 4 pairs of transverse commissural vessels, in II-V. Clitellum in ½V-½VIII; spermathecal ampulla round, duct distinct; vasa deferentia almost straight; atrial ampulla globular, duct well-defined; penial setae present, 3-5 per bundle, with a simple hook, occasionally with a vestigial distal prong.

Asia, Africa, Australia.

Specimens I had attributed to *A. inaequalis* were attributed to *A. pectinata* by Sperber (pers. comm.) on the basis that, although the needle teeth were slightly unequal, the upper tooth was the longer not the lower, and the teeth diverged somewhat.

Allonais lairdi Naidu, 1965
Fig. 7. 20H-K

Allonais lairdi NAIDU, 1965: 18, Fig. 1a-d.

l=7-9 mm, *s*=54-87. Eyes absent. Dorsal bundles from VI, 1 hair and 1 simple pointed needle. Ventral setae of II-V 5-6 per bundle, distal nodulus upper tooth longer than lower, in rest 4-8 per bundle, shorter, with upper tooth thinner and shorter than lower. Pharynx in II-IV with diverticulum, no stomach, no septal glands, ? no coelomocytes. Clitellum in ½V-VII. Penial setae present. Atrium spherical.

Southern India.

The needles resemble a *Nais* species more than they do any *Allonais*.

GENUS **Stephensoniana** CENOSVITOV, 1938

TYPE SPECIES: *Naidium* (?) *trivandranum* AIYER

No eyes. Dorsal setae from II onwards, hairs and straight, simple-pointed needles; ventral setae all of one type. No coelomocytes. Sexual organs one segment in front of the usual situation; atria provided with prostate gland cells; penial setae present. Four segments formed at anterior end on budding.

Asia, Africa.

Naidu (1963) attributed this genus to a separate sub-family Stephensonianinae defined, of course, in exactly the same way as the genus. The reasons for the decision were not discussed by Naidu, except in so far as a key to the sub-families was presented (Naidu, 1962 a), in which the separation from the Naidinae was based on the provision of dorsal setae from II onwards, not IV, V or VI. This ignores the (presumed non-Indian) genus *Homochaeta* which has dorsal setae from II but is yet in the Naidinae. The separation from the Pristininae involves the number of segments budded at the anterior end and the position of the gonads, this separation being also implied by the inclusion of *Stephensoniana* in the Naidinae. The only significant difference between the genus *Stephensoniana* and the rest of the Naidinae is the failure to develop a fifth segment at the front end following asexual reproduction, which places the gonads one segment forward of the position in the other Naidinae, and in the same position as in the Paranaidinae.

Stephensoniana trivandrana (Aiyer, 1926)
Fig. 7. 21B-F

Naidum (?) trivandranum AIYER, 1926: 139, Pl. V, VI.

Stephensonia trivandrana (Aiyer). AIYER, 1929: 27, Fig. 7, Pl. I, figs. 2-4.

Stephensoniana trivandrana (Aiyer). CERNOSVITOV, 1938 a: 539, Figs. 1-4. SPERBER, 1948: 208, Fig. 28c. NAIDU, 1963: 201, Fig. 28a-d.

l=2-6 mm, s=21-43. Body-wall with cutaneous glands, foreign matter adhering. Dorsal setal bundles containing anteriorly 3-4 hairs, and 3-4 needles, suddenly tapering towards the distal end, posteriorly 1-2 hairs and needles; ventral setae anteriorly 4 per bundle, decreasing to 1 posteriorly, all with proximal nodulus, and upper tooth longer than lower, the difference between the teeth growing stronger backwards. Pharynx in II-part of III; in VI a sudden dilatation of the gut (stomach?). Dorsal vessel situated to the left of the gut as far forwards as IV; 1 pair of contractile vessels in IV. Clitellum in $\frac{1}{2}$IV-$\frac{1}{2}$VI; spermathecae small, ampulla spherical or oval, duct distinct; vasa deferentia apparently devoid of prostate, atria small, pear-shaped, with thick, glandular epithelium, opening directly outwards; penial setae 4-5 per bundle, with distal nodulus and two blunt, short prongs.

Swims by brisk wriggling movements.

India, Palestine, Africa.

SUBFAMILY PRISTININAE LASTOCKIN, 1924

Type genus: Pristina EHRENBERG

Prostomium often forming a proboscis. No eyes. Dorsal setae usually beginning in II, consisting of hairs and needles. Pharyngeal and septal glands present; stomach of special structure, with intra-cellular canals. Commissural vessels mostly only in preovarial segments. Coelomocytes present. Nephridia sometimes invested with bladder-like peritoneal cells. Testes and spermathecae in VII, ovaria and atria in VIII. Seven segments formed at anterior end, by budding.

Cosmopolitan.

GENUS **Pristina** EHRENBERG, 1828

TYPE SPECIES: **Pristina longiseta** EHRENBERG

Dorsal setae hairs and simple, bifid or trifid, needles, with or without a nodulus. Dorsal vessel in the middle-line; commissural transverse vessels simple, anastomosing, or forming a plexus. Stomach, beginning in VI, VII, or VIII.

Chloragogen beginning in IV or V. Vasa deferentia with or without prostate cells joining atria on top; atria without prostate, most often weak, tubular; genital setae of various shapes, often accompanied by special glands.
Cosmopolitan.

Pristina

1.	Dorsal setae start in III or IV	*P. macrochaeta*
—	Dorsal setae start in II	2
2.	Needles simple-pointed. (Proboscis present)	3
—	Needles bifid, with at least a minute upper tooth, or pectinate. (Proboscis present or absent)	5
3.	No hair setae especially elongate	*P. proboscidea*
—	Some hairs longer than others	4
4.	Hairs of III longer than the rest	*P. longiseta* (2 subspecies)
—	Hairs of VIII onward (at least) very long	*P. biserrata*
5.	Prostomium with proboscis	6
—	Prostomium without proboscis	14
6.	Hairs of III longer than the rest	*P. longiseta bidentata*
—	Hairs of III not longer than the rest	7
7.	Needle teeth fine, equal teeth	8
—	Needle teeth long equal or unequal in length	10
8.	No giant ventral setae	*P. foreli*
—	Giant ventral setae in IV or V or both	9
9.	Giant ventral setae in IV or V or both, upper tooth at least twice as long as lower or lower absent. Genital setae not reported	*P. aequiseta*
—	Giant ventrally setae in V, enlarged genital setae in VII in some	*P. evelinae*
10.	Needle teeth unequal in length	11
—	Needle teeth equal in length	13
11.	Genital setae on VI and VIII. Upper tooth of needles slightly longer and thicker than lower	*P. plumaseta*

— No genital setae. Upper tooth of needles shorter than lower   12

12. Upper tooth of needles slightly shorter than lower, hairs non-serrate   *P. synclites*

— Upper tooth of needles much shorter than lower, hairs serrate   *P. americana*

13. Needle teeth diverging, dorsal bundles 1 non-serrate hair and 1 needle. Stomach in VII   *P. breviseta*

— Needle teeth parallel, dorsal bundles 2-4 serrate hairs, 1-3 needles. Stomach in VIII   *P. peruviana*

14. Needle teeth almost simple-pointed, upper tooth very fine   *P. menoni*

— Needles distinctly bifid or pectinate   15

15. Needles pectinate   *P. sima*

— Needles bifid   16

16. Needle teeth very short and about equal, or upper longer   17

— Needle teeth long, upper shorter than lower   20

17. Needle teeth parallel   *P. bilobata*

— Needle teeth diverging   18

18. Needle teeth diverge widely (hairs non-serrate)   *P. osborni*

— Needle teeth diverge at an acute angle (hairs serrate)   19

19. Upper tooth of needle longer than lower   *P. acuminata*

— Upper tooth of needle as long as lower   *P. notopora*

20. Ventral setae all with teeth equally long   21

— Ventral setae not all with teeth equally long   22

21. Needle teeth long, parallel, upper slightly shorter than lower   *P. idrensis*

— Needle teeth diverging, upper about half as long as lower   *P. jenkinae*

22. Anterior ventral setae with upper tooth slightly longer than lower, the difference less posteriorly   *P. rosea*

— Anterior ventral setae with upper tooth as long as or longer than lower, shorter posteriorly   *P. amphibiotica*

The distinctions between many species, especially those from couplet 18 onward, are extremely fine. When some of the more recently described species that are not accepted herein are considered, the distinctions between some of the species appear trivial, but the reproductive organs of many have not been described.

In only a few instances do we have a firm indication of variation in the form of the setae within a single specimen, variation which is elsewhere considered to be significant in the separation of species. The whole genus is in need of review following a study of the variability of characters such as those used in the above key.

Pristina rosea (Piguet, 1906)
Fig. 7. 22L, M

Naidium roseum PIGUET, 1906: 223, Pl. IX, figs. 22-23.

Naidium roseum Piguet, PIGUET, 1909: 175, Pl. III, fig. 18. PIGUET AND BRETSCHER, 1913: 25.

Pristina rosea (Piguet). MICHAELSEN, 1909 a: 28, Fig. 47. SCHUSTER, 1915: 50, 76-77. SVETLOV, 1924: 197. MALEVICH, 1927: 9. UDE, 1929: 30, Fig. 29. SPERBER, 1948: 209, Fig. 22a; 1950: 74, Fig. 26a, b. CEKANOVSKAYA, 1962: 216, Fig. 130. BOTEA, 1963: 338. BRINKHURST, 1963 b: 144. NAIDU, 1963: 205.

(?) *Pristina lutea* (Schmidt). SVETLOV, 1924: 197.

(? non) *Pristina rosea* (Piguet). MICHAELSEN AND BOLDT, 1932: 596. KONDO, 1936: 386, Pl. XXIV, fig. 16.

(non) *Naidium roseum* Piguet. MARCUS, 1943: 130, Pl. XXV, fig. 105, Pl. XXVI, fig. 106.

$l=4$-5.5 mm, $s=21$. No proboscis. Hair setae very finely serrated, 1-2 per bundle, needles 1-2 per bundle, slightly curved, with nodulus slightly distal, and moderately long, fine teeth, the upper being shorter than the lower; ventral setae increasing in length posteriorly in the first segments, all with nodulus slightly distal, and upper tooth longer than lower, the difference growing smaller towards the rear. Stomach beginning in VII or VIII; septal glands in III-V or IV-VI. Transverse commissural vessels in I-III forming a net-work, in IV-VII simple.
Europe.

Pristina menoni (Aiyer, 1929)
Fig. 7. 22A-C

(?) *Naidium mosquensis* UDALICOV, 1907.

(?) *Naidium heteroseta* UDALICOV, 1907.

Naidium menoni AIYER, 1929: 21, Fig. 3.

Pristina menoni (Aiyer). SPERBER, 1948: 213, Fig. 226e, Pl. XXI, fig. 3; 1950: 74, Fig. 26c-e; 1958: 50, Fig. 13-16. DAHL, 1957: 1162. BRINK-HURST, 1962: 320; 1963 a: 19, Fig. 36. CEKANOVSKAYA, 1962: 217, Fig. 131. NAIDU, 1963: 204.

$l=7$ mm, $s=28$. No proboscis. Hairs 1-2 per bundle, non-serrated; needles 1-2 per bundle, stout, simple-pointed or with a small upper tooth, and with bayonet-shaped distal half; ventral setae 2-5 per bundle, increasing in length backwards in anterior segments, in II nodulus median, in the rest distal, in anterior segments upper tooth longer than lower, in posterior segments equally long. Stomach beginning in VII, with intra-cellular canals; septal glands in III-V. Ventral vessel dividing in V; simple commissural vessels in II-VII, or those of II branching.
Europe, Asia, Africa.

Pristina bilobata (Bretscher, 1903).
Fig. 7. 22D, E

(?) *Naidium uniseta* BRETSCHER, 1900: 15.

Naidium bilobatum. BRETSCHER, 1903: 11, Pl. I, fig. 1.

(?) *Naidium uniseta* Bretscher. PIGUET, 1906: 216. PIGUET AND BRETSCHER, 1913: 26.

Naidium bilobatum Bretscher. PIGUET, 1906: 217, Pl. IX, figs. 24, 29. PIGUET AND BRETSCHER, 1913: 27.

(?) *Pristina uniseta* (Bretscher). MICHAELSEN, 1909 a: 28. UDE, 1929: 32.

Pristina bilobata (Bretscher). MICHAELSEN, 1909 a: 28, Figs. 48-49. SCHUSTER, 1915: 51. UDE, 1929: 31, Figs. 30-31. CERNOSVITOV, 1930: 9; 1938 a: 539. SPERBER, 1948: 216; 1950: 75, Fig. 26f. DAHL, 1957: 1159, Fig. 3-4. TIMM, 1959: 25. CEKANOVSKAYA, 1962: 215, Fig. 129. BOTEA, 1963: 338. BRINKHURST, 1963 b: 144. NAIDU, 1963: 205.

(?) *Pristina sp.* MICHAELSEN, 1926 c: 3.

(?) *Pristina sp.* MALEVICH, 1927: 9.

(?) *Pristina lutea* (O. Schmidt) ? KNOLLNER, 1935: 433, Fig. 5.

l (preserved)=about 4 mm, $s=34$. No proboscis. Hair setae serrated, 1-2 per bundle, bifid, with teeth short, equal, parallel; ventral setae 3-8 per bundle, bifid with teeth equally long, upper thinner than lower. Stomach beginning in VIII, slowly dilating. Commissural vessels in I-V forming a plexus. Posterior brain lobes deeply bifid (?).
Europe, Palestine, Turkestan, Africa.

Pristina amphibiotica Lastockin, 1927
Fig. 7. 22F-H

Pristina amphibiotica LASTOCKIN, 1927: 69, Figs. 1-2, SPERBER, 1948: 217,

Fig. 23a-c, Pl. XIX; 1950: 75, Fig. 27a-c, Pl. III, fig. 6. DAHL, 1957:
1161. CEKANOVSKAYA, 1962: 215, Fig. 128. NAIDU, 1963: 205.

Pristina amphibiotica changtuensis LIANG, 1963: 562, Fig. 2a-f.

l=3-6 mm, s=12-23. No proboscis. Hair setae 1-2 per bundle, non-serrated;
needles 1-2 per bundle, ? longer and thicker in IV (and V) than in other segments,
bifid, with upper tooth much shorter than lower, and nodulus distal; ventral
setae 3-6 per bundle, with nodulus distal, in anterior segments upper tooth
slightly longer than, or equal to lower, posteriorly much shorter. Stomach in
$\frac{1}{2}$VI-$\frac{1}{2}$VII, slowly dilating, with intracellular canals; septal glands in III-V.
Anastomosing commissural vessels in II-V. Nephridia paired in IX, unpaired or
missing in other segments. Coelomocytes numerous. Clitellum in $\frac{1}{2}$VII-$\frac{1}{2}$IX;
no spermathecae; male funnels very small, vasa deferentia very short and
narrow; no differentiated atria; probably no prostate; penial setae occasionally
present, in IX, 1 per bundle, curved, with two enormous converging distal
prongs; an ovisac formed; no sperm-sac.

Europe, Africa, China.

Dahl (1957) did not find any difference between the ventral setae of IV
and V and the rest.

The sub-species described by Liang (1963) had large, more or less simple
pointed needles in IV with those of V less modified, and had a developed
atrium.

Pristina idrensis Sperber, 1948
Fig. 7. 22I-K

Pristina idrensis SPERBER, 1948: 220, Fig. 23d-e, Pl. XX, XXI, fig. 1.

Pristina idrensis Sperber. SPERBER, 1950: 76, Fig. 27d-f. BRINKHURST, 1962:
320; 1963 a: 19, Fig. 3e. NAIDU, 1963: 205.

? *Pristina taita* STOUT, 1956: 99, Figs. 5-7.

? *Pristina nothofagi* STOUT, 1958: 289, Fig. 1-3.

? *Pristina longidentata* HARMAN, 1965 b: 28, Fig. 1, 2a-e.

l=3-4 mm, s=14-18. Colour whitish. No proboscis. Hair setae 1-2 per
bundle, non-serrated; needles 1-2 per bundle, curved distally, with nodulus
distal, bifid, and with teeth long and parallel, lower tooth slightly longer than
upper; in IV needles longer and stouter than in other segments; ventral setae
3-7 per bundle, increasing in length backwards in anterior segments, with distal
nodulus, and teeth in all segments equally long; in anterior segments setae and
teeth finer than behind. Stomach in $\frac{1}{2}$VI-$\frac{1}{2}$VII, slowly dilating, with intra-cellular
canals; septal glands in III-V. Dorsal vessel situated mid-dorsally. Nephridia in
IX paired, posteriorly unpaired. Coelomocytes present. Clitellum in $\frac{1}{2}$VII-IX;
no spermathecae; male funnels small, vasa deferentia short; atria very small,
thin-walled, atrial ducts distinct, uniting before opening together mid-ventrally;
no penial setae; ovisac formed; no sperm-sac.

Europe (? New Zealand? N. America).

This species is close to, but reasonably distinct from *P. amphibiotica*. Sperber (1948) mentions the problem of maintenance of forms within a species by the persistence of asexual reproduction or self-fertilization. In the description of *P. nothofagi* no real attempt was made to distinguish this supposed species from other *Pristina* species, the discussion centering largely around *P. minuta* (q.v.). *Pristina taita* differs from *idrensis* and *amphibiotica* by the number of segments anterior to the budding zone, and the dorsal segment bearing the longest setae. *Pristina longidentata* of Harman (1965 b) appears to be closely related to *P. idrensis*. The fact that serrations could be found on the hair setae using phase contrast at a magnification of X 1000 may not mean that this is a difference between this species and *P. idrensis*. Harman (1965) is concerned to show the distinction between *longidentata* and both *rosea* and *jenkinae*, *rosea* being the only other species with serrate hair setae and needles with unequal teeth. The position of the stomach was not described by Harman. Liang (1963) regarded *idrensis* as a sub-species of *P. amphibiotica*.

Pristina osborni (Walton, 1906)
Fig. 7. 22N-P

Naidium osborni WALTON, 1906 : 703, Fig. 12.

Naidium osborni Walton. GALLOWAY, 1911 : 302. SMITH, 1918; 640. HARMAN AND PLATT, 1961 : 93.

Naidium minutum STEPHENSON, 1914 : 327, Fig. 3-5.

Naidium minutum Stephenson. STEPHENSON, 1923 : 68, Fig. 22. MARCUS, 1943 : 129, Pl. XXV, figs. 103-104.

Pristina minuta (Stephenson). SPERBER, 1948 : 222. STOUT, 1958 : 291. NAIDU, 1963 : 206, Fig. 29a-c. BRINKHURST, 1964 : 217.

$l=2$ mm. No proboscis. Hair setae 1 per bundle, non-serrated; needles 1 per bundle, with distal nodulus, and short teeth, equally long, separated at a wide angle; ventral setae 3-5 per bundle anteriorly, 2-3 posteriorly, increasing in length in anterior segments, those of II being finer than the rest, with upper tooth longer than lower, and nodulus proximal; behind II nodulus distal, and prongs growing equally long posteriorly. Stomach in VII or VIII, sudden, with canals; septal glands in IV-V, occasionally also in III and VI.
India, Brazil, N. America, Africa.

The arguments of Stout (1958) and Naidu (1963) regarding the separation of *N. osborni* from *N. minutum*, and of *minutum* of Stephenson from *minutum* Stephenson of Marcus, are based on very slight differences. The drawing of a nodulus in needle setae that lack them occurs elsewhere in illustrations by Marcus (Sperber 1948, p. 232).

Pristina notopora Cernosvitov, 1937

Pristina notopora CERNOSVITOV, 1937 b: 140, Figs. 13-16.

Pristina notopora Cernosvitov. SPERBER, 1948: 223. NAIDU, 1963: 205.

$l=1.5-2.5$ mm, s=28. No proboscis. Hair setae 1-2 per bundle, serrated, needles 1-2 per bundle, bifid, with short, fine, equal teeth, diverging at an acute angle; ventral setae 4-7 per bundle anteriorly, 2-4 posteriorly, increasing in length backwards in anterior segments, with upper tooth longer than lower in II and III, equally long in the middle segments, and shorter posteriorly. Stomachal dilatation sudden, in VIII. Nephridia opening to the exterior in front of the dorsal setal bundles.
    S. America (Argentine).

The difference between this and *P. osborni* are slight.

Pristina acuminata Liang, 1958
Fig. 7. 22Q-S

Pristina acuminata LIANG, 1958: 42, Pl. III, fig. 26-32.

$l=3.3-5$ mm, $s=20-53$. Body whitish. Prostomium triangular, pointed apex in life but no proboscis. No eyes. Dorsal setae from II, 2-5 finely serrate hairs, 2-5 bifid needles with teeth narrowly diverging, upper longer than lower and thicker. Ventral setae 4-6 per bundle, upper tooth longer and thinner than lower. Septal glands in III-IV. Stomach in VIII, with intracellular canals. Dorsal vessel on left, commissurals in II-VII, winding. Nephridia paired in IX, single or missing posteriorly. Coelomocytes present.
    Nanking, China.

The species is close to *P. notopora* and perhaps *P. peruviana* but the number of setae and their form differs and *acuminata* lacks a proboscis. Few *Pristina* species have as many dorsal setae.

Pristina sima (Marcus, 1944)
Fig. 7. 22T-W

Naidium simum MARCUS, 1944: 68, Pl. XIII, figs. 58-59.

Pristina sima (Marcus). SPERBER, 1948: 223. NAIDU, 1963: 205.

$l=2$ mm. No proboscis. Hair setae 1 per bundle, non-serrated; needles 1 per bundle, bifid, with fairly short teeth, and usually 2-3 intermediate teeth; nodulus weak, distal; ventral setae 3-5 per bundle, increasing in length in anterior segments, in II-VII upper tooth longer, in the rest shorter than lower, nodulus in II proximal, in the neighbouring segments median, posteriorly distal. Stomach in VIII, with canals.
    S. America (Brazil).

Pristina jenkinae (Stephenson, 1931)
Fig. 7. 23A, B

(?) *Naidium lutem* Schmidt. MICHAELSEN, 1905 b: 306.

Naidium jenkinae STEPHENSON, 1931 b: 39, Fig. 1.

Naidium jenkinae Stephenson. STEPHENSON, 1932: 237.

(?) *Pristina rosea* (Piguet). MICHAELSEN AND BOLDT, 1932: 596. KONDO, 1936: 386, Pl. XXIV, fig. 16. YAMAGUCHI, 1953: 286.

Naidium roseum Piguet. MARCUS, 1943: 130, Pl. XXV, fig. 105, Pl. XXVI, fig. 106.

Pristina jenkinae (Stephenson). SPERBER, 1948: 224; 1958: 51, Fig. 17. (?) BOTEA, 1963: 338. NAIDU, 1963: 210, Fig. 31a-b.

l (preserved)=2·5-3 mm. No proboscis. Dorsal bundles containing usually 1 non-serrated hair, and 1 fairly stout needle, with distal nodulus, and long teeth, the upper tooth being shorter and thinner than the lower; ventral setae 4-6 per bundle anteriorly, 2-3 posteriorly, in anterior segments nodulus median, farther back distal, all with teeth about equally long. Stomach in VII.

Europe (?), S. America, Africa, Asia.

The specimens identified by Botea (1963) were not described.

Pristina synclites Stephenson, 1925.
Fig. 7. 23C-E

Pristina synclites STEPHENSON, 1925 c: 45, Pl. III, fig. 1.

Pristina synclites Stephenson. SPERBER, 1948: 225. NAIDU, 1963: 208, Fig. 30a-d.

l (preserved)=5-7 mm. *s*=35-61. Prostomium forming a short proboscis. Hair setae 1-2 per bundle, non-serrated; needles 1-2 per bundle, stout, with a weak nodulus, and long teeth, the lower slightly the longer and stouter; ventral setae 4 per bundle in anterior segments, decreasing in number backwards, with teeth equally long; in II-VII nodulus median or slightly proximal, in the following segments distal. Stomach in VIII; septal glands in IV-V. Free commissural loops at least in VI-VII.

Southern Asia (India), Africa.

Pristina breviseta Bourne, 1891
Fig. 7. 21G, 7. 23F-H

Pristina breviseta BOURNE, 1891: 353, Pl. XXVII, figs. 11-15.

Pristina breviseta Bourne. BEDDARD, 1895: 292. MICHAELSEN AND BOLDT, 1932: 595. SPERBER, 1948: 225. CEKANOVSKAYA, 1962: 214. NAIDU, 1963: 205. BRINKHURST, 1964: 217.

Naidium breviseta (Bourne). MICHAELSEN, 1900: 23. PIGUET, 1906: 216. STEPHENSON, 1923: 67; 1926 a: 290, Figs. 1-4. AIYER, 1925: 32, Figs. 1-2; 1929: 23. CHEN, 1940: 42, 1944: 5. MARCUS, 1943: 128, Pl. XXV, figs. 101-102. MARCUS, E. 1949: 2.

l=10-20 mm, *s*=35-120. Prostomium usually forming a short proboscis. Hair setae 1 per bundle, non-serrated; needles 1 per bundle, stout, curved distally, with a weak nodulus, and long, equal teeth; ventral setae 3-5 per bundle, increasing in length backwards in anterior segments, those of II straighter, with nodulus median, and upper tooth longer than lower, in posterior segments setae more curved, with nodulus distal, and lower tooth longer than upper. Stomachal dilatation in VII, with intracellular canals. Transverse commissural loops in II-VII, stronger in V-VII. Clitellum in ½VII-IX dorsally, absent or weak mid-ventrally; 1 single genital seta to the right of the spermathecae in VII, or in VIII, on the left side, both setae being strongly curved proximally, with two enormous converging prongs, united by a membrane, and both enclosed within genital glands; spermathecal ampullae small, thick-walled, ducts narrow, both opening into one pit, in common or apart; male funnels cup-shaped, vasa deferentia narrow, surrounded by gland cells; atrial ampullae fairly high, with narrow lumen, ducts narrow, thin-walled, both opening in common into a median pit.

S. and E. Asia, S. and N. America.

Pristina americana Cernosvitov, 1937
Fig. 7. 21I, 23I-M

Pristina americana f. *typica* CERNOSVITOV, 1937 b: 136, Figs. 1-5, 10.

Pristina americana var. *loretana* CERNOSVITOV, 1937 b: 139, Figs. 6-9, 11-12.

Pristina americana Cernosvitov. MARCUS, 1943: 106, Pl. XIX, fig. 82, Pl. XX, fig. 80, Pl. XXI, fig. 83, Pl. XXIV, fig. 98. SPERBER, 1948: 226. NAIDU, 1963: 205.

(?) *Pristina americana* f. *typica* Cernosvitov. MARCUS, E., 1947: 8, Fig. 7.

$l = 10$-30 mm. Prostomium forming a proboscis. Hair setae 1-2 per bundle, serrated; needles 1-2 per bundle, stout, slightly curved distally, with distal nodulus, and long teeth, the upper shorter than the lower, or occasionally absent; ventral setae in anterior segments 3-6 per bundle, diminishing posteriorly to 1-2, increasing in length backwards in anterior segments; in II upper tooth longer than lower, and nodulus proximal; in III teeth equally long, and nodulus median; the rest with upper tooth shorter than lower, and distal nodulus. Stomach beginning in VII, with intra-cellular canals. Commissural vascular loops in II-V branching and anastomosing, in VI-VII simple, pulsating. Clitellum in ½VII-½IX dorsally and ventrally; genital setae in VI 3-5 per bundle, curved, elongated, with upper tooth long and straight, lower tooth sometimes bifid; penetrating projections from the body-wall; in VIII 1 per bundle, very long and thick, straight, slightly curved and thickened distally, with a longitudinal groove; situated in a deep pocket, branching proximally, the anterior branch surrounded by a glandular mass; spermathecal ampulla long, duct more or less narrow; male funnels large, cup-shaped, vasa deferentia narrow, in their median part surrounded by gland cells; atria very long, tubular, sometimes curved, duct short, surrounded by gland cells.

S. America.

The specimens described by E. du B. R. Marcus (Marcus, E. 1947) had needles with teeth of equal length or others (in the same worm) with the upper tooth shorter than the lower. The hair setae were non-serrate (both conditions being observed in specimens of *P. proboscidea*).

Pristina peruviana Cernosvitov, 1939
Fig. 7. 23N-P

Pristina peruviana CERNOSVITOV, 1939: 83, Figs. 1-7.

Pristina peruviana Cernosvitov. MARCUS, E., 1947: 9. SPERBER, 1948: 227. NAIDU, 1963: 205.

l=4-4·5 mm (preserved). Prostomium forming a short proboscis. Hair setae 2-4 per bundle, stiff, closely serrated; needles 1-3 per bundle, stout, with a weak nodulus, and two long, equal, parallel teeth; ventral setae 4-6 per bundle, increasing in length in anterior segments, all approximately similar, with slightly longer upper tooth. Stomachal dilatation sudden, in VIII. Septal glands in IV-V. Anterior commissural vessels branching.

S. America (Peru).

Pristina foreli (Piguet, 1906)
Fig. 7. 24A-C

(?) *Pristina inaequalis* EHRENBERG, 1828: 112, foot-note.

(?) *Pristinais inaequalis* (Ehrenberg). GERVAIS, 1838: 17.

(?) *Pristina inaequalis* Ehrenberg. GRUBE, 1851: 105. VEJDOVSKY, 1884: 24. VAILLANT, 1890: 360.

Naidium foreli PIGUET, 1906: 222, Pl. IX, figs. 21, 25, 27-28.

Naidium foreli Piguet. PIGUET, 1909: 175.

Pristina foreli (Piguet). MICHAELSEN, 1909 a: 27, Figs. 45-46. PIGUET AND BRETSCHER, 1913: 53. SCHUSTER, 1915: 50. SVETLOV, 1924: 197. UDE, 1929: 32, Figs. 32-33. CERNOSVITOV, 1930: 9. SPERBER, 1948: 229, Pl. XXI, fig. 4; 1950: 76, Fig. 28a, Pl. III, fig. 7; 1958: 52. TIMM, 1959: 25. BRINKHURST, 1962: 320; 1963 a: 21, Fig. 3f; 1963 b: 144; 1964: 218, fig. 6a-d. CEKANOVSKAYA, 1962: 211, Fig. 125. MOSZYNSKA, 1962: 15. BOTEA, 1963: 338. NAIDU, 1963: 205.

Pristina palustris SCHUSTER, 1915: 51, Figs. 33-35.

Pristina palustris Schuster. HEMPELMANN, 1923: 434. UDE, 1929: 32. SZARSKI, 1936 b: 392, Pl. XVIII, figs. 6-8, 11, 13-15, 17-18.

(?) *Pristina capiliseta* KONDO, 1936: 388, Pl. XXIV, fig. 19.

(?) *Pristina nasalis* KONDO, 1936: 388, Pl. XXIV, fig. 20.

Pristina schmiederi CHEN, 1944: 4, Fig. 2.

Pristina sperberae NAIDU, 1963: 219, Fig. 35a-d.

Pristina schmiederi Chen. HARMAN, 1961: 94.

Pristina schniederi Chen. NAIDU, 1963: 204.

l=3-6·5 mm, s=20-26. Prostomium forming a proboscis. Hair setae very finely serrated, 1-4 per bundle; needles finely bifid, slightly curved distally, without a nodulus, 1-4 per bundle; ventral setae 2-8 per bundle, in II slightly longer than the rest, with nodulus proximal, in the rest nodulus distal; in II-VII upper tooth longer than lower, behind VII teeth equally long, the upper growing shorter backwards. Transverse commissural loops in II-VII, occasionally those of II and III anastomosing.

Europe, Turkestan, Japan?, Africa.

P. schmiederi of Chen (1944) seems to belong here, despite the (apparent) absence of serrations on the hair setae, and not in *P. plumaseta* as claimed by Marcus, E., 1947 (Brinkhurst, 1964). The differences between *P. foreli* and *P. sperberae* Naidu do not seem to be significant to me, again adding up to little more than the absence of serrations on the hairs, said to be very fine in *P. foreli*, and hence very difficult to resolve with most optical equipment.

Pristina plumaseta, Turner, 1935
Fig. 7. 21H

(?) *Pristina brevisieta* Bourne. HAYDEN, 1912: 531.

(?) *Pristina variabilis* HAYDEN, 1914: 136.

Pristina plumaseta TURNER, 1935: 253, Figs. 1-2.

Pristina plumaseta Turner. SPERBER, 1948: 228. NAIDU, 1963: 205. BRINK-HURST, 1964: 218.

(non) *Pristina schmiederi* Chen. MARCUS, E., 1947: 9.

l=8-12 mm, s=about 52. Prostomium forming a short proboscis. Hair setae 1-2 per bundle, serrated, with serrations in two rows; needles 1-2 per bundle, bifid, with distal half bayonet-curved, and with upper tooth slightly longer and thicker than lower, nodulus about median; ventral setae 4-8 per bundle anteriorly, with nodulus slightly distal, and teeth subequal. Stomachal dilatation beginning in VII (?), slow. Clitellum in VII-½X dorsally, in IX-½X ventrally; genital setae in VI and VIII, more than twice as long as the ordinary setae, slightly curved, simple-pointed, terminating distally in a knob, with a longitudinal groove along one side, in VI 4 per bundle, situated on lobular projections, in VIII situated behind the atria, in pockets, surrounded by gland cells; spermathecal ampulla large, duct fairly long, thin-walled; the ducts of both sides opening into a common, transverse groove; vasa deferentia very long, winding, covered with prostate cells; atria short, wide, duct inconspicuous.

N. America.

If the identity of *P. variabilis* Hayden could be established, the name would not be available in any case, as it is preoccupied (Friend 1912, ?=*Nais pseudobtusa* q.v.).

Pristina aequiseta Bourne, 1891.
Fig. 7. 24D-G

Pristina equiseta BOURNE, 1891: 352.

Pristina equiseta Bourne. BEDDARD, 1895: 291. MALEVICH, 1927: 9.

Pristina affinis GARBINI, 1898: 562, Fig. 1.

Pristina aequiseta Bourne. MICHAELSEN, 1900: 34 (partim); 1905 b: 309 (partim); 1926 c: 5; 1933: 341. PIGUET, 1909: 211. STEPHENSON, 1909 a: 269, Pl. XVIII, figs. 34-39, Pl. XIX, fig. 40; 1916: 304; 1922 a: 253; 1923: 71; 1931 b: 42; 1932: 237. PIGUET AND BRETSCHER, 1913: 52. SCHUSTER, 1915: 50. HAYDEN, 1922: 168. HEMPLEMANN, 1923: 380. SVETLOV, 1924: 197. UDE, 1929: 30, Figs. 27-28. CERNOSVITOV, 1930: 9. PASQUALI, 1938: 20. SCIACCHITANO, 1938: 257. CHEN, 1940: 48. MARCUS, 1943: 104, Pl. XIX, fig. 81. SPERBER, 1948: 230, Fig. 24, Pl. XXI, fig. 5; 1950: 77, Fig. 286, Pl. III, fig. 7; 1960: 158. BRINKHURST, 1962: 320; 1963 a: 191, Fig. 3h; 1963 b: 144; 1964: 218, Fig. 66. CEKANOVSKAYA, 1962: 213, Fig. 127. MOSZYNSKA, 1962: 15. NAIDU, 1963: 212, Fig. 32a-d.

Naidium tentaculatum PIGUET, 1906; 219, Pl. IX, figs. 18-20, 26.

Pristina tentaculata (Piguet). MICHAELSEN, 1909 a: 26, Fig. 44; 1909 b: 134.

Pristina aequiseta Bourne, var.? MICHAELSEN, 1913: 209.

l=2-8 mm, s=18-23. Prostomium forming a proboscis. Dorsal setae 1-2 finely serrated hairs and 1-2 finely bifid needles, slightly curved distally, but without a nodulus; ventral setae in most segments 5-8 per bundle, those of II longer and thinner than the rest, with nodulus slightly proximal, and with upper tooth twice as long as lower; in III-VII shorter and slightly thicker, with nodulus distal, and with upper tooth slightly longer than lower, or, usually, in IV, V, or VI or some of these, enlarged setae, much thicker than the rest, with upper tooth more than twice as long as lower, or the latter missing; behind VII setae thicker, more curved, with equally long teeth. Septal glands in III-V. Commissural vascular loops in II-VII.

Cosmopolitan. Also in oligohaline water (the W. Indies).

The distinction between this and *P. evelinae* is now reduced to the presence of genital setae in *evelinae*, and the absence of the lower tooth in the giant setae in the same. The proximity between the two is emphasized by the recognition of *P. aequiseta* Bourne of Hemplemann (1923), Aiyer (1929) and the var? of Michaelsen (1913) as *P. evelinae* by Naidu (1963).

Pristina evelinae Marcus, 1943.
Fig. 7. 24H, 7. 25A-D

Pristina evelinae MARCUS, 1943: 112, Pl. **XX**, fig. 80, Pl. XXII, figs. 89-91, Pl. XXIV, fig. 99.

Pristina evelinae Marcus, SPERBER, 1948: 232, Fig. 25. MARCUS, E., 1949: 2. NAIDU, 1963: 214, Fig. 332d.

(?) *Pristina aequiseta* Bourne. HEMPLEMANN, 1923: 380. AÏYER, 1929: 25, Fig. 5.

(?) *Pristina aequiseta* Bourne var? MICHAELSEN, 1913: 209.

$l=2$-5 mm, $s=13$-20. Prostomium forming a proboscis. Dorsal setal bundles containing 1 hair, and 1 finely bifid needle, slightly curved distally; ventral setae 4-6 per bundle, with nodulus distal in II longer than the rest, with upper tooth longer than lower; towards the posterior both teeth growing equally long; in V, 1 giant seta per bundle, very thick, twice as long as ordinary setae, with simple distal end. Septal glands in III-V. Transverse commissural vessels in II-VIII, enlarged in VI-VIII. Clitellum dorsally in VIII-IX, ventrally in IX; in VI, and sometimes in VII, strongly enlarged genital setae with bifid distal end; behind the setae of VI a pair of genital glands, and, medially in VIII, behind the male pores, an unpaired glandular mass; spermathecal ampulla fairly large, pear-shaped, duct well-defined, very narrow; male funnels large, tubular, vasa deferentia very thick, glandular; atrial ampulla infinitely small, duct very long, curved, with narrow lumen.

S. America (Brazil), India, ? Europe.

The species is probably no more than a form of *Pr. aequiseta*.

Pristina longiseta Ehrenberg, 1828
Fig. 7. 21J; 7. 25E-I

Prostomium forming a proboscis. Hair setae serrated, in III extremely elongated, non-serrated; needle setae fine, straight, without nodulus, simple-pointed or very finely bifid; ventral setae of II and III differing in shape from the rest, with nodulus median, and upper tooth 2-3 times as long as lower; in the rest nodulus distal, and upper tooth less than twice as long as lower. Stomach in VIII, septal glands in III-V or IV-VI. Simple transverse commissural vessels usually in II-VII, stronger in VI-VII, occasional anastomoses in anterior segments. Clitellum in $\frac{1}{2}$VII-$\frac{1}{2}$IX, weaker ventrally; genital setae in VI, in connection with a pair of genital glands; male funnels cup-shaped, vasa deferentia glandular, atria elongate, of varying size.

Pristina longiseta longiseta Ehrenberg, 1828

Pristina longiseta EHRENBERG, 1828: 112, foot-note.

Pristinais longiseta (Ehrenberg). GERVAIS, 1838: 17.

Nais longiseta (Ehrenberg). D'UDEKEM, 1855: 552; 1859: 22. TIMM, 1883: 153.

Stylaria longiseta (Ehrenberg). TAUBER, 1879: 73.

Pristina longiseta Ehrenberg. VEJDOVSKY, 1883: 219; 1884: 31, Pl. II, figs. 13-15. VAILLANT, 1890: 359 (partim). BEDDARD, 1892: 350; 1895: 290 (partim). BRETSCHER, 1896: 508. MICHAELSEN, 1900: 34; 1903 b: 186. PIGUET, 1906: 290, Pl. X, figs. 22-23, Pl. XII, figs. 21-25. STEPHENSON, 1909 a: 264, Pl. XVII, fig. 25; 1910 b: 235; 1920: 199; 1922 a: 254. PIGUET AND BRETSCHER, 1913: 50. SCHUSTER, 1915: 49. CUNNINGTON, 1920: 574. HEMPLEMANN, 1923: 395. MALEVICH, 1927: 9. UDE, 1929: 29, Figs. 25-26. STEPHENSON, 1931 b: 41, Fig. 2. KONDO, 1932: 388, Pl. XXIV, fig. 18. BERG, 1938: 45. CHEN, 1940: 46, Fig. 12a. CERNOSVITOV, 1942 a: 198. SPERBER, 1958: 52, Figs. 18, 19.

Nais proboscidea O. F. Müll. LEVINSEN, 1884: 219 (partim).

Pristina leidyi Smith. MICHAELSEN, 1905 a: 357 (partim).

Pristina longiseta Ehrenberg f. *typica* MICHAELSEN, 1905 b: 308.

Pristina longiseta Ehrenberg f. *typica* Michaelsen. MICHAELSEN, 1909 a: 25, Figs. 41-42. STEPHENSON, 1923: 70, Fig. 23. SVETLOV, 1924: 197. AIYER, 1929: 24, Fig. 4.

(?) *Pristina longiseta* Ehrenberg. JACKSON, 1931: 74.

(?) *Pristina longiseta* Ehrenberg f. *typica* Michaelsen. MICHAELSEN AND BOLDT, 1932: 595.

Pristina longiseta sinensis SPERBER, 1948: 237.

Pristina longiseta sinensis Sperber. NAIDU, 1963: 204.

Pristina longiseta longiseta Ehrenberg. SPERBER, 1948: 236, Pl. XXI, figs. 2, 6; 1950: 77, Fig. 28c, Pl. III, fig. 9. NAIDU, 1963: 216, Fig. 34a-k.

$l=3\cdot5$-$5\cdot5$ mm, $s=20$-33. Hair setae 1-4 per bundle, serration close and fine; needles 2-5 per bundle, simple-pointed; ventral setae 3-9 per bundle, in II slightly longer than the rest, in III slightly longer and thicker than in the following segments, with upper tooth much longer than lower, in the rest upper tooth $\frac{1}{3}$-$\frac{1}{2}$ longer than lower. Genital setae curved, simple (?) or with long, converging prongs, not enclosed in glands; spermathecal ampulla long, duct short; atria well-defined, glandular, duct narrow.

Europe, Asia, Africa, Australia.

Specimens reported from Australia by Johnson (1931) cannot be attributed to any of the forms for certain. The distinction between the sub-species *sinensis* and the above seems to be largely ignored by Sperber (1958), who referred to them as "races" but gave them sub-specific names.

Pristina longiseta leidyi Smith, 1896.

Pristina longiseta Ehrenberg. LEIDY, 1850: 44, Pl. II, fig. 3. VAILLANT, 1890: 359 (partim). BEDDARD, 1895: 350 (partim). HAYDEN, 1922: 167. ALTMAN, 1936: 14.

Stylaria longiseta (Ehrenberg). MINOR, 1863: 38.

Pristina leidyi SMITH, 1896: 396, Pl. XXV, figs. 1-6.

Pristina leidyi Smith. SMITH, 1900: 443. MICHAELSEN, 1900: 35; 1905 a: 357 (partim). MOORE, 1906: 166. GALLOWAY, 1911: 302. CHEN, 1940: 47, Fig. 12B, b; 1944: 4.

Pristina longiseta Ehrenberg var. *leidyi* Smith. MICHAELSEN, 1905 b: 309. SMITH, 1918: 640.

Pristina tangiseta HAYDEN, 1914: 136.

Pristina antenniseta HAYDEN, 1914: 137.

Pristina longiseta leidyi Smith. SPERBER, 1948: 237. HARMAN AND PLATT, 1961: 93. NAIDU, 1963: 204. BRINKHURST, 1964: 219, Fig. 6c.

l=4-8 mm, s (mature)=30. Hair setae 2-3 per bundle, teeth of serration far apart (6-16 μ); needles simple-pointed; ventral setae 5-9 per bundle, in II and III slightly stouter than the rest, with upper tooth 3 times as long as lower, and with weak nodulus. Genital setae long, straight with distal end bifid, resembling that of ordinary setae; enclosed within genital glands; spermathecal ampulla short, duct short; atria not distinguishable from vasa deferentia, muscular (?). N. America.

The separation of this form as a sub-species may not be justified. Sperber (1948) indicates that we are here classifying an Artenkreis, and if *P. longiseta sinensis* is no longer recognized as a sub-species distinct from the typical form (q.v.) it is almost impossible to separate *leidyi* from the combination of the two.

Pristina longiseta bidentata Cernosvitov, 1942
Fig. 7. 25J-M

(?) *Pristina leidyi* Smith. MICHAELSEN, 1905 a: 357 (partim).

(?) *Pristina longiseta* Ehrenberg f. *typica* Michaelsen. MICHAELSEN, 1913: 208; 1921 b: 2. CORDERO, 1931: 350.

(?) *Pristina longiseta* Ehrenberg. PIGUET, 1928: 91.

Pristina longiseta Ehrenberg var. *bidentata* CERNOSVITOV, 1942 a: 198, Figs. 1-15.

Pristina longiseta Ehrenberg f. *typica* Michaelsen. MARCUS, 1943: 107, Pl. XX, fig. 80, Pl. XXI, fig. 84, Pl. XXII, fig. 93, Pl. XXIII, Pl. XXIV, fig. 95, Pl. XXV, fig. 100. MARCUS, E., 1947: 9.

Pristina longiseta bidentata Cernosvitov. SPERBER, 1948: 238. NAIDU, 1963: 204.

l=2-12 mm, s=20-33. Hair setae 1-3 per bundle, closely and finely serrated;

needles 1-5 per bundle, extremely finely bifid; ventral setae 3-5 per bundle in anterior segments, up to 13 in posterior, in II considerably longer and thicker than the rest, with upper tooth twice as long as lower, in III slightly longer and thicker than in the following segments, with upper tooth 1½ as long as lower; in the other segments teeth about equally long. Genital setae long, straight, bifid like ordinary setae, not enclosed within glands; spermathecal ampulla small, duct long; atria very small and narrow; behind atria, a pair of glandular masses. S. America.

Pristina biserrata Chen, 1940
Fig. 7. 25N-P

Pristina biserrata CHEN, 1940: 49, Figs. 13-14.

Pristina biserrata Chen. SPERBER, 1948: SOKOLSKAYA, 1961: 67, Figs. 10, 11. NAIDU, 1963: 204.

$l=$up to 12 mm, $s=25$-30. Prostomium forming a proboscis. Hair setae 1-3 per bundle, serrated or notched, 2-4 hair-like needles, ventral setae in II-VII 4-6 per bundle, with upper tooth longer than lower, thinner than the rest, in II slightly longer and thicker than in III-VII, in the following segments 10 per bundle, thick. Septal glands in IV-V. Stomach in VIII, another swelling in X. Commissural vessels in II-IX. Clitellum in ½VII-IX; genital setae in VI, 3 per bundle, long, proximal prong disappearing; a pair of glandular masses behind the genital setae; spermathecal ampulla large, duct short; male funnels large, vasa deferentia glandular, with prostate covering, atria small but distinct, without prostate cells, duct short, narrow.
Swims with horizontal undulations.
E. Asia.

Chen (1940) failed to record needle setae, mistaking them for short hair setae. Consequently, the posterior hair setae are said to be very elongate, as the hair seta are, in fact, absent or represented by one moderately long hair setae anteriorly in some specimens (Sokolskaya, 1961). The species is at least close to *P. longiseta*.

Pristina proboscidea Beddard, 1896
Fig. 7. 21N-Q

Pristina proboscidea Beddard, 1896: 4, Fig. 18.

Pristan aequiseta Bourne. MICHAELSEN, 1900: 34 (partim); 1905 b: 309 (partim).

Pristina proboscidea Beddard f. *typica* MICHAELSEN, 1905 a: 359.

Pristina proboscidea Beddard var. *paraguayensis* MICHAELSEN, 1905 a: 360.

Pristina aequiseta Bourne var. *paraguayensis* (Michaelsen). MICHAELSEN, 1905 b: 309.

(?) *Pristina serpentina* WALTON, 1906: 701, Fig. 11.

Pristina proboscidea Beddard f. *typica* Michaelsen. MICHAELSEN, 1909 b: 133. STEPHENSON, 1911: 211; 1923: 73. MICHAELSEN AND BOLDT, 1932: 595. CERNOSVITOV, 1937 b: 136. CHEN, 1940: 49.

Pristina proboscidea Beddard var. *paraguayensis* Michaelsen. MICHAELSEN, 1909 b: 134. STEPHENSON, 1923: 73. AIYER, 1929: 26, Fig. 6. MICHAELSEN AND BOLDT, 1932: 596.

Pristina proboscidea Beddard. MARCUS, 1943: 111, Pl. XXI, fig. 87, Pl. XXII, fig. 88. MARCUS, E., 1947: 9. SPERBER, 1948: 239. NAIDU, 1963: 204; 1965: 20. BRINKHURST, 1964: 219.

l=2-5 mm, s=18-36. Prostomium forming a proboscis. Hair setae (?) serrated, 1-4 per bundle, none especially elongated; needles 1-4 per bundle, simple-pointed, straight and fine, without nodulus; ventral setae in anterior segments 2-4 per bundle, up to 9 posteriorly, all of one type, with upper tooth longer than lower, in II longer and thicker than the rest. Septal glands in III-V. Free commissural vessels in II-VII, enlarged in VI-VII.

S. America, Zanzibar, S. and E. Asia, Australia.

Marcus, E. (1947) described specimens with and without serrate hairs.

Pristina macrochaeta Stephenson, 1931
Fig. 7. 21K-M

Pristina macrochaeta STEPHENSON, 1931 c.: 299.

Pristina macrochaeta Stephenson. MARCUS, 1943: 109, Pl. XXI, fig. 85, Pl. XXII, fig. 86. SPERBER, 1948: 240; 1960: 158. NAIDU, 1963: 204.

l=3·8-7 mm, s=21. Prostomium forming a proboscis. Dorsal setae beginning in III or IV, hair setae very stout, serrated, 1-4 per bundle; needles simple-pointed, 1-3 per bundle; ventral setae 5-10 per bundle, in II longer than the rest, all of similar type, with upper tooth longer than lower. Septal glands in III-V or IV-VI. Stomach in VII-VIII. Enlarged vascular commissures in VI-VII.

S. America, Afghanistan.

The following names have also occurred in the literature but are not attributable to *Pristina*.

Tubificidae
Naidium luteum Schmidt, 1847.
Naidium palmeni Munsterhjelm, 1905 (or *Allodero*—q.v.).
Pristina papillosa Cernosvitov, 1935=*Peloscolex swirenkowi*.

Enchytraeidae
Naidium breviceps Schmidt, 1847.

Opistocystidae
Pristina flagellum Leidy, 1880.

Nomina dubia
  Naidium dadayi Michaelsen, 1905.
  Naidium bilongatum Chen, 1944—? *Allodero*—q.v.

OTHER NOMINA DUBIA

Papillonais orientalis Lastockin, 1949.

This species was never adequately described. In a key to the Russian Naididae, there appears a brief differentiation of this taxon from *Slavina appendiculata*, special attention being drawn to the abundance of papillae. The worm in question was, most probably, a tubificid of the genus *Peloscolex*.

Pristina elegans Finegenova, 1966, is also a papillate worm, with the dorsal setae (both hair and bifid setae, the latter with long parallel teeth) beginning in III. It is impossible to locate this species in the Naididae, except in so far as it would not seem to belong to *Pristina*.

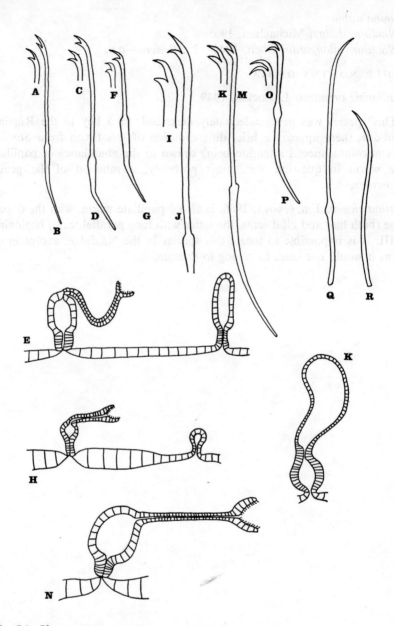

Fig. 7.1. *Chaetogaster.*

C. diastrophus. A,C—setae of II, B,D—setae of VI, E—reproductive organs.
C. langi. F—setae of II, G—seta of VI, H—reproductive organs.
C. diaphanus. I—setae of II, J—seta of VI, K—spermatheca.
C. cristallinus. L—seta of II, M—seta of VI, N—reproductive organs.
C. limnaei. O—seta of II, P—seta of VI.
C. setosus. Q—seta of II, R—posterior seta.

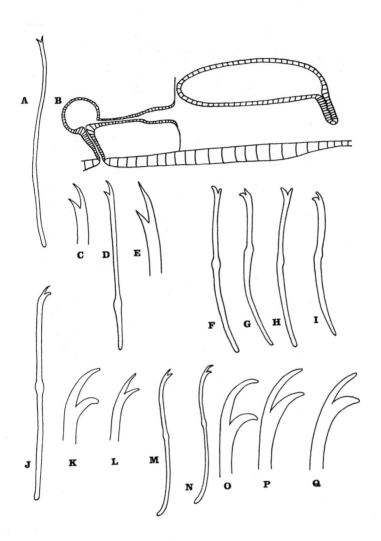

Fig. 7.2. *Amphichaeta* and *Paranais*.
A. sannio. A—seta, B—reproductive organs.
A. americana. C,D—seta of II, E—median seta.
A. magna. F—dorsal seta of III, G—same of VII, H—ventral seta of III, I—same of XII.
P. macrochaeta. J,K—ventral seta of II, L,M—setae of V and VI.
P. salina. N,O—ventral seta of II, P—same of III, Q—dorsal seta of VII.

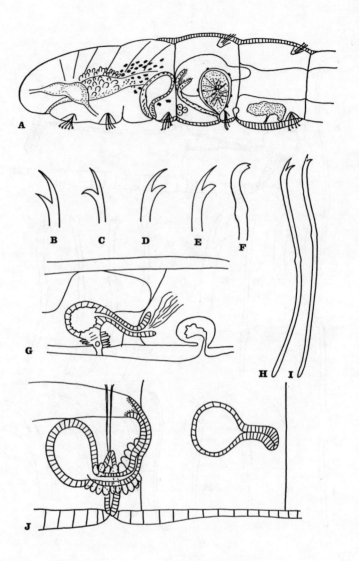

Fig. 7.3. *Waspa* and *Specaria.*

W. evelinae. A—sagital section including reproductive organs, B—ventral seta of II, C—last ventral seta.

W. mobilis. D—ventral seta of II, E—same of V, F—penial seta, G—reproductive organs.

S. josinae. H—ventral seta, I—dorsal seta, J—reproductive organs.

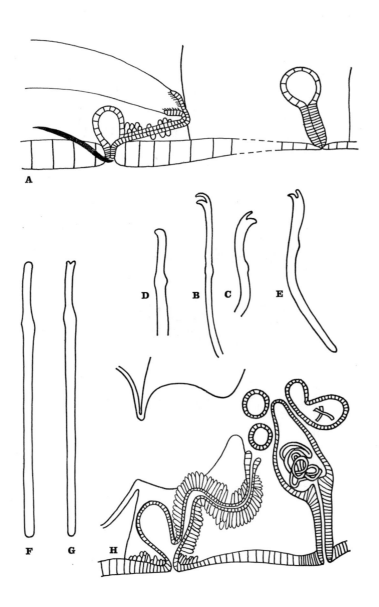

Fig. 7.4. *Uncinais* and *Ophidonais*.

U. uncinata. A—reproductive organs.
U. minor. B—ventral seta of II, C—same of X, D—penial seta, E—dorsal seta of VI.
O. serpentina. F,G—dorsal setae, H—reproductive organs.

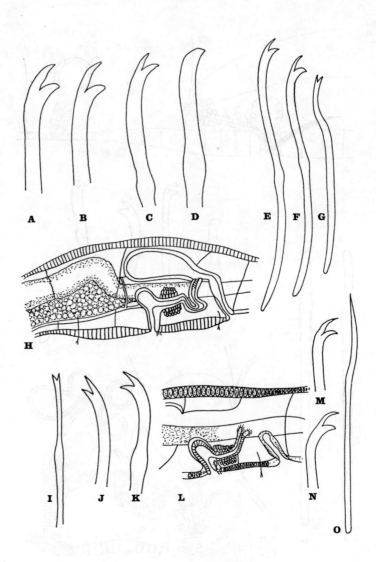

Fig. 7.5. *Nais.*

N. borutzkii. A—ventral seta of II, B—same of XII, C—dorsal seta, D—penial seta.
N. communis. E—ventral seta of II, F—same of posterior bundle, G—dorsal seta,
H—reproductive organs.
N. variabilis. I—dorsal seta of VI, J—ventral seta of II, K—same of VI, L—reproductive
organs.
N. simplex. M—ventral seta of II, N—posterior ventral seta, O—dorsal seta.

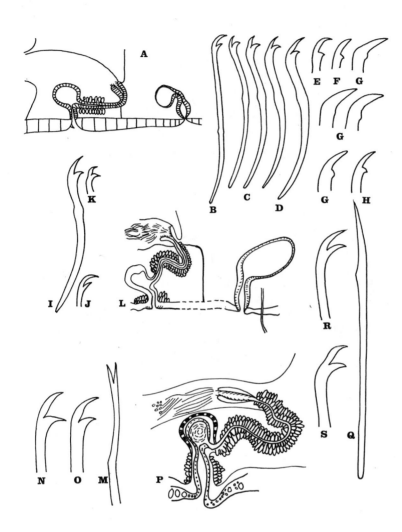

Fig. 7.6. *Nais* (continued).

N. simplex. A—reproductive organs.

N. bretscheri. B—ventral seta of II, C—same of XV, D—thick, non-giant seta from VIII, E—thick, ventral seta from VI, F—same from VII, G—giant seta from VIII-XI, H—thick seta of XIII from specimen with giant seta.

N. pardalis. I—thick seta from VI, J—ventral seta of II, K—same of XVIII, L—reproductive organs.

N. elinguis. M—dorsal seta, N—ventral seta of II, O—same of posterior segment, P—reproductive organs.

N. alpina. Q—dorsal seta, R—ventral seta of II, S—posterior ventral seta.

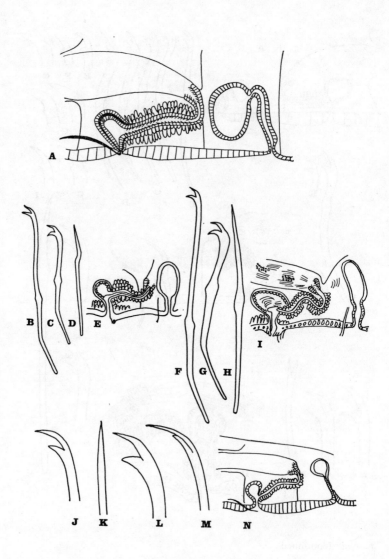

Fig. 7.7. *Nais* (continued).

N. alpina. A—reproductive organs.

N. pseudobtusa. B—ventral seta of II, C—same of posterior bundle, D—dorsal seta, E—reproductive organs.

N. barbata. F—ventral seta of II, G—same of posterior bundle, H—dorsal seta, I—reproductive organs.

N. behningi. J—ventral seta of II, K—dorsal seta of VIII, L—ventral seta of XIV, M—same of IV, N—reproductive organs.

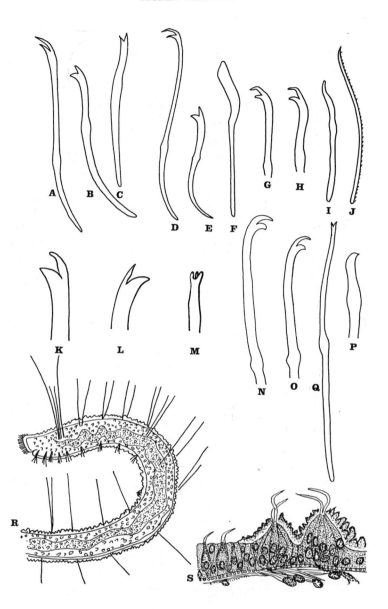

Fig. 7.8. *Nais* (continued) and *Slavina*.

N. raviensis. A—anterior ventral seta, B—posterior ventral seta, C—dorsal seta.

N. schubarti. D—ventral seta of II, E—same of X, F—dorsal seta.

N. baicalensis. G—ventral seta of II, H—same of VIII, I,J—dorsal setae of X.

N. africana. K—anterior ventral seta, L—posterior ventral seta, M—dorsal seta.

N. bekmani (?=*variabilis*). N—ventral seta of II, O—ventral seta of VII, P—dorsal seta of X, Q—penial seta.

S. appendiculata. R—anterior of body, S—papillae.

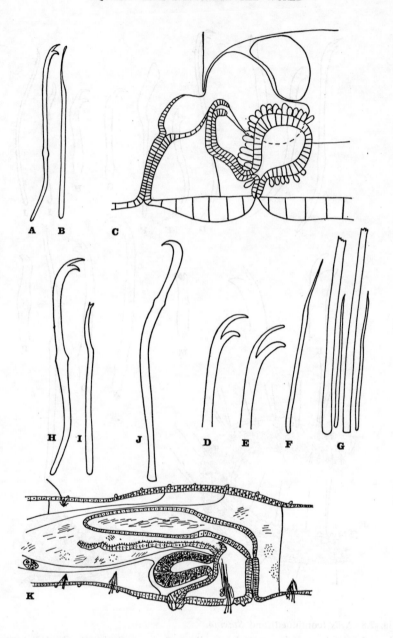

Fig. 7.9. *Slavina* (continued).

S. appendiculata. A—ventral posterior seta, B—dorsal seta, C—male efferent apparatus.

S. isochaeta. D—ventral seta of II, E—same of XI, F—dorsal seta, G—dorsal bundle.

S. sawayi. H—ventral seta, I—dorsal seta.

S. evelinae. J—penial seta, K—male efferent apparatus.

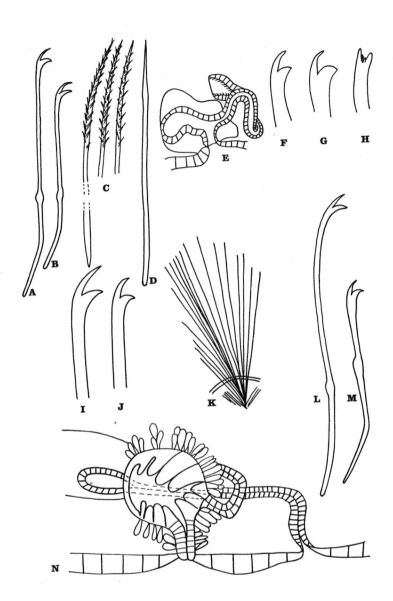

Fig. 7.10. *Vejdovskyella, Arcteonais* and *Ripistes*.

V. comata. A—ventral seta of II, B—ventral posterior seta, C—hair seta, D—dorsal seta, E—male duct.

V. hellei. F—ventral anterior seta, G—ventral posterior seta, H—dorsal seta.

A. lomondi. I—ventral seta of II, J—ventral seta of XXVIII, K—dorsal seta.

R. parasita. L—ventral anterior seta, M—ventral posterior seta, N—male efferent apparatus.

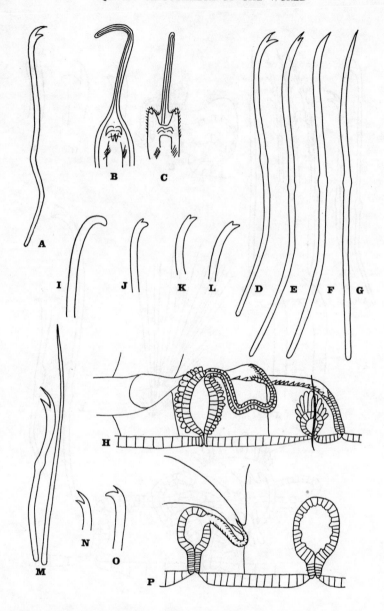

Fig. 7.11. *Stylaria, Piguetiella, Neonais* and *Haemonais*.
S. lacustris. A—ventral seta, C—prostomium.
S. fossularis. B—prostomium.
P. blanci. D—dorsal seta, E to G—spermathecal setae, H—reproductive organs.
N. elegans. I—ventral seta of III, J—same of XI, K—dorsal seta of VI, L—same of X.
H. waldvogeli. M—dorsal setae, N—ventral seta of II, O—same of X, P—male efferent ducts.

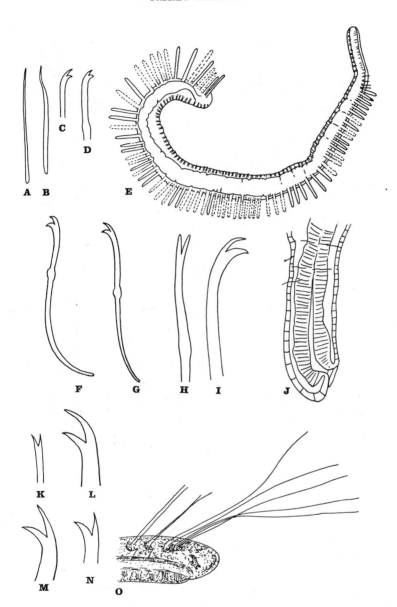

Fig. 7.12. *Branchiodrilus* and *Allodero*.

B. semperi. A—anterior dorsal seta, B—posterior dorsal seta, C—ventral seta of II, D—ventral seta of XV.

B. cleistochaeta. E—outline of body, right side, showing gills, F—ventral seta, anterior to XLVII, G—ventral seta, posterior to XLVII.

A. prosetosa. H—dorsal seta, I—ventral seta of II, J—posterior of body.

A. bilongata. K—dorsal seta, L—anterior ventral seta, M—mid ventral seta, N—post ventral seta, O—anterior of body.

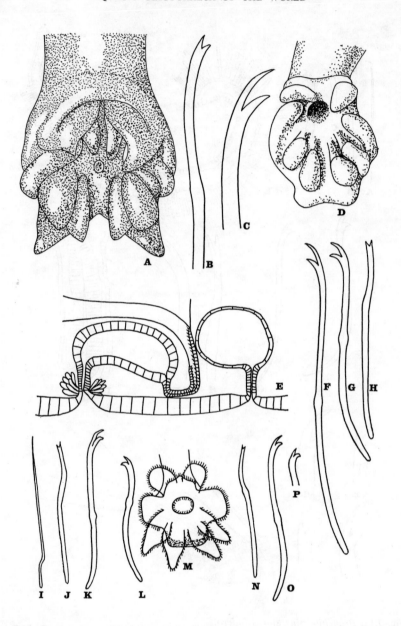

Fig. 7.13. *Dero.*

D. dorsalis. A—branchial fossa, dorsal aspect, B—dorsal seta, C—ventral anterior seta.

D. digitata. D—retracted branchial apparatus, E—male efferent apparatus, F—ventral seta of II, G—same of VI, H—dorsal seta.

D. indica. I—hair seta, J—dorsal seta, K—ventral seta of II, L—posterior ventral seta, M—branchial fossa.

D. zeylanica. N—dorsal seta, O—ventral seta of II, P—posterior ventral seta.

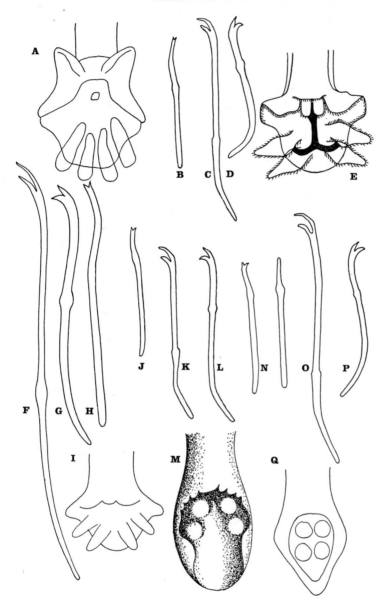

Fig. 7.14. *Dero* (continued).

D. zeylanica. A—branchial fossa.

D. cooperi. B—dorsal seta, C—anterior ventral seta, D—posterior ventral seta, E— branchial fossa.

D. obtusa. F—anterior ventral seta, G—posterior ventral seta, H—dorsal seta, I— branchial fossa.

D. nivea. J—dorsal seta, K—anterior ventral seta, L—posterior ventral seta, M—branchial apparatus.

D. sawayi. N—dorsal seta, side and face view, O—anterior ventral seta, P—posterior ventral seta, Q—branchial fossa.

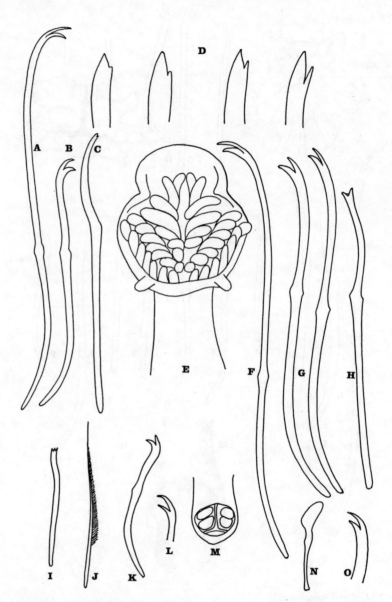

Fig. 7.15. *Dero* (continued).

D. multibranchiata. A—ventral seta of II, B—same of XIV, C—dorsal seta, D—tips of dorsal seta from various segments.

D. botrytis. E—branchial apparatus, F—ventral seta of II to V, G—ventral seta of X, H—dorsal seta.

D. pectinata. I—dorsal seta, J—hair seta, K—posterior ventral seta, L—anterior ventral seta, M—branchial apparatus.

D. palmata. N—dorsal seta, O—ventral seta.

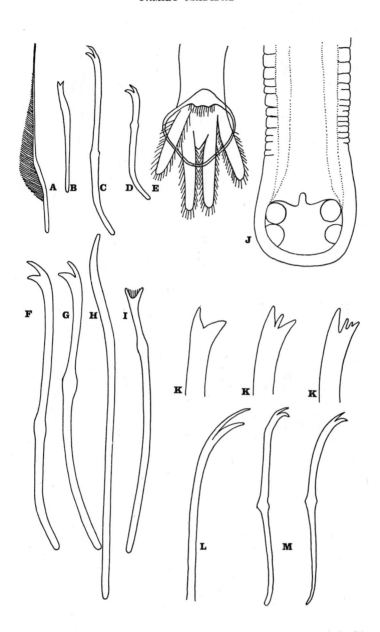

Fig. 7.16. *Dero* (continued).

D. plumosa. A—hair seta, B—dorsal seta, C—anterior ventral seta, D—posterior ventral seta, E—branchial apparatus.

D. asiatica. F—anterior ventral seta, G—posterior ventral seta, H—penial seta, I—dorsal seta, J—branchial apparatus.

D. evelinae. K—tips of dorsal setae, L—ventral seta of II, M—ventral setae of XII.

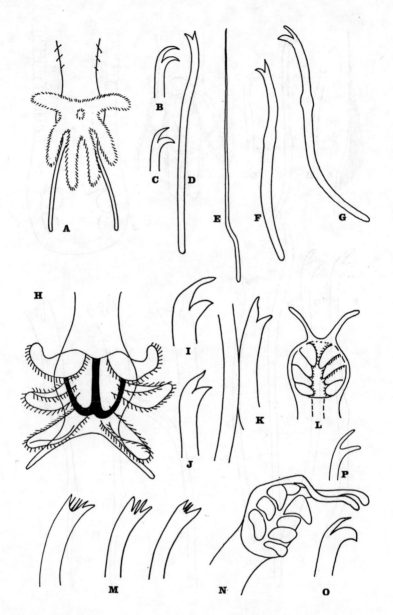

Fig. 7.17. *Aulophorus.*

A. furcatus. A—branchial fossa, B—ventral seta of VIII, C—ventral posterior seta, D—dorsal seta.

A. hymanæ. E—hair seta, F—dorsal seta, G—ventral seta of II, H—gills.

A. borellii. I—ventral seta of II, J—ventral posterior seta, K—dorsal setae, L—gills.

A. pectinatus. M—dorsal setae, N—gills, O—ventral seta from a posterior segment, P—same from an anterior segment.

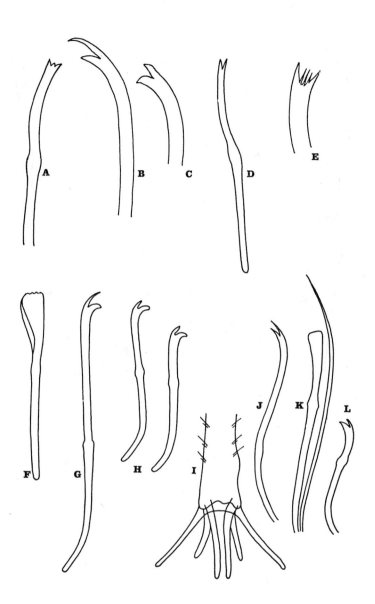

Fig. 7.18. *Aulophorus* (continued).

A. indicus. A—dorsal seta, B—ventral seta of II, C—ventral posterior seta.

A. gravelyi. D—dorsal seta.

A. beadlei. E—dorsal seta.

A. tonkinensis. F—dorsal seta, G—ventral seta of II, H—setae of posterior segments, I—gills.

A. carteri. J—anterior ventral seta, K—dorsal setae, L—posterior ventral seta.

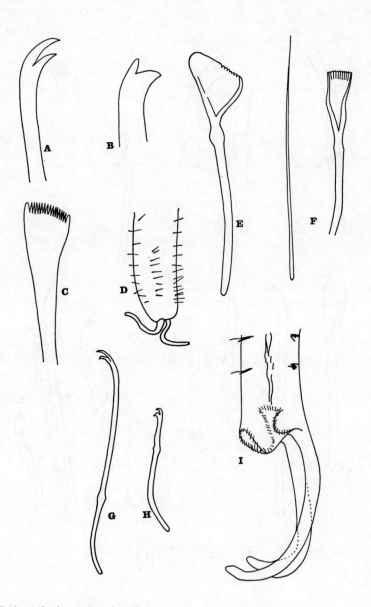

Fig. 7.19. *Aulophorus* (continued).

A. vagus. A—ventral anterior seta, B—ventral posterior seta, C—dorsal seta, D—posterior end of worm.

A. flabelliger. E—dorsal seta.

A. schmardai f. costata. F—dorsal seta, G—anterior ventral seta, H—posterior ventral seta, I—gills, side view.

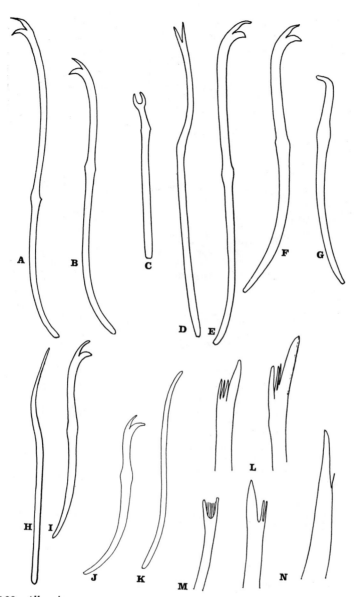

Fig. 7.20. *Allonais.*

A. chelata. A—ventral seta of II, B—same of X, C—dorsal seta.
A. gwaliorensis. D—dorsal seta, E—ventral seta of II, F—same of XIII, G—penial seta.
A. lairdi. H—dorsal seta, I—ventral seta of III, J—same of middle segments, K—penial seta.
A. inaequalis. L—two dorsal setae.
A. pectinata. M—dorsal seta.
A. paraguayensis. N—two dorsal setae.

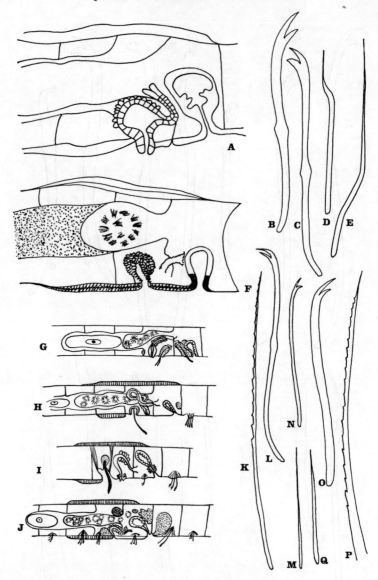

Fig. 7.21. *Allonais* (continued), *Stephensoniana* and *Pristina*.

A. paraguayensis. A—reproductive organs.
S. trivandrana. B—ventral seta of II, C—same of V, D—dorsal seta, E—hair seta, F—reproductive organs.
P. breviseta. G—reproductive organs.
P. plumaseta. H—reproductive organs.
P. americana. I—reproductive organs.
P. longiseta. J—reproductive organs.
P. macrochaeta. K—hair seta, L—ventral seta of III, M—dorsal seta.
P. proboscidea. P—hair seta, Q—dorsal seta.

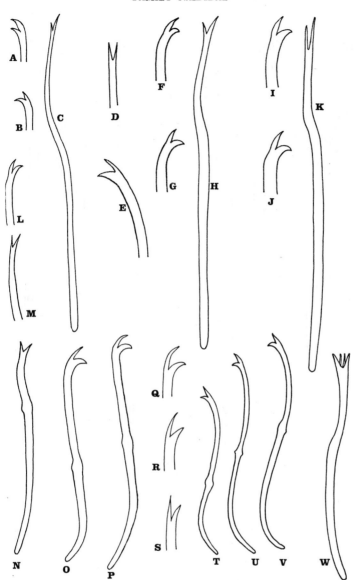

Fig. 7. 22. *Pristina* (continued).

P. menoni. A—ventral seta of II, B—posterior ventral seta, C—dorsal seta.
P. bilobata. D—dorsal seta, E—ventral seta.
P. amphibiotica. F—ventral seta of II, G—posterior ventral seta, H—dorsal seta.
P. idrensis. I—ventral seta of II, J—posterior ventral seta, K—dorsal seta.
P. rosea. L—ventral seta, M—dorsal seta.
P. osborni. N—dorsal seta, O—ventral seta of II, P—ventral seta of a median segment.
P. acuminata. Q—ventral seta of II, R—same of V, S—dorsal seta.
P. sima. T—ventral seta of II, U—same of III, V—same of VIII, W—dorsal seta.

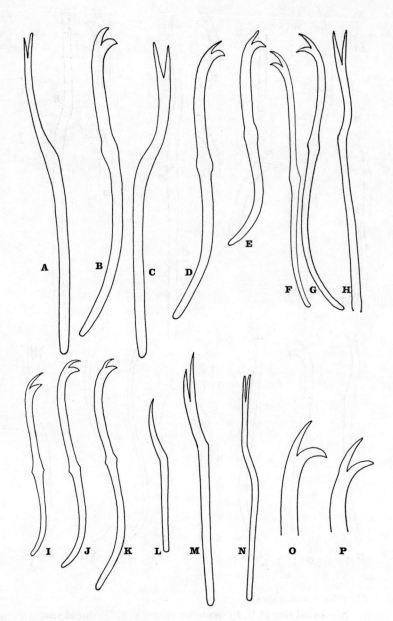

Fig. 7.23. *Pristina* (continued).

P. jenkinae. A—dorsal seta, B—ventral seta.

P. synclites. C—dorsal seta, D—ventral seta of II, E—ventral posterior seta.

P. breviseta. F—ventral seta of II, G—ventral seta of XX, H—dorsal seta.

P. americana. I—ventral seta of II, J—same of IV, K—same of VII, L—dorsal seta of II, M—dorsal seta of X.

P. peruviana. N—dorsal seta, O—ventral seta of IX, P—same of II.

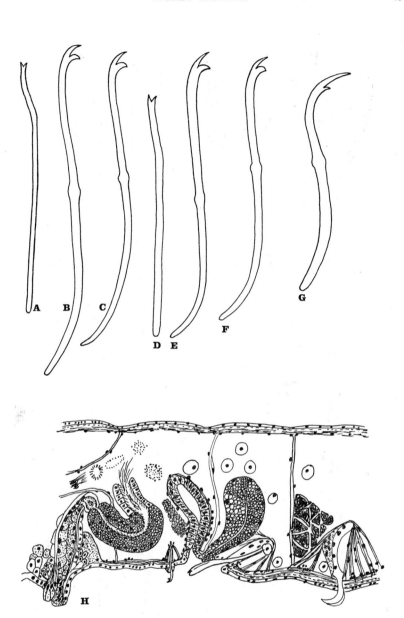

Fig. 7.24. *Pristina* (continued).

P. foreli. A—dorsal seta, B—anterior ventral seta, C—posterior ventral seta.

P. aequiseta. D—dorsal seta, E—ventral seta of II, F—same of IX, G—giant ventral seta of V.

P. evelinae. H—reproductive organs

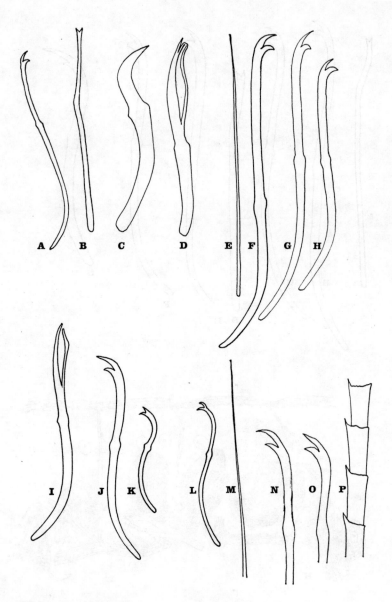

Fig. 7.25. *Pristina* (continued).

P. evelinae. A—ventral seta, B—dorsal seta, C—giant seta of V, D—giant genital seta of VI.

P. longiseta. E—dorsal seta, F—ventral seta of II, G—same of III, H—ventral posterior seta, I—genital seta.

P. bidentata. J—ventral seta of II, K—same of IV, L—same of XII, M—dorsal seta.

P. biserrata. N—ventral seta of IV, O—mid ventral seta, P—hair seta.

REFERENCES

AIYER, K. S. P. 1925. Notes on the aquatic Oligochaeta of Travancore I. *Ann. Mag. nat. Hist.*, (9) **16**, 31.

— 1926. Notes on the aquatic Oligochaeta of Travancore II. *Ann. Mag. nat. Hist.*, (9) **18**, 131.

— 1929. An account of the Oligochaeta of Travancore. *Rec. Indian Mus.*, **31**, 13.

ALTMAN, L. C. 1936. Oligochaeta of Washington. *Univ. Wash. Publs. Biol.*, **4**, 1.

ANNANDALE, N. 1905. Notes on an Indian worm of the genus *Chaetogaster*. *J. Asiat. Soc. Beng.*, **1**, 117.

— 1906. Notes on the freshwater fauna of India. V. Some animals found associated with *Spongilla carteri* in Calcutta. *J. Asiat. Soc. Beng.*, **2**, 187.

BAER, K. von. 1827. Beiträge zur Kenntnis der niedern Thiere III. *Nova Acta phys. med. Acad. Leop. Carol. Nat. Cur. Bonn*, **13**, 605.

BEDDARD, F. E. 1889. Contributions to the natural history of an Annelid of the genus *Dero*. *Proc. zool. Soc. Lond.*, **1889**, 440.

— 1890. On the homology between genital ducts and nephridia in the Oligochaeta. *Nature, Lond.*, **43**, 116.

— 1892. On some aquatic oligochaetous worms. *Proc. zool. Soc. Lond.*, **1892**, 349.

— 1895. *A monograph of the order of Oligochaeta.* Oxford.

— 1896. Naiden, Tubificiden und Terricolen. *Ergebn. Hamb. Magalh. Sammelr.*, **1892/93**, 3.

BEHNING, A. L. 1929. Materialien zur Hydrofauna des Kamaflusses. *Rab. volzh. biol. Sta. S.*, **9**, 163 (Russian with German summary).

BENHAM, W. B. 1892. Notes on some aquatic Oligochaeta. *Q. Jl microsc. Sci. (N.S.)*, **33**, 187.

— 1893. Note on a new species of the Genus *Nais*. *Q. Jl microsc. Sci. (N.S.)*, **34**, 383.

BERG, K. 1938. Studies on the bottom animals of Esrom Lake. *K. danske Vidensk. Selsk. Skr.*, (9) **8**, 1.

BLAINVILLE, H. de. 1825. Naïade, Naïde. *Dict. Sci. Nat.*, **34**, 127. Strasbourg et Paris.

— 1828. Vers. *Dict. Sci. Nat.*, **57**, 365.

BONNET, C. Observations sur quelques espèces de Vers d'eau douce. *Oeuvres d'histoire naturelle et de la philosophie.* **1**, 115. Paris.

BOSC, L. A. G. 1802. *Histoire naturelle des Vers.* 1. Paris.

BOTEA, F. 1963. Contributions to the study of Oligochaeta found in the phreatic water. *Rev. Biol. Acad. Roumaine*, **3**, 335.

BOURNE, A. G. 1890. On *Chaetobranchus*, a new genus of oligochaetous Chaetopoda. *Q. Jl microsc. Sci. (N.S.)*, **31**, 83.

— 1891. Notes on the naidiform Oligochaeta. *Q. Jl microsc. Sci. (N.S.)*, **32**, 335.

BOUSFIELD, E. C. 1886 a. On *Slavina* and *Ophidonais*. *J. Linn. Soc. Lond.*, **19**, 264.

— 1886 b. On the Annelids of the Genus *Dero*. *Rep. Br. Ass. Advmt. Sci. Lond.*, **1885**, 1097.

— 1887. The natural history of the genus *Dero*. *J. Linn. Soc. Lond.*, **20**, 91.

BRETSCHER, K. 1896. Die Oligochaeten von Zürich. *Revue suisse Zool.*, **3**, 499.

— 1899. Beitrag zur Kenntnis der Oligochaeten-Fauna der Schweiz. *Revue suisse Zool.*, **6**, 369.

— 1900. Mitteilungen über die Oligochaetenfauna der Schweiz. *Revue suisse Zool.*, **8**, 1.

— 1903. Beobachtungen über die Oligochaeten der Schweiz VII. *Revue suisse Zool.*, **11**, 1.

BRINKHURST, R. O. 1962. A check list of British Oligochaeta. *Proc. zool. Soc. Lond.*, **138**, 317.

— 1963 a. A guide for the identification of British aquatic Oligochaeta. *Scient. Publs. Freshwat. biol. Ass.*, **22**, 1.

— 1963 b. The aquatic Oligochaeta recorded from Lake Maggiore with notes on the species known from Italy. *Mem. Ist. Ital. Idrobiol.*, **16**, 137.

BRINKHURST, R. O. 1964. Studies on the North American acquatic Oligochaeta. I: Naididae and Opistocystidae. *Proc. Acad. nat. Sci. Philad.*, **116**, 195.

BRODE, H. S. 1898 A contribution to the morphology of *Dero vaga*. *J. Morph.*, **14**, 141.

CAIN, A. J. 1959. Oligochaeta new to Britain. *Ann. Mag. nat. Hist.*, **2**, 193.

CARTER, H. J. 1858. On the spermatology of a new species of *Nais*. *Ann. Mag. nat. Hist.*, (3) **2**, 20.

CEKANOVSKAYA, O. V. 1962. The aquatic Oligochaeta of the U.S.S.R. *Opred. Faune. S.S.S.R.*, **78**, 1 (in Russian).

CERNOSVITOV, L. 1930. Oligochaeten aus Turkestan. *Zool. Anz.*, **91**, 7.

— 1935. Über einige Oligochaeten aus dem See-und Brackwasser Bulgariens. *Bull. Inst. Roy. Hist. nat. Sofia*, **8**, 186.

— 1937 a. Die Oligochaetenfauna Bulgariens. *Izv. tsarsk. prirodonauch Inst. Sof.*, **10**, 69.

— 1937 b. Notes sur les Oligochaeta (Naididées et Enchytraeidées) de l'Argentine. *An Mus. nac. B. Aires*, **39**, 135.

— 1938 a. Oligochaeta. In WASHBOURN and JONES. Report of the Percy Sladen Expedition to Lake Huleh. *Ann. Mag. nat. Hist.*, (11) **2**, 535.

— 1938 b. Oligochaeta. *Mission scientifiques de l'Omo*. **4**, 255. Paris.

— 1939. Oligochaeta. The Percy Sladen Trust expedition to Lake Titicaca in 1937. *Trans. Linn. Soc. Lond.*, (3) **1**, 81.

— 1942 a. Oligochaeta from various parts of the world. *Proc. zool. Soc. Lond. (B)*, **111**, 197.

— 1942 b. Revision of Friend's types and descriptions of British Oligochaeta. *Proc. zool. Soc. Lond. (B)*, **111**. 237.

— 1942 c. Oligochaeta from Tibet. *Proc. zool. Soc. Lond. (B)*, **111**, 281.

CHEN, Y. 1940. Taxonomy and faunal relations of the limnitic Oligochaeta of China. *Contr. biol. Lab. Sci. Soc. China*, **14**, 1.

— 1944. Notes on naidomorph Oligochaeta of Philadelphia and vicinity. *Notul. Nat.*, **136**, 1.

CHIAJE, S. DELLE, 1827. *Memorie sulla storia e anaromia degli animali senza vertebri*. 2. Napoli.

COGNETTI DE MARTIIS, L. 1900. Contributo alla conoscenza degli oligocheti neotropicali. *Boll. Musei. Zool. Anat. comp. Torino*, **15**, 269.

COLLINS, D. S. 1937. The aquatic earthworms (Microdrili) of Reelfoot Lake. *J. Tenn. Acad. Sci.*, **12**, 188.

CORDA, A. J. C. 1837. *Copopteroma*, eine neue Gattung der Ringelwürmer. In WEITENWEBER. *Beiträge zur gesamten Natur-und Heilwissenschaft. Prag.* **1**, 390.

CORDERO, E. H. 1931. Notas sobre los Oligoquetos del Uruguay (primera serie). *An. Mus. nac. B. Aires*, **36**, 343.

CORI, C. I. 1928. Über die Art der Nahrungsaufnahme bei *Nais, Stylaria* und *Ripistes*. *Lotos*, **71**, 67.

CRAGIN. F. W. 1887. First contribution to knowledge of lower Invertebrata of Kansas. *Bull. Washburn Coll. Topeka*, **2**, 8.

CUNNINGTON, W. A. 1920. The fauna of the African lakes. *Proc. zool. Soc. Lond.*, **1-4**, 507.

CZERNIAVSKY, V. 1880. Materialia ad zoographiam Ponticam comparatum III. *Vermes. Bull. Soc. Imp. nat. Moscou*, **4**, 213.

DAHL, I. I. 1957. Results from the Danish Expedition to the French Cameroons. *Bull. Inst. fr. Afr. noire*, **19**, 1154.

— 1960. The oligochaete fauna of three Danish Brackish water areas. *Meddr. Kommn Danm. Fisk. -og Havunders.*

DALYELL, J. G. 1853. *The powers of the Creator displayed in the creation*. London.

DAVIES, O. B. 1913. On two new species of *Chaetogaster*. *Proc. R. Soc. Vict. (N.S.)*, **26**, 88

DEHORNE, L. 1916. Les Naidimorphes et leur reproduction asexuée. *Arch. Zool. exp. gén.*, **56**, 25.

DIEFFENBACH, O. 1885. Anatomische und systematische Studien an Oligochaetae limnicolae. *Diss. Giessen. Ber. Oberhess. Ges. Giessen*, **24**, 63.

DITLEVSEN, A. 1904. Studien an Oligochäten. *Z. wiss. Zool.*, **77**, 398.

Dugès, A. 1837. Nouvelles observations sur la zoologie et l'anatomie des Annèlides abranches sétigères, III. *Z. wiss. Zool.*, (2) **8**, 30.

Dujardin, F. 1839. Observations sur quelques Annélides marines. *Z. wiss. Zool.*, (2) **9**, 287.

— 1842. *Ripistes*, n. g. des Naidines. *P.-v. Soc. philomath. Paris*, **1842**, 93.

Dutrochet, H. 1819. Note sur un nouveau genre d'Annélides. *Bull. Soc. philomath. Paris*, **8**, 155.

Ehrenberg, C. G. 1828. Symbolae physicae —. *Animalia evertebrata Phytozoa*. Berlin.

Ekman, S. 1915. Die Bodenfauna des Vättern, qualitativ und quantitativ untersucht. *Int. Revue Hydrobiol. Leipzig*, **7**, 146.

Ercolini, A. 1956. Prime recerche sistematiche e biologiche sopia i Naididi die Piemonte. *Boll. Ist. Mus. zool. Univ. Torino*, **5**, 3.

Fabricius, O. 1780. *Fauna groenlandica*. Hafniae et Lipsiae.

Ferronnière, G. 1899. III: e contribution à l'etude de la daune de la Loire-inferieure (Annelides oligochaetes). *Bull. Soc. Sci. nat. Ouest Fr.*, **9**, 229.

Finegenova, N. P. 1962. *Studies on the Oligochaeta of the River Usa basin in fishes of the River Usa basin and their nutritional resources*. Moscow and Leningrad. 1962.

Floericke, C. 1892. Vorläufige Mitteilung über einige anscheinend neue Naidomorphen. *Zool. Anz.*, **15**, 468.

Friend, H. 1912. Some Annelides of the Thames Valley. *J. Linn. Soc. Lond.*, **32**, 95.

— 1916. Notes on Irish Oligochaetes. *Ir. Nat.*, **25**, 22.

Fuchs, K. 1907. Die Topographie des Blutgefässystems der Chätopoden. *Jena. Z. Naturw.*, **42**, 375.

Galloway, T. W. 1899. Observations on non-sexual reproduction in *Dero vaga*. *Bull. Mus. comp. Zool. Harv.*, **35**, 115.

— 1911. The common fresh-water Oligochaeta of the United States. *Trans. Am. microsc. Soc.*, **30**, 285.

Garbini, A. 1898. Una nuova specie di Pristina (*P. affinis* n. sp.). *Zool. Anz.*, **21**, 562.

Gay, C. 1849. Historia fisica y politica de Chile —. *Zoologia*. 3. Paris.

Gervais, P. 1838. Note sur la disposition systematique des Annélides chétopodes de la famille des *Nais*. *Bull. Acad. r. Belg. Cl. Sci.*, **5**, 13.

Gmelin, J. F. 1788-93. *Caroli a Linné systema naturae* (Ed. 13). **1**, VI, 3120. Lipsiae.

Golanski, J. 1911. Przyczynek do znajomosci fauny skaposzczetow wodnych (*Oligochaeta limicola*) Galicji. Ksiega Pam. *Prof. J. Nusbauma-Hilarowicza*. Lwow. 57.

Goodchild, C. G. 1951. A new endo parasitic oligochaete (*Naididae*) from a North American tree toad *Hyla squirella* Latreille, 1802, *J. Parasit.* **37**, 205.

Gould, A. A. 1841. *Report on the invertebrate animals of Massachusetts*. Cambridge.

Grebnitzky, N. 1873. Materialy dlja fauni Novorossiiskago kraja. *Zap. Novoross. Obshch. Odessa*, **2**.

Grube, A. E. 1851. *Die Familien der Anneliden*. Berlin.

Gruffydd, L. D. 1965. Evidence for the existence of a new subspecies of *Chaetogaster limnaei* (Oligochaeta), in Britain. *J. Zool.*, **146**, 175.

Gruithuisen, F. V. P. 1823. Anatomie der gezüngelten Naide. *Nova Acta phys.-med. Acad. Leop. Carol. Nat. Cur. Bonn*, **11**, 233.

— 1828. Über die *Nais diaphana* und *Nais diastropha* mit dem Nerven-und Blutsystem derselben. *Nova Acta phys.-med. Acad. Leop. Carol. Nat. Cur. Bonn*, **14**, 409.

Harman, W. J. 1965 a. A key to the genus *Slavina* (Oligochaeta, Naididae) with a description of a new species from Costa Rica. *Ann Mag. nat. Hist.*, **8**, 565.

— 1965 b. A new species of the genus *Pristina* (Oligochaeta, Naididae) from Louisiana. *Proc. La. Acad. Sci.*, **28**, 28.

Harman, W. J., and Platt, J. H., 1961. Notes on some aquatic Oligochaeta from Louisiana. *Proc. La. Acad. Sci.*, **24**, 90.

Hayden, H. E. 1912. Preliminary note on *Pristina* and *Naidium*. *Science (N.S.)*, **36**, 530.

— 1914. Further notes on *Pristina* and *Naidium*, with description of three new species. *Trans. Am. microsc. Soc.*, **33**, 135.

— 1922. Studies on American Naid Oligochaetes. *Trans. Am. microsc. Soc.*, **41**, 167.

HEMPELMANN, F. 1923. Kausalanalytische Untersuchungen über das Auftreten vergrösserter Borsten und die Lage der Teilungszone bei *Pristina*. *Arch. mikrosk. Anat. Entw Mech.*, **98**, 379.

HOUGHTON, W. 1860. On the occurrence of the fingered *Nais* (*Proto digitata*) in England. *Ann. Mag. nat. Hist.*, (3) **6**, 393.

HRABE, S. 1936. Zur Kenntnis der Oligochaeten des Aral-Sees. *Izv. Akad. Nauk. SSSR.*, **6**, 1265 (Russian with German summary).

— 1939. Über die ontogenetische Entwicklung des männlichen Ausführungsganges bei einigen Oligochäten. *Sb. Klubu přír. Brne*, **3**, 56 (Czech with German summary).

— 1941. Zur Kenntnis der Oligochaeten aus der Donau. *Acta Soc. Sci. nat. moravosiles*, **13** (12), 1 (Czech with German summary).

HYMAN, L. 1938. The fragmentation of *Nais paraguayensis*. *Physiol. Zöol.*, **11**, 126.

IMHOF, O. E. 1888. Ein neues Mitglied der Tiefseefauna der Süsswasserbecken. *Zool. Anz.*, **11**, 48.

ISOSSIMOV, V. V. 1949. Lumbriculidae of Lake Baikal. (A monographic treatment.) Doctorate Dissertation. Kazan.

JACKSON, A. 1931. The Oligochaeta of south-western Australia. *J. Proc. R. Soc. West. Aust.*, **17**, 71.

JOHNSTON, G. 1845. An index to the British Annelides. *Ann. Mag. nat. Hist.*, **16**, 433.

— 1865. *A catalogue of the British non-parasitical worms*. London.

KALLSTENIUS, E. 1892. Eine neue Art der Oligochaetengattung *Amphichaeta* Tauber. *Biol. Fören. Stockholm*, **4**, 42.

KESSLER, K. 1868. Materialy dlja poznanija onegskago ozera —. *Trudy I. Cyezda Russk. Est.* St. Petersburg. *Oligochaeta*, **102**.

KLEIBER, O. 1911. Die Tierwelt des Moorgebietes von Jungholz im südlichen Schwarzwald. *Arch. Naturgesch.*, 77, 1, **3**, 1.

KNÖLLNER, F. H. 1935. Okologische und systematische Untersuchungen über litorale und marine Oligochäten der Kieler Bucht. *Zool. Jb. Syst.*, **66**, 423.

KONDO, M. 1936. A list of naidiform Oligochaeta from the waterworks plant of the city of Osaka. *Annotnes. zool. jap.*, **15**, 382.

KOWALEWSKI, M. 1910. Materials for the fauna of Polish aquatic Oligochaeta I. *Bull. int. Acad. Sci. Lett. Cracovie* (B), **1910**, 804.

— 1917 a. Przyczyněk do lepszej znajomosci skaposzczeta *Amphichaeta leydigi* (Tauber 1879) M. Kowalewski 1910. *Rozpr. Wydz. mat.-przyr. pol. Akad. Umiejet. K.*, (3) **16**, 39 (Polish).

— 1917 b. A contribution to the knowledge of the Oligochaete: *Amphichaeta Leydigi* (Tauber 1879) M. Kowaleski 1910. *Rozpr. Wydz. mat.-przyr. pol. Akad. Umiejet K.*, (3) **16**, 77. *Bull. Int. Acad. Sci. Lett. Cracovie* (B), **1916**, 77.

LAMARCK, J. de 1816. *Histoire naturelle des animaux sans vertèbres*. 3. Paris.

LANKESTER, E. R. 1869 a. A contribution to the knowledge of the lower Annelides. *Trans. Linn. Soc. Lond.*, **26**, 631.

— 1869 b. The sexual form of *Chaetogaster Limnaei*. *Q. Jl microsc. Sci.*, **9**, 272.

— 1869 c. On the existence of distinct larval and sexual forms in the gemmiparous Oligochaetous worms. *Ann. Mag. nat. Hist.*, (4) **4**, 102.

— 1871. Outline of some observations on the organization of oligochaetous Annelids. *Ann. Mag. nat. Hist.*, (4) **7**, 90.

LASTOCKIN, D. A. 1918. Materialy do faune vodnych Oligochaeta Petrograd Gub. *Trudý leningr, obshch. Estest.*, **44**, 57.

— 1922. In BEKLEMICHEV, W. N. Nouvelles contributions à la faune du lac Aral. *Gidrobiol. Zh.*, **1**, 276 (Russian with French summary).

— 1924a. A New and rare Copepoda and Oligochaeta from Central Russia. *Ixv. gosud. gidrol. Inst.*, **9**, 1 (Russian with English summary).

— 1924 b. "Oligochaeta" Valdaiskogo ozera. *Tr. Ivan.-Vozn. Gub. Nauchn.*, **2**.

— 1927. Beiträge zur Oligochaetenfauna Russlands III. Fauna von *Oligochaeta limicola* in Gouvernements Iwanowo-Woznesensk und Vladimir. *Izv. Ivan-Voznesensk. Polytekh. Inst..* **10**, 65.

— 1937. New species of *Oligochaeta limicola* in the European part of the U.R.S.S. *Dokl. Akad. Nauk. SSSR. (N.S.)*, **17**, 233.

LEDERMÜLLER, M. F. 1763. *Mikroskopische Gemüths-und Augen-Ergötzung.* 1. Nürnberg.

LEIDY, J. 1850. Descriptions of some American *Annelida abranchia. J. Acad. nat. Sci. Philad.,* (2) **2**, 43.

— 1852 a. Description of new genera of Vermes. *Proc. Acad. nat. Sci. Philad.,* **5**, 124.

— 1852 b. Helminthological contributions II. *Proc. Acad. nat. Sci. Philad.,* **5**, 224.

— 1852 c. Corrections and additions to former papers on helminthology—. *Proc. Acad. nat. Sci. Philad.,* **5**, 284.

— 1852 d. Contributions to helminthology. *Proc. Acad. nat. Sci. Philad.,* **5**, 349.

— 1880. Notice of some aquatic worms of the family Naides. *Am. Nat.,* **14**, 421.

LEVANDER, K. M. 1901. Übersicht der in der Umgebung von Esbo-Löfö im Meereswasser vorkommenden Thiere. *Acta Soc. Fauna Flora fenn.,* 20, **6**, 1.

— 1904. Om en för Finland ny limicol oligochaet. *Meddn. Soc. Fauna Flora fenn.,* **29**, 199.

LEVINSEN, G. M. R. 1884. Systematisk-geografisk Oversigt over de nordiske Annulata, Gephyrea, Chaetognathi og Balanoglossi II. *Vidensk. Meddr dansk. naturh. Foren.,* **1883**, 92.

LEYDIG, F. 1857. *Lehrbuch der Histologie des Menschen und der Thiere.* Frankfurt a. M.

— 1864. *Tafeln zur vergleichenden Anatomie.* 1 Heft. Tübingen.

LIANG, Yan-Lin. 1958. On some new species of Naididae from Nanking including remarks on certain known species. *Acta hydrobiol. sin.,* **7**, 41 (in Chinese, English summary).

— 1963. Studies on the aquatic Oligochaeta of China. 1. Description of new naids and branchiobdellids. *Acta zool. sin.,* **15**, 560 (in Chinese, English summary).

— 1964. Studies on the aquatic Oligochaeta of China. II: On some species of Naididae from Sinkiang with description of a new species *Allodero prosetosa. Acta zool. sin.,* **16**, 643 (in Chinese, English summary).

LINNE, C. VON. 1767. *Systema naturae.* 1, II. Ed. 12. Holmiae.

LUTZ, A. 1927. Sur la *Schmardaella lutzi* Michaelsen. *C. r. hebd. Séanc. Acad. Sci. Paris,* **96**, 485.

MALEVICH, I. I. 1925. Bemerkung über die Oligochaeten der Schaturski-Seen (Gouv. Moskwa. *Trudý kosin biol. Sta.,* **3**, 65 (Russian with German summary).

— 1927. Oligochaeten der Wasserbecken zu Kossino. *Trudy kosin biol. Sta.,* **5**, 3 (Russian with German summary).

— 1929. Die Oligochaeten der Gewässer der Meschtschera-Niederung. *Trudy kosin biol. Sta.,* **9**, 41 (Russian with German summary).

MARCUS, E. 1942. Sôbre algumas Tubificidae do Brasil. *Bolm. Fac. Filos. ciênc. Univ. S. Paulo,* 25, *Zool.,* **6**, 153 (Portuguese with English summary).

— 1943. Sôbre *Naididae* do Brasil. *Bolm. Fac. Filos. ciênc. Univ. S. Paulo,* 32, *Zool.,* **7**, 3 (Portuguese with English summary).

— 1944. Sôbre Oligochaeta límnicos do Brasil. *Bolm. Fac. Filos. ciênc. Univ. S. Paulo,* 43, *Zool.,* **8**, 5 (Portuguese with English summary).

— 1965. Naidomorpha aus brasilianischem Brackwasser. *Beitr. neotrop. Fauna,* **4**, 61.

MARCUS, E. DU BOIS-REYMOND, 1944. Notes on freshwater Oligochaeta from Brazil. *Commun. zool. Mus. Hist. nat. Montev.,* **1** (20), 1.

— 1947. Naidids and tubificids from Brazil. *Commun. zool. Mus. Hist. nat. Montev.,* **2** (44), 1.

— 1949. Further notes on naidids and tubificids from Brazil. *Commun. zool. Mus. Hist. nat. Montev.,* **3** (51), 1.

MARTIN, C. H. 1907. Notes on some Oligochaetes found on the Scottish Loch survey. *Proc. R. Soc. Edinb.,* **28**, 21.

MAULE, V. 1906. Über die *Vejdovskyella comata* Mich. und "*Nais hamata* Timm". *Zool. Anz.,* **30**, 302.

MAYER, C. 1859. Reproductionsvermögen und Anatomie der Naiden. *Niederrhein. Ges. Natur. Heilk. (B). Verh. Naturh. Ver. preuss. Rheinl.,* **16**, 43.

MAYHEW, R. L. 1922. The anatomy of some sexually mature specimens of *Dero limosa* Leidy. *Trans. Am. microsc. Soc.,* **41**, 159.

MEHRA, H. R. 1920. On the sexual phase in certain Naididae (Oligochaeta). *Proc. zool. Soc. Lond.*, **1920**, 1-4, 457.
— 1924. The genital organs of *Stylaria lacustris*, with an account of their development. *Q. Jl microsc. Sci.*, **68**, 147.
— The atrium and the prostate gland in the Microdrili. *Q. Jl microsc. Sci.*, **69**, 399.
MICHAELSEN, W. 1899. Beiträge zur Kenntnis der Oligochäten. Oligochäten aus der Schweiz. *Zool. Jb. Syst.*, **12**, 119.
— 1900. Oligochaeta. *Das Tierreich*, 10. Berlin.
— 1902. Oligochäten des Zoologischen Museums zu St. Petersburg. und Kiew. *Izv. Akad. Nauk SSSR*, **15**, 137.
— 1903 a. Oligochäten. Hamburgische Elb-Untersuchung, IV. *Jb. hamb. wiss. Anst.*, **19**, 169.
— 1903 b. *Die geographische Verbreitung der Oligochaeten*. Berlin.
— 1905 a. Zur Kenntnis der Naididen. In E. V. DADAY. Die Süsswassermikrofauna Paraguays. *Zoologica, Stuttg.*, 18, **44**, 350.
— 1905 b. Die Oligochäten Deutsch-Ostafrikas. *Z. wiss. Zool.*, **82**, 288.
— 1905 c. Die Oligochaeten des Baikal-Sees. *Wiss. Ergebnisse Zool. Exped. Baikal-See Prof. Alexis Korotneff J. 1900-1902.* **1**, 1. Kiew und Berlin.
— 1909 a. Oligochaeta. In A. BRAUER. *Die Süsswasserfauna Deutschlands*, **13**, 1. Jena.
— 1909 b. The Oligochaeta of India, Nepal, Ceylon, Burma and the Andaman Islands. *Mem. Indian Mus.*, **1**, 103.
— 1910. Die Oligochäten der vorderindisch-ceylonischen Region. *Verh. naturw. Ver. Hamb.*, **19**, 1.
— 1912. Über einige zentralamerikanische Oligochäten. *Arch. Naturgesch.*, **78 A**, 9, 112.
— 1913. Die Oligochaeten Columbias. FUHRMANN and MAYR: Voyage d'exploration scientifique en Colombie. *Mém. Soc. neuchât. Sci. nat.*, **5**, 202.
— 1914. Oligochaeta. Beitr. Kennt. Land-u. *Süsswasserfauna Dt.-SüdwAfr.*, **1**, 137.
— 1923. Die Oligochäten der Wolga. *Rab. volzh. biol. Sta.*, **7**, 30.
— 1926 a. Oligochaeten aus dem Ryck bei Greifswald und von benachbarten Meeresgebieten. *Mitt. zool. St. Inst. Hamb.*, **42**, 21.
— 1926 b. Zur Kenntnis einheimischer und ausländischer Oligochäten. *Zool. Jb. Syst.*, **51**, 253.
— 1926 c. Oligochäeten aus dem Gebiet der Wolga und der Kama. *Rab. volzh. biol. Sta.*, **9**, 1.
— 1926 d. Schmarotzende Oligochäten nebst Erörterungen über verwandtschaftliche Beziehungen der Archioligochäten. *Mitt. zool. St. Inst. Hamb.*, **42**, 91.
— 1927 a. Oligochaeta. In G. GRIMPE. *Tierwelt N-u. Ostsee.* VI c, 1.
— 1927 b. Die Oligochaetenfauna Brasiliens. *Abh. senckenb. naturforsch. Ges.*, 40, **3**, 1.
— 1933. Süss-und Brackwasser-Oligochäten von Bonaire, Curacao und Aruba. *Zool. Jb. Syst.*, **64**, 327.
— and BOLDT, W. 1932. Oligochaeta der deutschen limnologischen Sunda-Expedition. In A. THIENEMANN. Tropische Binnengewässer II. *Arch. Hydrobiol. Suppl.*, **9**, 587.
— and VERESCHAGIN, G. 1930. Oligochaeten aus dem Selenga-Gebiete des Baikalsees. *Trudy Kom. Izuch. Ozera Baikala*, **3**, 213.
MINOR, W. C. 1863. Upon natural and artificial section in some chaetopod Annelids. *Amer. J. Sci. Arts*, (2) **35**, 35.
MODEER, A. 1798. Beskrifning af et slägte ibland maskkräken, kalladt slinga, *Nais*. *K. svenska Vetensk Akad. Nya Handl. Stockholm*, **19**, 107.
MOORE, J. P. 1905. Some marine Oligochaeta of New England. *Proc. Acad. nat. Sci. Philad.*, **57**, 373.
— 1906. Hirudinea and Oligochaeta collected in the Great Lakes region. *Bull. Bur. Fish. Wash.*, **25**, 153.
MOSZYNSKA, M. 1962. Skaposzezety Oligochaeta. *Kat. Fauny polski*, **11** (2), 1.
MOSZYSKI, A. 1933. Description d'une nouvelle espèce d'Oligochètes *Paranais setosa* n. sp. *Archwn. Hydrobiol. Ryb.*, **7**, 141.
MÜLLER, O. F. 1771. *Von Würmern des süssen und salzigen Wassers.* Kopenhagen.

MÜLLER, O. F. 1773. *Vermium terrestrium et fluviatilium II.* Hafniae et Lipsiae.
— 1776. *Zoologiae Danicae prodromus.* Havniae.
— 1784. *Zoologia Danica II.* Lipsiae.
MUNSTERHJELM, E. 1905. Verzeichnis der bis jetzt aus Finnland bekannten Oligochaeten. *Festschrift für Palmén.* N:013 Helsingfors.
NAIDU, K. V. 1962 a. Studies on the freshwater Oligochaeta of South India. 1. Aelosomatidae and Naididae. Part 2. *J. Bombay nat. Hist. Soc.,* **59**, 131.
— 1962 b. Studies on the freshwater Oligochaeta of South India. 1. Aelosomatidae and Naididae, Part 3. *J. Bombay nat. Hist. Soc.,* **59**, 520.
— 1962 c. Studies on the freshwater Oligochaeta of South India. 1. Aelosomatidae and Naididae Part 4. *J. Bombay nat. Hist. Soc.,* **59**, 897.
— 1963. Studies on the freshwater Oligochaeta of South India. 1. Aeolosomatidae and Naididae. Part 5. *J. Bombay nat. Hist. Soc.,* **60**, 201.
— 1965. Some freshwater Oligochaeta of Singapore. *Bull. Natn. Mus. St. Singapore,* **33**, 13.
NICHOLLS, G. E. 1921. On a new species of naidiform worm, Dero roseola. *J. Proc. Soc. West. Aust.,* **7**, 90.
OKEN, L. 1815. *Lehrbuch der Naturgeschichte* 3 : *Zoologie* 1 : *Fleischlose Thiere.* Leipzig.
ORSTED, A. S. 1842. Conspectus generum specierumque Naidum ad daunam Danicam pertinentium. *Naturhist. Tidsskr. Kjøbenhavn,* **4**, 128.
PASQUALI, A. 1938. Note sistematiche sugli Oligocheti aquicoli di Padova. *Zoologia, Padova,* **9**, 19.
PATARIDZE, A. J. 1957. A new species of Oligochaete *Nais ioriensis* from the River Iorda (Eastern Georgia). *Commun. Acad. Nauk. Georgia CCP,* **18**, 91 (in Russian).
PERRIER, E. 1872. Histoire naturelle du *Dero obtusa. Archs. Zool. exp. gén.,* **1**, 65.
PESSON, P. 1938. Notes de faunistique Armoricaine (4:e Note). *Bull. Soc. scient. Bretagne,* **15**, 39.
PIERANTONI, U. 1909. Sul genere *Paranais* e su di una nuova specie del golfo di Napoli (*Paranais elongata*). *Mitt. zool. Stn. Neapel. Berlin,* **19**, 445.
— 1911. Oligocheti del laghetto craterico do Astroni I. *Naididae. Fauna degli Astroni. Annuar. R. Mus. zool. R. Univ. Napoli (N.S.) Suppl.,* **3**, 1.
PIGUET, E. 1906. Observations sur les Naididées et revision systématique de quelques espèces de cette famille. *Revue suisse Zool.,* **14**, 185.
— 1909. Nouvelles observations sur les Naididés. *Revue suisse Zool.,* **17**, 171.
— 1919 a. Wasserbewohnende Oligochaeten der nordschwedischen Hochgebirge. *Naturw. Unters. Sarekgebirg. in Schwedisch-Lappland. Stockholm,* **4**, 779.
— 1928. Sur quelques Oligochètes de l'Amerique du Sud et de l'Europe. *Bull. Soc. neuchât. Sci. nat. (N.S.),* **52**, 78.
— and BRETSCHER, K. 1913. Oligochètes. *Catalogues des Invertébrés de la Suisse.* 7. Genève.
PLOTNIKOV, V. 1901. Nematoda, Oligochaeta i Hirudinea, Naidennyja v Bologovskom ozerje. *Ber. biol. Süsswasserstat. St. Petersburg,* **1**, 244 (in Russian).
POINTNER, H. 1911. Beiträge zur Kenntnis der Oligochaetenfauna der Gewässer von Graz. *Wiss. Zool.,* **98**, 269.
— 1914 a. Über einige neue Oligochaeten der Lunzer-Seen. *Arch. Hydrobiol.,* **9**, 606.
— 1914 b. Über Oligochaetenbefunde der Lunzer-Seen. *Arch Hydrobiol.,* **10**, 91.
POPESCU-MARINESCU, U., BOTEA, F., and BREZEANU, G. 1966. Untersuchungen über die Oligochaeten im romänischen Sektor des Donau Basins. *Arch. Hydrobiol. Suppl.,* **20**, 161.
PREU, T. 1937. Bemerkungen zur Histologie der Naiden. *Z. wiss. Zool.,* **149**, 258.
REAUMUR, R.-A. F. DE. 1748. *Mémoires pour servir à l'histoire des Insectes.* 6, 1. Amsterdam.
REDEKE, H. C., and VOS, A. P. C. DE. 1932. Beiträge zur Kenntnis der Fauna niederländischer oligotropher Gewässer. *Int. Revue Hydrobiol. Leipzig,* **28**, 1.
REIGHARD, J. 1885. On the anatomy and histology of *Aulophorus vagus. Proc. Amer. Acad. Sci. Boston,* **20**, 88.

Rösel Von Rosenhof, A. J. 1755. *Der monatlich-herausgegebenen Insecten- Belustigung dritter Theil.* Nürnberg.

Schmarda, L. 1861. *Neue wirbellose Thiere 1, II.* Leipzig.

Schmidt, O. 1847. Drei neue Naiden. *Froriep's Notizen der Heilkunde.* Weimar. (3) **3**, 320.

Schuster, R. W. 1915. Morphologische und biologische Studien an Naiden in Sachsen und Böhmen. *Int. Revue Hydrobiol. Leipzig, 7, Biol. Suppl.,* **2**, 1.

Sciacchitano, I. 1934. Sulla distribuzione geografica degli oligocheti in Italia. *Arch. zool. Torino,* **20**, 1.

— 1938. Oligocheti del Veneto. *Bull. Zool. Torino,* **9**, 253.

Semper, C. 1877. Beiträge zur Biologie der Oligochaeten. *Arb. Inst. Würzburg,* **4**, 65.

Smith, F. 1896. Notes on species of North American Oligochaeta. I. *Bull. Ill. St. Lab. nat. Hist. U.,* **4**, 397.

— 1900. Notes on species of North American Oligochaeta. III. *Bull. Ill. St. Lab. nat. Hist. U.,* **5**, 441.

— 1918. Aquatic earthworms and other bristle-bearing worms. In Ward and Whipple. *Freshwater biology.* New York.

Sokolskaya, N. L. 1958. Freshwater microdrili in the region of the Amur basin. *Tr. Amursk. ikhtiolog. eksped. 1945-1949,* **4**, 287.

— 1961. Material on the naidid fauna (Naididae, Oligochaeta) of the Maritime province. *Sbornik Trudov Zoologicheski Musee Mgu.,* **8**, 47 (in Russian).

— 1962. New data on the Naididae (Oligochaeta) of Lake Baikal. In *Oligochaeta and Planaria of Lake Baikal.* Ed. Jalazii, Moscow-Leningrad (in Russian). (Also in New species of Naididae (Oligochaeta) from Lake Baikal. *Zool. Zh. 1962,* **41**, 660 (in Russian).

— 1964. New species and subspecies in Family Naididae (Oligochaeta) from brackish reservoir in Kamchatka and South Sakhalin. *Byull. mosk. Obshch. Ispyt. Prir. (Biol.),* **69**, 57 (in Russian).

Southern, R. 1909. Contribution towards a monograph of the British and Irish Oligochaeta. *Proc. R. Ir. Acad.,* **27B**, 119.

— 1913. Oligochaeta. Clare Island survey. Part 48. *Proc. R. Ir. Acad.,* **31**, 3.

Sperber, C. 1939. Preliminary notes on the genital organs of *Nais blanci* and *Nais josinae.* *Ark. Zool.,* **31B**, 7.

— 1948. A taxonomical study of the Naididae. *Zool. Bidr. Upps.,* **28**, 1.

— 1950. A guide for the determination of European Naididae. *Zool. Bidr. Upps.,* **29**, 45.

— 1958. Über einige Naididae aus Europa, Asien und Madagascar. *Arkiv. Zool. K. Svenska. Vetenskaps.,* **12**, 45.

— 1960. Contribution à letude de la fauna d'Afghanistan 41. *Arkiv. Zool. K. Svenska. Vetenskaps.,* **13**, 155.

Stephenson, J. 1907 a. Description of an Oligochaete worm, allied to *Chaetogaster. Rec. Indian Mus.,* **1**, 133.

— 1907 b. Description of two freshwater Oligochaete worms from the Punjab. *Rec. Indian Mus.,* **1**, 233.

— 1909 a. Studies on the anatomy of some aquatic Oligochaeta from the Punjab. *Mem. Indian Mus.,* **1**, 255.

— 1909 b. Report on a collection of the smaller Oligochaeta made by Capt. F. H. Stewart, s.m.s., in Tibet. *Rec. Indian Mus.,* **3**, 105.

— 1910 a. Studies on the aquatic Oligochaeta of the Punjab. *Rec. Indian Mus.,* **5**, 59.

— 1910 b. On some aquatic Oligochaete worms commensal in *Spongilla carteri. Rec. Indian Mus.,* **5**, 233.

— 1911. On some aquatic Oligochaeta in the collection of the Indian Museum. *Rec. Indian Mus.,* **6**, 203.

— 1912. On a species of *Branchiodrilus* and certain other aquatic Oligochaeta, with remarks on cephalization in the *Naididae. Rec. Indian Mus.,* **7**, 219.

— 1913. On a collection of Oligochaeta, mainly from Ceylon. *Spolia zeylan.,* **8**, 251.

STEPHENSON, J. 1914. On a collection of Oligochaeta, mainly from northern India. *Rec. Indian Mus.*, **10**, 321.

— 1915 a. On *Haemonais laurentii* n. sp., a representative of a little known genus of Naididae. *Trans. R. Soc. Edinb.*, **50**, 769.

— 1915 b. On a rule of proportion observed in the setae of certain *Naididae. Trans. R. Soc. Edinb.*, **50**, 783.

— 1915 c. On the sexual phase in certain of the Naididae. *Trans. R. Soc. Edinb.*, **50**, 789.

— 1916. On a collection of Oligochaeta belonging to the Indian Museum. *Rec. Indian Mus.*, **12**, 299.

— 1917. Aquatic Oligochaeta from Japan and China. *Mem. Asiat. Soc. Beng.*, **6**, 85.

— 1918. Aquatic Oligochaeta of the Inlé Lake. *Rec. Indian Mus.*, **14**, 9.

— 1920. On a collection of Oligochaeta from the lesser known parts of India and from Eastern Persia. *Mem. Indian Mus.*, **7**, 191.

— 1921. Oligochaeta from Manipur, the Laccadive Islands, Mysore and other parts of India. *Rec. Indian Mus.*, **22**, 745.

— 1922 a. On the septal and pharyngeal glands of the *Microdrili* (Oligochaeta). *Trans. R. Soc. Edinb.*, **53**, 241.

— 1922 b. On some Scottish Oligochaeta, with a note on encystment in a common freshwater Oligochaete, *Lumbriculus variegatus* (Müll). *Trans. R. Soc. Edinb.*, **53**, 277.

— 1922 c. Contributions to the morphology, classification, and zoogeography of Indian Oligochaeta, IV. On the diffuse production of sexual cells in a species of *Chaetogaster* (fam. *Naididae*). *Proc. zool. Soc. Lond. 1922*, **1-2**, 109.

— 1923. Oligochaeta. The Fauna of British India. London.

— 1925 a. Oligochaeta from various regions —. *Proc. zool. Soc. Lond.*, **3-4**, 879.

— 1925 b. The Oligochaeta of Spitsbergen and Bear Island, some additions and a summary. *Proc. zool. Soc. Lond.*, **3-4**, 1293.

— 1925 c. On some Oligochaeta mainly from Assam, South India and the Andaman Island. *Rec. Indian Mus.*, **27**, 43.

— 1926. The sexual organs of the freshwater Oligochaete *Naidium breviseta* (A. G. Bourne). *Ann. Mag. nat. Hist.*, (9) **18**, 290.

— 1930. An Oligochaete worm parasitic in frogs of the genus *Phrynomerus*. *Ann. Mag. nat. Hist.*, (10) **6**, 367.

— 1931 a. Oligochaeta from the Malay Peninsula. *J. fed. Malay St. Mus.*, **16**, 261.

— 1931 b. Oligochaeta from Burma, Kenya, and other parts of the world. *Proc. zool. Soc. Lond. 1931*, **1-2**, 33.

— 1931 c. Oligochaeta from Brasil and Paraguay. *J. Linn. Soc. Lond.*, **37**, 291.

— 1932. Report on the Oligochaeta: Mr. Omer-Cooper's investigation of the Abyssinian fresh waters. *Proc. zool. Soc. Lond. 1932*, **1-2**, 227.

STIEREN, A. 1892. Über einige *Dero* aus Trinidad. *Protok, Obshch. Estest., Yurev.*, **10**, 103.

STOLC, A. 1886a. *Dero digitata* O. F. Müller. Anatomická a histologická studie. *Sber. K. böhm. Ges. Wiss.*, **1885**, 310 (Czech with French summary).

— 1886 b. Beiträge zur Kenntnis der Naidomorphen. *Zool. Anz.*, **9**, 502.

STOLTE, H. A. 1921. Untersuchungen über experimental bewirkte Sexualität bei Naiden. *Biol. Zbl.*, **41**, 535.

— 1927. Studien zur Histologie des Altersprozesses; —. *Z. wiss. Zool.*, **129**, 1.

STOUT, J. D. 1956. Aquatic Oligochaetes occurring in forest litter. 1. *Trans. R. Soc. N.Z.*, **84**, 97.

— 1958. Aquatic Oligochaetes occurring in forest litter. II. *Trans. R. Soc. N.Z.*, **85**, 289.

STRAND, E. 1928. Miscellania nomenclatorica zoologica et palaeontologica. *Arch. Naturgesch.*, **92A**, 8, 30.

STUHLMANN, F. 1891. Beiträge zur Fauna centralafrikanischer Seen. I. Süd-creek der Victoria-Niansa. *Zool. Jb. Syst.*, **5**, 924.

SVETLOV, P. 1924. Beobachtungen über Oligochaeta des Gouvernements Perm. II. *Izv. biol. nauchnoiss led. Inst. biol. Sta. perm. gosud. Univ.*, **3**, 187 (Russian with German summary).

SVETLOV, P. 1925. Einige Angaben über die Oligochaetenfauna des Tscherdynbezirkes (Uralgebiet). *Izv. biol. nauchnoiss led. Inst. biol. Sta. perm. gosud. Univ.*, **3**, 471 (Russian with German summary).

— 1926a. Zur Kenntnis der Oligochaetenfauna des Gouv. Samara. *Izv. biol. nauchnoiss led. Inst. biol. Sta. perm. gosud. Univ.*, **4**, 249 (Russian with German summary).

SZARSKI, H. 1936 a. Contribution to the physiology of Oligochaeta belonging to the genus *Chaetogaster*. *Bull. Acad. pol. Sci. Cl. B. Sév. Sci biol.*, **1936**, 101.

— 1936 b. Studies on the anatomy and physiology of the alimentary canal of worms belonging to the *Naididae family Bull. Acad. pol. Sci. Cl. B. Sév. Sci. biol.*, **1936**, 387.

TABLEAU ENCYCLOPÉDIQUE ET MÉTHODIQUE des trois règnes de la nature. *Vers, coquilles, mollusques et polypiers.* 1. 1827. Paris.

TAUBER, P. 1873. Om Naidernes Bygning og Kjønsforhold. *Naturhist. Tidsskr. Kjøbenhavn*, (3) **8**, 379.

— 1879. *Annulata Danica.* I. Kjøbenhavn.

TIMM, R. 1883. Beobachtungen an *Phreoryctes Menkeanus* Hoffmr. und *Nais. Arb. Inst. Würzburg*, **6**, 107.

TIMM, T. 1959. A survey of the freshwater Oligochaeta of Estonia. *Naturalists Soc. Acad. Sci. Esthonian S.S.R. (separaat)*, **1**, 23 (in Esthonian).

— 1962. Über die Fauna Ökologie und Verbreitung der Susswasser-Öligochaeten der Estnischen S.S.R. Zooloogia-Alaseia Töid, **2** (120), 84.

— 1965. Die Oligochaeten in Kustennahen Meere der Estnischen S.S.R. Easti Teaduste Akadaemia Toimetised XIV. *Biol. Ser.* **1**, 36-48 (Russian, Esthonian and German summary).

TREMBLEY, A. 1744. *Mémoires pour servir à l'histoire d'un genre de Polypes d'eau douce.* Leyden.

TRYBOM, F. 1901. Bexhedasjön, Norrasjön och Näsbysjön i Jönköpings län. *Meddn. K. LantbrStyr.*, 9, **76**, 1.

TURNER, C. 1935. A new species of freshwater Oligochaete from the Southeastern United States with a description of its sexual organs. *Zool. Anz.*, **109**, 253.

UDALICOV, A. 1907. Zur Fauna der *Naididae* des Glubokje-Sees und seiner Umgebung. *Trudy Otd. ikhtiol. Obshch. Akklim. Mosk.*, **6**, 144.

UDE, J. 1929. Oligochaeta. In F. DAHL. *Die Tierwelt Deutschlands*, **15**, 1. Jena.

D'UDEKEM, J. 1855. Nouvelle classification des Annélides sétigères abranches. *Bull. Acad. r. Belg. Cl. Sci.*, **22**, 548.

— 1856. Développement du Lombric terrestre. *Mém. Acad. r. Belg.*, **27**, 5.

— 1859. Nouvelle classification des Annélides sétigégéres abranches. *Mém. Acad. r. Belg.*, **31**, 1.

— 1861. Notices sur les organes génitaux des *Oelosoma* et des *Choetogaster*. *Bull. Acad. r. Belg.*, (2) **12**, 243.

VAILLANT, L. 1890. *Histoire naturelle des Annélés marins et d'eau douce.* 3. Paris.

VEJDOVSKY, F. 1883. Revisio Oligochaetorum Bohemiae. *Sber. Böhm. Ges. Naturw.*, **1883**, 215 (Czech).

— 1884. *System und Morphologie der Oligochaeten.* Prag.

— 1894. Description du *Dero tonkinensis* n. sp. *Mém. Soc. zool. Fr.*, **7**, 244.

VOGT, C. 1841. Zur Anatomie der Parasiten. *Müller's Archiv für Anatomie.* Berlin. **1841**, 33.

VOS, N. DE, 1922. Oligochaeten. *Flora en Fauna der Zuiderzee.* Rotterdam. 1922. 276.

WALTON, L. B. 1906. Naididae of Cedar Point, Ohio. *Am. Nat.*, **40**, 683.

WESENBERG-LUND, E. 1938. A peculiar occurrence of *Nais communis* Piguet in Denmark. *Vidensk. Meddr. dansk. naturh, Foren.*, **102**, 7.

WILLCOX, M. A. 1901. A parasitic or commensal Oligochaete in New England. *Am. Nat.*, **35**, 905.

WOLF, W. 1928. Über die Bodenfauna der Moldau im Gebiete von Prag im Jahreszyklus. Oligochaeta. *Int. Revue Hydrobiol. Leipzig*, **20**, 377.

YAMAGUCHI, H. 1938. Studies on the aquatic oligochaeta of Japan, V. The description of a new variety of *Branchiodrilus hortensis* (Stephenson). *Annotnes zool. jap.*, **17**, 530.

YAMAGUCHI, H. 1953. Studies on the Aequatic oligochaeta of Japan. VI. A systematic report, with some remarks on the classification and phylogeny of the Oligochaeta. *J. Fac. Sci. Hokkaido Univ. Ser. VI, Zool.,* **11,** 277.

YOSHIZAWA, H. 1928. On the aquatic Oligochaete, *Stylaria lacustris* L. *Scient. Rep. Tôhoku Univ.,* (4) **3,** 587.

8

FAMILY TUBIFICIDAE

Type genus: Tubifex LAMARCK

Prostomium without a proboscis. No eyes. Ventral setae, an indefinite number per bundle, beginning in II, bifid or (seldom) simple-pointed. Dorsal setae from II, hair setae present or absent; otherwise bifid crotchets, or any number of intermediate teeth forming obscurely to completely pectinate setae or (rarely) palmate setae, even simple-pointed setae but these mostly in posterior bundles. Spermathecal or penial setae or both modified or absent in mature specimens. Clitellum in a few segments in the region of the gonads. Testes and ovaries one pair each, in X and XI. Spermathecae paired in X, or single, or absent, spermatophores or free sperm masses in spermathecae. Male funnels in testes segment, atrium and male pores in succeeding segment together with female funnels. Female pores in furrow behind segment bearing male pores. All reproductive organs may be shifted from one to several segments forward. Sperm sacs and ovisacs usually present, unpaired. Asexual reproduction by fragmentation.

Cosmopolitan.

Various attempts have been made in the past to divide the Tubificidae into sub-families, but as yet no acceptable scheme has been proposed (Cekanovskaya, 1962; Brinkhurst, 1966). In some recent papers, Professor S. Hrabe has made several attempts to clarify the position, but thus far several genera have been omitted from the discussion.

The main characters used by Professor Hrabe to define sub-families are:

1. the form of the prostate gland,
2. the presence of spermatophores,
3. the presence of coelomocytes,
4. the position of the spermathecal pores,

444

5. the nature of the nephridia,
6. the mode of formation of the eggs.

Taking these characters in reverse order, the last appears to be of dubious value owing to the great scarcity of descriptions of the ovaries and egg formation. If the use of this character is based on personal observation by Professor Hrabe, this has not been documented. The nephridia of genera constituting one of the larger sub-families are said to be characterized by a glandular post-septal section. While this appears true of some species, once again a lack of evidence hampers evaluation of this character. Furthermore, the illustration of the nephridium of *Limodrilus hoffmeisteri* (Tubificinae) published by Naidu (1965) shows a narrow post-septal ampulla of exactly the same gross morphological appearance as that of *Bothrioneurum americanum* (Rhyacodrilinae), illustrated by Cernosvitov (1939). The position of the spermathecal pore relative to the longitudinal axis may give additional weight to a classification based on other characters, but again too much reliance on inadequate data might be involved.

The first three characteristics have the advantage that they are almost invariably mentioned in descriptions of tubificids, and that the family may be readily divided into three groups based on their use. This classification, considering these characters only, might be summarized as follows:

Telmatodrilinae Eisen, 1885

Type genus: *Telmatodrilus Eisen*

Prostate glands small and numerous on each atrium. Coelomocytes absent. Spermatophores present.

Tubificinae Eisen, 1885

Type genus: *Tubifex Lamarck*

Solid prostate gland on each atrium. Coelomocytes absent. Spermatophores present.

Rhyacodrilinae Hrabe, 1963

Type genus: *Rhyacodrilus* Bretscher

Prostate diffuse. Coelomocytes present. Spermatophores absent.

There remain some problems, however, which become apparent the minute one considers all of the genera in the family. The first of these is the monotypic genus *Branchiura*, in which the prostate glands are diffuse, there are no spermatophores, but there are no coelomocytes, possibly because of the unusual reduction of the coelom by muscle strands and other tissue. A new sub-family, Branchiurinae, was erected by Hrabe (1966) in response to correspondence on the affinities of this monotypic genus, but the footnote in which the definition is given mentions the presence of the diffuse prostate glands and "the absence of other typical signs of subfam. Rhyacodrilinae".

The unique nature of the atrial diverticulae may also be used as evidence in favour of the separation of this genus from the rest of the Tubificidae.

The genera *Aulodrilus*, *Siolidrilus* and *Limnodriloides* may also be separated from the rest as the prostate glands are attached to the atria by a wide base unlike the narrow stalk found in the Tubificinae, they lack coelomocytes, but the sperm are aggregated into loose bundles in the spermathecae, and are not formed into spermatophores. Hence a separate sub-family for these genera must be created in order to be consistent.

The position of *Macquaridrilus* in the Telmatodrilinae is also open to question. The prostate gland may consist of a series of discrete prostates, or it may represent a structure in some ways intermediate between that of the Rhyacodrilinae and the Telmatodrilinae. Such a condition is fore-shadowed in *Monopylephorus lacteus*, in which the diffuse prostate involves only part of the cross-section of the male duct on which it is developed (homologies of this male duct being difficult to ascertain). The prostate of *Macquaridrilus* seems to be attached by a series of discrete narrow stalks along one longitudinal axis of the male duct.

In previous attempts to clarify the tubificids into sub-families the position of the poorly-known marine genera has generally been ignored. The genus *Clitellio* fitted neatly into the Tubificinae when only *C. arenarius* was attributed to it, but the nature of the prostate could not be evaluated as it is absent. Having allied *C. arenicolus* with the genus we are forced to recognize the prostate as diffuse (or absent) in the genus. None of the other marine genera, with the possible exception of *Limnodriloides* and *Jolydrilus* fit the existing classification. They cannot be accepted as a single sub-family as they have little in common other than the marine habit and rather simple bifid setae with reduced teeth. There are, however, two groups which can be defined according to the three characters available for consideration. *Clitellio* and *Smithsonidrilus* lack coelomocytes but have spermatophores, and have diffuse prostates or none. The four genera* *Spiridion*, *Thalasso-drilus*, *Phallodrilus* and *Adelodrilus* have solid prostates with stalks, no spermatophores, and few, if any, coelomocytes. The genus *Jolydrilus* is difficult to place as the prostates are absent and so are the spermathecae, but there are abundant coelomocytes, and hence the genus is located within the Rhyacodrilinae. The form of the genital setae, when present, seems to be reasonably consistent within sub-families.

The genera of the Tubificidae may now be arranged in sub-families as follows:

Tubificinae

Tubifex, Limnodrilus, Isochaeta, Peloscolex Potamothrix, Psammoryc-tides, Ilyodrilus, Antipodrilus.

---

* The genus *Spiridion,* and possibly even the rest of its subfamily, may well be incorporated within the Aulodrilinae.

Rhyacodrilinae
Rhyacodrilus, Monopylephorus, Bothrioneurum, Epirodrilus, Paranadrilus, Jolydrilus.

Aulodrilinae
Aulodrilus Siolidrilus, Limnodriloides.

Telmatodrilinae
Telmatodrilus, Macquaridrilus.

Clitellinae
Clitellio, Smithsonidrilus.

Branchiurinae
Branchiura

Phallodrilinae
Phallodrilus, Adelodrilus, Spiridion, Thalassodrilus

These will be defined in the following systematic section.

KEY TO THE SUBFAMILIES AND GENERA OF THE FAMILY TUBIFICIDAE*

1.	Coelomocytes abundant	Rhyacodrilinae—2
—	Coelomocytes sparce or absent	7
2.	Spermathecae absent, sperm attached to body wall in sperm bearers	3
—	Spermathecae present, sperm loose within them	5
3.	Paratria present. Prostomium with ciliated pit dorsally	*Bothrioneurum*
—	Paratria absent. Prostomium without ciliated pit	4
4.	Vasa deferentia short, prostates absent	*Jolydrilus*
—	Vasa deferentia moderately long, prostates present	*Paranadrilus*
5.	Prostates absent, atria each of two wide tubular portions connected by a narrow neck	*Epirodrilus*
—	Prostates present, absent in *R. simplex* where atria broadly cylindrical with rounded apices	6

* Most regional keys are drawn up without keys to genera as the distinctions are difficult to describe and depend on the examination of mature worms. Compiling keys directly to species makes possible the identification of many immature specimens.

6.   Vasa deferentia moderately long, distinct from
      globular, pear-shaped or cylindrical atria
      which they join sub-apically   *Rhyacodrilus*
—   Vasa deferentia short, scarcely distinguishable
      from tubular atria bearing prostate cells   *Monopylephorus*

7.   Dorsal and ventral unpaired gill filaments
      present   Branchiurinae *Branchiura*
—   Gills absent   8

8.   Atria with single solid prostates attached by
      a narrow stalk, spermatophores* in sperma-
      thecae   Tubificinae—9
—   Atria with prostates not of this form, or,
      where prostates of similar form, no sperma-
      tophores formed   16

9.   Vasa deferentia much shorter than atria   *Potamothrix*
—   Vasa deferentia as long as or longer than atria   10

10.   Narrow, elongate tubular atria little wider
      than vasa deferentia. Prostates small   *Isochaeta*
—   Atria distinctly wider than vasa deferentia   11

11.   Thick walled cuticular penis sheaths cylindri-
      cal, usually much longer than broad, sur-
      rounded by spiral muscles   *Limnodrilus*
—   Penis sheaths tub shaped, thimble shaped,
      sometimes cylindrical, never heavily thickened,
      without spiral muscles   12

12.   Body wall papillate, first few segments non-
      papillate and rectractable within the papillate
      segments, papillation varying from a complete
      armoured sheath to a slight modification of the
      body wall detectable in sections. Atria mostly
      horse-shoe shaped with vasa deferentia enter-
      ing apices, narrowing to ejaculatory ducts.
      Penis sheaths thin, cylindrical when present   *Peloscolex*
—   Body wall non-papillate. Penis sheaths when
      present, usually tub-shaped or conical. Atria
      not horse-shoe shaped.   13

* The term spermatophores may be confused with the externally attached structures
found in *Bothrioneurum* and megadriles, and should be replaced by the term
spermatozeugma—see Chapter 1.

13. Atria with globular ental section, narrowing abruptly to a tubular section leading to a naked penis. *Antipodrilus*

— Atria not of this form. Cuticular penis sheaths normally present. 14

14. Atria with narrow duct-like portion between small atria bearing prostate glands and terminal expensions leading to penes with thin cuticular penis sheaths *Psammoryctides*

— Atria terminate in penial bulbs without distinct tube-like narrowings or ejaculatory ducts 15

15. Prostates attached to anterior face of upright atria that are broad apically and narrow gradually towards the penial bulbs. Vasa deferentia frequently very long *Tubifex*

— Prostates attached to posterior face of upright atria. Vasa deferentia relatively wide and short. Atria more tubular than elongate pear-shaped *Ilyodrilus*

16. Atrial walls penetrated at several points by stalks of the prostate gland or glands Telmatodrilinae—17

— Prostates one or two per atrium or diffuse 18

17. Spermathecae with diverticulae *Macquaridrilus*

— Spermathecae without diverticulae *Telmatodrilus*

18. Prostates diffuse. Spermatophores present Clitellinae—19

— Prostates solid. Spermatophores absent 20

19. Paratria present, with a cap of prostate cells *Smithsonidrilus*

— Paratria absent, prostates diffuse on atria or absent *Clitellio*

20. Prostates attached to atria by broad bases. Sperm in bundles in spermathecae. Coelomocytes absent Aulodrilinae—21

— Prostates attached to atria by narrow stalks. Sperm free in spermathecae. Coelomocytes sparse or absent Phallodrilinae—23

21. Vasa deferentia at least as long as atria. Gut with pair of preclitellar diverticulae *Limnodriloides*

—   Vasa deferentia short. No gut diverticulae   22

22.   Prostates solid. Penial setae present or absent.
       No spermathecal setae   *Aulodrilus*

—   Prostates tufts of cells attached to atria. Penial
       setae absent, spermathecal setae present   *Siolidrilus*

23.   Vasa deferentia as long as or longer than atria   24

—   Vasa deferentia shorter than atria   25

24.   Each atrium bears one prostate gland opening
       almost apically but just posterior to vas
       deferens   *Spiridion*

—   Each atrium bears two prostates, one opening
       near the vas deferens, the other near male
       pore   *Phallodrilus*

25.   Prostates small, attached to vasa deferentia
       near union with atria   *Thalassodrilus*

—   Prostates large, paired on each male duct and
       seemingly attached to penial bulbs   *Adelodrilus*

SUB-FAMILY TUBIFICINAE EISEN, 1879

Type genus: Tubifex LAMARCK

Solid prostates with stalk-like attachments to atria when present. Sperm as spermatophores within spermathecae. Coelomocytes absent.

GENUS **Tubifex** LAMARCK, 1816

Type species: Lumbricus tubifex Müller

Male efferent ducts with vasa deferentia at least as long as the pear-shaped to cylindrical atria, which are frequently turned-over apically, otherwise standing vertically in XI, no distinct ejaculatory ducts; prostate glands large, closely connected to the atria subapically on anterior side. Penes present, with tub-shaped or conical to elongate penis sheaths. Spermathecae sometimes absent, spermatophores present unless spermathecae absent. No genital setae. No coelomocytes.

Cosmopolitan.

The relative lengths of the vas deferens and the atrium vary through wide limits in the genus, in some species the vas deferens being inordinately long. The vas deferens may join the atrium apically or on either side of the apical end, adjacent to or opposite to the union of the prostate gland and atrial wall. The distinction between the typical genus *Tubifex* and the obviously closely related genus *Peloscolex*, and, to a lesser extent, *Psammoryctides*,

Limnodrilus and *Isochaeta* is not always clear. Typical species of each genus are readily distinguishable by a variety of characteristics, but in each instance there are one or two species which seem to be intermediate between two genera, one of which is usually *Tubifex*. Consequently, it is simpler to write regional keys that proceed directly to species than to key-out the genera.

According to Hrabe (1962 a) the genus contained 13 species plus 3 sub-species of *T. tubifex*. In a later account (Hrabe 1966) the list was shortened to 5 sub-species of *T. tubifex* (*T. chacoensis, T. kleerekoperi, T. siolii* having been omitted from the first account without comment) plus 11 other species. Of the latter I regard *T. davidis* of Benham as belonging to the distinct genus *Antipodrilus*.

I regard *Tubifex speciosus* of Hrabe (1931 a) as correctly located in *Peloscolex* because of the nature of the body wall and male efferent ducts. I regard *T. smirnovi* and *T. minor* as inadequately described and best regarded as a species inquirendae, and *T. flabellisetosus* as not certainly in *Tubifex*.

Two species formerly accepted as part of *Tubifex* by Hrabe (1962 a) and transferred by him to *Isochaetides* (=*Isochaeta*) I would retain in *Tubifex* because of the form of the male efferent ducts, despite the lack of hair setae and pectinate setae, which I do not regard as a valid generic character (i.e. *T. newaensis, T. pseudogaster*). Two other species originally located in *Tubifex* but shifted to *Isochaeta* by Brinkhurst (1963 a) have now been recognized as distinct from *Tubifex* by Hrabe (1966) (i.e., *hamatus, lacustris*) but *dojranensis* was left within *Tubifex* despite its over-all morphological relationship to other *Isochaeta* species, again because of the nature of the dorsal setae.

Tubifex templetoni was referred to the little-known genus *Ilyodrilus* by Hrabe (1966), a decision supported by an examination of type material and newly discovered specimens (vide *Ilyodrilus*).

Hence, 6 of Hrabe's latest list of *Tubifex* species are rejected, 2 are retained in *Tubifex* though they are not on that list, and 2 other species have been added by subsequent description. The total count therefore becomes 11 with 1 species (*T. kessleri*) divided into 2 sub-species, and 3 forms of *T. tubifex* being recognized, while *Tubifex flabellisetosus* is included as a species incerta sedis, and *T. smirnovi, T. minor* and *T. kryptus* are regarded as species inquirendae.

Tubifex

1.   Ventral setae completely pectinate, with
     several intermediate teeth   *T. flabellisetosus* (sp. i. sedis).

—   Ventral setae simple pointed, bifid, or with a
     small intermediate tooth on the lower lateral
     tooth   2

2.   Hair setae in at least some dorsal bundles   3
—   Hair setae absent   10

3.   Few hair setae in a restricted number of
        bundles   4
—   Hair setae generally present, more than one in
        most bundles   5

4.   Penis sheaths thin, tub-shaped.   *T. tubifex* ("bergi" form)
—   Penis sheaths elongate, wide proximally, thin
        distally with lateral openings   *T. kessleri kessleri*

5.   Pectinate setae absent.   *T. natalensis*
—   Pectinate setae present   6

6.   Unusually thin worms, with obviously serrate
        hair setae, elongate in some segments in the
        clitellar region.   *T. ignotus*
—   Worms not especially thin, serration on hairs
        close and fine or? absent, none especially
        elongate   7

7.   Pectinate setae practically palmate, little dis-
        tinction between lateral and intermediate
        teeth   *T. montanus*
—   Pectinate setae with lateral teeth distinct from
        intermediate teeth   8

8.   Penis sheaths thin, short, tub-shaped.   9
—   Penis sheaths distinct, elongate.   *T. kessleri americanus*

9.   Vasa deferentia long, atria comma-shaped,
        pectinate setae with several, irregular inter-
        mediate teeth   *T. tubifex* ("tubifex" *form*)
—   Vasa deferentia short, atria short and wide,
        pectinate setae with few thin intermediate
        teeth   *T. nerthus*

10.   Some dorsal setal bundles with palmate setae   *T. costatus*
—   No palmate setae   11

11.   Penis sheaths short and tub-shaped   12
—   Penis sheaths conical to elongate   13

12.   Vasa deferentia entering concave side of

comma-shaped atria at anatomical apex

        T. tubifex ("blanchardi" form)

— Vasa deferentia entering convex side of coma-
shaped atria   *T. newaensis*

13. Penis sheaths thimble-shaped with terminal
opening   *T. pseudogaster*

— Penis sheaths bent, elongate with a lateral
notch, no opening   *T. longipenis*

Tubifex tubifex (Müller, 1774)
Fig. 8. 1A-D. Fig. 8. 3G-J

Lumbricus tubifex MÜLLER, 1774: 27 (in part).

Nais tubifex (Müller). OKEN, 1815: 364.

Tubifex rivulorum LAMARK, 1816: 225.

Nais filiformis DUGÈS, 1828: 336.

Tubifex filiformis (Dugès). DUGÈS, 1837: 33.

Tubifex filiformis (Dugès). BRETSCHER, 1900 b: 447, Pl. XXXIII, fig. 4.

Blanonais filiformis (Dugès). GERVAIS, 1838: 16.

Saenuris variegata HOFFMEISTER, 1842: 9, Pl. II, figs. 19, 20, 22 (in part).

Saenuris tubifex (Müller). WILLIAMS, 1852: 264. JOHNSTON, 1865: 64.

Tubifex rivulorum Lamark. UDEKEM, 1853: 3, Pl. I-IV. LANKESTER, 1871 a:
 93, Fig. 2. EISEN, 1879: 14; 1886: 892. LEVINSEN, 1884: 224. VEJDOVSKY,
 1884 b: 46, Pl. VIII, figs. 1-8, Pl. IX, figs. 2-19, Pl. X, figs. 1-5, 7-16.
 STOLC, 1886: 643; 1888: 39. BENHAM, 1892: 209. BEDDARD, 1895: 244.
 BRETSCHER, 1896: 506. FRIEND, 1897 c: 296. SMITH, 1900: 444. CHEN,
 1934: 99, Figs. 1-7.

Nais sanguinea DOYÈRE, 1856: 306 (in part).

Tubifex bonnetti CLAPARÈDE, 1862: 230, Pl. II, figs. 1-6, Pl. IV, fig. 5.

Tubifex bonneti Claparède. EISEN, 1879: 15; 1886: 893.

?*Tubifex campanulatus* EISEN, 1879, 16.

Saenuris diversisetosa forma *charcoviensis* CZERNIAVSKY, 1880: 334.

Saenuris diversisetosa forma *suchurmica* CZERNIAVSKY, 1880: 332.

Saenuris peculiaris CZERNIAVSKY, 1880: 330.

Saenuris taurica CZERNIAVSKY, 1880: 332.

?*Tubifex campanulatus* Eisen VEJDOVSKY, 1884 b: 45. EISEN, 1886: 893, Pl. VIII, fig. 7. MICHAELSEN, 1900: 49.

Tubifex blanchardi VEJDOVSKY, 1891: 596.

Tubifex blanchardi Vejdovsky. BEDDARD, 1895: 245. NAIDU, 1966: 219. MICHAELSEN, 1900: 49. HRABE, 1930: 2, Figs. 10, 11; 1931 b: 314. BRINK-HURST, 1963 a: 22.

Saenuris variegata Hoffmeister, FUHRMAN, 1897: 492.

Tubifex alpinus BRETSCHER, 1900: 13.

Tubifex tubifex (Müller). MICHAELSEN, 1900: 48. BRETSCHER, 1901: 196. DITLEVSEN, 1904: 422, Pl. XVI, figs. 5-6, Pl. XVIII, fig. 9. PIERANTONI, 1904 b: 2; 1905: 231. SOUTHERN, 1909: 138. GALLOWAY, 1911: 313. LASTOCKIN, 1918: 62; 1924: 5; 1927 a: 67. CERNOSVITOV, 1926: 321; 1929: 145; 1930: 321; 1931: 87; 1935 a: 2; 1936: 226; 1938 b: 266. SVETLOV, 1962 a: 252; 1962 b: 344; 1936: 145; 1946: 104. HRABE, 1929: 1; 1930: 2; 1939 a: 222, Figs. 1-2; 1941: 14; 1942: 26; 1952: 6; 1962 a: 300. UDE, 1929: 85, Figs. 81, 99-104. ALTMAN, 1936: 15. MARCUS, 1942: 198. BERG, 1948: 50. CEKANOVSKAYA, 1952: 294; 1956: 660; 1962: 272, Fig. 177. CAUSEY, 1953: 423. MALEVICH, 1956: 403. BÜLOW, 1957: 99. JUGET, 1957: 2; 1958: 88, Figs. 13, 14h. BRINKHURST, 1960: 402, Fig. 6; 1962 c: 362, Figs. 1a, b, l; 1963 a: 22, Figs. 2, 3, 3d; 1963 c: 37, Fig. 9c; 1965: 123, Fig. 2a-d. SOKOLSKAYA, 1961 b: 86. BRINKHURST and KENNEDY, 1962 a: 184; 1962 b: 375. MOSZYNSKA, 1962: 28. KENNEDY, 1964: 227. NAIDU, 1965: 464, Fig. 1a-h; 1966: 219. POPESCU-MARINESCU et al., 1966: 163.

Tubifex (Tubifex) tubifex (Müller). MICHAELSEN, 1909: 37. POINTNER, 1911: 637; Figs. 1-3, Pl. XXIX, fig. 31. PIGUET and BRETSCHER, 1913: 63. PIGUET, 1919: 791. STEPHENSON, 1921: 753; 1923: 106. LASTOCKIN, 1927 a: 67. CERNOSVITOV, 1928: 5, Pl. I, fig. 2; (?) 1938 a: 543; 1939 a: 36; 1942 a: 213. HRABE, 1935 b: 79; 1962 a: 308. MARCUS, 1942: 198, figs. 24, 27, 29. NAIDU, 1965: 464.

Tubifex (Tubifex) fontaneus POINTNER, 1914: 607.

Tubifex tubifex var. *heterochaeta* CERNOSVITOV, 1926: 321.

Tubifex hattai NOMURA, 1926: 193.

Limnodrilus chacoensis STEPHENSON, 1931 b: 309.

Tubifex (Tubifex) bergi HRABE, 1935 b: 76, figs. 1-6; 1962 a: 307.

Tubifex (Tubifex) blanchardi Vejdovsky. CERNOSVITOV, 1942 a: 213.

Limnodrilus kleerekoperi MARCUS, 1944: 71, Figs. 62-65.

Limnodrilus siolii MARCUS, E., 1947: 11, Figs. 10-14.

Tubifex tubifex Michaelsen. JAROSCHENKO, 1948: 69.

Tubifex ignotus (Stolc). BRINKHURST, 1960: 404, Fig. 5.

Tubifex hattai Nomura. YAMAGUCHI, 1953: 295. HIRAO, 1964: 439, Figs. 1-3, Pl. XVIII, figs. 20-27, Pl. XV, figs. 4-10, Pl. 16, figs. 11-19.

Tubifex bergi Hrabe. CEKANOVSKAYA, 1962: 273, fig. 172. BRINKHURST, 1963 a: 23.

Tubifex tubifex tubifex (Müller). HRABE, 1962 a: 308; 1966: 74.

Tubifex tubifex blanchardi Vejdovsky. HRABE, 1962 a: 308; 1966: 74.

Tubifex tubifex bergi Hrabe. HRABE, 1962 a: 308; 1966: 74.

Tubifex siolii (Marcus). BRINKHURST, 1963 a: 25.

Tubifex kleerekoperi (Marcus). BRINKHURST, 1963 a: 25.

Tubifex tubifex chacoensis (Stephenson). HRABE, 1966: 74.

Tubifex tubifex kleerekoperi (Marcus). HRABE, 1966: 74.

Tubifex tubifex siolii (Marcus). HRABE, 1966: 74.

$l=20$-200 mm, $s=34$-120. Anterior dorsal bundles with 3-5 pectinate setae and 1-4 or 6 serrate or non-serrate hair setae, pectinate setae with several irregular intermediate teeth, lateral teeth more or less equal; posterior dorsal crotchets with upper tooth longer and thinner than the lower, intermediate teeth reduced or absent or nothing but bifid crotchets in all dorsal bundles. Ventral bundles with 3-6 or 10 setae anteriorly, falling to 2 posteriorly; upper tooth thinner and more or less as long as the lower, becoming relatively shorter posteriorly rarely with a single intermediate tooth. Spermathecae with elongate ampullae, or single, or absent. Spermatophores often elongate. Vasa deferentia elongate when fully formed, divided into thin and thick sections, atria comma-shaped, vasa deferentia and prostates entering together on the concave side. Penis sheaths often indistinct, tub-shaped, often granular.
Cosmopolitan.

This species has been found in a variety of forms, originally described as species, then varieties, most recently as sub-species (Hrabe, 1966) in order to conform to the dictates of the International Code of Zoological Nomenclature of 1958 (published 1961). The difficulty in accepting these variants as sub-species is that they are often found together in the same locality. The difference is one we shall meet in other genera, and consists solely of the loss of hair setae and of pectinate setae and their replacement by bifid setae. The simplest solution seems to be to give these forms names by which they may be recognized but which have no systematic standing according to the code. We may then recognize three main forms, the "tubifex" form with a full complement of hair setae and pectinate setae, the "bergi" form with some specimens having a few hair setae, most with a few meagrely pectinate setae (apparently only found in L. Issyk-kul, U.S.S.R.) and the "blanchardi" form with bifid setae only. The last form includes the former species *chacoensis, kleerekoperi, siolii* as well as *blanchardi* as the differences between them seem to be trivial, consisting chiefly of a difference in the form of the spermatophores.

The vas deferens enters the convex side of the atrium, but the atrium is bent over the union with the prostate so that the vas deferens actually enters its proximal extremity. The spermathecae are sometimes missing, sometimes there is only one.

Tubifex ignotus (Stolc, 1886)
Fig. 8. 1E-G

Lophochaeta ignota Stolc., 1886: 646.

Lophochaeta ignota Stolc. STOLC, 1888: 41, Pl. I, figs. 4-6, Pl. II, figs. 6, 10, Pl. III, figs. 7-9, 13, Pl. IV, fig. 136. BEDDARD, 1895: 270, Figs. 38, 41. MICHAELSEN, 1900: 53.

Tubifex filum MICHAELSEN, 1901 a: 3.

Tubifex filum Michaelsen. MICHAELSEN 1903 b: 194. LASTOCKIN, 1927 a: 67; 1953: 181; 1955: 62. JUGET, 1958: 88. POPESCU-MARINESCU et al., 1966: 163.

Tubifex longiseta BRETSCHER, 1905: 670.

Tubifex longiseta Bretscher. PIGUET, 1906 a: 392.

Tubifex (Tubifex) filum Michaelsen. MICHAELSEN, 1909: 37, fig. 72. PIGUET and BRETSCHER, 1913: 62.

Tubifex (Tubifex) ignotus (Stolc). MICHAELSEN, 1909: 36, Fig. 69. CERNOSVITOV, 1928: 6.

Tubifex ignotus (Stolc). DITLEVSEN, 1904: 421. HRABE, 1929: 1; 1939 a: 227; 1962 a: 301, Figs. 16-21. UDE, 1929: 90. BERG, 1948: 51. SOKOLSKAYA, 1953: 89. MALEVICH, 1956: 425. JUGET, 1958: 88, fig. 14f. CEKANOVSKAYA, 1962: 271. MOSZYNSKA, 1962: 28. BRINKHURST, 1963 a: 23, Fig. 5a-e; 1963 c: 35, Fig. 96. KENNEDY, 1964: 229. BRINKHURST and COOK, 1966, 13.

Tubifex nerthus Michaelsen. BRINKHURST, 1960: 404, Fig. 8.

(non) *Tubifex ignotus* (Stolc). BRINKHURST, 1960: 404, Fig. 5.=*Tubifex tubifex.*

Tubifex ignota (Stolc). BRINKHURST, 1962 c: 328, Fig. 1d-l. BRINKHURST and KENNEDY, 1962 b: 375.

l=80 mm, *s*=170. Very thin, elongate worms. Dorsal bundles with 1-3 pectinate setae with U-shaped lateral teeth and few intermediates, hair setae beset with long, relatively sparse lateral hairs, very long in some segments around the clitellum at least. Ventral setae 3-5 per bundle anteriorly, upper tooth longer and thinner than lower unless worn down, only 2 per bundle posteriorly with the upper tooth as long as the lower. Vasa deferentia long, divided into thin and thick portions, entering small, rather globular atria apically, penes with short cuticular sheaths, weakly developed.

Europe, N. America, S. Africa.

The species is generally scarce, and few mature individuals have been described. The illustration of the male efferent ducts in Hrabe (1962 a) lacks detail.

Tubifex newaensis (Michaelsen, 1903)
Fig. 8. 1H-J

Limnodrilus newaensis MICHAELSEN, 1903 a : 3, Figs. 1, 2.

Limnodrilus crassus ANDRUSOW, 1914: 92, Figs. 1-3.

Limnodrilus newaensis Michaelsen. LASTOCKIN, 1927 a : 67. JAROSCHENKO,

1948 : 60. CEKANOVSKAYA, 1962: 251, Fig. 157.

Tubifex newaensis (Michaelsen). BRINKHURST, 1962 b: 307, Fig. 1-3; 1963 a : 23; 1963 c: 42, Fig. 9c. HRABE, 1962 a : 310; (?) 1966 a : 737. BRINKHURST AND COOK, 1966 : 13.

Isochaetides newaensis (Michaelsen). HRABE, 1966: 75.

l=4-65 mm, s=60. Setae all bifid crotchets with the upper tooth thinner than but as long as the lower, or worn down to appear rudimentary. Vasa deferentia long, divided into thin and thick portions, entering the comma-shaped atria sub-apically on the convex side opposite the massive prostate gland. Penis sheaths distinct, tub-shaped.

Europe, N. America.

The American specimens attributed to this species differ from the Russian forms only in size, the latter being amongst the largest, thickest tubificids ever described. The species is very close to *T. tubifex*, differing chiefly in total absence of hair and pectinate setae (the "tubifex" form of the latter being most commonly encountered) and point of union of the vas deferens with the atrium. The fact that the species was first described as a *Limnodrilus* and has been placed in *Isochaeta* (as *Isochaetides*) by Hrabe (1966) indicates the degree of relationship of these genera, but as the atria are distinctly those of a *Tubifex* as defined above it is difficult to see the reasoning behind the last decision. There is a division of the vas deferens into thin and thick parts in my specimens, but not, apparently, in those studied by Professor Hrabe.

Tubifex nerthus Michaelsen, 1908
Fig. 8. 1K-M

Tubifex insignis (Eisen). MICHAELSEN in THIENEMANN, 1907: 88.

Tubifex nerthus MICHAELSEN, 1908 a: 155, Pl. III, figs. 15-20.

Tubifex (*Tubifex*) *nerthus* Michaelsen. MICHAELSEN, 1909: 35, Figs. 66-88.

Tubifex nerthus Michaelsen. KNÖLLNER, 1935: 425, Fig. 39. BÜLOW, 1955:
261; 1957: 99. JUGET, 1958: 88, fig. 14D (a, b). BRINKHURST, 1962 c:
328, Fig. 1f, q; 1963 a: 24; 1963 b: 712; 1963 c: 35, Fig. 9a. BRINKHURST
and KENNEDY, 1962 b: 375. CEKANOVSKAYA, 1962: 276, Fig. 174. HRABE,
1962 a: 307.

(*non*) *Tubifex nerthus* Michaelsen. BRINKHURST, 1960: 404.

$l=15$-30 mm, $s=52$. Dorsal bundles with 3-4 pectinate setae and 1-2 serrate
hairs, pectinate setae thin with lateral teeth somewhat U-shaped, with few fine
intermediate teeth. Ventral bundles mostly with 4 setae, with the upper tooth
twice as long as the lower, the lower being reduced and ? absent in XI. Vasa
deferentia short and broad, opening into the narrowly pear-shaped atria apically,
prostate glands large, penis sheaths short, thin.
Europe, brackish water.

This species has short, broad vasa deferentia according to the original
account. It may be that these were not fully developed, as there is some
suggestion that elongation of the vas deferens takes place late in development, or the character is subject to intra-specific variation (Dzwillo, pers.
comm.). This feature will be discussed in relation to the genus *Ilyodrilus*
(q.v.). *Tubifex kryptus* Bülow may belong here, the description being too
brief to enable a decision to be made.

Tubifex costatus (Claparède, 1863)
Fig. 8. 2A-D

Heterochaeta costata CLAPARÈDE, 1863: 25, Pl. XIII, figs. 16-19.

Heterochaeta costata Claparède. VEJDOVSKY, 1884 b: 45. BENHAM, 1892:
188, Pl. V, figs. 1-17, 32, Pl. VI, figs. 18-31. BEDDARD, 1895: 258. FRIEND,
1897 a: 63; 1897 c: 297.

Psammoryctes costatus (Claparède). MICHAELSEN, 1900: 52; 1927: 18.

Tubifex costatus (Claparède). MICHAELSEN, 1900: 525; 1926 b: 22; 1927:
18, Fig. 21. DITLEVSEN, 1904: 419. STEPHENSON, 1911: 33, Fig. 1a-b.
UDE, 1929: 89. STAMMER, 1932: 578. KNÖLLNER, 1935: 478, Fig. 41.
BERG, 1948: 51. JAROSCHENKO, 1948: 60. MALEVICH, 1951: 181. BÜLOW,
1955: 261; 1957: 99. DAHL, 1960: 18. BRINKHURST, 1962 c: 327, Fig.
1h, j; 1963 a: 24, Fig. 6a, b; 1963 b: 712; 1963 c: 38, Fig. 10c. BRINK-
HURST and KENNEDY, 1962 b: 375. CEKANOVSKAYA, 1962: 274, fig. 173.
HRABE, 1962 a: 308. KENNEDY, 1964: 228. MOSZYNSKA, 1962: 28. POPESCU-
MARINESCU, 1966: 163.

Tubifex costata (Claparède). SOUTHERN, 1909 : 139.

Tubifex thompsoni SOUTHERN, 1909 : 140, Pl. IX, fig. 7a-c.

Heterochaeta thompsoni (Southern) FRIEND, 1912 b : 284.

Tubifex thompsoni Southern. HRABE, 1962 a : 307.

$l=16$ mm, $s=40$. Dorsal bundles of anterior segments from about V to about XIV with 5-11 palmate setae, more anterior bundles with pectinate setae, intermediate teeth broad, upper tooth longer than the lower, posteriorly simple bifid setae with equal teeth; ventral bundles with 1-4 bifid setae with the upper tooth longer and thinner than the lower anteriorly. Vasa deferentia moderately long, entering the comma-shaped atria sub-apically on the convex side, penis sheaths short, tub-shaped.

Europe, brackish water.

Tubifex montanus Kowalewski, 1919
Fig. 8. 2E-I

Tubifex montanus KOWALEWSKI, 1919 : 131, Pl. IX, figs. 3-5.

Tubifex montanus Kowalewski. HRABE, 1939 a : 223, Figs. 3-8; 1926 a : 307. BRINKHURST, 1963 a : 24, Fig. 7.

$l=8$-12 mm, $s=40$-50. Anterior dorsal bundles with 2-3 pectinate crotchets almost palmate with lateral teeth scarcely distinguishable from broad intermediate teeth, lateral teeth equally long anteriorly, upper tooth longer than lower posteriorly, 2-3 serrate hair setae. Ventral bundles with 3-4 or 5 setae anteriorly with the upper tooth thinner than and twice as long as the lower, posteriorly 1-3 with equal teeth, the upper only slightly thinner than the lower. Vasa deferentia moderately long, all one width, entering narrowly pear-shaped atria apically, prostate glands large. Penis sheaths conical, narrowing distally. Spermathecae absent.

Tatra Mountains, Europe.

Tubifex kessleri Hrabe, 1962
Fig. 8. 2J-O

Tubifex kessleri HRABE, 1962 a : 305, Figs. 22-24.

Penis sheaths broad proximally, elongate narrow distally with lateral openings.

Until further European material is available, two sub-species must be recognized because of the differences in the form of the setae. The original description by Hrabe (1962) was based on a single fragment.

sub-species *kessleri* subsp. nov.
Fig. 8. 2J-K

$l=8$ mm, $s=32$. Dorsal bundles with 3-5 pectinate setae, but in V, VI, VIII, X, single short hair setae, pectinate setae with 1-2 intermediate teeth; ventral bundles with 3-5 bifid setae, rarely with an intermediate tooth, upper tooth longer than the lower. Vasa deferentia divided into proximal ciliated part and distal wider non ciliated part, twice as long as atria ? spindle-shaped. Penis sheaths wide proximally, narrow distally with lateral openings.

Onega Lake, U.S.S.R.

In the description (translated summary) the vasa deferentia are described as above, but in the following key they are said to be not divided into two distinct parts. Put together, these statements seem at first sight to be contradictory. The explanation may be that, while the duct widens and loses the lining of cilia distally, the distinction is not apparent superficially as an abrupt narrowing. A single specimen was available for study, and that was incomplete. The setal differences between this and the American material seem to me to be most likely to be due to loss of the hair setae, as they are otherwise exactly the same, the ventral setae even showing the reasonably frequent intermediate tooth noted by Professor Hrabe.

sub-species *americanus* BRINKHURST and COOK, 1966

Fig. 8. 2L-O

Tubifex kessleri americanus BRINKHURST and COOK, 1966: 13, Fig. 6b-d.

$l=22$ mm, $s=c.$ 75. Dorsal bundles with 3-4 hair setae, 3-4 pectinate setae with several intermediate teeth, posteriorly 1 of each, ventral bundles with 3-5 bifid setae, upper tooth longer than the lower which frequently bears a small spine. Vasa deferentia about three times the length of the atria, enter atria apically, atria spindly-shaped with large prostate glands. Penis sheaths broad proximally, narrow distally with lateral openings.

Lake Superior, Michigan, U.S.A.

This sub-species is probably the same as the typical form, but Professor Hrabe did not accept it as the same when sent specimens for comparison.

Tubifex pseudogaster (Dahl, 1960)

Limnodrilus pseudogaster DAHL, 1960: 13, Fig. 3, Pl. 1.

Tubifex pseudogaster (Dahl). BRINKHURST, 1962 c: 328, Fig. 1k, m; 1963 a: 25, Figs. 2, 9a-c; 1963 b: 712; 1963 c: 42, Fig. 12f; 1965: 124, Fig. 2h-i. BRINKHURST AND COOK, 1962 b: 375.

Isochaetides pseudogaster (Dahl). HRABE, 1966: 75.

Tubifex sp. Cernosvitov. [In Mss.—manuscript at the B.M. (N.H.)]

$l=12$-13 mm, $s=50$-85. Dorsal bundles with 3-4 or 6 bifid setae the upper tooth thinner than and as long as or, more often, shorter than the lower, fewer setae of the same form ventrally. Vasa deferentia moderately long, atria narrow, coiled in life. Penes with thimble-shaped penis-sheaths. Spermathecae with spermatophores.

Europe, N. America, brackish water.

This species resembles *Peloscolex gabriellae* fairly closely, and a detailed study of the male efferent ducts is needed to establish the generic affiliation for certain.

Tubifex longipenis (Brinkhurst, 1965)

Fig. 8. 2P-R

Tubifex longipenis BRINKHURST, 1965: 124, fig. 2j-1.

Tubifex longipenis Brinkhurst. BRINKHURST and COOK, 1966: 14.

l=25-30 mm, s=c. 100. Setae all bifid crotchets, 2-3 anteriorly, broad with upper teeth shorter than lower, posteriorly setae simple-pointed. Atria pear-shaped, penes with elongate, complex sheaths.
   Maine, U.S.A., brackish water.

   Only a single specimen was obtained, and later forms may prove to shed at least the distal portion of the penis sheaths.

Tubifex natalensis Brinkhurst, 1967
Fig. 8. 3A-C

Tubifex natalensis BRINKHURST, 1967: 146, Fig. 3.

l=c. 10 mm. Dorsal anterior bundles with 2 bifid crotchets and 1-2 short serrate hair setae, bifid setae with teeth short, broad and about equally long, posteriorly 1 of each. Ventral setae 1-3 per bundle anteriorly, 1 posteriorly, as dorsal setae but straight beyond nodulus. Vasa deferentia short, atria widely comma-shaped, with union of vasa deferentia on convex side opposite stalk of prostate. Penis sheaths broadly conical. Spermathecae with short broad spermatophores.
   N. of Durham, S. Africa.

SPECIES INQUIRENDAE

Tubifex smirnowi (Lastockin, 1927)

Tubifex (*Tubifex*) *smirnowi* LASTOCKIN, 1927 a: 71, Fig. 2.

Tubifex smirnovi Lastockin. CEKANOVSKAYA, 1962: 275. BRINKHURST, 1963 a: 26 (sp. dub.).

Tubifex smirnowi Lastockin. HRABE, 1962 a: 308; 1966: 74.

l=10 mm, s=37-42. Dorsal bundles with 1-2 serrate hair setae, 1-4 pectinate setae with 1-3 fine intermediate teeth. Ventral bundles with 5-6 ventral setae, 2-4 posteriorly, upper tooth shorter than lower. Vasa deferentia long, prostates large.
   U.S.S.R.

   The species is inadequately described and it is impossible to place it in the genus for certain. In a recent personal communication, Dr N. Sokolskaya stated "I suppose *Tubifex smirnovi* is a real species because it has the typical *Tubifex* male duct and the original chaetae. But I have never met worms of this species."

Tubifex kryptus Bülow, 1955
Fig. 8. 3K, L

Tubifex kryptus BÜLOW, 1955 : 261, Figs. 7-13.

Tubifex kryptus Bülow. BÜLOW, 1957: 99. BRINKHURST, 1963 a: 26 (sp. dub.); 1963 b: 712 (sp. dub.).

$l=18$ mm, $s=43$. Dorsal bundles with 1-2 short, smooth hair setae, pectinate setae present, ventral setae 3-4 per bundle, the upper tooth slightly longer and thinner than the lower. Vasa deferentia long. Spermathecae with spermatophores. Germany (brackish water?).

The absence of a clear description of the atria and penes makes it impossible to identify this species.

Tubifex minor Sokolskaya, 1961

Tubifex minor SOKOLSKAYA, 1961 b: 87, Fig. 1.

Edmondsonia minor (Sokolskaya). BRINKHURST, 1965 : 154.

Anterior dorsal bundles with 2-3 hair setae with lateral hairs, 3 pectinate setae with thin intermediate teeth, posterior bundles with 1 hair seta and 2 pectinate setae, anterior ventral bundles with 3-5 setae with long upper teeth, no genital setae. Penes skittle-shaped.
Amur basin, Asia.

In the original account abundant coelomocytes were described, but in a recent communication to the author, Dr Sokolskaya stated that *"Tubifex minor* has no real coelomocytes", and hence the genus *Edmondsonia* Brinkhurst is no longer valid (*E. montana* having been recognized as attributable to the genus *Rhyacodrilus* q.v.). The species is supposed to differ from *T. tubifex* because of the length of the upper teeth of the ventral setae and the shape of the penes, but as these are inadequately illustrated, it is not possible to make a detailed comparison.

SPECIES INCERTA SEDIS
Fig. 8. 3D-F

Tubifex flabellisetosus Hrabe, 1966

Tubifex flabellisetosus Hrabe. STAMMER, 1932 : 578, (nomen nudum). BRINKHURST, 1963 a: 26, (nomen nudum). HRABE, 1966 : 65, Figs. 21-26.

Tubifex (Tubifex) flabellisetosus Hrabe. HRABE, 1936 : 404, (nomen nudum). CERNOSVITOV, 1939 a: 35, (nomen nudum).

$l=?$ 10 mm. Anterior dorsal bundles 2-4 finely serrate hairs, 3-4 or 5 pectinate setae with long lateral teeth and two or three broad intermediate teeth, posteriorly 3 hairs and 4-5 narrower pectinate setae of similar form. Ventral bundles with up to 6 pectinate setae, with upper tooth longer than the lower, separated at an acute angle, with about three fine intermediate teeth. Vasa deferentia elongate, gradually widening posteriorly, enter atria sub-apically (?), near prostates; atria almost cylindrical, slightly widened apically. No penis sheaths. Free sperm in the spermathecae.

Jugoslavia (cave near source of R. Rjeka).

This species was described by Hrabe (1966) although it had been referred to (as a nomen nudum) in earlier publications. The anterior end of a single specimen constitutes the entire material available. While the setal characteristics differ from those of any known tubificid, the absence of penis sheaths and the (apparent) absence of spermatophores renders its position in the genus *Tubifex* doubtful.

GENUS **Limnodrilus** CLAPARÈDE, 1862

Type species: Limnodrilus hoffmeisteri CLAPARÈDE

Male efferent ducts with long vasa deferentia, small bean-shaped atria bearing large prostate glands, ejaculatory ducts long, penes more or less elongate with thick cylindrical sheaths. Genital setae, hair setae and pectinate setae absent. Coelomocytes absent. Spermathecae with spermatophores.

Cosmopolitan.

The genus is one of the most clearly defined, but the species are not always easy to separate owing to the simplicity of the setae. There are often distinct forms within any one locality which may be held to represent valid species, but comparison with material from other sites soon blurs the picture. The following account represents the current state of knowledge, while at the same time it is clear that intermediate forms have been described. The distinction between *L. claparedeianus* and *L. cervix* is marked in Britain, where the latter has been introduced, but is much less obvious in N. American specimens, particularly those from heavily polluted localities where many thousands of *Limnodrilus* specimens can be found, often to the exclusion of most other benthic animals.

The spiral muscles around the penis sheath are an unusual characteristic (Fig. 5E).

Limnodrilus

1.	Penis sheaths short, broad, spatulate	*L. silvani*
—	Penis sheaths more or less cylindrical	2
2.	Penis sheaths more than 14 times longer than broad when fully developed	3
—	Penis sheaths up to 14 times longer than broad when fully developed	5

3. Penis sheaths with walls two layers thick from base to near hood, outer layer disappears distally to leave a narrow neck below the hood, which typically bears two triangular processes ... *L. cervix*

— Penis sheaths with wall thin, if thick then not two layers thick, hood roughly pear-shaped ... 4

4. Hood small, pear-shaped, shaft enters hood centrally ... *L. claparedeianus*

— Hood large, irregularly pear-shaped, shaft enters hood eccentrically ... *L. maumeensis*

5. Upper tooth of anterior setae much longer than lower ... 6

— Upper tooth of anterior setae at most a little longer than lower ... 7

6. Upper tooth of anterior setae thicker than lower, penis sheaths with narrow flange-like hood ... *L. udekemianus*

— Upper tooth of anterior setae about as thick as the lower, penis sheaths with swollen distal ends with lateral openings ... *L. neotropicus*

7. Penis sheaths with thin-walled broad basal section, narrowing abruptly into shaft, hood reflected over shaft ... *L. angustipenis*

— Penis sheaths narrowing gradually from base to apex, or cylindrical ... 8

8. Penis sheaths up to 7 times longer than broad, more or less straight, with hood reflected back over shaft unless forced forward ... *L. profundicola*

—· Penis sheaths up to fourteen times longer than broad, hood characteristically with opening at right-angles to shaft, or hood a plate with scalloped edges ... *L. hoffmeisteri*

Limnodrilus hoffmeisteri Claparède, 1862
Fig. 8. 3M, O; 8. 4C, H, I; 8. 5E

Limnodrilus hoffmeisteri CLAPARÈDE, 1862: 226, Pl. I, figs. 1-3, Pl. III, fig. 12, Pl. IV, fig. 6.

Camptodrilus californicus EISEN, 1879: 24, Fig. 6.

Camptodrilus spiralis EISEN, 1879: 22, Fig. 5.

Camptodrilus corallinus EISEN, 1879: 23.

Clitellio hoffmeisteri (Claparède). CZERNIAVSKY, 1880: 325.

Limnodrilus corallinus (Eisen). VEJDOVSKY, 1884 b: 45. BEDDARD, 1895: 254. MICHAELSEN, 1900: 46.

Limnodrilus californicus (Eisen). VEJDOVSKY, 1884 b: 45.

Limnodrilus spiralis (Eisen). VEJDOVSKY, 1884 b: 45.

Limnodrilus claparedeanus Ratzel VEJDOVSKY, 1884 b: 48, Pl. VIII, figs. 22, 23. GALLOWAY, 1911: 315. ALTMAN, 1936: 16. CAUSEY, 1953: 423. TETER, 1960: 193.

Limnodrilus hoffmeisteri Claparède. LEVINSEN, 1884: 225. VEJDOVSKY, 1884 b: 47, Pl. VIII, figs. 13-17, Pl. XI, fig. 4. STOLC, 1888: 41. BEDDARD, 1895: 252. BRETSCHER, 1896: 506; 1899: 371; 1902, 4. RYBKA, 1898: 390. PIGUET, 1899: 73; 1906 a: 390; 1913: 137, Fig. 10. MICHAELSEN, 1900: 43. LINDER, 1904: 250. POINTER, 1911: 634. PIGUET and BRETSCHER, 1913: 81, Fig. 19c. FRIEND, 1912 b: 271. LASTOCKIN, 1927 a: 67. SVETLOV, 1926 a: 252; 1926 b: 345; 1946: 104. HRABE, 1929: 2; 1941: 4; 1958: 338; 1962 a: 310. UDE, 1929: 82, Figs. 30, 97c. CERNOSVITOV, 1928: 6; 1930: 321; 1938 a: 548; 1939 b: 103, Figs. 76-85; 1945: 530. MALEVICH, 1937 b: 133. CHEN, 1940: 109. MARCUS, E., 1947: 11, Fig. 8. BERG, 1948: 49. JAROSCHENKO, 1948: 69. GAVRILOV and PAZ, 1950: 533. CEKANOVSKAYA, 1952: 296; 1956: 659. BÜLOW, 1957: 99, Pl. XXX, fig. 12. JUGET, 1957: 3; 1958: 90, Fig. 14a, c. BRINKHURST, 1960: 401, Fig. 4a, b, f-i, k; 1962 c: 321; 1963 a: 36, Fig. 21a, b; 1963 c: 41, Fig. 12b; 1965: 127, Fig. 4a. SOKOLSKAYA, 1961 b: 86. BRINKHURST and KENNEDY, 1962 a: 184; 1962 b: 375. MOSZYNSKA, 1962: 27. KENNEDY, 1964: 229. NOMURA, 1965: 477. BRINKHURST and COOK, 1966: 16. NAIDU, 1965: 477, Fig. 6; 1966: 220. POPESCU-MARINESCU et al., 1966: 63.

Camptodrilus corallinus Eisen. EISEN, 1886: 900, Pl. XVI, fig. 14a-h., Pl. XVIII, fig. 14i-k, Pl. XVIII, fig. 14.

Camptodrilus californicus Eisen. EISEN, 1886: 901, Pl. XVIII, fig. 16.

Camptodrilus spiralis Eisen. EISEN, 1886: 899, Pl. XVII, fig. 15.

Clitellio (Limnodrilus) hoffmeisteri (Claparède). VAILLANT, 1890: 424.

Clitellio (Limnodrilus) corallinus (Eisen). VAILLANT, 1890: 431.

Clitellio (Limnodrilus) spiralis (Eisen). VAILLANT, 1890: 429.

Clitellio (Limnodrilus) californicus (Eisen). VAILLANT, 1890: 432.

Limnodrilus dugesi RYBKA, 1898: 389, Pl. V, figs. 1-17.

Limnodrilus gotoi HATAI, 1899: 5, Fig. 3 (in part).

Limnodrilus dugesi Rybka. MICHAELSEN, 1900: 45. POPESCU-MARINESCU, 1966: 163.

Limnodrilus lucasi BENHAM, 1903: 216, Pl. XXV, figs. 18-22.

Limnodrilus vejdovskyanus BENHAM, 1903: 213, Pl. XXV, figs. 10-17.

Tubifex hoffmeisteri (Claparède). DITLEVSEN, 1904: 422.

Limnodrilus subsalus MOORE, 1905 b: 392, Pl. XXXIII, figs. 19-22.

Limnodrilus aurostriatus SOUTHERN, 1909: 136, Pl. II, fig. 3a-g.

Limnodrilus parvus SOUTHERN, 1909: 137, Pl. VIII, fig. 5a-c.

Limnodrilus hoffmeisteri forma *parvus* Southern. SOUTHERN, 1909: 137.

Limnodrilus aurantiacus FRIEND, 1911: 414.

Limnodrilus aurostriatus Southern. FRIEND, 1912 b: 274.

Limnodrilus parvus Southern. FRIEND, 1912 b: 274. CERNOSVITOV, 1939 b: 103; 1945: 528, Figs. 13-19.

Limnodrilus socialis STEPHENSON, 1912 a: 294, Figs. 9-16.

Limnodrilus socialis Stephenson. STEPHENSON, 1912 b: 237; 1913 a: 740; 1913 b: 260; 1917 b: 93, Pl. IV, figs. 6-7; 1923: 96, Fig. 36; 1925 a: 48; 1926: 250. MICHAELSEN and BOLDT, 1932: 598. YAMAGUCHI, 1953: 296. INOUE and KONDO, 1962: 97.

Limnodrilus gotoi Hatai. NOMURA, 1913: 3, Fig. 1; 1929: 131.

Limnodrilus parvus var. *biannulatus* LASTOCKIN, 1927 a: 67.

Limnodrilus pacificus CHEN, 1940: 118, Fig. 33a-e.

Limnodrilus hoffmeisteri forma *divergens* MARCUS, 1942: 169, Pl. II, figs. 6-9, 11-15.

Limnodrilus hoffmeisteri forma *parva* Southern. MARCUS, 1942: 167, Figs. 4-5.

Limnodrilus auranticus Friend. MALEVITCH, 1956: 423.

Limnodrilus subsalus Moore. MARCUS, 1944: 73, Fig. 66.

Limnodrilus hoffmeisteri forma *socialis* Stephenson. GAVRILOV and PAZ, 1950: 563.

Limnodrilus hoffmeisteri forma *parvus* Southern. CEKANOVSKAYA, 1962: 254.

Limnodrilus hoffmeisteri forma *typica* CEKANOVSKAYA, 1962: 253, Figs. 153, 154.

CORRIGENDA

The publishers and authors regret the necessity of inserting this list of corrections of printing errors which escaped their close scrutiny at final page proof stage, especially in view of the detailed taxonomic nature of the book. It is hoped that no other misprints remain uncorrected but readers are invited to send details to the publisher. In this list 'p. 43/11' means, for example, 'page 43, line 11'.

TEXT

Preliminary pages: *for* 'D. V. Anderson' *and* 'D. V. A.' *read* 'D. T. Anderson' *and* 'D. T. A.'.
p. 43/11: *for* 'Fig. 15A-C' *read* 'Fig. 1.5A-C'.
p. 44/32: *for* 'Fig. 15.6G' *read* 'Fig. 1.6G'.
p. 46/1: *for* 'Fig. 15' *read* 'Fig. 15.9'.
p. 46/3: *for* 'Fig. 15.9' *read* 'Fig. 15.10'.
p. 56/15: *for* 'p. 000' *read* 'p. 771'.
p. 146/26: *for* 'p. 000' *read* 'p. 714'.
p. 147/18: *for* 'TRIBE GLOSSOSCOLECINI Neotropical: as for the subfamily but' *read* 'TRIBE MICROCHAETINI Ethiopian Region: S. Africa, *Microchaetus*'.
p. 155/28: *for* 'p. 000' *read* 'p. 815'.
p. 181/17: *for* 'acanthodrile' *read* 'megascolecid'.
p. 181/24: *for* 'an acanthodrilid' *read* 'a megascolecid'.
p. 183/8: *for* 'organs' *read* 'origin'.
p. 716/14: *for* 'Fig. 14.1E' *read* 'Fig. 14.1E, F'.
p. 730/bottom: *for* 'Fig. 1.3' *read* 'Fig. 1.7'.
p. 739/5: *for* 'east' *read* 'last'.
p. 755/8: *for* 'Fig. 13.3A-C' *read* 'Fig. 15.3A-C'.
p. 758/12: *for* 'Fig. 15.1H' *read* 'Fig. 15.2G, H'.

ILLUSTRATIONS

Fig. 1.6 Legend: *add* '1.m, longitudinal muscle'.
Fig. 1.8 Legend: *add* '1n, lateroneural; v.lp, ventral lateroparietal'.
Fig. 1.9 Labels: *in A for* 'pe' *read* 'oe'; *for* 'sup 5/6' *read* 'sep 5/6'.
          in B for 'n.ts' *read* 'n.t3'; *for* 's.q.' *read* 's.g.'
Legend: *for* 'h 1, 2' *read* 'h 1, 5'.
Fig. 14.1 Labels: *in A for* 'VII' *read* 'VIII'; *scale of* D = 100μ.
Fig. 15.1 Legend: *for* 'C—cerebral ganglia' *read* 'c—cerebral ganglia';
          for 's.v.2' *read* 's.v.4'.
Fig. 15.2 Labels: *in C and F for* 'b.u.' *read* 'b.v';
          in H for 'd' *read* 'd.v'; *for* 's.u.4' *read* 's.v.4';
          and for 'sp' *read* 'sn'.
    Legend: *add* 'c.g, chromophil glands; sin, sinus; sn, subneural vessel;
          su, suboesophageal ganglion'.
Fig. 15.3 Labels: *in H for* 'i.g.m.' *read* 'l.g.m.'.
Fig. 15.6 Labels: *in C for* 'c.a.' *read* 'cla'.
    Legend: *for* 'sin.ve' *read* 'sin.oe'.
Fig. 15.8: Legend *is correct. Illustration above* F *is* G; G-J *are* H-K *respectively*, K *and* L *are both* L.
Fig. 15.10 Labels: *in F for top* 'Sem.f 2' *read* 'sem.f 1; *for* 's.u1' *and* 's.u.2' *read* 's.v.1' *and* 's.v.2'; *for* 'Sp.4' *read* 'sp.5'.
          in G for 'd.u' *read* 'd.v'.
    Legend: *last (6th) heart is in* 13; *for* 'sp.2' *read* 'sp.5'; *for* 'end.sin.ve' *read* 'end.sin.oe'.
    Add 'mus, intestinal gizzards'.
Fig. 15.11 Legend: *in line* 7 *for* 'o, male pore' *read* '$\male$, male pore'.
Fig. 15.12 Labels: *in C for* 'h.c' *read* 'n.c'.
    Legend: *for* 'co,' *read* 'co.g,'.
Fig. 15.13 Legend: *add* 'tub, tuberculum pubertatis'.

R. O. Brinkhurst & B. G. M. Jamieson, *Aquatic Oligochaeta of the World*

$l=20$-35 mm, $s=55$-95. Anterior bundles with 3-10 (mostly 7) setae, posteriorly fewer, all bifid with the upper tooth shorter or longer than the lower, usually thinner than the lower. Vasa deferentia long, atria small, prostates large, ejaculatory ducts lead to penes. Penis sheaths 1-14 times longer than broad, straight with plate-like extremity, sometimes with scalloped edges, or somewhat bent with a hood set at an angle to the shaft, the lumen of the sheath turning through 90°, the latter often slightly shorter than the former, both types narrowing from base to head.
Cosmopolitan.

Specimens with the "long, straight, plate-topped" penis sheaths and those with "typical" penis sheaths (more sinuous, shorter, lumen turning through 90° in complex distal hood) may frequently be recognized, the former apparently quite common in the western states of the United States, but this distinction should not be made the basis for the erection of a new species unless it can be shown to differ from the "typical" form in some other way.

This species is the commonest, most widely distributed tubificid.

Limnodrilus udekemianus Claparède, 1862
Fig. 8. 4A, B

Limnodrilus udekemianus CLAPARÈDE, 1862: 243, Pl. I, figs. 4, 5, Pl. III, figs. 13, 13a, Pl. VII, fig. 1.

Limnodrilus ornatus EISEN, 1879: 17.

Limnodrilus steigerwaldi EISEN, 1879: 18.

Clitellio udekemianus (Claparède). CZERNIAVSKY, 1880: 325.

Limnodrilus udekemianus Claparède. VEJDOVSKY, 1884 b: 47, Pl. VIII, figs. 18-21, Pl. IX, fig. 20, Pl. X, figs. 19-20, Pl. XI, figs. 1-3. LEVINSEN, 1884: 225. STOLC, 1885: 143; 1888: 41. BEDDARD, 1895: 252. FRIEND, 1896: 127; 1897 b: 207; 1897 c: 297; 1912 b: 292; 1913: 171. BRETSCHER, 1896: 505; 1899: 371; 1900: 5; 1902: 5; 1903: 2. RYBKA, 1898: 390. MICHAELSEN, 1900: 45. PIGUET, 1906 a: 390, Fig. 10a. SOUTHERN, 1909: 135. BAUMANN, 1910: 675. PIGUET and BRETSCHER, 1913: 78, Fig. 19a. LASTOCKIN, 1927 a: 67. CERNOSVITOV, 1928: 6, 1942 a: 214; 1942 c: 285. HRABE, 1929: 2; 1962 a: 310; UDE, 1929: 82, Fig. 97a. MARCUS, 1942: 175, Figs. 10, 16. BERG, 1948: 49. JAROSCHENKO, 1948: 69. BÜLOW, 1955: 261, Pl. XLIII, figs. 4-6; 1957: 98, Pl. XXX, figs. 1-3. JUGET, 1958: 90, Fig. 14a. BRINKHURST, 1960: 402, Fig. 4c, d, j; 1962 c: 321; 1963 a: 38, Fig. 23a, b; 1963 c: 41, Fig. 12d; 1965: 128, Fig. 4b-g. BRINKHURST and KENNEDY, 1962 b: 375. SOKOLSKAYA, 1961 a: 55; 1961 b: 86. CEKANOVSKAYA, 1962: 252. MOSZYNSKA, 1962: 27. NAIDU, 1966: 220. POPESCU-MARINESCU et al., 1966: 163.

Limnodrilus ornatus Eisen. (?) VEJDOVSKY, 1884 b: 45. EISEN, 1886: 894,

Pl. IX, fig. 8, Pl. X, fig. 9. BEDDARD, 1895: 253. MICHAELSEN, 1900: 43. GALLOWAY, 1911: 315.

Limnodrilus steigerwaldii Eisen. (?) VEJDOVSKY, 1884 b: 45. EISEN, 1886: 895.

Limnodrilus steigerwaldi Eisen. BEDDARD, 1895: 253. MICHAELSEN, 1900: 46.

Clitellio (Limnodrilus) udekemianus (Claparède). VAILLANT, 1890: 425.

Clitellio (Limnodrilus) steigerwaldi (Eisen). VAILLANT, 1890: 427.

Clitellio (Limnodrilus) ornatus (Eisen). VAILLANT, 1890: 426.

Limnodrilus wordsworthianus FRIEND, 1898 a: 120.

Limnodrilus gotoi HATAI, 1899: 5, Fig. 3 (in part).

Limnodrilus gotoi Hatai. MICHAELSEN, 1900: 44.

Tubifex udekemianus (Claparède). DITLEVSEN, 1904: 422.

Limnodrilus udekemianus var. *wordsworthianus* Friend. FRIEND, 1912 b: 292.

Limnodrilus willeyi NOMURA, 1913: 34, Figs. 25-34.

Limnodrilus inversus GAVRILOV and PAZ, 1949: 537, Figs. 1-5, Pl. I-III.

Limnodrilus virulentus? det. by "Cernosvitov, B.M. (N.H.), Reg. No. 1949. 3.1.382 (immature specimen)?

Limnodrilus willeyi Nomura. INOUE and KONDO, 1962: 97, Figs. 152, 153.

$l=20$-90, $s=c$. 160. Anterior bundles with 3-8 setae, decreasing to two posteriorly, anteriorly with the upper tooth much longer than the lower, and thicker. Vasa deferentia long, atria bean-shaped, prostate glands large, ejaculatory ducts lead to penes. Penis sheaths short, 1-4 times longer than broad, seldom longer, with a simple plate-like hood.
    Cosmopolitan.

Hrabe (1966) claimed that *Isochaeta virulenta* Pointner was nothing but an immature form of *L. udekemianus*. After examining the type specimen I cannot agree completely with this decision (vide *Isochaeta*).

Limnodrilus claparedeianus Ratzel, 1868.
Fig. 8. 3Q, R

Tubifex rivulorum (non Lamark 1816!). BUDGE, 1850: 1, Pl. I (in part).
Limnodrilus claparedianus RATZEL, 1868: 590, Pl XLII, fig. 24.

Captodrilus igneus EISEN, 1879: 23.

Clitellio claparedianus (Ratzel). CZERNIAVSKY, 1880: 325.

Limnodrilus igneus (Eisen). VEJDOVSKY, 1884b: 45. EISEN, 1886: 900, Fig. 13. BEDDARD, 1895: 255. RYBKA, 1898: 390. MICHAELSEN, 1900: 46.

(non) *Limnodrilus claparedianus* Ratzel. VEJDOVSKY, 1884 b: 48, Pl. XI, fig. 5-8. GALLOWAY, 1911: 315. ALTMAN, 1936: 16. CAUSEY, 1953: 899. TETER, 1960 (=*L. hoffmeisteri*).

Limnodrilus claparedianus Ratzel. STOLC, 1886: 647; 1888: 42, Pl. I, fig. 7, Pl. II, fig. 11, Pl. III, fig. 5, 6. BEDDARD, 1895: 251. RYBKA, 1898: 390. SMITH, 1900: 441. JAROSCHENKO, 1948: 69.

Camptodrilus igneus Eisen. EISEN, 1886: 909, Pl. XV, fig. 13a-f.

Clitellio (Limnodrilus) igneus Eisen. VAILLANT, 1890: 430.

Limnodrilus claparedeanus Ratzel. BRETSCHER, 1896: 504; 1900 a: 445; 1901: 204; 1902: 2; 1903: 5; 1905: 664. PIGUET, 1913: 113, Fig. 10d. PIGUET AND BRETSCHER, 1913: 82. LASTOCKIN, 1927 a: 67. CERNOSVITOV, 1928: 6. UDE, 1929: 82, Fig. 97d. BERG, 1948: 49. HRABE, 1962 a: 310. JUGET, 1958: 90, Fig. A(d). BRINKHURST, 1963 a: 35, Fig. 21c; 1963 c: A1, Fig. 12c; 1965: 129, Fig. 4h, i. BRINKHURST AND KENNEDY, 1962 b: 368. CEKANOVSKAYA, 1962: 255, Fig. 153. MOSZYNSKA, 1962: 27. KENNEDY, 1964: 231. POPESCU-MARINESCU *et al.*, 1966: 166.

Limnodrilus claparèdeianus Ratzel. MICHAELSEN, 1900: 45; 1909: 41, Fig. 81. HRABE, 1929: 2.

Limnodrilus longus BRETSCHER, 1901: 204, Pl. XIV, fig. 2, 3.

Tubifex claparedianus (Ratzel). DITLEVSEN, 1904: 422.

? *Limnodrilus gracilis* MOORE, 1905 a: 169, Fig. V.

Limnodrilus subsalsus MOORE, 1905 b: 392, Pl. XXXIII, fig. 19-22 (in part).

Limnodrilus longus Bretscher. PIGUET, 1906 a: 392; 1913: 138. SOUTHERN, 1909: 136. FRIEND, 1912 b: 274. BRINKHURST, 1962 c: 321. BRINKHURST AND KENNEDY, 1962 a: 183. 1962 b: 368.

Limnodrilus motomurai NOMURA, 1929: 131, Fig. 5, Pl. 1, fig. 4.

Limnodrilus hoffmeisteri Claparède. BRINKHURST, 1960: 401 (in part).

(non) *Limnodrilus claparèdeanus* Ratzel. BRINKHURST, 1960: 402, Fig. 4e, k, i; 1962 c: 321. BRINKHURST AND KENNEDY, 1962 a: 188.

l=30-60 mm, *s*=50-120. Anterior bundles with 4-9 setae per bundle, a few with upper tooth much longer than lower but not noticeably thicker, mostly upper tooth somewhat longer and thinner than the lower, fewer setae posteriorly. Vasa deferentia long, atria small, prostates large, ejaculatory ducts present, leading to very long penes. Penis sheaths up to 43 times longer than broad. Head of penis sheaths with narrow pear-shaped hoods set at an angle to the shaft (viewed laterally), walls one layer thick, or with a thin, cellular (?) layer externally. Cosmopolitan.

A few specimens may be found that seem almost exactly intermediate between *L. claparedeianus* and *L. cervix*, but typical specimens of both may be separated with ease.

The spelling of this name has been thoroughly confused. Having been named for Claparède the simple addition of the suffix "ianus" is presumeably the correct spelling, as used by Michaelsen (1900).

Limnodrilus profundicola (Verrill, 1871)
Fig. 8. 4D, E

Tubifex profundicola VERRILL, 1871: 451.

Tubifex profundicola Verrill. SMITH, 1874: 699.

Limnodrilus alpestris EISEN, 1879: 10.

Limnodrilus monticola EISEN, 1879: 18.

Limnodrilus alpestris Eisen. VEJDOVSKY, 1884 b: 45. EISEN, 1886: 896, Pl. XII, fig. 11a-h; Pl. XVII, fig. 11i-k; Pl. XIX, fig. 18. MICHAELSEN, 1900: 44; 1914: 16, Pl. V, fig. 5. BEDDARD, 1895: 254. RYBKA, 1898: 390. GALLOWAY, 1911: 315.

Limnodrilus monticola Eisen. VEJDOVSKY, 1884 b: 45. EISEN, 1886: 896, Pl. XI, fig. 10a-h. BEDDARD, 1895: 254. MICHAELSEN, 1900: 46. GALLOWAY, 1911: 315.

Clitellio (Limnodrilus) alpestris (Eisen). VAILLANT, 1890: 428.

Clitellio (Limnodrilus) monticola (Eisen). VAILLANT, 1890: 427.

Limnodrilus helveticus PIGUET, 1913: 134, Fig. 8-10.

Limnodrilus helveticus Piguet. PIGUET AND BRETSHER, 1913: 79, Fig. 19b. HRABE AND CERNOSVITOV, 1929: 212. HRABE, 1954: 306. JUGET, 1957: 3; 1958: 90, Fig. 14a, b. MALEVITCH, 1957: 82. SOKOLSKAYA, 1958: 310; 1961 a: 56; 1961 b: 85. BRINKHURST AND KENNEDY, 1962: 185. MOSZYNSKA, 1962: 27. BRINKHURST, 1963 a: 38, Fig. 24a-c; 1963 c: 41, Fig. 12g.

Limnodrilus profundicola (Verrill). BRINKHURST, 1965: 130, Fig. 4k-m.

$l=20$-45 mm, $s=50$-90. Anterior bundles with 5-9 setae, all bifid crotchets with the upper tooth a little longer and thinner than lower unless worn. Vasa deferentia long, atria small, ejaculatory ducts lead to penes. Penis sheaths 2-7 times longer than broad, short and more or less straight with hood reflected back over shaft (unless forced forward).
Cosmoplitan.

Limnodrilus neotropicus Cernosvitov, 1939.
Fig. 8. 4J, K; 8. 5A-D

Limnodrilus neotropicus CERNOSVITOV, 1939 b: 106, Fig. 86-101.

Limnodrilus neotropicus Cernosvitov. BRINKHURST, 1963 a: 38, Fig. 22.

l=50 mm, *s*=45-170. Dorsal anterior bundles with 3-4 setae, 5-6 in ventral bundles, fewer posteriorly, anterior setae with the upper tooth much longer than lower, teeth about equally thick, posteriorly upper tooth as long as or just longer than the lower. Vasa deferentia long, atria bean-shaped, ejaculatory ducts lead to penes. Penis sheaths 3-4 times longer than broad below the distal dilatation, narrowing from base to union of shaft with distal swollen part, openings mostly lateral.
S. America (L. Titicaca).

The distal ends of the penis sheaths are more regular in shape than in *L. hoffmeisteri* ("typical" form), and only in these two does the hood form a chamber on the end of the shaft. According to Cernosvitov (1939) there are no spiral muscles around the penis sacs, but in the illustrated section of the penis there are some thick, darkly stained structures outside the wall of the penis sac, which were not mentioned in the text.

Limnodrilus cervix Brinkhurst, 1963
Fig. 8. 3S, T

Limnodrilus claparedeanus Ratzel. BRINKHURST, 1960: 402, Fig. 4e, k, l; 1962 c: 321. BRINKHURST AND KENNEDY, 1962 a: 188.

Limnodrilus cervix BRINKHURST, 1963 a: 36, Fig. 2a-b; 1963 c: 41, Fig. 12e; 1965: 129, Fig. 4j. BRINKHURST AND COOK, 1966: 15, Fig. 7a-b.

l=30-80 mm, *s*=55-160. Anterior bundles with 4-9 or 10 setae, posteriorly fewer, all bifid crotchets with upper tooth a little longer and thinner than the lower. Vasa deferentia long, atria small, prostate glands large, ejaculatory ducts lead to long penes. Penis sheaths up to 48 times longer than broad, typically with wall of two thick layers basally, narrowing abruptly near the head, head with two solid triangular lobes.
N. America, introduced to England.

Limnodrilus silvani Eisen, 1879
Fig. 8. 4F, G; 8. 5F

Limnodrilus silvani EISEN, 1879: 19, Fig. 4.

Limnodrilus silvani Eisen. VEJDOVSKY, 1884 b: 45. EISEN, 1886: 897, Pl. XIII, fig. 12, Pl. XIV, fig. 12. BEDDARD, 1895: 254. RYBKA, 1898: 390. MICHAELSEN, 1900: 44. GALLOWAY, 1911: 315. BRINKHURST, 1965: 131, Fig. 4n-o.

Clitellio (*Limnodrilus*) *silvani* (Eisen). VAILLANT, 1890: 428.

Limnodrilus sp. STEPHENSON, 1929: 227.

Limnodrilus grandisetosus NOMURA, 1932: 511, Fig. 1-5, Pl. XIII-XVII.

Limnodrilus grandisetosus Nomura. SOKOLSKAYA, 1958: 310. CEKANOV-SKAYA, 1962: 250. BRINKHURST, 1963 a: 39, Fig. 25. NAIDU, 1966: 219.

l=60-180 mm, s=85-95. Setae mostly 3 per bundle, posteriorly only 1, upper tooth as long as, or longer than the lower, and thicker. Giant ventral setae may be present. Vasa deferentia long, atria small, prostate glands large, ejaculatory ducts lead to penes. Penis sheaths short, broad, spade-shaped.

California, U.S.A. and Asia (Japan, Amur basin, Burma and Malaya (new record)).

Limnodrilus maumeensis Brinkhurst and Cook, 1966
Fig. 8. 3P

Limnodrilus maumeensis BRINKHURST AND COOK, 1966: 15, Fig. 7F.

l=20 mm, s=100-135. Setae all bifid crotchets, up to 10 per bundle anteriorly, fewer posteriorly, with the upper tooth a little longer than the lower, but only a little thinner. Vasa deferentia long, atria small, prostate glands large, ejaculatory ducts leading to penes. Penis sheaths up to 48 times longer than broad, lumina wide, wall thicker one side than the other near the head, causing a bend in the shaft, typically heads broad, triangular, set asymmetrically on the shaft.

N. America, Great Lakes.

The lumen of each penis sheath is about $1\frac{1}{2}$ times wider than in either *L. claparedeianus* or *L. cervix*.

Limnodrilus augustipenis Brinkhurst and Cook, 1966
Fig. 8. 3N

Limnodrilus augustipenis BRINKHURST AND COOK, 1966: 15, 7H.

l=10-15 mm, s=25-50. Setae all bifid crotchets with rather short teeth, the lower often broad, the upper longer or shorter than lower. Vasa deferentia long, atria small, ejaculatory ducts lead to penes. Penis sheaths with broad base containing penes, narrow distal section, hood relected over shaft distally, about 10 times longer than broad. Spermathecae with spermatophores about twenty times longer than broad.

N. America, Great Lakes.

GENUS **Isochaeta** POINTNER, 1911

Type species: Isochaeta virulenta POINTNER

Male efferent ducts with elongate, narrow atria scarcely distinguishable from vasa deferentia, prostate glands small, proximal on atria, penes long, cuticular sheaths very thin or cuticle not thickened over penes. Genital setae often modified. Spermathecae with spermatophores. No coelomocytes.

Eurasia, N. and S. America.

The name *Isochaetides* was proposed by Hrabe (1966) for an assemblage which included the species included here by Brinkhurst (1963 a) (i.e. *baica-lensis, hamatus, arenarius, michaelseni, lacustris*) together with *suspectus*, which was described subsequent to that review and is accepted here. Also included were *Tubifex newaensis* and *T. pseudogaster*, which were excluded from *Tubifex* by virtue of the absence of hair and pectinate setae in dorsal bundles. *I. dojranensis* was retained in *Tubifex* in the account by Hrabe, again largely because of the nature of the dorsal setae despite the form of the atria and the penial seta found in one specimen. These decisions are rejected here, according to the principals to be used in the separation of genera in this family. *Peloscolex nomurai* was placed in *Isochaeta* by Brinkhurst (1963 a) but may be better placed in *Peloscolex* as suggested by Hrabe (1966), though there is some evidence that the body wall of some *Isochaeta* species may secrete material to which foreign matter may adhere.

The name *Isochaeta* was rejected by Hrabe (1966) following an examination of the type material, traced by Brinkhurst (1963 a) to the University of Graz. As the penes of *I. virulenta* are moderately long but lack penis sheaths, it is obvious that sectioned material would appear very similar to incompletely developed specimens of *Limnodrilus udekemianus*. The setae of *virulenta* do not resemble those of *L. udekemianus* as the upper tooth is thinner than the lower, not thicker, and there are fewer setae in anterior dorsal bundles than in anterior ventral bundles. I do not accept the synonymy claimed.

The type species was recently identified in Roumania by Popescu-Marinescu *et al.* (1966) in material from the R. Danube, into which drains the river system on which Graz, the type locality, stands, but these specimens are no longer available for study.

Species in which the setae are all bifid crotchets are difficult to distinguish, particularly as the relative lengths of the two teeth are subject to variation, some of which is caused by wear. Hence many of the distinguishing characteristics are rather inadequate, but until a large number of specimens are examined from Lake Baikal, Lake Ochrid, and Lake Titicaca, it might be unwise to regard the various forms as comprising a single species. The genus is closely related to *Potamothrix* via *I. suspectus* and *I. nevadana*. The form of atrium and the spermathecal setae is also reminiscent of those in the genus *Psammoryctides*.

Isochaeta

1.	Hair setae and pectinate setae present	2
—	Bifid setae in all bundles	4
2.	Pectinate setae absent.	*I. dojranensis*

—	Pectinate setae present	3
3.	Ventral setae bifid, vas deferens short	*I. nevadana*
—	Ventral setae with a single intermediate tooth, vas deferens long	*I. israelis*
4.	Posterior dorsal setae strongly recurved with rudimentary upper teeth	*I. hamata*
—	Posterior dorsal setae not recurved, upper tooth at most shorter than the lower	5
5.	Ventral setae of X, XI, XIII bifid with upper tooth much longer than lower, more anterior setae with upper tooth shorter than lower, no ventral setae in XII	*I. michaelseni*
—	Ventral setae of X, XII, and XIII unmodified, those of XI may be long, straight and hollow tipped, or normal, or absent	6
6.	Spermathecae setae long, thin hollow tipped	7
—	No modified genital setae	*I. virulenta*
7.	Cuticular penis sheaths distinct	*I. suspectus*
—	Cuticular penis-sheaths rudimentary	*I. baicalensis, I. arenarius, I. lacustris*

The distinctions between all or most of these last three species may be thought to be trivial, and several may represent local forms of a single species. A long series of specimens of one or two of these must be examined in order to establish the variability within any one locality, but most of the type localities are inaccessible.

Isochaeta virulenta POINTNER, 1911
Fig. 8. 7A

Isochaeta virulenta POINTNER, 1911: 637. Pl. XXVIII, fig. 6-16, Pl. XXIX, fig. 17-30.

Isochaeta virulenta Pointner. BRINKHURST, 1963 a: 31.

Limnodrilus virulentus (Pointner). POPESCU-MARINESCU, 1966: 163.

? dimensions. Dorsal anterior bundles with 3-4 setae, ventral bundles with 4-8, all with upper tooth longer and thinner than lower, median segments with 2-3 setae, posterior bundles with 0-2, all with teeth equally long. Vasa deferentia long, atria scarcely distinguishable from the vasa deferentia; penes long, slender, without penis sheaths. Spermathecae spherical with short, wide glandular ducts.
    R. Danube system, Europe.

Isochaeta baicalensis (Michaelsen, 1901)
Fig. 8. 7D, F, I, P-S

Limnodrilus baicalensis MICHAELSEN, 1901: 140, Pl. II, fig. 11, 12.

Lymnodrilus baicalensis Michaelsen, 1905 a: 22.

Limnodrilus baicalensis Michaelsen. CEKANOVSKAYA, 1962: 248, Fig. 149.
HRABE, 1966: 75.

Tubifex lacustris CERNOSVITOV, 1939 b: 101, Fig. 68-75.

Tubifex lacustris Cernosvitov. HRABE, 1962 a: 307.

Isochaeta baicalensis (Michaelsen). BRINKHURST, 1963 a: 32, Fig. 2, 16.

Isochaetides lacustris (Cernosvitov). HRABE, 1966: 75.

$l=25$-38 mm, $s=90$-190. Anterior bundles up to 8 setae, most bundles with 4-6, upper tooth as long as the lower or up to twice as long; posterior setae with teeth equally long. Spermathecal setae in ventral bundles of X straight and hollow-tipped. Vasa deferentia long, atria about seven times longer than broad, twice as wide as vasa deferentia, heart-shaped or lobed prostate glands have very short stalks. Penes egg-shaped, rounded distally, and as thick as atria, no cuticular sheaths. Spermathecae with more or less egg-shaped ampullae, short ducts, spermatophores present. Posterior setae with teeth equally long.
Lake Baikal, L. Titicaca.

The species described by Cernosvitov (1939) from Lake Titicaca under the name *Tubifex lacustris* does not appear to be separable from *I. baicalensis*. Having regarded the species as a *Tubifex*, Cernosvitov was not obliged to distinguish it from any other *Tubifex* species in detail owing to the very distinct characteristics displayed by the species, which even rendered its position in the genus anomalous. The setae of *lacustris* are rather thin, and the prostate a little large, but the description of *baicalensis* is so lacking in detail that the matter can only be clarified by study of new material from Lake Baikal.

Isochaeta arenaria (Michaelsen, 1926)
Fig. 8. 7H

Limnodrilus arenarius MICHAELSEN, 1926 a: 155, Pl. II. fig. 1.

Limnodrilus arenarius Michaelsen. MICHAELSEN AND VERESCHAGIN, 1930: 219. CEKANOVSKAYA, 1962: 247, Fig. 148.

Limnodrilus arenaria forma *inaequalis* MICHAELSEN AND VERESCHAGIN, 1930: 213.

Isochaeta arenaria (Michaelsen). BRINKHURST, 1963 a: 32.

Isochaetides arenarius (Michaelsen). HRABE, 1966: 75.

$l=35$ mm, $s=100$. Anterior bundles with mostly 5 setae per bundle, upper tooth

very little longer and thinner than lower, or distinctly longer and thinner, with lower tooth recurved and thick, posteriorly only 2 setae. Spermathecal setae single, ? modified in shape. Vasa deferentia long, entering thickened proximal ends of atria, atria tubular, shorter than vasa deferentia, with prostates on proximal ends, penes without true penis sheaths, merely covered with cuticle.
Lake Baikal.

The relationship between this and the other *Isochaeta* species from Lake Baikal (*I. baicalensis*) was not discussed by Michaelsen (1926 a) although *baicalensis* had also been described as a *Limnodrilus*. The setal form of *I. arenaria* forma *inaequalis* of Michaelsen and Vereschagin (1930) seems to be that of *I. baicalensis*, but neither are adequately illustrated.

Isochaeta hamata (Moore, 1905)
Fig. 8. 7B, J, K, O

Tubifex hamatus MOORE, 1905 b: 389, Pl. XXXII, fig. 12-18.

Tubifex hamatus Moore. GALLOWAY, 1911: 314. HRABE, 1962 a: 307.

Isochaeta hamata (Moore). BRINKHURST, 1963 a: 32, fig. 17; 1963 b: 713; 1965: 126, Fig. 3a-c.

Isochaetides hamatus (Moore). HRABE, 1966: 75.

l=35-40 mm, *s*=85-110. Anterior bundles with 1-4 setae with the upper tooth thinner than the lower, teeth diverging at a right-angle, the upper tooth progressively shortened in posterior setae, posterior dorsal setae characteristically bent over with upper tooth markedly reduced. Long, straight, hollow-tipped spermathecal setae single in ventral bundles of X. Vasa deferentia long, atria tubular, prostate small, penes long, without cuticular penis sheaths.
Massachusetts, U.S.A. In brackish water.

Isochaeta suspectus Sokolskaya, 1964
Fig. 8. 6A-D

Limnodrilus suspectus SOKOLSKAYA, 1964: 1071, Figs. 1-7.

Isochaetides suspectus (Sokolskaya). HRABE, 1966: 75.

l=8-13 mm, *s*=*c*. 60. Dorsal and ventral anterior setae with upper tooth longer than the lower, 3-5 per bundle ventrally, falling to 2 posteriorly. Single straight, hollow-tipped spermathecal setae in ventral bundles of X. Vasa deferentia long, prostate glands large, atria wider than vasa deferentia, narrowing distally virtually to an ejaculatory duct. Penes long with distinct cylindrical penis-sheaths. Spermathecae small, spherical, with short spermatophores.
U.S.S.R. (South Sakhalin). Brackish lakes.

The narrowing of the atrium to form an ejaculatory duct and the thick penis sheaths are reminiscent of *Limnodrilus* species, but the spermathecal setae are like those of other *Isochaeta* species.

Isochaeta dojranensis (Hrabe, 1958)

Isochaeta dojranensis (Hrabe, 1958)
Fig. 8. 6E, F; 8. 7C

Tubifex dojranensis HRABE, 1958: 342, Fig. 1-5.

Tubifex dojranensis Hrabe. HRABE, 1962 a: 307.

Isochaeta dojranensis (Hrabe). BRINKHURST, 1963 a: 33, Fig. 16.

$l=20\text{-}25$ mm, $s=140\text{-}150$. Dorsal bundles with 3-5 setae with the upper tooth much shorter and thinner than the lower, together with 1 long serrate hair seta in some bundles, at any point along the body, always very long, four times the width of the thin body; ventral bundles with 3-6 setae as anterior dorsal setae, posteriorly all bundles with 2, then 1 seta. 1 specimen with a single penial seta in XI, long, straight, hollow distally. Vasa deferentia long, atria scarcely wider than vasa deferentia, prostate glands small, penes short, in long sacs, without cuticular sheaths. Spermathecal ampullae long, ducts short, spermatophores present.

Lake Dojran, Jugoslavia.

Hrabe (1966) maintained that this species is a member of the genus *Tubifex*, largely on the basis of the setal characteristics.

Isochaeta michaelseni (Lastockin, 1937)
Fig. 8. 6G-J, K-N; 8. 7E

Limnodrilus michaelseni LASTOCKIN, 1936: 188 (nomen nudum).

Limnodrilus michaelseni Lastockin. LASTOCKIN, 1937: 233, Fig. 1. HRABE 1941: 28, Fig. 29-41; 1950: 257. JAROSCHENKO, 1948: 60. CEKANOV-SKAYA, 1962: 249, Fig. 150. POPESCU-MARINESCU, 1966: 163.

Isochaeta michaelseni (Lastockin). BRINKHURST, 1963 a: 32, Fig. 2, 18.

Isochaetides michaelseni (Lastockin). HRABE, 1966: 75.

$l=25\text{-}33$ mm, $s=90\text{-}118$. Anterior setae 4-6 per bundle, upper tooth shorter and somewhat thinner than the lower, in median segments upper tooth more or less as long as lower but clearly thinner, posteriorly both teeth shorter than anteriorly, upper tooth thinner and a little shorter than the lower. Setae of ventral bundles in X, XI and XIII single, with upper tooth much longer than lower. Vasa deferentia long, atria about twice as wide as vasa deferentia apically, where small prostate glands attached, atria about seven times as long as average width, penes long without cuticular sheaths.

Europe. Fresh and brackish water.

Isochaeta nevadana Brinkhurst, 1965
Fig. 8. 7L-N

Isochaeta nevadana BRINKHURST, 1965: 127, fig. 3d-g.

Small worms, dimensions unknown. Dorsal anterior bundles with 1-3 hair setae,

2-3 pectinate setae with well separated lateral teeth and a few distinct intermediate spines. Anterior ventral bundles with 3-6 setae, the upper tooth longer
and thinner than the lower, median setae about the same, ventrals with the
upper tooth about as long as the lower. No genital setae. Vasa deferentia short,
atria long and tubular, prostate small, penes with delicate cuticular sheaths.
Spermathecae with short spermatophores.

Lake Tahoe, California/Nevada, U.S.A.

This species could well be placed in *Potamothrix* but for the noticeable
cuticular penis-sheaths (see *Potamothrix* (?) *dniprobugensis*). The relative
length of the vasa deferentia and atria varies within the genus *Tubifex*, and
also in *Peloscolex*, so that objection to its inclusion here on that basis may
not be valid. The setal characteristics are not regarded as valid criteria on
which to base a generic definition.

Isochaeta israelis sp. nov.

Fig. 8. 5G-I

$l=c$. 7 mm, $s=c$. 40. Body wall with some foreign matter adhering to it. Dorsal
anterior bundles with 1 serrate hair seta and 2 pectinate setae with upper tooth
longer than the lower, a few distinct teeth between laterals, posteriorly no hair
setae, pectinates with several very short teeth. Ventral bundles with 2-3 setae
anteriorly, upper tooth longer than the lower, and with a single intermediate
tooth. No genital setae. Vasa deferentia long, atria long and tubular, prostate
glands small at union of vasa deferentia and atria, penis sheaths distinct but
thin.

.A salt spring by Lake Tiberias, Israel.

Holotype: Hebrew University, Jerusalem.

The foreign matter found adhering to the body wall has been noted in
American specimens of other species in the genus. In this respect they
resemble species of *Peloscolex* that are devoid of papillae but actively secrete
a substance which binds foreign material into a cuticular sheath. Other
species of *Isochaeta* have been recorded from brackish water localities.

GENUS **Psammoryctides** HRABE, 1964

Type species: Saenuris barbata GRUBE

Male efferent ducts with small, more or less globular atria with solid prostate
glands, narrow ejaculatory ducts between atria and distal swollen sections preceeding penes; eversible penes bearing thin cuticular sheaths. Spermathecal setae
often modified, long, narrow with hollow distal end. No coelomocytes.

Europe, N. America.

The genus was originally proposed by Vejdovsky (1876) for the reception
of the single species *Saenuris barbata* Grube. It was subsequently regarded
as part of the genus *Tubifex*, then recognized as a sub-genus of *Tubifex*, but
was revived from synonymy by Hrabe (1950). The name had been used as
a generic name in earlier papers by the same author, but the justification
for this re-erection did not appear until the complete survey of the genus in

1950. In a later paper (Hrabe, 1964) the name was amended to *Psammoryctides* owing to the prior use of the earlier form of the name for a genus of Rodentia. The limits of the genus have remained stable from 1950 until 1965 but 3 new species were described recently from N. America. Another species, originally described under the generic name *Tubificoides*, was associated with *Psammoryctides* by Cekanovskaya (1962) but has been shown to be synonymous with *Peloscolex swirenkowi*. The genus now contains 7 species, and one placed here incerta sedis.

The structure of the penis in this genus could bear further elucidation. It would seem that there is a permanent penis within a folded voluminous penis-sac. It has been illustrated in the retracted condition in some species, everted in others.

Psammoryctides

1.	Bifid setae in all bundles	2
—	Hair setae and pectinate setae in at least some anterior dorsal bundles	4
2.	Anterior setae with upper tooth shorter than lower	*P. lastoschkini*
—	Anterior setae with upper tooth as long as or longer than lower	3
3.	Posterior ventral setae with distal end strongly recurved, upper tooth rudimentary, not diverging strongly from lower	*P. curvisetosus*
—	Posterior ventral setae with lower tooth set at a right angle to the distal end of the shaft, upper tooth short, diverging widely from the lower tooth	*P. ochridanus* ("variabilis")
4.	Anterior dorsal setae broadly palmate	*P. barbatus*
—	Anterior dorsal setae pectinate	5
5.	Spermathecal setae present	6
—	Spermathecal setae absent	9
6.	Hair setae in anterior and posterior bundles	7
—	Hair setae restricted to anteriormost bundles, scarce as far as XX, absent posteriorly	8
7.	Anterior pectinate setae with parallel lateral teeth, almost palmate	*P. californianus*

— Anterior pectinate setae with long lower tooth, short upper tooth, few short intermediate teeth .... *P. albicola*

8. Pectinate setae with teeth diverging at an acute angle, intermediate teeth shorter than laterals, very fine, sometimes only rudimentary intermediate teeth .... *P. ochridanus* ("typica")

— Pectinate setae with teeth diverging widely, several broad intermediate teeth as long as lower lateral tooth .... *P. moravicus*

9. Thick penial setae present, of normal shape. *P. minutus* (sp i. sedis)

— No penial setae .... *P. deserticola*

Psammoryctides barbatus (Grube, 1861)
Fig. 8. 8A-F

Saenuris barbata GRUBE, 1861: 152, Pl. IV, fig. 10.

Saenuris umbellifer KESSLER, 1868: 107, Fig. 1.

Tubifex umbellifer (Kessler). LANKASTER, 1871: 93, Fig. 1a-c.

Psammoryctes umbellifer (Kessler). VEJDOVSKY, 1875: 194; 1876: 137, Fig. 1-5, 7-12. EISEN, 1879: 13; 1886: 891.

Psammoryctes barbatus (Grube). VEJDOVSKY, 1884 a: 224; 1884 b: 46, Pl. VIII, fig. 9-12, Pl. IX, fig. 1, Pl. X, fig. 17, 18. STOLC, 1886: 644; 1888: 39, Pl. III, fig. 14-17. BENHAM, 1892: 208, Pl. VII, fig. 33. BEDDARD, 1895: 260. BRETSCHER, 1896: 505. FRIEND, 1897 a: 102; 1897 c: 297. MICHAELSEN, 1900: 52. DITLEVSEN, 1904: 415, Fig. 12. MALEVITCH, 1937 b: 133. BERG, 1948: 51. HRABE, 1950: 269; 1954: 308. CEKANOVSKAYA, 1962: 267, Fig. 166; 167. BRINKHURST, 1963 a: 27, Fig. 10a-d, 1963 c: 37, Fig. 8f. KENNEDY, 1964: 234.

?*Clitellio lemani* IMHOF, 1888: 48.

Tubifex barbatus (Grube). MICHAELSEN, 1909: 36, Fig. 71. SOUTHERN, 1909: 139. HRABE, 1929: 1 UDE, 1929: 88. STAMMER, 1932: 578. JAROSCHENKO, 1948: 69 BÜLOW, 1957: 100. JUGET, 1958: 88, Fig. 14g. BRINKHURST, 1960: 405, fig. 7. POPESCU-MARINESCU, et al., 1966: 163.

Tubifex (Tubifex) barbatus (Grube). MICHAELSEN, 1909: 36, Fig. 71. PIGUET AND BRETSCHER, 1913: 65. LASTOCKIN, 1927 a: 67. MOSZYNSKA, 1962: 28.

Psammoryctes barbata (Grube). FRIEND, 1912 b: 291. BRINKHURST, 1962 c: 321. BRINKHURST AND KENNEDY, 1962 a: 184.

Tubifex (Psammoryctes) barbatus (Grube). HRABE, 1934 a : 50. CERNOSVITOV, 1931 a : 36; 1942 a : 214.

Psammorcyctides barbatus (Grube). HRABE, 1964 : 107.

l=30-50 mm, *s*=90. Dorsal anterior bundles with 7-8 broad palmate setae, from X posteriorly 2-3 bifid setae, sometimes a few setae intermediate in form immediately behind X, otherwise with upper tooth shorter and thinner than the lower, also 2-3 serrate or non-serrate hair setae; ventral bundles with 3-5 setae anteriorly, with the upper tooth longer and thinner than the lower, only 2-3 posteriorly, with the upper tooth shorter and thinner than the broad lower tooth. The ventral setae of X long, thin, straight distally with the teeth elongate to form a hollow distal end. Vasa deferentia long, atria small and globular, ejaculatory ducts narrow, then wide proximal to penes, penes with long cuticular sheaths.
Europe.

Psammoryctides albicola (Michaelsen, 1901)
Fig. 8. 8G-I

?Hemitubifex insignis EISEN, 1879 : 13.

Hemitubifex insignis Eisen. LEVINSEN, 1884 : 224. VEJDOVSKY, 1884 b : 45. EISEN, 1886 : 890, Pl. VIII, fig. 6a-f, Pl. VIII, fig. 6g, h. BEDDARD, 1895 : 261.

?Psammoryctes insignis (Eisen). MICHAELSEN, 1900 : 52.

?Tubifex insignis Eisen. MICHAELSEN, 1900 : 525. MOSZYNSKA, 1962 : 29.

Lophochaeta ignota MICHAELSEN, 1900 : 75 (in part).

Lophochaeta albicola MICHAELSEN, 1901 a : 4.

Lophochaeta ignota Michaelsen. MICHAELSEN, 1903 a : 36, (in part).

Lophochaeta albicola Michaelsen. MICHAELSEN, 1901 b : 137; 1903 b : 202, Fig. 5-7.

Psammoryctes illustris DITLEVSEN, 1904 : 416, Pl. XVI, fig. 18-20.

non *Tubifex insignis* (Eisen). MICHAELSEN, 1907 : 22.

Tubifex albicola Michaelsen. MICHAELSEN, 1908 a : 155. JAROSCHENKO, 1948 : 69. JUGET, 1958 : 88, Fig. 14e. BRINKHURST, 1960 : 404, Fig. 8g, h.

Tubifex (Tubifex) albicola Michaelsen. MICHAELSEN, 1909 : 36, Fig. 70. LASTOCKIN, 1927 a : 67.

Psammoryctes albicola Michaelsen. HRABE, 1931 b : 309; 1950 : 269; 1964 : 104. CERNOSVITOV, 1935 b : 186. MALEVITCH, 1937 b : 133. BRINKHURST, 1962 c : 321; 1963 a : 28, Fig. 11a, b; 1963 c : 33, Fig. 8g. CEKANOVSKAYA, 1962 : 266.

Tubifex (Psammoryctes) albicola Michaelsen. CERNOSVITOV, 1938 a: 543, Fig. 11-15; 1945: 526, Fig. 1-12.

l=25-35 mm, s=70-98. Dorsal anterior bundles with 1-3 serrate hair setae, 1-3 pectinate setae with the upper tooth shorter and thinner than the lower, intermediate teeth short and thin, posteriorly hair setae and bifid setae with lower tooth broad, re-curved. Ventral bundles with 1-3 broad setae, upper tooth thinner and shorter than lower especially where worn, lower tooth broad and re-curved, especially in posterior segments, ventral setae of X long, thin straight distally with elongate narrow teeth forming a hollow tip. Vasa deferentia long, atria small, globular, ejaculatory ducts narrow, then broad proximal to penis sacs, penes elongate, penis sheaths short.

Europe, Western Asia.

Specimens of *P. albicola* described by Cernosvitov (1938a) have the lower tooth of the pectinate setae longer than the upper, but it is not so broad and recurved as in more typical specimens such as those illustrated by Brinkhurst (1963 c). Specimens of this type have been observed from Lake Tiberias, Israel (near L. Huleh that was the source of the material described by Cernosvitov) and in Esrom Lake, Denmark (where they had been identified as *Tubifex nerthus*).

A similar specimen was described in a collection from Windermere, England by Cernosvitov (1945). The difference between these forms and the typical specimens is the same as the difference between *P. deserticola* and *P. oligosetosus*, which are therefore regarded as variants of a single species, which is recorded from the Balkans lakes together with *P. albicola*. The difference between *P. albicola*, *P. deserticola* and *P. oligosetosus* seems merely to consist of the absence of spermathecal setae in the latter, a fact which may come to be regarded as insufficient evidence on which to base the separation of these two species.

Psammoryctides deserticola (Grimm, 1877)
Fig. 8. 8J-R

Tubifex deserticola GRIMM, 1877: 108, Pl. V, fig. 8-12

Tubifex deserticola Grimm. VEJDOVSKY, 1884 b: 45. MICHAELSEN, 1900: 56.

Tubifex (Psammoryctes) oligosetosus HRABE, 1931 a: 21, Fig. 3a-c.

Psammoryctes deserticola (Grimm). LASTOCKIN, 1937: 233. HRABE, 1950: 264, Fig. 6-17. CEKANOVSKAYA, 1962: 269, Fig. 169. BRINKHURST, 1963 a: 28, Fig. 12.

Ilyodrilus raduli JAROSCHENKO, 1948: 61, Fig. 1-3.

(?) *Limnodrilus lastoschkini* JAROSCHENKO, 1948: 64, Fig. 7, 8.

Psammoryctes oligosetosus (Hrabe). HRABE, 1950: 269; 1964: 107. BRINKHURST, 1963 a: 30, Fig. 14.

l=25-32 mm, s=150. Anterior dorsal bundles with 1-2 hair setae, 1-3 pectinate setae with upper tooth as long as or shorter than lower, which may be somewhat thicker than the upper, or broad and recurved, usually several short intermediate teeth; ventral bundles with 2-3 setae anteriorly, with the upper tooth shorter and thinner than the recurved lower tooth, both dorsal and ventral posterior bundles with a single bifid seta with lower tooth recurved, upper tooth markedly shorter and thinner. No spermathecal setae. Penes conical with simple, small sheaths. Spermathecal ampullae long, spermatophores present.

Caspian Sea, Dnestr estuary, Black Sea, Lakes Ochrid and Dojran, Jugoslavia.

Having established *P. oligosetosus* prior to the re-discovery of *P. deserticola*, it was initially separable from all other species owing to the absence of spermathecal setae. The distinction between this and *deserticola* is, however, very slight, particularly in view of the variation in form of the pectinate setae in *P. albicola*.

Psammoryctides ochridanus (Hrabe, 1931)
Fig. 8. 8S, T; 8. 9A-G

Tubifex (Psammoryctes) ochridanus HRABE, 1931 a : 16, Fig. 2a-i.

Tubifex (Psammoryctes) ochridanus forma *typica* HRABE, 1931 a : 16, Fig. 2.

Tubifex (Psammoryctes) ochridanus forma *variabilis* HRABE, 1931 a : 16.

Psammoryctes ochridanus (Hrabe). HRABE, 1950 : 269; 1964 : 107. BRINKHURST, 1963 a : 30, Fig. 2, 13.

Psammoryctes ochridanus forma *typica* Hrabe. HRABE, 1950 : 269.

Psammoryctes ochridanus forma *variablilis* Hrabe. HRABE, 1950 : 269.

l=30-40 mm, s=120-160. Anterior dorsal bundles with 2-3 pectinate setae and 1-2 short serrate hair setae, or bifid setae only, bifid and pectinate setae with teeth about equally thick, upper tooth as long as or longer than the lower, intermediate teeth absent, or a few short spines on lower tooth or several short distinct intermediate teeth; post clitellar bundles with bifid setae, with short upper tooth as thick as, recurved lower tooth; ventral anteclitellar bundles 3-4 setae with upper tooth clearly longer than the lower but about as thick, posterior ventral setae single, with short upper tooth and thick, recurved lower tooth. Spermathecal setae long, straight, distally, with long narrow teeth forming a hollow distal end. Vasa deferentia long, atria short, ovoid to pear-shaped, ejaculatory duct first narrow then broad proximal to penes, penis sheaths long. Spermathecae with long sac-like ampullae, spermatophores present.

Lakes Ochrid and Prespa, Jugoslavia.

Two forms were described, the "typical" form with hair setae and pectinate setae and the "variabilis" form lacking these and having bifid setae only. Intermediate forms were found with a few hair setae and traces of intermediate teeth in anterior dorsal bundles. The figure of the inverted penis-sac in the original account is difficult to interpret in relation to the illustration of the similar structure in *P. barbatus* shown in the everted condition by Hrabe (1939 b), the fact that these two penes are much the same being indicated in the key by Hrabe (1950).

Psammoryctides moravicus (Hrabe, 1934)
Fig. 8. 9H-P

Tubifex sp. LASTOCKIN, 1927 b: 18.

Tubifex moravicus HRABE, 1929: 1 (nudum).

Tubifex (Psammoryctes) moravicus Hrabe.HRABE, 1934 a: 48.

Tubifex (Psammoryctes) moravicus moravicus HRABE, 1934: 33, Fig. 1-10.

Tubifex (Psammoryctes) moravicus fontinalis HRABE, 1934 b: 34.

Psammoryctes moravicus Hrabe. HRABE, 1938: 6; 1950: 269; 1954: 308; 1964: 107. CEKANOVSKAYA, 1962: 268, Fig. 168. BRINKHURST, 1963 a: 30.

Psammoryctes moravicus fontinalis Hrabe. HRABE, 1950: 269.

Psammoryctes moravicus moravicus Hrabe. HRABE, 1950: 269.

l=20-25 mm, s=80-95. Dorsal anteclitellar bundles with 1-5 finely serrate hair setae, 2-3 pectinate setae with teeth equally long and thick, diverging widely, with a series of broad, intermediate teeth; posteriorly hair setae scarce, absent from XX on, with 1 seta with a short thin upper tooth and a recurved broad lower tooth. Ventral anterior bundles of II-III with 2-4 setae, the upper tooth 2-4 times longer than the lower, the two teeth enclosing an almost right-angled space, rest with upper tooth a little longer but as thick as lower, posteriorly ? with upper tooth reduced. Spermathecal setae long, straight distally, with two long teeth forming hollow distal end. Vasa deferentia long, atria globular, ejaculatory ducts thin, then thick. Penes small, reduced penis sheaths. Spermathecal ampullae elongate.
Europe.

In the original account, the posterior dorsal setae are said to have the upper tooth longer than the lower, but it is clearly illustrated as shorter than the lower as in most species in this genus. Two forms have been named, "moravicus" from Czechoslovakia with the ventral setae of II-III being thinner than the rest and having very long upper teeth (3-4 x as long as lower), whereas the "fontanalis" form from near Lake Ochrid had shorter upper teeth.

Psammoryctides curvisetosus Brinkhurst and Cook, 1966
Fig. 8. 10K-R

Psammoryctides curvisetosus BRINKHURST and COOK, 1966: 14, Fig. 6f-m.

l=20-25 mm, s=50-90. All setae bifid crotchets, anterior dorsal setae and ventral setae of III-IX 2-5 per bundle, upper tooth as long as or a little longer than lower, ventral setae of II 2-3 per bundle, with upper tooth shorter and thinner than the lower. Postclitellar dorsal bundles and ventral bundles of XI to about XXX with 2-3 setae with the upper tooth shorter, very short in posterior ventral bundles, where the lower tooth is strongly recurred. Spermathecal setae long, straight distally, with elongate teeth forming a hollow distal end. Vasa deferentia long, atria small, somewhat bean-shaped, ejaculatory ducts wider above penis sacs, penis sheaths scarcely thicker than normal cuticle. Spermathecal ampullae small, not clearly delineated from short ducts. Spermatophores short and broad.

Great Lakes, N. America.

Psammoryctides californianus Brinkhurst, 1965
Fig. 8. 10A-E

Psammoryctides californianus BRINKHURST, 1965: 125, Fig. 2 R.V.

$l=c.$ 75 mm, $s=c.$ 40. Anterior dorsal bundles with 2-3 serrate hair setae, 3-4 pectinate setae, each with a somewhat palmate distal end, posteriorly with 2 hair setae and 2 pectinate setae with about three fine intermediate teeth. Anterior ventral bundles with 3-4 setae, the upper tooth longer and thinner than the lower, posteriorly 2-3 per bundle, upper tooth thinner but little longer than the lower. Ventral bundles of X with modified spermathecal setae, one per bundle, straight distally, with long teeth forming a hollow distal end. Atria cylindrical but short, ejaculatory ducts thin, thicker proximal to penis sacs, penis sheaths bluntly conical, very thin. Spermathecal with globular ampullae and distinct ducts.

California, U.S.A.

Only a single specimen was found in a collection from Coyote Creek, Santa Clara County, California. The atria are a little larger than in other species but the characteristic ejaculatory ducts, together with the presence of thin spermathecal setae suggests that the species is correctly assigned to the genus. With the description of *P. curvisetosus* the presence of the genus in N. America was confirmed.

Psammoryctides lastoschkini (Jaroschenko, 1948)

Limnodrilus lastoschkini JAROSCHENKO, 1948: 64, Fig. 7-8.

Limnodrilus lastockini Jaroschenko. CEKANOVSKAYA, 1962: 249.

Psammoryctes lastockini (Jaroschenko). BRINKHURST, 1963 a: 78.

$l=19$-20 mm, $s=60$-72. All setae with the upper tooth shorter and thinner than the lower, 1-2 per bundle in all but anterior ventral bundles, where 2-3. No genital setae. Vasa deferentia short, atria small, ejaculatory duct narrow, then widening, penes short. Spermathecae with elongate, spermatophores thin, spindle-shaped.

Dnieprobug estuary, U.S.S.R.

This species may well be no more than a form of *P. deserticola* in which the hair and pectinate setae are absent.

SPECIES INCERTA SEDIS

Psammoryctides (?) *minutus* Brinkhurst, 1965
Fig. 8. 10F-J

Psammoryctides (?) *minutus* BRINKHURST, 1965: 125, Fig. 2m-q.

l=7-12 mm, s=40-84. Dorsal bundles with 3-5 hair setae, 3-5 pectinate setae, intermediate teeth mostly fine, median bundles with 0-2 hair setae, 2-3 pectinates, posteriorly 2 bifid crotchets. Ventral bundles with 4-6 setae, teeth more or less equal, 3-4 in median segments, 2 posteriorly with upper tooth shorter than lower. Ventral setae of XI thicker than the rest but not modified in shape. Thin cuticular penis sheaths present, usually crumpled in whole mounts, much shorter than penes. Spermathecae with short spermatophores.
Lake Tahoe, California—Nevada, U.S.A.

The penes with short, thin sheaths are reminiscent of those described (but not illustrated) in other species, but the shortage of mature individuals makes it impossible to confirm the generic identity by studying the male ducts. Such parts of the male ducts that have been observed seem to resemble those of *Psammoryctides* species.

GENUS **Potamothrix** VEJDOVSKY AND MRAZEK, 1902

Type species: Potamothrix moldaviensis VEJDOVSKY AND MRAZEK

Male efferent ducts with short vasa deferentia, long wide tubular atria with small discrete prostate glands, no ejaculatory ducts, penes short without cuticular sheaths. Spermathecae with spermatophores. No coelomocytes. Spermathecal setae modified or unmodified.

Holarctic, ?Africa, Australia.

The distinction between *Ilyodrilus* Eisen and *Potamothrix* as herein defined is marked. Two of the three species described by Eisen (1879) remain in *Ilyodrilus*, together with some recently described species, all of which are known from western N. America, but *templetoni* (formerly in *Tubifex*) has a wider distribution. The third species to be described was *Ilyodrilus hammoniensis* (Michaelsen, 1901), which differed from Eisen's species in several fundamental ways (the male ducts are totally different, the penis lacks a sheath, and spermathecal setae are present), but all subsequently described species were allied to *Ilyodrilus* (*sensu* Michaelsen), ignoring Eisen's species.

The only exception to this occurred with the description of *Potamothrix moldaviensis* by Vejdovsky and Mrazek (1902) and so the name of the genus should be *Potamothrix*.

This name was not utilized by Brinkhurst (1963 a *et seq.*) owing to an erroneous attempt to conserve some trace of the original name, the prefix "Eu" being added to *Ilyodrilus*, and *E. hammoniensis* (Michaelsen) being adopted as the generic type. As it is inevitable that this name will eventually be rejected in favour of *Potamothrix* in order to conform with the code of

Zoological Nomenclature, stability will be achieved by rejecting the name *Euilyodrilus* here.

Further study of some of these species may result in the recognition of a small number of variable species in which the progressive reduction of the hair setae and pectinate setae, and the presence or absence of spermathecal setae may be acknowledged by the recognition of "forms". At present it is not possible to associate the species in this way with any degree of certainty, but the possible relationships will be mentioned in each instance.

The presence or absence of the minute prostate gland is used as a specific character with considerable hesitation but its rejection must await evidence of variation within a discrete population of a single species.

Potamothrix

1.	Hair setae present	2
—	Hair setae absent	7
2.	Pectinate setae present	3
—	Pectinate setae absent	5
3.	Spermathecal setae with broad, triangular distal end, much wider than shaft basally	*P. bavaricus*
—	Spermathecal setae more or less parallel sided, hollow distally	4
4.	Spermathecal setae broad, gutter-shaped distally with upper tooth broader than acuminate lower tooth	*P. hammoniensis*
—	Spermathecal setae narrow, hollow, sharply pointed distally, upper tooth not broader than lower or only so at distal end, where it forms a recurved tooth	*P. heuscheri*
5.	Hair setae present in anterior dorsal bundles	*P. vejdovskyi*
—	Hair setae present only in postclitellar bundles	6
6.	Spermathecal setae modified	*P. prespaensis*
—	Spermathecal setae absent or unmodified	*P. ochridanus*
7.	Spermathecal setae absent or unmodified	*P. caspicus*
—	Spermathecal setae modified	8
8.	Prostate glands absent	*P. moldaviensis*
—	Prostate glands present	*P. isochaetus*

Potamothrix moldaviensis Vejdovsky and Mrázek, 1902
Fig. 8. 11A-C

Potamothrix moldaviensis VEJDOVSKY and MRÁZEK, 1902: 1, Fig. 1-10.

Potamothrix okaensis LASTOCKIN, 1927 b: 9, Fig. 1-3.

Tubifex (Ilyodrilus) moldaviensis (Vejdovsky and Mràzek). UDE, 1929: 90, Fig. 113-114.

Potamothrix okaensis Lastockin. LASTOCKIN, 1936: 167.

Ilyodrilus moldaviensis (Vejdovsky and Mrázek). PIGUET, 1928: 98, Fig. 9. HRABE, 1941: 23, Fig. 26. JAROSCHENKO, 1948: 60; 1957: 54. MALEVICH, 1956: 424.

Ilyodrilus moldaviensis moldaviensis (Vejdovsky and Mrázek). HRABE, 1950: 275, Fig. 35-39. CEKANOVSKAYA, 1962: 261, Fig. 160.

Ilyodrilus moldaviensis mitropolskiji HRABE, 1950: 273, Fig. 26-34. CEKANOVSKAYA, 1962: 262.

Ilyodrilus grimmi HRABE, 1950: 277, Fig. 40-44.

Ilyodrilus (Potamothrix) moldaviensis JUGET, 1957: 2.

Ilyodrilus sp. (*I. moldaviensis?*) JUGET, 1957: 3.

Ilyodrilus isochaetus BRINKHURST, 1960: 405, Fig. 10.

Euilyodrilus moldaviensis (Vejdovsky and Mrázek). BRINKHURST, 1962 c: 329; 1963 a: 51, Fig. 34d; 1963 c: 42, Fig. 12a; 1965: 140, Fig. 7A-B. BRINKHURST and KENNEDY, 1962 a: 189. BRINKHURST and COOK, 1966: 17.

Ilyodrilus grimmi Hrabe. CEKANOVSKAYA, 1962: 262, Fig. 161.

Euilyodrilus grimmi (Hrabe). BRINKHURST, 1963 a: 52, Fig. 34c.

$l = 15$-40 mm. Anterior dorsal bundles with up to 9 setae, all bifid crotchets. Spermathecal setae modified, thick with elongate teeth forming a gutter-shaped distal end. Penial setae slightly different to other ventral setae, retained in mature specimens. Vasa deferentia very short, atria long, tubular, no prostate glands, penes short, without cuticular sheaths.
Holarctic.

According to Hrabe (1950) the sub-species *mitropolskiji* is different from sub-species *moldaviensis* in that the upper tooth of the bifid setae was a little shorter than the lower in the former, longer in the latter. In *grimmi* (described as a distinct species by Hrabe 1950) the upper tooth is fractionally shorter than the lower. The spermathecal setae, as illustrated, differ slightly in each instance, but not more than has been found to occur within *P. hammoniensis*. Differences in the detailed structure of the spermathecal setal sacs could easily be attributed to fixation and normal variation. The dimensions of the spermathecal seta of *grimmi* (138-160 μ) are of the same order of magnitude as those of *mitropolskiji* (104-166 μ). Hence it is difficult to see the two entities from the Caspian Sea (*mitropolskiji, grimmi*) as more than local forms of *moldaviensis*. ·

The American specimens were checked by Prof. S. Hrabe.

Potamothrix isochaetus (Hrabe) 1931
Fig. 8. 12X-Z

Ilyodrilus isochaetus HRABE, 1931 a : 31, Fig. 5g-i.

Ilyodrilus mrazeki HRABE, 1941 : 25, Fig. 27-28.

Ilyodrilus danubialis HRABE, 1941 : 33.

Ilyodrilus mrazeki Hrabe. HRABE, 1950 : 279. CEKANOVSKAYA, 1962 : 262.

Ilyodrilus isochaetus Hrabe. HRABE, 1950 : 279.

Ilyodrilus danubialis Hrabe. HRABE, 1950 : 279. CEKANOVSKAYA, 1962 : 261.

(non) *Ilyodrilus isochaeta* Hrabe. BRINKHURST, 1960 : 405, Fig. 10.

Euilyodrilus mrazeki (Hrabe) BRINKHURST, 1963 a : 52.

Euilyodrilus isochaetus (Hrabe). BRINKHURST, 1963 a : 52.

Euilyodrilus danubialis (Hrabe). BRINKHURST, 1963 a : 52.

$l=25$ mm, $s=$ 70-165. Anterior bundles with up to 10 setae, upper tooth thinner than but as long as the lower. Spermathecal setae long and parallel sided, broad, with gutter-shaped distal end. Vasa deferentia short, atria tubular, prostate glands small, penes small without cuticular sheaths. Spermathecae with large ampullae, short distinct ducts, spermatophores present.

Europe.

Ilyodrilus mrazeki resembles *isochaetus*, but was said to lack spermathecal setae, to have only four or five bifid setae anteriorly, and to have twice as many segments. A hitherto inadequately described species, *Ilyodrilus danubialis* resembled *I. mrazeki* (Hrabe 1941). If the latter be no more than a form of *mrazeki*, then there is nothing significant separating these from *isochaetus* described by Hrabe from L. Ochrid (Hrabe, 1931 a). The following species may also be little more than another deviant form of this same species.

Potamothrix caspicus (Lastockin, 1937)
Fig. 8. 11D-G

Ilyodrilus caspicus LASTOCKIN, 1937 : 234.

Ilyodrilus caspicus Lastockin. HRABE, 1950 : 271, Fig. 21-25. CEKANOVSKAYA, 1962 : 263, Fig. 162.

Euilyodrilus caspicus (Lastockin). BRINKHURST, 1963 a : 52

$l=26$ mm, $s=130$. Anterior bundles with about 5 setae, the upper tooth twice as long as the lower, about equally thick anteriorly but the upper thinner posteriorly. Spermathecal setae absent. Vasa deferentia short, atria tubular, prostate glands small, penes without sheaths.

Caspian Sea.

This species closely resembles *isochaetus* (save for the absence of spermathecal setae) but differs from both this and *mrazeki* in the form of the bifid setae. The distinction is narrow.

Potamothrix heuscheri (Bretscher, 1900)
Fig. 8. 11H-I

Tubifex heuscheri BRETSCHER, 1900 a : 11, Pl. I, fig. 1-4.

Tubifex heuscheri Bretscher. BRETSCHER, 1905 : 664. PIGUET, 1906 a : 391; 1913 : 127, Fig. 4a-b.

Ilyodrilus heuscheri (Bretscher). PIGUET, 1913 : 127. STAMMER, 1932 : 578. JAROSCHENKO, 1948 : 57. HRABE, 1950 : 280. JUGET, 1958 : 89, Fig. 141. CEKANOVSKAYA, 1962 : 260, Fig. 159.

Tubifex (Ilyodrilus) heuscheri Bretscher. PIGUET AND BRETSCHER, 1913 : 69, Fig. 14a, b.

Ilyodrilus orientalis CERNOSVITOV, 1938 a : 545, Fig. 16-23.

Ilyodrilus orientalis Cernosvitov. HRABE, 1950 : 279.

Euilyodrilus orientalis (Cernosvitov). BRINKHURST, 1963 a : 51.

Euilyodrilus heuscheri (Bretscher). BRINKHURST and KENNEDY, 1962 a : 184. BRINKHURST, 1963 a : 49, Fig. 34.

$l=6$-15 mm, $s=60$-150. Dorsal anterior bundles with 3-5 straight hair setae, 3-5 pectinate setae; ventral setae 4-5 per bundle, the upper tooth somewhat longer and thinner than the lower. Spermathecal setae long, narrow, hollow-tipped with upper tooth broadening to a hook distally. Vasa deferentia short, atria tubular, no prostate glands. Spermathecal ampullae large, ducts short, with spermatophores.

Europe, Israel, and a new record—Belgian Congo.

The discovery of undoubted specimens of *P. heuscheri* in Lake Tiberias, which lies in the Jordan Valley drainage area as did the former L. Huleh (the type locality of *I. orientalis*) led to a re-examination of the type specimens of the latter. It was immediately apparent that pectinate setae were present dorsally, not simple bifid setae as described by Cernosvitov (1938). No mention of *heuscheri* occurs in the description by Cernosvitov, who discussed *orientalis* in relation to *bavaricus*, in which the spermathecal setae differ considerably from those of *heuscheri*. Specimens of *P. hammoniensis* were also discovered amongst the type material by Mr. A. Gitay, who also discovered this species in Lake Tiberias.

Potamothrix hammoniensis (Michaelsen, 1901)
Fig. 8. 11J; 8. 12A, B

Ilyodrilus hammoniensis MICHAELSEN, 1901 a : 1.

Tubifex camarani de VISART, 1901 : 1, Fig. 1-5.

Tubifex cameranoi de Visart. BRETSCHER, 1903 : 12. PIGUET, 1913 : 128

Ilyodrilus hammoniensis Michaelsen. MICHAELSEN, 1903 b : 188; 1923 : 38
1926 b : 22; 1927 : 19, Fig. 22-24. LASTOCKIN, 1927 a : 67. HRABE, 1929 :
1; 1931 a : 27; 1950 : 279. BERG, 1948 : 52. JUGET, 1958 : 89, Fig. 14K (a, b).
SAPKAREV, 1959 : 1. BRINKHURST, 1960 : 406, Fig. 9. CEKANOVSKAYA, 1962 :
257, Fig. 156. MOSZYNSKA, 1962 : 29. POPESCU-MARINESCU et al. 1966 : 163.

Tubifex (Ilyodrilus) hammoniensis Michaelsen. MICHAELSEN, 1903 b : 188,
Fig. 10; 1908 a : 154; 1909 : 38, Fig. 74, 75. PIGUET, 1913 : 125, Fig. 5.
PIGUET AND BRETSCHER, 1913 : 70, Fig. 15, 16.

Psammoryctes fossor DITLEVSEN, 1904 : 417, Pl. XVI, fig. 15-17.

Ilyodrilus hammoniensis forma *lacustris*. LASTOCKIN, 1927 a : 71, Pl. LXVII.

Ilyodrilus hammoniensis forma *hammoniensis* HRABE, 1958 : 352.

Ilyodrilus hammoniensis forma *subdorsalis* HRABE, 1958 : 352.

Ilyodrilus hammoniensis forma *supralinearis* HRABE, 1958 : 352.

Euilyodrilus hammoniensis (Michaelsen). BRINKHURST, 1962 c : 321; 1963 a :
50, Fig. 34b; 1963 c : 37, Fig. 10a. KENNEDY, 1964 : 235.

$l=15$-45 mm, $s=75$. Dorsal anterior bundles with 1-5 straight hair setae and
3-5 bifid setae with the upper tooth longer and thinner than the lower, rarely with
a small intermediate tooth, spermathecal setae broad with parallel sides, gutter-
shaped distally. Vasa deferentia short, atria tubular, prostate glands small, penes
without sheaths. Spermathecae with large ampullae, spermatophores present.
Holarctic.

The species is scarce in N. America, but a typical specimen was obtained
from Green Bay, Lake Michigan, Wisconsin, U.S.A. This may be one of
those predominantly European species introduced to N. America, as it is a
very common species in Britain.

Various forms of this species have been described. Lastockin (1927 a)
described a forma *lacustris* in which the hair setae drop out in mature
worms. This implies that the pectinate setae remain and are not replaced
by bifid setae, but this is not made clear in the brief description. Hrabe
(1958) named four forms (*hammoniensis*, *subdorsalis*, *sublinearis*, and
supralinearis) all depending on the position of the spermathecal pores.
Apparently the position of the pores varies considerably, and elsewhere (as
in *Peloscolex*) Professor Hrabe uses this character as part of the definition
of species. As such wide variations are displayed by one of the commonest

European species some doubt may be cast upon the value of the position of the spermathecal pores as a significant anatomical feature.

The spermathecal setae vary considerably. When partially developed they resemble those of *bavaricus* and there are specimens with broad, straight spermathecal setae that are no thicker than those of most specimens of *heuscheri*. The only other major characteristic separating these three species is that the prostate glands are present in *hammoniensis* but absent in both *heuscheri* and *bavaricus*. There is no resemblance between the spermathecal setae of *bavaricus* and *heuscheri* although those of *hammoniensis* may resemble both.

Potamothrix bavaricus (Öschmann, 1913)
Fig. 8. 12H, I

Tubifex (Ilyodrilus) bavaricus ÖSCHMANN, 1913: 559, Fig. 1-5.

Ilyodrilus bedoti PIGUET, 1913: 124, Fig. 4c-e.

Tubifex (Ilyodrilus) bedoti Piguet. PIGUET AND BRETSCHER, 1913: 67, Fig. 14c-e.

Ilydorilus bavaricus Öschmann. MICHAELSEN, 1926 b: 22; 1927: 20, Fig. 25, 26. HRABE, 1935 a: 11; 1950: 280. CEKANOVSKAYA, 1959: 1152; 1962: 258, Fig. 157. BRINKHURST, 1960: 406, Fig. 11.

Ilyodrilus bedoti Piguet. PIGUET, 1928: 97, Fig. 8. UDE, 1929: 93. JARO-SCHENKO, 1948: 58. HRABE, 1950: 279; 1954: 305. JUGET, 1958: 88, Fig. 14J (a, b.). SOKOLSKAYA, 1959: 1154. CEKANOVSKAYA, 1962: 257, Fig. 155.

Euilyodrilus bedoti (Piguet). BRINKHURST AND KENNEDY, 1962 a: 186. BRINKHURST, 1963 a: 50.

Euilyodrilus bavaricus (Öschmann). BRINKHURST, 1962 c: 321; 1963 a: 50; 1963 c: 37, Fig. 106; 1965: 140, Fig. 7c. KENNEDY, 1964: 234.

$l=15$-35 mm, $s=55$-80. Dorsal anterior bundles with 1-5 straight hair setae, 2-5 pectinate setae; ventral bundles with 3-4 bifid setae with teeth equally long, but the upper thinner than the lower. Spermathecal setae relatively short, with broad triangular blade-like tips. Vasa deferentia short, atria tubular, prostate glands absent, penes without cuticular sheaths.

Holarctic, Australia, New Zealand.

In the variant originally described as *Ilyodrilus bedoti* the reproductive organs are in VII-IX not X-XII. This variation is caused by regeneration following damage or fragmentation.

Potamothrix prespaensis (Hrabe, 1931)
Fig. 8. 12J-R

Ilyodrilus prespaensis HRABE, 1931 a: 30.

Ilyodrilus prespaensis typica CERNOSVITOV, 1931: 315, Fig. 2, 4, 5, 11-14.

Ilyodrilus prespaensis scutarica CERNOSVITOV, 1931: 317, Fig. 1, 3, 6-10.

Ilyodrilus prespaensis Hrabe. HRABE, 1950: 279.

Euilyodrilus prespaensis (Hrabe). BRINKHURST, 1963 a: 50.

l=20-33 mm, s=63-120. Preclitellar segments with 5-8 dorsal, 5-10 ventral setae, posterior to the clitellum 1-2 short hair setae with 5-6 bifid setae, ventral bundles with 4-5 setae, posterior segments with 3-4 setae, spermathecal setae modified, long broad, with parallel sides and a gutter-shaped end. Vasa deferentia short, atria tubular, prostate glands absent, penes without cuticular sheaths.
Jugoslavia.

This species may prove to be no more than a form of *P. hammoniensis* as it differs from that species in much the same way as the "bergi" form differs from *Tubifex tubifex*, in which case the *P. isochaetus* complex may represent the final stage in the reduction of the hair setae and pectinate setae. As all of these forms are so limited geographically, it is probably best to leave any further reduction in their status until similar variability in *P. hammoniensis* is established elsewhere. Cernosvitov (1931) found typical specimens of this species in lakes and streams near Scutari, and described a new form (*scutarica*) from Scutari lake itself. In the so-called typical specimen the upper tooth of the setae is thicker but shorter than the lower, not thinner and as long as the lower as described by Hrabe (1931 a). The new form was smaller (7-10 mm), the upper tooth of the setae a little thinner and shorter than the lower, the hair setae at most twice the length of the bifid setae. The remaining differences were trivial and are concerned with variations in the annulation of the body, size, and size of the chloragogen cells.

Potamothrix ochridanus (Hrabe, 1931)

Fig. 8. 12S-W

Ilyodrilus ochridanus HRABE, 1931 a: 27, Fig. 5a-i.

Ilyodrilus ochridanus Hrabe. HRABE, 1950: 279.

Euilyodrilus ochridanus (Hrabe). BRINKHURST, 1963 a: 51.

l=20-25 mm, s=95-122. Anterior bundles with 4-7 setae dorsally 5-7 ventrally, upper tooth thinner and shorter than the lower, post clitellar bundles with 1-2 short hair setae with 5-7 bifid setae dorsally, 6-8 setae ventrally, 2-3 setae posteriorly, spermathecal setae absent. Vasa deferentia short, atria tubular, prostate glands small, penes without cuticular sheaths.
Lake Ochrid, Jugoslavia.

The species differs from *prespaensis* in lacking spermathecal setae, and while this species is found in Lake Ochrid *prespaensis* is known from the Scutaria area and Lake Prespa (close to Lake Ochrid). It probably represents no more than a form of the *hammoniensis-prespaensis-isochaetus* complex.

Potamothrix vejdovskyi (Hrabe, 1941)
Fig. 8. 12C-E

Ilyodrilus vejdovskyi HRABE, 1941 : 22, Fig. 20-25.

Ilyodrilus lastockini (nomen nudem et in litt.). JAROSCHENKO, 1948: 60.

Ilyodrilus vejdovskyi Hrabe, HRABE, 1950: 279. CEKANOVSKAYA, 1962: 259, Fig. 158.

Ilyodrilus sp. (*Aulodrilus pluriseta* Piguet?). JUGET, 1958: 89.

Euilyodrilus vejdovskyi (Hrabe). BRINKHURST, 1963 a : 51. BRINKHURST AND COOK, 1966: 17.

$l=10$-21 mm, $s=93$. Dorsal bundles with 2-4 short, bent hair setae, 4-6 bifid setae with upper tooth a little shorter and thinner than the lower, ventral setae up to 10 per bundle similar to dorsal setae, posteriorly 1 hair and 2 bifid setae dorsally, 2 bifid setae ventrally, spermathecal setae long, broad, parellel sided, with gutter-shaped end. Vasa deferentia short, atria tubular, prostate glands small, penes without penis sheaths. Spermathecae present or absent.
River Danube, Europe and Great Lakes, N. America.

The identity of the N. American specimens has been confirmed by Prof. S. Hrabe.

SPECIES INQUIRENDAE

Potamothrix svirenkoi (Lastockin, 1937)

Ilyodrilus svirenkoi LASTOCKIN, 1937: 235.

Ilyodrilus svirenkoi Lastockin. HRABE, 1950: 278.

Ilyodrilus swirenkoi Lastockin. CEKANOVSKAYA, 1962: 260.

Euilyodrilus svirenkoi (Lastockin). BRINKHURST, 1963 a : 50.

$l=38$ mm, $s=180$. Dorsal bundles with 4-7 setae, ventral bundles with 5-8, the upper tooth reduced, postclitellar dorsal bundles with 1 long hair setae (1 mm long) per bundle; spermathecal setae relatively short, thin, parallel sided, with a straight tip. ?Prostate gland.
European, U.S.S.R.

The species was inadequately described, and has not been re-examined.

Potamothrix (?) *dniprobugensis* (Jaroschenko, 1948)

Limnodriloides dniprobugensis JAROSCHENKO, 1948: 66, Fig. 9, 10.

(?) *Clitellio dniprobugensis* (Jaroschenko). BRINKHURST, 1963 a : 78.

(?) *Isochaeta dniprobugensis* (Jaroschenko). BRINKHURST, 1966 a : 739.

Ventral setae 4-5 per bundle, the upper tooth longer than the lower, dorsal setae similar, genital setae in X and XI small, nearly straight distal teeth. Vasa deferentia short, prostate glands small and stalked, at union of vasa deferentia and long tubular atria.

Dniprobug estuary, Black Sea.

As with most of the species described by Jaroschenko (1948) it is almost impossible to tell what the true identity of this worm is. The illustration of the male efferent ducts suggests an affinity either with *Potamothrix* species, although the atria are unusually long, or with the American species *Isochaeta nevadana*, which was included in *Isochaeta* with some hesitation.

GENUS **Ilyodrilus** EISEN, 1879

Type species: Ilyodrilus perrieri EISEN

Male efferent ducts with vasa deferentia broad, about as long as atria, which are thin-walled, saccular when fully developed, with large prostate glands on the upper posterior walls, ejaculatory ducts separated from atria, but broad. Penes with conical cuticular sheaths. Spermathecae with spermatophores. No coelomocytes.

Europe, Asia, N. America (? S. Africa).

The male efferent ducts resemble those of *Tubifex* species but careful examination reveals several basic differences. The vasa deferentia are very broad in *Ilyodrilus*, narrowing abruptly before entering the atria. The prostate glands are situated on the posterior walls of the atria, whereas those of *Tubifex* all seem to lie on the anterior face. The irregular, sac-like atria of fully mature *Ilyodrilus* species are apparently separated from broad tubes just proximal to the penis-sacs by an abrupt narrowing which was taken to be an artefact when first observed. This narrowing seems to be present in most specimens however, and is frequently the site at which the atria are accidentally broken away from the distal tube during dissection. The part of the efferent system beyond the narrowing is therefore termed the ejaculatory duct. The atria of *Tubifex* are usually of a regular shape, broad proximally, gradually narrowing towards the penis sacs. There is a distinction between the narrow part of the atrium and the wide proximal part of the duct immediately preceding the penis sac in *Tubifex*, but it is not so pronounced as in *Ilyodrilus* species. Since it appears that the centres of distribution of *Ilyodrilus* and *Tubifex* are western N. America and western Europe respectively, there does seem to be some value in retaining the separation of these two rather similar genera.

The separation of species in the genus *Ilyodrilus* presents some difficulty. In all probability, the various forms described constitute no more than a single species demonstrating the same range of setal variation observed in *Tubifex tubifex*. Differences in the degree of maturity probably account for the observed differences in the form of the atria and the penis sheaths, because most of the forms with conical penis sheaths and narrow atria have empty spermathecae. The penis sheaths of the widely distributed species *I. templetoni* are very long prior to copulation, but are narrowly conical with a distal tear at full maturity. The distal portion seems to be shed prior to copulation. The short conical penis sheaths in *I. perrieri* seem to lack a distal opening, but may resemble those of *I. frantzi* after copulation. The tube lacking a distal opening may line the penis sac, as a shorter, broader sheath seems to cover the penis itself in some preparations. The genus is very poorly known, and the following must not be regarded as more than one possible division of the available forms into species.

Ilyodrilus

1. Most specimens without hair setae, no pectinate setae. (Spermatophores with globular heads, narrow tails.)   *I. frantzi*
— Most specimens with hair setae and pectinate setae   2

2. Penis sheaths narrowly conical   *I. templetoni*
— Penes broadly conical, truncated   *I. perrieri* (? *I. fragilis*)

Ilyodrilus perrieri Eisen, 1879
Fig. 8. 13H-J; 8. 14A-F

Ilyodrilus perrieri EISEN, 1879: 11.

Ilyodrilus perrieri Eisen. VEJDOVSKY, 1884: 45. MICHAELSEN, 1900: 47. GALLOWAY, 1911: 314. HRABE, 1966: 65.

Ilyodrilus perrierii Eisen. EISEN, 1886: 887, Pl. IV, fig. 3a-k. BRINKHURST, 1963 a: 54; 1965: 141, Fig. 7d-f.

$l=10$-12 mm, $s=27$. Dorsal anterior bundles, with 2-4 hair setae, 2-4 pectinate setae with teeth equally long, or the upper the longer, rather parallel, and a series of fine intermediate teeth; ventral bundles with 4-5 setae, the upper tooth longer and thinner than the lower; posteriorly only 2 setae with equal teeth. Vasa deferentia as long as atria, atria rather tubular (semi-mature), penes short, conical, ? penis sheaths. Prostate glands attached to atria opposite union with vasa deferentia. Spermathecal ampullae as long as their ducts.
Fresno, California, U.S.A. (below 300'). ? British Columbia.

The atria, as figured by Eisen (1886) are irregularly winding, suggesting a thin-walled sac like that found in *I. fragilis* (q.v.). The pectinate setae of the lectotype lack the elongate upper tooth figured by Eisen (1886), and are pectinate, not palmate. There are no cuticular penis sheaths in the semimature type specimens.

The name of this species is spelt both as *perrieri* and *perrierii* by Eisen (1879, 1886), but the former is clearly the correct version as the species is named for a Mr Perrier.

Specimens from the late D. S. Rawson's collection from Okanagan Lake, British Columbia, are larger than the average *I. templetoni* and have rather broadly conical penis sheaths. The hair setae are serrate, dorsal setae pectinate, and the anterior ventral setae sometimes have a small spine on the inner side of the lower tooth, which is thicker and shorter than the upper tooth. The head of the atrium is exactly the same as that seen in *I. frantzi*. These specimens may be fully mature *I. perrieri*, in which case the two forms of *I. frantzi* are probably no more than forms of *I. perrieri*. Alternatively, the Okanagan specimens may represent the third form of *I. frantzi*, which would then include the full spectrum of variants observed in *T. tubifex* and others, but it would be very difficult to separate *I. frantzi* from both *perrieri* and *templetoni* if this interpretation were accepted.

Ilyodrilus templetoni (Southern, 1909)
Fig. 8. 13A-F

Tubifex templetoni SOUTHERN, 1909: 140, Pl. VIII, fig. 6a-e.

Tubifex templetoni Southern. FRIEND, 1912 b: 292. MALEVITCH, 1956: 425. BRINKHURST, 1962 c: 326, Fig. 1n; 1963 a: 23; 1965: 124, Fig. 2e-g. CEKANOVSKAYA, 1962: 275; HRABE, 1962 a: 308. KENNEDY, 1964: 228.

Tubifex thempletoni Southern. LASTOCKIN, 1927 a: 67.

Tubifex templetoni var *typica* BRINKHURST, 1963 c: 37, Fig. 9d.

Tubifex templetoni var *walshi* BRINKHURST, 1963 c: 42, Fig. 9c; 1966: 736, Fig. 1j-l.

Ilyodrilus templetoni (Southern). HRABE, 1966: 61, Fig. 9-20.

$l = 10$-30 mm, $s = 120$. Dorsal bundles with 3-4 pectinate setae and 1-4 hair setae or all bifid setae; ventral bundles with 3-4 setae per bundle, unworn anteriormost setae with upper tooth much longer and thinner than lower. Vasa deferentia short and broad, moderately long in some, entering atria at recurved apical end, prostate large, atria cylindrical to pear-shaped. Penis sheaths conical tapering distally, with irregular opening. Spermathecae present or absent.
Europe, Asia, N. America, ? S. Africa.

This apparently widely distributed species may be no more than a variant of *I. perrieri* or *I. fragilis*, or both. Most specimens are relatively small, and the penis sheaths are decidedly narrow cones. The penis sheaths of *perrieri* and *fragilis* may be similar, as illustrated by Eisen (1879, 1886), but this cannot be established from the type of material available.

Ilyodrilus frantzi Brinkhurst, 1965
Fig. 8. 13O-Q

Ilyodrilus frantzi BRINKHURST, 1965: 142, Fig. 7k-m, 8a, b.

Ilyodrilus frantzi Brinkhurst. BRINKHURST AND COOK, 1966: 17.

Ilyodrilus frantzi typica BRINKHURST AND COOK, 1966: 18.

Ilyodrilus frantzi capillatus BRINKHURST AND COOK, 1966: 18.

$l=12$-20 mm, $s=60$-100. Bifid setae with upper tooth larger than the lower in anterior bundles, where 4-7 setae per bundle, hair setae absent, or 2-3 short, bent, hair setae in some dorsal bundles behind V at least. Vasa deferentia about as long as atria, broad forming cap-like proximal ends of atria; rest of atria at first narrowly tubular, becoming broad thin-walled sacs at maturity; narrow connections between atria and broad ejaculatory ducts, penes with truncated cone-shaped sheaths. Spermathecae globular with spermatophores having broad "heads" and recurved narrow "tails".
Western U.S.A.

Before complete maturation the male ducts of this species are very similar to those of *perrieri* and *fragilis*. They resemble those of some *Tubifex* species, but at full maturity they are distinctive. The form with a few hair setae was found in the estuary of the San Joaquin-Sacramento rivers, which is subject to incursions of salt-water. Initially described as a variety of *frantzi*, this form was then elevated to the status of a sub-species in order to comply with the International Code of Zoological Nomenclature (Article 45 (e) (ii)). Because this would seem to give undue weight to a trivial variation, these varieties or sub-species are herein called forms, and are given names in parentheses, so that in this instance specimens with hair setae will be attributed to the "capillatus" form.

Material from Okanagan Lake, B.C., Canada, has atria strongly reminiscent of *I. frantzi*, but the penis sheaths are longer than those illustrated for *I. frantzi* and the spermatophores are not of the characteristic form described in this species. I have tentatively ascribed these specimens to *I. perrieri* (q.v.).

SPECIES INQUIRENDAE

Ilyodrilus fragilis Eisen, 1879
Fig. 8. 13K-N

Ilyodrilus fragilis EISEN, 1879: 12.

Ilyodrilus fragilis Eisen. VEJDOVSKY, 1884: 45. EISEN, 1886: 888, Fig. 4a-g
MICHAELSEN, 1900: 47. GALLOWAY, 1911: 314. BRINKHURST, 1963 a: 54;
1965: 142, Fig. 7g-j. HRABE, 1965: 65.

$l=c.$ 10 mm, $s=c.$ 30. Dorsal anterior bundles with 2-4 hair setae, 2-4 pectinate setae, the latter with the upper tooth longer than the lower, diverging quite widely, with a few short fine intermediate teeth; ventral setae up to 5 in anterior bundles, upper tooth longer and thinner than the lower. Vasa deferentia as long as atria, prostate glands attached opposite vasa deferentia, atria ?

separated from ejaculatory ducts, penes conical, with thin cuticular sheaths. Spermathecal ampullae globular, ducts short.

Sierra Nevada, California, U.S.A. (5,000'-7,000').

Having discovered that the dorsal setae of the type specimens are pectinate, the distinction between this species and *I. perrieri* is not clear. The pectinations between the teeth of the dorsal setae may be finer than in *I. perrieri*, but with so few specimens available for study, it is impossible to determine the variability of *I. perrieri*.

NOMEN DUBIUM

Ilyodrilus asiaticus of Chen (1940) cannot be retained in the genus, nor can it be attributed to any known genus in the family.

GENUS **Peloscolex** LEIDY, 1851

Type species: Peloscolex variegatus LEIDY

Male efferent ducts with vasa deferentia long to fairly long, atria broad tubes, mostly curved horse-shoe shaped over connection with large prostate glands situated $\frac{1}{3}$ to $\frac{1}{2}$ way down atria from union with vasa deferentia, narrowing to short ejaculatory ducts which enter true penes with or without penis sheaths. Penis sheaths commonly cylindrical, about 3 times longer than broad. Spermathecal setae modified or unmodified, penial setae rarely modified. Spermathecae with spermatophores. No coelomocytes. Body wall variously covered with secretion with included foreign particles forming either papillae or a sheath covering all or some of the body, or as little as a thin granular tube; may be shed periodically or apparently absent. Sensory papillae present or absent, usually associated with dense sheath or papillate covering. Prostomium and first segment commonly retractable within other anterior segments.

Eurasia, N. and S. America.

The definition of this genus is perhaps one of the most difficult to establish. While species with a definite sheath or obvious papillae may clearly be placed here, related species without these characteristics, or individuals in which the sheath or covering of papillae has been shed (Brinkhurst, 1964) are difficult to place with certainty. Since the redescription of the type species, *P. variegatus* (Brinkhurst, 1962 a), which had not been recorded since its original discovery by Leidy (1851), it has been possible to consider the form of the male efferent duct in defining the genus, when it becomes apparent that there is considerable uniformity in the form of the atria, the penes, the penis sheaths when present, and the spermathecal or penial setae when modified. In many species with hair setae and pectinate setae, these are of a distinctive form, the former being broad at the base but narrowing abruptly (sabre shaped), and the distal ends of the latter are frequently lyre-shaped. There is a strong tendency towards the development of simple-pointed setae in both dorsal and ventral bundles.

Most of the species in the genus were included in the key to *Peloscolex* in a recent paper by Hrabe (1964)with the exception of a number of

American species described subsequently, *P. ringulatus* (placed in *Alexandrovia*-vide *Telmatodrilus*), and *P. speciosus*, which was placed in *Tubifex*. The decision regarding *P. speciosus* is rejected here owing to the synonymy of *P. zavreli*, *P. simsi*, and *P. speciosus monfalconensis* with *P. speciosus*. The terminalia of the male ducts are very similar in these entities and they share the unique feature of having both penial and spermathecal setae modified as well as distinct penis sheaths. The body wall is scarcely modified but shows distinct evidence that a secretion is produced, and its genital setae and dorsal pectinate setae resemble those of other species in the genus. There does not seem to be sufficient evidence to create a separate genus for this species, and the inclusion of *speciosus* in *Tubifex* only serves to widen the limits of that genus to the point of confusion with *Peloscolex*.

The only other species with a similarly modified body-wall is *Alexandrovia onegensis*, here regarded as a species within *Telmatodrilus* because of the form of the male ducts. There is a series of prostate glands on each atrium in this genus, a characteristic shared with no other species in the genus. According to Hrabe (1964), the same characteristic is shared by *Peloscolex ringulatus*, although in the original description of that species the male ducts were described and figured and clearly resemble those of other *Peloscolex* species with a horse-shoe shaped tubular atrium with a distinct prostate gland. The union of the prostate with the atrium was not illustrated, however, and the gland was not mentioned in the text. In a recent personal communication, Dr. N. Sokolskaya admitted that the original description was erroneous and the species is correctly located in *Telmatodrilus* (or *Alexandrovia*, which is included in *Telmatodrilus*).

While the genus is the largest in the family, many of the species were described in ignorance of the form of the male ducts, and several may prove to belong to *Telmatodrilus* or even genera as yet unrecognized.

The modified genital setae resemble those of *Psammoryctides* species, being long and narrow with two elongate teeth forming a hollow distal end.

Peloscolex

1.   Hair setae present   2
—   Hair setae absent   26

2.   Hair setae together with thin, often hair-like
     setae in some or all dorsal bundles, the tips
     of the latter palmate, pectinate, bifid or simple
     but very difficult to observe   3
—   Hair setae together with bifid or pectinate
     setae in anterior dorsal bundles, sometimes
     simple pointed setae with hairs posteriorly   10

3.   Body wall covered by a cuticular sheath composed of foreign matter embedded in a secretion. Few, if any, discrete papillae.   *P. nomurai*

—   Body wall characteristically papillate (papillae may be shed, when worms appear naked or with a sheath)   4

4.   Exceptionally large papillae in 2 or 3 rows along the body. Up to 16 or 19 hair setae per bundle behind the clitellum   *P. stankovici*

—   Papillae not exceptionally large, body more or less evenly covered with papillae   5

5.   Ventral bundles all with bifid crotchets   *P. yamaguchii* nom. nov.

—   Ventral bundles of at least some anterior segments with one or both setae simple pointed   6

6.   Ventral bundles of many segments simple pointed   *P. velutinus*

—   Ventral bundles of all but the pre-clitellar segments with bifid setae   7

7.   Spermathecal setae present   8

—   Spermathecal setae absent (not observed in *oregonensis*)   9

8.   Ventral setae of II-IX. 1 bifid seta with teeth narrowly diverging, and 1 simple pointed   *P. carolinensis*

—   Ventral setae with teeth diverging widely, simple pointed setae restricted to one or two anterior segments   *P. nikolskyi*

9.   Posterior ventral setae with teeth equally long or upper longer than lower   *P. variegatus*

—   Posterior ventral setae with upper tooth shorter and thinner than broad lower tooth   *P. oregonensis*

10.   Hair setae and bifid setae in anterior dorsal bundles
(Brackish or salt water species).   11

—   Hair setae and pectinate setae in anterior dorsal bundles
(Mostly freshwater species, except *P. nerthoides*, *P. aculeatus*).   16

11. A few short hair setae only, otherwise two
     bifid setae with reduced upper teeth or simple
     pointed setae in all bundles. (Papillae
     prominent, covering body wall thinly but
     evenly)   *P. benedeni*
—   Hair setae in all anterior bundles at least,
     other dorsal setae unlike ventral setae   12

12. Upper tooth of all bifid setae shorter and
     thinner than the lower, posterior dorsal bifid
     setae with thickened lower tooth, posterior
     ventral setae with somewhat thickened lower
     tooth. (Body wall densely papillate)   *P. dukei*
—   Bifid setae with upper tooth at most some-
     what thinner than the lower, teeth usually
     subequal. (Body wall sparsely to finely
     papillate)   13

13. Hair setae straight. Posterior dorsal setae
     hairs plus bifid setae   *P. euxinicus*
—   Hair setae bent. At least some posterior dorsal
     setae hair like, simple pointed or with closely
     applied parallel teeth   14

14. Hair setae heavily serrate. Some posterior
     dorsal setae with thin parallel teeth closely
     applied, some probably simple pointed   *P. apectinatus*
—   Hair setae sparsely serrate or naked. Posterior
     dorsal setae with hair-like simple pointed
     setae as well as hair setae   15

15. Hair setae with sparse lateral hairs. Vasa
     deferentia length/length of atrium=4·5, penis
     length/diameter=2·5   *P. intermedius*
—   Hair setae smooth. Vasa deferentia length/
     length of atrium=2·5, penis length/dia-
     meter=1·0   *P. swirencowi*

16. With modified genital setae replacing ventral
     setae of X or X and XI.   17
—   With no modified genital setae   19

17. With spermathecal and penial setae modified.
     Body wall glandular, or grooved, ? in thin
     sheath   *P. speciosus*

— With spermathecal setae modified, penial setae absent. Body wall closely papillate 18

18. Anterior ventral bundles of II-III, with a simple-pointed and a bifid seta, pectinate setae with thin, short intermediate teeth ..... *P. kurenkovi*

— Anterior ventral bundles of II-VI with one or both setae simple pointed, pectinate setae virtually palmate ..... *P. beetoni*

19. Body wall characteristically papillate (unless papillae shed, when may appear naked or with a sheath) 20

— Body wall naked, or with a sheath composed of foreign matter embedded in a secretion 22

20. Exceptionally large papillae in rings around the body in line with the setae, other rings of smaller papillae. Up to 14 hair setae in dorsal bundles ..... *P. multisetosus*

— Papillae not exceptionally large, densely covering the body when fully developed, less than 8 hair setae per bunde 21

21. Hair setae 2-4 per bundle (? teeth of postclitellar ventral setae diverge at an acute angle) ..... *P. inflatus*

— Hair setae up to 7 per bundle (teeth of postclitellar ventral setae diverge at an obtuse angle) ..... *P. ferox*

22. Sheath forming a complete armour plating over the body (recorded only from L. Baikal). ..... *P. werestschagini**

— Sheath less completely enveloping body or absent. 23

23. Pectinate setae with lateral teeth both curved inward, two or three distinct intermediate teeth, the distal end lyre-shaped, broader than the shaft. Posterior ventral setae strongly sigmoid with recurved distal end ..... *P. tenuis*

— Pectinate setae not lyre-shaped, distal end not broader than shaft. Posterior ventral setae with distal end not strongly recurved 24

* Description of setae inadequate—probably belongs here.

24. Pectinate setae with short teeth, penis sheath cylindrical with a basal flange (freshwater— Great Lakes)   *P. superiorensis*

— Pectinate setae with elongate teeth, penis sheath elongate thimble-shaped (brackish and salt water)   25

25. Penis sheath ornamented with spines. Hair setae with sparse lateral hairs   *P. aculeatus*

— Penis sheath not ornamented. Hair setae smooth   *P. nerthoides*

26. Spermathecal setae modified   *P. freyi*

— Spermathecal setae not modified
(? in *P. pigueti*—immature material only)   27

27. Simple pointed setae present in some bundles   28

— No simple pointed setae present   29

28. Most bundles with 2 setae, upper tooth short or absent. Body thinly but evenly covered with papillae.   *P. benedeni*

— Most anterior bundles with more than two bifid setae, most mid and hind bundles with 1-2 sharply pointed setae   *P. heterochaetus*

29. Body wall with a sheath of foreign matter in a secretion fused into rings on IX-XII, papillae on the rest of the body   *P. pigueti*

— Body wall with papillae in some specimens   *P. gabriellae*

Peloscolex nomurai (Yamamoto and Okado, 1940)
Fig. 8. 21G-M

Tubifex (*Peloscolex*) *nomurai* YAMAMOTO and OKADO, 1940: 427, Fig. 1-20.

Isochaeta nomurai (Yamamoto and Okado). BRINKHURST, 1963 a: 33, Fig. 2, 19.

Peloscolex nomurai (Yamamoto and Okado). HRABE, 1964: 109.

l=70-80 mm, s=90-140. Dorsal anterior bundles with up to 5 hair setae, 2-3 narrowly palmate setae with a web between two lateral teeth. Ventral bundles of II-IX with 2-3 bifid setae, the upper tooth longer than the lower. Diverticulum with gland in spermathecal duct, may be associated with spermathecal setae, no spermathecal setae observed. No ventral setae beyond IX in mature worms. Vasa deferentia long, atria saccular with large prostates, ejaculatory ducts lead to long penes, no penis sheaths. Spermathecal ampullae long. Body wall enclosed in a sheath with foreign particles in a secretion, papillae around male pores and in clitellum.

Lake Tazawa, Japan.

The sheath on the body-wall of this species forms the so-called tubes of the original account, and hence Hrabe (1964) attributed the species to *Peloscolex* not *Isochaeta* (vide Brinkhurst, 1963 a). The distinct atria in the illustration in the original account suggest that this location is, in fact, more reasonable if one accepts the interpretation of the true nature of the tube in which the worms dwell.

Peloscolex stankovici Hrabe, 1931
Fig. 8. 20X; 2. 21E, F

Peloscolex stankovici HRABE, 1931 a: 33, Fig. 6a-b, Pl. I, fig. 3-6.

Peloscolex stankovici forma *typica* HRABE, 1931 a: 34, Fig. 6, Pl. I, fig. 3, 4, 6.

Peloscolex stankovici forma *sublitoralis* HRABE, 1931 a: 38, Pl. I, fig. 5.

Peloscolex stankovici forma *litoralis* HRABE, 1931 a: 33.

Peloscolex stankovici forma *typica* Hrabe. MARCUS, E., 1950: 4.

Peloscolex stankovici Hrabe. BRINKHURST, 1963 a: 45.

Peloscolex stankovici stankovici HRABE, 1964: 110.

Peloscolex stankovici sublitoralis HRABE, 1964: 110.

Peloscolex stankovici litoralis HRABE, 1964: 110.

l=6-24 mm, s=30-115. Dorsal anterior bundles with up to 12 hair setae and few hair-like bifid setae, postclitellar bundles with up to 19 hair setae, longer than in anterior bundles, posteriorly reduced to 7 per bundle. Anterior ventral bundles with 2 setae, the outermost in most anterior bundles bifid, the rest simple pointed with bent tips, spermathecal setae long, straight, hollow tipped, in glandular sac opening via spermathecal pore. Vasa deferentia long, atria saccular with moderately large prostate glands, ejaculatory ducts short, penes long with thin sheaths covering penes and penis-sacs. Spermathecae with long spermatophores. Two or three rows of very large papillae on body wall, other small papillae numerous.

Lake Ochrid, Jugoslavia.

There are 3 different forms (varieties *typica, litoralis, sublitoralis,*— Hrabe, 1931 a, sub-species—Hrabe, 1964) that differ in size, number of hair setae and number of large papillae on the body wall. The exceptionally large papillae are also found in *P. multisetosus.*

Peloscolex yamaguchii nom. nov.

Peloscolex sp. YAMAGUCHI, 1953: 295.

Dorsal bundles with 5-7 hair setae. 4-6 crotchets with parallel teeth, ventral bundles with 1-4 setae, teeth equally thick, posteriorly upper tooth much thinner than lower, spermathecal setae long and thin, hollow tipped, in glandular sac. Vasa deferentia long, penes conical without penis sheaths. Sensory papillae in two rows per segment, ? papillate with secretion plus foreign matter. Japan.

While this species was inadequately described and not illustrated by Yamaguchi (1953) it has a unique combination of characters and seems certainly to belong to a hitherto undescribed species as that author suggested. The mention of *P. ferox* and *P. velutinus* in the description implies that the species may be closely papillate, but this feature is not mentioned.

Peloscolex velutina (Grube, 1879)
Fig. 8. 20T-W

Saenuris velutina GRUBE, 1879 : 116.

Tubifex velutinus (Grube). IMHOF, 1888 : 48. MICHAELSEN, 1903 b : 196. BRETSCHER, 1905 : 670. PIGUET, 1906 a : 390.

? *Saenuris velutina* Grube. ZSCHOKKE, 1891 : 120.

Embolocephalus velutinus (Grube). RANDOLPH, 1892 : 463, Fig. 17-18. BEDDARD, 1895 : 272.

Psammoryctes velutinus (Grube) MICHAELSEN, 1900 : 50. BRETSCHER, 1903 : 2.

Tubifex (Peloscolex) velutinus (Grube). MICHAELSEN, 1903 b : 196; 1909 : 39, Fig. 76. PIGUET, 1913 : 130. PIGUET AND BRETSCHER, 1913 : 72, Fig. 17a-c.

Tubifex sarnensis PIERANTONI, 1905 : 228, Fig. 1-4.

Peloscolex velutinus (Grube). UDE, 1929 : 95. CERNOSVITOV, 1930 : 321; 1935 a : 2; 1939 a. 37; 1942 a : 214. MICHAELSEN AND VERESCHAGIN, 1930 : 213. STAMMER, 1932 : 578. MALEVITCH, 1947 : 15. JAROSCHENKO, 1948 : 69. MARCUS, E., 1950 : 3. JUGET, 1958 : 89, Fig. 14B. BRINKHURST, 1960 : 407. Fig. 12a, b; 1962 c : 321; 1963 a : 45, Fig. 33a-d; 1963 c : 31, Fig. 8e. BRINKHURST AND KENNEDY, 1962 a : 184. CEKANOVSKAYA, 1962 : 283, Fig. 181. HRABE, 1964 : 110. POPESCU-MARINESCU et al, 1966 : 163.

(?) *Peloscolex cernosvitovi* HRABE, 1958 : 349, Fig. 24-35.

(?) *Peloscolex cernosvitovi* Hrabe. BRINKHURST, 1963 a : 42, Fig. 29. HRABE, 1964 : 110.

Peloscolex fontinalis HRABE, 1964 : 107, Fig. 22-24.

l=25-50, s=40-70. Dorsal bundles with 1-4 hair setae, up to 4 hair-like crotchets. Ventral setae 1-2 per bundle, simple-pointed or bifid with reduced upper tips, ventral setae of modified spermathecal setae, long, straight with hollow ends.

No cuticular penis sheaths. Body wall closely covered with papillae. Europe.

The separation of *P. fontinalis* from *P. velutinus* on the basis of the absence of bifid setae in all ventral bundles from IV does not seem justified (Brinkhurst 1966 a). The poorly described *P. cernosvitovi* probably belongs here. The key to *Peloscolex* in Hrabe (1964) seems to be misleading in the separation of *cernosvitovi* from *velutinus*, claiming that all ventral setae from XIII are simple-pointed (couplet 13), whereas in the original account many posterior *setae* have short but distinct upper teeth (fig. 29-35, p. 351, Hrabe, 1958).

The species seems to be rare, confined to deep lakes and small streams.

Peloscolex carolinensis Brinkhurst, 1965.
Fig. 8. 20P-S

Peloscolex carolinensis BRINKHURST, 1965: 137, Fig. 6K-N.

l=15 mm+, s=60+. Prostomium and segment I retractile. Dorsal anterior bundles with 3-6 hair setae, up to 8 posteriorly, and 1-2 bifid setae with minute tips, ventral setae of II-VII or IX with simple-pointed seta, 1 bifid with upper tooth longer than the lower, rest with 1 long seta, upper tooth progressively reduced, setae progressively more sigmoid, posterior setae smaller than median setae, modified spermathecal setae present, hollow distally in glandular sacs. Atria horse-shoe shaped, large prostate glands, penes with very thin cuticular sheaths. Body wall closely covered with papillae.
N. Carolina, U.S.A.

The species is close to *P. velutinus, P. nikolskyi,* and possibly *P. cernos-vitovi.*

Peloscolex nikolskyi Lastockin and Sokolskaya, 1953
Fig. 8. 20K-M

Peloscolex nikolskyi LASTOCKIN AND SOKOLSKAYA, 1953: 409, Fig. 1.

Peloscolex nikolskyi Lastockin and Sokolskaya. SOKOLSKAYA, 1961 b: 92.
CEKANOVSKAYA, 1962: 282, Fig. 180. BRINKHURST, 1963 a: 45. HRABE,
1964: 107, Fig. 20.

l=20-40 mm, s=?. Anterior dorsal bundles with up to 7 sabre-shaped hair setae, a few bifid or pectinate setae with small distal ends. Ventral bundles with some simple-pointed setae in anteriormost bundles, otherwise bifid with upper tooth twice as long as the lower, posteriorly setae enlarged, upper tooth shorter and thinner than lower but not strongly sigmoid, spermathecal setae long, thin and hollow ended. Atria horse-shoe shaped, large prostate glands, ejaculatory ducts distinct, lead to penes lacking cuticular sheaths. Spermathecal ampullae large, spermatophores long. Body wall papillate, sensory papillae 2 rows per segment.
Asiatic Russia, R. Amur basin.

The dorsal setae were shown to be bifid in the original figure, pectinate in the re-drawn figure in Cekanovskaya (1962). If spermathecal setae are discovered in the rare species *P. variegatus* it will become practically impossible to separate *nikolskyi* from it.

Peloscolex variegatus Leidy, 1851
Fig. 8. 20G-J

Peloscolex variegatus LEIDY, 1851: 124.

Peloscolex variegatus Leidy. VEJDOVSKY, 1884 b: 45. BEDDARD, 1895: 258. MICHAELSEN, 1900: 53. MARCUS, E., 1950: 3. BRINKHURST, 1962 a: 301, Fig. 1-3; 1963 a: 45; 1965: 132, Fig. 5a-e. HRABE, 1964: 110.

l=10-20 mm?, *s*=? Prostomium retractile. Dorsal anterior bundles with 2-4 hair setae and 2-4 hair-like setae, ? bifid or pectinate minute tips; ventral setae of II-IV simple-pointed or bifid or both, remaining pre-clitellar setae bifid with upper tooth longer and thinner than the rest, rarely with intermediate tooth, posteriorly setae single, teeth equally thick, upper slightly longer than lower. Vasa deferentia long and thin, atria horse-shoe shaped, prostate glands large, attached to concave side of atria medially, ejaculatory ducts not distinct, penes long with cuticular sheaths. Body wall closely covered with papillae.
N. America.

Peloscolex oregonensis BRINKHURST, 1965
Fig. 8. 20D-F

Peloscolex oregonensis Brinkhurst, 1965: 138, Fig. 6O-Q.

l=7+ mm, *s*=32+. Dorsal bundles with 1-2 hair setae and about 2 setae with minute tips. Ventral setae of II and III, a single sharply pointed seta and 1 bifid seta with long, thin upper tooth, posteriorly setae progressively broader with shorter upper tooth and thicker lower tooth, not strongly sigmoid. Body wall closely covered with papillae.
Oregon, U.S.A.

A single immature specimen was obtained, and the description requires amplification, especially in view of the similarity to the *variegatus-nikolskyi* complex, which may well include *carolinensis* and *velutinus* (c.f. *P. tenuis*). In *P. kurenkovi* the pectinate setae are narrow-tipped, and this species is also related to this assemblage, but in *P. beetoni* the dorsal setae are practically palmate.

Peloscolex benedeni (Udekem, 1855)
Fig. 8. 15U; 8. 20A-C

Tubifex benedii UDEKEM, 1855: 544.

? *Nais pusulosa* WILLIAMS, 1859: 96.

Clitellio ater CLAPARÈDE, 1862: 253, Pl. IV, fig. 7-11.

Pachydermon elongatum CLAPARÈDE, 1862: 254, Fig. 12.

Tubifex papillosus CLAPARÈDE, 1863: 25, Pl. XIII, fig. 14-15.

Clitellio benedii (Udekem). VAILLANT, 1868: 251.

Peloryctes inguitina ZENGER, 1870: 221.

Limnodrilus benedeni (Udekem). TAUBER, 1879: 71. VEJDOVSKY, 1884 b: 45.

Clitellio inguitinus (Zenger). CZERNIAVSKY, 1880: 356.

Clitellio ater Claparède LEVINSEN, 1884: 225. VEJDOVSKY, 1884 b: 45.

Lumbricillus verrucosus ÖRSTED, 1884: 68 (nomen nudem).

Hemitubifex ater (Claparède). BEDDARD, 1889: 485, Fig. A.

Hemitubifex benedii (Udekem). BEDDARD, 1889: 487, Pl. XXIII, fig. 6-9; 1895: 261.

Hemitubifex salinarium FERRONNIÈRE, 1889: 272, Pl. XIX, fig. 5-7.

Tubifex papillosus Claparède. BEDDARD, 1889: 487.

Clitellio (Clitellio) benedeni (Udekem). VAILLANT, 1890: 418.

(non) *Hemitubifex benedii* (Udekem). FRIEND, 1896: 128; 1897 c: 297.

(non) *Hemitubifex benedeni* var. *pustulatus*. FRIEND, 1898 a: 120.

Psammoryctes benedeni (Udekem). MICHAELSEN, 1900: 51.

Tubifex benedeni (Udekem). MICHAELSEN, 1900: 524; 1927: 18. SOUTHERN, 1909: 139; 1911: 5. GALLOWAY, 1911: 314.

Clitellio irrorata Verrill. MOORE, 1905 b: 384 (in part).

(non) *Hemitubifex benedeni* (Udekem). FRIEND, 1912 b: 290.

(non) *Hemitubifex pustulatus* Friend. FRIEND, 1912 b: 290.

Tubifex (Peloscolex) Benedeni Udekem. MICHAELSEN, 1912: 95.

Tubifex (Peloscolex) benedeni Udekem. STEPHENSON, 1922: 289.

Tubifex (Peloscolex) insularis Stephenson. STEPHENSON, 1922: 290.

Peloscolex insularis STEPHENSON, 1922: 290.

Peloscolex benedeni (Udekem). DITLEVSEN, 1904: 421. MICHAELSEN, 1927: 18, Fig. 20, 35. UDE, 1929: 96. STAMMER, 1932: 578. KNÖLLER, 1935: 480, Fig. 42. MARCUS, E., 1950: 3. MALEVITCH, 1951: 181. BÜLOW, 1957: 101. DAHL, 1960: 17. BRINKHURST, 1962 c: 321; 1963 a: 44; 1963 b: 712; 1963 c: 31, Fig. 8c; 1965: 133, Fig. 51K. BRINKHURST AND KENNEDY, 1962 b: 375. CEKANOVSKAYA, 1962: 280, Fig. 178. MOSZYNSKA, 1962: 29. HRABE, 1964: 110.

Peloscolex insularis Stephenson. MARCUS, E., 1950: 4. BRINKHURST, 1962 c:

321; 1963 a: 44. HRABE, 1964: 111.

$l=35$-55 mm, $s=75$-100. All setae indistinctly bifid with reduced upper teeth or simple-pointed, hair setae short in a few bundles of some (small) specimens. Vasa deferentia enter somewhat pear-shaped atria apically, large prostate glands attached near apex, ejaculatory ducts not distinct, penis sheaths cylindrical with recurved distal end. Body wall thinly but evenly covered with papillae.

Europe, Atlantic coast of N. America, brackish water and marine.

Peloscolex dukei Cook, 1970

Peloscolex dukei COOK, 1970 a: 492, Fig. 1.

$l=8$ mm, $s=36$. Prostomium and first segment retractile. Papillae densely covering the body apart from the prostomium and anteriormost segments. Dorsal setae from II, one or two short, bent hair setae with one or two bifid setae with curved tips and short, thin upper tooth. Hair setae non-serrate. Posterior dorsal setae one very short, thin hair seta scarcely longer than the single bifid seta, which has a very thin short upper tooth but a thickened lower tooth. Ventral setae one or two per bundle with upper tooth shorter and thinner than lower, posterior bundles with upper tooth very short, lower tooth somewhat thickened but less so than dorsal posterior setae. No genital setae. Vas deferens about one and a half times the length of the atrium, entering atrium on the opposite side to the large prostate gland; atrium cylindrical, bulging apically, bent over itself; penis with a somewhat elongate thimble-shaped penis sheath. Spermathecae in X, with spermatophores.

Beaufort Shelf Transect, N. Carolina, U.S.A. 19-200 m. (from 34° 19-34° 34·5 N and 75° 53-76·25·5 W).

Holotype: U.S.N.M. 35522.

Paratypes: U.S.N.M. 35523.

None of the specimens were sectioned, and so the internal anatomy other than the form of the male efferent ducts cannot be described. The male ducts and the spermathecae were in the normal position, however, and there is no reason to suppose that there is anything unusual about the rest of the anatomy.

The other marine and brackish water *Peloscolex* species with hair setae and bifid dorsal setae are *P. swirencowi* (Jaroschenko) and *Peloscolex apectinata* Br. The latter and *P. nerthoides* Br. were both described as varieties of *P. gabriellae* by Brinkhurst (1965) but are now regarded as separate species. *Peloscolex swirencowi* has posterior dorsal setae which all resemble hairs because the teeth of the bifid setae have been completely lost, but the hair setae are non-serrate (Hrabe, 1964). There is little difference between this species and *P. apectinata*, except that some of the posterior dorsal setae of the latter still have two short, fine parallel teeth which are very difficult to see, so that some of the apparently simple-pointed posterior dorsal setae of *apectinata* may in fact be bifid setae looked at from the side. In addition, the hair setae of the anterior bundles of *apectinata* bear a definite line of lateral hairs giving the setae a plume-like appearance when viewed at the appropriate angle. This is such an obvious feature once a hair seta is examined from the appropriate angle that I doubt if Professor Hrabe would have overlooked it in *P. swirencowi*. Hence *P. dukei* is easily distinguished from both *apectinata* and *swirencowi*.

A few specimens of *P. benedeni* (Udekem) have short hair setae in the dorsal bundles, but the other setae are rather characteristically bent, and are bifid with a reduced upper tooth, or simple-pointed without any broadening of the lower tooth in posterior bundles. The papillae are also rather widely spaced in *P. benedeni*.

Somewhat nearer to *P. dukei* may be the recently described species *P. euxinicus* Hrabe, in which the hair setae are short posteriorly but are straight, lacking the characteristic bent hairs of *dukei*, and the upper tooth of both dorsal and ventral setae in all bundles is never as reduced as it is in *dukei* and may be almost as long as the lower tooth anteriorly. There are no discrete papillae in *euxinicus* specimens examined so far (Hrabe, 1966), but this characteristic is subject to some variation. The male efferent ducts of *euxinicus* seem to differ from those of *dukei* in various features described, but not illustrated, in the original account.

Peloscolex euxinicus Hrabe, 1966
Fig. 8. 19H-K

Peloscolex euxinicus HRABE, 1966 : 57, Fig. 1-6.

$l=8$ mm. $s=53$. Prostomium non retractile. Dorsal bundles with 2-3 non-serrate straight hair setae and 2 bifid setae with teeth equally long, upper somewhat thinner than lower, posteriorly 1 hair and 1 bifid; ventral bundles with 2-3 bifid setae anteriorly, upper tooth thinner but about as long as the lower, posteriorly 1-2 bifid setae, upper tooth diverging from lower. Vasa deferentia long, enter apical ends of pear-shaped atria, compact prostate glands present, penes with thin cuticular sheaths. Body wall with a sheath.

Black Sea.

Peloscolex swirencowi (Jaroschenko, 1948)
Fig. 8. 18G-H; 8. 19A-G

Pristina papillosa CERNOSVITOV, 1935 b : 186.

Tubificioides heterochaetus LASTOCKIN, 1937 : 234, Fig. 2.

Pristina papillosa Cernosvitov. SPERBER, 1948 : 241. CEKANOVSKAYA, 1962 : 217, Fig. 132.

Tubificioides swirencowi JAROSCHENKO, 1948 : 62, Fig. 4-6.

Psammoryctes heterochaetus Lastockin. CEKANOVSKAYA, 1962 : 270, Fig. 170. BRINKHURST, 1963 a : 78.

Paranais papillosa Cernosvitov. BRINKHURST, 1963 a : 79 (laps pro *Pristina*).

Tubificoides heterochaeta Lastockin. BRINKHURST, 1963 a : 26.

Peloscolex svirenkoi HRABE, 1964: 101, Fig. 1-12.

$l=13$ mm, $s=50$. Prostomium non retractile. Dorsal anterior bundles with up to 4 smooth hair setae, and 1-3 bifid setae with upper tooth somewhat longer than lower, posteriorly (behind XIV at least) bifid setae replaced by simple-pointed setae; ventral anterior bundles with 2-4 bifid setae, teeth equally long, posteriorly 2 per bundle with upper tooth $1\frac{1}{2}$ times longer than lower, all ventral setae with teeth equally thick. Vasa deferentia little longer than atria, entering atria sub-apically opposite large prostates, atria largely cylindrical, penes elongate with cuticular sheaths, no ejaculatory ducts. Papillae from XII-XV, readily detached.

Black Sea.

Peloscolex apectinatus Brinkhurst, 1965

Peloscolex gabriellae var *apectinata*. BRINKHURST, 1965: 133, Fig. 5o.

Peloscolex gabriellae apectinata Brinkhurst. BRINKHURST AND COOK, 1966: 17.

$l=5$-20 mm. $s=$up to 70. Anterior dorsal bundles with 2-3 bent serrate hair setae and 2-3 bifid setae with short teeth, the upper tooth much thinner than but about as long as the lower, posteriorly 1 hair and 1 hair-like seta, most of the latter with long closely parallel teeth, many possibly simple-pointed. Ventral anterior setae with teeth short, upper tooth thinner than lower, teeth about equally long, posteriorly single with upper tooth thin, projecting straight along the line of the shaft, lower tooth thick, curved away from the shaft, teeth more or less equally long. Vasa deferentia about twice as long as atria, entering posterior aspect of atria subapically, penis sheaths indistinct, truncate cones. Body wall with a thin covering of foreign matter, no papillae.

N. America. Brackish water.

Peloscolex intermedius Cook, 1969

Peloscolex intermedius COOK, 1969 a: 11, Fig. 2.

$l=8$ to 10 mm, $w=0.28$ to 0.45 mm, $s=42$. Prostomium small, conical, with small papilla anteriorly. Body wall non papillate to densely, but very finely, papillate. Dorsal setae from II to VI (VII) 3 bifids plus 1 to 3 hairs per bundle; bifids 50 to 80 μ long with equal teeth which become increasingly shorter in more posterior segments or whose upper tooth becomes increasingly reduced; hairs 110 to 160 μ long bearing a few very short, indistinct lateral hairs which are often not detectable; from VII (VIII) to terminal segment 3 simple-pointed, hair-like crotchets, 75 to 100 μ long plus 3 true hair setae, 110 to 160 μ long present per bundle. Ventral setae anteriorly 3 to 4 per bundle, posteriorly 2 to 3 per bundle; ventral bifids 60 to 80 μ long with approximately equal teeth. 1 unmodified ventral bifid per bundle on X and XI. One pair spermathecal pores situated just anterior to and in line with ventral setae. One pair male pores.

Coiled vasa deferentia, 15 μ diameter, 1 mm long, joins atria subapically and dorsally. Vasa deferentia 5 to 7 times as long as atria. Atria elongate with long axis directed posteriorly with muscle bands arranged circularly. Atria 150 to 220 μ long, 60 to 90 μ wide, joined to cylindrical, cuticularized penes, 75 to

100 μ long, 31 to 37 μ wide, by short discrete proximal ducts. Large prostate glands join atria ventrally and subapically to medially. Spermathecae with sacciform ampullae and long discrete ducts which have a bulbous swelling proximally. Spermatophores spindle-shaped 130 μ long, 35 μ diameter at the median swelling.

Types: USNM.

Locality: Cape Cod Bay, Massachusetts, USA; 41° 55′ 5″ N, 70° 21′ 1″ W.

Depth: 30 metres.

Distribution: Massachusetts.

The following table illustrates differences between *Peloscolex intermedius* and its two most closely related species, *P. apectinatus* and *P. swirencowi*.

Character	P. intermedius	P. apectinatus	P. swirencowi
Hair setae	with sparse lateral hairs	Serrate	Smooth
Anterior dorsal setae	Bifid to VI	Bifid	Bifid to VIII
Posterior dorsal setae	Single-pointed elongate	At least some bifid with parallel teeth	Single pointed elongate
Ventral setae; upper tooth compared to lower	Longer	Same or shorter	Longer
Length vas deferens / length atrium	4·5	less than 3·0	2·5
Length penis / diameter penis	2·4-2·6	1·5-1·6	1·0

P. intermedius is closely related to *P. apectinatus* Brinkhurst, 1965 and *P. swirencowi* (Jaroschenko, 1948). These three species, together with *P. euxinicus* Hrabe, 1966, *P. gabriellae* Marcus, 1950 and *P. nerthoides* Brinkhurst, 1965, form a complex of small, sparsely or finely papillate, marine worms whose male genitalia bear a strong resemblance to those of *Tubifex*. It is suspected that future work on this group may reveal more intermediate types in this complex of species which will invalidate many of the entities or reduce them to sub-specific rank, and that such work will make necessary a critical re-examination of the generic limits of *Peloscolex* and *Tubifex*.

Peloscolex nerthoides BRINKHURST, 1965

Peloscolex gabriellae var *nerthoides* BRINKHURST, 1965: 134.

Peloscolex gabriellae nerthoides Brinkhurst. BRINKHURST AND COOK, 1966: 17.

l=5-70 mm, s=35-50. Dorsal anterior bundles with 2-3 smooth bent hair setae, 2-3 pectinate setae with long somewhat parallel teeth, the upper thinner than the lower, hair and pectinate setae single posteriorly with shorter teeth. Ventral anterior bundles with 1-3 setae with teeth long, upper tooth thinner than lower, about as long as or shorter than lower, teeth diverging narrowly or more or less parallel, posteriorly single setae, upper tooth thinner and shorter than lower. Vasa deferentia about twice as long as atria, entering posterior aspect of atria subapically, almost opposite to prostate glands, atria largely cylindrical, penis sheath elongate thimble-shaped. Body wall with a fine coating of foreign matter, no papillae.

N. America. Brackish water.

Peloscolex speciosus (Hrabe, 1931)
Fig. 8. 17H-S; 8. 18A-F

Tubifex (Tubifex) speciosus HRABE, 1931 a: 24, Fig. 4a-f.

Tubifex speciosus Hrabe. STAMMER, 1932: 578. HRABE, 1962 a: 307.

Peloscolex zavreli HRABE, 1942: 23, Fig. 1-10.

Peloscolex zavreli Hrabe. MARCUS, E., 1950: 4. BRINKHURST, 1963 a: 41, Fig. 28a-d. HRABE, 1964: 109.

Peloscolex speciosus (Hrabe). BRINKHURST, 1963 a: 42, Fig. 15a-c.

Peloscolex simsi BRINKHURST, 1966: 737, Fig. 6a-e.

Tubifex speciosus speciosus Hrabe. HRABE, 1966: 68, Fig. 27.

Tubifex speciosus monfalconensis HRABE, 1966: 68, Fig. 28.

l=6-12 mm, s=40-70. Dorsal bundles anteriorly with 1-3 hair setae, 1-3 pectinate setae with narrow lateral teeth with fine intermediate teeth. Posteriorly 1 short hair seta and one bifid seta, or hair setae missing in most posterior segments; ventral bundles with up to 5 setae anteriorly, upper tooth longer and thinner than the lower, posteriorly 1-2 with teeth equally long, ventral setae of both X and XI long, straight, with hollow distal ends. Vasa deferentia long, divided into thin and thick positions, atria rather narrow, tubular with small prostate glands attaching one third the way from the proximal end, ejaculatory ducts lead to long penes with cuticular sheaths, penis sacs compounded with penial setal sacs with glands. Cuticular penis sheaths cylindrical. Spermathecae with elongate ampullae, spermatophores elongate. Body wall non-papillate, with glandular belts or a slight sheath, appearing grooved.

Europe.

The placing of *speciosus* in *Peloscolex* rather than *Tubifex* is rejected by Hrabe (1964), as there was no sign of a covering derived by secretion on the body wall of the original specimens although the epidermis is clearly glandular. According to Hrabe (1964) "*Tubifex speciosus* with its sub-species is a typical representative of the genus *Tubifex*. It possesses hair and pectinate setae in the anteclitellar bundles. Its bipartite sperm duct is much longer than the fusiform atrium." The differences between this and any other *Tubifex* species are extensive, and include the form of the atrium, the prostate, the penis sheath and the presence of genital setae. The genital setae of *speciosus* are unique, in that both spermathecal and penial setae are modified in a species in which a cuticular penis sheath is well developed. The penial setae lie in large setal sacs with specialized glands, these setal sacs being united with the penis sacs distally. *Peloscolex zavreli* was described by Hrabe in a later paper (1942), but the differences between this and *speciosus* are scant. According to the original illustrations, the male efferent ducts and setae of these two are very similar. Only the details of the form of the pectinate setae and the exact nature of the body wall can separate the two, and these differences are at a level subject to variation within species. A single British specimen was described as a separate species (*simsi*) by Brinkhurst (1966 a) as the form of the pectinate setae appeared to be intermediate between those of *P. speciosus* and *zavreli*. As the original limits of *speciosus* have been widened by the description of the sub-species *monfalconensis* (the pectinate setae of which are more like those of *simsi* or *zavreli* than those of *speciosus speciosus*) the distinction is no longer valid, and all of these entities are merged within *P. speciosus*.

Peloscolex kurenkovi Sokolskaya, 1961
Fig. 8. 17A-G

Peloscolex kurenkovi SOKOLSKAYA, 1961 a: 59, Fig. 4.

Peloscolex kurenkovi Sokolskaya. BRINKHURST, 1963 a: 46. HRABE, 1964: 107, Fig. 21.

$l=20$ mm, $s=63-70$. Prostomium and I retractile. Dorsal anterior bundles with 3-4 smooth, sabre-shaped hair setae, up to 4 pectinate setae with parallel teeth and fine intermediate teeth, posteriorly 2 hair setae with ? pectinate setae. Ventral setae of II-III with sharply pointed seta with one bifid seta with long upper tooth, other bundles with 2-4 of the latter, which are markedly sigmoid, posterior ventral setae 2 per bundle, the upper tooth gradually becoming shorter until equal to or even shorter than the lower, but thinner, spermathecal setae long, thin, hollow ended, in a glandular sac. Vasa deferentia very long, atria strongly horse-shoe shaped. Prostate glands very large, ejaculatory ducts discrete, penes without cuticular sheaths. Body wall papillate, also 2 rows of sensory papillae anteriorly.

Asiatic Russia, Kamchatka.

Peloscolex beetoni Brinkhurst, 1965
Fig. 8. 16G-M

Peloscolex beetoni BRINKHURST, 1965: 136, Fig. 6c-j.

$l=25$ mm, $s=47$. Prostomium and I retractile. Dorsal bundles with 2-3 hair setae and 1-2 palmate/pectinate setae, ventral bundles of II-VI either 1 with upper tooth longer than the lower, the other with teeth equally long, or worn until 1 simple-pointed, posterior bundles with 1 seta with short thin upper tooth or bluntly simple-pointed, spermathecal setae long, straight, hollow ended. Atria large, penes with conical penis sheaths, thicker proximally than distally. Spermathecae with extremely elongate spermatophores. Body wall closely covered with papillae.

Lake Tahoe, U.S.A.

Peloscolex multisetosus (Smith, 1900)
Fig. 8. 16B-F

Embolocephalus multisetosus SMITH, 1900: 452, Pl. XXXIX, fig. 1-5.

Tubifex multisetosus (Smith). MICHAELSEN, 1900: 525. COLLINS, 1937: 199.

Embolocephalus multisetosus Smith. GALLOWAY, 1911: 314.

Peloscolex multisetosus (Smith). MARCUS, E., 1950: 4. BRINKHURST, 1963 a: 42; 1965: 138, Fig. 6 (r-u). HRABE, 1964: 110.

Peloscolex sp. WURTZ and DOLAN, 1960: 470.

Peloscolex multisetosus typica BRINKHURST AND COOK, 1966: 16.

Peloscolex multisetosus longidentus BRINKHURST AND COOK, 1966: 16.

l=19-35 mm, s=49-106. Prostomium and I retractile. Dorsal anterior bundles with 3-14 hair setae, 1-5 pectinate setae, somewhat lyre-shaped with fine intermediate teeth; anterior ventral bundles with 2-3 bifid crotchets, upper tooth long and thin, often very long in II, posteriorly setae broader, the upper tooth reduced, distal end curved over shaft, or all ventral setae with upper tooth longer than and at least as thick as lower. Vasa deferentia long, atria horse-shoe shaped, large prostate glands, no distinct ejaculatory duct, penes without penis sheaths. Body wall covered with a sheath anteriorly, extremely large sensory papillae protrude through the sheath in 2 rings on each segment with other scattered papillae.

N. America.

The only other species with widely spaced papillae is *P. stankovici* from Jugoslavia.

Sub-species *multisetosus*

Posterior ventral setae with upper tooth short, lower tooth strongly recurved.

Sub-species *longidentus*

Posterior ventral setae with upper tooth long, as long and as thick as lower, not recurved.

Peloscolex aculeatus Cook, 1970

Peloscolex aculeatus COOK, 1970 a: 493, Fig. 2.

l=up to 15 mm, w=0·17 to 0·24 mm, s=about 70. Prostomium as long as it is broad at peristomium, with very small papilla anteriorly. Posteriorly body wall bears a number of concentric ridges in the furrows of which substrate particles often accumulate. Dorsal setae anteriorly, 2 bifids per bundle with slender, only slightly divergent teeth and 1 to 2 thin intermediate teeth, major upper tooth thinner and shorter than lower, also 2 hair setae per bundle with short, very sparse, lateral hairs, posteriorly usually 1 bifid and 1 short hair seta per bundle. Ventral setae anteriorly, 2 bifids with slender, very slightly diverging teeth, upper of which very thin and a little shorter than lower; posteriorly 1 or 2 bifids of same form. Setae reach maximum length in about VI to VIII, then become shorter in more posterior segments. Paired spermathecal pores in line with, and anterior to ventral setae of X. Paired male pores in place of ventral setae of XI.

Pharyngeal glands in IV and V. Pair of very long, coiled vasa deferentia, extend into XII and join atria subapically. Atria cylindrical, curved into comma-shape apically. Prostate glands penetrate atria opposite the vasa deferentia, (inside the comma). Atria terminating in cuticularized penis sheaths which bear series of cuticular hooks near their terminal ends. Spermathecae with discrete ducts with proximal dilation, and pear-shaped ampullae. Spermatophores thick-walled, cocoon shaped.

Locality: 39° 46·5′ N, 70° 43·3 W, 1330-1470 metres, Atlantic ocean, 39° 4·20′ N, 70° 39′ W, 2000 metres.

Peloscolex inflatus (Michaelsen, 1902)
Fig. 8. 16A

Tubifex inflatus MICHAELSEN, 1901 b: 141, Pl. I, fig. 8-10.

Tubifex inflatus Michaelsen. MICHAELSEN, 1902: 43.

Tubifex (Peloscolex) inflatus Michaelsen. MICHAELSEN, 1905 a: 23.

Peloscolex inflatus (Michaelsen). MICHAELSEN, 1926 a: 157. MARCUS, E., 1950: 4. CEKANOVSKAYA, 1962: 284. BRINKHURST, 1963 a: 41. HRABE, 1964: 110.

$l=40$ mm, $s=120$-140. Prostomium and I retractile. Dorsal anterior bundles with 2-4 sabre-shaped hair setae, 2-4 ribbed or pectinate setae. 2 of each posteriorly. Anterior ventral bundles with 3-4 bifid setae on II-IV, with upper tooth shorter than lower, the rest with 2 bifid setae or rarely simple-pointed setae, usually upper tooth as long and thick as lower or upper somewhat shorter. Vasa deferentia long, coiled, atria conical with egg-shaped prostate glands, ejaculatory ducts lead to penes without cuticular sheaths. Body wall papillate, ? fused to rings on XV-XVIII.

Lake Baikal.

The description of this species lacks detail, and it has always been assumed to be distinct because of its presence in L. Baikal, where many endemic species are found. Most of the tubificids described from the lake by Prof. Michaelsen from time to time urgently require further study, but I have not been able to gain access to collections. As the Naididae and Lumbriculidae of Lake Baikal have been reviewed recently, it may well be that a revision of the tubificids will be available before too long.

Peloscolex ferox (Eisen, 1879)
Fig. 8. 15Q-T

? *Nais filiformis* (part?). WILLIAMS, 1852: 182, Pl. III, fig. 8, Pl. VIII, fig. 72; 1859: 93.

? *Nais papillosa* KESSLER, 1868: 105, Pl. VI, fig. 2.

Spirosperma ferox EISEN, 1879: 10.

Spirosperma ferox Eisen. LEVINSEN, 1884: 224. VEJDOVSKY, 1884 b: 45. STOLC, 1886: 644. 1888, 40, Pl. I, fig. 3, Pl. III, fig. 4, 10-12, Pl. IV, fig. 1-3, 13a. EISEN, 1886: 884, Pl. II, fig. 2a-g, Pl. III, fig. 2g-h. BENHAM, 1892: 207, Pl. VII, fig. 36c, d. FRIEND, 1912 b: 291.

Embolocephalus plicatus RANDOLPH, 1892: 469, Pl. XIX.

Embolocephalus plicatus Randolph. BEDDARD, 1895: 273.

Spirosperma papillosus BEDDARD, 1895: 263. BRETSCHER, 1900 a: 5.

(?) *Hemitubifex benedii* (Udekem). FRIEND, 1896: 128; 1897 c: 297; 1912 b: 290.

(?) *Hemitubifex benedeni* var *pustulatus* FRIEND, 1898 a: 120.

Peloscolex ferox (Eisen). MICHAELSEN, 1900: 51. DITLEVSEN, 1904: 420, Pl. XVI, fig. 8. HRABE, 1929: 2; 1938 a: 227; 1962 a: 308. Fig. 25-35; 1964: 110. UDE, 1929: 94. CERNOSVITOV, 1942 a: 214; 1945: 526. BERG, 1948: 53. JAROSCHENKO, 1948: 69. MARCUS, E., 1950: 3. JUGET, 1958: 89, Fig. 14C. BRINKHURST, 1960: 407, Fig. 12c-f; 1962 c: 321; 1963 a: 42, Fig. 26a. c, d; 1963 b: 31, Fig. 8d; 1965: 133, Fig. 5F-H. SOKOLSKAYA, 1961 a: 64, Fig. 6. BRINKHURST AND KENNEDY, 1962 a: 185; 1962 b: 375. CEKANOVSKAYA, 1962: 281, Fig. 179. MOSZYNSKA, 1962: 29. KENNEDY, 1964: 233. POPESCU-MARINESCU et al; 1966: 163.

Tubifex ferox (Eisen). MICHAELSEN, 1900: 252; 1901 b: 141. BRETSCHER, 1905: 664. SOUTHERN, 1909: 139.

Tubifex plicatus (Randolph). MICHAELSEN, 1900: 524.

Embolocephalus plicatus var *pectinata* Randolph. BRETSCHER, 1900 b: 446, Pl. XXXIII, fig. 2, 3.

Psammoryctes plicatus (Randolph). MICHAELSEN, 1900: 50.

Psammoryctes plicatus var *pectinatus* (Randolph). BRETSCHER, 1901: 196.

Tubifex (Peloscolex) ferox (Eisen). MICHAELSEN, 1903: 196. BRETSCHER, 1904: 260. PIGUET, 1906: 391; 1913: 131. PIGUET AND BRETSCHER, 1913: 74, Fig. 18. LASTOCKIN, 1927 a: 67. CERNOSVITOV, 1928: 6.

Tubifex (Peloscolex) ferox var *pectinatus* (Eisen). PIGUET, 1906 a: 396; 1913: 132.

(?) *Hemitubifex pustulatus* Friend. FRIEND, 1912 b: 290.

(?) *Hemitubifex benedeni* (Udekem). FRIEND, 1912 b: 290.

l=15-40 mm, s=c. 50. Prostomium and I retractile. Dorsal bundles with up to 7 sabre-shaped hair setae, up to 5 type-shaped pectinate setae; anterior ventral bundles with up to 7 bifid setae, occasionally with a rudimentary intermediate tooth, upper tooth a little longer than the lower, posteriorly fewer with shorter upper tooth. Vasa deferentia long, atria horse-shoe shaped, prostate glands large, ejaculatory ducts lead to penes with cuticular sheaths. Spermethecal ampullae long, spermatophores thin and very long. Body wall closely covered with papillae when developed.
Holarctic.

Peloscolex werestschagini Michaelsen, 1933

Peloscolex werestschagini MICHAELSEN, 1933: 327, Fig. 1-3.

Peloscolex werestschagini Michaelsen. MARCUS, E., 1950: 4. CEKANOVSKAYA, 1962: 284. BRINKHURST, 1963 a: 41. HRABE, 1964: 109.

l=3-5 mm, s=29. Dorsal bundles with 1 hair setae and some (?) crotchets. Body completely enclosed in a heavy sheath.
Lake Baikal.

The species needs re-describing. The male efferent ducts were said to resemble those of *P. inflatus* (q.v.).

Peloscolex tenuis Hrabe, 1931
Fig. 8. 15F-P

Peloscolex tenuis HRABE, 1931 a: 40, Fig. 7a-9.

Peloscolex tenuis Hrabe. MARCUS, E., 1950: 4. BRINKHURST, 1963 a: 41, Fig. 27. HRABE, 1964: 107, Fig. 13.

Peloscolex apapillatus LASTOCKIN AND SOKOLSKAYA, 1953: 411.

(?) *Peloscolex scodraensis* HRABE, 1958: 347, Fig. 13-17.

Peloscolex kamtschaticus SOKOLSKAYA, 1961 a: 56, Fig. 2.

Peloscolex apapillatus Lastockin. SOKOLSKAYA, 1961 a: 57, Fig. 3; 1961 b: 92. CEKANOVSKAYA, 1962: 278, Fig. 175. BRINKHURST, 1963 a: 42. HRABE, 1964: 109.

Peloscolex kamtschaticus Sokolskaya. BRINKHURST, 1963, 1963 a: 46. HRABE, 1964: 107, Fig. 14.

(?) *Peloscolex scodraensis* Hrabe. BRINKHURST, 1963 a: 44, Fig. 30. HRABE, 1964: 107, Fig. 15.

l=20 mm, s=55-85. Prostomium and segment 1 retractile. Dorsal bundles with 3-7 sabre-shaped hair setae, up to 4 type-shaped pectinate setae, ventral bundles with up to 5 setae anteriorly, mostly 2, with upper tooth longer and thinner than lower, posterior setae single, thick, with upper tooth shorter than lower. No modified genital setae. Vasa deferentia long, atria saccular, horse-shoe shaped, with large prostates attached to concave wall of atrium medially, ejaculatory duct leads to penis with cuticular sheaths. Spermathecae with long sacs, with spermatophores. Body wall with a sheath, sensory papillae present. Eurasia.

Sub-species *tenuis* sub-species novum.
Fig. 8. 15L-P

Ventral posterior setae strongly recurved.
L. Ochrid, Prespa, Jugoslavia.

Sub-species *kamtschaticus* sub-species novum
Ventral posterior setae with teeth diverging widely.
Kamchatka.

Sub-species *apapillatus* sub-species novum
Fig. 8. 15F-K

Ventral posterior setae bent, not fully recurved, teeth at a narrow angle to each other.
Amur basin, U.S.S.R. (? L. Skadar, Jugoslavia-*scodraensis*).

The differences between *P. tenuis, P. apapillatus, P. kamtschaticus* and the dubious *P. scodraensis* are few, and those that have been reported seem open to question. For instance, Sokolskaya (1961) did not perceive the cuticular penis sheaths of *P. kamtschaticus*, reported later in cotypes by Hrabe (1964). The latter did not mention the apparent absence of the penis sheaths in *P. apapillatus*, a species described by Lastockin in a paper published posthumously.

Hrabe (1964) published a key to the species of *Peloscolex* and these four species (together with *P. zavreli*) are treated in a group of related couplets, some of which are incomplete (i.e. one half of the couplet describes a characteristic not mentioned in the other half) and most concern trivial differences such as setal lengths and the position of spermathecal pores. The description of the posterior ventral setae of *P. kamtschaticus* by Hrabe (1964) does not tally with that of the original account, and the degree of recurvature of the lower tooth of posterior ventral setae has been found to vary within what I consider to be the single species *P. multisetosus*, the distinction being recognized at the sub-specific level merely to ensure that the characteristic be recorded by subsequent workers.

Peloscolex superiorensis Brinkhurst and Cook, 1966
Fig. 8. 15A-E

Peloscolex superiorensis BRINKHURST AND COOK, 1966: 16, Fig. 7J-N.

$l=c.$ 7 mm. Anterior dorsal bundles with 4-5 serrate hair setae, c. 4 pectinate setae with broad lateral teeth and few distinct intermediate teeth or latter somewhat fused together, posteriorly hair setae with bifid setae with teeth about equal, some with small intermediate tooth; ventral setae 4-6 per bundle, upper tooth as long as but thinner than lower, no modified genital setae. Atria horseshoe shaped, large prostate glands, penes with long cylindrical sheaths covering penis sac walls basally. No papillae.

N. America—Lake Superior, and St. Marys River.

The similarity between this species and others in *Peloscolex* is slight, and its placing here depends largely on the form of the penis sheaths and atria.

Peloscolex freyi Brinkhurst, 1965
Fig. 8. 14J-M

Peloscolex freyi Brinkhurst, 1965: 139, Fig. 6v-y.

l=20-25 mm, *s*=140. Setae up to 8 per bundle anteriorly, upper tooth thinner than lower but teeth equally long, sometimes both worn down, only 2-3 posteriorly, spermathecal setae straight, long, hollow-tipped. Vasa deferentia short, atria strongly curved, short and broad, with large prostates on concave side, penes with thin cuticular sheaths covering walls of penis sacs.

N. America, Great Lakes and N. Carolina.

Peloscolex heterochaetus (Michaelsen, 1926)
Fig. 8. 14G-I

Limnodrilus heterochaetus MICHAELSEN, 1926 b: 22, Fig. A (a-d); 1927: 17, Fig. 19.

Limnodrilus heterochaetus Michaelsen. BÜLOW, 1957: 99. POPESCU-MARINESCU et al., 1966: 163.

Peloscolex heterochaetus (Michaelsen). de VOS, 1936: 86, Fig. 1, 2. MARCUS, 1942: 156. MARCUS, E., 1950: 4. CEKANOVSKAYA, 1962: 279, Fig. 176, 177. BRINKHURST, 1963 a: 44, Fig. 31; 1963 b: 713. HRABE, 1964: 109.

l=7-9 mm, *s*=46. Setae of anterior bundles up to 5 bifid setae, postclitellar bundles with 1-2 sharply pointed setae. Vasa deferentia short, atria more or less tubular, prostate glands small, compact, penes with swollen bases, no cuticular penis sheaths. Scattered papillae on body wall of some specimens.

Europe, N. Sea coasts, Black Sea estuary of Danube (?).

Peloscolex pigueti Michaelsen, 1933

Trachydrilus plicatus (Randolph). PIGUET, 1928: 93, Fig. 6.

Peloscolex pigueti MICHAELSEN, 1933: 326.

Peloscolex pigueti Michaelsen. MARCUS, E., 1950: 4. BRINKHURST, 1963 a: 41. HRABE, 1964: 109.

l=3 mm, *s*=24. 3 bifid setae per bundle, set on pads. Body wall with a sheath forming rings on IX-XII, papillae also present.

Lake Naticocha, Peru.

The species was included in *Peloscolex* following the description of *P. werestchagini*, but badly needs re-describing. *P. freyi* is the only other freshwater species lacking hair setae, but there are considerable differences between the two although it should be remembered that no mature specimen of *P. pigueti* has been seen.

Peloscolex gabriellae Marcus, 1950

Peloscolex gabriellae MARCUS, E., 1950: 1, Pl. I, fig. 1-4.

"Oligochaeta S, L, S.L., S.H., P." JONES, 1961: 245.

Peloscolex gabriellae Marcus. BRINKHURST, 1963 a: 44; 1963 b: 173; 1965: 133, Fig. 5L-T. HRABE, 1964: 110. BRINKHURST AND COOK, 1966: 17.

(non) *Peloscolex gabriellae* var *apectinata* BRINKHURST, 1965: 133, Fig. 50(T).

(non) *Peloscolex gabriellae* var *nerthoides* BRINKHURST, 1965: 34.

l=10-20 mm, s=40-60. ? Prostomium retractile. Anterior bundles of setae with up to 6 bifid setae, teeth more or less equal, often fewer ventrally with upper tooth shorter and thinner than lower. Vasa deferentia short entering atria opposite large prostate glands, no ejaculatory ducts. Penes with thin conical sheaths. Spermathecae with short spermatophores. Body wall variable from naked to papillate from III posteriad.
Atlantic coasts, N. and S. America, Pacific coast, N. America.

The varieties of this species described by Brinkhurst (1965) are here regarded as species.

SPECIES INQUIRENDAE

Peloscolex marinus (Ditlevsen, 1904)

Tubifex marinus DITLEVSEN, 1904: 421, Pl. XVI, Fig. 9-11.

Peloscolex marinus (Ditlevsen). MARCUS, E., 1950: 4. BRINKHURST, 1963 a: 46; 1963 b: 713 (sp. dub.).

1-3 short bent hair setae, 2-3 bifid setae in anterior dorsal bundles, 3-5 setae in ventral bundles, upper tooth longer and thinner than the lower. Vasa deferentia long, atria broad, prostate glands large, cuticular penis sheaths present.

This species presumably has a papillate body wall, as it was said to be related to *P. ferox* (as *Embolocephalus plicatus*) by Ditlevsen (1904). It is unidentifiable from the original description. Several brackish water and marine *Peloscolex* species have these characteristically bent hair setae.

Peloscolex evelinae Marcus, 1942=*Slavina evelinae* (Naididae—q.v.).

Spilodrilus stellatus Piguet, 1928. Gen. et sp. dubia. This species probably belongs in *Peloscolex* but it is impossible to determine from the very brief original account. The single specimen was found in L. Huaron, S. America.

Antipodrilus gen. nov.

Type species: Tubifex davidis BENHAM

Male efferent ducts with long narrow vasa deferentia, small globular atria narrowing to tubular ejaculatory ducts terminating in non-cuticular penes. Prostate glands solid, joining atria anteriorly just below union of vasa deferentia and atria. Spermathecae present. Coelomocytes absent.
Australia, New Zealand.

There is considerable similarity between *Tubifex, Ilyodrilus* and *Antipodrilus*, but the small globular atria of the latter, which narrow so abruptly that the term ejaculatory ducts can be applied to the ectal end of the tube, are rather different to the atria of the other two genera. The absence of penis sheaths and the presence of modified spermathecal setae also serve to distinguish the new genus from these two. The differences between *Antipodrilus* and other genera of the Tubificidae may be inferred from the key to the genera of the Tubificidae.

Antipodrilus davidis (Benham, 1907) comb. nov.

Tubifex davidis BENHAM, 1907: 252, Pl. XLVI, fig. 1-6.

Tubifex davidis Benham. HRABE, 1962 a: 307; 1966: 74.

Euilyodrilus heuscheri (Bretscher). BRINKHURST, 1963 a: 49 (partim).

l=25-40 mm. Dorsal setae with 3-7 hair setae, 2-4 pectinate setae, ventral bundles with 3-5 setae, anteriorly the upper tooth longer and thinner than the lower. Spermathecal setae in X modified, hollow ended or with short lower tooth. Vasa deferentia long, atria small, globular with anterior prostates, atria narrowing abruptly to form ejaculatory ducts, penes small without thickened cuticular sheaths. Spermathecae present. No coelomocytes.
Australia, New Zealand.

Having examined the original material from the Australian Museum and some specimens collected at Rotorua, New Zealand, by Dr. G. R. Fish, it is now necessary to recognize the existence of this monotypic genus, the single species having earlier been regarded as synonymous with *Potamothrix* (then *Euilyodrilus*) *heuscheri* (Brinkhurst, 1963). The New Zealand specimens were rather small (10 mm. *c.* 40 segments) and there is no record of the number of segments in the larger worms described by Benham (1907).
(A second species has been found in Australia.)

SUB-FAMILY AULODRILINAE SUBFAM. NOV.

Prostate glands more or less solid but attached to the atria by broad bases. Sperm in bundles in spermathecae, not as spermatophores. Coelomocytes absent.

Type genus: Aulodrilus BRETSCHER

GENUS **Aulodrilus** BRETSCHER, 1899

Type species: Aulodrilus limnobius BRETSCHER

Male efferent ducts with short to very short vasa deferentia, globular or bean-shaped to elongate cylindrical atria with solid prostate, large eversible pseudopenes. Spermathecae with sperm masses, or spermathecae absent. Penial setae spoon-shaped or absent. No coelomocytes. Inhabit tubes, using unsegmented posterior end as respiratory organ.
    Cosmopolitan.

Species of *Aulodrilus* reproduce asexually, and are commonly found to have the reproductive organs located more anteriorly than usual. Degenerate testes often appear in maturing worms in the segment in front of the fully developing testes, but no extra male ducts have been described. The posterior end of the body is usually thin, without setae, and more or less unsegmented. This may be a region of regeneration but has usually been described as functioning as a respiratory organ. The hair setae, when present, are short and bent.

Aulodrilus

1.	Hair setae present	2
—	Hair setae absent	4
2.	Hair setae present from II, dorsal setae bifid or pectinate from II, no penial setae	*A. pluriseta*
—	Hair setae often absent from II or III, dorsal setae change in form around VI, penial setae modified	3
3.	Dorsal setae behind VII or so oar-shaped	*A. pigueti*
—	Dorsal setae behind VII with several equal-sized teeth	*A. pectinatus*
4.	Setae bifid with lateral webs	*A. limnobius*
—	Setae of at least mid and hind region broad with many teeth	*A. americanus*

Aulodrilus limnobius Bretscher, 1899
Fig. 8. 23G, H

Aulodrilus limnobius BRETSCHER, 1899: 388.

Aulodrilus limnobius Bretscher. MICHAELSEN, 1900: 55. PIGUET AND BRETSCHER, 1913: 58. KOWALEWSKI, 1914: 25. IASTOCKIN, 1927 a: 67;

1927 b: 9. UDE, 1929: 97. CERNOSVITOV, 1930: 321; 1935 a: 1; 1935 a: 35; 1942 a: 213. JAROSCHENKO, 1948: 69. HRABE, 1952: 5. JUGET, 1958: 89. SOKOLSKAYA, 1961 b: 93. CEKANOVSKAYA, 1962: 224, Fig. 134. MOSZYNSKA, 1962: 30. BRINKHURST, 1963 a: 66, Fig. 2; 1965: 150, Fig. 90; 1966: 732, Fig. 4H. BRINKHURST AND COOK, 1966: 19, Fig. 7(T-U).

Paranais tenuis CERNOSVITOV, 1937: 143, Fig. 21-24.

(non) *Aulodrilus limnobius* Bretscher. MARCUS, 1144: 78, Fig. 69-76.

Aulodrilus tenuis (Cernosvitov). MARCUS, E., 1947: 13.

$l = 12$-15 mm, $s = 80$. Up to 10 setae per bundle anteriorly, all bifid setae with the upper tooth shorter and thinner than the lower, those of the first bundles shorter and thicker and more strongly sigmoid than the rest, setae of mid and posterior segments with lateral wings. Vasa deferentia long, atria long, cylindrical, eversible pseudopenes.
Cosmopolitan.

Specimens described as *Paranais tenuis* by Cernosvitov (1937) were recognized as *A. limnobius* by Marcus (1944). This synonymy was rejected by Cernosvitov in correspondence with E. du B.-R. Marcus (quoted in E. Marcus, 1947) but the differences are trivial.

The lateral wings on the setae are easily overlooked when they are examined laterally in order to observe the teeth. When viewed from in front or behind, the teeth project as a single tooth on top of a wide palm. This characteristic has been observed on material from most parts of the world.

Aulodrilus pluriseta (Piguet, 1906)
Fig. 8. 23J-N

Naidium pluriseta PIGUET, 1906 b: 218.

Aulodrildus pluriseta (Piguet). PIGUET AND BRETSCHER, 1913: 118, Fig. 1, 2. KOWALEWSKI, 1914: 600. LASTOCKIN, 1927 a: 67; 1927 b: 7. CERNOSVITOV, 1928: 4, Pl. I, fig. 1; 1930: 321; 1939 a: 35 UDE, 1929: 97. MALEVITCH, 1930: 83. STAMMER, 1932: 579. HRABE, 1937 b: 3; 1939 a: 221; 1952: 5; 1954: 302. BERG, 1948: 53. JAROSCHENKO, 1948: 69. JUGET, 1958: 89. BRINKHURST, 1962 c: 322; 1963 a: 66, Fig. 2, 50-53; 1963 c: 31, Fig. 8h; 1965: 150, Fig. 9 (R, S). BRINKHURST AND KENNEDY, 1962 a: 187; 1962 b: 375. CEKANOVSKAYA, 1962: 225, Fig. 135. MOSZYNSKA, 1962: 30. KENNEDY, 1964: 235. NAIDU, 1965: 466, Fig. 2(a-c); 1966: 221.

Aulodrilus trivandranus AIYER, 1925: 36, Fig. 6-9.

Aulodrilus trivandranus Aiyer. STEPHENSON, 1925 b: 884. Aiyer, 1929 a: 43. MICHAELSEN AND BOLDT, 1932: 598. MARCUS, 1944: 76.

? *Aulodrilus* sp. CERNOSVITOV, 1938 b: 290, Fig. 47-52.

? *Aulodrilus japonicus* YAMAGUCHI, 1953: 298, Fig. 12, Pl. VII, fig. 5-7.

Aulodrilus pleuriseta (Piguet). BRINKHURST, 1960: 408, Fig. 13 (laps pro *pluriseta*).

? *Aulodrilus japonicus* Yamaguchi. BRINKHURST, 1963 a: 70.

l=10-25 mm, *s*=65-105. Anterior dorsal bundles with up to 8 short bayonet-shaped hair setae, and up to 10 setae with 1 to several upper teeth which are thinner and shorter than the lower, anterior ventral setae up to 16 per bundle, upper tooth shorter and thinner than the lower. Vasa deferentia fairly long, atria globular, large eversible pseudopenes.
? Cosmopolitan (not yet recorded from S. America).

According to Naidu (1965) there are two pairs of testes, found in V and VI although there is no male efferent duct opening into V. The majority of mature specimens found in temperate regions have the genital pores in the usual position (male pores in XI for example). Replication and apparent forward shifting of the reproductive organs are frequently associated with asexual reproduction, and this phenomenon has been reported for several species in the genus (Hrabe, 1937; E. Marcus, 1944).

A. japonicus was at first thought to be close to *A. pectinatus* but the short thin upper teeth of the pectinate setae are quite similar to those seen more recently in *A. pluriseta*. In addition, the hair setae begin in II and so do the pectinate setae, there being no change in the dorsal setae at about VII as in *pigueti* and *pectinatus*. Gonads were seen in VIII, IX and X but no fully mature specimens have been described.

Aulodrilus pigueti Kowalewski, 1914
Fig. 8. 23I

Aulodrilus pigueti KOWALEWSKI, 1914: 25, Fig. 12.

Aulodrilus remex STEPHENSON, 1921: 753, Fig. 2-6, Pl. XXVIII.

Aulodrilus kashi MEHRA, 1922: 946, Fig. 1-7, Pl. I-III, fig. 1-12.

Aulodrilus stephensoni MEHRA, 1922: 963, Fig. 8, 9, Pl. III, fig. 13.

Aulodrilus remex Stephenson. STEPHENSON, 1923: 107, Fig. 42-45; 1930: 272. AIYER, 1925: 35, Fig. 5; 1929 b: 81, Pl. IV, fig. 1-9. NAIDU, 1965: 470, Fig. 3a-e; 1966: 221.

Aulodrilus pigueti Kowalewski. HRABE, 1937: 3. CEKANOVSKAYA, 1962: 227, Fig. 137. BRINKHURST, 1963 a: 69, Fig. 51; 1965: 150, Fig. 9 (P, Q).

Aulodrilus prothecatus CHEN, 1940: 68, Fig. 22a-f.

Aulodrilus cernosvitovi MARCUS, E., 1947: 14, Fig. 15-21.

Aulodrilus kashi Mehra. NAIDU, 1966: 221.

Aulodrilus stephensoni Mehra. NAIDU, 1966: 221.

l=2-28 mm, *s*=100. Dorsal anterior bundles with 4-5 or up to 10 simple-pointed or bifid setae, the upper tooth shorter and thinner than the lower, and 2-5 hair setae which often start in IV-VII, beyond VII bifid setae become oar-shaped, anterior ventral bundles with 4-7 or 11 bifid setae with upper tooth shorter and thinner than the lower. Penial setae modified, 2 hollow spoon-shaped setae per bundle. Vasa deferentia short, atria bean-shaped, large eversible pseudopenes open via a median inversion of the body wall.

Cosmopolitan.

The absence of hair setae in the first few bundles and the presence of oar-shaped setae are common to all of the descriptions cited above. The male efferent ducts are also consistent in form. The differences between the various forms that have been described as species are trivial, concerning chiefly the position of the spermathecal pores and the form of the expanded web of the oar-shaped setae. The former character is known to vary intra-specifically, and the web of the oar setae is difficult to see and easy to interpret in a variety of ways.

In specimens described by Chen (1940) under the name *A. prothecatus* and by Aiyer (1929 b) and Naidu (1965) under the name *A. remex*, a pair of testes was found in V as well as VI, but these degenerated at full maturity. In most accounts the presence of a mass of spongy tissue is described as filling a crescent-shaped area of segment VI, being thickest ventrally, and extending into V and VII in some specimens. The function of this structure is unknown. Specimens described by E. du B.-R. Marcus (1947) under the name *A. cernosvitovi* lack spermathecae, a characteristic observed as an uncommon deviation in several tubificid species, including *T. tubifex*.

Aulodrilus pectinatus Aiyer, 1928
Fig. 8. 23C, D

Aulodrilus pectinatus AIYER, 1928: 345, Fig. 1, 2, Pl. X, fig. 1-6.

Aulodrilus pectinatus Aiyer. CHEN, 1940: 72. CEKANOVSKAYA, 1962: 225, Fig. 136. BRINKHURST, 1963 a: 70. NAIDU, 1966: 221.

$l=$50-60 mm, $s=$32-43. Hair setae from IV, curved, bayonet-shaped, 1-2 per bundle, anterior setae bifid with teeth equally long, from about VII pectinate with 2-3 or 4 intermediate teeth of the same thickness and length as the lower, 3-5 or 7 per bundle, penial setae modified, 1 or 2 per bundle, with spoon-like distal ends. Vasa deferentia very short, atria ? bean-shaped, eversible pseudo-penes large.

Asia.

In the absence of the hair setae in anteriormost bundles, the change in the form of the dorsal setae from VII, and the presence of modified penial setae, *A. pectinatus* closely resembles *A. pigueti*. The male ducts do seem to differ from those of *pigueti* in that the vasa deferentia are shorter, and the inner ends of the eversible pseudopenes form thick-walled muscular ducts, but these differences might be more apparent than real. The pectinate setae appear to be almost exactly intermediate in form between those of *A. pluriseta* and *A. pigueti*.

Aulodrilus americanus Brinkhurst and Cook, 1966
Fig. 8. 23A, B

Aulodrilus americanus BRINKHURST AND COOK, 1966: 19, Fig. 7 (R, S).

$l=10$-15 mm, $s=c$. 100. Anterior setae up to 10 per bundle, more or less simple-pointed or with a rudimentary tooth or teeth. From about VI onwards all setae palmate, with a series of small teeth at the end of a broad lobe, mostly 8 per bundle, fewer posteriorly. Vasa deferentia short, atria small, rather globular.

N. America.

GENUS **Siolidrilus** MARCUS, 1949

Type species: Siolidrilus adetus MARCUS

Male ducts with moderately long vasa deferentia, atria elongate, ovoid to tubular, prostates tufts of cells at junction of vasa deferentia and atria, pseudopenes present. Spermathecae present, spermatophores absent. Coelomocytes absent.

The definition of the genus is almost identical to that of *Aulodrilus*, in which the prostates are generally described as solid, not formed of a "tuft" of cells. According to E. Marcus (1949) *Siolidrilus* lacks the posterior "gill" of *Aulodrilus*, but that structure in the latter hardly merits the name. While *A. pigueti* has penial setae, the genital setae of *Siolidrilus* are located in IX, ahead of the spermathecal pores.

Siolidrilus adetus Marcus, 1949
Fig. 8. 23O-Q

Siolidrilus adetus MARCUS, E., 1949: 6, Pl. II. fig. 12-18.

Siolidrilus adetus Marcus. BRINKHURST, 1963 a: 70, Fig. 2, 58.

$l=20$-30, $s=100$-140. Setae all bifid with teeth equally long in anterior bundles, the upper slightly shorter in median and posterior segments, 5-8 anteriorly, 2-4 posteriorly. Vasa deferentia moderately long, joining ovoid-tubular atria apically along with tufts of prostate cells. Pseudopenes present. Spermathecae long, no spermatophores. Hollow-ended. Spermathecal setae in IX, setae of X unmodified. No coelomocytes.

GENUS **Limnodriloides** PIERANTONI, 1904

Type species: Limnodriloides appendiculatus PIERANTONI

Gut in immediate preclitellar region with a pair of elongated diverticulae. Vasa deferentia, as long as or slightly longer than atria, join latter more or less apically. Each atrium bears a discrete prostate gland, broadly attached ventral or anterior to vas deferens. Atria with thin muscle layer. Penes present or absent. Spermatophores absent but sperm often aggregates into more or less discrete, oriented bundles. Coelomocytes absent.

Europe, N. America, Africa, Black Sea.

Limnodriloides was included in *Clitellio* by Brinkhurst (1963 a) but was reinstated by Hrabe (1967), who included *Thalassodrilus prostatus* (Knollner, 1935) within it. *Limnodriloides* is considered to be distinct from *Clitellio* by virtue of three contrasting criteria, thus: *Limnodriloides* has gut diverticulae, broadly attached, discrete prostate glands, no spermatophores while *Clitellio* possesses no gut diverticulae, no prostate gland or a diffuse one, and has well developed spermatophores. *T. prostatus* (Knollner, 1935) is exluded from *Limnodriloides* as it has no gut diverticulae, possesses a series of penial setae, and has a peculiarly thick muscular atrium with a pedunculate prostate gland.

In his original description of the genus, Pierantoni (1904) included the species *L. roseus* and *L. pectinatus* in *Limnodriloides*. These two species have no gut diverticulae, the former have pedunculate prostate glands joining the atria dorsal or posterior to the vasa deferentia, and the latter possesses a series of modified penial setae. Neither species have apparently been seen by other workers who have regarded them as species dubiae of *Limnodriloides*. Since *Spiridion* Knollner, 1935 has penial setae and pedunculate prostate glands which join the atria dorsal or posterior to the vasa deferentia, it is proposed that *L. roseus* and *L. pectinatus* should be included as species inquirendae of this genus. This action clarifies the definitions of this group of marine genera as *Limnodriloides*, excluding *L. roseus* and *L. pectinatus*, becomes a homogenous group, while *Spiridion*, even with its new species inquirendae, retains its cohesion as a genus.

Limnodriloides

1.	Male and spermathecal pores contained within a median ventral fold in the body wall; i.e. external apertures appear as elongate median slits arranged transversely	*L. medioporus*
—	Male and spermathecal pores paired, in line of ventral setae; i.e. external apertures appear as simple, paired pores	2
2.	Ventral setae of X modified and contained within a muscular sac	*L. winckelmanni*
—	Ventral setae of X unmodified or absent	3
3.	Gut diverticulae in VIII 2 setae per bundle in most posterior segments	*L. appendiculatus*
—	Gut diverticulae in IX. 1 seta per bundle in most posterior segments	*L. agnes*

Limnodriloides winckelmanni Michaelsen, 1914
Fig. 8. 24D

Limnodriloides winckelmanni MICHAELSEN, 1914: 155-160, Pl. V, fig. 6, 7.

Limnodriloides winckelmanni Michaelsen. BOLDT, 1928: 146-148, Fig. 1. HRABE, 1967: 339, 345-347, Fig. 25-29. COOK, 1969 a: 24.

Clitellio winckelmanni (Michaelsen). BRINKHURST, 1963 a: 73; 1963 b: 713; 1966: 153.

l=12 to 18 mm, w=0·2 to 0·25 mm, posteriorly and 0·6 mm in clitellar region. 3 setae per bundle anteriorly, 2 per bundle posteriorly, smaller than middle setae. 1 ventral modified seta per bundle of X; this spermathecal seta hollow, contained in large, vacuolated gland cells and surrounded by thick muscle layer. Male and spermathecal pores paired in line of ventral setae.

Paired intestinal diverticulae present in IX. Vasa deferentia as long as atria, join latter apically. Atria ovoid with long, narrow, proximal ducts, terminating in small penes (Hrabe, 1967 states that it has no penes but illustrates a penis-like structure). Large prostate gland joins each atrium ventrally on a broad base. Spermathecal ampullae sacciform, with thick, discrete ducts. Sperm oriented into long, narrow bundles.

Known only from type locality, Swakopmund, S.W. Africa, under stones, intertidal.

Limnodriloides medioporus Cook, 1969
Fig. 8. 24A-C

Limnodriloides medioporus Cook, 1969 a: 21, Fig. 7.

l=8 mm, w=0·2 mm, s=40. Prostomium usually longer than it is broad at peristomium, with a small, thin-walled papilla on its tip. Setae 2 to 4 per bundle anteriorly, 2 per bundle posteriorly. Setae bifid with teeth of about equal length, 30 to 50 μ long. Ventral bundles absent on X and XI. Clitellum on IX and XII. One pair male pores open inside a dumbell-shaped, median bursa on XI. Pharyngeal glands penetrate into V. Chloragogen cells begin in VI. A pair of diverticulae are joined to the gut in the posterior part of IX and extend anteriorly to 8/9. A pair of vasa deferentia, 11 to 14 μ diameter and as long as, or slightly shorter than the atria, join the atria apically. Atria cylindrical, 110 to 130 μ long, 40 to 55 μ diameter, which narrow to a pair of ducts 70 to 80 μ long (relaxed condition), 15 μ diameter. These ducts terminate in small conical penes, 25 μ long, which open into a median bursa medio-laterally. Long axis of atria directed anteriorly. Large compact prostate glands open into atria medially and ventrally. One pair of spermathecae present in X, with ovoid ampullae and short, indistinct ducts. Spermatophores not developed although not often sperm in spermathecae oriented in a definite manner.

Type: USNM.
Cape Cod Bay, 41° 54·9′N, 70° 15·2′W. Depth 36·5 metres.
Massachusetts, USA.
Also Bermuda Transect (W.H.O.I.) 40° 20·5′ N, 70° 47′ W, 97 m.

Limnodriloides appendiculatus Pierantoni, 1904
Fig. 8. 24E, F

Limnodriloides appendiculatus PIERANTONI, 1904: 187-188, Fig. 1.
Limnodriloides appendiculatus Pierantoni. BOLDT, 1928: 145-151, Fig. 2-3.
HRABE, 1967: 339, 344. COOK, 1969 a: 23.

Clitellio appendiculatus (Pierantoni). BRINKHURST, 1963 a: 73; 1963 e: 713; 1966 a: 300; 1967: 115.

l=10-18 mm, w=0·2-0·4 mm, s=40-50. Clitellum $\frac{1}{2}$X-$\frac{1}{2}$XII. Setae bifid with upper teeth thinner, and posteriorly shorter, than lower. 2(3) setae per bundle anteriorly, 2 per bundle posteriorly. No modified genital setae. One pair of male and spermathecal pores just anterior to ventral setae of X and XI respectively.

Pharyngeal glands present in III to V. Chloragogen cells begin in VI. A pair of oesophageal diverticulae present in VIII, extending to septum 7/8. Paired vasa deferentia, 13 μ diameter, as long as, or slightly shorter than atria which they join apically. Atria cylindrical with a median constriction and narrowing proximally. Atria open into a pair of pear-shaped eversible, pseudo-penes laterally. Large, discrete, non-petiolate prostate glands join atria subapically. Paired spermathecae in X with very short, indistinct ducts.

Gulf of Naples, 3 metres depth, Mediterranean.

Limnodriloides agnes Hrabe, 1967
Fig. 8. 24G-K

Limnodriloides agnes HRABE, 1967: 331, 339-334, Fig. 13-24. COOK, 1969 a: 23.

l=10 mm, w=0·4 mm, s=68. Prostomium rounded. Clitellum on $\frac{1}{2}$X-XII. Setae bifid with upper teeth shorter and thinner than lower, 2(3) per bundle anteriorly, 1 per bundle posteriorly, 77 to 112 μ long. Ventral setae of X unmodified, of XI absent. One pair spermathecal pores anterior, and a little lateral to ventral setae of X. One pair male pores in place of ventral setae of XI.

Pharyngeal glands in III to IV. Chloragogen cells begin in VI. A pair of oesophageal diverticulae present in IX which extend to septum 8/9. Vasa deferentia shorter than atria which they join apically. Diameter of vasa deferentia near small male funnels, 16 μ widening to 25 μ near atrial junction. Atria elongate and tapering proximally, 38 μ diameter near vasa deferentia narrowing to 11 μ about half way along its length. Atria open into a pair of large eversible pseudo-penes, 208 μ long, 64 μ diameter. Large, compact, non-petiolate prostate glands join atria near vasa deferentia. Paired spermathecae with long, cylindrical ampullae and very short, inconspicuous ducts.

Nesebar (Mesembria), near fisherman's pier, Black Sea.

SUB-FAMILY TELMATODRILINAE EISEN, 1879

Type genus: Telmatodrilus EISEN

Prostates a series of discrete, small bodies. Spermatophores present. Coelomocytes absent.

Two characteristics are reminiscent of the Lumbriculidae. A series of discrete small prostates is found in several representatives of that family as well as in *Telmatodrilus*, as are the spermathecal diverticulae of *Macquaridrilus*. These two genera, of very limited geographical distribution, may represent survivors of an ancient line of tubificids.

GENUS **Telematodrilus** EISEN, 1879

Type species: Telematodrilus vejdovskyi EISEN

Male efferent ducts with vasa deferentia moderately long, atria globular or tubular, with several small discrete prostate glands on each. Penes present or absent. Spermatophores present. No coelomocytes.

California, Kamchatka, Onega Lake, Tasmania.

The genus was erected by Eisen (1879), based upon a single species found at 7,000′ in the Sierra Nevada, California, U.S.A. A second species from the same general area was described by the same author in 1900. Neither of these species was found again and there remained only a few specimens in the British Museum (Natural History) and the U.S. National Museum, one of which was selected as lectotype by Brinkhurst (1965). Shortly before this, Hrabe (1962 a) described *Alexandrovia*, a monotypic genus in which the male ducts are of the same form as Eisen's *Telmatodrilus* but the body wall is covered with a sheath consisting of a secretion together with foreign matter which is not, however, aggregated into discrete papillae. While it is true that the separation of *Peloscolex* from *Tubifex* is primarily based on the nature of the body wall, the male ducts, where described, do differ from those of other genera. It is possible that too much reliance on the form of the body wall has led to the inclusion within Peloscolex of species which should be recognized as belonging to other, possibly new, genera. This fact was given special emphasis in the recent discovery by Hrabe (1964) that the atria of *Peloscolex ringulatus* had been incorrectly described by Sokolskaya (1961 a), a finding confirmed recently by Dr. Sokolskaya in a personal communication. The male ducts of *ringulatus* are like those of other species considered here, all of which are included in *Telmatodrilus* as the characters used to separate *Alexandrovia* from *Telmatodrilus* are thought to be inadequate (Brinkhurst, 1965). Two new species were discovered in a collection from Tasmania, and the type species was recently re-discovered at a site near the type locality.

Spermatophores are described in the two species formerly attributed to *Alexandrovia*. Their presence has yet to be confirmed in other species, although the sperm does seem to be in organized bundles in at least one of the Tasmanian species. Eisen (1886) described the sperm as being "agglomerated together in pear-shaped or globular balls" not forming spermatophores. If other differences in characters, which are normally constant within a genus, are found to separate *Alexandrovia* from *Telmatod-*

rilus, the decision to disregard the body wall character may have to be reviewed.

The scattered distribution of the species in this genus (California, Karelia, Kamchatka, Tasmania) may indicate the antiquity of these species, which are unique among tubificids in respect to the form of the prostate glands. Such discrete, numerous prostates are found on the atria of some members of the Lumbriculidae, and this fact alone was sufficient to cause Eisen (1886) and Hrabe (1963 b) to erect the sub-family Telmatodrilinae containing the single genus *Telmatodrilus* (or two genera with *Alexandrovia* according to Hrabe).

Telmatodrilus

1.	Hair setae present	2
—	Hair setae absent	3
2.	3-6 hair setae and some short hair-like setae in anterior dorsal bundles	*T. onegensis*
—	Up to 10 serrate hairs and minutely bifid or bifid setae in anterior dorsal bundles	*T. ringulatus*
3.	Posterior setae pectinate	*T. pectinatus*
—	Posterior setae simple pointed or bifid	4
4.	Anterior setae simple pointed, the rest bifid with rudimentary upper tooth. No spermathecal setae	*T. vejdovskyi*
—	All setae simple pointed. Spermathecal setae large	*T. multiprostatus*

Telmatodrilus vejdovskyi Eisen, 1879
Fig. 8. 25I-K

Telmatodrilus vejdovskyi EISEN, 1879 : 8, Fig. 1.

Telmatodrilus vejdovskyi Eisen. VEJDOVSKY, 1884 b : 45. EISEN, 1886 : 880, Pl. I, fig. 1(a-i), Pl. II, fig. 1(k); 1900 : 243. BEDDARD, 1895 : 264. MICHAELSEN, 1900 : 42. GALLOWAY, 1911 : 316. BRINKHURST, 1963 a : 72; 1965 : 152, Fig. 10(A).

? *Telmatodrilus mcgregori* EISEN, 1900 : 244.

l=35-60 mm, s=at least 130. Setae 8-15 per bundle, bluntly simple pointed or bifid with small upper tooth. Spermathecae with small ampullae. Vasa deferentia moderately long, atria tubular, with numerous small, discrete, prostate glands.

Penes long, narrow, conical, without thickened sheaths. No spermathecal setae, penial setae present, unmodified.

Sierra Nevada, California, U.S.A.

Some worms were recently received from Dr G. E. Gates which had been collected in Whitakers Forest, Tulare County, California, U.S.A., in June, 1966, and in Sagehen Creek nr. Truckee, California, at 6300′ in July 1966 by E. V. Komarek. These proved to belong to *Telmatodrilus*, and as the differences between *T. vejdovskyi* and *T. mcgregori* are trivial, they have been identified as *T. vejdovskyi*. The setae are bifid, but appear simple pointed in the first eight segments, probably owing to wear. Penial setae were observed in sections, but were unmodified. The small spermathecae open laterally in the new material, but it is difficult to tell whether spermatophores are present or not.

Telmatodrilus ringulatus (Sokolskaya, 1961)
Fig. 8. 26A-D

Peloscolex ringulatus SOKOLSKAYA, 1961 a: 61, Fig. 5.

Peloscolex ringulatus Sokolskaya. BRINKHURST, 1963 a: 47.

Alexandrovia ringulatus (Sokolskaya). HRABE, 1964: 108.

l=14-15 mm, *s*=50-54. Prostomium and first segment retractile. Anterior dorsal bundles with up to 10 serrate hair setae, thin hair-like crotchets minutely bifid, with one tooth also bifid in some. Ventral setae 3-5 anteriorly, 2 posteriorly, at first upper tooth much longer than the lower, upper tooth gradually reducing posteriorly. Spermathecal setae modified, hollow distally. Atria horse-shoe shaped with a series of prostate glands. Penes without cuticular sheaths. Sensory papillae present, body covered with a sheath. Spermatophores elongate.

Kamchatka.

The species was placed in *Alexandrovia* by Hrabe (1964) who examined the co-types and demonstrated the presence of numerous prostate glands on the atria. In a personal communication, Dr Sokolskaya confirmed that the original description was erroneous in respect to the prostate glands.

Telmatodrilus onegensis (Hrabe, 1960)
Fig. 8. 25A-C

Alexandrovia onegensis HRABE, 1960: 277.

Alexandrovia onegensis Hrabe. HRABE, 1962 a: 291, Fig. 1-15.

i=10-20 mm, *s*=c. 35. Anterior dorsal bundles with 3-6 long hair setae and some short hair-like setae, ventral bundles with 3-4 bifid setae, the upper tooth pointed and twice as long as the lower in II, the upper tooth progressively shorter in more posterior bundles until the teeth are equally long posteriorly, and equally thick at the base. Vasa deferentia long, atria short and cylindrical with numerous prostate glands. Penes rudimentary in eversible penis sacs. Spermathecae with ovoid ampullae short ducts. Spermatophores long and thin. Spermathecal setae modified, hollow ended. Body wall covered with a secretion mixed with mineral particles, sensory papillae present.

Onega lake, U.S.S.R.

It is interesting to note that the head and first two segments can be inverted within segment III in this species, just as in many *Peloscolex* species in which the body wall is similarly modified.

Telmatodrilus multiprostatus sp. nov.

Fig. 8. 27A-F

$l = 16$ mm, $s = 65$. Prostomium broadly conical. Setae all simple pointed, sigmoid, up to eight setae in anterior bundles, as few as three in posterior bundles; large spatulate spermathecal setae singly in the ventral bundles of X. Superficial pores crescentic, within the line of the ventral setae on XI, spermathecal pores outside the line of the ventral setae, anterior and lateral on X. Female pores not observed.

Pharynx with thickened roof bearing a cuticular lining and a large number of tall gland cells, narrow foregut telescoped into wider portion of gut through V-VI. Testes in X elongate, protruding into anterior sperm-sac from 9/10 into IX. Sperm funnels in posterior sperm-sacs on X/XI, vasa deferentia less than twice as long as atria, entering atria apically. Atria elongate pear-shaped, discharging into inverted sacs from the body-wall or eversible pseudopenes. Minute rudiments of true penes at the point of union of the atria and pseudo-penes, being the distal end of the atria lined with cuticle and protruding slightly into the pseudopenis. Several prostate glands on each atrium, each entering the atrium separately in the uppermost third of that organ; the lower third of the atrium sheathed in tissue, possibly a post prostate gland. Ovaries in XI, attached to 10/11 near the midline on the posterior wall of the sperm-sac. Spermathecae large, lobed in X, containing spermatophores (?). The large spermathecal setae in huge sacs and their associated glands fill the posterior sperm-sacs protruding into XI.

Lake Pedder, S.W. Tasmania. Collected: W. D. Williams, March 2-4, 1966.

Holotype and Paratype: The Australian Museum, Sydney.

Telmatodrilus pectinatus sp. nov.*

Fig. 8. 25D-H

Dimensions unknown. Dorsal anterior bundles with 9-12 setae, ventral bundles with 11-14, upper tooth much longer than the lower and slightly hooked. Posterior bundles with 5-8 pectinate setae. Vasa deferentia moderately long, opening at the summit of the pear-shaped atria, which bear numerous prostate glands. Atria open into eversible chamber on the ventral body wall. Spermathecae capacious, containing sperm at least organized into bundles if not spermatophores. Spermathecal setae modified. Penial setae present, probably unmodified.

L. Pedder, S.W. Tasmania, coll. W. D. Williams, 2/3/66, Dartnall 11/2/67. Three specimens and fragments.

Holotype and Paratype: The Australian Museum, Sydney.

This second species from L. Pedder is characterized by the remarkable

* See Brinkhurst (1971), listed in Appendix, p. 838.

development of pectinate setae in posterior bundles. Pectinate setae are usually found in anterior dorsal bundles, and are frequently replaced by ordinary bifid setae posteriorly.

GENUS **Macquaridrilus** JAMIESON, 1968

Type species: Macquaridrilus benettae JAMIESON

Male efferent ducts with long vasa deferentia, tubular atria with numerous prostate glands, stout muscular ejaculatory ducts, small conical penes in elaborate glandular penis sacs. Spermathecae with diverticulae. Coelomocytes absent.
    S. Pacific, Macquarie Island.

    The single species *M. bennettae* is known only from Macquarie Island in the South Pacific Ocean. The form of the prostate seems to resemble that of *Telmatodrilus* in which *T. pectinatus* has a complex male pore, but no species of *Telmatodrilus* has the complex muscular ejaculatory duct found in this species. The penial bulb of *T. vejdovskyi* is heavily muscular, but is globular rather than tubular. It seems advisable to retain the genus *Macquaridrilus* because of the presence of the unique muscular ejaculatory duct and the presence of spermathecal diverticulae, which are not found in other tubificids. The terminology used to describe the male ducts is discussed below.

Macquaridrilus bennettae Jamieson, 1968
Fig. 8. 26E-I

Macquaridrilus bennettae JAMIESON, 1968 : 55, Figs. 1-9

$l=6.5-12.5$ mm, $s=40-49$. Lateral muscular groove present. Setae all bifid crotchets with the upper tooth shorter and thinner than the lower, 3-5 per bundle, setae absent on XI. Vasa deferentia long, atria tubular with numerous prostates, ejaculatory ducts with tightly spiral muscle layers, elongate tubes terminating in short conical penes. The linings of the ectal third of the ejaculatory ducts heavily cuticularized, continuous with spout-like penes, which are surrounded with cuticular rings basally. The penis sacs surrounding the penes open into ventral chambers which bear post prostate glands. Spermathecae with short tubular diverticulae ending in small spherical bulbs, spermathecal pores lateral on X in the line of the unmodified setae.
    Macquarie Island.

    As indicated by Jamieson (1968), the terminology to be applied to the regions of the male duct is by no means settled. The lack of muscle tissue on the part of the duct termed atrium (which bears the prostate glands) may indicate that this is, in fact, part of the vas deferens. The presence of the elongate muscular tube termed ejaculatory duct may indicate that this is the true atrium, to use the terminology normally applied in the Tubificidae. The same problems arise in considering the male ducts of *Monopylephorus*, where the distal section of the duct is heavily muscular and the diffuse prostate enters an elongate non-muscular tube. These problems can only be solved by a study of the development of the male ducts during maturation.

SUB-FAMILY RHYACODRILINAE HRABE, 1963

Type genus: Rhyacodrilus BRETSCHER

Prostates diffuse when present. Spermatophores absent. Coelomocytes present, usually abundant.

GENUS **Paranadrilus** GAVRILOV, 1955

Type species: Paranadrilus descolei GAVRILOV

Male efferent ducts with moderately long vasa deferentia, atria tubular with diffuse prostates, elongate penes. Spermathecae absent, sperm in sperm bearers* attached to the body wall. Coelomocytes present.
Argentina.

In this monotypic genus uniparental reproduction, possibly by partheno-genesis, seems to be the rule rather than the exception. Copulation is a rare event, culminating in the attachment of sperm in sperm-bearers to the body wall, as in *Bothrioneurum*.

Paranadrilus descolei Gavrilov, 1955
Fig. 8. 27G

Paranadrilus descolei GAVRILOV, 1955: 295, Fig. 1-20

Paranadrilus descolei Gavrilov. GAVRILOV, 1958: 149, Fig. 1-6, Pl. I-III.
BRINKHURST, 1963 a: 71, Fig. 2.

$l=16\text{-}40$ mm, $s=83\text{-}130$. Setae 2-8 per bundle anteriorly, falling to 2-3 posteriorly, all bifid with teeth more or less equal. Vasa deferentia moderately long, thin, atria irregular tubes at first thick and then thinning to enter capacious penis sacs with long penes which bear only thin cuticular sheaths. Diffuse prostate cells cover atria. No spermathecae, sperm in sperm bearers attached to the body walls. Coelomocytes present.
Argentina.

GENUS **Epirodrilus** HRABE, 1930

Type species: Epirodrilus michaelseni HRABE

Male efferent ducts with fairly short vasa deferentia, atria wide tubes, consisting of two main sections connected by a narrower tube, the thickness of the wall varying from region to region, no prostate glands. Eversible pseudopenes present. Spermathecae present, no spermatophores. Coelomocytes present.
Europe, S. America, S. Africa.

---

* See Chapter 1—spermatophores retained as term for these externally attached structures.

Epirodrilus

1. All setae bifid with reduced upper tooth, penial
   setae modified — E. antipodum

— Hair setae and pectinate setae present, penial
   setae unmodified or absent — 2

2. Pectinate setae with broad intermediate teeth,
   anterior ventral setal bundles with simple
   pointed and bifid setae — E. allansoni

— Pectinate setae with fine intermediate teeth,
   ventral setae all bifid — 3

3. Proximal part of atrium swollen, globular
   distal part markedly narrower — E. michaelseni

— Proximal and distal parts of atrium wide,
   irregularly tube like, possibly separated by
   a constriction — E. pygmaeus

Epirodrilus michaelseni Hrabe, 1930
Fig. 8. 28H-L

Epirodrilus michaelseni HRABE, 1930: 2, Fig. 1-9.

Epirodrilus michaelseni Hrabe. HRABE, 1931 b: 309, Fig. 1. BRINKHURST,
1963 a: 71, Fig. 2, 54.

l = 10 mm, *s* = 50-55. Dorsal anterior bundles with 1-4 hair setae and 2-5 pectinate
setae, ventral bundles with 4-8 setae, upper tooth slightly longer than the lower,
median bundles with 4 bifid setae, posterior bundles with only 2 setae, the upper
tooth shorter than the lower. Vasa deferentia short, entering elongate atria at
expanded distal ends, the tubular continuation of each atrium at first wide, then
narrow, and ending in eversible pseudopenes. Prostate glands absent. Spermathecae present, no spermatophores. Coelomocytes present.
   Greece.

The distal end of the male duct presumably constitutes an eversible
pseudopenis.

Epirodrilus pygamaeus (Hrabe, 1935)
Fig. 8. 28D-G

Moraviodrilus pygmaeus HRABE, 1935 a: 4, Fig. 1-9.

Epirodrilus pygmaeus (Hrabe). CEKANOVSKAYA, 1962: 229, Fig. 138. BRINK-
HURST, 1963 a: 78.

Moraviodrilus pygmaeus Hrabe. BRINKHURST, 1963 a: 71, Fig. 2.

$l=15$-20 mm, $s=52$-55. Dorsal anterior bundles with 1-3 short hair setae, 3-4 pectinate setae, ventral bundles with 3-5 setae with teeth more or less equal in length and width, posterior bundles with 2-3 setae (3-4 ventrally) with upper tooth shorter than the lower. Vasa deferentia short, atria wide elongate tubes, width of lumen and thickness of walls varying. No prostate glands. Spermathecae present, no spermatophores. Coelomocytes present.

Czechoslovakia.

The difference between this species and *E. michaelseni* seems trivial, although Hrabe (1935 a) originally designated the genus *Moraviodrilus* for it on the basis of differences of the atria. The almost globular proximal atrial chamber of *E. michaelseni* is not found in *E. pygmaeus* or *E. antipodum*.

Epirodrilus antipodum Cernosvitov, 1939
Fig. 8. 28A-C

Epirodrilus antipodum CERNOSVITOV, 1939 b: 94, Fig. 50-56.

Epirodrilus antipodum Cernosvitov. BRINKHURST, 1963 a: 71, Fig. 55.

$l=8$-16 mm, $s=70$. Dorsal and ventral bundles with 5-7 setae, the upper tooth rudimentary, 3-5 posteriorly. Vasa deferentia short, atria bulky tubes, thickness of lumen and walls varying, terminating in eversible pseudopenes. No prostate glands. Penial setae simple pointed, 5-6 per bundle. Spermathecae present, no spermatophores. Coelomocytes present.

Lake Titicaca.

Epirodrilus allansoni Brinkhurst, 1966
Fig. 8. 27H-L

Epirodrilus allansoni BRINKHURST, 1966: 152, Fig. 6.

$l=10$ mm, $s=50$. Dorsal anterior bundles with stout hair setae and about 3 pectinate setae with wide intermediate teeth, anterior ventral bundles with 2-4 setae, an equal number of broad simple pointed setae and bifid setae with the upper tooth thinner than but as long as the lower, fewer setae posteriorly, where all ventral setae bifid with short upper tooth. Vasa deferentia apparently rather short, atria bilobed, no distinct prostate gland. Spermathecae present, without spermatophores. Coelomocytes present.

S. Africa.

The male ducts were not studied in detail in the single mature specimen obtained, and the position of this species may be subject to amendment. The terminal portion of the male duct presumably forms an eversible pseudopenis.

GENUS **Bothrioneurum** STOLC, 1888

Type species: Bothrioneurum vejdovskyanum STOLC

Male efferent ducts with short vasa deferentia, atria tubular and covered proximally with prostate cells and partially coiled in the segment bearing testes, distal part of atria naked, opening into voluminous eversible pseudopenes bearing glandular paratria and accessory glands. Penial setae sometimes modified. Prostomium with a sensory pit dorsally. Coelomocytes present. Spermathecae absent, sperm bearers attached externally. Body wall richly vascular anteriorly and posteriorly. Reproducing by fragmentation.

Cosmopolitan.

The separation of species in the genus depends on rather questionable characteristics. It is quite possible that variations in the reproductive organs are related to the predominance of reproduction by fragmentation. The variation within what has been recognized as *B. iris* by several authors further confuses the issue, and it seems quite possible that there is, in reality, only one variable species in this genus. The scarcity of descriptions of the type species, and the small number of individuals upon which some of the other species descriptions are based, all contribute to the doubts.

Many of the authors failed to make clear the distinction between new and existing species in their accounts. Under the circumstances, each species will be described separately, but no attempt to key them out will be made.

The unique paratrium probably functions during copulation to provide an insertion and possibly some cementing material by which the sperm bearers *(the term spermatophores being retained for the organized sperm bundles of *Tubifex, Psammoryctides,* etc.) are attached to the body wall of the con-copulant. These bodies must be firmly attached by a burrowing animal if they are not to be rubbed off. The sperm bearers have a solid stem and discharge the sperm through the unattached end, probably into the cocoon as it is formed and passes over the sperm bearer although this has never been established.

Bothrioneurum vejdovskyanum Stolc, 1888
Fig. 8. 29A-D

Bothrioneuron vejdovskyanum STOLC, 1886: 647

Bothrioneuron vejdovskyanum Stolc. STOLC, 1888: 43, Pl. I, fig. 8, 9, Pl. II, fig. 5, 9, Pl. IV, fig. 6-11, 13e. BEDDARD, 1895: 269, Fig. 40.

? *Clitellio arenarius* (part). VAILLANT, 1890: 415.

Bothrioneurum vejdovskyanum Stolc. MICHAELSEN, 1900: 54; 1909: 30, Fig. 50-52. HRABE, 1929: 2; 1934 c: 1; 1935 a: 12. CHEN, 1940: 102, Fig. 29a-c. BRINKHURST, 1962 c: 329; 1963 a: 64, Fig. 46a, b; 1963 c: 39, Fig. 11a-d; 1965: 149, Fig. 9L-N. CEKANOVSKAYA, 1962: 289, Fig. 183.

l=28-35 mm, s=10-140. Setae 4-6 anteriorly, fewer posteriorly, upper tooth longer than the lower anteriorly, teeth equally long posteriorly, penial setae 4 per bundle, bluntly hooked. Sperm funnels large, vasa deferentia long, atria tubular, covered with diffuse prostate gland, except for terminal section leading to eversible pseudopenes, the latter with paratria bearing glands, eversible

* See footnote p. 537.

pseudopenes open via a median chamber. Integumental plexus present.
Europe, Asia, N. America, Africa.

The species often reproduces asexually (Hrabe 1934 b, 1935 a) and the reproductive organs may be shifted forward in some specimens. Specimens from many parts of the world seem to be referable to this species, but the S. American fauna is supposed to contain three different species in the absence of *B. vejdovskyanum*. As the type species is so poorly known the situation would seem to require confirmation.

Bothrioneurum iris Beddard, 1901.
Fig. 8. 29H-L

Bothrioneuron iris BEDDARD, 1901: 81, Fig. 8-16.

Bothrioneurum iris Beddard. MICHAELSEN, 1908 b: 135; 1909: 135; 1910: 8. STEPHENSON, 1910: 241, Fig. 1-2; 1923: Fig. 40; 1924: 322. AIYER, 1925: 39. CERNOSVITOV, 1939 b: 100, Fig. 63-67. CHEN, 1940: 107. MARCUS, 1942: 205, Fig. 35-38. BRINKHURST, 1963 a: 64, Fig. 48. NAIDU, 1965: 475, Fig. 5a-g; 1966: 220.

$l=8$-50 mm, $s=38$-170. 2-4 setae in anterior dorsal bundles, sometimes 5 or 6 ventrally, fewer posteriorly, anterior setae with upper tooth longer than the lower. Male pore single, median. Penial setae present or absent ? Integumental plexus present.
Asia, S. America.

There has been so much confusion about the anatomy of this species that it is impossible to be certain of its validity. When first described, the single male pore was said to be in XII, there were no genital setae, the integumental plexus was absent, and so was the cap of gland cells on the paratrium.

Stephenson (1910) found the male pore to be either on XI or XII, but confirmed the lack of an integumental plexus and paratrial gland. The male ducts were described as coiled in a protrusion of XI (or XIII) into X (or XI), indicating that they were as extensive as those described in *B. americanum*. Cernosvitov (1939) also noted the similarity between the male ducts of *B. iris* and *B. americanum*, and at the same time established the presence of the integumental plexus and paratrial glands. The setae of *iris* were also stated to resemble those of *americanum*, the differences between the two species being reduced to the unification of the male pore and the thick muscle layer of the paratrium in *iris*. Chen (1940) found a single mature specimen that had penial setae and the integumental plexus, but lacked the paratrial glands, and suggested that *iris* was, in fact, in general identical to *vejdovskyanum*. Specimens described by Naidu (1965) have a vascular body wall and penial setae are present, but the paratria were not described.

In the account by Marcus (1942) the paratria have gland cells but there are no penial setae and the integumental plexus is absent. The male ducts are illustrated from what must be a partially mature specimen (only two mature specimens were obtained) because the ducts are much shorter than in descriptions of other species, being wholly confined to the eleventh segment. In earlier accounts it is clear that the male ducts distend septum 10/11 forwards or have broken through into X as in other species in the genus.

All of this variability in what were originally taken to be significant characteristics throws considerable doubt on the separation of *B. iris* from *B. vejdovskyanum* in particular. These two forms are supposed to differ from the three S. American species *americanum*, *brauni* and *pyrrhum* in that the eversible pseudopenes (unmodified inversions of the body wall into which open the atria, and which bear the paratria) open via a single median depression in the former, but via separate openings in the latter. This characteristic may well prove to be subject to variation and hence to be of little significance in a species which probably reproduces mostly by fragmentation.

Bothrioneurum americanum Beddard, 1894
Fig. 8. 29E-G

Bothrioneuron americanum BEDDARD, 1894: 206.

Bothrioneurum americanum Beddard. BEDDARD, 1895: 269; 1896: 6, Fig. 16, 20

Bothrioneurum americanum Beddard. MICHAELSEN, 1900: 54. CERNOSVITOV, 1936 b: 96, Pl. X, fig. 57-62. BRINKHURST, 1963 a: 64, Fig. 47.

$l=$up to 50 mm, $s=$up to 190. Setae 4-8 per bundle in anterior dorsal bundles, more ventrally, falling to 2-3 posteriorly, upper tooth longer than the lower anteriorly, teeth equally long posteriorly, no penial setae. Sperm funnel small, vasa deferentia distinct from glandular atria, coiled in X and XI, eversible pseudopenes receive large paratria with thick glandular cap, distal part cuticular, muscle layers thin. Male pores separate. Integumental network present.
S. America.

Cernosvitov (1939) saw several mature specimens, Beddard saw many worms but did not state the number of mature worms available. Penial setae are lacking, and the male pores are separate, these being apparently the only characters by which this species may be separated from *B. vejdovskyanum*.

Bothrioneurum brauni Marcus, 1949
Fig. 8. 30E-G

Bothrioneurum brauni MARCUS, E., 1949: 4, Pl. I, fig. 7-10, Pl. II, fig. 11.

Bothrioneurum brauni Marcus. BRINKHURST, 1963 a: 65, Fig. 2.

$l=14$ mm, $s=c$. 70. Setae bifid, the upper tooth longer but thinner than the lower anteriorly, upper tooth thinner than, but as long as, the lower posteriorly. Setal glands unicellular. Testes in IX and X, ovaries in X and XI. Male funnels greatly enlarged, vasa deferentia long, atria long, tubular with non-glandular tube and eversible pseudopenis, paratria with muscular walls, glands, but non-cuticular tips. Accessory glands entering eversible pseudopenes basally. Male pores paired.

Brazil, 1 mature specimen.

Most of the male ducts are protruded through 10/11 into XI, as in several other species. There is no duplication of the male ducts or female ducts to match the duplication of the gonads. Penial setae were not found, but in the absence of a long series of specimens this may not be significant.

Bothrioneurum pyrrhum Marcus, 1942
Fig. 8. 30A-D

Bothrioneurum pyrrhum MARCUS, 1942: 201, Pl. IX-X, fig. 30-34.

Bothrioneurum pyrrhum Marcus. BRINKHURST, 1963 a: 65, Fig. 49a, b.

$l=20\text{-}30$, $s=50\text{-}70$. Anterior setae 2-5 per bundle, upper tooth longer than lower, 2-3 in median bundles, 1 posteriorly, teeth about equally long penial setae in IX, 4-7 per bundle with thick blunt hooked heads. Testes in VIII ovaries in IX, sometimes rudimentary gonads in other segments, sperm funnels large, ventral. First part of atria narrow, coiled in VIII, wider non-glandular parts leading to eversible pseudopenes with glands, paratria muscular, with glands, male pores separate. Cutaneous blood vessels well developed.

Brazil, 10 mature specimens.

The position of the male pores seems to be the chief reason for the separation of this from other *Bothrioneurum* species, despite the known variation in the position of the male pores in *B. iris* and *B. vejdovskyanum*. No mention of the difference between *B. pyrrhum* and *B. brauni* was made in the description of the latter species by E. Marcus (1949). The coiled vasa deferentia are shown as being involved in the glandular covering of the atrium in the original illustrations, but they are probably devoid of gland cells as in other species. Brinkhurst (1963 a) noted that the male ducts were short, as *B. iris* of Marcus (1942), but in actual fact the detailed drawing (Fig. 33, Marcus, 1942) shows the ducts to be very much like those of *B. vejdovskyanum*.

SPECIES INQUIRENDAE

Bothrioneurum aequatorialis (Michaelsen, 1935)

Limnodrilus (?) *aequatorialis* MICHAELSEN, 1935 b: 34.

Bothrioneurum aequatorialis (Michaelsen). CERNOSVITOV, 1938 b: 266.
BRINKHURST, 1963 a: 64.

$l=25$ mm, $s=75$. Setae bifid, 3-6 in ventral bundles, 2-4 in dorsal bundles, both teeth relatively short, bent, more or less equally long. No spermathecae. Coelomocytes present.

Congo.

The original description is too brief to permit an identification of this species to be made.

GENUS **Rhyacodrilus** BRETSCHER, 1901

Type species: Rhyacodrilus falciformis BRETSCHER

Male efferent ducts with moderately long vasa deferentia joining atria subapically, atria usually covered with prostate cells, pseudopenes present or absent. Penial setae often modified. Spermathecae paired, no spermatophores. Coelomocytes present.

Cosmopolitan (rare in S. America).

Michaelsen (1909) recognized the relationship between Vejdovsky's *Tubifex coccineus* and *Rhyacodrilus falciformis* of Bretscher (1901), thus ending the confusion concerning the true position of *coccineus*. Before that publication Michaelsen had contributed to the confusion by allying *coccineus* with *Branchiura* (Michaelsen, 1900) and then with *Taupodrilus* of Benham (Michaelsen, 1908 a). Stolc (1885) allied *R. falciformis* and *coccineus* with Eisen's *Ilyodrilus*, a decision that was supported by Beddard (1895) and Ditlevsen (1904). One of the most characteristic features of the male efferent ducts of *Rhyacodrilus* species is the way in which the vasa deferentia enter the atria sub-apically, a rather unusual characteristic in the Tubificidae.

Rhyacodrilus

1.	Prostate glands absent	*R. simplex*
—	Prostate glands present	2
2.	Hair setae present	3
—	Hair setae absent	10
3.	Pectinate setae present	4
—	Pectinate setae absent	9
4.	Anterior ventral setae often simple-pointed	*R. lepnevae*
—	Anterior ventral setae bifid	5
5.	Upper tooth of anterior pectinate setae mostly much longer than the lower	6
—	Upper tooth of pectinate setae little longer than or as long as the lower	7
6.	Hair setae of II much longer than the others, ventral setae sometimes minutely pectinate, upper tooth never shorter than the lower	*R. montana*

—	Hair setae of II not longer than the others, ventral setae bifid, upper tooth shorter than the lower posteriorly	*R. subterraneus*
7.	Ventral setae pectinate	*R. punctatus*
—	Ventral setae rarely if ever with a single intermediate tooth	8
8.	Pectinate setae with long parallel teeth, intermediate teeth fine	*R. sodalis*
—	Pectinate setae with short, diverging teeth, intermediate teeth distinct	*R. coccineus*
9.	Anterior ventral setae often simple pointed, dorsal setae bifid	*R. brevidentatus*
—	Ventral setae bifid, posterior dorsal setae simple pointed	*R. lindbergi*
10.	Penial setae sickle-shaped	*R. falciformis*
—	Penial setae straight	11
11.	Ventral setae with fine webs between teeth	*R. korotneffi* (I. sedis)
—	Ventral setae bifid	12
12.	Upper tooth of anterior setae longer than lower	*R. stephensoni*
—	Upper tooth of anterior setae shorter than lower	*R. balmensis*

Rhyacodrilus simplex (Benham, 1903)
Fig. 8. 31I-L

Taupodrilus simplex BENHAM, 1903: 209, Pl. XXIV, fig. 1-9.

Branchiura coccinea ? *simplex* var (Benham). MICHAELSEN, 1905: 11; 1908 a: 141.

Rhyacodrilus simplex (Benham). HRABE, 1931 a: 11; 1963 a: 252; 1963 b: 55. BRINKHURST, 1963 a: 58.

$l=15$ mm, $s=70$. Dorsal anterior bundles with 3-4 pectinate setae, 1-2 hair setae on most bundles or reduced in number or perhaps absent, ventral bundles with 4-6 setae, the upper tooth thinner than the lower and longer than, or as long as, the lower, some with intermediate spines. Penial setae up to 8 per bundle, simple pointed and hooked. Vasa deferentia coiled around atria, atria broadly cylindrical with rounded apices, no prostate glands. Coelomocytes present. Spermatophores absent. Spermathecae not communicating with gut.

New Zealand.

Hrabe (1931 a) confirmed the absence of the prostate and the lack of connection between the spermathecae and the gut by examining a type specimen. Brinkhurst (1963 a) mistranslated the account by Hrabe (1931 a) but most of the description has now been confirmed from an examination of several specimens from Lake Coleridge, New Zealand. In these specimens the vasa deferentia are coiled several times (usually thrice) around the atria, but they clearly resemble the original specimens. The form of the atria and absence of prostates might be construed as sufficiently distinctive characters to warrant separation of *simplex* from *Rhyacodrilus*, in which case the name *Taupodrilus* would be available as the genus was created by Benham (1903) for its reception. The lack of prostate glands in some species of *Clitellio* and *Potamothrix* has not been taken as evidence to justify the division of these genera, and so this difference is not accepted as justifying the separation of *Taupodrilus* from *Rhyacodrilus*. A more detailed examination of fresh material might provide more significant evidence, but for the moment the more conservative attitude will be adopted. In most other respects the species resembles those in the genus *Rhyacodrilus*.

Rhyacodrilus falciformis Bretscher, 1901
Fig. 8. 30H, I

Rhyacodrilus falciformis BRETSCHER, 1901: 204, Pl. XIV, fig. 4, 5.

Rhyacodrilus falciformis Bretscher. BRETSCHER, 1903: 13, Pl. I, fig. 2, 3; 1904: 260. PIGUET AND BRETSCHER, 1913: 59, Fig. 11-13. UDE, 1929: 79. HRABE, 1935 a, Fig. 10-14; 1954: 300; 1963 a: 252. CERNOSVITOV, 1942 a: 212, Fig. 51, 52. BERG, 1948: 53. BÜLOW, 1957: 97. JUGET, 1958: 90, Fig. 14L. BRINKHURST, 1962 c: 321; 1963 a: 57, Fig. 37. CEKANOVSKAYA, 1962: 240, Fig. 145. MOSZYNSKA, 1962: 26.

Rhyacodrilus lemani PIGUET, 1906 a: 396, Fig. a-c.

Ilyodrilus filiformis DITLEVSEN, 1904: 408, Pl. XVI, fig. 3-5.

Taupodrilus lemani (Piguet). MICHAELSEN, 1908 a: 146; 1909: 32, Fig. 62, 63.

$l=8$-10 mm, $s=48$. Anterior dorsal bundles with 3-4 bifid setae, anterior ventral setae 3-5 setae with the upper tooth thinner and longer than the lower, lower sometimes divided. Penial setae twice as long and six times as thick as the other setae, with sickle-shaped ends. Spermathecae with or without diverticulae connecting spermathecal ampullae to intestine. Atria elongate pear-shape.

Europe.

Cernosvitov (1942 a) and Hrabe (1935) claimed that there was no communication between the spermathecae and intestine in material from several sites, but Piguet (1906 a) and Piguet and Bretscher (1913) quite definitely stated that such a communication existed. This discrepancy may be further evidence to support the view that *R. ekmani* and *R. palustris* are no more than variants of *R. coccineus* (q.v.).

Rhyacodrilus balmensis Juget, 1959
Fig. 8. 32K-N

Rhyacodrilus balmensis JUGET, 1959: 399, Fig. 4 (A-D).

Rhyacodrilus balmensis Juget. BRINKHURST, 1963 a: 58, Fig. 39.

$l=2.7$-3 mm, $s=c$. 20. 2-5 bifid setae per bundle anteriorly, upper tooth thinner and shorter than the lower, 1-3 posteriorly with teeth equally long, upper thinner than lower. Penial setae 7 per bundle, with long straight shafts and slightly sickle-shaped tips. Atria globular.

Cave in Jura mountains, France.

Two mature specimens were obtained.

Rhyacodrilus stephensoni Cernosvitov, 1942.
Fig. 8. 31E-H

Limnodrilus sp? STEPHENSON, 1909: 112, Fig. 2-3, Pl. VIII, fig. 3-4.

Rhyacodrilus stephensoni CERNOSVITOV, 1942 c: 284.

Rhyacodrilus stephensoni Cernosvitov. CEKANOVSKAYA, 1962: 241, Fig. 146. BRINKHURST, 1963 a: 58; 1966: 148. HRABE, 1963 a: 252. NAIDU, 1966: 220.

$l=8$-20 mm, $s=40$-64. Dorsal and ventral bundles with 4-7 bifid setae, all thin, upper tooth slightly longer than the lower. Penial setae 3-4 per bundle, heads close together, upper tooth reduced, distal ends hooked. Vasa deferentia enter ovoid or pear-shaped atria near summit.

Asia, S. Africa.

The species may be no more than the ultimate variant of *R. coccineus*, lacking hair setae and pectinate setae. In all other respects this species closely resembles *R. coccineus*.

Rhyacodrilus coccineus (Vejdovsky, 1875)
Fig. 8. 31A-D

Tubifex rivulorum MACINTOSH, 1870: 253 (in part).

Tubifex coccineus VEJDOVSKY, 1875: 193.

Tubifex coccineus Vejdovsky. EISEN, 1879: 14; 1886: 892. MICHAELSEN, 1909: 32, Fig. 55-58.

Tubifex rivulorum var *coccineus* (Vejdovsky). VEJDOVSKY, 1884 b: 46.

Ilyodrilus coccineus (Vejdovsky). STOLC, 1885: 656; 1886: 642; 1888: 38,

Pl. I, fig. 1, 2, Pl. II, fig. 1-4, 7, 8, Pl. III, fig. 1-3, Pl. IV, fig. 4, 5, 12, 13c, d. BEDDARD, 1895: 266. Fig. 39. DITLEVSEN, 1904: 408, Pl. XVI, fig. 2(D-F).

Branchiura coccinea (Vejdovsky). MICHAELSEN, 1900: 40; 1903 b: 187, Pl. XVIII.

? *Branchiura coccineus* forma *inaequalis* MICHAELSEN, 1905 a: 3.

Branchiura pleurotheca BENHAM, 1907: 256, Pl. XLVI, fig. 7-12.

Taupodrilus coccineus (Vejdovsky). MICHAELSEN, 1908 a: 141.

Rhyacodrilus coccineus (Vejdovsky). MICHAELSEN, 1909: 31; 1929 a: 320. MICHAELSEN AND VERESCHAGIN, 1930: 219. LASTOCKIN, 1927 a: 67; 1927 b: 9. CERNOSVITOV, 1928: 4. UDE, 1929: 78. HRABE, 1929: 1; 1931 a: 11; 1963 a: 252. JUGET, 1958: 89. BRINKHURST, 1960: 399, Fig. 2; 1962 c: 321; 1963 a: 55, Fig. 35; 1963 c: 35, Fig. 9e. SOKOLSKAYA, 1961 a: 55. CEKANOVSKAYA, 1962: 234, Pl. V, fig. 140. MOSZYNSKA, 1962: 26. KENNEDY, 1964: 235. BRINKHURST AND COOK, 1966: 18, Fig. 7O-Q.

? *Tubifex (Taupodrilus) lunzensis* POINTNER, 1914: 614, Pl. XVIII, fig 10-11.

? *Rhyacodrilus riabuschinski* MICHAELSEN, 1929 a: 320.

? *Rhyacodrilus coccineus* forma *inaequalis* Michaelsen. CEKANOVSKAYA, 1962: 235.

Rhyacodrilus riabuschinskii Michaelsen. HRABE, 1931 a: 11; 1963 a: 252. CERNOSVITOV, 1942 c: 281. CEKANOVSKAYA, 1962: 231. BRINKHURST, 1963 a: 56.

l=10-35 mm, s=60-110. Anterior dorsal bundles with 3-5 hair setae, up to 5 pectinates with a series of fine intermediate teeth, hair setae absent from a variable number of postclitellar bundles, 3-5 ventral setae per bundle, upper tooth thinner and a little longer than the lower, penial setae 3-5 per bundle, knobbed with the heads close together. Atria globular.
Eurasia, N. America, Australasia.

Branchiura pleurotheca of Benham (1907) differs little from *R. coccineus*. The spermathecae open laterally, a fact easily confirmed from an examination of the original sections. The innermost parts of the spermathecal ampullae appear to be connected to the wall of the intestine, as reported in *R. ekmani* and *R. palustris*. However, the faded condition of the slides made it impossible to be certain of this. A similar connection has been sought in undoubted *R. coccineus* specimens. Although the spermathecae have been seen to narrow posteriorly into a duct-like prolongation in sections, no connections with the gut has yet been established. The systematic position of *R. ekmani* and *R. palustris* depends entirely on the outcome of this issue. The hair setae of var *inaequalis* of Michaelsen (1905 a) are restricted to segments II-III or IV, and the pectination of the dorsal setae is indistinct. No mention of a connection between the spermathecae and gut appeared in the original description so that this may represent a second record of *R. palustris* as noted by Michaelsen (1908), itself a possible variant of *R. coccineus*.

R. riabuschinskii has the spermathecal pores in VIII, the male pores in IX and possibly also in X, the female pores in 10/11. Testes were located only in VIII, but ovaries in IX and X, which does not correspond with the description of the ducts. This clearly constitutes an abnormality, as noted by Michaelsen (1929), and as there was only one mature and one partially mature specimen found, and *R. coccineus* is also found in the same general area, there seems to be little reason for maintaining *riabuschinskii* as a valid species. In the original account, Michaelsen (1929) drew attention to the parallel example of *Potomothrix bavaricus* and *P. bedoti* and pointed out the close similarity in most respects between *R. coccineus* and *R. riabuschinskii*. Assuming the differences in the position of the reproductive organs to be unimportant, two other features were used to differentiate these species. The upper tooth of the ventral setae of *riabuschinskii* were said to be longer than the lower, whereas those of *coccineus* are of equal length, but the description of *coccineus* has now been amended as most specimens have ventral setae with the upper tooth at least a little longer than the lower anteriorly. Differences in the proportions between the vasa deferentia, atria and terminal ducts of the atria are impossible to substantiate as the reproductive organs have seldom been described.

Tubifex (Taupodrilus) lunzensis of Pointner (1914) was included here by Brinkhurst (1963 a) largely on the basis of statements in the original description indicating the close relationship between that species and *R. coccineus*. The form of the male efferent ducts of *lunzensis* as described by Pointner bear little resemblance to those of *Rhyacodrilus* species, and there was no mention of coelomocytes. Penial setae were described by Pointner, but not figured. The relationship between *lunzensis* and *coccineus* (or any other *Rhyacodrilus*) has not been satisfactorily established.

Rhyacodrilus sodalis (Eisen, 1879)
Fig. 8. 33A-H

Ilyodrilus sodalis EISEN, 1879: 11.

Ilyodrilus sodalis Eisen. VEJDOVSKY, 1884: 45. EISEN, 1886: 887, Pl. IV, fig. 5. MICHAELSEN, 1900: 47. GALLOWAY, 1911: 314.

Rhyacodrilus altaianus MICHAELSEN, 1935 a: 298, Fig. 1.

Tubifex sinicus CHEN, 1940: 75, Fig. 23.

Rhyacodrilus altaianus Michaelsen. CERNOSVITOV, 1942 c: 281. CEKANOV-SKAYA, 1962: 232. HRABE, 1963 a: 251.

Rhyacodrilus sinicus (Chen). SOKOLSKAYA, 1961 b: 84. CEKANOVSKAYA, 1962: 235, Fig. 141. BRINKHURST, 1963 a: 57. HRABE, 1963 a: 252.

Rhyacodrilus altainus Michaelsen. BRINKHURST, 1963 a: 56.

Rhyacodrilus sodalis (Eisen). BRINKHURST, 1963 a: 59; 1965: 144, Fig. 8E-K. BRINKHURST AND COOK, 1966: 23.

$l=$5-10 mm, $s=$33-100. Anterior dorsal bundles with 1-2 hair setae, 2-3 pectinate setae with long teeth and very fine intermediate teeth, posteriorly 1 hair and 1-2 pectinate setae with shorter teeth, anterior ventral bundles with up to 6 setae per bundle, the upper teeth longer and thinner than the lower, becoming shorter in mid and posterior bundles, penial setae blunt with curved tips, 2-4 per bundle. Atria globular to ovoid.

Asia, N. America.

The synonymy of this species was discussed by Brinkhurst (1965), who claimed that the supposed absence of intermediate teeth in the dorsal setae of specimens described by Eisen (1879) and Chen (1940) was probably erroneous, as the very fine intermediate teeth in American specimens were very difficult to see with the most modern microscopes.

Rhyacodrilus brevidentatus Brinkhurst, 1965
Fig. 8. 32O, P

Rhyacodrilus brevidentatus BRINKHURST, 1965: 144, Fig. 8C-D.

Rhyacodrilus brevidentatus Brinkhurst. BRINKHURST AND COOK, 1966: 18.

$l=$c. 12 mm, $s=$36-45. Dorsal bundles with 1-2 hair setae, 2-3 bifid setae with the upper tooth shorter and thinner than the lower, ventral bundles with up to 4 setae, either simple-pointed or with the upper tooth shorter and thinner than the lower. Atria tall, narrow, pear-shaped, male ducts in a median invagination at full maturity.

L. Tahoe, U.S.A.

The description of the atria of *R. falciformis* was overlooked when this species was first described, so that the elongate shape of the atria of this American species was thought to be unusual. The presence of coelomocytes has not been firmly established, but in all other aspects the species is clearly attributable to *Rhyacodrilus*. Penial setae have yet to be recorded in *brevidentatus*.

Rhyacodrilus punctatus Hrabe, 1931
Fig. 8. 31Q-U

Rhyacodrilus punctatus HRABE, 1931 a: 11, Fig. 1(a-g).

Rhyacodrilus punctatus Hrabe. BRINKHURST, 1963 a: 56, Fig. 36. HRABE, 1963 a: 252.

$l=$30 mm maximum, $s=$85-112. Anterior dorsal bundles with 2-5 long and 1 short hair seta, 3-6 or 8 pectinate setae with somewhat U-shaped tips and a series of fine intermediate teeth, posterior pectinate setae with upper tooth thinner than the lower, both curved, anterior ventral setae pectinate, with the upper tooth thinner and slightly longer than the lower, posteriorly still pectinate but upper tooth shorter than lower. Penial setae 4-5 per bundle, blunt with rudimentary upper tooth.

L. Ochrid, Europe.

Rhyacodrilus montana (Brinkhurst, 1965)
Fig. 8. 31M-P

Edmondsonia montana BRINKHURST, 1965: 155, Fig. 10(F-J).

Rhyacodrilus montana (Brinkhurst). BRINKHURST AND COOK, 1966: 23.

l=15-30 mm, s=100-150. Anterior dorsal bundles with 4-6 serrate hair setae and 3-5 pectinate setae with the upper tooth much longer than the lower in many instances, hair setae of II often very long, median bundles with 2 hairs, pectinate setae with upper tooth as long as but thinner than the lower, posterior bundles with 1 hair and 1 pectinate seta, ventral anterior bundles with up to 6 bifid setae, the upper tooth much longer than the lower but teeth equally thick, sometimes minutely pectinate, posteriorly fewer in number, with teeth equally long but the upper thinner than the lower. Penial setae with heads close together. Coelomocytes often very large. Body square in cross section posteriorly.

N. America.

The species was originally placed in the genus *Edmondsonia*, the type species of which was the species described as *Tubifex minor* by Sokolskaya (1961). The latter was said to have male ducts like those of *Tubifex* but with coelomocytes in the body cavity. This latter observation has now been retracted by Dr. Sokolskaya (vide *Tubifex*) and, as *E. montana* has now been recognized as being attributable to *Rhyacodrilus*, the ill-fated genus *Edmondsonia* no longer exists.

Rhyacodrilus lepnevae Malevitch, 1949
Fig. 8. 33M-O

Rhyacodrilus lepnevae MALEVITCH, 1949: 119.

Rhyacodrilus lepnevae Malevitch. CEKANOVSKAYA, 1962: 233, Fig. 139. BRINKHURST, 1963 a: 78. HRABE, 1963 a: 252.

Dorsal bundles with 2-3 hair setae and 2 pectinate or palmate setae, anterior ventral setae 2-4 per bundle, bifid or simple pointed, often mixed in one bundle. Penial setae straight, simple pointed.

U.S.S.R.

Rhyacodrilus subterraneus Hrabe, 1963
Fig. 8. 32A-F

Rhyacodrilus subterraneus HRABE, 1963 a: 249, Fig. 1-7.

l=c. 10 mm, *s*=70. Anterior dorsal bundles with 1-2 hair setae and 1-2 pectinate setae with upper tooth much longer than lower, postclitellar bundles with 1 hair and 2 bifid setae, posterior bundles with 2-3 bifid setae, with the upper tooth shorter and thinner than the lower, anteriorly 2-4 bifid setae, posteriorly 2-3, again, anteriormost setae with upper tooth much shorter and thinner than the lower. Penial setae single, blunt with recurved tips. Small globular gland-covered part of atria at summit of long non-glandular section.

No connection between spermathecae and gut.

E. Germany, in springs. Five specimens, only two mature.

Rhyacodrilus lindbergi Hrabe, 1963
Fig. 8. 32G-J

Rhyacodrilus lindbergi HRABE, 1963 b: 54, Fig. 1-4.

l=6.5 mm, *s*=63. Dorsal anterior bundles with 1 or 2 hair setae, 2-3 bifid setae with almost parallel long teeth, the upper slightly shorter than the lower, posterior bundles with 3-4 simple pointed setae, ventral bundles with 4-5 setae with the upper tooth shorter than the lower. Penial setae present ? Atria globular. Spermathecae not communicating with gut.

Portugal, in a cave.

In *Peloscolex swirenkowi* the postclitellar dorsal setae were all said to be simple pointed by Hrabe (1964). In fact there are both hair setae and simple pointed setae, the latter being somewhat shorter and thicker than the hair setae. To judge by the number of hair setae in posterior bundles compared with those of anterior bundles, the same may be true of *R. lindbergi*. One mature specimen was described.

SPECIES INCERTA SEDIS

Rhyacodrilus korotneffi (Michaelsen, 1905)
Fig. 8. 33I-L

Clitellio korotneffi MICHAELSEN, 1905 a: 6, Fig. 1.

Taupodrilus korotneffi (Michaelsen). MICHAELSEN, 1908 a: 148, Pl. III, fig. 21-23; 1909: 31.

Rhyacodrilus korotneffi (Michaelsen). HRABE, 1931 a: 11; 1963 a: 252. CEKANOVSKAYA, 1962: 238, Fig. 143. BRINKHURST, 1963 a: 57, Fig. 40.

l=55-65 mm, *s*=140-190. Anteclitellar bundles with 6-11 bifid setae, postclitellar bundles with 4-6 setae, upper tooth a little longer and thinner than the lower, ventral setae webbed. Penial setae straight, bent over distally, sharp pointed. Atria bilobed, distinct from ejaculatory ducts?

L. Baikal.

The form of the atria, as illustrated by Michaelsen (1905 a) is reminiscent of that of *Rhyacodrilus* species, but detailed comparison shows that the species cannot be regarded as a *Rhyacodrilus* for certain. The species was included first in *Taupodrilus* by Michaelsen (1908 a) and then in *Rhyaco-*

drilus in later accounts by Hrabe (1931 a, 1963 a) and Brinkhurst (1963 a) and there seems to be no other place for it, other than in a distinct genus. The vasa deferentia of *R. korotneffi* open into the base of the most proximal of two ovoid parts of the atrium, joined side to side. Both chambers are covered in a diffuse layer of prostate cells. The terminal ducts leading from these to the male pores are rather long, but they are also elongate in the more recently described *R. subterraneus*.

SPECIES INQUIRENDAE

Rhyacodrilus palustris (Ditlevsen, 1904)
Fig. 8. 33P, Q

Ilyodrilus palustris DITLEVSEN, 1904: 408, Pl. XVI, fig. 1-2.

? *Branchiura coccinea* var *inequalis* MICHAELSEN, 1905 a: 10.

Branchiura coccinea var *palustris* (Ditlevsen). MICHAELSEN, 1905 b: 9.

Taupodrilus palustris (Ditlevsen). MICHAELSEN, 1908 a: 144; 1909: 32, Fig. 59-61.

Rhyacodrilus palustris (Ditlevsen). MICHAELSEN, 1909: 31. UDE, 1929: 80. JUGET, 1958: 90. CEKANOVSKAYA, 1962: 234. BRINKHURST, 1963 a: 56. HRABE, 1963 a: 251.

l=6 mm. Dorsal bundles of II-V with 1 hair seta and bifid setae, other dorsal bundles with bifid setae only, the upper tooth longer and thinner than the lower, ventral setae of the same form. Penial setae straight, simple pointed. Spermathecae connected to the gut by short ducts.
Europe, ? Asia, S. Georgia.

The dispute about the presence or absence of the connection between the spermathecae and the gut of *R. falciformis* suggests that there may be interspecific variation of this character, so *R. palustris* and *R. ekmani* may be no more than variants of *R. coccineus*. The series *coccineus-ekmani-palustris* and perhaps even *stephensoni* may demonstrate the same sort of intraspecific variation of the setae observed in *Tubifex tubifex* and other tubificid species.

Rhyacodrilus ekmani Piguet, 1928
Fig. 8. 33R-W

Rhyacodrilus ekmani PIGUET, 1928: 94, Fig. 7a-d.

Rhyacodrilus ekmani Piguet. UDE, 1929: 80. BRINKHURST AND KENNEDY, 1962 a: 188, Pl. VII. BRINKHURST, 1963 a: 56. HRABE, 1963 a: 251.

Rhyacodrilus ekmani forma *profundalis* LASTOCKIN, 1937: 234 .

Rhyacodrilus ekmani forma *typica* CEKANOVSKAYA, 1962: 237.

Rhyacodrilus ekmani forma *profundalis* Lastockin. CEKANOVSKAYA, 1962:
238, Fig. 142.

$l=20$ mm, $s=106$. Dorsal bundles with 4-5 bifid setae, 2-3 posteriorly, and hair
setae from XIX-LXIX, membrane between prongs of dorsal setae in some,
ventral setae 4-7 per bundle. Penial setae straight and blunt. ? Connection
between spermathecae and gut.
Sweden, U.S.S.R.

The Russian form *profundalis* was said to lack the connection between
the spermathecae and the gut. If this observation is valid, it further supports
the contention that *ekmani* is part of the variable species *R. coccineus.*
Detailed examination of the type collection of *R. ekmani* revealed the
presence of a connection between the spermathecae and between them and
the gut, as illustrated by Brinkhurst and Kennedy (1962 a). Careful study
of these illustrations indicates that sperm from both spermathecae can be
traced to the dorsal side of the gut, but that there is no sign of connections
between the walls of the spermathecal ampullae and the gut. There is some
indication that, in this instance, the spermathecae have burst, and that
sperm is being mopped up by the chloragogen tissue. More material must
be studied before this situation can be clarified.

Rhyacodrilus multispinus (Michaelsen, 1905)

Clitellio multispinus MICHAELSEN, 1905 a : 8.

Taupodrilus multispinus (Michaelsen). MICHAELSEN, 1908 a : 149, Pl. III,
fig. 12-14.

Clitellio multispinus. Michaelsen. MICHAELSEN, 1926 a : 153.

Rhyacodrilus multispinus (Michaelsen). HRABE, 1931 a : 11; 1963 a : 252.
BRINKHURST, 1963 a : 57, Fig. 40.

Rhyacodrilus multispinus var *multiovis* BUROV, 1936 : 20.

Rhyacodrilus multispinus forma *multiovata* (Burov). CEKANOVSKAYA, 1962 :
240.

Rhyacodrilus multispinus forma *typica* CEKANOVSKAYA, 1962 : 239, Fig. 144.

$l=57$ mm, $s=110$. Up to 16 setae per bundle anteriorly, 5-6 posteriorly, in the
dorsal bundles the lower tooth at right angles to the thick upper tooth. Ventral
setae with the upper tooth much longer than the lower, and bent over it, a thin
web set deep in the angle of the setae.
Lake Baikal.

Mature forms were not available when the species was first described by
Michaelsen (1905). The *multiovis* form differs so little from the typical form
that Cekanovskaya (1962) merged the two after correcting the name of the
variant. The separation of *multispinus* from *korotneffi* seems difficult to
justify on the available evidence.

GENUS **Monopylephorus** LEVINSEN, 1884

Type species: Monopylephorus rubroniveus LEVINSEN

Male efferent ducts, with vasa deferentia entering apically, tubular atria, covered with prostate cells; modified protrusible penes in some species, male pores and spermathecal pores frequently included in median inversions of the body-wall. Spermathecal and/or penial setae sometimes modified. Spermathecae without spermatophores. Coelomocytes abundant.

Cosmopolitan.

The original description of *Monopylephorus* by Levinsen (1884) was inadequate, and some confusion about the name of the genus has persisted right up to the present day. The genus was re-defined by Ditlevsen (1904) who claimed to have rediscovered Levinsen's original species, but this claim has been disputed by some authors, who prefer to retain the name *Rhizodrilus* first used by Smith (1900). The earlier synonym *Vermiculus* of Goodrich (1892) is not available, having been used previously in zoological literature. The genera *Postiodrilus* of Boldt (1926) and *Littodrilus* of Chen (1940) can now be shown to be attributable to *Monoplyephorus*, apparent differences in the structure of the male ducts being due solely to inadequate descriptions in earlier studies.

The identification of parts of the male duct is quite difficult. The narrow tubular male ducts are usually sheathed in prostate gland cells from the sperm funnel to the ventral sac-like body which usually forms a protrusible pseudopenis. Goodrich (1895) described muscle cells lying beneath the prostate cells of the tubular male duct, which probably represents the atrium. If this interpretation is true, the vas deferens is normally rudimentary, and the vertical chamber is the protrusible or eversible penis sac. Furthermore, the median inversion of the body wall would then be a tertiary inversion, the first being the atrium, the second being the penis sac. The view is supported by the description of the male ducts of *M. montanus* by Hrabe (1962 b), in which there are distinct vasa deferentia. Part of the atrium may be devoid of prostate cells, as in *M. frigidus*.

Monopylephorus

1.	Hair setae present	2
—	Hair setae absent	3
2.	Hair setae twisted distally, other dorsal setae bifid or with a single intermediate tooth. Brackish water species	*M. irroratus*
—	Hair setae straight, other dorsal setae distinctly pectinate. Fresh water species	*M. montanus*

3. Spermathecae single with median ventral pore.
    Simple-pointed setae usually present posteri-
    orly   *M. parvus*

—   Spermathecae paired. Few, if any, simple-
    pointed setae   4

4. Ventral setae of IX and XI modified, spatu-
    lated and single in IX (rarely also X), 4-5
    knobbed setae in ventral bundles of XI   *M. lacteus*

—   Ventral setae unmodified, or modified only in
    XI   5

5. Up to 13 penial setae per bundle   *M. frigidus*

—   Penial setae absent   6

6. All setae with the upper tooth shorter and
    thinner than the lower   *M. limosus*

—   At least anterior setae with upper tooth as long
    as or a little longer than lower tooth   *M. rubroniveus*

Monopylephorus rubroniveus Levinsen, 1884
Fig. 8. 35E

Monopylephorus rubroniveus LEVINSEN, 1884: 225.

Vermiculus pilosus GOODRICH, 1892: 47, Fig. 1, 2.

Vermiculus pilosus Goodrich. GOODRICH, 1895: 253, Fig. 26-28. BEDDARD,
    1895: 271. MICHAELSEN, 1900: 41.

Vermiculus fluviatilis FERRONNIÈRE, 1899: 278, Pl. XIX, fig. 10-12.

(?) *Vermiculus glotini* FERRONNIÈRE, 1899: 250, Pl. XX, fig. 13-15.

Rhizodrilus pilosus (Goodrich). MICHAELSEN, 1900: 523; 1927: 14, Fig. 15,
    16. UDE, 1929: 76, Fig. 84. KNÖLLNER, 1935: 460. MOSZYNSKI AND
    MOSZYNSKA, 1957: 44. MOSZYNSKA, 1962: 26. BÜLOW, 1955: 260; 1957:
    96.

Monopylephorus rubroniveus Levinsen. DITLEVSEN, 1904: 423. CHEN, 1940:
    86. BRINKHURST, 1962 c: 328; 1963 a: 60, Fig. 2; 1963 b: 713; 1963 c:
    43; 1965: 147, Fig. 9A-D. CEKANOVSKAYA, 1962: 287, Fig. 182. KENNEDY,
    1964: 235.

Monopylephorus glaber MOORE, 1905 b: 378, Pl. XXXII, fig. 1-6.

Rhizodrilus kermadecenis BENHAM, 1915: 180, Fig. 8-10.

Monopylephorus corderoi MARCUS, E., 1949: 1, Fig. 1-6.

Rhizodrilus kermadecenis Benham. HRABE, 1962 b: 344.

Rhizodrilus fluviatilis (Ferronnière). HRABE, 1962 b: 344.

Rhizodrilus glaber (Moore). HRABE, 1962 b: 344.

Rhizodrilus ponticus HRABE, 1967: 332, Fig. 1-12.

l=10·40 mm, s=48-74. Setae mostly bifid, with the upper tooth thinner than the lower, longer than the lower or teeth equally long, 4-6 anteriorly, 2 posteriorly, most may be simple-pointed. Vasa deferentia rudimentary, atria long, mostly covered with prostate cells, short section without prostate cells separated from protrusible pseudopenis by a constriction, pseudopenes weakly developed, cuticular lining present. Male ducts and spermathecae paired.

Cosmopolitan, brackish water.

Recent examination of sections and dissections of freshly caught material (as well as type specimens of *M. glaber*) indicates that the vertical chamber of the male duct becomes modified into a protrusible pseudopenis at full maturity, and that the cells lining the lower third may be heavily cuticularized. The spermathecae may be closely paired, or may unite before opening into the median inversion of the body wall. This variation was observed within the type collection of *M. glaber* found in the Academy of Natural Sciences, Philadelphia. The male pores may also be separate in immature worms, linked by a bridge of tissue in partially mature specimens (vide *glotini*), and may open separately or together into a median inversion when fully developed.

Some simple pointed setae were found by Ferronnière (1899) in his *V. fluviatilis*, and the schematic illustration of the male ducts of *R. kermadecensis* by Benham (1915) exaggerated the prostate-free section of the atrium at the expense of the pseudopenis. Other differences used by Hrabe (1962 b) to separate these species are trivial or based on unreliable characteristics such as the hairy nature of the cuticle and the detailed structure of the nephridia.

Monopylephorus parvus Ditlevsen, 1904

Monopylephorus parvus DITLEVSEN, 1904: 427, Pl. XVI, fig. 25, 26.

Monopylephorus parvus Ditlevsen. MOORE, 1905 b: 383, Pl. XXXIII, fig. 29-34. STEPHENSON, 1917 a: 485, Fig. 1a, b; 1923: 104, Fig. 41a, b. BRINKHURST, 1962 c: 321; 1963 a: 61; 1963 b: 713; 1965: 147, Fig. 9e. MARCUS, E., 1965: 75, Fig. 25, 26. NAIDU, 1966: 220.

Rhizodrilus parvus (Ditlevsen). HRABE, 1962 b: 343.

l=8-15 mm, s=38-64. Anterior bundles with 3-4 or 5 setae, fewer posteriorly, anterior setae bifid, the upper tooth as long as lower or longer, teeth more or less equally thick or upper thinner than lower, some with reduced upper teeth or single-pointed setae often present posteriorly. Vasa deferentia rudimentary, atria tubular, covered with prostate cells, protrusible pseudopenis present. Male ducts paired, entering median bursa through separate or combined pores, sperma-thecae single (on left) with median pore.

Cosmopolitan, brackish water.

There is some variation in the proximity of the two male pores where they join the median inversion of the body wall. Material described by Ditlevsen (1904) and Marcus (1965) had the male pores entering the ventral chamber separately, whereas the ducts were united before entering the ventral chamber in specimens described by Moore (1905 b) and Stephenson (1917 a). Speci-mens of *T. tubifex* with one spermathecae or even none are considered to be no more than variants of that species. It may be that *M. parvus* is no more than a local variant of *M. rubroniveus*.

Monopylephorus limosus (Hatai, 1898)
Fig. 8. 34H

Vermiculus limosus HATAI, 1898: 103, Fig. 1-5.

Vermiculus limosus Hatai. MICHAELSEN, 1900: 41.

Rhizodrilus limosus (Hatai). MICHAELSEN, 1900: 523. YAMAGUCHI, 1953: 297. HRABE, 1962 b: 344.

Monopylephorus limosus (Hatai). NOMURA, 1915: 1, Fig. 1-30. CHEN, 1940: 83, Fig. 25, 26. CEKANOVSKAYA, 1962: 288, Fig. 182. BRINKHURST, 1963 a: 61; 1963 b: 713.

$l=15$-70 mm, $s=43$-130. Anterior setal bundles with 3-5 or 6 setae, fewer posteriorly, upper tooth shorter and thinner than lower, simple pointed setae in ventral bundle of II. Vasa deferentia short, atria tubular, two thirds of each covered with prostate cells, short naked section then enters protrusible pseudo-penis on each side. Male ducts and spermathecae paired.

Asia, saline city ditches.

The terminal portion of the male duct appears to have a folded lining layer according to the illustrations in the account by Nomura (1915), but I have not examined the type material in order to establish the form of the pseudopenis in relation to recent re-interpretations of the structure in the genus.

Monopylephorus frigidus sp. nov.
Fig. 8. 35A-D

$l=6$-10 mm, $s=40$-51. Anterior bundles with 4-6 bifid setae, the upper tooth thinner than the lower, teeth more or less equally long, 3-4 setae posteriorly with upper tooth shorter and thinner than the lower. Penial setae 11-13 per bundle, simple pointed. Vasa deferentia rudimentary, atria long, protruding through 10/11 into X, just over half covered with prostate cells along one side, the rest naked, with an enlarged lumen, long vertical chambers opening via rudimentary penes into separate eversible penis sacs, the ventral chambers sur-rounded by the crescentic inversions of the body-wall.

Alaska, brackish water. (Olsen Creek, Prince William Sound, Coll. J. Hele.) Types: USNM, 35520, 35521.

This species resembles *M. limosus* except for the presence of penial setae and the degree of development of the atria, which are so large they cannot be accommodated in XI.

Monopylephorus lacteus (Smith, 1900)

Rhizodrilus lacteus SMITH, F., 1900: 444, Pl. IIIa, fig. 45, Pl. XL, fig. 6-8.

Rhizodrilus lacteus Smith. MICHAELSEN, 1900: 523; GALLOWAY, 1911: 316. HRABE, 1962 b: 343.

Monopylephorus lacteus (Smith). BRINKHURST, 1963 a: 62, Fig. 44a, b; 1965: 148, Fig. 9J-K.

$l=70$-100 mm, $s=215$-365. Setae bifid, up to 6 anteriorly, fewer posteriorly, a single spatulate seta in each ventral bundle of IX, 4-5 knobbed setae in XI in which the teeth are enlarged and fused, setae of X usually normal, rarely like those of IX. Vasa deferentia short, atria tubular and covered with prostate cells attached to one side. Male ducts and spermathecae paired. Elongate tubular glands associated with ventral setae of IX or IX and X.

N. America, freshwater.

There is no detailed account of the terminal section of the male efferent duct of this apparently scarce species.* The genital setae are quite distinct.

Monopylephorus irroratus (Verrill, 1873)
Fig. 8. 34A-B

Clitellio irrorata VERRILL, 1873: 324 (in part).

?*Vermiculus glotini* FERRONNIÈRE, 1899: 250, Pl. XX, fig. 13-15.

?*Monopylephorus trichochaetus* DITLEVSEN, 1904: 421, Fig. 21-23.

Tubifex irrorata (Verrill). MOORE, 1905 b: 384, Pl. XXXII, fig. 7-11.

Rhizodrilus auklandicus BENHAM, 1909: 258, Pl. X, fig. 2-7.

Tubifex irorratus (Verrill). GALLOWAY, 1911: 314.

Rhizodrilus auklandicus Benham. MICHAELSEN, 1924: 199. HRABE, 1962 b: 343.

Postiodrilus sonderi BOLDT, 1926: 177, Fig. 1-3.

Monopylephorus irroratus (Verrill). BOLDT, 1926: 177. BRINKHURST, 1963 a: 61, Fig. 43; 1965: 147, Fig. 9F-I.

---

* The type specimens have been re-examined by D. G. Cook, and a re-appraisal of the anatomy of the male ducts will appear elsewhere.

Postiodrilus sonderi Boldt. MICHAELSEN, 1927: 15, Fig. 17. BÜLOW, 1957: 96.

Littodrilus auklandicus (Benham). CHEN, 1940: 96.

Monopylephorus irrorata (Verrill). BRINKHURST, 1962 c: 329; 1963 b: 713; 1963 c: 33.

Rhizodrilus irroratus (Verrill). HRABE, 1962 b: 343.

Rhizodrilus sonderi (Boldt). HRABE, 1962: 343.

Rhizodrilus trichochaetus (Ditlevsen). HRABE, 1962 b: 343.

Monopylephorus auklandicus (Benham). BRINKHURST, 1963 a: 63; 1963 b: 713.

$l = 15$-35 mm, $s = 70$-90. Anterior dorsal bundles with 1 or 2 thin hair setae with twisted distal ends in most bundles, with up to 4 crotchets, the upper tooth longer and thinner than the lower, sometimes with a small intermediate tooth, ventral bundles with 3 or rarely up to 7 anteriorly, fewer posteriorly, upper tooth large and thinner than the lower, penial setae rarely modified. Vasa deferentia rudimentary, atria mostly covered with prostate cells, protrusible pseudopenes with cuticular lining in outer-third. Male ducts and spermathecae paired.

Cosmopolitan, brackish water.

The examination of type specimens, including freshly made serial sections, has made possible the recognition of *M. auklandicus* as a synonym of *M. irroratus*. The dorsal bundles of setae include the characteristically twisted hair-setae, which were described by Benham (1909) as "long, fine hairs— entangled with the chaetae, which at first examination were mistaken for capilliform chaetae". The male ducts closely resemble those of *M. irroratus* specimens from Britain, France and the east and west coasts of N. America, and it is clear that there is a progressive increase in complexity of the male ducts with advancing maturation. Hrabe (1966 b) found penial setae in a specimen from the original material used by Boldt (1926) in describing *P. sonderi*, but no details are available. Ventral setae were observed in segments X and XI in sections of mature specimens from Alaska, but did not appear to be modified. The *sonderi-trichochaetus-glotini-auklandicus* complex was separated by Hrabe (1966 b) using characters depending on the degree of maturation of the reproductive organs, and he separated these from *irroratus* using the erroneus description of the elongate penes illustrated by Moore (1905 b). Re-examination of the type material of *M. irroratus* as well as more mature material from Alaska has made possible a complete description of the terminal portion of the male ducts.

Monopylephorus montanus (Hrabe, 1962)
Fig. 8. 34C-G

Rhizodrilus montanus HRABE, 1962 b: 339, Fig. 1-13.

$l=14$ mm, $s=86$. Anterior dorsal bundles with 2-5 smooth hair setae, 2-5 pectinate setae, median segments with 1 or 2 hairs, 2-3 bifid setae, posteriorly only 1 or 2 bifid setae, ventral bundles with 3-6 bifid setae, the upper tooth longer and thinner than the lower anteriorly, equally long posteriorly. Up to 8 simple-pointed setae in each ventral bundle of XI. Vasa deferentia long, atria very long, tubular, closely covered with prostate cells, protrusible pseudopenis present. Male ducts and spermathecae paired.

S. Europe, freshwater.

Both the male efferent ducts and setae are different from those of any other *Monopylephorus* species.

SPECIES INQUIRENDAE

Monopylephorus africanus Michaelsen, 1913
Fig. 8. 35F

Monopylephorus africanus MICHAELSEN, 1913: 143, Fig. 1.

Rhizodrilus africanus (Michaelsen). HRABE, 1962 b: 343.

Monopylephorus africanus Michaelsen. BRINKHURST, 1963 a: 62, Fig. 45; 1966: 149.

$l=16$ mm, $s=84$. Anterior bundles with up to 4 bifid setae with equal teeth, posteriorly upper tooth thinner than ventral lower tooth. Penial setae 4-5 per bundle. Vasa deferentia long, entering atria sub-apically, atria tubular, covered with prostate cells except for short section leading to upright chamber receiving penial setae. Spermathecal pores just anterior to 9/10 (?).

S. Africa, freshwater.

This species may well be correctly located in *Rhyacodrilus* because the long vas deferens joins the atrium sub-apically,* the atria being elongate as in *R. brevidentatus* and *R. falciformis*. The proximal part of each atrium and the vas deferens is supposedly coiled within X having penetrated 10/11, while the spermathecae are wholly in IX with their pores exactly in 9/10. This arrangement is unique, but as only two specimens were obtained, one incomplete, one being sectioned, this interpretation of the positioning of the reproductive organs requires confirmation. The way in which the atria of *africanus* open into inverted portions of the body wall including the penial setae is, however, reminiscent of the male openings in other species. The genital setae are situated in the same position as those of *M. lacteus* (Smith).

Specimens with similar genital setae have been found off the coast of N. America and in the harbour of Anvers Island, Antarctica.† These specimens lack coelomocytes, and it is of interest to note that the coelomocytes are not mentioned in the original description of *M. africanus*.

GENUS **Jolydrilus** MARCUS, 1965

Type species: Jolydrilus paulus MARCUS

* Now observed in other species in the genus—D. G. Cook.
† See Appendix, p. 838.

Male ducts with short vasa deferentia, cylindrical atria lacking prostates, small penes opening into a common bursa. Spermathecae absent. Coelomocytes present.

S. America.

The absence of the prostate and of the spermathecae makes it difficult to place this monotypic genus, as the nature of the prostate and the presence or absence of spermatophores are two of the most important characters upon which generic distinctions are normally based.

Jolydrilus jaulus Marcus, 1965
Fig. 8. 36A-C

Jolydrilus jaulus MARCUS, 1965 : 77-80, Fig. 27-30.

Jolydrilus jaulus Marcus. BRINKHURST, 1966 a : 302.

$l=35$ mm, $w=0.6-1.0$ mm, $s=70$. Prostomium short. Clitellum in $\frac{1}{2}$X to XIII. Dorsal setae anteriorly 4 to 6 bifids with upper tooth longer than lower, 120 μ long; posteriorly 2 to 3 bifids of the same form and length; in middle segments up to 10 straight, smooth hair setae, 650 μ long. Ventral setae anteriorly 4 to 6 bifids per bundle, posteriorly 2 to 3 per bundle, all with upper teeth shorter and thinner than lower, 120 μ long. In immature worms dorsal bifids shorter than ventrals (90 μ and 200 μ respectively). Male pore single, median, ventral. Spermathecal pores absent.

Pharyngeal glands in IV and V. Chloragogen cells begin in V. A pair of short vasa deferentia join a pair of cylindrical atria. These terminate in a pair of small penes (not pseudopenes) which open laterally into a common median chamber. Prostates absent. Spermathecae absent. Cocoons ovoid, containing about 11-eggs. Cocoons 1·4 to 2· 6mm long, 0·8 to 1·3 mm diameter.

Brazil (16-20‰ salinity, in mangrove swamp near Sao Paulo).

SUB-FAMILY BRANCHIURINAE HRABE, 1966

Type genus: Branchiura BEDDARD

Prostate glands diffuse. No spermatophores. No coelomocytes.

The only species (*B. sowerbyi*) has gills and atrial diverticulae that are omitted from the above as being generic characteristics.

GENUS **Branchiura** BEDDARD, 1892

Type species: Branchiura sowerbyi BEDDARD

Male efferent ducts with short vasa deferentia, atria covered with prostate cells, atrial diverticulae present, ejaculatory ducts enter eversible pseudopenis. Spermathecae paired, no spermatophores. No coelomocytes. Dorsal and ventral gill filaments on each segment posteriorly.

Cosmopolitan.

This monospecific genus is quite unlike any other in the Tubificidae, the diffuse prostate gland and lack of spermatophores usually being associated with the presence of abundant coelomocytes. No other tubificid has gills, although these occur in some species belonging to other families, and no other species has atrial diverticulae (formerly termed paratria, a term here reserved for the structures so named in *Bothrioneurum*).

Branchiura sowerbyi Beddard, 1892
Fig. 8. 36D-F

Branchiura sowerbyi BEDDARD, 1892: 325, Pl. XIX, fig. 1-15.

Branchiura sowerbii Beddard. BEDDARD, 1895: 271.

Branchiura sowerbyi Beddard. MICHAELSEN, 1900: 40; 1908 a: 134, Pl. III, fig. 1-6; 1909: 30, Fig. 53, 54; 1934: 494; 1936: 95, Pl. I. SOUTHERN, 1909: 135. FRIEND, 1912 b: 290. STEPHENSON, 1912 a: 285, Pl. I, II, fig. 1-8; 1912 b: 234, Pl. XII, fig. 1-5; 1913 a: 741; 1917 b: 89; 1918: 12, Fig. 1-3; 1920: 200; 1921: 752; 1923: 99; 1924: 321; 1929: 226; 1930: 272. KEYL, 1913: 199, Fig. 1, 2, 17-19, 28-30, 36-56, Pl. IX, fig. 2, 5-7, 9, Pl. X, fig. 10-15, Pl. XI, fig. 16, 17. MEHRA, 1920: 457. SPENCER, 1932: 267. MALEVITCH, 1937 a: 131. CHEN, 1940: 90. CAUSEY, 1953: 423. COLE, 1954: 127. STRECHER, 1954: 280. WURTZ AND ROBACK, 1955: 183. JUGET, 1957: 2. EVANS, 1958: 223. MANN, 1958: 732. BRINKHURST, 1960: 398, Pl. IX, fig. 1a-c; 1962 c: 321; 1963 a: 59, Fig. 2, 4; 1963 c: 29, Fig. 8a, b; 1965: 146, Fig. 8L-M. WURTZ AND DOLAN, 1960: 470. SOKOLSKAYA, 1961 b: 84; 1961 c: 605. CEKANOVSKAYA, 1962: 291, Fig. 184. INOUE and KONDO, 1962: 97. STAMMER, 1963: 390. HARO, 1964: 137.

Kawamuria japonica STEPHENSON, 1917 b: 89, Fig. 1-5.

$l=38$-185 mm, $s=74$-270. Dorsal anterior bundles with 1-3 short hair setae, 11-12 setae with bifid tips, rarely a single intermediate tooth, often the upper tooth rudimentary or absent, ventral bundles with 10-11 similar bifid setae. Male efferent ducts with atrial diverticulae. Gills present posteriorly.

Cosmopolitan.

The species is now known to be widespread in N. America and Europe, and has been recorded from S. America, S. Africa, Mauritius, and Australia. Earlier records were restricted to Asia and to botanical gardens in Europe, and it was thought to have been introduced to temperate localities. It is certainly of sporadic occurrence in temperate habitats, being frequently observed in numbers in artificially warmed sites and man-made situations, but by no means absolutely restricted to them.

In the illustration of the atrial diverticulae by Chen (1940) the diffuse "prostate" cells were omitted, whereas they were included in the diagrammatic figure illustrating the supposed evolution of male ducts in the Tubificidae which appeared in the same paper.

Sections of mature specimens collected by Dr. Jamieson near Nairobi (Athi River) enabled this point to be clarified. The structure formerly called "paratrium" is in fact a glandular diverticulum on each atrium.

The gland cells investing the outside of each diverticulum differ histo-logically from the more vacuolated prostate cells that cover the tubular atria. The wall of the atrium bears the usual muscle layer but this is missing on the diverticulum. The lumen of the diverticulum is wider than that of the atrium and, in my sections, was filled with a secretion. The atrium narrows proximally, and the muscle layer becomes separated from the epithelial lining at the beginning of the pseudopenis, creating a space between the two layers which has been referred to as a coelomic sac (Hrabe, 1967). The atrial lining forms a thin tube within this muscular sac, or distal part of the pseudopenis. This tube winds about in the sac until it becomes con-fluent with the wide chamber of the pseudopenis. The diverticulum arises from the narrow tubular atrium within the confines of the pseudopenis, and the two tubes are closely associated at this point. They remain in close proximity even when invested with a common muscular sheath at the innermost end of the pseudopenis, but the diverticulum then separates from the atrium, losing all trace of a muscular sheath, at the point where the glandular layer becomes apparent. This final separation is obscured by the mass of gland cells surrounding the two ducts.

The vas deferens is very wide immediately behind the funnel on each side.

SUB-FAMILY PHALLODRILINAE SUBFAM. NOV.

Type genus: Phallodrilus PIERANTONI

Prostates solid with stalk-like attachments to the male ducts when present. No spermatophores. Coelomocytes sparse or absent.

The genus *Phallodrilus* has stalked prostates on both the anterior and posterior ends of the atria. The prostates of *Adelodrilus* are also four in number, two being attached to a small chamber on the male ducts, the homologies of which are difficult to determine.

GENUS **Phallodrilus** PIERANTONI, 1902

Type species: Phallodrilus parthenopaeus PIERANTONI

Vasa deferentia as long as, or longer than, pear-shaped to cylindrical atria. Vasa deferentia join atria apically. Each atrium bears two, discrete, petiolate prostate glands, one of which joins near vas deferens, the other near the basal end of the atrium. Spermatophores not developed. Coelomocytes sparse to absent.
Europe, N. America.

Phallodrilus

1.   Penial setae absent. Spermatheca single with
       mid-dorsal pore   *P. monospermathecus*

—   Penial setae present. Spermathecae paired
     with lateral to ventral pores   2

2.   Up to 13 hooked penial setae, shorter than
     body setae present. Spermathecal pores situ-
     ated between lines of dorsal and ventral setae   *P. coeloprostatus*

—   Up to 7 bifid to simple-pointed penial setae,
     longer than body setae, present. Spermathecal
     pores situated near to ventral setae   3

3.   Spermathecal setae with upper teeth 2·5
     times longer than lower   *P. parthenopaeus*

—   No modified spermathecal setae   4

4.   4-6 (rarely 7) unmodified to single-pointed
     penial setae present. Spermathecal ducts
     short and narrow   *P. aquaedulcis*

—   2-3 (rarely 4) single-pointed penial setae
     present. Spermathecal ducts long and thick   *P. obscurus*

Phallodrilus coeloprostatus Cook, 1969
Fig. 8. 38D-F

Phallodrilus coeloprostatus Cook, 1969 a: 16, Fig. 5.

l=6-10 mm, w=0·17-0·30 mm, s=58-60. Prostomium rounded, as long as or a
little longer than it is broad at peristomium. Clitellum well developed on
$\frac{1}{2}$X-XII. Setae bifid with upper teeth shorter and thinner than lower, 48 to 55 μ
long; 4 to 5 per bundle anteriorly, 3 to 4 posteriorly. Setae of X unmodified. 10
to 13 penial setae, 40 to 45 μ long, hooked distally, present on XI. Paired male
pores situated just lateral to penial setae. Paired spermathecal pores situated in
anterior part of X, mid-way between lines of dorsal and ventral setae.
    Pharyngeal glands extend into VI. Chloragogen cells begin in VI. Atria very
small, cylindrical, curved towards anterior end of worm and very closely applied
to body-wall. Vasa deferentia, 6 μ diameter, longer than atria, join latter apically.
Atria 95 to 130 μ long, 19 to 26 μ diameter, with very thin musculature and
thick lining cells, terminating in small, truncated cone-shaped penes. Two pairs
very large prostate glands join atria by thick discrete ducts near vasa deferentia
and posteriorly near the penial end of the atria. Prostates, which lie medially to
and completely cover atria, have distinct boundaries but the cells are loosely
packed and cavities are usually present in them. Paired spermathecae have short
discrete ducts and large ovoid to elongate ampullae.
    Cape Cod Bay. 41° 53·5′ N, 70° 10·65′ W. Depth 18·3 metres.
    Massachusetts, USA.
    Holotype: USNM.

Phallodrilus obscurus Cook, 1969
Fig. 8. 38A-C

Phallodrilus obscurus Cook, 1969 a : 17, Fig. 6.

$l=7$ mm, $w=0.16-0.20$ mm, $s=40$. Prostomium longer than it is broad at peristomium. Setae, except ventrals of XI, bifid with upper tooth equal to, or shorter and thinner than lower tooth. Setae 45 to 55 μ long, 4 to 6 setae per bundle anteriorly, 4 to 5 per bundle posteriorly. Spermathecal setae unmodified. 2 to 3(4) slightly curved, simple-pointed penial setae present, 55 to 70 μ long. One pair spermathecal pores in anterior part of X, in line of ventral setae. Male pores paired, just lateral to penial setae.

Pharyngeal glands extend to V. Chloragogen cells begin in VI. Atria pear to comma-shaped, shorter than vasa deferentia which join atria apically. Atria 70 to 120 μ long, 30 to 35 μ diameter. Vasa deferentia 100-130 μ long, 5.5-7.5 μ diameter. Anterior prostate enters atrium near vas deferens, posterior prostate 10-40 μ from the male pores. Penes absent. One pair spermathecae with long, thick ducts and cylindrical ampullae open in the anterior part of X, near septum 9/10.

Cape Cod Bay, near Ellisville, Massachusetts, USA, 41° 51' N, 70° 31.1' W. Depth 8.5 metres.

Holotype: USNM.

Phallodrilus aquaedulcis Hrabe, 1960
Fig. 8. 37A-C

Phallodrilus aquaedulcis HRABE, 1960: 248, Fig. 1-4.

Phallodrilus aquaedulcis Hrabe. BRINKHURST, 1963 a : 74, 77. COOK, 1969 a : 19.

$l=3-4$ mm, $w=0.15$ mm at X, $s=28-32$. Prostomium rounded, as long as it is broad at peristomium. Setae, except ventrals of XI, bifid with upper tooth thinner and shorter than lower, about 42 μ long. 3-4(5) setae per bundle anteriorly, 2 per bundle posteriorly. Spermathecal setae unmodified. 4-6(7) penial setae present; unmodified bifids to single-pointed setae. Spermathecal pores in anterior part of X, in line with ventral setae. Male pores slightly anterio-lateral to penial setae.

Pharyngeal glands present in III to V. Chloragogen cells begin in V or VI. Atria cylindrical, longer than vasa deferentia which open into atria apically. Anterior prostate opens near vas deferens junction, posterior prostate near basal part of atrium. Penes absent. A pair of spermathecae present with long cylindrical ampullae and short narrow ducts.

Germany (R. Weser).

Phallodrilus parthenopaeus Pierantoni, 1902
Fig. 8. 37F-I

Phallodrilus parthenopaeus PIERANTONI, 1902: 114, Fig. 1, 2.

Phallodrilus parthenopaeus Pierantoni. HRABE, 1960: 251. BRINKHURST, 1963 a : 74; 1963 e : 714; 1967: 115. COOK, 1969 a : 18.

$l=$ up to 12 mm, $w=0.2$ mm, $s=40-60$. Setae, except ventrals of X and XI, bifid with equal teeth. 4 setae per bundle anteriorly, 2 per bundle posteriorly. 2

spermathecal setae per bundle anteriorly, 2 per bundle posteriorly. 2 spermathecal setae per bundle with upper teeth about $2\frac{1}{2}$ times longer than lower. 2 straight, blade-shaped penial setae per bundle with simple rounded ends. Male and spermathecal pores paired, near line of ventral setae.

Atria elongate, pear-shaped, shorter than vasa deferentia which join atria apically. Anterior prostate opens near vas deferens junction, posterior prostate near base of atrium. Penes absent. Spermathecae paired in X.

Italy: Gulf of Naples, 4 metres depth.

Phallodrilus monospermathecus (Knöllner, 1935) n. comb.
Fig. 8. 37D, E

Aktedrilus monospermathecus KNOLLNER, 1935: 482, Fig. 43-50.

Aktedrilus monospermathecus Knöllner. BÜLOW, 1955: 262; 1957: 102. HRABE, 1960: 251-254, Fig. 5-12. CEKANOVSKAYA, 1962: 286. BRINKHURST, 1963 a: 75; 1963 c: 1203; 1963 e: 714; 1964 b: 12; 1967: 115.

Phallodrilus monospermathecus (Knöllner). COOK, 1969 a: 19

$l=3$-8 mm, $w=0.11$-0.23mm, $s=25$-35. Prostomium longer than it is broad at peristomium. Setae bifid with upper tooth shorter and thinner than lower, 32 to 45 μ long (2)3 to 4(5) setae per bundle anteriorly, (1)2 to 3(4) per bundle posteriorly. Ventral setae of XI absent. Spermathecal pore single, mid-dorsal on X; cuticle thickened in region of pore. Male pores paired in line of ventral setae.

Pharyngeal glands in IV to VI. Chloragogen cells begin in VI. Atria narrow, cylindrical, about as long as vasa deferentia. Atria 16 to 17 μ in diameter, terminating in ovoid penes 25 to 30 μ long, 17 to 19 μ wide, contained in penial chamber. Anterior prostate opens into atrium apically with vas deferens, smaller posterior prostate joins near base of atrium. Spermathecae single, cylindrical with thick duct opening mid-dorsally.

Kiel Bay, Baltic, Mediterranean, N.-E. Atlantic.

GENUS **Spiridion** KNÖLLNER, 1935

Type species: Spiridion insigne KNÖLLNER

Vasa deferentia about as long as atria. One discrete, pedunculate prostate gland joins each atrium almost apically but dorsal or posterior to junction of vas deferens. Atrial muscle thin. True penes absent. Spermatophores not developed. Coelomocytes absent.

Europe, N. America.

No key to species is provided as 3 out of the 4 species here attributed to *Spiridion* are species inquirendae.

Spiridion insigne Knöllner, 1935
Fig. 8. 38G-J

Spiridion insigne KNÖLLNER, 1935: 427, Fig. 35-38.

Spiridion insigne Knöllner. BÜLOW, 1957: 98. HRABE, 1960: 255, Fig. 13, 14.

CEKANOVSKAYA, 1962: 243. BRINKHURST, 1963 a: 74, Fig. 2, 57; 1967: 115. COOK, 1969 a: 24.

$l=$5-10 mm, $w=$0·34 mm at XI, 0·13 mm in pre-clitellar segments, $s=$34. Setae bifid with upper teeth shorter and thinner than lower, 35 to 45 μ long. 3 to 5 setae per bundle anteriorly, 1-2(3) posteriorly. 4 to 6 single-pointed, hooked penial setae, 77 to 90 μ long, present on XI. Ventral setae absent from X. Male pores paired anterior to and in line with ventral setae. One pair spermathecal pores, in line with ventral setae, present in anterior part of X.

Vasa deferentia a little longer than atria, join latter apically. Atria cylindrical, 140 μ long, 20 μ diameter. Prostate glands enter atria near junction with vasa deferentia. Spermathecae with long cylindrical ampullae and short discrete ducts, opening near septum 9/10.

Strander Bach; Schilkseebucht; Kiel Bay. Northern Europe, U.S.A. (?).

SPECIES INQUIRENDAE

Spiridion scrobicularae Lastockin, 1937

Spiridion scrobicularae LASTOCKIN, 1937: 234.

Spiridion scrobiculare Lastockin. BRINKHURST, 1963 a: 75; 1967: 115. COOK, 1969 a: 25.

Clitellum rudimentary, developed only in region of spermathecal pores and genital cavity. Male pore unpaired, median. Penial setae present.

Paired atria muscular, consisting of a narrow anterior part and pear-shaped posterior part. Prostate gland joins atrium apically. Both atria and bundles of penial setae open into the median genital chamber.

Gulf of Finland, U.S.S.R.

Spiridion roseus (Pierantoni, 1904)

Limnodriloides roseus PIERANTONI, 1904: 188, Fig. 2.

Limnodriloides roseus Pierantoni. BOLDT, 1928: 146. HRABE, 1967: 347. *Clitellio roseus* (Pierantoni). BRINKHURST, 1963 a: 73; 1963 b: 713 *Spiridion roseus* (Pierantoni). COOK, 1969 a: 25.

Setae bifid, 4 per bundle anteriorly, 3 per bundle posteriorly. Vasa deferentia short, join atria apically. Atria elongate, pear-shaped to cylindrical, erect, terminating in short protrusible penes. Large prostate gland joins each atrium just posterior to junction of vas deferens.

Gulf of Naples, 3-4 metres.

Spiridion pectinatus (Pierantoni, 1904)

Limnodriloides pectinatus PIERANTONI, 1904: 190, Fig. 3.

Limnodriloides pectinatus Pierantoni. BOLDT, 1928: 146. HRABE, 1967: 347.

Clitellio pectinatus (Pierantoni). BRINKHURST, 1963 a: 73; 1963 b: 713.

Spiridion pectinatus (Pierantoni). COOK, 1969 a: 26.

l=12-15 mm, w=0·25 mm, s=50. Clitellum on ½X to ½XII. Setae bifid, 4 per bundle up to about XIV, then 2-3 per bundle. Ventral setae of XI modified to 12 small penial setae situated on tubercles. Paired male pores lateral to penial setae. Spermathecal pores in line with and anterior to ventral setae. Short vasa deferentia and large prostate glands join atria apically.

Gulf of Naples.

GENUS **Adelodrilus** COOK, 1969

Type species: Adelodrilus anisosetosus COOK

Vasa deferentia very short, about 0·2 the length of atria, join latter apically. Atria thin walled, cylindrical, without prostate cells, terminating in pear-shaped penial bulbs. Penial bulbs each bear two large, thickly stalked, prostate glands. Spermatophores not developed. Coelomocytes only sparsely distributed and small.

N. America.

Adelodrilus anisosetosus Cook, 1969
Fig. 8. 39A-F

Adelodrilus anisosetosus Cook, 1969 a: 13, Fig. 3.

l=4-6·5 mm, w=0·18-0·4 mm, s=30-45. Prostomium broadly rounded, longer than it is wide at peristomium. Clitellum well developed on X to XII. Segments, especially posteriorly, deeply annulated with body wall nuclei concentrated in rows on the crests of the ridges formed by the annulation. Setae 3, sometimes 4, per bundle in all body regions. Anterior, and posterior ventral setae bifid with widely diverging teeth, the upper of which become thinner and shorter in more posterior segments. Posterior dorsal setae single pointed and strongly curved distally, 85 to 110 μ long. Anterior setae 80 to 100 μ long and mid-body setae 70 to 75 μ long. Ventral setae of XI highly modified. Each penial bundle contains one giant, simple-pointed, strongly curved setae, 135 to 150 μ long, 6·5 to 8·8 μ thick, and 8 to 12 small, thin, straight seta, 70 to 90 μ long, 1·5 μ thick which are clubbed distally and which bear a thin, hooked tooth, originating apically and curving around the club. Male and spermathecal pores paired in line of ventral setae.

Pharyngeal glands extend into VI. Male ducts consist of a pair of vasa deferentia, 40 μ long, 15 μ diameter, which join a pair of cylindrical, thin-walled, atrial ampullae apically. Each atrial ampullae, 250 μ long, 28 to 37 μ diameter, joins a pear-shaped penial bulb. This structure, 50 to 65 μ long, 30 to 45 μ diameter, bears two thickly stalked, discrete prostate glands, one apically near atrial junction, and one near the proximal end. Penial bulbs open into a pair of spherical, petiolate chambers about 70 μ diameter, into which also protrude the penial setae. Penes are formed from the protruded ends of elongate lining cells of the penial bulb. Spermathecae with sacciform ampullae 200 μ long, up to 55 μ wide and ill-defined ducts 45 μ long, 25 μ diameter which open near septum 9/10.

Holotype: USNM.

Cape Cod Bay, Massachusetts. 41° 53·5′ N, 70° 10·65′ W. Depth 18·3 metres. USA.

GENUS **Thalassodrilus** BRINKHURST, 1963

Type species: Rhyacodrilus prostatus KNÖLLNER

Male ducts with large funnels, short broad vasa deferentia entering pear-shaped atria laterally. Small solid prostates on vasa deferentia close to atria. No penes. No spermatophores. No coelomocytes.
Europe.

Thalassodrilus prostatus (Knöllner, 1935)
Fig. 8. 39G-K

Rhyacodrilus prostatus KNÖLLNER, 1935: 427, Fig. 30-34.

Rhyacodrilus prostatus Knöllner BÜLOW, 1957: 97. CEKANOVSKAYA, 1962: 242. BRINKHURST, 1963 a: 58.

Thalassodrilus prostatus (Knöllner). BRINKHURST, 1963 a: 55; 1963 e: 711, 714, Fig. 1; 1967: 115.

Limnodriloides prostatus (Knöllner). HRABE, 1967: 347.

l=6-10 mm, w=0.16-0.23 mm, s=34. Dorsal setae anteriorly (3) 5-6 (7) bifids per bundle, posteriorly 4-(7) simple-pointed setae per bundle. Ventral setae anteriorly 5-7 bifids per bundle, posteriorly 4-6 bifids per bundle. Setae 50 to 63 μ long, bifids with upper teeth thinner and shorter than lower. 15 to 16 single-pointed, slightly hooked, penial setae per bundle present on XI. Penials 72 to 92 μ long, 3 to 4 μ thick. One pair male pores situated anterior to and in line with penial setae, enclosed within a median invagination of body wall. One pair spermathecal pores situated in line with ventral setae, in anterior part of X.

Pharyngeal glands present in V. Chloragogen cells begin in VI. One pair of vasa deferentia (plus atria ?), 60 μ long, 15 μ diameter, join one pair of thick walled, pear-shaped atria (pseudo-penes ?) subapically; thickness of atrial muscle 12 μ. Small prostate glands join vasa deferentia (plus atria ?) near proximal end. One pair sacciform spermathecae with short, thick ducts. Coelomocytes absent.

Germany, Britain.

SUB-FAMILY CLITELLINAE SUBFAM. NOV.

Type genus: Clitellio SAVIGNY

Prostates diffuse when present. Spermatophores present. Coelomocytes absent.

Whilst the paratria of *Smithsonidrilus* resemble those of *Bothrioneurum* (Rhyacodrilinae), the resemblance ceases there, the presence of spermatophores and the absence of coelomocytes, indicating profound differences.

GENUS **Smithsonidrilus** BRINKHURST, 1966

Type species: Smithsonidrilus marinus BRINKHURST

Male efferent ducts tubular, opening into a common chamber in association with two paratria. Prostate glands cap the paratria. Spermathecae present, with spermatophores. No coelomocytes.
N. America.

Smithsonidrilus marinus Brinkhurst, 1966
Fig. 8. 40A-E

Smithsonidrilus marinus BRINKHURST, 1966 a: 300, Fig. 2.

$l=10$-15 mm, $w=0.75$ mm, $s=40$-70. Worm wound in a thin coil, with long narrow segments. Prostomium short and blunt. Setae bifid with upper teeth shorter and thinner than lower, often upper teeth minute or absent in more posterior segments; setae 2 to 5 per bundle anteriorly, 65 to 90 μ long, 4·5 to 8·5 μ thick. 1 to 2 per bundle posteriorly, 60 to 65 μ long. 3 to 4 blunt spermathecal setae in ventral bundles of X. 3 broad simple pointed penials on XI. Male pore unpaired, mid-ventral. One pair spermathecal pores in line of ventral setae (?).

One pair vasa deferentia, 27 μ diameter, continuous with a pair of thin walled "atria", 42 μ diameter, 285 μ long, which are without prostate glands. The pair of "atria", and a pair of thick-walled pear-shaped "paratria", 100 μ in diameter, 135 μ long, join an ovoid chamber whose long axis is directed across the body (left to right) and which communicates with the exterior by a single median duct. "Paratria" bear prostate glands apically. Spermathecae containing spermatophores 10 to 15 times longer than broad. ? genital setae.
Florida, N. Carolina, U.S.A.

The genital setae described in specimens from N. Carolina were not observed in the holotype, but resemble those of *Phallodrilus*. Their presence in specimens with male ducts as described from Florida specimens requires confirmation.

GENUS **Clitellio** SAVIGNY, 1820

Type species: Lumbricus arenarius MÜLLER

Male efferent ducts tubular, prostate glands absent or diffuse. Penes present. Spermathecae present, with spermatophores. No coelomocytes.
Europe, N. America, Australia.

Clitellio arenarius (Müller, 1776)
Fig. 8. 40F-H

Lumbriculus arenarius MÜLLER, 1776: 216.

Clitellio arenarius (Müller). SAVIGNY, 1820: 104, Pl. I, III; 1826: 443, Pl. XXI. CLAPARÈDE, 1861: 106, Fig. 1-18. VEJDOVSKY, 1884: 45. LEVINSEN, 1884: 228. BEDDARD, 1889: 485, Fig. b, Pl. XXIII, fig. 1-5; 1895: 247. (Partim) VAILLANT, 1890: 414. FERRONNIÈRE, 1899: 246. MICHAELSEN, 1900: 41; 1908: 150; 1912: 94; 1927: 16, Fig. 18. DITLEVSEN, 1904: 422. MOORE, 1905: 377. SOUTHERN, 1909: 135, Pl. III, fig. 4; 1911: 4. UDE, 1929: 81, Fig. 96. KNÖLLNER, 1935: 427, 475. HRABE, 1952: 5; 1967: 349-351, Fig. 41-44. BÜLOW, 1957: 98, Pl. XXIX, fig. 1. BRINKHURST, 1960: 400, Fig. 3; 1962 a: 322; 1963 a: 72, 73; 1963 e: 713; 1965 c: 152; 1966 a: 299; 1966 c: 734; 1967: 115. BRINKHURST AND KENNEDY, 1962 a: 373. KENNEDY, 1964: 234.

(?) *Nais littoralis* VINCENT, 1824: 134, Fig. 4, 5, 7, 8, 10.

Peloryctes arenarius (Müller). LEUKART, 1849: 161.

Tubifex hyalinus UDEKEM, 1855: 544.

Tubifex lineatus (non *Saenuris lineata* Hoffmeister, 1843) UDEKEM, 1859: 11.

Pachydermon acuminatum CLAPARÈDE, 1861: 88.

Limnodrilus hyalinus (Udekem) TAUBER, 1879: 71. VEJDOVSKY, 1884: 45.

Psammobius hyalinus LEVINSEN, 1884: 224.

Psammoryctes hyalinus (Levensen) MICHAELSEN, 1900: 53.

Limnodrilus arenarius var *inaequalis* MICHAELSEN AND VERESCHAGIN, 1927: 219.

Peloscolex canadensis BRINKHURST, 1965 c: 135, Fig. 6a, b.

l=20-65 mm, w=0·5-0·8 mm, s=64-120. Setae faintly bifid with upper tooth reduced, or single-pointed, 90 to 130 μ long. 2 to 3 setae per bundle anteriorly, 1 to 2 per bundle posteriorly. One seta per bundle ventrally on X. Ventral setae absent on XI. Male and spermathecal pores in line of ventral setae, the latter anterior to setal bundle.

Vasa deferentia longer than atria; 1200 to 1500 μ long, 40 to 50 μ diameter. Atria long, tubular, with thick internal lining cells. Atria, 990 to 1250 μ long, 80 to 110 μ diameter, terminate in protrusible penes 42 to 55 μ long, 38 to 60 μ wide. Spermathecae with discrete ducts, 180 μ long and long sacciform to ovoid ampullae. Long, thin tapering spermatophores (1000 μ to 45 μ).

Europe, N. America.

Clitellio arenicolus (Pierantoni, 1902)
Fig. 8. 40I-S

Heterodrilus arenicolus PIERANTONI, 1902: 116, Fig. 3.

Clitellio subtilis PIERANTONI, 1917: 83, Pl. IV, fig. 1-5.

Heterodrilus arenicolus Pierantoni. PIERANTONI, 1916: 87, Pl. IV, fig. 6-11.
BRINKHURST, 1963 a: 74; 1963 e: 714; 1967: 115. HRABE, 1967: 351.

Clitellio subtilis Pierantoni. BRINKHURST, 1963 a: 73; 1963 e: 713; 1967: 115.

Clitellio arenicolus (Pierantoni). BRINKHURST, 1963 a: 299, Fig. 1.

l=7-10 mm, w=0·16-0·40 mm, s=45-60. Anterior end of worm markedly narrow. Prostomium elongate, conical. Anterior setae 2 per bundle, trifid to blunt ended, 77 to 108 μ long, 6·5 μ thick. Posterior setae 1 per bundle, bifid with short teeth, the upper tooth usually the shorter or rudimentary, 70 to 100 μ long. Ventral setae of X, 1 per bundle, unmodified; of XI, 2 per bundle, simple-pointed, 110 to 120 μ long. One pair male and spermathecal pores situated just posterior to ventral setae.

One pair tightly coiled vasa deferentia, 13 to 20 μ diameter, join one pair cylindrical atria, 740 to 1000 μ long, 35 to 48 μ diameter, which narrow to 22 μ proximally and terminate in swellings which contain internal penes. Atria bear a diffuse covering of prostate cells. One pair spermathecae with short discrete ducts and pear-shaped ampullae contain spermatophores.
Mediterranean, N. America (Florida, N. Carolina).

SPECIES INQUIRENDAE

Clitellio abjornseni Michaelsen, 1907
Clitellio abjornseni MICHAELSEN, 1907: 124.

$l = 3$ mm, $s = 24$. 3-4 dorsal, 4-5 ventral bifid setae with upper tooth shorter and thinner than the lower, no ventral setae on XI. Short vasa deferentia, tube-like atria. No prostate gland.
S.W. Australia, marine.

The type specimens of this poorly described material are in Perth but they have yet to be re-described.

GENERA ET SPECIES DUBIAE

Ilydrilus meganymphus Friend, 1912

Clitellio heterosetosus Czerniavsky, 1880

Ilyodrilus meganymphus Friend, 1912

Ilydorilus pallescens Friend, 1912

Ilyodrilus robustus Friend, 1910 (nomen nudum), 1912.

Ilydorilus glandulosus Friend, 1910 (nomen nudum).

Ilyodrilus asiaticus Chen, 1940

Limnodrilus bogdanovi Grimm, 1877

Limnodrilus novaezelandiae Beddard, 1895

Limnodrilus galeritus Friend, 1912

Limnodrilus inaequalis Friend, 1912

Limnodrilus nervosus Friend, 1912

Limnodrilus trisetosus Friend, 1912

Limnodrilus papillosus Friend, 1910

Lumbricus lineatus Rathke, 1843

Lumbricus littoralis Dalyell, 1853

Pododrilus Czerniavsky, 1880

Rhyacodrilus dichaetus Friend, 1912

Rhyacodrilus bisetosus Friend, 1912

Saenuris batillifera Schmankewitsch, 1873

Saenuris remifera Schmankewitsch, 1873

Saenuris neurosoma Leukhart, 1847

Saenuris variabilis Friend, 1912

Strephuris agilis Leidy, 1850

Tubifex serpentinus Orsted, 1844

Tubifex elongatus Udekem, 1855

Tubifex diaphanus Tauber, 1879

Tubifex contrarius Giard, 1893

Tubifex globulatus Friend, 1912

Tubifex (Taupodrilus) lunzensis Pointner, 1914

Tubifex gentilinus Dugès, 1837 (Gen. et. sp. dub. Lumbriculidae—Michaelsen, 1900—also *T. gentilianus* Dugès).

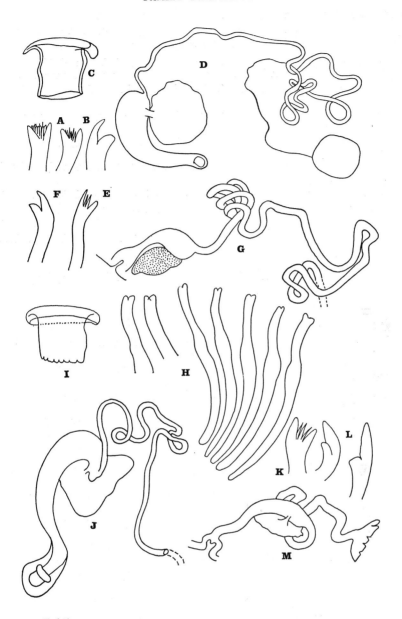

Fig. 8.1. *Tubifex.*
T. tubifex. A—anterior dorsal setae, B—anterior ventral seta, C—penis sheath, D—male duct.
T. ignotus. E—anterior dorsal seta, F—anterior ventral seta, G—male duct.
T. newaensis. H—setae, I—penis sheath, J—male duct.
T. nerthus. K—anterior dorsal setae, L—anterior ventral setae, M—male duct.

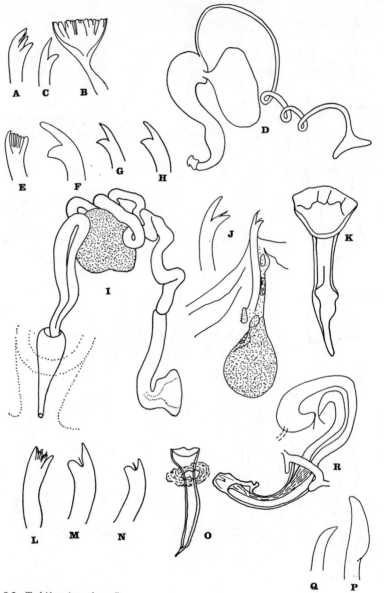

Fig. 8.2. *Tubifex* (continued).

T. costatus. A—anterior dorsal seta, B—palmate dorsal seta, C—anterior ventral seta, D—male duct.

T. montanus. E—dorsal seta, F,G,H,—ventral setae of V, X and XXX respectively, I—male duct.

T. kessleri kessleri. J—ventral seta from IX, showing intermediate tooth and setal gland, K—penis sheath.

T. kessleri americanus. L—anterior dorsal seta, M,N—ventral setae of anterior and posterior segments, O—penis sheath.

T. longipenis. P,Q—ventral setae of median and posterior segments, R—atrium and penis.

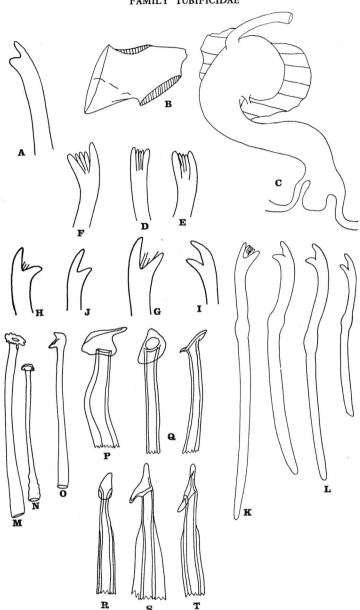

Fig. 8.3. *Tubifex* (continued) and *Limnodrilus.*
T. natalensis. A—seta, B—penis sheath, C—male duct.
T. flabellisetosus. D—dorsal seta of VI, E—dorsal seta of XVII, F—ventral seta of VI.
T. tubifex (bergi). G,H—dorsal setae of VII and XX, I,J—ventral setae of II and XVIII.
T. kryptus. K—dorsal seta, L—ventral setae.
Limnodrilus species M to T—penis sheaths, M—*L. hoffmeisteri* (straight-headed variant), N—*L. angustipenis,* O—*L. hoffmeisteri* (typical), P—*L. maumeensis,* Q—*L. claparedeianus,* R—*claparedeianus/cervix* (rare intermediate), S,T—*L. cervix* (typical).

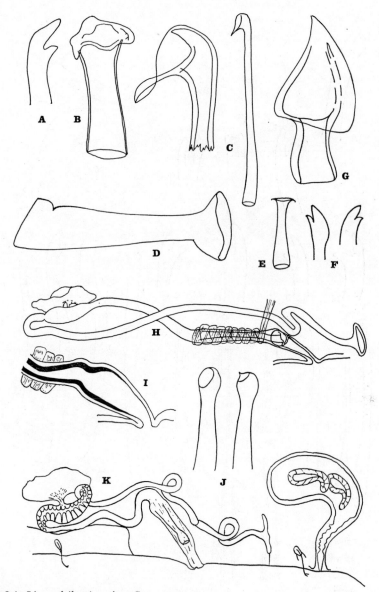

Fig. 8.4. *Limnodrilus* (continued).
L. udekemianus. A—seta, B—penis sheath,
L. hoffmeisteri. C—penis sheath, H—male duct, I—male pore, section (sheath in solid black).
L. profundicola. D,E—penis sheath.
L. silvani. F—setae, G—penis sheath.
L. neotropicus. J—penis sheaths, K—male duct and spermatheca.

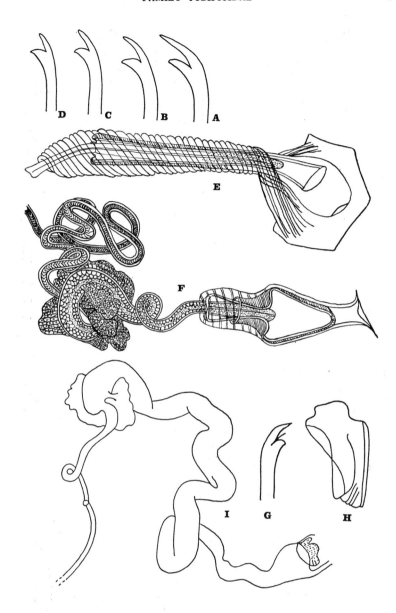

Fig. 8.5. *Limnodrilus* (continued) and *Isochaeta*.
L. neotropicus. A to D—setae from III, VIII, XV and a posterior segment respectively.
L. hoffmeisteri. E—spiral muscles around penis sac acc. to Eisen.
L. silvani. F—male duct.
I. israelis. G—anterior ventral seta, H—penis sheath, I—male duct.

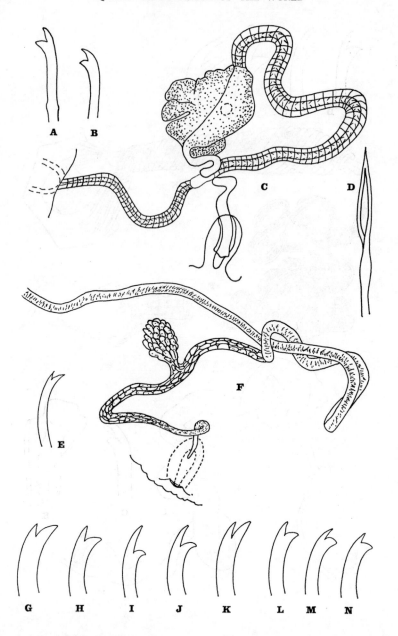

Fig. 8.6. *Isochaeta* (continued).
I. suspectus. A—ventral seta from VI, B—dorsal seta from V, C—male duct, D—spermathecal seta.
I. dojranensis. E—seta, F—male duct.
I. michaelseni. G to J—ventral setae of V, VIII, X, XIII, K,L—dorsal setae of III, VIII, M,N—ventral setae of median and posterior segments.

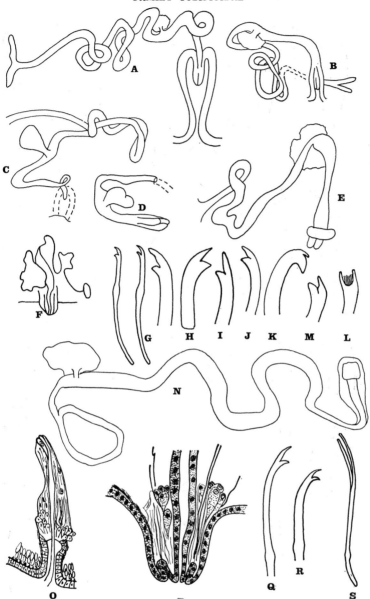

Fig. 8.7. *Isochaeta* (continued).
I. virulenta. A—male duct, G—setae.
I. hamata. B—male duct, J—anterior dorsal seta, K—posterior dorsal seta, O—spermathecal seta.
I. dojranensis. C—male duct.
I. baicalensis. D—male duct, I—seta.
I. michaelseni. E—male duct.
I. baicalensis. F—male duct, P—penis, Q—anterior dorsal seta, R—posterior dorsal seta, S—spermathecal seta.
I. arenaria. H—seta.
I. nevadana. L—dorsal seta, M—ventral seta, N—male duct.

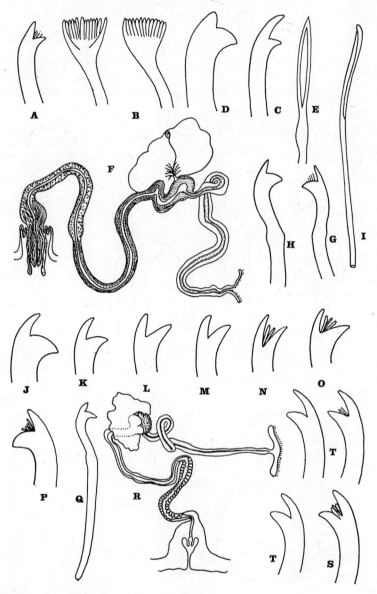

Fig. 8.8. *Psammoryctides.*

P. barbatus. A—anterior dorsal seta, B—palmate setae, C—anterior ventral seta, D—posterior ventral seta, E—spermathecal seta, F—male duct.

P. albicola. G—dorsal seta, H—ventral seta, I—spermathecal seta.

P. deserticola. J—ventral seta of VIII, K—ventral seta of III, L—dorsal seta of VIII, M—dorsal seta of IV, N,O—dorsal setae of IX and II, pectinate, P—dorsal seta, Q—ventral seta, R—male duct.

P. ochridanus. S—dorsal seta of III, T—various setae from dorsal bundle of IV.

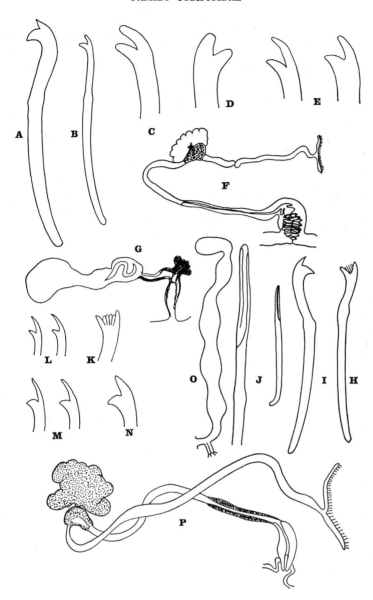

Fig. 8.9. *Psammoryctides* (continued).

P. ochridanus. A—posterior dorsal seta, B—dorsal seta of VII, C—ventral seta of IV, D—dorsal seta of II, E—the same another worm, F—male duct, G—spermatheca.
P. moravicus. H—dorsal seta of VI, I—posterior seta, J—spermathecal seta, K—dorsal seta of IV, L to N—ventral seta of II, III and IV, O—spermatheca, P—male duct.

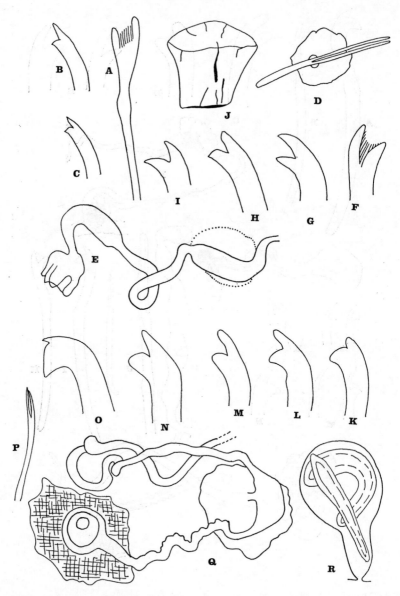

Fig. 8.10. *Psammoryctides* (continued).

P. californianus. A—dorsal seta, B,C—anterior and posterior ventral setae, D—spermathecal seta, E—male duct.

P. minutus. F—dorsal seta, G to I—ventral setae of progressively more posterior segments, J—penis sheath.

P. curvisetosus. K to O—setae of II, III, IV median and posterior segments (O to a smaller scale), P—spermathecal seta, Q—male duct, R—spermatheca.

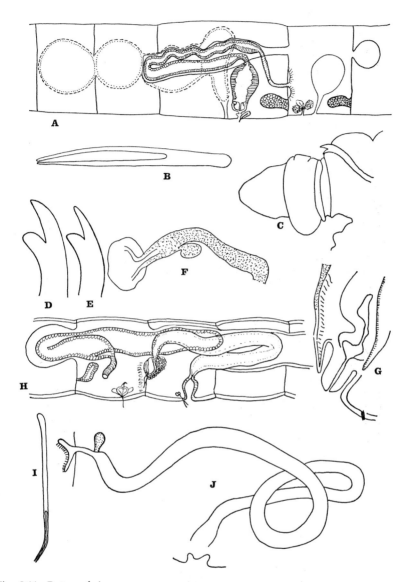

Fig. 8.11. *Potamothrix.*
P. moldaviensis. A—reproductive organs, B—spermathecal seta, C—penis.
P. caspicus. D,E—setae of V and XI, F—sperm funnel and prostate gland on vas deferens, G—penis.
P. heuscheri. H—reproductive organs, I—spermathecal seta.
P. hammoniensis. J—male duct.

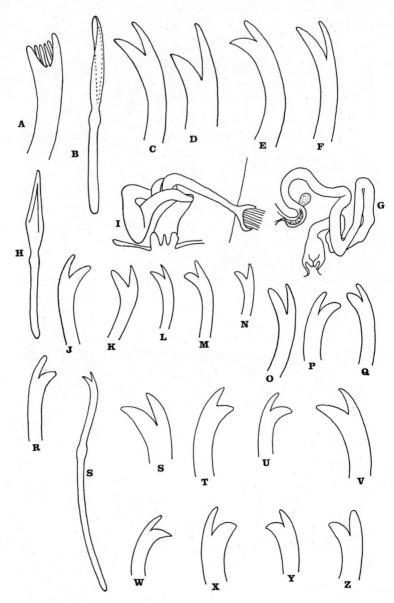

Fig. 8.12. *Potamothrix* (continued).

P. hammoniensis. A—dorsal seta, B—spermathecal seta.

P. vejdovskyi. C to E—setae of dorsal bundles of II and VI and median segments, F—ventral seta of IV, G—male duct.

P. bavaricus. H—spermathecal seta, I—male duct.

P. prespaensis scutarica. J to N—setae from dorsal IV, ventral IV, ventral XI, dorsal XXV, ventral XIV respectively.

P. prespaensis prespaensis. O to R—setae from dorsal IV, ventral IV, dorsal XXV, ventral XIV respectively.

P. ochridanus. S to W—setae from dorsal VII, dorsal mid region and posterior bundle, ventral of VII, ventral posterior bundle respectively.

P. isochaetus. X to Z—setae of dorsal VII, ventral VII, dorsal posterior bundle respectively.

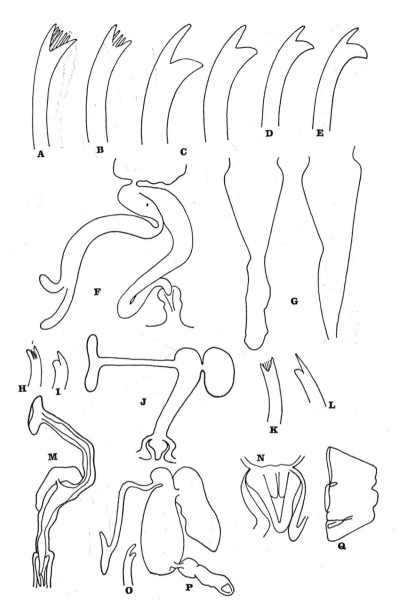

Fig. 8.13. *Ilyodrilus.*
I. templetoni. A,B—dorsal setae of V, VIII, C to E—ventral setae of VII, XV, and XXIII, F—male duct, G—penis sheaths (with complete distal ends).
I. perrieri. H—dorsal seta, I—ventral seta, J—male duct.
I. fragilis. K—dorsal seta, L—ventral seta, M—male duct, N—penis.
I. frantzi. O—seta, P—male duct, Q—penis sheath.

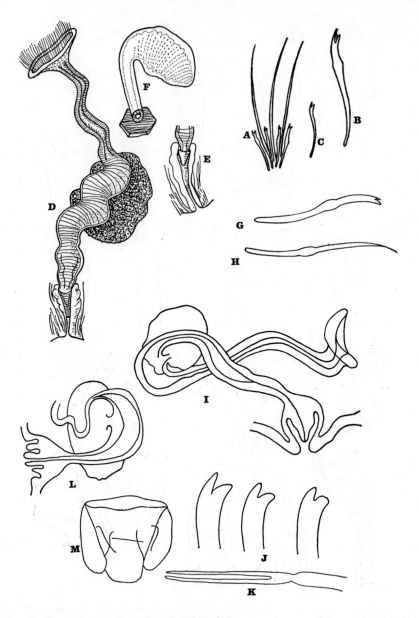

Fig. 8.14. *Ilyodrilus* (continued) and *Peloscolex.*
I. perrieri. A,B—dorsal setae, C—ventral seta, D—male duct, E—penis, F—spermathecae.
P. heterochaetus. G—anterior seta, H—posterior seta, I—male duct.
P. freyi. J—setae, K—spermathecal seta, L—male duct, M—penis sheath.

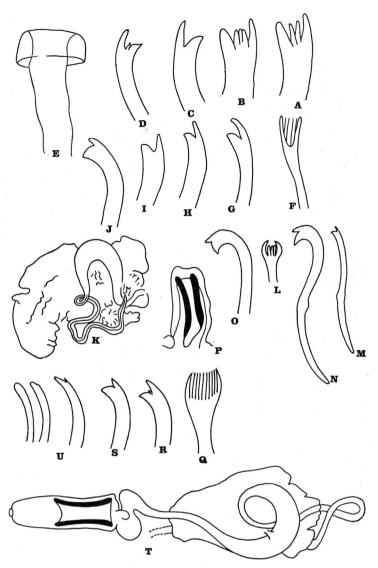

Fig. 8.15. *Peloscolex* (continued).

P. superiorensis. A,B—dorsal seta, C,D—ventral setae of anterior and posterior bundles, E—penis sheath.

P. tenuis apapillatus. F—dorsal seta of III, G to J—ventral setae of II, V, VII, XXI, K—male duct.

P. tenuis tenuis. L—dorsal seta, M to O—ventral setae of progressively more posterior segments, P—penis and sheath (section).

P. ferox. Q—dorsal setae, R,S—anterior and posterior ventral setae, T—male ducts.

P. benedeni. U—setae.

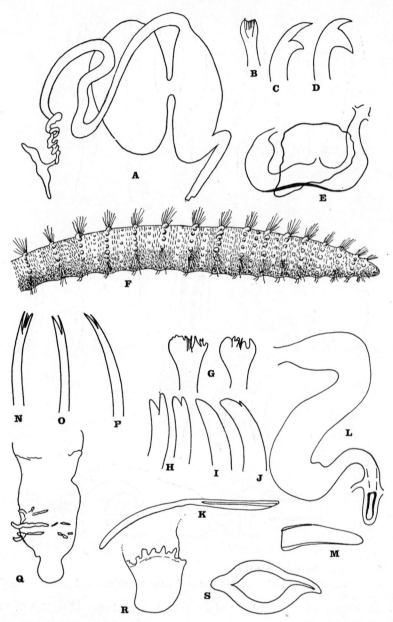

Fig. 8.16. *Peloscolex* (continued).

P. inflatus. A—male duct.

P. multisetosus. B—dorsal seta, C,D—anterior and posterior ventral setae, E—male duct, F—anterior end, whole worm.

P. beetoni. G—dorsal setae, H to J—ventral setae of II, VIII, posterior bundle respectively, K—spermathecal seta, L—male duct, M—penis sheath.

P. aculeatus. N—anterior dorsal seta, O,P—anterior ventral setae, Q,R—penis sheaths, with hooks, S—spermatophore.

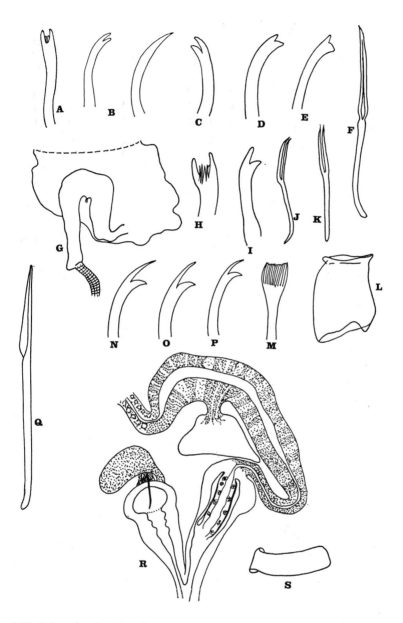

Fig. 8.17. *Peloscolex* (continued).

P. kurenkovi. A—dorsal seta, B to E—ventral setae of II, V, LV, LVIII respectively, F—spermathecal seta, G—male duct.

P. speciosus "simsi". H—dorsal seta. I—ventral seta, J,K—spermathecal and penial setae, L—penis sheath.

P. speciosus "zavreli". M—dorsal seta, N to P—ventral setae of III, VIII, posterior bundle, Q—spermathecal seta, R—male duct and penial setal sac, S—penis sheath.

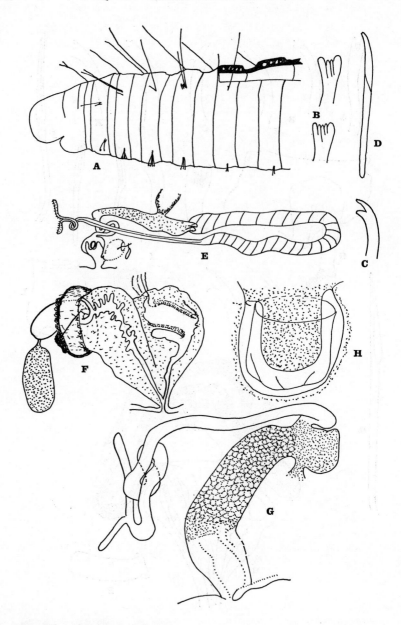

Fig. 8.18. *Peloscolex* (continued).

P. speciosus. A—anterior end of worm—body wall partly sectioned, B—dorsal setae, C—ventral seta, D—spermathecal seta, E—male duct, F—penis and penial setal sac.

P. swirenkovi. G—male duct, H—penis sheath.

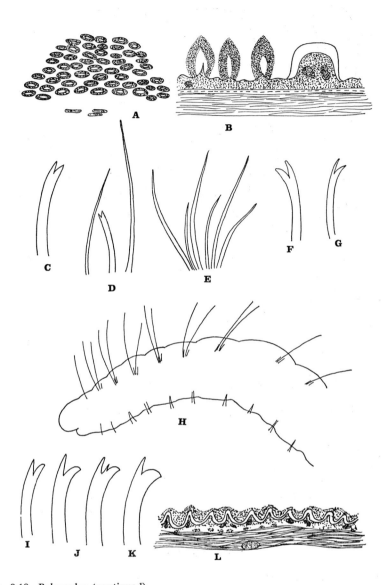

Fig. 8.19. *Peloscolex* (continued).
P. swirenkovi. A,B papillae, sections, C to E—dorsal setae of VI, VIII, posterior bundle, F,G—anterior and posterior ventral setae.
P. euxinicus. H—anterior end, I—dorsal seta from VIII, J,K—ventral setae from V, VIII, L—section of body wall.

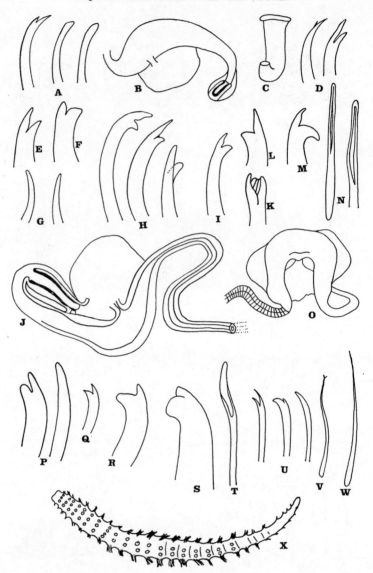

Fig. 8.20. *Peloscolex* (continued).

P. benedeni. A—setae, B—male duct, C—penis sheath.

P. oregonensis. D to F—setae from anterior bundle, V and posterior bundle respectively.

P. variegatus. G to I—ventral setae from II-IV, II-X, posterior bundle respectively, J—male duct.

P. nikolskyi. K—dorsal seta, L,M—ventral setae of VI, XIX, N—spermathecal setae, O—male duct.

P. carolinensis. P to S—ventral setae of VII, X, XII, XV.

P. velutinus. T—spermathecal seta, U—ventral setae, V—anterior dorsal seta, W—rare spermathecal seta.

P. stankovici. X—whole worm.

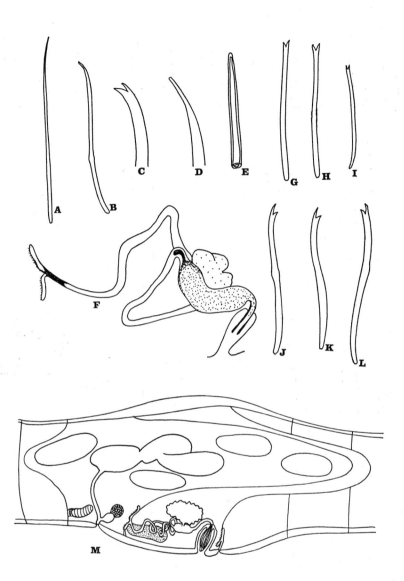

Fig. 8.21. *Peloscolex* (continued).
P. stankovici. A—hair seta of IV, B to D—ventral setae of III, E—spermathecal seta,
F—male duct.
P. nomurai. G to I—dorsal setae of III, VII, XXIX, J to L—ventral setae of III, V, IX,
M—reproductive organs.

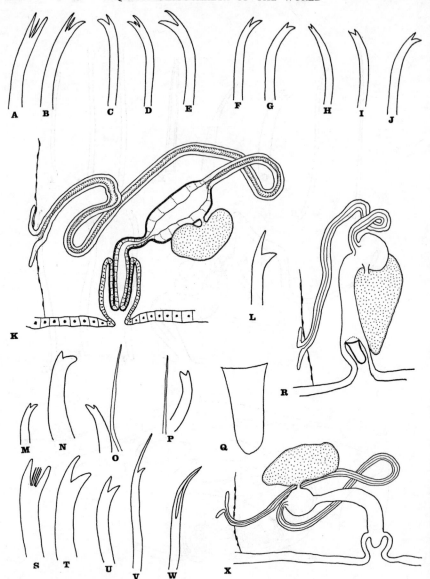

Fig. 8.22. *Peloscolex* (continued) and *Antipodrilus.*

P. nerthoides. A to C—dorsal setae of II, VI and posterior bundle, D,E—anterior and posterior ventral setae.

P. apectinatus. F to H—dorsal setae of II, VIII and posterior bundle, I,J—anterior and posterior ventral setae.

P. intermedius. K—male duct, L—seta.

P. dukei. M,N—anterior and posterior ventral setae, O,P—anterior and posterior dorsal setae, Q—penis sheath, R—male duct.

A. davidis. S—dorsal seta, T,U—anterior and posterior ventral setae, V,W—spermathecal setae, X—male duct.

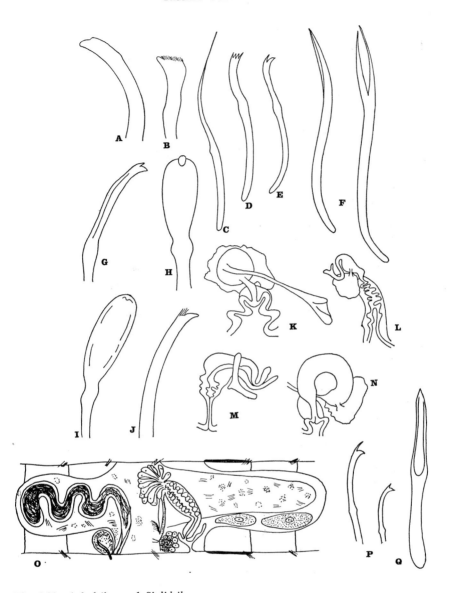

Fig. 8.23. *Aulodrilus* and *Siolidrilus*.

A. americanus. A—anterior seta, B—mid posterior seta.
A. pectinatus. C—hair seta, D—pectinate dorsal seta, E—ventral seta, F—penial seta.
A. limnobius. G,H—side and face view of seta of median segment.
A. pigueti. I—dorsal seta, median segment.
A. pluriseta. J—dorsal seta, median segment, K to N—male ducts: K—*A. pluriseta*,
   L—*A. pigueti* (as *A. prothecatus*), M—*A. pectinatus*, N—*A. limnobius*.
S. adetus. O—reproductive organs, P—anterior and posterior setae, Q—spermathecal
   seta.

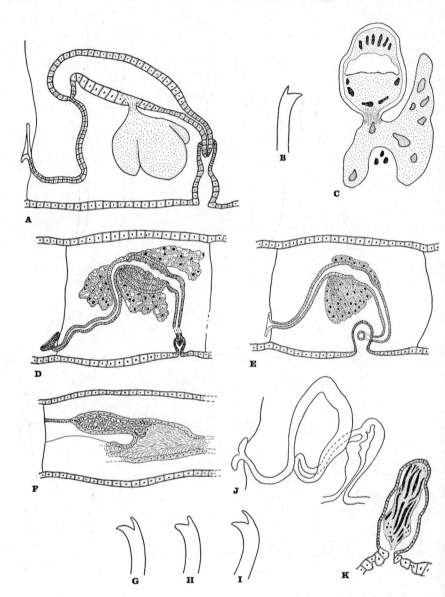

Fig. 8.24. *Limnodriloides.*
L. medioporus. A—male duct, B—seta, C—section through union of atrium and prostate.
L. winckelmanni. D—male duct.
L. appendiculatus. E—male duct, F—gut diverticulum.
L. agnes. G,H,I—dorsal setae of IV, V and a posterior segment, J—male duct (minus prostate), K—sperm bundles in spermatheca.

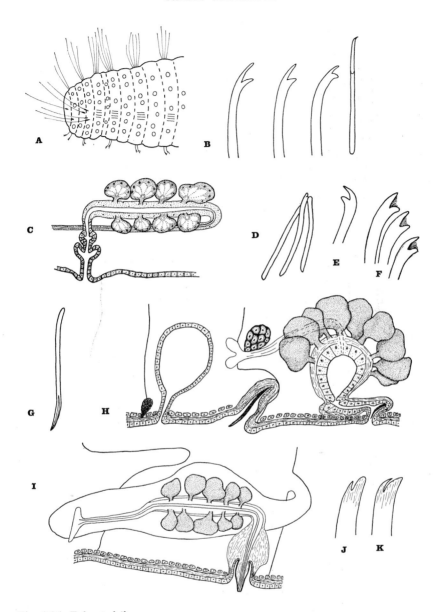

Fig. 8.25. *Telmatodrilus.*
T. onegensis. A—papillate body wall, B—ventral setae of II, III posterior segment and spermathecal segment, C—male duct.
T. pectinatus. D—penial setae, E—anterior seta, F—posterior setae, G—spermathecal seta, H—reproductive organs.
T. vejdovskyi. I—male duct, J,K—setae with and without wear.

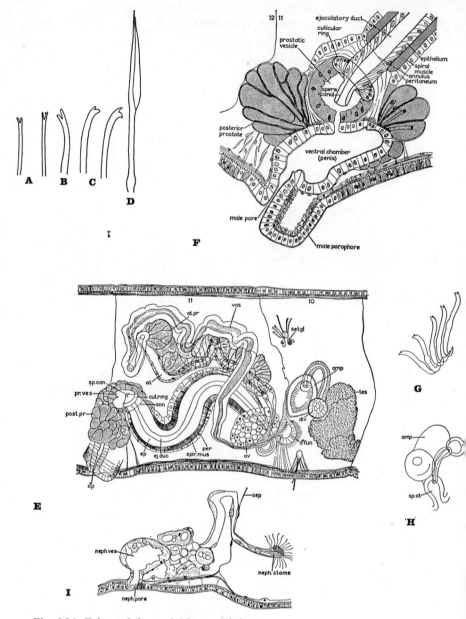

Fig. 8.26. *Telmatodrilus* and *Macquaridrilus*.

T. ringulatus. A—dorsal setae, B—ventral seta of V, C—posterior ventral setae, D—spermathecal seta.

M. bennettae. E—reproductive organs (amp.—spermathecal ampulla, ann—annulus, at—atrium, at. pr.—prostate, cut. ring.—cuticular ring, div—spermathecal diverticulum, ej. duc.—ejaculatory duct, ep.—epithelium, ♂ f—male funnel, ♂ p— male pore, ov—ovary, per—peritonium, pr. ves.—prostate vesicle, post. pr.—posterior prostate, set. gl.—setal gland, sp. cna.—sperm canal, spir. mus.—spiral muscle, tes.—testis, vas.—vas deferens). F—penial apparatus, G—ventral setae of XXXVIII. H—spermatheca (amp.—ampulla; div.—diverticulum; sp. at.—ectal chamber or atrium of spermatheca). I—nephridium—longitudinal section (sep.—septum; neph. stome—nephrostome; neph. ves.—vesicle; neph. pore—nephropore).

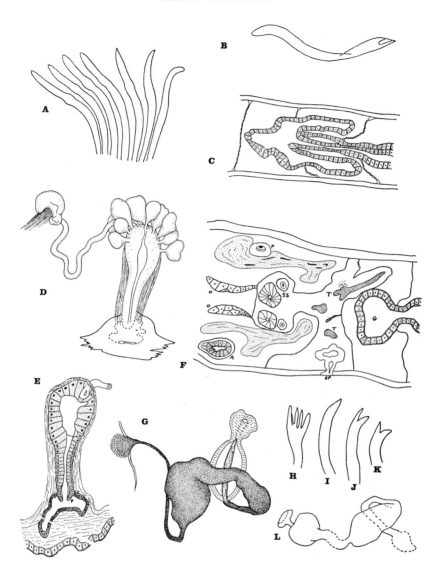

Fig. 8.27. *Telmatodrilus, Paranadrilus* and *Epirodrilus.*

T. multiprostatus. A—anterior setae, B—spermathecal seta, C—gut introvertion in V to VII, D—male duct, E—horizontal longitudinal section through reproductive organs. (ss.—spermathecal setal sac and gland, sp.—spermathecal pore, T.—testis, O.—ovary, A.—atrium, G.—gut, P.—penis), F—atrium, penis and penis sac, with part of vas deferens.

P. descolei. G—male duct.

E. allansoni. H—pectinate dorsal seta, I,J—anterior ventral seta, K—posterior ventral seta, L—male duct (reconstruction).

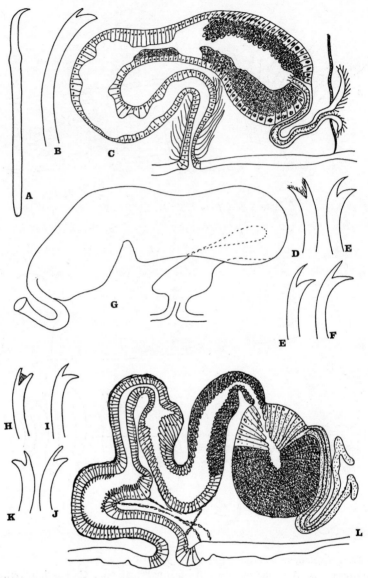

Fig. 8.28. *Epirodrilus.*
E. antipodum. A—penial seta, B—anterior seta, C—male duct.
E. pygmaeus. D—dorsal seta of V, E—anterior ventral setae, F—posterior ventral seta,
   G—male duct.
E. michaelseni. H to K—setae from dorsal VII, XXXVI, ventral VII, XXXVI
   respectively, L—male duct.

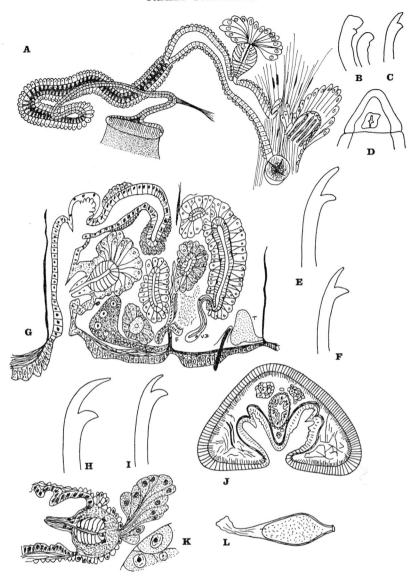

Fig. 8.29. *Bothrioneurum.*

B. vejdovskyanum. A—male duct, B—penial setae, C—normal seta, D—prostomial pit, dorsal view.

B. americanum. E—anterior seta, F—posterior seta, G—male duct (note vas deferens and atrium coiled within testes-bearing segment, T—testes, VD—vas deferens, F—funnel).

B. iris. H—anterior seta, I—posterior seta, J—section through superficial male pore, K—paratrium, L—sperm bearer.

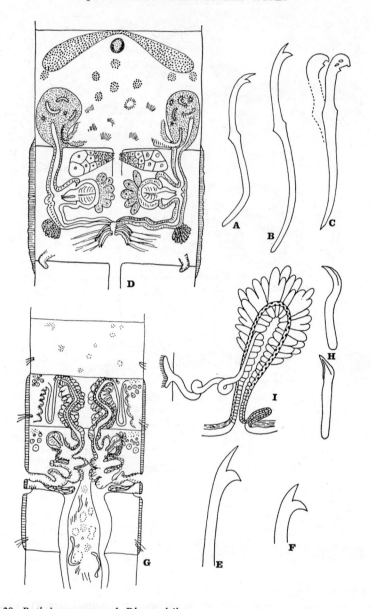

Fig. 8.30. *Bothrioneurum* and *Rhyacodrilus*.
B. pyrrhum. A,B—setae of anterior and posterior segments, C—penial setae, D—horizontal section, through reproductive organs.
B. brauni. E,F—anterior and posterior setae, G—horizontal section through reproductive organs.
R. falciformis. H—penial setae, I—male duct.

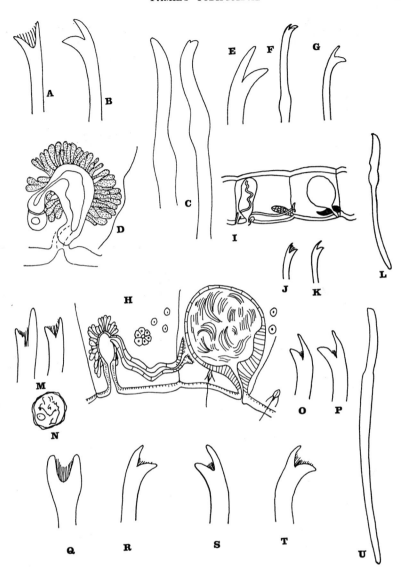

Fig. 8.31. *Rhyacodrilus* (continued).

R. coccineus. A—dorsal seta, B—ventral seta, C—penial seta, D—male duct.

R. stephensoni. E—dorsal seta of II, F—penial seta, G—posterior seta, H—male duct and spermatheca.

R. simplex. I—reproductive organs, J—anterior dorsal seta, K—anterior ventral seta, L—penial seta.

R. montana. M—pectinate dorsal setae, N—coelomocyte, O,P—anterior and posterior pectinate ventral setae.

R. punctatus. Q—anterior dorsal seta, R—posterior dorsal seta, S—anterior ventral seta, T—posterior ventral seta, U—penial seta.

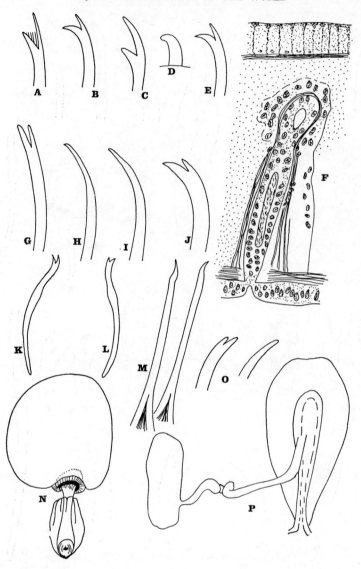

Fig. 8.32. *Rhyacodrilus* (continued).

R. subterraneus. A—anterior dorsal seta, B—posterior dorsal seta, C—anterior ventral seta, D—penial seta, E—posterior ventral seta, F—section of atrium.

R. lindbergi. G—anterior dorsal seta, H—median dorsal seta, I—posterior dorsal seta, J—median ventral seta.

R. balmensis. K—anterior ventral seta, L—median ventral seta, M—penial seta, N—spermatheca.

R. brevidentatus. O—ventral setae, P—male duct.

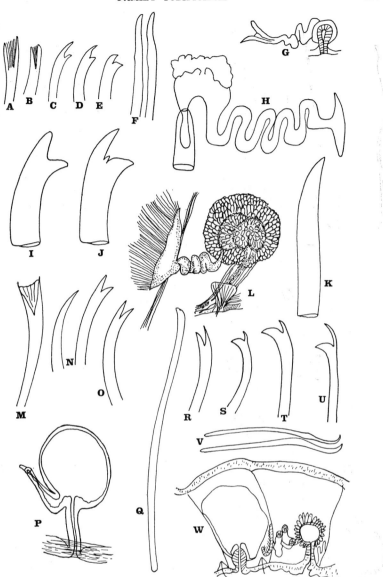

Fig. 8.33. *Rhyacodrilus* (continued).
R. *sodalis*. A—anterior dorsal seta, B—posterior dorsal seta, C—anterior ventral seta,
 D—median ventral seta, E—posterior ventral seta, F—penial seta, G—male duct
 (*Tubifex sinicus*), H—male duct (*Ilyodrilus sodalis*).
R. *korotneffi*. I—dorsal seta, J—ventral seta, K—penial seta, L—male duct.
R. *lepnevae*. M—anterior dorsal seta, N—ventral setae of VII, O—ventral seta of X.
R. *palustris*. P—spermatheca, showing canal connecting to gut, Q—penial seta.
R. *ekmani*. R—anterior dorsal seta, S—posterior dorsal seta, T—anterior ventral seta,
 U—posterior ventral seta, V—penial setae, W—reproductive organs.

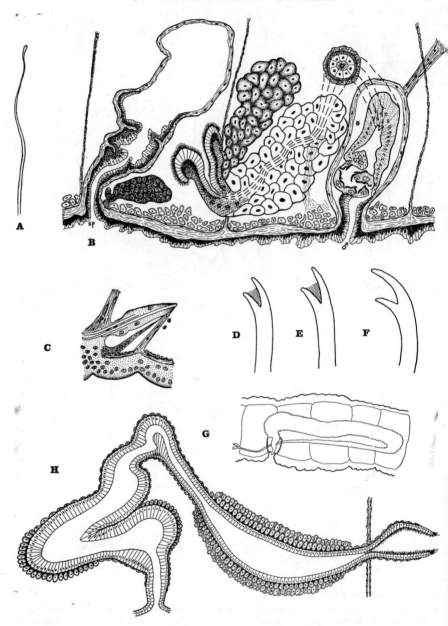

Fig. 8.34. *Monopylephorus.*

M. irroratus. A—hair seta, B—reproductive organs (*—gap between muscle layer and lining of terminal portion of male duct, ♂ —male pore, sp.—spermathecal pore).

M. montanus. C—penial setae, D—anterior dorsal seta, E—median dorsal seta, F— anterior ventral seta, G—male duct.

M. limosus. H—male duct.

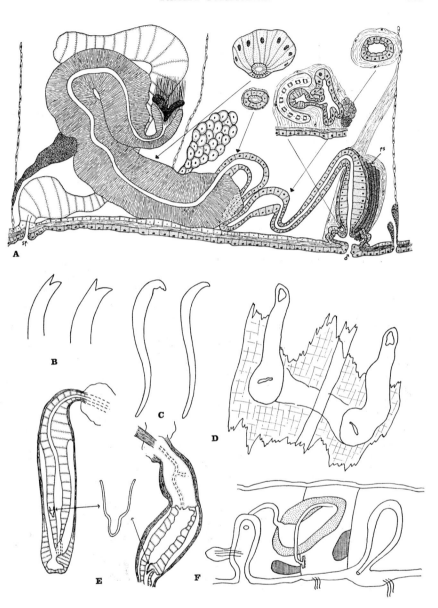

Fig. 8.35. *Monopylephorus* (continued).
M. frigidus. A—reproductive organs (ps.—penial setae, ♂—male pore, sp.—spermathecal pore), B—anterior and posterior setae, C—penial setae, D—spermathecae, prior to copulation.
M. rubroniveus. E—terminal portions of two male ducts, and detail of cuticular process.
M. africanus. F—reproductive organs.

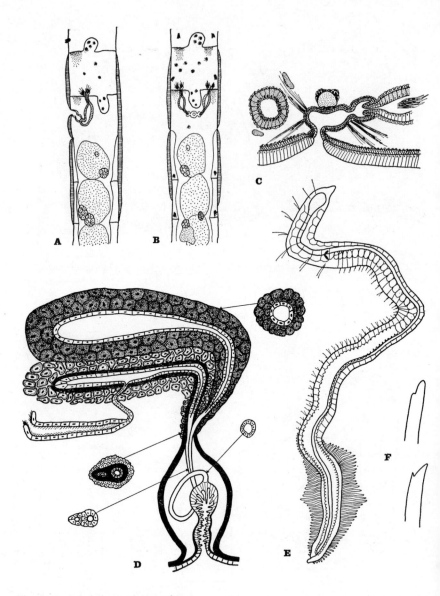

Fig. 8.36. *Jolydrilus* and *Branchiura*.
Jolydrilus jaulus. A,B—reproductive organs, side and dorsal views, C—section of common male pore.
B. sowerbyi. D—male duct, E—whole animal to show gills, F—bifid setae.

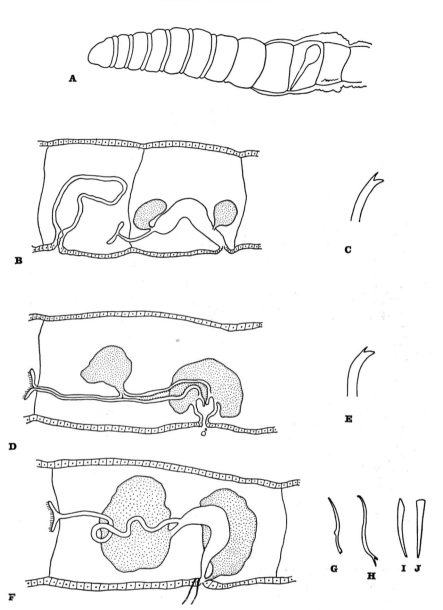

Fig. 8.37. *Phallodrilus.*
P. aquaedulcis. A—anterior of body and spermatheca, B—reproductive organs, C—anterior ventral seta.
P. monospermathecus. D—male duct, E—seta.
P. parthenopaeus. F—male duct, G—anterior seta, H—spermathecal seta, I,J—penial setae.

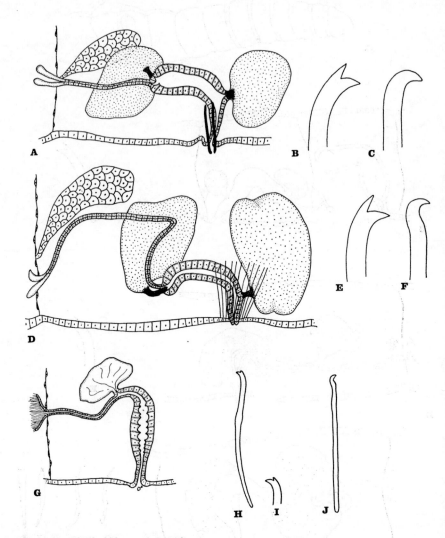

Fig. 8.38. *Phallodrilus* and *Spiridion*.
P. obscurus. A—male duct, B—seta, C—penial seta.
P. coeloprostatus. D—male duct, E—seta, F—penial setae.
S. insigne. G—male duct, H,I—seta, J—penial seta.

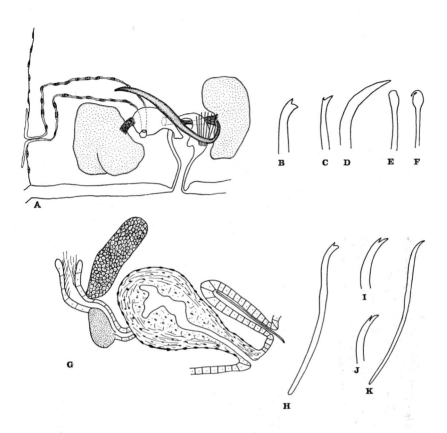

Fig. 8.39. *Adelodrilus* and *Thalassodrilus*.
A. anisetosus. A—male duct, B—anterior seta, C—posterior seta, D—posterior seta, E,F—penial setae.
T. prostatus. G—male duct, H—ventral seta, I—anterior dorsal seta, J—median dorsal seta, K—posterior dorsal seta.

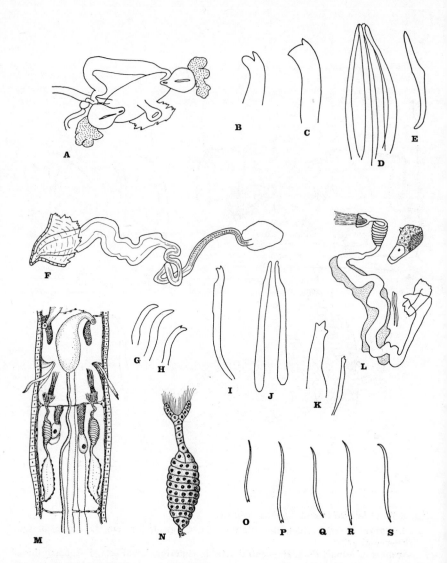

Fig. 8.40. *Smithsonidrilus* and *Clitellio*.

S. marinus. A—male duct, B—anterior seta, C—posterior seta, D—penial setae, E—spermathecal seta.

C. arenarius. F—male duct, G,H—setae.

C. arenicolus. I—posterior seta, J—penial setae, K—anterior trifid setae, L—male duct (and ovary).

C. subtilis. M—horizontal section, reproductive organs, N—vas deferens, coiled, O,P—anterior setae, Q—seta of VIII, R—setae of IX, S—posterior seta.

REFERENCES

AIYER, K. G. P. 1925. Notes on the aquatic Oligochaeta of Travancore. I. *Ann. Mag. nat. Hist.,* **16**, 31.
— 1928. On a new species of the Oligochaete genus *Aulodrilus* Bretscher. *Rec. Indian Mus.,* **30**, 345.
— 1929a. An account of the Oligochaeta of Travancore. *Rec. Indian Mus.,* **31**, 13.
— 1929b. On the sexual organs of the Tubificid worm *Aulodrilus remex* Steph. *Rec. Indian Mus.,* **31**, 81.
ALTMAN, L. C. 1936. Oligochaeta of Washington. *Univ. Wash. Publs. Biol.,* **4**, 1.
ANDRUSOV, L. 1914. Beiträge zur Oligochaetenfauna der Umgebung von Kiew. *Zap. kiev. Obshch. Estest.,* **33** (4), 91.
BAUMANN, F. 1910. Beiträge zur Biologie der Stockhornseen. *Revue suisse Zool.,* **18**, 647.
BEDDARD, F. E. 1889. On certain points in the structure of *Clitellio* (Claparède). *Proc. zool. Soc. Lond.,* **1888**, 485.
— 1892. A new branchiate Oligochaete (*Branchiura sowerbyi*). *Q. Jl. Micros. Sci.,* **33**, 325.
— 1894. Preliminary notice of S. American Tubificidae collected by Dr. Michaelsen, including a description of a branchiate form. *Ann. Mag. nat. Hist.,* **13**, 205.
— 1895. *A monograph of the order Oligochaeta.* Clarendon Press, Oxford.
— 1896. Naiden, Tubificiden und Terricolen. *Ergebnisse Hamburger Magalhaensischen Sammelreise.* I. Hamburg.
— 1901. On a freshwater Annelid of the genus *Bothrioneuron* obtained during the Skeat Expedition to the Malay Peninsula. *Proc. zool. Soc. Lond.,* **1**, 81.
BENHAM, W. B. 1892. Notes on some aquatic Oligochaeta. *Q. Jl Micros. Sci.,* **33** (1891), 187.
— 1903. On some new species of aquatic Oligochaeta from New Zealand. *Proc. zool. Soc. Lond.,* **2**, 202.
— 1907. On the Oligochaeta from the Blue Lake, Mount Kosciusko, Sydney, N. S. W. *Rec. Aust. Mus.,* **6**, 251.
— 1909. *Report on the Oligochaeta in the Subantarctic Islands of New Zealand.* Wellington. Ed. C. Chilton. **1** (12), 251.
— 1915. Oligochaeta from the Kermadec Isles. *Trans. Proc. N. Z. Inst.,* **47**, 174.
BERG, K. 1948. Biological studies on the river Susaa. *Folio limnol. scand.,* **4**, 1.
BOLDT, W. 1926. Vorläufige Mitteilungen über die Oligochaeten des Oldesloer Salzgebietes. *Mitt. Georgr. Ges. Lübeck.,* **2** (31), 177.
— 1928. Mitteilung über Oligochaeten der Familie Tubificidae. *Zool. Anz.,* **75**, 145.
BORY de ST. VINCENT, J. 1824. Encyclopédie méthodique. *Histoire naturelle des zoophytes on animaux rayonnes,* **2**, 134.
BRETSCHER, K. 1896. Die Oligochaeten von Zürich. *Revue Cuisse Zool.,* **3**, 499.
— 1899. Beitrag zur Kenntniss der Oligochaetenfauna der Schweiz. *Revue suisse Zool.,* **6**, 369.
— 1900a. Mitteilungen über die Oligochaetenfauna der Schweiz. *Revue suisse Zool.,* **8**, 1.
— 1900b. Südschweizerische Oligochaeten. *Revue suisse Zool.,* **8**, 435.
— 1901. Beobachtungen über die Oligochaeten der Schweiz. *Revue suisse Zool.,* **9**, 189.
— 1902. Beobachtungen über die Oligochaeten der Schweiz. *Revue suisse Zool.,* **10**, 1.
— 1903. Beobachtungen über die Oligochaeten der Schweiz. *Revue suisse Zool.,* **11**, 1.
— 1904. Beobachtungen über die Oligochaeten der Schweiz. *Revue suisse Zool.,* **13**, 259.
— 1905. Beobachtungen über die Oligochaeten der Schweiz. *Revue suisse Zool.,* **13**, 663.
BRINKHURST, R. O. 1960. Introductory studies on the British Tubificidae (Oligochaeta). *Arch. Hydrobiol.,* **56**, 395.

BRINKHURST, R. O. 1962a. A re-description of *Peloscolex variegatus* Leidy (Oligochaeta, Tubificidae) with a consideration of the diagnosis of the genus *Peloscolex*. *Int. Revue ges. Hydrobiol.*, **47**, 301.
— 1962b. A redescription of *Tubifex newaensis* (Michaelsen). (Oligochaeta, Tubificidae) with a consideration of its taxonomic position in the genus. *Int. Revue ges. Hydrobiol.*, **47**, 307.
— 1962c. A check-list of British Oligochaeta. *Proc. zool. Soc. Lond.*, **138**, 317.
— 1963a. Taxonomical studies on the Tubificidae (Annelida, Oligochaeta). *Int. Revue ges. Hydrobiol.*, **48**, 7.
— 1963b. Notes on the brackish-water and marine species of Tubificidae (Annelida, Oligochaeta). *J. mar. biol. Ass. U.K.*, **43**, 709.
— 1963c. A guide for the identification of British aquatic Oligochaeta. *Scient. Publs Freshwat. biol. Ass.*, **22**, 29.
— 1964. Observation on the biology of Lake-dwelling Tubificidae. *Arch. Hydrobiol.*, **60**, 385.
— 1965. Studies on the North American aquatic Oligochaeta. II. Tubificidae. *Proc. Acad. nat. Sci. Philad.*, **117**, 117.
— 1966a. Taxonomical studies on the Tubificidae (Annelida, Oligochaeta). Supplement. *Int. Revue ges. Hydrobiol.*, **51**, 727.
— 1966b. A contribution to the systematics of the marine Tubificidae (Annelida, Oligochaeta). *Biol. Bull. mar. biol. Lab., Woods Hole*, **130**, 297.
— 1966c. A contribution towards a revision of the aquatic Oligochaeta of Africa. *Zoologica Africana*, **2**, 131.
BRINKHURST, R. O., AND COOK, D. G. 1966. Studies on the North American aquatic Oligochaeta. III. Lumbriculidae and additional notes and records of other families. *Proc. Acad. nat. Sci. Philad.*, **118**, 1.
BRINKHURST, R. O., AND KENNEDY, C. R. 1962a. A report on a collection of aquatic Oligochaeta deposited at the University of Neuchâtel by Dr. E. Piguet. *Bull. Soc. neuchâtel. Sci. nat.*, **85**, 184.
— 1962b. Some aquatic Oligochaeta from the Isle of Man with special reference to the Silver Burn Estuary. *Arch. Hydrobiol.*, **58**, 367.
BUDGE, J. 1850. Uber die Geshlechtsorgane von *Tubifex rivulorum*. *Arch. Naturgesch.*, **16**, 1.
BÜLOW, T. 1955. Oligochaeten aus den Endgebieten der Schlei. *Kieler Meeresforsch.*, **11**, 253.
— 1957. Systematische Studien an enulitoralen Oligochaeten der Kimbrischen Halbinsel. *Kieler Meeresforsch.*, **13**, 69.
BUROV, W. S. 1936. Die Oligochaeta Ost-Sibiriens. Zur Systematik und Biologie der Gattung *Clitellio* aus dem Baikalsee. *Izy. Biologo-geogre. nauchnoissled. Inst., Irkutsk*, **7**, 17.
CAUSEY, D. 1953. Microdrili in artificial lakes in Northwest Arkansas. *Am. Midl. Nat.*, **50**, 420.
CEKANOVSKAYA, O. V. 1952. K faune maloshchetinkoveikh chervei Zapadnogo Kazakhstana. *Trudy zool. Inst., Leningr.*, **11**, 293.
— 1956. K faune maloshchetinkoveikh chervei basseina Eniseya. *Zool. Zh.*, 35, 657.
— 1959. On Oligochaeta from the bodies of water in Central Asia. *Zool. Zh.*, **38**, 1152.
— 1962. The aquatic Oligochaete fauna of the USSR. *Opred. Faune SSSR*, **78**, 1 (in Russian).
CERNOSVITOV, L. 1926. Eine neue Variation des *Tubifex* Müll. *Zool. Anz.*, **65**, 321.
— 1928. Die Oligochaetenfauna der Karpathen. *Zool. Jb. Syst.*, **55**, 1.
— 1929. Communications preliminaire sur les Oligochètes récoltés par M. P. Remy pendant la croisière artique effectuée par le "Pourquoi-Pas?" en 1926 sous la direction du Dr. J. B. Charcot. *Bull. Mus. Hist. nat. Paris*, (2) **1**, 144.
— 1930. Zur Kenntnis der Oligochätenfauna des Balkans I. Über die Oligochäten aus Bosnien. *Zool. Anz.*, **86**, 319.
— 1931. Zur Kenntnis der Oligochaeten-Fauna des Balkans. III. *Zool. Anz.*, **95**, 312.
— 1935a. Exploration biologique des Cavernes de la Belgique et du Limbourg

Hollandais. XXIII. Contribution: Oligochètes. *Bull. Mus. r. Hist. nat. Belg.*, **11** (22), 1.

CERNOSVITOV, L. 1935b. Resultats zoologiques du voyage de M. le Dr. Storkan au Mexique V. *Vest. csl. Spol. zool.*, **3**, 80.

— 1937. Notes sur les Oligochaeta (Naidides et Enchytraeidés) de l'Argentine. *An. Mus. argent. Cienc. nat.*, **39**, 135.

— 1938a. The Oligochaeta (in Studies on the freshwater of Palestine). *Ann. Mag. nat. Hist.*, **2**, 535.

— 1938b. Oligochaeta. Mission scient. Omo, P. IV, Zoologie. *Mèm. Mus. Hist. nat. Paris*, **38**, 255.

— 1939a. Études biospéologiques. X (1): Catalogue des Oligochètes hypogés. *Bull. Mus. r. Hist. nat. Belg.*, **15** (22), 35.

— 1939b. Oligochaeta from the Percy Trust Expd. to Lake Titacaca. *Trans. Linn. Soc. Lond. (Zool.)*, **1**, 81.

— 1942a. Oligochaeta from various parts of the world. *Proc. zool. Soc. Lond.*, **3**, 197.

— 1942b. A revision of Friend's types and descriptions of British Oligochaeta. *Proc. zool. Soc. Lond.*, **3**, 327.

— 1942c. Oligochaeta from Tibet. *Proc. zool. Soc. Lond.*, **3** (1941), 281.

— 1945. Oligochaeta from Windermere and the Lake District. *Proc. zool. Soc. Lond.*, **114**, 523.

CHEN, P. S. 1934. Regenerationsversuche an Embryonen und jungen Würmchen von *Tubifex rivulorum* Lam. *Z. wiss. Zool.*, **145**, 99.

CHEN, Y. 1940. Taxonomy and faunal relations of the limnic Oligochaeta of China. *Contr. biol. Lab. Sci. Soc. China, Zool.*, **14**, 1.

CLAPARÉDE, E. R. 1862. Recherches anatomiques sur les Oligochètes. *Mèm. Soc. Phys. Hist. nat. Genève*, **16**, 217.

— 1863. *Beobachtungen über Anatomie und Entwicklungsgeschichte wirbelloser Tiere.* Leipzig.

COLE, G. A. 1954. An occurance of *Branchiura sowerbyi* Beddard in Kentucky. *Trans. Ky Acad. Sci.*, **15**, 127.

— 1966. *Branchiura sowerbyi* Beddard (Annelida: Oligochaeta) in Arizona. *J. Ariz. Acad. Sci.*, **4**, 43.

COOK, D. G. 1969. The Tubificidae (Annelida, Oligochaeta) of Cape Cod Bay, with a taxonomic revision of three genera, *Phallodrilus* Pierantoni, 1902, *Limnodriloides* Pierantoni, 1903, and *Spiridion* Knöllner, 1935. *Biol. Bull. Woods Hole*, **136**, 9.

— 1970. *Peloscolex dukei* n. sp. and *P. aculeatus* n. sp. (Oligochaeta, Tubificidae) from the North-west Atlantic, the latter being from abysall depth. *Trans. Amer. microsc. Soc.* **88**, 492.

COLLINS, D. S. 1937. The aquatic earthworms (microdrili) of Reelfoot Lake. *J. Tenn. Acad. Sci.*, **12**, 188.

CZERNIAVSKY, V. 1880. Materialia ad zoographiam ponticam comparatam. III. Vermes. *Bull. Soc. imp. Nat. Moscow*, **55** (4), 213.

DAHL, I. O. 1960. The Oligochaeta fauna of 3 Danish brackish water areas. *Meddr. Danm. Fisk. -og. Havundens*, **2** (26), 1.

DITLEVSEN, A. 1904. Studien an Oligochaeten. *Z. wiss. Zool.*, **77**, 398.

DOYÉRE, M. P. 1856. Essai sur l'anatomie de la *Naïs sanguinea. Mèm. Soc. linn. Normandie*, **10** (1884-5), 306.

DUGÉS, A. 1828. Recherches sur la circulation, la respiration et la reproduction des Annélides abranches. *Annls. Sci. nat.*, I, **15**, 284.

— 1837. Nouvelles observations sur la zoologie et l'anatomie des Annélides abranches sétigères. *Annls. Sci. nat.*, II, **8**, 15.

EGGLETON, F. E. 1931. A limnological study of the profundal bottom fauna of certain freshwater lakes *Ecol. monogr.*, **1**, 231.

EISEN, G. 1879. Preliminary report on genera and species of Tubificidae. *Bih. K. svenska VetenskAkad. Handl.*, **5** (16), 1.

— 1886. *Oligochaetological researches.* XX. Rep. U.S. Commnr. Fish.

— 1900. Researches in American Oligochaeta. *Proc. Calif. Acad. Sci. Zool.*, III, **2**, 243.

EVANS, K. J. 1958. A record of the Oligochaete *Branchiura sowerbyi* from Oklahoma. *S. West. Nat.*, **3**, 223.

FABRICIUS, O. 1780. *Fauna Grønlandica*. Hafniae et Lipsiae.

FERRONNIÉRE, G. 1899. Contribution a l'étude de la faune de la Loire-Inférieur. *Bull. Soc. Sci. nat. Ouest. Fr.*, **9**, 229.

FRIEND, H. 1896. Irish freshwater worms. *Ir. Nat.*, **5**, 125.

— 1897a. Field days in Ulster. *Ir. Nat.*, **6**, 61, 101.

— 1897b. Annelids new to Ireland. *Ir. Nat.*, **6**, 206.

— 1897c. The tube-forming worms. *Ir. Nat.*, **6**, 294.

— 1898a. Note on British Annelids. *Zoologist*, IV, **2**, 119.

— 1898b. New Irish Annelids. *Ir. Nat.*, **7**, 195.

— 1911. New records for British Annelids. *Naturalist, Hull*, **1911**, 411.

— 1912a. Some Annelids of the Thames Valley. *J. Linn. Soc.*, **32**, 95.

— 1912b. British Tubificidae. *Jl R. microsc. Soc.*, **1912**, 265.

— 1912c. Irish Oligochaetes. *Ir. Nat.*, **21**, 171.

— 1912d. Oligochaetes of Great Britain and Ireland. *Naturalist, Hull*, **1912**, 76.

— 1913. Notes on Dublin Oligochaetes. *Ir. Nat.*, **22**, 169.

FUHRMANN, O. 1897. Recherches sur la faune des lacs alpins du Tessin. *Revue suisse Zool.*, **4**, 489.

GALLOWAY, T. W. 1911. The common freshwater Oligochaeta of the United States. *Trans. Am. microsc. Soc.*, **30**, 285.

GAVRILOV, K. 1955a. Ein neuer Spermathekenloser Vertreter der Tubificiden. *Zool. Anz.*, **155**, 294.

— 1955b. Über die uniparentale Vermehrung von *Paranadrilus*. *Zool. Anz.*, **155**, 303.

— 1958. Notas adicionales sobre *"Paranadrilus"*. *Acta zool. lilloana*, **16**, 149.

GAVRILOV, K., AND PAZ, N. G. 1949. *Limnodrilus inversus* n. sp. y su reproduccion uniparental. *Acta zool. lilloana*, **8**, 537.

— 1950. Nota adicional sobre la reproduccion de *Limnodrilus*. *Acta zool. lilloana*, **9**, 533.

GERVAIS, P. 1838. Note sur la disposition systématique des Annélides chétopodes de la famille des Nais. *Bull. Acad. r. Belg. Cl. Sci.*, **5**, 13.

GOODRICH, E. S. 1892. Note on a new Oligochaete. *Zool. Anz.*, **15**, 474.

— 1895. On the structure of *Vermiculus pilosus*. *Q. Jl micros. Sci.*, **37**, 253.

GRUBE, A. E. 1861. *Ein Ausflug nach Trieste und dem Quarnero*. (Beiträge zur Kenntniss der Thierwelt dieses Gebietes.) Berlin.

GRUBE, E. 1879. Untersuchungen über die physikalische Beschaffenheit und Flora und Fauna der Schweizer Seen. *Jber. schles. Ges. vaterl. Kult.*, **58**, 115.

HARO, A. de, 1964. *Branchiura sowerbyi* Beddard (1892), foruna cosmopolita, encontrado en España. *Boln. R. Soc. esp. Hist. nat. (Biol.)*, **62**, 137.

HATAI, S. 1898. *Vermiculus limosus*, a new species of aquatic Oligochaeta. *Annot. zool. Jap.*, **2**, 103.

— 1899. On *Limnodrilus gotoi*, n. sp. *Annot. zool. Jap.*, **3**, 5.

HIRAO, Y. 1964. Reproductive system and oogensis in the freshwater Oligochaete, *Tubifex hattai*. *J. Fac. Sci., Hokkaido Univ.*, IV, **15**, 439.

HOFFMEISTER, W. 1842. De vermibus quibusdam ad genus *Lumbricorum* pertinentibus. Dissertatio inaugur. 4. Berlini.

HRABE, S. 1929. Prispevek k poznání moravských Tubificid a Lumbriculid. *Biol. Listy*, **14**, 1.

— 1930. Príspevek k poznání oligochaet z jezeva Janiny, jeho okoli a z ostrova Korfu. *Mém Soc. r. Sci. Bohème*, **2**, 1.

— 1931a. Die Oligochaeten aus den Seen Ochrida und Prespa. *Zool. Jb. (syst.)*, **61**, 1.

— 1931b. Über eine neue Tubificiden-Gattung. *Epirodrilus* (Oligochaeta) nebst Beiträgen zur Kenntniss von *Tubifex blanchardi*. *Zool. Anz.*, **93**, 309.

— 1934a. *Tubifex (Psammorcytes) moravicus* n. sp. (Tschechisch). *Sb. Klubu prìr. Brne*, **16**, 48.

— 1934b. *Tubifex (Psammoryctes) moravicus* n. sp. *Zool. Anz.*, **107**, 33.

— 1934c. O nepohlavním rozmnozování nítenky *Bothrioneurum vejdovskýanum* Stolc. *Sb. Klubu přír. Brne*, **17**, 1.

HRABE, S. 1935a. Über *Moraviodrilus pygmaeus* n. g., n. sp., *Rhyacodrilus falciformis* Br., *Ilyodrilus bavaricus* Öschm. und *Bothrioneurum vejdovskyanum* St. *Spisy vydáv. přír. Fak. Masaryk. Univ.*, **209**, 4.

— 1935b. Die Oligochaeten des Issykkulsees. *Trudy Kirgiz. Kompleks. Ekxped.*, **3** (2), 73.

— 1936. *Trichodrilus strandi* n. sp., ein neuer Vertreter der Höhlen-Lumbriculiden. *Festschrift Prof. E. Strand. Riga*, **1**, 404.

— 1937. Zur Kenntnis der *Lamprodrilus mrazeki, Aulodrilus*, etc. *Sb. Klubu přír Brne*, **19**, 3.

— 1939a. Oligochètes aquatiques des Hautes Tatras. *Vest. csl. Lék.*, **6-7**, 209.

— 1939b. Über die ontogenetische Entwicklung des männlichen Ausführungsganges bei einigen Oligochäten. *Sb. Klubu přír. Brne*, **3**, 56 (Czech. with German summary).

— 1941. Zur Kenntnis der Oligochaeten aus der Donau. *Acta Acad. Sci. nat. moravosiles.*, **22**, 1 (reprint).

— 1942. Zur Kenntnis der Brunnen und Quellenfauna aus der Slowakei. *Sb. Klubu přír. Brne*, **24**, 23.

— 1950. Oligochaeta from the Caspian Sea. *Acta Acad. Sci. nat. moravo-siles.*, **22**, 251.

— 1952. Oligochaeta. 2 Limicola. In *The Zoology of Iceland*, **2** (20b), 1.

— 1958. Die Oligochaeten aus den Seen Dojran und Skadar. *Spisy vydav. přír. Fak. Masaryk Univ.*, **397**, 337.

— 1960. Oligochaeta limicola from the collection of Dr. Husmann. *Spisy vydav., přír. Fak. Masaryk Univ.*, **415**, 245.

— 1962a. Oligochaeta limicola from Onega lake collected by Mr. B. M. Alexandrov. *Spisy přír Fac. Univ. Brne*, **435**, 291.

— 1962b. *Rhizodrilus montanus* n. sp. from the glacial lake in the Perister Mountains in South Macedonia. *Spisy přír Fac. Univ. Brne*, **435**, 335.

— 1963a. *Rhyacodrilus subterraneus* n. sp., eine neue Tubificiden-Art aus den Brunnen in der Umgebung von Leipzig. *Zool. Anz.*, **170**, 249.

— 1963b. On *Rhyacodrilus lindbergi* n. sp., a new cavernicolous species of the fam. Tubificidae (Oligochaeta) from Portugal. *Bolm Soc. port. Cienc. nat.*, **10**, 52.

— 1964. On *Peloscolex svirenkoi* (Jaroschenko) and some other species of the genus *Peloscolex*. *Spisy přír Fac. Univ. Brne*, **450**, 101.

— 1966. New or insufficiently known species of the family Tubificidae. *Spisy přír Fac. Univ. Brne*, **470**, 57.

— 1967. Two new species of the family Tubifiidae from the Black Sea, with remarks about various species of the subfamily Tubificinae. *Spisy přír Fac. Univ. Brne*, **485**, 331.

HRABE S. AND CERNOSVITOV, L. 1926. Ueber eine neue Lumbriculiden—Art (*Rhychelmis vejdovskii* n. sp.). *Zool. Anz.*, **65**, 265.

IMHOF, O.-E. 1888. Ein neues Mitglied der Fauna der Süsswasserbecken. *Zool. Anz.*, **11**, 48.

INOU, T., AND KONDO, K. 1962. Susceptibility of *Branchiura sowerbyi, Limnodrilus socialis* and *Limnodrilus willeyi* for several agricultural chemicals. *Inst. Insect Control, Kyoto Univ.*, **27**, 97.

JAMIESON, B. G. M. 1968. *Macquaridrilus*: A new genus of Tubificidae (Oligochaeta) from Macquarie Island. *Pap. Dep. Zool. Univ. Qd*, **3** (5), 55.

JAROSCHENKO, M. F. 1948. Oligochaeta. Dneprobugskogo limana. *Nauch Zap. Moldavsk. Nauchnoissled. bazy Akad. Nauk. SSSR*, **1**, 57.

JOHNSTON, G. 1865. *A catalogue of the British non-parasitical worms*. London.

JONES, M. L. 1961. A quantitative evaluation of the benthic fauna off Point Richmond, California. *Univ. Calif. Publs Zool.*, **67**, 219.

JUGET, J. 1957. Quelques aspects de la faune limicole des environs de Saint-Jean-de-Losue. *Trav. Lab. Zool. Stn. aquic. Grimaldi Dijon*, **22**, 1 (extrait).

— 1958. Recherche sur la faune de fond du Léman et du Lac d'Annecy. *Annls Stn cent. Hydrobiol. appl.*, **7**, 9.

— 1959. Recherches sur la faune aquatique de deux grottes du Jura Méridional Français: la grotte de la Balme (Isère) et la grotte de Corveissiat (Ain). *Annls Spéléol.*, **14**, 391.

KENNEDY, C. R. 1964. Studies on the Irish Tubificidae. *Proc. R. Irish Acad.*, **63** (B), 225.

KESSLER, K. 1868. Materialy dlja poznanija onegskago ozera. *Trudý I. Cyezda Russk. Est. Oligochaeta,* **102**.

KEYL, F. 1913. Beiträge zur Kenntniss von *Branchiura sowerbyi* Bedd. *Z. wiss. Zool.*, **107**, 199.

KNÖLLNER, F. H. 1935. Ökologische und systematische Untersuchungen über litorale und marine Oligochäten der Kieler Bucht. *Zool. Jb.* (*Syst.*), **66**, 425.

KOWALEWSKI, M. 1914. Rodzaj *Aulodrilus* Bretscher 1899 i jego przedstawiciele. *Bull. int. Acad. Sci. Lett. Cracovie,* **54**, 598.

— 1919. Zbadan nad skaposzczetami. *Rozpr. Wydz. mat-przyr. Akad. Vmiejet, III,* 18.

LAMARK, J. B. P. 1816. *Histoire naturelle des animaux sans vetèbres.* 3. Paris.

LANKASTER, E. R. 1871a. Some observations on the organization of the aquatic Oligochaete annelids *Ann. Mag. nat. Hist.*, IV, 7, 90.

— 1871b. On the structure and origin of the spermatophors or sperm-ropes of two species of *Tubifex. Q. Jl micros. Sci.*, **11**, 181.

LASTOCKIN, D. A. 1918. Materialy po faune vodnykh Oligochaeta Rosii, 1. *Tr. Petrogr. obshch. estestvoisp.*, **49**, 57.

— 1927a. Beiträge zur Oligochaetenfauna Russlands: 3. *Izv. Ivan.-Voznesensk. Politekh. Inst.*, **10**, 65.

— 1927b. Oligochaeta limicola des Oka-Flusses. *Rab. nauchno-promýsl. Éksped. Izveh. Reki Obi.,* **5**, 34.

— 1936. Gidrobiologicheskie issledovanija rek Volgi i Mologie. *Tr. Ivanovsk. s.-kh. Inst.,* **2**, 167.

— 1937. New species of Oligochaeta limicola in the European part of the U.S.S.R. *Dokl. Akad. Nauk SSSR,* **17**, 233.

LASTOCKIN, D. A., AND SOKOLSKAYA, N. L. 1953. New species of the genus *Peloscolex* (Tubificidae) from the Amur basin. *Zool. Zh.*, **32**, 409 (in Russian).

LEIDY, J. 1850. Description of some American Annelids abranchia. *J. Acad. nat. Sci. Philad.,* **2**, 43.

— 1851. Description of new genera of Vermes. *Proc. Acad. nat. Sci. Philad.,* **5** (6), 124.

LEUKART, R. 1849. Beitrag zur Kenntnis der Fauna von Island. *Arch Naturgesch.,* **15**, 149.

LEVINSEN, G. M. R. 1884. Systematisk—geofrafisk—oversigt over de nordiske Annulata. Gephyren, Choetognathi og Balenoglossi. *Vidensk. Meddr dansk naturh. Foren.,* **4** (1883), 192.

LINDER, C. 1904. Etude de la faune pelagique du lac de Bret. *Revue suisse Zool.,* **12**, 149.

MACINTOSH, W. C. 1870. On some points in the structure of *Tubifex. Trans. roy. Soc. Edinb.,* **26**, 253.

MALEVITCH, J. J. 1930. Zur Kenntnis der Oligochaetenfauna des Kaukasus. *Rab. sev.-kavk. gidrol. Sta. gorsk. sel. khoz. Inst.,* **3**, 81 (in Russian).

— 1937a. Notes on the geographical distribution of *Branchiura sowerbyi* Bedd. *Sb. Trud. gos. zool. Muz.,* **4**, 131 (in Russian).

— 1937b. Note on the Oligochaeta of the Azow-Sea. *Sb. Trud. gos. zool. Muz.,* **4**, 133, (in Russian).

— 1947. Oligochaeta proper to caves of the Caucasus. *Byull. mosk. Obshch. Ispýt. Prir.* (*biol.*). *N. S.,* **52** (4), 11 (in Russian with English summary).

— 1949. K faune oligokhet Teledkogo ozera. *Trudý zool. Inst., Leningr.,* **7**, 119.

— 1951. Materialy k poznaniio fauny maloshchetinkoveikh iervei (Oligochaeta) poberezheja Belogo Morja. *Sb. Trud. zool. Mus.,* **7**, 171.

— 1956. Maloshchetinkokye chervi (Oligochaeta) moskovskoi oblasti. *Uchen. Zap. mosk. gor. ped. inst. Potomkina,* **61**, 403.

— 1957. Nekotorye novye dannye o raspostranenii maloshchetinkveikh chervei (Oligochaeta) v CCCP. *Trudý leningr. Obshch. Estest.,* **73** (4), 81.

MANN, K. H. 1958. Occurrence of an exotic oligochaeta *Branchiura sowerbyi* Beddard, 1892, in the River Thames. *Nature, Lond.,* **182**, 732.

MARCUS, E. 1942. Sobre algumas Tubificidae do Brasil. *Bolm Fac. Filos. Cienc. Univ. S. Paulo,* **25**, 153.

Marcus, E. 1944. Sobre Oligochaeta limicos do Brasil. *Bolm Fac. Filos. Cienc. Univ. S. Paulo*, **43** (8), 5.

— 1965. Naidomorpha aus brasilianischem Brackwasser. *Beitrage zur Neotropischen Fauna*, **4**, 61.

Marcus, E. du B.-R. 1947. Naidids and Tubificids from Brazil. *Comun. zool. Mus. Hist. nat. Montev.*, **2** (44), 1.

— 1949. Further notes of Naidids and Tubificids from Brazil. *Comun. zool. Mus. Hist. nat. Montev.*, **3** (51), 1.

— 1950. A marine Tubificid from Brazil. *Comun. zool. Mus. Hist. nat. Montev.*, **3** (59), 1.

— 1952. A new Tubificid from the Bay of Montevideo. *Comun. zool. Mus. Hist. nat. Montev.*, **3** (56, 1949), 1.

Mehra, H. R. 1920. On the sexual phase in certain Indian Naididae (Oligochaeta). *Proc. zool. Soc. Lond.*, **1920**, 457.

— 1922. Two new Indian species of the little known genus *Aulodrilus* Bretscher. *Proc. zool. Soc. Lond.*, **59**, 943.

Michaelsen, W. 1900. Oligochaeta. In *Das Tierreich*. 10. Berlin.

— 1901a. Neue Tubificiden des Niederelbgebiets. *Verh. naturw. Ver. Hamb.*, **3** (8), 1.

— 1901b. Oligochaeten der Zoologischen Museen zu St. Petersburg und Kiew. *Izv. Acad. Nauk SSSR*, **5** (15), 137.

— 1902. Die Oligochaeten-Fauna des Baikal-Sees. *Verh. naturw. Ver. Hamb.*, **3** (9), 43.

— 1903a. Neue Oligochaeten und neue Fundorte alt-bekannter. *Mitt. Naturh. Mus. Hamb.*, 1901, **19**, 1.

— 1903b. Oligochaeten *Mitt. Naturh. Mus. Hamb.*, 1901, **19**, 169.

— 1905a. Die Oligochaeten des Baikal-Sees. *Wiss. Ergebn. Zool. Exped. Baikal-See* . . . 1900-1902. 1. Kiew und Berlin.

— 1905b. Die Oligochaeten der deutschen Südpolar-expedition 1901-3. *Deutsche Südpolar—Exp. IX, Zool.*, **1**, 7.

— 1907. Oligochaeta. In Michaelsen and Hartmeyer. *Die Fauna Südwest-Australiens. Ergebnisse der Hamburger Südwest-Australischen Forchungsreise* 1905. 1 (2): 117.

— 1908a. Zur Kenntnis der Tubificiden. *Arch. Naturgesch.*, **74**, 129.

— 1908b. Oligochaeta of India, Nepal, Ceylon, Burma and the Andaman Islands. *Mem. Indian Mus.*, **1**, 103.

— 1909. Oligochaeta. In Brauer, *Die Süsswasserfauna Deutschlands*, **13**, 1.

— 1910. Die Oligochätenfauna der vorderindisch-ceylonischen Region. *Abh. Geb. Naturw. Hamburg*, **19** (5), 1.

— 1912. Litorale Oligochäten von der Nordküste Russlands. *Trudy Imperatorskago S.-Peterburgsko obshchestva estestvois tatelei*, **42** (1911), 94. (in Russian and German).

— 1913. Oligochäten vom tropischen und südlichsubtropischen Afrika. *Zoologica, Stuttg.*, **67** (B 26), 141.

— 1914. *Beiträge zur Kenntnis der Land-und Süsswasserfauna Deutsch-Südwest-afrikas: Oligochaeta.* Hamburg. 139.

— 1923. Die Oligochaeten der Wolga. *Rab. Volzhsk. biol. st.*, **7**, 30.

— 1924. Oligochaeten von Neuseeland und den Auckland-Campbell Inseln, nebst einigen anderen Pacifischen Formen. *Vidensk. Meddr dansk naturh. Foren.*, **75**, 197.

— 1926a. Zur Kenntnis der Oligochaeten des Baikal-Sees. *Russk. gidrobiol. Zh.*, **5**, 153.

— 1926b. Oligochaeten aus dem Ryck bei Greifswald und von benachbarten Meeres-gebieten. *Mitt. zool. StInst. Hamb.*, **42**, 21.

— 1926c. Schmarotzende Oligochäten nebst Erörterungen über verwandtschaftliche Beziehungen der Archiologochäten. *Mitt. zool. StInst. Hamb.*, **42**, 91.

— 1927. Oligochaeta. In Grimp and Wagler, *Die Tierwelt der Nord-und Ostsee*. 6c. Leipzig.

— 1929a. Oligochäten der Kamtschatka-Expedition. *Ezheg. zool. Muz.*, **30**, 315.

— 1929b. Zur Stammesgeschichte der Oligochäten. *Zeit. wiss. Zoologie*, **134**, 693.

MICHAELSEN, W. 1933. Ein Panzenoligochät aus dem Baikal-See. *Zool. Anz.*, **102**, 326.
— 1934. Oligochäten von Französisch-Indochina. *Archs Zool. exp. gén.*, **76**, 493.
— 1935a. Oligochaeten aus den Seen des Zentral-Altai. *Serv. Hydro. Meteor. U.R.S.S.*, *Inst. Hydrol.*, **8**, 298.
— 1935b. Oligochäten von Belgisch Kongo. *Rev. Zool. Bot. afr.*, **27**, 34.
— 1936. *Branchiura sowerbyi* Bedd. and its synonomy. *Rec. Indian Mus.*, **38**, 95.
MICHAELSEN, W., AND BOLDT, W. 1932. Oligochaeta der Deutschen Limnologischen Sunda-Expedition. *Arch. Hydrobiol.*, **9**, 587.
MICHAELSEN, W., AND VERESCAGIN, G. 1930. Oligochaeten aus dem Selenga-Gebiete des Baikalsees. *Trudý Kom. Izuch. Ozera Baikala*, **3**, 213.
MOORE, J. P. 1905a. Hirudinea and Oligochaeta collected in the Great Lakes Region. *Bull. U.S. Bur. Fish.*, **25**, 153.
— 1905b. Some marine Oligochaeta of New England. *Proc. Acad. nat. Sci. Philad.*, **57**, 373.
MOSZYNSKA, M. 1962. Skaposzczety Oligochaeta. *Kat. Fauny polski*, **11** (2), 1.
MOSZYNSKI, A., AND MOSZYNSKA, M. 1957. Skaposzczety (Oligochaeta) Polski i niektórych krayów sasiednich. *Poznan. tow. Přir zyjac. nauk.*, **18** (6), 1.
MÜLLER, O. F. 1774. *Vermium terrestrium et fluviatilium.* Havniae et Lipsiae. 1773-74.
NAIDU, K. V. 1965. Studies on the fresh-water Oligochaeta of South India. II. Tubificidae. *Hydrobiologia*, **26**, 463.
— 1966. Checklist of freshwater Oligochaeta of the Indian subcontinent and Tibet. *Hydrobiologia*, **27, 208**.
NOMURA, E. 1913. On two species of aquatic Oligochaeta, *Limnodrilus gotoi* Hatai and *Limnodrilus willeyi*, n. sp. *J. Coll. Sci. imp. Univ. Tokyo*, **35** (4), 1.
— 1915. On the aquatic Oligochaete, *Monopylephorus limosus* (Hatai). *J. Coll. Sci-imp. Univ. Tokyo*, **35** (9), 1.
— 1926. On the aquatic Oligochaete, *Tubifex hattai*, n. sp. *Scient. Rep. Tohoku imp. Univ.* (*Biol.*), **1**, 193.
— 1929. On *Limnodrilus grandisetosus* sp. n., an aquatic Oligochaete. *Annotnes zool. jap.*, **12**, 131.
— 1932. *Limnodrilus grandisetosus*, nov. sp., a freshwater Oligochaete. *Scient. Rep. Tôhoku imp. Univ.*, (4) **7**, 511.
ORSTED, A. S. 1844. De regionibus marinis. Elementa topographiae historico-naturalis freti Oeresund. Diss. inaug. Havniae.
ÖSCHMANN, A. 1913. Über eine neue Tubificiden—Art *Tubifex* (*Ilyodrilus*) *bavaricus*. *Zool. Anz.*, **42**, 559.
OKEN, L. 1815. *Lehrbuch der Naturgeschichte.* 3. *Zoologie*, 1. *Oligochaeta.* Leipzig.
PIERANTONI, U. 1902. Due nuovi generi di Oligocheti marini vinvenuti nel Golfo di Napoli. *Boll. Soc. Nat. Napoli*, **16**, 113.
— 1904a. Altri nuovi Oligocheti del Golfo di Napoli. (*Limnodriloides* n. gen.).—Il nota sui Tubificidae. *Boll. Soc. Nat. Napoli*, **17**, 185.
— 1904b. Sopra alcuni Oligocheti raccolti nel fiume Sarno. *Annuar. R. Mus. zool. R. Univ. Napoli*, **1** (26), 1.
— 1905. Oligocheti del fiume Sarno. *Archo zool. ital.*, **2**, 227.
— 1916. Sull' *Heterodrilus arenicolus* Pierant. e su di una nuova specie del genere *Clitellio*. *Boll. Soc. Nat. Napoli*, **9**, 82.
PIGUET, E. 1899. Notice sur la réparition de quelques Vers Oligochètes dans le lac Léman. *Bull. Soc. vaud. Sci. nat.*, **35**, 71.
— 1906a. Oligochètes de la Suisse français. *Revue suisse Zool.*, **14**, 389.
— 1906b. Observations sur les Naididées. *Revue suisse Zool.*, **14**, 185.
— 1913. Notes sur les Oligochètes. *Revue suisse Zool.*, **21**, 111.
— 1919. Oligochètes communs aux Hautes Alpes suisse et scandinaves. *Revue suisse Zool.*, **27**, 1.
— 1928. Sur quelques Oligochètes de l'Amerique du Sud et d'Europe. *Bull. Soc. neuchâtel Sci. nat.*, **52** (1927), 78.
PIGUET, E., AND BRETSCHER, K. 1913. Oligochètes. In *Fasc. 7 Catalogue des Invertébrés de la Suisse.* Mus. Hist. nat. Genève.
POINTNER, H. 1911. Beitràge Zur Kenntnis der Oligochaetenfauna der Gewässer von Graz. *Z. wiss. Zool.*, **98**, 626.

POINTNER, H. 1914. Über einige neue Oligochaeten der Lunzer Seen. *Arch. Hydrobiol.*, **9**. 606.
POPESCU-MARINESCU, V., BOTAE, F. AND BREZEANU, G. 1966. Untersuchungen über die Oligochaeten im rumänischen Sektor des Donaubassins. *Arch. Hydrobiol.*, **20**, 161.
RANDOLPH, H. 1892. Ein Beitrag zur Kenntnis der Tubificiden. *Vjschr. naturf. Ges. Zürich*, **1892**, 1.
RATZEL, F. 1868. Beiträge zur anatomischen und systematischen Kenntnis der Oligochaeten. *Z. wiss Zool.*, **18**, 563.
RYBKA, J. 1898. Contribution a la morphologie et la classification du genre *Limnodrilus* Claparède (1). *Mèm. Soc. zool. Fr.*, **11**, 376.
SAPKAREV, J. 1959. *Ilyodrilus hammionensis* Mich (Oligochaeta) in the large lakes of Macedonia (Ohrid, Prespa and Dojran). *Section des Sci. Nat. Univ. Skopje*, **7**, 1.
SAVIGNY, 1820. *Système des Annélides principalement des Côtes d'Egypt et de la Syrie*. Paris.
SCHAEFFER, K. F., HARREL, R. C., AND MATHIS, B. J. 1965. *Branchiura sowerbyi* (Tubificidae, Annelida) in Oklahoma. *Proc. Okla. Acad. Sci.*, **45**, 71.
SMITH, F. 1900. Notes on species of North American Oligochaeta. III. *Bull. Ill. St. Lab. nat. Hist.*, **5**, 441.
SMITH, S. I. 1874. Sketch of the invertebrate fauna of Lake Superior. *Rep. U.S. Commnr. Fish.* *(1872-1873)*, **2**, 690
SMITH, S. I., AND VERRILL, A. E. 1871. Notice of the Invertebrata dredged in Lake Superior in 1871 by the U.S. Lake Survey under direction of Gen. C. B. Comstock, S. I. Smith, naturalist. *Am. J. Sci.*, III, **2**, 448.
SOKOLSKAYA, N. L. 1953. Maloshchetinkovye chervi ozer Belorvsskoi CCCP. *Uchen. Zap. beloruss. gos. Univ.*, **17**, 88.
— 1958. Presnovodnye maloshchetinkovye eervi basseina Amvra. *Tr. Amvrsk. ikhtiolog. eksped. 1945-1949*, **4**, 287.
— 1961a. Material dealing with the fresh-water Oligochaete fauna of Kamchatka. *Byull. mosk. Obshch. Ispyt. Přír*, **66**, 54 (in Russian).
— 1961b. Material on the fauna of freshwater Microcrili in the Amur Basin based on the joint Soviet-China Amur Expedition 1957-58. *Sb. Trud. zool. Mus.*, **8**, 79 (in Russian).
— 1961c. Novye dannye o geograficheskom rasprostranenii oligokhety *Branchiura sowerbyi* Bedd. I. nekotorye svedeniya po ekologii vida. *Zool. Zh.*, **40**, 605.
— 1964. A new species of the genus *Limnodrilus* Claparède (Tubificidae, Oligochaeta) from brackish lakes of South Sakhalin. *Zool. Zh.*, **43**, 1071.
SOUTHERN, R. 1909. Contribution towards a monograph of the British and Irish Oligochaeta. *Proc. R. Irish Acad.*, **27** (B, 8), 119.
— 1911. Oligochaeta in the Clare Island Survey. *Proc. R. Irish Acad.*, **31** (48), 1.
SPENCER, W. P. 1932. A gilled Oligochaete *Branchiura sowerbyi* new to America. *Trans. Am. micros. Soc.*, **51**, 267.
SPERBER, C. 1948. A taxonomical study of the Naididae. *Zool. Bidr. Upps.*, **28**, 1.
STAMMER, H. -J. 1932. Die Fauna des Timavo. *Zool. Jb. (Syst.)*, **63**, 521.
— 1963. *Branchiura sowerbyi* Beddard in Franken. *Zool. Anz.*, **171**, 390.
STEPHENSON, J. 1910. On *Bothrioneurum iris*, Beddard. *Rec. Indian Mus.*, **5**, 241.
— 1911. On some littoral Oligochaeta of the Clyde. *Trans. roy. Soc. Edinb.*, **48**, 6.
— 1912a. On *Branchiura sowerbyi* Bedd. and on a new species of *Limnodrilus* with distinctive characters. *Trans. roy. Soc. Edinb.*, **48**, 285.
— 1912b. On a new species of *Branchiodrilus* and certain other aquatic Oligochaeta, with remarks on cephalization in the Naididae. *Rec. Indian Mus.*, **7**, 219.
— 1913a. On intestinal respiration in Annelids with consideration of the origin and evolution of the vascular system in that group. *Trans. roy. Soc. Edinb.*, **49**, 735.
— 1913b. On a collection of Oligochaeta mainly from Ceylon. *Spolia zeylan*, **8**, 251.
— 1917a. Oligochaeta. In: the fauna of Chilka Lake. *Mem. Indian Mus.*, **5**, 485.
— 1917b. Aquatic Oligochaeta from Japan and China. *Mem. Asiat. Soc. Beng.*, **6**, 83.
— 1918. Aquatic Oligochaeta of the Inlé Lake. *Rec. Indian Mus.*, **14**, 9.

STEPHENSON, J. 1920. On a collection of Oligochaeta from the lesser known parts of India and from Eastern Persia. *Mem. Indian Mus.*, **7**, 191.

— 1921. Oligochaeta from Manipur, the Laccadive Islands, Mysore, and other parts of India. *Rec. Indian Mus.*, **32**, 745.

— 1922. On some Scottish Oligochaeta with a note on encystment in a common freshwater Oligochaeta. *Trans. roy. Soc. Edinb.*, **53**, 277.

— 1923. Oligochaeta. In *The fauna of British India.* London.

— 1924. On some Indian Oligochaeta, with a description of two new genera of Ocnerodrilinae. *Rec. Indian Mus.*, **26**, 317.

— 1925a. On some Oligochaeta mainly from Assam, South India and the Andaman Islands. *Rec. Indian Mus.*, **27**, 43.

— 1925b. Oligochaeta from various regions, including those collected by the Mount Everest Expedition 1924. *Proc. zool. Soc. Lond.*, **1925**, 879.

— 1926. Description of Indian Oligochaeta. *Rec. Indian Mus.*, **28**, 249.

— 1929. The Oligochaeta of the Indawgyi Lake (Upper Burma). *Rec. Indian Mus.*, **31**, 225.

— 1930. *The Oligochaeta.* Oxford.

— 1931a. Oligochaeta from the Malay Peninsula. *J. fed. Malay St. Mus.*, **16**, 261.

— 1931b. The Oligochaeta. (Reports of an Expedition to Brazil and Paraguay.) *J. Linn. Soc. (Zool.)*, **37**, 291.

STOLC, A. 1885. Vorlaufiger Bericht über *Ilyodrilus coccineus.* (Ein Beitrag zur Kenntnis der Tubificiden). *Zool. Anz.*, **8**, 638.

— 1886. Přehled ceskych Tubificidu. *Sber. K. böhm. Ges. Wiss.*, **1885**, 640.

— 1888. Monographie ceskych Tubificidu. *Sber. K. böhm. Ges. Wiss.*, VII, **2**, 1.

STRECKER, R. L. 1954. A new Ohio locality record for the gilled Oligochaete Branchiura *sowerbyi. Ohio J. Sci.*, **54**, 280.

SVETLOV, P. G. 1926a. Zur Kenntnis der Oligochaetenfauna des Gouv. Samara. Isv. biol. Nauchno-issled. *Inst. biol. Sta. perm. gosud. Univ.*, **4**, 249.

— 1926b. Beobachtungen über die Oligochaeten des Gouv. Perm. III. Die Familien Tubificidae, Lumbriculidae und Discodrilidae. *Inst. biol. Sta. perm. gosud. Univ.*, **4**, 343 (in Russian with German resumé).

— 1936. The Oligochaeta of the Kama Expedition, 1935. *Izv. fiz.-mat. Obshch. (imp.) kazan. Univ.*, **10** (4), 145 (in Russian with English summary).

— 1946. K faune Oligochaeta Tomskoi ovlasti. *Trudý tomsk. gos. Univ.*, **97**, 103.

TAUBER, P. 1879. *Annulata Danica.* Kopenhagen.

TEAL, J. M. 1957. Community metabolism in a temperate cold spring. *Ecol. Monogr.*, **27**, 283.

TETER, H. E. 1960. The bottom fauna of Lake Huron. *Trans. Am. Fish. Soc.*, **89**, 193.

THIENEMANN, A. 1907. Die Tierwelt der kalten Bäche und Quellen auf Rügen. *Mitt. natw. Ver.*, **38**, 74.

UDE, H. 1929. Oligochaeta. In Dahl, *Die Tierwelt Deutschlands*, **15** (1), 1.

UDEKEM, J. D.' 1853. Histoire naturelle du *Tubifex* des Ruisseaux. *Mem. cour. Acad. r. Sci. Belg.*, **26**, 3.

— 1855. Nouvelle classification des Annélides sétigères abranches. *Bull. Acad. r. Belg. Cl. Sci.*, **22** (10), 1.

— 1859. Nouvelle classification des Annélides sétigères abranches. *Mèm. Acad. r. Belg. Cl. Sci.*, **31**, 1.

VAILLANT, L. 1868. Note sur l'anatomie de deux espèces du genre *Perichaeta*, et essai de classification des Annélides Lombricines. *Annls. Sci. nat. V, Zool.*, **10**, 225.

— 1890. *Histoire naturelle des Annelés marins et d'eau douce.* III. *Lombriciniens, Hirudiniens, Bdellomorphes, Térétulariens et Planariens.* Paris.

VEJDOVSKY, F. 1875. Beiträge zur Oligochaetenfauna Böhmens. *Sber. böhm. Ges. Wiss. Prag.*, **1874**, 191.

— 1876. Über *Psammorcytes umbellifer* und die ihm verwandten Gattungen. *Z. wiss. Zool.*, **27**, 137.

— 1884a. Reviso Oligochaetorum Bohemiae. *Sber. K. böhm. Ges. Wiss.*, **1883**, 215.

— 1884b. *System und Morphologie der Oligochaeten.* Prag.

— 1891. Note sur un *Tubifex* d'Algerie. *Mem. Soc. zool. Fr.*, **4**, 596.

VEJDOVSKY, F., AND MRAZEK, A. 1902. Ueber *Potamothrix (Clitellio?) moldaviensis* n. g., n. sp. *Sber. K. böhm. Ges. Wiss. Prag.*, **24**, 1.

VERRILL, A. E. 1873. Report on the invertebrate animals of Vineyard Sound. *Rep. U.S. Commnr. Fish.*, **324**, 622.

VISART, E. de. 1901. Res Italicae. *Boll. Musei. Zool. Anat. comp. R Univ. Torino,* **16** (387), 1.

VOS, A. P. C. de 1936. Chaetopoda. In: Flora Fauna Zuiderzee. *Helder Suppl.,* **85**, 134.

WILLIAMS, T. 1852. Report on the British Annelida. *Rep. Br. Ass. Advmt. Sci.,* **1851**, 159.

— 1859. Researches on the structure and nomology of the reproductive organs of the Annelids. *Phil. Trans. R. Soc.,* **148** (1858), 93.

WURTZ, C. B., AND DOLAN, T. 1960. A biological method used in the evaluation of effects of thermal discharge in the Schuylkill River. *Proc. 15th Ind. Waste Conference Purdue*, 461.

WURTZ, C. B., AND ROBACK, S. S. 1955. The invertebrate fauna of some Gulf Coast Rivers. *Proc. Acad. nat. Sci. Philad.,* **107**, 167.

YAMAGUCHI, H. 1953. Studies on the aquatic Oligochaeta of Japan. VI. *J. Fac. Sci. Hokkaido Univ. (Zool.),* **11**, 277.

YAMAMOTO, G. AND OKADO, K. 1940. *Tubifex (Peloscolex) nomurai* sp. nov. *Sci. Rep. Tohoku Univ. (Biol.),* **15**, 427.

ZENGER, N. 1870. *Peloryctes inquilina. Byull. mosk. Obshch. Ispyt. Přir,* **43**, 221.

ZSCHOKKE, F. 1891. Weiterer Beitrag zur Kenntnis der Fauna von Gebirgsseen. *Zool. Anz.,* **14**, 119.

9

FAMILY PHREODRILIDAE

Type genus: Phreodrilus BEDDARD

Dorsal setae from III or IV, hairs and very small needles, or restricted to a few posterior segments. Ventral setae two per bundle, simple-pointed or bifid with a rudimentary upper tooth or (commonly) one of each. Spermathecal setae modified or unmodified. Testes in XI, ovaries in XII, sperm sacs and egg sacs mostly present, male pores on XII, female pores in 12/13, spermathecal pores in XII. All reproductive organs may be shifted one or two segments forward, spermathecae may be duplicated. Pharynx eversible, with tall ciliated cells. Septal glands in IV or V to VII or VIII, loose cellular aggregations. Nephridia restricted to a few anteclitellar segments, or a single large nephridium in VII. No spermatophores, no spermathecal diverticulae.

Southern hemisphere and Ceylon.

Phreodrilus (Astacopsidrilus) novus is supposed to have two pairs of testes. In *P. (Ph.) zeylanicus* an extra segment is supposed to exist between the male pores and supermathecal pores. Neither description is sufficiently detailed to give full weight to these observations. A single specimen of *P. (A) niger* (as *albus*) has been reported with the reproductive organs one segment in advance of the usual position. The dorsal setae are usually referred to as hair setae with short "reserve" setae buried in the body wall. Stout (1958) showed that the shorter setae differ from the hairs and should perhaps be termed needles. Their form has not been adequately described in most species.

GENUS **Phreodrilus** BEDDARD, 1891

Fig. 9. 1

Type species: Phreodrilus subterraneus BEDDARD

Defined and distributed as the family.

626

The characteristics displayed by the various taxa that have been described do not show any clear-cut groupings that might be recognized as constituting genera (Brinkhurst, 1965). Any single character (presence or absence of gills, penes, coelomocytes, genital setae or spermathecae, position of spermathecal pores, nature of spermathecal pores or male pores *et al.*) could be selected to form the basis of separation of genera, but this is clearly undesirable. Six groups of species can be recognized on the basis of the association of a series of characters, any one of which would not suffice to separate one of these groups from at least one another. Therefore, *Phreodrilus, Phreodriloides, Astacopsidrilus, Hesperodrilus, Gondwanaedrilus, Schizodrilus* and *Tasmaniaedrilus*, were united into the single genus *Phreodrilus* and six sub-genera were recognized (Brinkhurst, 1965).

Phreodrilus

1.	Living on crayfish *Astacopsis*	2
—	Free living	3
2.	Dorsal setae from **XXXI-XLIII**	*P. (As.) goddardi*
—	Dorsal setae from **IV**	*P. (As.) fusiformis*
3.	With gills	*P. (P.) branchiatus*
—	Without gills	4
4.	Spermathecae opening jointly with male or female pores	5
—	Spermathecae opening independently of other genital pores, or absent	6
5.	Spermathecae opening with female pores	*P. (As.) novus*[*]
—	Spermathecae opening into penis sacs	*P. (G.) africanus*
6.	Spermathecae absent	7
—	Spermathecae present	8
7.	No reproductive organs present. Coelomocytes present	*P. (S.) major*
—	Other reproductive organs present. No coelomocytes	*P. (Ph.) notabilis*
8.	Testes in IX, male pores in X, spermathecae in XI or XI and XII	*P. (S.) nothofagi*
—	Testes in XI male pores in XII, spermathecal pores in XIII, (once seen 1 segment anteriad)	9

* And a new species from L. Tanganyika to be described. (See Appendix, p. 838).

9.   Spermathecal pores dorsal   10
—   Spermathecal pores ventro-lateral or ventral   13

10.   True penis present   *P. (A.) niger*
—   Protrusible pseudopenis present   11

11.   Inner duct of pseudopenis coiled once or twice
within muscular sac   *P. (P.) beddardi*
—   Inner duct of pseudopenis coiled many times
within muscular sac   12

12.   Ventral setal bundles with one bifid and one
simple pointed seta   *P. (P.) mauienensis*
—   Ventral setal bundles with setae dissimilar
but both simple pointed   *P. (P.) subterraneus*

13.   Spermathecal pores fused, median ventral   *P. (Ph.) kerguelenensis*
—   Spermathecal pores ventro-lateral   14

14.   At least rudimentary vestibulae on spermathecal pores   15
—   No vestibulae on spermathecal pores   17

15.   Vestibulae rudimentary, paired spermathecal
setae   *P. (I.) lacustris*
—   Vestibulae large, spermathecal setae single
or absent   16

16.   Spermathecal setae single   *P. (I.) litoralis*
—   Spermathecal setae absent   *P. (I.) cambellianus*

17.   Vasa deferentia enter atria apically   *P. (Ph.) zeylanicus*
—   Vasa deferentia enter atria basally.   *P. (Ph.) crozetensis*

SUB-GENUS **Phreodrilus** BEDDARD, 1891

SUB-GENERIC TYPE : **Phreodrilus subterraneus** BEDDARD

Spermathecal pores dorsal, with muscular vestibulae formed from an inversion of the body wall. Pseudopenes present, with muscular sacs. Vasa deferentia open into atria basally.

New Zealand, Chile.

Phreodrilus (Phreodrilus) subterraneus Beddard, 1891

Phreodrilus subterraneus BEDDARD, 1891 : 273, Pl. I, fig. 1-15, Pl. II, fig. 16, 18, 19, 30-33, Pl. III, fig. 34, 37.

Phreodrilus subterraneus Beddard. BEDDARD, 1895: 275. MICHAELSEN, 1900: 37, 1924: 207. STEPHENSON, 1930: 758.

Phreodrilus subterraneous Beddard. STOUT, 1958: 298.

Phreodrilus (Phreodrilus) subterraneus Beddard. BRINKHURST, 1965: 377, Fig. 2.

$l=50$ mm. Ventral setae both simple-pointed but one thicker and more curved than the other. No spermathecal setae. Pseudopenes extremely long, distal part of male duct and pseudopenis enclosed in muscular sac, true male pore deep in the penis sac.
New Zealand (S. Island).

It is a strange fact that the most unique, mostly extremely developed form of the male efferent ducts was observed in this, the first species to be described. It is not surprising, therefore, that the interpretation of the complex male ducts presented some difficulty, just as it has until recently in the tubificid genus *Monopylephorus*. When the series *Phreodrilus branchiatus, beddardi mauienensis* and *subterraneus* is considered, it will be seen that the progressive development of the protrusible pseudopenis, involving as it must the separation of the lining layer of the system from the muscular layers (the system all being derived from the body wall, be it atrium plus ejaculatory duct or the secondary involution of a penis sac, eversible and with or without a penis), leads to a situation where the lining layer forms a coiled pipe within a muscular sac, the whole being capable of considerable elongation when everted and protruded through the superficial male opening. The lining layer has to be much longer than the muscular wall in order to accomplish this extension, during which it forms a double-walled structure.

The testes were said to appear as one pair, lying on the posterior face of X and a second on the anterior face of XI (Beddard, 1895). There is probably only a single pair.

Phreodrilus (Phreodrilus) mauienensis Brinkhurst, 1971*

Dimensions unknown. Dorsal setae from III, one hair and two short setae in follicles, ventral setae one broad, clearly bifid with upper tooth shorter than the lower, the other thinner, straighter and simple pointed. Vasa deferentia long, atria extremely long and tubular, these uniting just with muscular sacs. Lining of male duct much convoluted within muscular sacs. Small globular organs (? gland) attached to muscular sacs. Distal part of pseudopenes long. Spermathecal setae absent. Spermathecae present, spermathecal pores dorsal, with atria.

Type locality: Lake Okataina, New Zealand (150'). Holotype & Paratype: Otago Museum, Dunedin NZ. Also Lake Coleridge, New Zealand, specimens in The Australian Museum, Sydney, Australia.

The male ducts of this species are rather like those of *P. subterraneus*, but the little accessory organ, probably a gland, has not been reported in

* See Appendix on p. 838 for reference.

that species. Unlike the other species in the sub-genus, the ventral setae are distinctly dissimilar, not only in width and curvature, but also by virtue of the constantly bifid end of the broader setae. It is quite conceivable that immature forms of this species were described under the name *P. mauianus* (Benham, 1903) especially as most of the other species described in that paper were found in the collection from Lake Coleridge. The truth of this cannot be ascertained until the number of phreodrilid species in New Zealand which have dissimilar setae can be decided. Hence there is no alternative to the application of a new name for this species, but a name has been chosen which approximates the two names used by Benham (1903, 1904), which are *P. mauianus* and *P. mauiensis* respectively.

Phreodrilus (Phreodrilus) beddardi Benham, 1904

Phreodrilus beddardi BENHAM, 1904: 281.

Phreodrilus beddardi Benham. MICHAELSEN, 1924: 207. STEPHENSON, 1930: 758, Fig. 143.

Phreodrilus (Phreodrilus) beddardi Benham. BRINKHURST, 1965: 377, Fig. 2.

$l=40$ mm, $s=78$. Ventral setae both simple-pointed, spermathecal setae unmodified. Pseudopenes long, lining layer coiled once or twice within muscular sac, short penis sac.
New Zealand.

New material from Lake Coleridge has confirmed most of the earlier description of the species. The ventral setae are slightly dissimilar in form but they are both simple pointed, one set is thick, sigmoid and has a nodulus, the other is thin, nearly straight and has no nodulus.

Phreodrilus (Phreodrilus) branchiatus Beddard, 1891

Hesperodrilus branchiatus BEDDARD, 1894: 207.

Hesperodrilus branchiatus Beddard. BEDDARD, 1895: 257; 1896: 15. MICHAELSEN, 1900: 38; 1924: 207.

Phreodrilus branchiatus (Beddard). MICHAELSEN, 1903: 136.

Phreodrilus (Phreodrilus) branchiatus (Beddard). BRINKHURST, 1965: 377, Fig. 2, 5.

$s=53$. Ventral setae 1 bifid, 1 simple-pointed per bundle, spermathecal setae unmodified. Protrusible pseudopenes short. Gill filaments lateral, below dorsal setae, 13 pairs.
S. Chile.

SUB-GENUS **Antarctodrilus** BRINKHURST, 1965

SUB-GENERIC TYPE : **Hesperodrilus niger** BEDDARD

Spermathecal pores dorsal, with vestibulae. Vasa deferentia open into atria basally, true penes present.

Phreodrilus (Antarctodrilus) niger (Beddard, 1894)

Hesperodrilus niger BEDDARD, 1894: 208.

Hesperodrilus albus BEDDARD, 1894: 209.

Hesperodrilus pellucidus BEDDARD, 1894: 210.

Hesperodrilus albus Beddard. BEDDARD, 1895: 256; 1896: 11, Fig. 17-19. MICHAELSEN, 1900: 39.

Hesperodrilus pellucidus Beddard. BEDDARD, 1895: 256: 1896: 14. MICHAELSEN, 1900: 39.

Hesperodrilus niger Beddard. BEDDARD, 1895: 257; 1896: 16. MICHAELSEN, 1900: 38.

Phreodrilus niger (Beddard). MICHAELSEN, 1903: 136; 1924: 207.

Phreodrilus albus (Beddard). MICHAELSEN, 1903: 136; 1924: 207.

Phreodrilus pellucidus (Beddard). MICHAELSEN, 1903: 136.

Phreodrilus africanus GODDARD and MALAN, 1913 b: 242, Pl. XIV, fig. 1, 2, 4.

Phreodrilus albus var *pellucidus* (Beddard). MICHAELSEN, 1916: 5; 1924: 207.

Phreodrilus africanus Goddard and Malan. MICHAELSEN, 1924: 207.

Phreodrilus (Antarctodrilus) niger (Beddard). BRINKHURST, 1965: 378, Fig. 2, 6, 7.

$l=15$-30 mm. Ventral setae one bifid and one more slender simple-pointed seta per bundle, no spermathecal setae. Spermathecal pores dorsal, with vestibulae. True penes present, vasa deferentia open to atria basally.

S. Africa, S. America, Falkland Islands.

Most of the other species with dorsal spermathecal pores lack true penes, most species with true penes have ventral spermathecal pores or have them opening into the penis sac.

A single specimen of *albus* had the gonads displaced one segment in front of the usual position (Beddard, 1894).

SUB-GENUS **Phreodriloides** BENHAM, 1907

SUB-GENERIC TYPE: **Phreodriloides notabilis** BENHAM

Spermathecal pores ventra-lateral, median or ? absent, without vestibulae. No penes. Position of union of vasa deferentia with atria variable.

The sub-genus may be considered to be the stem form from which all the other conditions may have arisen by the development of true penes or pseudo-penes, the development of vestibulae on the spermathecal pores and the shifting of the pores to a dorsal position in many instances.

Phreodrilus (Phreodriloides) notabilis (Benham, 1907)

Phreodriloides notabilis BENHAM, 1907 : 260, Pl. XLVII, fig. 13-17.

Phreodriloides notabilis Benham. MICHAELSEN, 1924 : 208. STEPHENSON, 1930 : 759.

Phreodrilus (Phreodriloides) notabilis (Benham). BRINKHURST, 1965 : 379, Fig. 2.

$l=8$ mm. Ventral setae 1 or 2 simple-pointed, no spermathecal setae. Spermathecae absent. No penes, vasa deferentia open into atria basally.
Australia (1 specimen only).

The muscular sac into which the vas deferens leads is apparently the atrium devoid of its lining layers, possibly having exhausted its function of feeding the sperm. The spermathecae may have been overlooked, as the ducts are often obscure and the ampullae lie several segments behind the pores. Species lacking spermathecae are known to occur in other families however. The type specimen is in the Australian Museum, Sydney.

Phreodrilus (Phreodriloides) kerguelenensis Michaelson, 1903

Phreodrilus kerguelenensis MICHAELSEN, 1903 : 136, Fig. 1-5.

Phreodrilus kerguelenensis Michaelsen. MICHAELSEN, 1905 a : 5.

Hesperodrilus kerguelenensis (Michaelsen). MICHAELSEN, 1924 : 208.

Hesperodrilus kerguelenensis (Michaelsen). STEPHENSON, 1930 : 757.

Phreodrilus (Phreodriloides) kerguelenensis Michaelsen. BRINKHURST, 1965 : 379, Fig. 2.

$l=10$-20 mm, $s=40$-70. Ventral setae simple-pointed, modified spermathecal setae 1 per bundle. Spermathecal pores fused, mid-ventral, no vestibulae. Male pores mid-ventral, vasa deferentia open into atria medially (?).
Kerguelen Island.

The form of the male duct is not adequately described, but it is assumed to be similar to that of *crozetensis*. Apart from this doubt, the possession of spermathecal setae, and the median position of the pores, the species is about the most generalized, representing the least amount of specialization from a presumed ancestral form. The median position of the opening of the vas deferens may be more apparent than real, as the duct may run within the atrial wall for some distance.

Phreodrilus (Phreodriloides) crozetensis Michaelsen, 1905

Phreodrilus crozetensis MICHAELSEN, 1905 a : 5.

Phreodrilus crozetensis Michaelsen. MICHAELSEN, 1905 b : 2, Pl. I, fig. 8; 1916 : 5.

Hesperodrilus crozetensis (Michaelsen). MICHAELSEN, 1924 : 207. CERNOSVITOV, 1934 : 3. STOUT, 1958 : 298.

Phreodrilus (Phreodriloides) crozetensis Michaelsen. BRINKHURST, 1965 : 379, Fig. 2.

$l=15$-18 mm, $s=60$. Ventral setae bifid, same size, no spermathecal setae. Spermathecal pores closely associated with female pores. Small eversible pseudopenes, vasa deferentia open into atria sub-medially.
Crozet Islands, S. Georgia, S. America.

With the development of a penis and possibly a further elaboration of the relationship between the spermathecal and male pores the *Astacopsidrilus* condition is reached.

Phreodrilus (Phreodriloides) zeylanicus (Stephenson, 1913)

Hesperodrilus zeylanicus STEPHENSON, 1913 : 257, Pl. I, fig. 6.

Hesperodrilus zeylanicus Stephenson. MICHAELSEN, 1924 : 208 (sp. inquir.) STEPHENSON, 1930 : 257.

Hesperodrilus zelanicus Stephenson. STOUT, 1958 : 298.

Phreodrilus (Phreodriloides) zeylanicus (Stephenson). BRINKHURST, 1965 : 380, Fig. 2.

$l=8$ mm. Ventral setae one bifid and one thinner simple-pointed, no spermathecal seta. Spermathecal pores ventro-lateral. Vasa deferentia open into atria apically, no penes.
Ceylon.

An additional septum was said to lie behind the male pore, penetrated by the atria and vasa deferentia. A single specimen was described, which requires more detailed description. The setal characteristics alone justify its position in the family.

SUB-GENUS **Insulodrilus** BRINKHURST, 1965

SUB-GENERIC TYPE : **Phreodrilus lacustris** BENHAM

Spermathecal pores ventral, with vestibulae. Penes present, vasa deferentia opening to atria basally.

The three species in this sub-genus are perhaps closer to each other than species in other sub-genera, and may be no more than local forms of a single species.

Phreodrilus (Insulodrilis) lacustris Benham, 1903

Phreodrilus lacustris BENHAM, 1903 : 204.

Phreodrilus lacustris Benham. BENHAM, 1904: 272, Pl. XIII, fig. 1-10.

Hesperodrilus lacustris (Benham). MICHAELSEN, 1924: 208.

Phreodrilus (Insulodrilus) lacustris Benham. BRINKHURST, 1965: 380, Fig. 2.

$l=20$ mm, $s=75$. Ventral setae both with nodulus, 1 minutely bifid, 1 simple-pointed, slightly dissimilar in shape anteriorly, (?) posteriorly both bifid, spermathecal setae paired. Spermathecal pores ventro-lateral, rudimentary vestibulae.
New Zealand, S. Island.

Specimens from Lake Coleridge, N.Z., sent to me by Dr. M. Flain were identifiable as *P. lacustris*, but most of the ventral setae were broken at the top. There was no clear sign of even a minute tooth on any of the setae.

Phreodrilus (Insulodrilus) campbellianus Benham, 1909

Phreodrilus campbellianus BENHAM, 1909 : 256, Pl. X, fig. 1.

Hesperodrilus campbellianus (Benham). MICHAELSEN, 1924: 208.

Phreodrilus (Insulodrilus) campbellianus Benham. BRINKHURST, 1965 : 380, Fig. 2.

$l=18$ mm, $s=60$. Ventral setae anteriorly both nodulate, 1 minutely bifid, 1 slightly thinner simple-pointed, some bundles both bifid, no spermathecal setae. Spermathecal pores anterior in segment, with vestibulae.
Campbell Island.

The male pores were said to be anterior in position (Benham, 1909) but were figured as lying posteriorly, close to 12/13.

Phreodrilus (Insulodrilus) litoralis (Michaelsen, 1924)

Hesperodrilus litoralis MICHAELSEN, 1924: 208, Fig. 1.

Hesperodrilus litoralis Michaelsen. STEPHENSON, 1930: 757. STOUT, 1958: 298.

Phreodrilus (Insulodrilus) litoralis (Michaelsen). BRINKHURST, 1965 : 380, Fig. 2.

$l=18$ mm, $s=55$-65. Ventral setae 1 bifid, 1 simple-pointed, spermathecal setae present. Spermathecal pores with vestibulae close to female pores, male pores median on preceeding segment.
Campbell Island.

The median position of the male pores is not borne out by the figure in the original description.

SUB-GENUS **Astacopsidrilus** GODDARD, 1909

SUB-GENERIC TYPE : **Astacopsidrilus notabilis** GODDARD

Spermathecal pore or pores united with female openings. Penes present. Vasa deferentia opening into bases of atria (where described).

Three very poorly described species have been placed here. The spermathecal pores may, in reality, have large dorsal vestibulae (the ducts turn upwards at the end) or the vestibulae and female pores may have been confused if the former open ventrally, possibly via a common opening to the exterior.

Phreodrilus (Astacopsidrilus) goddardi Brinkhurst, 1965

Astacopsidrilus notabilis GODDARD, 1909 a: 769, Pl. XXIX, XXXI, figs. 1-5, 8, 12, 13-17.

Hesperodrilus notabilis (Goddard). MICHAELSEN, 1924: 208.

Phreodrilus (Astacopsidrilus) goddardi BRINKHURST, 1965: 381, Fig. 2.

(non) *Phreodrilus notabilis* BENHAM, 1907.

l=5·5 mm, *s*=53. Ventral setae 1 bifid, 1 simple pointed, no spermathecal setae Dorsal setae from XXI-XLIII only. Spermathecal duct single, rudimentary (?) but ampullae paired.
Living on crayfish *Astacopsis*.
Australia.

The occluded spermathecal duct may be no more than the normal duct, which has a very narrow lumen in most species.

Phreodrilus (Astacopsidrilus) fusiformis (Goddard, 1909)

Astacopsidrilus fusiformis GODDARD, 1909 a: 781, Pl. XXIX, XXX, figs. 6, 7, 9-11.

Hesperodrilus fusiformis (Goddard). MICHAELSEN, 1924: 208.

Phreodrilus (Astacopsidrilus) fusiformis (Goddard). BRINKHURST, 1965: 381, Fig. 2.

l=2·8 mm, *s*=46. Ventral setae 1 bifid, 1 simple-pointed, no spermathecal setae. Hair setae from IV. One spermathecal ampulla. Body form modified, spindle-shaped.
Living on crayfish *Astacopsis*.
Australia.

This species seems to be no more than a more modified version of *P. (A.) goddardi*, but no final decision can be reached without examining new material.

Phreodrilus (Astacopsidrilus) novus (Jackson, 1931)

Astacopsidrilus novus JACKSON, 1931: 77, Pl. XVI, fig. 1-3.

Phreodrilus (Astacopsidrilus) novus (Jackson). BRINKHURST, 1965: 381, Fig. 2.

l=7-10 mm. Ventral setae may be both simple-pointed or 1 bifid, 1 simple pointed. Spermathecal pores paired, opening ventro-laterally with female pores. ?2 pairs of testes.
    Free living.
    Australia.

SUB-GENUS **Gondwanaedrilus** GODDARD AND MALAN, 1913

SUB-GENERIC TYPE : **Gondwanaedrilus africanus** GODDARD AND MALAN

Spermathecae open dorsally into enlarged penis-sacs (?). Penes present, vasa deferentia open into atria basally.

Phreodrilus (Gondwanaedrilus) africanus (Goddard and Malan, 1913)
Gondwanaedrilus africanus GODDARD AND MALAN, 1913 a : 232, Pl. XI-XIII, fig. 1-9.

Gondwanaedrilus africanus Goddard and Malan. MICHAELSEN, 1924 : 208. STEPHENSON, 1930 : 759.

Phreodrilus (Gondwanaedrilus) africanus (Goddard and Malan). BRINKHURST, 1965 : 382.

(non) *Phreodrilus africanus* GODDARD AND MALAN, 1913 b : (=*P. (A.) niger*)

l=20-22 mm. Ventral setae 1 bifid and 1 thinner simple-pointed, no spermathecal setae.
    S. Africa.

An error in one of the original figures was noted by Brinkhurst (1965) and this considerably clarified the interpretation of the genital pores. It is possible that the spermathecal pores actually open into large dorsal vestibulae, in which case this may merely be a synonym of *P. (A.) niger*.

SUB-GENUS **Schizodrilus** STOUT, 1958

SUB-GENERIC TYPE : **Schizodrilus nothofagi** STOUT

Spermathecae with ventro-lateral pores, ? with vestibule, in IX or in XI and XII. Testes in IX, penes present, vas deferens open into atria medially? Male pores on X.

Phreodrilus (Schizodrilus) nothofagi (Stout, 1958)

Schizodrilus nothofagi STOUT, 1958 : 292, Fig. 7-16.

Phreodrilus (Schizodrilus) nothofagi (Stout). BRINKHURST, 1965 : 382, Fig. 2.

l<10 mm, s=c. 50. Ventral setae 2 bifids of same size, spermathecal setae single. Reproduction by fragmentation.
    New Zealand, N. Island.

The atria of this species are very odd as illustrated in the very diagrammatic illustrations of the original account. According to these, the vas deferens runs straight to the penis with a discrete prostate gland close to the sperm funnel. I have interpreted this as being in reality a vas deferens entering an atrium medially, the distal part of the "vas deferens" of Stout (1958) being the distal part of the atrium, possibly narrowing into an ejaculatory duct, or perhaps even the detached lining of a protrusible penis with the muscle layer removed. The latter seems not a very plausible explanation, as surely the muscle layer would have been noted in the quite extensive description.

The spermathecal setae lie in a sac behind the spermathecal pore and are not, therefore, penial setae (Stout, 1958).

Phreodrilus (*Schizodrilus*) *major* (Stout, 1958)

Schizodrilus major STOUT, 1958: 296, Fig. 12-22.

Phreodrilus (*Schizodrilus*) *major* (Stout). BRINKHURST, 1965: 383, Fig. 2.

$l < 20$ mm, $s = 120$. Ventral setae 1 bifid, 1 simple pointed, same size, no spermathecal setae. Reproduces asexually by fragmentation. Coelomocytes present. Septal glands in V-VIII.
New Zealand, N. Island.

The presence of coelomocytes in this species is unique in the family.

SPECIES INQUIRENDAE

Phreodrilus mauianus Benham, 1903

(=*Phreodrilus mauiensis* Benham, Benham 1904, Michaelsen, 1924)

Tasmaniaedrilus tasmaniaensis Goddard and Malan, 1913 a.

Neither species was sufficiently described to permit a distinction to be made between these and other species with dissimilar ventral setae and gonads in XI and XII.

Two other phreodrilids have been seen in a collection from Lake Pedder, Tasmania. One had a proboscis, one bifid and one simple seta ventrally, three strongly serrate hair setae dorsally, but no mature specimens were found. They are mentioned in order to bring out the fact that these characteristics have now been reported in the family. Another new species from Africa has cuticular penis sheaths. (See Appendix, p. 838).

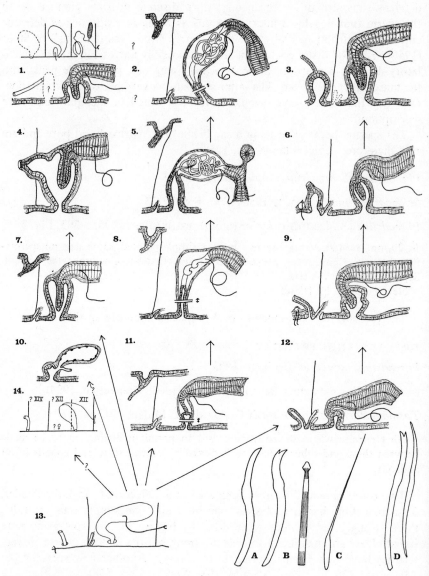

Fig. 9.1. Phreodrilidae.
1-13—possible evolutionary interrelationships of various types of phreodrilid male ducts as illustrated by known species (1. *Shizodrilus* spp., *Astacopsidrilus* spp; 2. *subterraneus*; 3. *campbellianus*; 4. *africanus*; 5. *mauienensis*; 6. *litoralis*; 7. *niger*; 8. *beddardi*; 9. *lacustris*; 10. *notabilis*; 11. *branchiatus*; 12. *crozetensis*; 13. *kerguelenensis*; 14. *zelanicus*).
A to D—setal diversity of *P. major*, A—simple pointed and bifid ventral setae, B—dorsal seta, C—hair seta.
D—typical simple pointed and bifid ventral setae from a single bundle of a phreodrilid.

REFERENCES

BEDDARD, F. E. 1891. Anatomical description of two new genera of aquatic Oligochaeta. *Trans. R. Soc. Edinb.*, **36**, 273.

— 1894. Preliminary notice of South American Tubificidae collected by Dr. Michaelsen, including the description of a branchiate form. *Ann. Mag. nat. Hist.*, (6) **13**, 205.

— 1895. *A monograph of the Order Oligochaeta.* Clarendon Press, Oxford.

— 1896. Naiden, Tubificiden und Terricolen I. Limicole Oligochaeten *Ergebn. Hamb. Magalh. Sammelr.*, **1**, 5.

BENHAM, W. B. 1903. On some new species of aquatic Oligochaeta from New Zealand. *Proc. zool. Soc. Lond.*, **2**, 202.

— 1904. On some new species of the genus *Phreodrilus. Q. Jl microsc. Sci.*, **48**, 271.

— 1907. On the Oligochaeta from the Blue Lake, Mount Kosciusko. *Rec. Aust. Mus.*, **6**, 259.

— 1909. *Report on Oligochaeta of the Subantarctic Islands of New Zealand.* Wellington (ed. C. Chilton). **1** (12), 251.

BRINKHURST, R. O. 1965. A taxonomic revision of the Phreodrilidae. *J. Zool.*, **147**, 363.

CERNOSVITOV, L. 1934. Oligochètes. *Résult. Voyage S.Y. Belgica.* (*Zool.*), **1935**, 1.

GODDARD, E. J. 1909a. Contribution to a further knowledge of Australasian Oligochaeta. Part 1. Description of two species of a new genus of Phreodrilidae. *Proc. Linn. Soc. N.S.W.*, **33**, 768

— 1909b. Contribution to a further knowledge of Australasian Oligochaeta. Part II. Description of a Tasmanian Phreodrilid. *Proc. Linn. Soc. N.S.W.*, **33**, 845.

— and Malan, D. E. 1913a. Contribution to a knowledge of South African Oligochaeta. Part I. On a Phreodrilid from Stellenbosch Mountain. *Trans. R. Soc. S. Afr.*, **3**, 231.

— and Malan, D. E. 1913b. Contribution to a knowledge of South African Oligochaeta. Part II. Description of a new species of *Phreodrilus. Trans. R. Soc. S. Afr.*, **3**, 242.

JACKSON, A. 1931. The Oligochaeta of South-Western Australia. *J. Proc. R. Soc. West Aust.*, **17**, 71.

MICHAELSEN, W. 1900. Oligochaeta. *Das Tierreich*, **10**, 1.

— 1903. Die Oligochäten der Deutschen Tiefsee-Expedition nebst Erörterung der Terricolenfauna oceanischer Inseln, insbesondere der Inseln des subantarktischen Meeres. *Wiss. Ergebn. dt. Tiefsee-Exped. "Valdivia"*, **3**, 131.

— 1905a. Die Oligochaeten der deutschen Südpolar-Expedition 1901-1903 nebst Erörterung der Hypothese über einen früheren grossen, die Südspitzen der Kontinente verbindenen antarktischen Kontinent. *Dt. Südpol.-Exped. 1901-1903.* **9** *Zool.*, I, 5.

— 1905b. Die Oligochaeten der Swedischen Südpolar-Expedition, 1901-03. *Wiss Ergebn. schwed. Südpolarexped.*, **5** (3), 1.

— 1916. Oligochäten aus dem Naturhistorischen Reichsmuseum zu Stockholm. *Ark. Zool.*, **10** (9). 1.

— 1924. Oligochaeten von Neuseeland und den Auckland-Campbell Inseln, nebst einigen anderen Pacifischen Formen. *Vidensk. Meddr. dansk. naturh. Foren.*, **75**, 197.

STEPHENSON, J. 1913. On a collection of Oligochaeta mainly from Ceylon. *Spolia zeylan*, **8**, 251.

— 1930. *The Oligochaeta.* Oxford University Press.

— 1932. Oligochaeta Part I. Microdrili, mainly Enchytraeidae. *"Discovery" Rep.*, **4**, 235.

STOUT, J. D. 1958. Aquatic Oligochaetes occurring in forest litter II. *Trans. R. Soc. N. Z.*, **85**, 289

10

FAMILY OPISTOCYSTIDAE

Type genus: Opistocysta CERNOSVITOV

Prostomium with proboscis. More than two setae per bundle, dorsal bundles with hairs, or hairs and needle-like setae, hairs serrate or non-serrate. One pair of ovaries, one pair of testes. Sperm funnels in testis segment, atria in segment with ovaries, spermathecae in segment behind atria. Coelomocytes present or absent. Meganephridia. Eggs large and yolky. Asexual reproduction by budding.
    N. and S. America, Africa.

GENUS **Opistocysta** CERNOSVITOV, 1936
Fig. 10. 1E-O

Type species: Opistocysta funiculus CORDERO
Defined as the family.

These small, obscure worms are known from very few localities. The nomenclature of the single genus is very badly confused, but has been reduced to a more orderly arrangement by W. J. Harman. The separation of species is based on rather narrow differences, but this is necessary at this stage of our knowledge. The inclusion of all species in a single genus is also a provisional situation.

No attempt will be made to present a key, as knowledge of the various species does not allow this to be done with any confidence.

Opistocysta funiculus Cordero, 1948
Fig. 10. 1E, F, H-O

---

* I am indebted to Professor W. J. Harman for access to his revision in Mss. form.

Opistocysta flagellum (Leidy). CERNOSVITOV, 1936: 75, Fig. 1-19. MARCUS, 1944: 69, Fig. 60, 61. MARCUS, E., 1947: 9.

Opisthocysta funiculus CORDERO, 1948: 3, Fig. 1-5.

Opistocysta funiculus Cordero. MARCUS, 1., 1949: 2 (?)BRINKHURST AND COOK: 21.

l=5-12 mm, *s*=40-106. Eyeless. Prostomium up to 785 μ. Dorsal setae from II, 2-3 serrate hair setae, 150-160 μ, maximum length 560-640 μ in mid body, 2-3 needles, 90-105 μ, smooth, straight with nodulus. Ventral bundles with 3-5 bifid setae, the upper tooth a little thinner and longer than the lower, setal lengths being 120 μ in II, 104 μ in III-V, 116-120 μ median segments, 128-130 μ posteriorly. Testes in XXI, ovaries in XXII, sperm funnels on 21/22, atria oval with diffuse prostates, peculiar eversible form of penes. Male pores on XXII, ? female pores in 22/23, spermathecal pores on anteriormost end of XXIII, ampullae separated from ducts within XXIII. Sperm sac from 21/22 into XXII, egg sac on 22/23 to XXVI. Coelomocytes present. Posterior end with two lateral processes (? gills) and one thin median process variously developed, processes ciliated. The dorsal pulsatile blood vessel divides beneath the dorsal ganglia, the two branches running ventrally and then posteriorly to unite in III, forming the non-pulsatile ventral vessel. In V the dorsal vessel runs into the left side of the gut, the commissurals in IV and V being the only ones running from dorsal to ventral position. A right lateral sinus and a dorsal sinus are present on the gut in median segments.

Argentina, Brazil, ? Africa.

Opistocysta corderoi Harman, 1970
Fig. 10. 1G

Opisthocysta flagellum (Leidy). CORDERO, 1948: 3, Fig. 1-5.

Opistocysta corderoi HARMAN, 1970

l=10-12 mm, *s*=56-76. Prostomium with proboscis *c.* 660 μ long. Dorsal setae from II, 2 non-serrate hair setae, 72 μ in III, 360 μ in XII and XIII, may be of unequal length. Ventral setae 4-5 per bundle, about 80 μ, upper tooth thinner than lower. Clitellum at least from XV-XVI. Testes in XIV or XV, ovaries in XV or XVI, spermathecae in XVI or XVII. Prostate glands diffuse on narrow portion of atria and/or vasa deferentia, main body of atria naked, narrowing to eversible pseudopenes. Spermathecae narrow, tubular. Postero-lateral gills 900 μ long, 150 μ broad at base, median process 190 μ by 72 μ.
Uruguay.

As suggested by Harman (1970), it may prove necessary to separate this species from *O. funiculus* at the generic level.

Opistocysta serrata Harman, 1970

Pristina flagellum Leidy. MICHAELSON, 1905: 350.

Opistocysta flagellum (Leidy). CORDERO, 1948: 1 (partim).

Opistocysta serrata HARMAN, 1970

$l=2.2$-10 mm, $s=17$-76. Proboscis present. Dorsal bundles from II with serrate hair setae, up to 0.25 mm long. Ventral setae bifid, teeth equally long, the upper thinner than the lower, 150 μ long, 3-5 per bundle. Two long lateral gills and a shorter median process posteriorly.

Paraguay.

Harman (1970) suggested that the species should be regarded as distinct from *O. corderoi*, from which it clearly differs in various details. Very little is known about this species, however, and the genital arrangement is unknown.

Opistocysta tribranchiata Harman, 1970

Opistocysta tribranchiata HARMAN, 1970

Proboscis averaging 102 μ (preserved). No eyes. Dorsal setae from II, 1-3 finely serrate hairs, the longest of II being 155 μ, median segment 337 μ, posterior 142 μ. Ventral setae 3-5 per bundle, upper tooth longer than the lower anteriorly, as long as or slightly longer posteriorly, those of II from 66-75 μ long. Two long caudal gills. 160 μ preserved) and one retractile median process.

Mississippi, Louisiana U.S.A.

Holotype—USNM 33094. Paratypes USNM 33093, BM(NH) 1965.24.1., Department of Zoology, Louisiana State U. 255.

The serrations on the hair setae separate this species from *O. corderoi*, together with differences in the lengths of the proboscis, setae and gills, and differences in the ventral setae. The species is shorter than *O. serrata* with shorter ventral setae, longer hair setae. It has no needle setae as found in *O. funiculus,* and the setae differ in size in the two species.

SPECIES INQUIRENDAE

Opistocysta flagellum (Leidy, 1880)

Pristina flagellum LEIDY, 1880: 421.

Pristina flagellum Leidy. SMITH, 1918: 640. PENNAK, 1953: 292, Fig. 186J.

Opistocysta flagellum (Leidy). GOODNIGHT, 1959: 528, Fig. 21. 7.

(non) *Opistocysta flagellum* (Leidy). MICHAELSEN, 1905: 350. CERNOSVITOV, 1936: 75, Fig. 1-19. MARCUS, 1944: 69. MARCUS, E., 1947: 9. CORDERO, 1948:

$l=6$-7 mm, $s=c$. 30. Prostomium prolonged. 3-6 dorsal setae, 4 ventral setae per bundle. Posterior end with lateral gills and median appendage, laterals 0.75 mm long, median one 0.375 mm.

N. America (New Jersey and Pennsylvania).

Both W. J. Harman and the author have sought the type specimens in vain, and a search of the type locality has failed to yield fresh material. In view of the diversity of characters displayed by S. American material and material from the southern states of the U.S.A., it is no longer possible to accept this as a valid species. I agree with Harman (1970) that the best course of action is to remove this contentious name from consideration in the current nomenclature of the family.

The citations by Brinkhurst (1964) and Brinkhurst and Cook (1966) should refer to *O. flagellum* (Leidy) but the descriptions of other species were confused with this.

An unidentifiable species was represented in a collection made by J. Rzoska in the White Nile, Africa (Brinkhurst, 1966). It is possibly identifiable as *O. funiculus*.

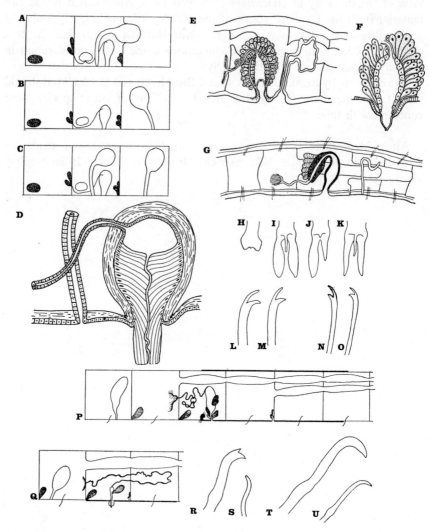

Fig. 10.1. *Dorydrilus*, *Lycodrilus* and *Opistocysta*.

A to D—*Dorydrilus*, A—*D. michaelseni*, B—*D. mirabilis*, C—*D. tetrathecus*, D—*D. michaelseni*.

E to O—*Opistocysta*, E—reproductive organs, *O. funiculus*, F—atrium and penis of the same, G—reproductive organs of *O. corderoi*, H to K—variation in posterior appendages *O. funiculus*, L,M—setae of anterior and posterior ventral bundles, *O. funiculus* (after Cernosvitov), N,O—ventral setae of II and XV, *O. funiculus* (after Marcus).

P to U—*Lycodrilus*, P—reproductive organs *L. parvus*, Q—the same, *L. schizochaetus*, R,S—ventral setae of VIII and XVII of the same, T,U—ventral setae of VIII and XVI, *L. dybowskii*.

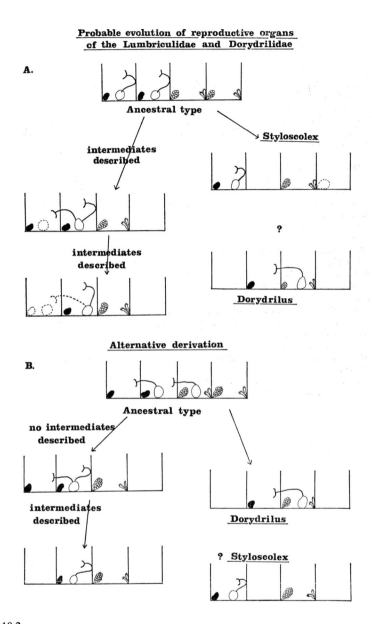

Fig. 10.2.
Possible derivation of *Dorydrilus* and *Styloscolex* from a lumbriculid ancestral form.

REFERENCES

BRINKHURST, R. O. 1964. Studies on the North American Aquatic Oligochaeta. 1. Naididae and Opistocystidae. *Proc. Acad. nat. Sci. Philad.*, **116**, 195.

— 1966. A contribution towards a revision of the Aquatic Oligochaeta of Africa. *Zool. Afr.*, **2**, 131.

BRINKHURST, R. O., AND COOK, D. G. 1966. Studies on the North American Aquatic Oligochaeta. III. Lumbriculidae and additional notes and records of other families. *Proc. Acad. nat. Sci. Philad.*, **118**, 1.

CERNOSVITOV, L. 1936. Oligochaeten au Südamerika, Systematische Stellung der *Pristina flagellum* Leidy. *Zool. Anz.*, **113**, 75.

CORDERO, E. H. 1948. Zur Kenntnis der Gattung *Opisthocysta* Cern. (Archiolgochaeta). *Comun. zool. Mus. Hist. nat. Montev*, **2** (50), 1.

GOODNIGHT, C. J. 1963. Oligochaeta. In Ward and Whipple. *Freshwater Biology.* Ed. Edmondson; New York, 2nd ed. 522.

HARMAN, W. J. 1970. Revision of the family Opistocystidae (Oligochaeta). *Trans. Amer. microsc. Soc.*, **88**, 472 (1969 year).

LEIDY, J. 1880. Notice on some Aquatic worms of the family Naides. *Amer. Nat.*, **14**, 421.

MARCUS, E. 1944. Sobre Oligochaeta limnicos do Brasil. *Bolm. Fac. Filos. Ciêne. Univ. S. Paulo*, **43**, *Zool*, **8**, 5.

MARCUS, E. du BOIS-REYMOND. 1947. Naidids and Tubificids from Brazil. *Comun. zool. Mus. Hist. nat. Montev.*, **2** (44), 1.

— 1949. Further notes on Naidids and Tubificids from Brazil. *Comun. zool. Mus. Hist. nat. Montev.*, **3** (51), 1.

MICHAELSEN, W. 1905. Zur Kenntnis der Naididen. *Zoologica*, **44**, 350.

PENNAK, R. W. 1953. *Fresh-water Invertebrates of the United States.* Ronald, New York.

SMITH, F. 1918. Aquatic earth-worms and other bristle bearing worms (Chaetopoda). In Ward and Whipple, *Freshwater Biology*, New York, 1st ed. 632.

11

FAMILY DORYDRILIDAE

D. G. COOK

Type genus: Dorydrilus PIGUET

Small unpigmented worms, with rounded prostomium. Setae four bundles, beginning in II, thin, sigmoid, single pointed. One pair male pores on X. One or two pairs spermathecal pores, on X, anterior to male pores but behind ventral setae, or on the post-atrial segment, or on both.

Blind posterior lateral blood vessels absent. Pharynx eversible. Pharyngeal glands present. One pair testes and one pair male funnels in IX. One pair vasa deferentia join one pair large, muscular atria, in X. One pair ovaries in X. One or two pairs spermathecae, in the atrial or the post-atrial segment, or in both.

Europe, ? Lake Baikal.

Dorydrilus, described as, and until the present considered as, a member of the Lumbriculidae, has been the subject of much misunderstanding. In the original description, Piguet wrongly assumed that two pairs of testes were present and that the ovaries were in the usual post-atrial position. Hrabe (1936), however, clarified this situation (see family definition) by redescribing Piguet's original material and at the same time describing *D. mirabilis* from among the same collection. To accommodate this species, Hrabe erected the sub-genus *Piguetia*. In 1933 Michaelsen described *Dorydrilus wiardi* and put this in a third sub-genus, *Guestphalinus*, which, however, has the lumbriculid arrangement of the male genitalia.

Yamaguchi (1953) rightly pointed out that *Dorydrilus* (sensu Piguet, emmended Hrabe) and *Dorydrilus* (sensu Michaelsen) were homonymous. He proposed, contrary to the laws of precedence, and presumably because he misinterpreted Hrabe (1936), that Michaelsen's view of *Dorydrilus* was correct, and that *Dorydrilus* (sensu Piguet and Hrabe) should be called *Piguetia*, and suggested that this was not a member of the Lumbriculidae.

With the description of *D. tetrathecus*, Hrabe (1960) himself states that there was no longer any justification for the sub-genera *Dorydrilus* and *Piguetia*, and elevated Michaelsen's sub-genus *Guestphalinus* to generic rank.

Hrabe (1936) argued that *Dorydrilus* is close to the ancestral lumbriculid, and that a *Lamprodrilus* type of organization may be derived directly from it by the forward movement of the atria one segment. He supported this contention by pointing out the fact that in many species of Lumbriculidae, e.g. *Lamprodrilus, Trichodrilus, Stylodrilus,* the posterior pair of vasa deferentia penetrate into the post-atrial segment before joining the atria. Although this view may be valid, to recognize *Dorydrilus* as a lumbriculid seems to be an extreme case of vertical classification and does not sufficiently reflect the fundamental morphological differences between the arrangement of the genitalia in *Dorydrilus*, and the lumbriculid type of arrangement (Fig. 10. 2). Hence in this account, *Dorydrilus* is placed in a separate family, the Dorydrilidae, characterized by the tubificid arrangement of the male genitalia and lumbriculid type setae and spermathecal position. The strong resemblance of *Dorydrilus,* to the lumbriculid genus *Trichodrilus* in the form and arrangement of the setae and in general similarity of body form (prostomium and biannulations) is, in the author's opinion, the result of convergent evolution of these two groups which are found in very similar habitats.

One genus, containing three species, is recognized, and the dubious genus *Lycodrilus* is placed here incertae sedis.

GENUS **Dorydrilus** PIGUET, 1913

Fig. 10. 1-2

Type species: Dorydrilus michaelseni PIGUET

Defined as the family.

  Europe.

Dorydrilus

1.	One pair spermathecal pores on X or XI	2
—	Two pairs spermathecal pores on X and XI	*D. tetrathecus*
2.	One pair spermathecal pores in same segment as male pores, situated in front of male pores, behind ventral setae	*D. michaelseni*
—	One pair spermathecal pores in segment behind male pores	*D. mirabilis*

Dorydrilus michaelseni Piguet, 1913
Fig. 10. 1A, D

Dorydrilus michaelseni PIGUET, 1913: 141, Fig. 12.

Dorydrilus michaelseni Piguet. PIGUET AND BRETSCHER, 1913: 158, Fig. 39b.
MICHAELSEN, 1933: 8. HRABE, 1936: 3, Fig. 1, 2; 1960: 246, Fig. 15, 16.
YAMAGUCHI, 1953: 312. BRINKHURST AND KENNEDY, 1962: 185. COOK,
1967: 353, Fig. 1a, 8; 1968: 275, Fig. 3a.

Dorydrilus (*Dorydrilus*) *michaelseni* Piguet. HRABE, 1936: 3, Fig. 1, 2.

$l=8-15$ mm, $w=0.25-0.5$ mm, $s=53-65$. Prostomium rounded, as long as it is
broad at peristomium. Secondary annuli present from IV. Clitellum developed
on X and XI. Setae single pointed, 93 μ long; ratio lengths, pt.-node/total$=0.45$.
Male pores paired on X, behind one pair spermathecal pores also on X, both
behind ventral setae.

Pharyngeal glands present in IV to VII. Chlorogogen cells begin in VI or VII.
Posterior lateral blood vessels absent. One pair testes in IX. Vasa deferentia join
atria apically. Atria spherical to pear-shaped, in X, 100-200 μ long, 90-180 μ
wide, with muscle layer 10-45 μ thick. Penis large, internal, usually protruded.
Ovaries in X. One pair spermathecae, in X, with spherical to elongate ampullae
and discrete, often twisted ducts, 250-400 μ long, 25-35 μ diameter.

Switzerland, Austria, British Isles.

Dorydrilus mirabilis Hrabe, 1936
Fig. 10. 1B

Trichodrilus pragensis Vejdovsky. PIGUET AND BRETSCHER, 1913: 157.

Dorydrilus mirabilis HRABE, 1936: 3, Fig. 3-7.

Dorydrilus (*Piguetia*) *mirabilis* HRABE, 1936: 3, Fig. 3-7.

Dorydrilus mirabilis Hrabe. HRABE, 1937: 2; 1938: 74; 1960: 259. YAMA-
GUCHI, 1953: 312. COOK, 1968: 275, Fig. 3b.

Piguetea mirabilis (Hrabe). YAMAGUCHI, 1953: 312.

$l=13-17$ mm, $w=0.37-0.43$ mm, $s=60$. Prostomium conical. Secondary annuli
present. Clitellum not well developed. Setae single pointed, 75-117 μ long. Male
pores paired on X. One pair spermathecal pores on XI.

Pharyngeal glands present in IV to VII. Posterior lateral blood vessels absent.
One pair testes in IX. Vasa deferentia join atria apically. Atria pear-shaped in
X, 145 to 215 μ long, 103-150 μ wide, with muscle layer 22 μ thick. Penes internal.
Ovaries in X. One pair spermathecae in XI, with ovoid ampullae, 200 μ long,
100 μ wide, and discrete ducts, 107 μ long.

Switzerland.

Dorydrilus tetrathecus Hrabe, 1960
Fig. 10. 1C

Dorydrilus tetrathecus HRABE, 1960: 259, Fig. 17-19.

Dorydrilus tetrathecus Hrabe. COOK, 1968: 275, Fig. 3c.

$l=15$ mm, $w=0.33$ mm. Prostomium rounded. Secondary annuli present from III. Clitellum not developed. Setae single pointed, 90-143 μ long. Male pores paired on X. Two pairs spermathecal pores on X and XI; first pair in front of male pores, but both behind ventral setae.

In IX, epidermis+circular muscle=8 μ thick, longitudinal musce layer, up to 26 μ thick. Pharyngeal glands present in IV to VI. Chlorogogen cells cover gut from VII. Posterior lateral blood vessels absent. Testes in IX. Vasa deferentia penetrate atria apically. Atria pear-shaped, in X, 155-167 μ long, 102-110 μ wide, with muscle layer up to 23 μ thick. Penes large, internal. Ovaries in X. Two pairs spermathecae in X and XI, with large ovoid ampullae and long discrete ducts.

Germany.

GENUS INCERTA FAMILIA

GENUS **Lycodrilus** GRUBE, 1873

Type species: Lycodrilus dybowskii GRUBE

Setae paired or single, simple pointed, rarely bifid, four or eight per segment, rarely more. Testes in X, ovaries in XI, male pore in XI, female pore in 11/12. Vasa deferentia moderately long, small atria with stalked prostates, penes present or absent. Spermathecae in VIII or X, spermatophores present. Pharynx with pharyngeal glands. No contractile blood vessel in VIII.

Lake Baikal.

Five species of *Lycodrilus* have been described, but all are poorly known and the descriptions need revision. A sixth species, *kraepelini*, based on some inadequately described material from the Congo, was attributed to this genus by Michaelsen (1914). Its true systematic position cannot be determined.

The genus was merged with *Limnodrilus* by Michaelsen (1926), but the combination of the setal characters and genital characters is almost unique. Cekanovskaya (1962) recognized this, and placed the genus in a new family, the Lycodrilidae, attributed to Svetlov. The family is not formally defined in Cekanovskaya's work, and the author has not found any subsequent publication by Svetlov. As the family Lycodrilidae is so poorly defined, it would seem wiser to place this assemblage of species in the newly defined Dorydrilidae, erected for the single genus *Dorydrilus* which also has setae like those of many lumbriculids but a genital arrangement like that of the tubificids. The genus *Lycodrilus* differs from *Dorydrilus* in the precise segments occupied by the gonads and in having the spermathecae in segments in front of that bearing the male pores, not in the male pore segment or the one behind it. The form of the atrium and penis in *Lycodrilus* appears to be different to that in *Dorydrilus* as well.

Lycodrilus

1.	More than 2 setae per bundle in most anterior segments	*L. grubei*
—	Setae single or paired, 4 or 8 per segment	2
2.	Anterior ventral setae greatly enlarged	3
—	No enlarged setae	4
3.	Enlarged setae with bifid tips	*L. schizochaetus*
—	Enlarged setae with simple pointed tips	*L. dybowskii*
4.	Single dorsal setae with thin, hair-like tips	*L. phreodriloides*
—	Paired dorsal setae sigmoid	*L. parvus*

Lycodrilus dybowskii Grube, 1873
Fig. 10. 1, T, U.

Lycodrilus dybowskii GRUBE, 1873: 67.

Rhynchelmis dybowskii (Grube). VAILLANT, 1889: 221.

Lycodrilus dybowskii Grube. MICHAELSEN, 1900: 65; 1901: 183, Pl. 1, fig. 7; 1905: 15. CEKANOVSKAYA, 1962: 385, Fig. 245. BRINKHURST, 1963: 76.

Limnodrilus dybowskii (Grube). MICHAELSEN, 1926: 154.

$l=75$-130 mm. Setae four pairs per segment, rarely single in anterior segments, dorsal setae fine, posteriorly especially so, ventral setae enlarged anteriorly, with hooked ends, smaller setae sigmoid.
Lake Baikal.

No details of the reproductive system are known.

Lycodrilus schizochaetus Michaelsen, 1901
Fig. 10. 1, Q-S

Lycodrilus dybowskii var *schizochaetus* MICHAELSEN, 1901: 184, Pl. I, fig. 6.

Lycodrilus schizochaetus Michaelsen. MICHAELSEN, 1903: 49.

Lycodrilus schizochaetus Michaelsen. MICHAELSEN, 1905: 12. CEKANOVSKAYA, 1962: 387, Fig. 247. BRINKHURST, 1963: 76.

Limnodrilus schizochaetus (Michaelsen). MICHAELSEN, 1926: 154.

$l=38$ mm, $s=150$. Ventral setae of II-III, dorsal setae of II-VII paired, other setae single. Anterior ventral setae enlarged, bifid, with the upper tooth half as long as the lower, smaller setae sigmoid, simple-pointed or indistinctly bifid. Male pores in XI, with penes; spermathecal pores in X, female pores 11-12. Vasa deferentia thin, then thick, joining small atria with long prostate glands, long ejaculatory ducts.

Lake Baikal.

Lycodrilus phreodriloides Michaelsen, 1905

Lycodrilus phreodriloides MICHAELSEN, 1905: 16.

Lycodrilus phreodriloides Michaelsen. CEKANOVSKAYA, 1962: 386. BRINK-HURST, 1963: 76.

$l=2.5$-3.0 mm, $s=21$-24. Ventral setae paired, strongly bent distally, simple pointed. Dorsal setae with hair-like tips. Vasa deferentia fairly long, coiled, enter elongate atria proximally, prostate gland elongate, ejaculatory duct short, narrow.

Lake Baikal.

Lycodrilus parvus Michaelsen, 1905
Fig. 10. 1, P.

Lycodrilus parvus MICHAELSEN, 1905: 18, Fig. 3.

Lycodrilus parvus Michaelsen. CEKANOVSKAYA, 1962: 386, Fig. 246. BRINK-HURST, 1963: 76.

$l=12$-22 mm, $s=65$-75. Setae strongly sigmoid, paired, simple-pointed throughout. Male pore and spermathecal pore median, male pore on X, spermathecal on VIII, female pores in 11/12. Atria long, slender, pear-shaped, prostates simple.

Lake Baikal.

Lycodrilus grubei Michaelsen, 1905

Lycodrilus grubei MICHAELSEN, 1905: 20.

Lycodrilus grubei Michaelsen. CEKANOVSKAYA, 1962: 388. BRINKHURST, 1963: 76.

$l=25$ mm, $s=86$. Up to six setae per bundle in II, less in III-X, single behind genital region, all sigmoid with blunt tips. Spermathecal setae modified, with hollow tips, spermathecae in X with lateral pores. Male pores in ventral setal line of XI, penes elongate cones. Vasa deferentia fairly long, coiled, enter pear-shaped atria distally, ejaculatory ducts present.

Lake Baikal.

NOMEN DUBIUM

Lycodrilus kraepelini Michaelsen, 1914: 81; 1932: 5, Fig. 1.

Pelodrilus kraepelini (Michaelsen). HRABE, 1931: 52; 1933: 225, Fig. 1.

The description of this Congolese worm is totally inadequate (Brinkhurst, 1966. p. 51).

REFERENCES

BRINKHURST, R. O. 1963. Taxonomical studies on the Tubificidae (Annelida, Oligochaeta). *Internat. Rev. ges Hydrobiol. Syst. Beih.*, **2**, 1.
— 1966. A taxonomic revision of the family Haplotaxidae (Oligochaeta). *J. Zool. Lond.*, **150**, 29.
BRINKHURST, R. O. AND COOK, D. G. 1966. Studies on the North American aquatic Oligochaeta. III. Lubcriculidae, etc. *Proc. Acad, nat. Sci. Philad,* **118**, 1.
BRINKHURST, R. O., AND KENNEDY, C. R. 1962. A report on a collection of Aquatic Oligochaeta deposited at the University of Neuchatel by Dr. E. Piguet. *Bull. Soc. neuchat. Sci. nat.*, **85**, 183.
CEKANOVSKAYA, O. V. 1962. The aquatic oligochaete fauna of the U.S.S.R. *Tabl. Anal. Faune USSR*, **78**, 1 (in Russian).
COOK, D. G. 1967. Studies on the Lumbriculidae (Oligochaeta) in Britain. *J. Zool. Lond.*, **153**, 353.
— 1968. The genera of the family Lumbriculidae and the genus *Dorydrilus* (Annelida, Oligochaeta). *J. zool. Lond.*, **156**, 273.
GRUBE, E. 1873. Über einige bischer noch unbekannte Bewohner des Baikal Sees. *Jber. Schles. Ges. vaterl. Kult.*, **50**, 66.
HRABE, S. 1931. Die Oligochaeta aus den Seen Ochrida und Prespa. *Zool. Jb. (Syst.),* **61**, 1.
— 1933. Zur Kenntnis der *Pelodrilus kraepelini* (Michaelsen). *Zool. Anz.*, **104**, 225.
— 1936. Über *Dorydrilus (Piguetia) mirabilis* n. sub. gen. n. sp. aus einem Sodbrunnen in der Umgebung von Basel sowie über *Dorydrilus (Dorydrilus) Michaelseni* Pig. und *Bichaeta sanguinea* Bret. *Spisy. vydav. přír. Fak. Univ. Brne*, **227**, 3.
— 1937. Contribution a l'étude du genre *Trichodrilus* (Olig. Lumbriculidae) et description de deux espèces nouvelles. *Bull. Mus. r. Hist. nat. Belg.*, **13** (32), 1.
— 1938. *Trichodrilus moravicus* und *claparedei*, neue Lumbriculiden *Zool. Anz.*, **15**, 73.
1960. Oligochaeta limicola from the collection of Dr. S. Husmann. *Spisy přír. Fak. Univ. Brne*, **415**, 245.
MICHAELSEN, W. 1900. Oligochaeta. *Tierreich*, **10**, 1.
— 1901. Oligochaeten der Zoologischen Museen zu St. Petersburg und Kiev. *Izv. imp. Akad. Nauk.*, **15**, 145.
— 1903. *Die Geographische Verbreitung der Oligochaeta.* Berlin.
— 1905. *Die Oligochaeten des Baikal Sees. Wissenschaftliche Ergebnisse einer Zoologischen Expedition nach dem Baikal-See unter Leitung des Profs. Alexis Korotneff in den Jahren 1900-1902. Kiev und Berlin.*
— 1914. Oligochaeten vom tropischen Afrika *Mitt. naturh. Mus. Hamb.*, **31**, 81.
— 1926. Zur Kenntnis der Oligochaeten des Baikal-Sees. *Russk. gidrobiol. Zh.*, **5**, 153.
— 1932. Ein neuer *Phreoryctes* von der Tropinsel Poeloe Berhala. *Miscnea Zool. sumatr.*, **71**, 1.
— 1933. Uber Höhlen Oligochäten. *Mitt. Höhl-u, Kartforsch.*, **1933**, 1.
PIGUET, E. 1913. Notes sur les Oligochaetes. *Revue suisse Zool.*, **21**, 111.
— and BRETSCHER, K. 1913. *Oligochètes in Catalogue des Invertébrés Suisse*, **7**, 1.
VAILLANT, L. 1889. *Histoire naturelle des annelés marins et d'eau douce. 3 Lombriciniens, Hirudiniens, Bdellomorphes, Teretulariens, et Planariens.* Paris.
YAMAGUCHI, H. 1953. Studies on the aquatic Oligochaeta of Japan VI. A systematic report with some remarks on the classification and phylogeny of the Oligochaeta. *J. Fac. Sci. Hokkaido. Univ.*, **11**, 277.

12

FAMILY ENCHYTRAEIDAE

Type genus: Enchytraeus HENLE

Prostomium without a proboscis. No eyes. Setae in four bundles per segment from II, dorsal and ventral setae usually the same, sigmoid or straight, simple pointed (bifid in *Propappus*) sometimes reduced in number or absent (*Achaeta*). Clitellum in a few segments in the region of the gonads. Testes and ovaries, one pair of each, in XI and XII. Spermathecae paired in V with pores in 4/5. Sperm funnels in XI long, glandular, vasa deferentia may be broad behind funnel, may be coiled in XII, atria mostly absent, glands and penial bulbs present or absent. Nephridia often with anteseptal, duct often in (? syncitial) interstitial cell mass.
   Cosmopolitan.

   The family has been partially reviewed by Nielsen and Christensen (1959, 1961, 1963) but even the following somewhat superficial review has indicated that the rulings of the I.C.Z.N. were not applied in that work. No attempt has been made here to deal with species in this family as few may be described as primarily aquatic (other than the freshwater species of *Propappus* and the marine species of several genera, including *Grania*). Descriptions of the European species may be found in reviews cited, together with lists of non-European species and doubtful entities. The senior authors have no experience with species of this family, and none of the current authorities on the Enchytraeidae were in a position to review the group for this publication. Hence most of this section is derived more or less directly from published reviews.

GENUS **Propappus** MICHAELSEN, 1905

Type species: Propappus glandulosus MICHAELSEN

Setae sigmoid, furcate; behind each bundle a setal gland. Head pore and dorsal pores absent. Sudden transition between oesophagus and intestine. Dorsal vessel arises at anterior end of intestine. Peptonephridia and intestinal diverticula absent. Anteseptale of nephridia consisting of funnel only; postseptale with reduced interstitial tissue. Copulation glands absent. Male pore in front of setal zone XII; spermathecal opening in front of setae IV. Sperm funnel completely or partly behind disseptimentum XI/XII, much thinner than the ectal part. Sperm duct with distal atrium-like enlargement. Epidermal fold and prostatic glands (accessory glands) absent. Spermatheca free, not communicating with the oesophagus, without diverticula and reaching backwards through a few segments. Chromosome numbers unknown.

In freshwater.

GENUS **Cernosvitoviella** NIELSEN AND CHRISTENSEN, 1959

Type species: Marionina atrata BRETSCHER

Setae sigmoid, with nodulus. Head pore on prostomium or at 0/1. Dorsal pores absent. Brain incised posteriorly. Gradual transition between oesophagus and intestine. Peptonephridia and oesophageal appendages absent. Intestinal diverticulae absent. Dorsal vessel arising in or behind the clitellar region. Blood colourless or coloured. Two or three pairs of primary and secondary septal glands present. Interstitial tissue of nephridia much reduced; anteseptale consisting of funnel only; efferent duct of nephridium arising terminally. Seminal vesicle present or absent. Vas deferens short, stout, and often enlarged ectally; no true penial bulb. Male pore with small atrial glands ectally. Spermatheca simple and not attached to the oesophagus. Ectal duct and aperture without glands.

Small, active and very contractile worms mostly occurring in limnic or to some extent in marine surroundings, or in permanently wet soil.

GENUS **Analycus** LEVINSON, 1883

(=*Mesenchytraeus* Eisen)

Type species: Analycus armatus LEVINSEN

Head pore at the apex of O. Dorsal pores absent. Setae sigmoid, with nodulus. Brain rather short slightly incised posteriorly and with stout circumpharyngeal connectives. Gradual transition between oesophagus and intestine. Oesophagus and intestine without diverticula or appendages. Peptonephridia absent. Dorsal vessel originating in XII or behind the clitellar region (appr. XVI-XX). Nephridia practically without interstitial tissue, hence the canal is much more conspicuous than in any other genus of the Enchytraeidae; anteseptale consisting of nephrostome and a thin stalk connecting with the posteptale which latter is bilobed, with a large dorsal and smaller ventral lobe; the long and narrow efferent duct originates between the lobes. Coelomocytes of one type only, colourless or, in some species strongly yellow. Egg sac present.

Considerable nomenclatural confusion was found to exist when an attempt was made to discover the identity of the type species of the genus known as *Mesenchytraeus*. According to the Law of Priority only three species were eligible, having been included in the original assemblage described by Eisen (1878). These species (*M. falciformis, M. mirabilis* and *M. primaevus*) were included in reviews of the Enchytraeidae by Michaelsen (1889, 1900) and Cernosvitov (1937), but were regarded as species dubiae by Nielsen and Christensen (1959). If this position is to be maintained, the generic name *Mesenchytraeus* cannot be utilized. Apstein (1915) sought to maintain the generic name as a Nomen Conservandum, incorrectly electing *M. beumeri* (Michaelsen, 1886) as the type species (the basis for the oligochaete section of this work apparently originating with Michaelsen himself however), and this suggests that the name was considered to require validification even then.

The next generic name to be considered is *Neoenchytraeus* Eisen, 1878, the type species of which is *N. fenestratus*, but this species was considered to be a species dubia by Cernosvitov (1937). Following this, Levinsen (1883) defined the genus *Analycus* with three species (*glandulosus, armatus, flavus*). All three are recognized as European species by Neilsen and Christensen.

Hence the correct generic name should be *Analycus* Levinsen, and *A. armatus* is available for designation as the type species. This species is selected because it is one of the three listed by Levinsen in his original designation of the genus, and it was not regarded as a sp. dub. by Cernosvitov (1937) as was *A. glandulosus*.

GENUS **Cognettia** NIELSEN AND CHRISTENSEN, 1959

Type species: Pachydrilus sphagnetorum VEJDOVSKY

Setae sigmoid, without nodulus and usually few in number. Head pore at O.1. Dorsal pores absent. Brain incised posteriorly. Gradual transition between oesophagus and intestine. Peptonephridia and oesophageal appendages absent. Dorsal vessel originating in or behind the clitellar region. Blood colourless or coloured. Often more than three pairs of primary septal glands; secondary septal glands may be present. Interstitial tissue of nephridia well developed; anteseptale consisting of funnel only; the slender efferent duct usually arising antero-ventrally. Sperm duct usually long and narrow, confined to XII. Penial bulb present. Testes compact. Testes and ovaries may be displaced forwards by up to four segments. Seminal vesicle present or absent. Atrium and atrial glands absent. Spermatheca simple and not attached to the oesophagus. Ectal duct usually with a single, compact gland at the ectal orifice.

Small to medium-sized, usually very slender, worms which mostly occur in limnic surroundings or in permanently wet soil.

GENUS **Hemienchytraeus** CERNOSVITOV, 1935

Type species: Enchytraeus stephensoni COGNETTI

Setae straight and few in number (2 in all species known so far). Head pore on prostomium. Dorsal pores absent. Brain slightly incised or rounded posteriorly, circumpharyngeal connectives powerful. Transition between oesophagus and intestine gradual or more or less sudden. Unpaired dorsal salivary gland. No other oesophageal or intestinal diverticulae. Dorsal vessel arising in the clitellar region (XII-XIII). Blood colourless. Three pairs of primary septal glands. Interstitial tissue of nephridia well developed; anteseptale large, consisting of funnel and coils of the nephridial canal; origin of efferent duct variable in position. Sperm duct long and often coiled into a spiral. Penial bulb compact but often much reduced in size. Spermatheca simple, and not communicating with the oesophagus.

GENUS **Achaeta** VEJDOVSKY, 1877

Type species: Achaeta eiseni VEJDOVSKY

Setae absent but ovoid or pear-shaped transparent bodies often occupy the position of some or all of the missing setal bundles (setal follicles, Borstendrüsen). Head pore on O. Dorsal pores absent. The transition between oesophagus and intestine more or less sudden. Intestinal diverticula absent; peptonephridia absent but oesophageal appendages usually present, (some of which have been called peptonephridia). Dorsal vessel originates in VI, VII or VIII. Coelomocytes of one kind. Nephridia with well developed interstitial tissue. Sperm sac compact when present. Egg sac absent. The spermathecae without diverticula and not communicating with the oesophagus.

GENUS **Enchytronia** NIELSEN AND CHRISTENSEN, 1959

Type species: Enchytronia parva NIELSEN AND CHRISTENSEN

Setae straight without nodulus. Head pore at 0/1. Dorsal pores absent. Brain longer than wide, convex anteriorly and deeply indented posteriorly, lateral margins converging towards the anterior end. Paired peptonephridia absent. Transition between oesophagus and intestine at VI/VII. A pair of lateral intestinal diverticula originating at the transition and extending forwards into VI; the diverticula communicating separately with the intestine through a canal which gives off finer branches into the body of the diverticula. Dorsal vessel arising in XIII. Blood colourless. Nephridia with well developed interstitial tissue; consisting of anteseptale with indistinctly demarcated nephrostome and coils of the nephridial canal, and an elongate postseptale with the efferent duct arising postero-ventrally. Seminal vesicle absent. Seminal funnel cylindrical. Vas deferens long and thin. Spermathecae unite entally and seem to be attached to the dorsal wall of the oesophagus; no open communication with the oesophagus has been observed. Genital organs in normal position.

Terrestrial.

GENUS **Stercutus** MICHAELSEN, 1888

Type species: Stercutus MICHAELSEN

Setae sigmoid, without nodulus. Head pore apparently absent. Dorsal pores absent. Oesophagus and intestine merging gradually. Intestinal diverticula absent. Peptonephridia and oesophageal appendages absent. Anteclitellar origin of dorsal vessel. Blood colourless. Interstitial tissue of nephridia well developed. Seminal funnel small, funnel-shaped. Spermatheca simple, without diverticula, and not communicating with the oesophagus. In the European species the posterior part of the intestine is reduced to a solid strand of tissue, and the anus is closed.

GENUS **Guaranidrilus** CERNOSVITOV, 1937

Type species: Guaranidrilus glandulosus CERNOSVITOV

Setae straight, without nodulus, two per bundle. Head pore near apex of O. Dorsal pores absent. Sudden transition between oesophagus and intestine; one pair of intestinal diverticula at oesophago-intestinal transition. Paired oesophageal appendages in VI (peptonephridia?). Dorsal vessel originating in front of or in the clitellar region. Blood colourless. Interstitial tissue of nephridia well developed; anteseptale large, consisting of funnel and coils of the nephridial canal. Seminal vesicle present. Funnel cylindrical; vas deferens long, thin and wound into a spiral. Penial bulb compact. Spermatheca simple, not communicating with the oesophagus.

GENUS **Aspidodrilus** BAYLIS, 1914

Type species: Aspidodrilus kesalli BAYLIS

Setae straight, without nodulus, two per bundle. Body cylindrical in front, dorso-ventrally flattened posteriorly. Head pore and dorsal pores absent. Sudden transition between oesophagus and intestine (located in VIII). Dorsal vessel arising in VIII. One pair of intestinal (oesophageal?) diverticula in VII. Other oesophageal appendages absent (?). Interstitial tissue of nephridia well developed; anteseptale large. Seminal funnel cylindrical; vas deferens thin. Penial bulb compact. Spermatheca simple, not communicating with the oesophagus. Epizoic on earthworms.

GENUS **Pelmatodrilus** MOORE, 1943

Type species: Pelmatodrilus planariformis MOORE

Setae straight to slightly sigmoid, no nodulus, dorsal setae smaller than ventrals, single, ventral setae numerous, 16-30 per bundle. Head pores and dorsal pores absent. Body flattened throughout. Transition from oesophagus to intestine abrupt in VIII, oesophagus with an abrupt sigmoid flexure in VII, with diverticulae. Dorsal vessel associated with intestinal plexus throughout. Nephridia with funnel and part of plexus as preseptal, the rest of the plexus postseptal. Seminal funnels on X/XI, preseptal small, postseptae large, broader than the gut in some, vasa deferentia long, coiled. Penial bulbs globular, compact, with gland. Testes and ovaries paired in X and XII, each fusing mid ventrally. Female pores XII/XIII. Spermathecae tubular, without ampullae, glands surround dorso-lateral orifices in IV/V, opening to oesophagus. Epizooic on earthworms.

A single species from Jamaica was described (without illustrations) by Moore (1943). The description seems to have been overlooked by Nielsen and Christensen (1959). The genus is included here with no attempt to justify its existence as a distinct entity.

GENUS **Henlea** MICHAELSEN, 1889

Type species: **Enchytraeus ventriculous** UDEKEM

Setae straight or slightly bent, unequal or equal in size within the bundle; nodulus absent. Head pore at 0/1. Dorsal pores absent. Oesophagus expanding abruptly into the intestine (possibly with the exception of the non-European genus (subgenus) *Hepatogaster* Cejka 1910). Peptonephridia absent but oesophageal appendages present in VI or IV-VI. Intestinal diverticula present or absent. Nephridia with well developed interstitial tissue. The dorsal vessel arises in the anteclitellar region. Blood colourless. The brain is slightly longer than wide and more or less incised posteriorly. Coelomocytes uniform in size and shape (discoid to slightly oval). Seminal vesicle absent. Egg sac absent. Spermathecae simple, usually without glands and diverticula (glands are reported from a few non-European species and diverticula occur in the Siberian species, *H. diverticulata* Cejka and possibly in the Californian *H. ehrhorni* Eisen). The ental ducts unite before the attachment to the oesophagus.

Small to large species inhabiting terrestrial or limnic environments.

GENUS **Buchholzia** MICHAELSEN, 1887

Type Species: **Enchytraeus appendiculatus** BUCHHOLZ

Setae sigmoid, without nodulus, within the bundles the setae grow gradually smaller towards the dorsal and ventral midlines of the body. Head pore at 0/1. Dorsal pores absent. Brain longer than wide, slightly concave posteriorly. Oesophagus expanding abruptly into the intestine at VII/VIII. Dorso-lateral, hollow oesophageal appendages or irregular outline present in IV where they communicate with the oesophagus. A single or a pair of dorsal intestinal diverticula at VII/VIII. Nephridia with well developed interstitial tissue. Postseptale elongate, gradually narrowing into the terminal efferent duct. Origin of dorsal vessel anteclitellar, at the summit of the intestinal diverticula. Blood colourless. Coelomocytes ovoid, two types: large ones of about the same length as the setae, finely granulated and with distinct nucleus, and much smaller hyaline ones, present in great numbers.

GENUS **Bryohenlea** CERNOSVITOV, 1934

Type species: **Bryodrilus udei** EISEN

Setae slightly curved without nodulus. Head pore at 0/1. Dorsal pores absent. Gradual transition between oesophagus and intestine. Two pairs of intestinal (oesophageal?) diverticula in VIII. Oesophageal appendages present (?). Dorsal vessel arising in XII. Interstitial tissue of nephridia well developed; anteseptale consisting of funnel only. Seminal vesicle absent. Seminal funnel cylindrical; vas deferens rather short (3-4 times longer than funnel). Penial bulb compact. Spermatheca simple, joining before communicating with oesophagus.

GENUS **Bryodrilus** UDE, 1892

Type species: Bryodrilus ehlersi UDE

Setae sigmoid or almost straight, without nodulus; those towards the dorsal and ventral midlines of the body often gradually smaller. Head pore at 0/1. Dorsal pores absent. Four oesophageal diverticula in VI. Peptonephridia and intestinal diverticula absent. Dorsal vessel originating in or behind the clitellar region. Blood colourless. Coelomocytes of uniform size and shape. Interstitial tissue of nephridia well developed; anteseptale consisting of funnel only. Spermatheca simple; ental ducts unite and communicate jointly with the dorsal region of the oesophagus.

GENUS **Hemifridericia** NIELSEN AND CHRISTENSEN, 1959

Type species: Hemifridericia parva NIELSEN AND CHRISTENSEN

Setae straight, without nodulus, of equal length within the bundle (only one species with two or three setae per bundle is known). Head pore at 0/1. Dorsal pores absent. The transition between oesophagus and intestine is gradual. Pepto-nephridia and oesophageal appendages absent. Intestinal diverticula absent. The dorsal vessel originates in XII-XIII. The coelomocytes are of two kinds. Egg sac absent. Seminal vesicle absent or poorly developed. The spermatheca is simple, without diverticula and without glands along the ectal duct. The ental ducts unite and communicate with the oesophagus mid-dorsally.

GENUS **Enchytraeus** HENLE, 1837

Type species: Enchytraeus albidus HENLE

Setae straight, without nodulus, sometimes missing on some segments. Head pore at 0/1. Dorsal pores absent. Brain rounded posteriorly. Gradual transition between oesophagus and intestine. Paired, unbranched peptonephridia present. Oesophageal appendages absent. Intestinal diverticula absent. Dorsal vessel arising in or behind the clitellar region. Blood colourless or faintly yellowish. Three pairs of primary septal glands; secondary glands absent. Interstitial tissue of nephridia well developed; anteseptale consisting of funnel only; the efferent duct usually of postero-ventral origin. Seminal vesicle present and usually well developed. Vas deferens long, sometimes extending backwards to XXI. Penial bulb present. Male pore without atrial glands. Spermatheca simple, communicating with the oesophagus. The ectal duct usually with glands.

Small to large species occupying marine, limnic and terrestrial habitats.

GENUS **Lumbricillus** ORSTED, 1844

Type species: Lumbricillus lineatus MÜLLER

Setae sigmoid in most species but almost straight in some, without nodulus; the number of setae per bundle shows wide interspecific variation. Head pore at 0/1. Dorsal pores absent. Brain usually incised posteriorly. Peptonephridia and oesophageal appendages absent. Intestinal diverticula absent. Gradual transition between oesophagus and intestine. Three pairs of primary septal glands; secondary glands absent. Dorsal vessel originating in or behind the clitellar region. Blood usually coloured. Interstitial tissue of the nephridia well developed; anteseptale consisting of funnel only; efferent duct terminal, subterminal or postero-ventral. Testes large, lobed, enclosed in large seminal vesicles. Sperm duct long and narrow. Penial bulb present. Atrial glands absent. Copulatory glands often present. The spermathecae simple, attached to the oesophagus, usually with prominent glands at the ectal aperture and sometimes also along the ectal duct.

Small to large species, the majority littoral forms but some also occuring in limnic or terrestrial environments.

GENUS **Enchylea** NIELSEN AND CHRISTENSEN, 1963

Type species: Enchylea heteroducta NIELSEN AND CHRISTENSEN

Setae straight, without nodulus of equal size within the bundle. Head pore at 0/1. Dorsal pores absent. Brain rounded posteriorly. Sudden expansion of intestine at VII/VIII. Paired unbranched peptonephridia present. Oesophageal appendages absent. Intestinal diverticula present in VII. Dorsal vessel arising in VII. Blood colourless. Three pairs of septal glands in normal position. Interstitial tissue of nephridia well developed; anteseptale consisting of funnel only; the efferent duct of postero-ventral origin. Seminal vesicle present. Penial bulb present. Male pore without atrial glands. Spermathecae communicating with the oesophagus.

GENUS **Distichopus** LEIDY, 1882

Type species: Distichopus silvestris LEIDY

Setae straight, without nodulus. The innermost setae of the bundles pairwise shorter than the outer ones; the number of setae per bundle ranges from two to sixteen. Head pore at 0/1. Dorsal pores present on all segments from about VII. Oesophagus merges gradually with the intestine. Peptonephridia present. Oesophageal appendages and intestinal diverticula absent. Brain rounded or truncate posteriorly. Three pairs of septal glands. Dorsal vessel arising in the postclitellar region. Interstitial tissue of the nephridia well developed; anteseptale usually large, oval, with coils of the nephridial canal; efferent duct arising ventrally or terminally on postseptale but point of origin often variable within the species. Blood colourless or (rarely) coloured. Two kinds of coelomocytes present: large nucleated and small hyaline, anucleate. Testes small and compact; seminal vesicle present or absent. Egg sac absent. Sperm funnel cylindrical. Vas deferens long and narrow, confined to XII. Penial bulb present. Spermathecae simple or with diverticula; the ectal duct usually long and well demarcated, often with glands at the ectal orifice; entally the spermathecae communicate with the oesophagus.

Small to large species mostly occurring in terrestrial habitats, sometimes in limnic surroundings.

Cernosvitov (1933) placed *D. silvestris* in the genus *Fridericia* Michaelsen,

1889, the type species of which is *F. hegemon* (Vejdovsky) according to Michaelsen in Apstein (1915). The latter author placed *Fridericia* on a list of Nomina Conservanda, and Cernosvitov (1933) decided to reject the name *Distichopus* in favour of *Fridericia* because of this. No application to the I.C.Z.N. has been made to place *Distichopus* on the Official List of Rejected Names, and as it clearly takes priority over *Fridericia* there would seem to be little purpose in refusing to apply the law of priority in this instance.

GENUS **Stephensoniella** CERNOSVITOV, 1934

Type species: Enchytraeus marina MOORE

Setae straight, without nodulus. Head pore at 0/1. Dorsal pores absent. Transition between oesophagus and intestine gradual. Peptonephridia and oesophageal appendages absent. Intestinal diverticula absent. Dorsal vessel arising in or behind the clitellar region. Blood colourless. Interstitial tissue of nephridia well developed; anteseptale consisting of funnel only. Testes enclosed in large seminal vesicles. Sperm funnel cylindrical. Sperm duct long, confined to XII. Penial bulb compact. Spermatheca attached to the oesophagus. Diverticula present or absent.

No type species was nominated by Cernosvitov (1934), but there are only two in the genus, *S. marina* (Moore) and *S. barkudensis* (Stephenson), and the former was re-described by Cernosvitov in the paper in which the genus was established. Hence that species has been selected as the type species in the apparent absence of any earlier citation of a type.

Nielsen and Christensen (1959) considered the status of the genus to be rather doubtful, and suggested possible synonymy with *Marionina*, but left the situation to be clarified by subsequent study of the species concerned.

GENUS **Marionina** MICHAELSEN, 1889

Type species: Pachydrilus georgianus MICHAELSEN

Setae straight in most species but distinctly sigmoid in some; nodulus absent; the number of setae per bundle is variable but usually low. Setal bundles may be missing on some segments. Head pore at 0/1. Dorsal pores absent. Brain incised posteriorly or rounded. Peptonephridia absent. Oesophageal appendages are usually absent but present in three species and then in segments IV or VI. Intestinal diverticula absent. Three pairs of primary septal glands; secondary glands absent. Dorsal vessel originating in the clitellar region or—more rarely—farther back. Blood mostly colourless but coloured in a few species. Interstitial tissue of the nephridia well developed; anteseptale consisting of funnel and a few coils of the nephridial canal, rarely of funnel only; efferent duct usually arising in the posterior region of the postseptale but the origin may be variable in some species according to the segments considered. Testes are small and compact, a well developed seminal vesicle is usually absent. Sperm duct usually long and narrow. Penial bulb present. Atrial glands absent. The spermatheca with or without diverticula and usually attached to the oesophagus, in a few species free. Glands are usually associated with the spermatheca along the ectal duct, at the ectal aperture or in either of these positions.

Usually small species inhabiting marine, limnic or terrestrial habitats.

No type species has been designated in this genus. The original assemblage of species placed in *Marionina* by Michaelsen (1889) is represented only by *M. georgiana* (Michaelsen), in the current list of species recognized as belonging to that genus by Nielsen and Christensen (1959) and so that species has been selected as the type. The current position of the other species first attributed to *Marionina* by Michaelsen (1889) is as follows:

sphagnetorum—type of *Cognettia.*
sphagnetorum var *glandulosa*—species of *Cognettia.*
semifusca—species of *Lumbricillus.*
crassa—
ebudensis— ⎫ species dubiae in *Lumbricillus.*
enchytraeoides— ⎭

The type species of *Pachydrilus* (*P. verrucosus* Clap., 1861) was first placed in *Lumbricillus* Ørsted, 1844 and was then synomized with *Lumbricillus lineatus* (Müller, 1774), and so the generic name does not compete for recognition with the later name *Marionina*. The only other generic name used for this assemblage that pre-dates *Marionina* is *Enchytraeoides* Roule. The type species of *Enchytraeoides* is *E. marioni* Roule, long regarded as a dubious species. It is difficult to see why this name should have been raised as a possible alternative to *Marionina* as mentioned by Nielsen and Christensen (1959, pp. 38, 42).

GENUS **Grania** SOUTHERN, 1913

Type species: Grania maricola SOUTHERN

Setae straight, stout, without nodule, slightly curved or enlarged at the base, completely absent anteriorly. Brain incised posteriorly. Peptonephridia present or absent. Septal glands separate on each side of the gut, more or less fragmented into secondary glands. Dorsal vessel originating in XV to XX. Nephridia with a short anteseptal lacking loops of the nephridial canal. Coelomocytes with large nuclei. Sperm sacs unequally developed, reaching to XV or XVIII. Male ducts reach from XII to XIII or XVIII. Large spermathecae attached dorsally to oesophagus with globular ampulla and cylindrical muscular canal devoid of the glands.

Marine.

There has been a revived interest in marine oligochaetes in the last few years, and the status of the various morphological forms described has changed several times (Kennedy, 1966; Lasserre, 1966, 1967). The latest revision recognizes three species, one represented by four sub-species.

NOMINA DUBIA

The following species (and genera) are inadequately described and require re-evalution. Additional lists may be found in Michaelsen (1900) and Cernosvitov (1937).

Enchytraeina lutheri Bulow, 1957.
Hydrenchytraeus stebleri Bretscher, 1901.
Hydrenchytraeus nematoides Bretscher, 1901.

REFERENCES

APSTEIN, C. 1915. Nomina Conservanda. *Sitz. ges. natur. Freunde Berlin*, **1915 (5)**, 119.

CERNOSVITOV, L. 1933. Revision der Enchytraeiden—Gattung *Distichopus* Leidy. *Zool. Anz.*, **104**, 73.

— 1934. Zur Kenntnis der Enchytraeiden 1. Über *Enchytraeus marinus* Moore 1902; Über die Beziehung zwischen den Gattungen *Enchytraeus* und *Pachydrilus*. *Zool. Anz.*, **105**, 233.

— 1937. System der Enchytraeiden. *Bull Ass. Russe Rech. Prague Sci.*, **5**.

EISEN, G. 1878. Redogörelse För Oligochaeter, samlade under de svenska expeditionerna till Arktiske trakter. *Öfvers, auf Kongl. Vet.-Akad. Förh 1878 nr*, **3**, 63.

KENNEDY, C. R. 1966. A taxonomic revision of the genus *Grania* (Oligochaeta: Enchytraeidae). *J. Zool.*, **148**, 399.

LASSERRE, P. 1966. Oligochètes Marins des côtes de France. I: Bassin d'Arcachon: Systematique. *Cahiers Biol. Marine*, **7**, 295.

— 1967. Oligochètes Marins des côtes de France. II: Roscoff, Penpoull, Etangs saumâtres de Concarneau: Systematique, Ecologie, *Cahiers. Biol. Marine*, **8**, 273.

LEVINSEN, G. M. R. 1884. Systematesk—geografisk oversigt over der nordiske Annulata, Gephyren, Choetognathi og Balenoglossi. *Vidensk. Meddr. dansk naturh. Foren*, **4(1883)**, 192.

MICHAELSEN, W. 1889. Synopsis der Enchytraeiden. *Abh. d. naturw. Ver. Hamburg*, **11**.

— 1900. Oligochaeta. *Das Tierreich*, **10**. Berlin.

MOORE, J. P. 1943. *Pelmatodrilus planariformis*, a new oligochaete (Enchytraeidae), modified for epizootic life on Jamaican earthworms *Not. Nat.*, **128**, 1.

NIELSEN, C. O. and B. CHRISTENSEN. 1959. The Enchytraeidae: Critical revision and taxonomy of European species. *Natura jutl.*, **8-9**, 1.

— 1961. The Enchytraeidae: Critical revision and taxonomy of European species. Suppl. 1. *Natura jutl.*, **10**, 1.

— 1963. The Enchytraeidae: Critical revision and taxonomy of European species. Suppl. 2. *Natura jutl.*, **10**, 1.

13

FAMILY AEOLOSOMATIDAE*

J. VAN DER LAND

Type genus: Aeolosoma EHRENBERG

Small, mostly asexually reproducing worms; up to 10 mm long, but generally much smaller. A small number of segments (up to 17 in the first zooid). Prostomium ciliated ventrally. Eyes absent. Epidermis with numerous glands, part of which generally have vacuoles with refractive and most often coloured contents. Intersegmental furrows present in some species only. Setae in four bundles per segment or absent (in *Rheomorpha*); each bundle with an indefinite number of hair setae and needle setae (the latter occur only in part of the species and they may be restricted to the ventral and/or posterior bundles). A muscular ciliated pharynx in I. A simple oesophagus from I to III, IV, or V. Anterior part of the intestine dilated. Septa absent in most species. Commissural vessels mostly wanting. Two nephridia may occur in each segment, but usually several are absent. Cerebral ganglion, pharyngeal commissures and ventral nervecord attached to the epidermis. Copulatory gland ventrally in the middle or posterior part of the body. Two to five pairs of spermathecae in the anterior half of the body or one median spermatheca just anterior to the copulatory gland (*Potamodrilus*). Functional testes in two to several segments (may be both anterior and posterior to ovaries); spermatogenesis free in body cavity; two or several nephridia which are not at all or only slightly differentiated, act as male ducts. One or two functional ovaries in mid-body; one median female pore in middle of copulatory gland.

Cosmopolitan. In fresh-water and brackish water, less often terrestrial..

* To be excluded from the Clitellata according to R.O.B. (p. 176).

KEY TO THE GENERA

1.	Setae absent	*Rheomorpha*
—	Setae present	2

2.	With a ventral pharyngeal pocket with a muscular tongue; with a short postanal adhesive tail; no asexual reproduction by budding (perhaps by fragmentation)	*Potamodrilus*
—	No pharyngeal pocket; anus terminal; asexual reproduction by budding	3

3.	Dorsal setae of each bundle arranged in two transverse rows or in an ellipse; epizoic on crayfishes	*Hystricosoma*
—	Setae not arranged in a special way; freeliving	*Aeolosoma*

NOTE: The poorly described, epizoic *Aeolosoma setosa* Georgévitch, 1957, with pharyngeal teeth, might be a representative of a separate genus.

GENUS **Aeolosoma** EHRENBERG, 1828

Type species: Aeolosoma hemprichi EHRENBERG

Very small to relatively very large species (chains about 1-10 mm). Refractive or coloured epidermal glands generally present. Intersegmental furrows mostly absent. Setae present including needle setae in several species. Septa mostly absent. Commissural vessels present in some species only. Two nephridia per segment in large species, usually several absent in small species. Two to five pairs of spermathecae, each consisting of one cell, in the anterior half of the body. Functional testes in several segments. Normal nephridia of the posterior half of the body act as male ducts. One functional ovary in the mid-body.

Cosmopolitan. Freeliving in fresh and brackish water, less often terrestrial.

Aeolosoma

1.	Bundles of setae consist of long, flexible hair setae and short, stiff sigmoid setae (the latter often only in the posterior bundles; rarely even absent in the first zooid)	2
—	Sigmoid setae completely absent	10

2.   Sigmoid setae with very thin distal point and two rows of 5-10 relatively large teeth on concave side of distal end, 30-45 μ; epidermal glands colourless   *A. travancorense*

—   Sigmoid setae smooth or with one row of small teeth near distal end (use oil-immersion of good microscope; setae in lateral view)   3

3.   Large, body width 120-200 μ; hair setae reaching a length of 150-200 μ; sigmoid setae 45-70 μ, with one or more teeth on convex side of distal end or smooth (?) (these teeth are easily overlooked!)   4

—   Small, body width 50-100 μ; hair setae up to 135 μ, but mostly considerably shorter; sigmoid setae up to 45 μ, rarely up to 50 μ 6

4.   Sigmoid setae may occur in all bundles, although they are mostly absent in II (epidermal glands green)   *A. leidyi*

—   Sigmoid setae only in posterior bundles (from V backwards, sometimes only in ventral bundles, rarely absent in first zooid)   5

5.   Epidermal glands orange   *A. japonicum*

—   Epidermal glands greenish yellow to olive green   *A. tenebrarum*

6.   Sigmoid setae with 5-10 small teeth on convex side of distal end (epidermal glands red)   *A. psammophilum*

—   Sigmoid setae smooth or with teeth on concave side of distal end   7

7.   Sigmoid setae only in ventral bundles; ciliated field with typical latero-dorsal continuations (epidermal glands orange-red)   *A. evelinae*

—   Sigmoid setae also in dorsal bundles   8

8.   Sigmoid setae smooth, distal end sharply bent; epidermal glands colourless or absent   *A. beddardi*

—   Sigmoid setae mostly with teeth (they may be invisible or absent), distal end not sharply bent; epidermal glands coloured   9

9. Sigmoid setae with 2 or 3 very small teeth
    (they may be invisible or absent), 28-35 μ;
    epidermal glands lemon   *A. sawayai*

—   Sigmoid setae with 6 or 7 teeth, 40-46 μ;
    epidermal glands red   *A. corderoi*

10. Very large, body width more than 350 μ;
    dissipiments present, resulting in evident ex-
    ternal segmentation; hair setae up to 600 μ;
    epidermal glands green   *A. marcusi*

—   Small to large, body width always less than
    300 μ; dissepiments lacking, no evident ex-
    ternal segmentation; hair setae up to 350 μ 11

11. Fission-zone after XIV to XVI; a large species
    (body width 150-200 μ); (epidermal glands
    orange-red)   *A. gertae*

—   Fission-zone after VII to XI   12

12. Hair setae very short, 25-75 μ, mostly con-
    siderably shorter than body width, the shortest
    setae resembling sigmoid setae; a small species
    body width 50-100 μ); (epidermal glands
    orange-red)   *A. quaternarium*

—   Hair setae mostly reaching a greater length,
    longer than body width, the shortest not
    resembling sigmoid setae   13

13. Longest hair setae more than 175 μ; body
    width about 150-300 μ; (epidermal glands
    green or yellow)   14

—   Longest hair setae shorter than 160 μ; body
    width 40-175 μ 16

14. Epidermal glands yellow; commissural vessels
    present   *A. tenuidorsum*

—   Epidermal glands green; no commissural
    vessels   15

15. Body width 200-300 μ; hair setae up to 250 μ;
    fission-zone after VII or VIII, rarely after
    IX; intestine dilated from IV to VI or VII   *A. viride*

—   Body width 150-200 μ; hair setae up to 400 μ;
    fission-zone after IX to XI, mostly after X;
    intestine dilated from IV to VIII, IX or X   *A. headleyi*

16. Refractive epidermal glands colourless or absent   17
— Coloured epidermal glands present   18

17. Small, body width about 80 μ; fission-zone after VI or VII, rarely after VIII; epidermal glands without satellite cells; hair setae subequal, 50-80 μ, mostly 2 or 3 per bundle   *A. niveum*
— Moderately small, body width 80 to 130 μ; fission-zone after IX to XI, rarely after VIII; epidermal glands with satellite cells; hair setae unequal, up to more than 100 μ, 3-5 in most bundles   *A. hyalinum*

18. Epidermal glands orange-red   19
— Epidermal glands green or yellow   20

19. Large, body width more than 100 μ, up to 175 μ; fission-zone after VIII-XII, rarely after VII, mostly after X; hair setae up to 150 μ, up to 10 per bundle   *A. litorale*
— Small, body width 40-100 μ; fission-zone after VI to VIII, mostly after VII; hair setae up to 120 μ, up to 6, rarely up to 8 per bundle   *A. hemprichi*

20. Fission-zone after XI; intestine dilated from III to IX; epidermal glands brilliant golden green   *A. aureum*
— Fission-zone after VI-IX; intestine dilated to V, VI, or VII; epidermal glands not brilliant golden green   21

21. Epidermal glands golden yellow; intestine dilated from III-V   *A. flavum*
— Epidermal glands green or greenish yellow   22

22. Epidermal glands faintly green, with homogenous contents, without satellite cells; fission-zone after VI or VII   *A. olivaceum*
— Epidermal glands greenish yellow, containing granules and globules, with satellite cells; fission-zone after VII-IX   *A. variegatum*

Notes: The location of the fission-zone can not be determined in sexually reproducing individuals.

The body widths given are those of uncontracted living specimens. In fixed specimens the body widths are up to 50% greater. The characters of the epidermal glands can only be observed in living specimens. Consequently fixed material of many species can not be identified with certainty.

Aeolosoma japonicum Yamaguchi, 1953
Fig. 13. 1B

Aeolosoma japonicum YAMAGUCHI, 1953: 279, Figs. 2-3.

Aeolosoma japonicum Yamaguchi. YAMAGUCHI, 1957: 162. NAIDU, 1961: 648. BUNKE, 1967: 192, Fig. 14.

Large, chains 2-3 mm × 130-180 μ. Fission-zone after X. Prostomium rounded, wider than following segments; tactile hairs not numerous; sensory pits connected with ventral field. Epidermal glands globular, diameter up to 12 μ, orange. Hair setae unequal, 0-5(8) per bundle; short hair setae 40-120 μ, 0-3(5) per bundle; long hair setae 130-200 μ, 0-2(3) per bundle. Sigmoid setae with some small teeth on convex side of distal end, 42-50 μ, 1 or 2 per bundle, only in 1 or 2 of the posteriormost segments, sometimes absent in the first zooid. Intermediate forms between hair setae and sigmoid setae, up to 70 μ long, may occur. Intestine dilated from posterior part of IV to VIII or IX. First pair of nephridia in II/III.

Aquarium, Sapporo, Japan., Marshy soil, near Leiden, The Netherlands,

Aeolosoma hemprichi Ehrenberg, 1828

Aeolosoma hemprichi EHRENBERG, 1828: Pl. Phytozoa 5, fig. 2.

Aeolosoma hemprichi Ehrenberg. EHRENBERG, 1831: [59, 61]. OKEN, 1832: 1286. DESMAREST, n.d.: 374. GRUBE, 1851: 105. VAN DER HOEVEN, 1859: 265. MAGGI, 1865: 4. LEYDIG, 1865: 360. VEJDOVSKY, 1885 b: 276. VAILLANT, 1890: 461, Pl. 22, fig. 26. BEDDARD, 1892 b: 352; 1895: 178, Fig. 35. BRETSCHER, 1896: 500; 1899: 370; 1900: 2; 1903: 31. FERRO-NIÈRE, 1899: 231. SMITH, 1900: 443. WALDVOGEL, 1900: 344. MICHAEL-SEN, 1900: 13; 1903 a: 41; 1905: 289; 1909 a: 6, Fig. 9; 1928: Fig. 23A. STOLC, 1903 a: 3, Figs. 5-6, 9; 1903 b: 1; 1903 c: 75; 1903 d: 153, Figs. 1-26; 1903 e: 638, Figs. 1-26. MUNSTERHJELM, 1904: 32; 1905: 4. PIGUET 1906: 391; 1913: 112. STEPHENSON, 1909: 277, Pl. 20, figs. 53-55; 1913: 743; 1923: 41. 1930: 29, Fig. 227. SOUTHERN, 1909: 121. CHINAGLIA, 1910: 2. POINTNER, 1911: 627. TOIVONEN, 1911: 16. PIGUET AND BRETSCHER, 1913: 11, Fig. 6. KORSCHELT, 1914: 553. DEHORNE, 1916 a: 122; 1916 b: 72. ROMIJN, 1919: xlii. HAMMERLING, 1924: 581, Pls. 16-19, figs. 1-29, figs. A-H; 1930: 350. MOSZYNSKI, 1925 a: 28; 1925 b: 4; 1933: 245; 1935: 81. VAN OYE, 1927: 359. BASKIN, 1928: 231. AIYER, 1929: 16. UDE, 1929: 11, Fig. 12. LAMEERE, n.d.: Fig. 170. CERNOSVITOV, 1930: 9. STOLTE, 1933-55: 128, Figs. 167a, 604-605, 610-612. KNÖLLNER, 1935: 427. LASTOCKIN, 1935 a: 643; 1949: 123. KONDO,

1936: 382, Pl. 23, fig. 1. COLLINS, 1937: 194, Pl. 1, fig. 1. WESENBERG-LUND, 1937: Fig. 404, Pl. 10, fig. 5. BERG, 1938: 44. PASQUALI, 1938 a: 19; 1938 b: 25, Fig. 27. KENK, 1941: 3. SCHAEFER, 1941: 160. KESSELYAK, 1942: 47. MARCUS, 1944: 10, Fig. 7A-B. VAN DER WERF, 1946: 253. LEENTVAAR, 1946: 3, Figs. 4-5. BUISAN, 1946: 166, Figs. 11-12, 22A, 24. BERG, 1948: 41. SPERBER, 1948: 265. ANON., 1949: 1378, Figs. 1-4. HERLANT-MEEWIS, 1950 a: 140; 1950 b: 211; 1950 c: 123, Figs. 3-4; 1950 d: 173, Pl. 1, figs. 1-9; 1951 a: 429, Pl. 1, figs. 1-3, Pl. 2, figs. 5-10, Pl. 3, figs. 11-13; 1951 b: 231, Fig. 4R; 1953: 119; 1954: 75. HERLANT-MEEWIS AND BOULANGER, 1950: 138. AUCLAIR ET AL., 1951: 162. PENNAK, 1953: 189, Fig. 183 a. ISTVAN, 1955: 8. COLE, 1955: 219. VON BÜLOW, 1955: 253. RIIKOJA, 1955: 11. KAESTNER, 1955, 1960: 422, 430: JÄRNE-FELT, 1956: 4. STOUT, 1956: 99. RUTTNER-KOLISKO, 1956: 384. CORBELLA ET AL., 1956: 216. KAMEMOTO AND GOODNIGHT, 1956: 219. MALEVIC, 1957: 112. AX, 1957: 430. VON BÜLOW, 1957: 71. MOSZYNSKI AND MOSZYNSKA, 1957: 333. YAMAGUCHI, 1957: 161, Fig. 1. SEMAL-VAN GANSEN, 1958: 241, Figs. 5, 10. AVEL, 1959: 369, Fig. 286. GOODNIGHT, 1959: 528, Fig. 21. 6. PARKER AND KAMEMOTO, 1959: 207. TIMM, 1959: 24. DELAMARE DEBOUTTEVILLE, 1960: 185. NAIDU, 1961: 643, Fig. 2. DIONI, 1961: 107. HARMAN AND PLATT, 1961: 90. CEKANOVSKAYA, 1962: 79, Figs. 69, 70a; 1964: 113. MOSZYNSKA, 1962: 7. BOTEA, 1962: 401. BRINKHURST, 1963 b: 15. DALES, 1963: 161. WACHS, 1963: 510. CORIC AND CORIC, 1965: 233, Figs. 1-2, 4-6. BUCHER, 1965: 97, Figs. 1-6. VAN DER LAND, 1965: 236. NAIDU, 1966: 209. HARMAN, 1966: 239. BRINKHURST, 1966: 132. PETERS, 1967: 51, Figs. 1-4. LAAKSO, 1967: 561. AX AND BUNKE, 1967: 222, Fig. 3. JUGET, 1967: 218. BUNKE, 1967: 192, Figs. 1-2, 62-63, 65.

Aeolonais hemprichi (Ehrenberg). GERVAIS, 1838: 14.

Aeolosoma ehrenbergi (pro *A. decorum* et *A. hemprichi*) ØRSTED, 1842: 137, Pl. 3, fig. 7. CARUS, 1863: 448. LEYDIG, 1865: 361. CLAUS, 1876: 420. TAUBER, 1879: 75. VEJDOVSKY, 1880: 505; 1884: 4; 1885 a: 16, Pl. 1, figs. 1-7; 1892: 171. BEDDARD, 1888: 214; 1889 a: 262; 1889 b: 55; 1892 a: 12. STOLC, 1890: 183, Pl. 7, figs. 1-9. LAMEERE, 1895: 200. PERRIER, 1897: 1716. STENROOS, 1898: 41. JANDA, 1901: 1. ISSEL. 1901: 4. MAUPAS, 1919: 159. STOLTE, 1933-55: 177, 495, 740. HERLANT-MEEWIS, 1954: 78.

Aelosoma ehrenbergi Ørsted. D'UDEKEM, 1858: 23; 1861: 244, Fig. 1.

Aeolosoma emprichi Ehrenberg. FERRONIÈRE, 1899: 232.

Aeolosoma ehrenbergi Ørsted. JANDA, 1901: 8.

Aeolosoma kashyapi STEPHENSON, 1923: 12.

Aeolosoma kashyapi Stephenson. AIYER, 1926: 138; 1929: 18. STEPHENSON, 1930: 1. MICHAELSEN AND BOLDT, 1932: 589-591. SCHAEFFER, 1941: 167.

MARCUS, 1944: 10. MARCUS, E., 1944: 5. SPERBER, 1948: 264. HERLAND-MEEWIS, 1951 b: 277; 1954: 80. STOUT, 1952: 97, Fig. 2a; 1956: 97, Fig. 2. YAMAGUCHI, 1957: 161. AVEL, 1959: Figs. 263, 306. NAIDU, 1961: 648; 1966: 210. BUNKE, 1967: 192, Fig. 11.

Aeolosoma hemprici Ehrenberg. SCIACCHITANO, 1934: 2. BRINKHURST, 1962: 318; 1963 a: 139; 1967: 112.

Aeolosoma hemprichi forma *typica*. CHEN, 1940: 23.

Aeolosoma hemprichi var. *kashyapi* CHEN, 1940: 17, Fig. 1A.

Aeolosoma hemprihi Ehrenberg.CORIC AND CORIC, 1965: 233.

Small, first zooid 0·3-1 mm × 40-100 μ; chains up to 6 zooids, up to 2 mm. Fission-zone after VII, rarely after VI or VIII. Prostomium rounded or slightly triangular, distinctly wider than following segments; tactile hairs up to 10 μ; ciliated field restricted to ventral surface or with small dorso-lateral continuations; sensory pits round, diameter up to 15 μ, often indistinct and not clearly separated from ventral field. Epidermal glands globular, diameter up to about 10 μ, orange to dark red, without satellite cell. Hair setae unequal, in small specimens sometimes subequal, 2-6 (8) per bundle; short setae 35-80 μ, 1-3 (4) per bundle; long setae 70-120 μ, 0-3 (4) per bundle. Nephridia irregularly distributed, up to 5 pairs in the first zooid; first nephridium in II/III or III/IV. Intestine dilated from posterior part of VI. Copulatory gland in IV/V or V/VI. Up to 3 pairs of spermathecae from II to IV.

Cosmopilitan, one of the most common species, in all kinds of habitats.

Most records in the literature are unreliable and may as well refer to one of the related species (*A. litorale*, *A gertae*, and *A. quaternarium*). On the other hand part of the records of *A. quaternarium* undoubtedly refers to *A. hemprichi*.

Very small forms of this species are not uncommon (length of chains less than 1 mm; body width about 50 μ, sensory pits often indistinct; often only one or two setae per bundle, 40-80 μ long). Some authors have considered the small form a separate species (*A. kashyapi*).

The animals move very well on a flat surface, regularly and often quite rapidly.

Aeolosoma gertae MARCUS, 1944
Fig. 13. 2E

Aeolosoma gertae MARCUS, 1944: 14, Figs. 5A-C, 18.

Aeolosoma gertae Marcus. HERLANT-MEEWIS, 1951 b: 277. AVEL, 1959: Fig. 305. DIONI, 1961: 112. NAIDU, 1961: 647. BUNKE, 1967: 192, Fig. 12.

Aeolosoma gestae Marcus. DIONI, 1961: 110.

Large, chains of 2 or 3 zooids 2-4 mm × 150-200 μ. Fission-zone after XIV to XVI. Prostomium much wider than following segments; tactile hairs numerous, 12 μ; ciliated field restricted to ventral surface; sensory pits round, diameter about 20 μ, separated or not separated from ventral field. Epidermal glands globular or somewhat irregular, diameter up to 12 μ, orange to dark red, with grey satellite cell; also glands with colourless granular contents. Hair setae unequal, (3) 5-7 (8) per bundle; short setae of 65-120 μ, 2-5 per bundle; long setae 150-250 μ, 0-2 (3) per bundle. Nephridia irregularly distributed; first pair in II/III or III/IV. Intestine dilated from posterior part of IV-IX or XI.

Brazil and The Netherlands.

There are some differences between the S. American and the European populations. In the specimens from Brazil the sensory pits are separated from the ventral field and the intestine is dilated to XI. In the specimens from the Netherlands the sensory pits are always connected with the ventral field and the intestine is only dilated to IX; the long hair setae are shorter, never much longer than 200 μ.

The animals move rapidly, with frequent sudden stops; the prostomium and the first segment form one large circle and the posterior part of the body is bent.

Aeolosoma tenebrarum Vejdovsky, 1880
Figs. 13. 1J, 13. 2G

Aeolosoma tenebrarum VEJDOVSKY, 1880: 505.

Aeolosoma tenebrarum Vejdovsky. VEJDOVSKY, 1884: 4; 1885 a: 18, Pl. 1, figs. 16-36; 1885 b: 279; 1905: 10. BEDDARD, 1888: 214; 1889 a: 262; 1889 b: 51, Pl. 5, figs. 1-8; 1892 b: 353; 1895: 19. STOLC, 1890: 191. VAILLANT, 1890: 460. BRETSCHER, 1896: 501. PERRIER, 1897: 1686. GRIFFITHS, 1898: 448. BRACE, 1898: 363. SMITH, 1900: 443. MICHAELSEN, 1900: 13; 1903 a: 41; 1909 a: 6, Figs. 8, 10; 1928: Fig. 24. ISSEL, 1901: 5, (?) BRACE, 1901: 177, Pl. 21, figs. 1-16. JANDA, 1901: 1, Figs. 1-22; 1902: 172. PIGUET, 1906: 391. MOORE, 1906: 164. SOUTHERN, 1909: 125. (?) KRIBS, 1910: 45. TOIVONEN, 1911: 15. PIGUET AND BRETSCHER, 1913: 12. DEHORNE, 1916 a: 122; 1916 b: 32. HAMMERLING, 1924: 585. BASKIN, 1928: 231. UDE, 1929: 20, Figs. 14-15. STEPHENSON, 1930: 42. STOLTE, 1933-55: 127, 177, 495, 546, 739, Figs. 133a-d, 171-172, 253. ALTMAN, 1936: 7. (?) COLLINS, 1937: 194. CHEN, 1940: 16. SCHAEFER, 1941: 165. KENK, 1941: 1. MARCUS, 1944: 11. LEENTVAAR, 1946: 10. BUISAN, 1946: 152, Figs. 14, 25. LASTOCKIN, 1949: Fig. 74.1. HERLANT-MEEWIS AND BOULANGER, 1950: 137. HERLANT-MEEWIS, 1950 c: 127; 1950 d: 173; 1954: 78. AUCLAIR ET AL., 1951: 163. YAMAGUCHI, 1953: 83. PENNAK, 1953: 289. ISTVAN, 1955: 9, Fig. 3. MOSZYNSKI AND MOSZYNSKA, 1957: 406. GOODNIGHT, 1959: 528. NAIDU, 1961: 648. CEKANOVSKAYA, 1962: 144, Figs. 70, 72. BRINKHURST, 1962: 318; 1963 b: 8; 1967: 112. BEGER, 1966: 113. BUNKE, 1967: 193, Fig. 25. LAAKSO, 1967: 561.

A large species, chains 2-10 mm × 120-200 μ, up to 5 zooids. Fission-zone after VII-IX. Prostomium much wider than following segments, slightly triangular; tactile hairs up to 10 μ; ciliated field with small triangular latero-dorsal continuations; sensory pits large, probably connected with ventral field. Epidermal glands globular, diameter up to 18 μ, with satellite cell, pale greenish yellow to olive green. Hair setae unequal, 0-6 (7) per bundle; long setae 110-200 μ, 0-3 per bundle; short setae 50-105 μ, 0-4 (5) per bundle. Sigmoid setae sickle shaped, with one or more small teeth on convex side of distal end, 50-60 μ, only in posterior bundles, from V, sometimes restricted to ventral bundles, rarely absent in first zooid. Nephridia regularly distributed, first pair in II/III. Intestine dilated from anterior part of IV-VII.

Many European countries, U.S.A., and Canada.

The sigmoid setae are stated to be bifid by Vejdovský (1885 a) and Moore (1906), but in fact there seems to be a row of teeth on the convex side of which only the first is large enough to be observed through most microscopes. Even this first one is rather small so it may have been overlooked by Beddard (1889 b) who states that the setae are smooth.

Movements are clumsy, with strongly bent body. There is a considerable mucous secretion.

Aeolosoma quaternarium Ehrenberg, 1831.
Figs. 13. 1H, 13. 2F, H

Aeolosoma quaternarium EHRENBERG, 1831: 60.

Aeolosoma quaternarium Ehrenberg. GRUBE, 1851: 105. MAGGI, 1865: 4. LEYDIG, 1865: 360, Pl. 8B, figs. 1-2. GREBNICKIJ, 1873: 268. CLAUS, 1876: 420. GRIMM, 1877: 116. VEJDOVSKY, 1880: 505; 1884: 4; 1885 a: 16, Pl. 1, figs. 8-15; 1885 b: 276; 1905: 9. TIMM, 1883: 155: KRAEPELIN, 1886: 8. LEUNIS, 1886: 774. BEDDARD, 1888: 214; 1889 a: 262; 1889 b: 55; 1892 b: 352; 1895: 14. VAILLANT, 1890: 461. STOLC, 1890: 191; 1903 a: 2, Fig. 10; 1903 c: 75. FRENZEL, 1891: 21. GARBINI, 1895: 106. LAMEERE, 1895: 200. MICHAELSON, 1900: 13; 1903 a: 41; 1909 a: 6, Fig. 11; 1928: 68, Fig. 3A. ISSEL, 1901: 3, Figs. 1-3, Pl. 1, fig. 3. JANDA, 1901: 1. DITLEVSEN, 1904: 441, Pl. 17, figs. 50-51. MUNSTERHJELM, 1905: 4. SOUTHERN, 1909: 125. KRIBS, 1910: 45. CHINAGLIA, 1910: 3. MINKIEWICZ, 1914: 126. KOWALEWSKI, 1914: 107. CLAUS AND GROBBEN, 1917: 438. ROMIJN, 1919: xxxix. BOVERI-BONER, 1920: 507. VAN OYE, 1927: 359. BASKIN, 1928, 240. AIYER, 1929: 16. UDE, 1929: 13, Fig. 13. STEPHENSON, 1930: 215; 1931: 297. CERNOSVITOV, 1931: 1. STOLTE, 1933-55: 742. MOSZYNSKI, 1933: 237. COLLINS, 1937: 194. WESENBERG-LUND, 1937: 332. CHEN, 1940: 16. KENK, 1941: 3. SCHAEFER, 1941: 159, Figs. 1-54. KESSELYAK, 1942: 47. MARCUS, 1944: 9. MARCUS, E., 1944: 2. LEENTVAAR, 1946: 10. HERLANT-MEEWIS, 1951 a: 430; 1954: 89. PENNAK, 1953: 289. ISTVAN, 1955: 8. MALEVIC, 1956: 401. MOSZYNSKI AND MOSZYNSKA, 1957: 333. YAMAGUCHI, 1957: 162. AVEL, 1959: 321. GOODNIGHT, 1959: 528. DIONI, 1961: 107. CEKANOVSKAYA, 1962: 145, Fig. 70b. MOSZYNSKA, 1962: 7. BRINKHURST, 1962: 318; 1963 a: 142; 1963 b: 15; 1967: 112. BEGER, 1966: 120. PETERS, 1967: 53. BUNKE, 1967: 192, Figs.

3-4, 64, 66-68, 74-75. AX AND BUNKE, 1967: 222. LAAKSO, 1967: 561. FINO-GENOVA, 1968: 233.

Aeolonais quaternarium (Ehrenberg). GERVAIS, 1838: 14.

Aeolosoma quaternarium Ehrenberg. DESMAREST, n.d.: 374.

Chaetodemus multisetosus (pro *Aeolosoma quaternarium*—Lankester, 1869) CZERNIAVSKY, 1880: 307.

Chaetodemus quaternarius (Ehrenberg). CZERNIAVSKY, 1880: 307.

Aeolosoma quaternarpium Ehrenberg. LEVINSEN, 1884: 221.

Aeolosoma thermale ISSEL, 1900: 55 (nom. nud.).

Aeolosoma quarternarium Ehrenberg. STOLTE, 1933-55: 177, 496, 546, Figs. 105a, 260. NAIDU, 1961: 647.

Aeolosoma quaternarum Ehrenberg. HERLANT-MEEWIS, 1954: 78, 89.

[non] *Aeolosoma quaternarium* Ehrenberg. CORDERO, 1931: 347 (=*A. headleyi*).

Small, first zooid 0·7-1·2 mm×50-100; chains up to 5 zooids, up to 2·5 mm. Fission-zone after VII or VIII. Prostomium rounded, slightly wider than following segments; tactile hairs up to 10 μ, numerous; ciliated field restricted to ventral surface; sensory pits round, diameter up to 15 μ, usually not distinctly separated from ventral field. Epidermal glands globular or somewhat irregular, diameter up to 12 μ, orange-red, without satellite cells. Hair setae unequal in most bundles, but also often subequal, especially in the ventro-posterior bundles, 1-5 (6) per bundle; short hair setae 25-50 μ, stiff proximal part 50-70% of total length, resembling sigmoid setae, 1-4 per bundle; long hair setae relatively short, 50-75 μ, mostly considerably shorter than body width, 1-3 per bundle. Intestine dilated from posterior part of III to posterior part of VI. Nephridia irregularly distributed; first nephridium usually in III/IV. Copulatory gland in V/VI or VI/VII, sometimes also in second zooid. Up to 5 pairs of spermathecae in II-VI.

Europe.

This species is only known with certainty from some European countries. Most records in the literature are unreliable. Many may as well refer to *A. hemprichi, A. gertae,* and *A. litorale*. Part of the records of *A. hemprichi* probably refer to *A. quaternarium*.

Sexual reproduction seems to be relatively common in this species. In fact is is the only species of *Aeolosoma* in which complete sexual reproduction has been observed by several students. In other species egg cells seem usually to atrophy.

Aeolosoma beddardi Michaelsen, 1900
Fig. 13. 1I

Aeolosoma niveum (non Leydig, 1865) BEDDARD, 1892 b: 351.

Aeolosoma niveum Beddard. BEDDARD, 1895 : 177. CORDERO, 1931 : 348.

Aeolosoma beddardi (nom. nov. pro. *A. niveum* Beddard, 1892, non Leydig, 1865). MICHAELSEN, 1900 : 13.

Aeolosoma beddardi Michaelsen. MICHAELSEN, 1903 a : 41. SOUTHERN, 1909 : 125. STEPHENSON, 1923 : 40; 1930 : 722, 726. STOLTE, 1933-55 : 128. MARCUS, 1944 : 12. CORDERO, 1951 : 231. PENNAK, 1953 : 289. NAIDU, 1961 : 648. DIONI, 1961 : 107. BRINKHURST, 1962 : 318; 1963 b : 8. BUNKE, 1967 : 194.

Aeolosoma lucidum CHEN, 1940 : 17, Fig. 2.

Chains of 2 zooids 1-2 mm × 50-100 μ. Fission-zone after VI (rarely after V). Prostomium slightly wider than following segments; tactile hairs about 8 μ; ciliated pits round, connected with ventral field. Epidermal glands inconspicuous, colourless (a few may be light orange or greenish yellow), or no refractive glands present. 0-5 hair setae per bundle; long setae 90-135 μ, 0-2 (3) per bundle; short setae 40-80 μ, 0-2 per bundle. Sigmoid setae smooth, only slightly sickle shaped, 30-50 μ, 0-2 (3) per bundle. First nephridia in II/III. Intestine dilated from posterior part of III to anterior part of VI.

England, China (Nanking and Chunking), Surinam (Lelydorp, among vegetable debris in swamp).

The original description was based on one specimen and lacks detail. The specimens from China (*A. lucidum*) are evidently smaller than the specimens from Surinam (body width 50-70 μ and about 100 μ respectively; hair setae up to 78 μ and up to 135 μ, respectively; sigmoid setae up to 35 μ and up to 50 μ respectively), but they have several important characters in common.

Aeolosoma headleyi Beddard, 1888

Aeolosoma headleyi BEDDARD, 1888 : 213, Pl. 12, figs. 1-8.

Aeolosoma headleyi Beddard. BEDDARD, 1889 a : 262; 1889 b : 51; 1892 b : 352; 1895 : 14. STOLC, 1890 : 192; (?)1903 a : 3; 1903 c : 77. SMIDT, 1896 161. MICHAELSEN, 1900 : 13; 1903 a : 41; 1909 a : 6; 1909 b : 104. JANDA, 1901 : 3. STEPHENSON, 1907 : 236; 1911 : 205; 1923 : 42; 1930 : 722. SOUTHERN, 1909 : 125. TOIVONEN, 1911 : 15. MEYER, 1926 : 344. BASKIN, 1928 : 231. UDE, 1929 : 20. (?)CORDERO, 1931 : 348. STOLTE, 1933-55 : 128. COLLINS, 1937 : 194. (?)WESENBERG-LUND, 1937 : 332. (?)PASQUALI, 1938 a : 19; (?)1938 b : 27, Fig. 2. KENK, 1941 : 3, Figs. 3-4. SCHAEFER, 1941 : 167. BUISAN, 1946 : 152, Figs. 1-4, 6, 13, 15-16, 20, 22b, 23. LASTOCKIN, 1949 : 123. HERLANT-MEEWIS, 1950 c : 127; 1951 a : 430; 1951 b : 277; 1954 : 78. PENNAK, 1953 : 289, Fig. 183C. MALEVIC, 1956 : 406. MOSZYŃSKI AND MOSZYŃSKA, 1957 : 406. GOODNIGHT, 1959 : 528. BRINKHURST, 1962 : 318; 1963 a : 142; 1963 b : 15; 1967 : 112. CEKANOVSKAYA, 1962 (p.p.) : 144. LAAKSO, 1967 : 561. BUNKE, 1967 (p.p.) : 239.

Aeolosoma bengalense STEPHENSON, 1911: 204.

Aeolosoma bengalense Stephenson. STEPHENSON, 1923: 16; 1930: 316.
AIYER, 1926: 136; 1929: 18. MICHAELSEN AND BOLDT, 1932: 589. STOLTE,
1933-55: 128. SCHAEFER, 1941: 166. MARCUS, 1944: 9. MARCUS, E., 1944:
1, Figs. 11-12. HERLANT-MEEWIS, 1954: 80. YAMAGUCHI, 1953: 279, Fig.
1; 1957: 163. DIONI, 1961: 112. NAIDU, 1961: 643, Fig. 1; 1966: 209.
BUNKE, 1967: 193, Figs. 17-18.

Aeolosoma headlei POINTNER, 1911: 627, Fig. 1.

Aeolosoma spec. AIYER, 1926: 131, Figs. 1-3.

Aeolosoma quaternarium (non EHRENBERG)—CORDERO, 1931: 347.

(?) *Aeolosoma* spec. I STEPHENSON, 1931: 291.

(?) *Aeolosoma* spec. II STEPHENSON, 1931: 291.

Aeolosoma headleyi Beddard. SCIACCHITANO, 1934: 2.

Aeolosoma variegatum (non VEJDOVSKY)—CHEN, 1940: 17, Fig. 1B.

Aeolosoma pointneri (pro *Aeolosoma headlei* POINTNER) MARCUS, 1944: 12.

Aeolosoma pointneri Marcus. CORDERO, 1951: 231. NAIDU, 1961: 648.
BRINKHURST, 1967: 112.

Aeolosoma haedleyi Beddard. HERLANT-MEEWIS, 1950 d: 173; 1951 b: 232,
Fig. 3.

Aeolosoma readleyi Beddard. CIONI, 1961: 116.

Large, first zooid up to 2 mm × 150-200 μ; chains up to 4 zooids, up to 4 mm.
Fission-zone after IX to XI. Prostomium somewhat triangular, slightly wider
than following segments; tactile hairs up to about 12 μ, not numerous; ciliated
field not reaching anterior tip, with small latero-dorsal continuations; large, oval
sensory pits, situated posterior to the level of the brain, separated from ventral
field. Epidermal glands mostly of irregular form, green, sometimes yellowish
or bluish green, with satellite cells; besides colourless refractive glands often
occur. Indistinct intersegmental furrows usually present. Posterior end often
two-lobed. Hair setae unequal, up to 6 (9) per bundle; short setae 75-200 μ,
2-5 per bundle; long setae 150-400 μ, 0-4 per bundle; ventral setae usually
shorter than dorsal setae. Intestine dilated from IV-VIII, IX, or X. Nephridia
mostly regularly distributed, up to 9 pairs; first pair in II/III. Clitellum in V,
VI, VII or VIII, with one gonopore.

Europe, America, Asia; tropical and temperate zones.

There has been some confusion about this species. Stephenson (1911)
described a new species, *A. bengalense*, although his description shows that
there are no essential differences with *A. headleyi*. Marcus (1944) recognized Stephenson's species, at the same time attaching the name *A. headleyi*
to another species, which clearly differs from Beddard's species and which

is here called *A. marcusi* n. sp. Several authors followed Marcus.

Aeolosoma marcusi n. sp.
Fig. 13. 2I

Aeolosoma headleyi (non BEDDARD) MARCUS, 1944: 7, Figs. 6, 19-34. MARCUS, E., 1944: 5. SPERBER, 1948: 264. AVEL, 1959: Figs. 206, 256, 292. ERCO-LINI, 1960: 9. NAIDU, 1961: 648. DIONI, 1961: 107, Figs. a-b, Pl. 1, figs. 1-4. CEKANOVSKAYA, 1962 (p.p.): 45, Figs. 30, 33, 41, 71. BUNKE, 1967 (p.p.): 191. AX AND BUNKE, 1967: 222.

(?) *Aeolosoma headleyi* var. *pointneri* DIONI, 1961: 107, Pl. 1. fig. 5.

Very large, first zooid 2-6 mm × 350-600 μ; chains up to 5 zooids, up to 10 mm. Fission-zone after XIII to XVIII. Prostomium rounded or somewhat truncate, slightly wider than following segments; tactile hairs numerous, up to 24 μ; ciliated field with large latero-dorsal continuations; sensory pits large, elliptical, behind level of brain, separated from ventral field. Epidermal glands of irregular form, green to geenish blue; besides colourless glands are present; satellite cells (?). Dissipiments present, resulting in evident external segmentation. Hair setae unequal, up to 600 μ, 4-12 per bundle, mostly 5-7. Intestine dilated from posterior part of IV to XI-XIII. A pair of commissural vessels in III-IV. Nephridia usually regularly distributed, present in nearly all segments; first nephridia in II/III, sometimes a rudimentary nephridium in I; occasionally the nephridia are double or multiple. Copulatory gland between IV and X, not sharply delimited, with several gonopores.
Brazil (Sao Paulo and surroundings), Uruguay (Montevideo).

Marcus first described this species, but he considered it to be identical with *A. headleyi* Beddard. However, it is obvious from the original des-cription that *A. headleyi* is a considerably smaller species and that it has no dissipiments.

A. marcusi n. sp. differs much from all other species of the genus; it has a comparatively large size; its body-cavity is divided by dissipiments; more than two nephridia may be present in one segment; commissural vessels are present in the oesophageal region; several ovocytes may develop simultaneously. It is much less reduced than the other species, so it is probably the most original one of the known species.

In his laboratory cultures of this species Dioni (1961) found a smaller form, with fission-zone after IX and intestinal dilatation from III/IV to VIII. It is not stated whether dissipiments are present or not. He considered this form to be a physiological race, which he called *A. headleyi* var *pointneri*. It is not impossible that this material belongs to *A. headleyi* sensu Beddard.

Aeolosoma hyalinum Bunke, 1967.

Aeolosoma hyalinum BUNKE, 1967: 190, Figs. 29-30.

Moderately small, first zooid 0·5-1 mm × 80-130 μ; chains up to 4 zooids, up to

2 mm. Fission-zone after (VIII)IX to X(XI). Prostomium somewhat triangular, evidently wider than following segments; tactile hairs up to 9 μ, numerous; ciliated field restricted to ventral surface; sensory pits round, diameter up to 12 μ, separated from the ventral field. Refractive epidermal glands colourless, diameter up to 14 μ, with satellite cell; sometimes no refractive glands present. Hair setae unequal, 2-5 (6) per bundle; short setae 40-95 μ, 1-4 (5) per bundle; long setae 100-130 (140) μ, 0-2 per bundle. Intestine gradually widening in anterior part of IV, gradually narrowing in VI-VIII. Nephridia irregularly distributed; first in II/III or III/IV. Copulatory gland in V/VI.

Several localities in Germany and The Netherlands.

This species has been recognized only quite recently. Part of the records of *A. niveum* in the literature undoubtedly refer to *A. hyalinum*.

This is a tube-building species. Movements are somewhat awkward, with contracted and bent hind body. The animals produce much mucous with which they stick to the stubstrate.

Aeolosoma niveum Leydig, 1865.

Aeolosoma niveum LEYDIG, 1865: 365, Pl. 8B, fig. 3.

Aeolosoma niveum Leydig. VEJDOVSKY, 1885 a: 17; 1885 b: 7. BEDDARD, 1888: 215; 1892 b: 351; 1895: 182. VAILLANT, 1890: 460. BRETSCHER, 1896: 501. SEKERA, 1896: 375. MICHAELSEN, 1900: 13; 1903 a: 41; 1909 a: 6. STOLC, 1903 a: 3, Figs. 3, 8; 1903 c: 77. MUNSTERHJELM, 1905: 7. POINTNER, 1911: 627. TOIVONEN, 1911: 16. PIGUET AND BRETSCHER, 1913: 10. ROMIJN, 1919: xxxix. STEPHENSON, 1923: 40; 1930: 722. UDE, 1929: 19. MOSZYŃSKI AND URBANSKI, 1932: 49. STOLTE, 1933-55: 128. MOSZYŃSKI, 1933: 237. COLLINS, 1937: 194. CHEN, 1940: 26. MARCUS, 1944: 11. LASTOCKIN, 1949: 123. STOUT, 1952: 97, Figs. 3-4; 1956: 97, Figs. 1, 3-4. PENNAK, 1953: 289. MAELVIC, 1956: 402. JÄRNEFELT, 1956: 5. MOSZYŃSKI AND MOSZYŃSKA, 1957: 341. YAMAGUCHI, 1957: 163. JUGET, 1959: 393; 1967: 218. GOODNIGHT, 1959: 528. DIONI, 1961: 107. NAIDU, 1961: 648. BRINKHURST, 1962: 318; 1963 b: 14; 1967: 112. MOSZYŃSKA, 1962: 7. CEKANOVSKAYA, 1962: 144. VAN DER LAND, 1965: 236. BUNKE, 1967: 194, Fig. 26. LAAKSO, 1967: 561. AX AND BUNKE, 1967: 222.

Aeolosoma lacteum (pro. *A. niveum* Leydig). TIMM, 1883: 155.

(non) *Aeolosoma niveum* Leydig. CORDERO, 1931: 348. KOND, 1936: 383, Pl. 23, fig. 2. LEENTVAAR, 1946: 10.

Small, first zooid 0·5-0·8×about 80 μ; chains up to 4 zooids, up to 1·5 mm. Fission zone after VI to VII(VIII). Prostomium somewhat triangular, slightly wider than the following segments; tactile hairs up to about 7 μ; ciliated field restricted to the ventral surface; sensory pits round, relatively large, diameter up to 10 μ, separated from the ventral field. Refractive epidermal glands colourless or with very faint greenish tinge, diameter up to 8 μ, without satellite cell; sometimes no refractive glands present. Hair setae subequal, 50-80 μ, 1-3 (4) per bundle. Intestine dilated from posterior part of III to posterior part of V. Nephridia irregularly distributed; first in II/III or III/IV.

Europe, temperate zone, rather common. New Zealand.

Part of the records in the literature may as well refer to *A. hyalinum*, which is probably not less common.

Aeolosoma variegatum Vejdovský, 1885
Fig. 13. 2D

Aeolosoma variegatum VEJDOVSKY, 1885 b: 275, Figs. 1-6.

Aeolosoma variegatum Vejdovsky. BEDDARD, 1888: 214; 1889 a: 263; 1889 b: 52; 1892 b: 352; 1895: 178. STOLC, 1890: 191; 1903 a: 3, Fig. 1; 1903 c: 76. VAILLANT, 1890: 464. SEKERA, 1896: 375. BRETSCHER, 1896: 501. MICHAELSEN, 1900: 13; 1903 a: 41; 1905: 289; 1909 a: 6. JANDA, 1901: 2. MUNSTERHJELM, 1905: 7. SOUTHERN, 1909: 125. TOIVONEN, 1911: 16. PIGUET AND BRETSCHER, 1913: 12. ROMIJN, 1919: xxxix. MOSZYŃSKI, 1925 a: 28; 1925 b: 4. BASKIN, 1928: 240. UDE, 1929: 19. STEPHENSON, 1930: 722. MOSZYŃSKI AND URBANSKI, 1932: 49. MOSZYŃSKI, 1933: 239. WESENBERG-LUND, 1937: 332. (?). COLLINS, 1937: 194, Pl. 1, fig. 2. SCHAEFER, 1941: 202. KENK, 1941: 5. MARCUS, 1944: 10. BERG, 1948: 40. LASTOCKIN, 1949: 123. HERLANT-MEEWIS, 1951 a: 431, Pl. 1, fig. 4, Pl. 3, figs. 14-16; 1954: 78. AUCLAIR et al., 1951: 162. PENNAK, 1953: 286. ISTVAN, 1955: 9. KAMEMOTO AND GOODNIGHT, 1956: 219. MALEVIC, 1956: 404. JÄRNEFELT, 1956: 30. MOSZYŃSKI AND MOSZYŃSKA, 1957: 341. GOODNIGHT, 1959: 528. NAIDU, 1961: 648 (p.p.). CEKANOVSKAYA, 1962: 144. MOSZYŃSKA, 1962: 7. BRINKHURST, 1962: 318; 1963 a: 143; 1963 b: 15; 1967: 112. BUCHER, 1965: 97. JUGET, 1967: 218. BUNKE, 1967: 193, Figs. 15-16, 68-69, 73. AX AND BUNKE, 1967: 222, Fig. 2b. LAAKSO, 1967: 561.

Aeolosoma varium BEDDARD, 1895: 177.

(non) *Aeolosoma variegatum* Vejdovsky. MICHAELSEN, 1903 b: 170. CHEN, 1940: 17, Fig. 1B. KESSELYAK, 1942: 47.

Small, first zooid 0·8-1·2 mm; chains up to 4 zooids, up to about 2 mm. Fissionzone after VIII, rarely after VII or IX. Prostomium very slightly triangular, somewhat wider than following segments; tactile hairs up to 8 μ, numerous; ciliated field restricted to ventral surface, not reaching anterior tip; sensory pits round, diameter 12-15 μ, separated from ventral field. Epidermal glands globular or of somewhat irregular form, faintly greenish yellow, with many motile granules and globules; with satellite cells; colourless cells may also be present. Hair setae unequal, 2-6 per bundle; short setae 30-80 μ, 1-4 per bundle; long setae 95-135 μ, 0-2 per bundle. Intestine dilated from anterior part of IV-VI. First nephridia usually in III/IV. Copulatory gland in IV/V. Spermathecae in III and IV.

Europe, temperate zone, not very common, in various habitats including peat-moors. Some less reliable records from N. America (Canada; U.S.A.) and the tropics (Zanzibar; Surinam).

Colourless epidermal glands are not always present (Bunke, 1967). On the other hand Štolc (1903) states that coloured glands may be absent. The coloured glands are often very few in number and the faint colour may disappear rapidly from the vacuoles under circumstances, e.g., when the animals are observed under a coverglass (the vacuoles are not emptied then as in other species!).

Aeolosoma viride Stephenson, 1911

Aeolosoma spec. STEPHENSON, 1907: 233, Fig. 1, Pl. 8, figs. 1-4.

Aeolosoma viride STEPHENSON, 1911: 205.

Aeolosoma viride Stephenson. STEPHENSON, 1913: 743; 1923: 12; 1930: 525. STOLTE, 1933-55: 466. SCHAEFER, 1941: 204. MARCUS, 1944: 10, Figs. 13-14. HERLANT-MEEWIS, 1951 b: 231, Figs. 1-2, 4-5, Pl. 1, figs. A-F; 1953: 120, Figs. 1-13, Pl. 1, figs. 1-9, Pl. 2, figs. 1-9; 1954: 75, Pl. 3, figs. 1-17, Pl. 4, figs. 18-24. AVEL, 1959: 369, Fig. 286. ERCOLINI, 1960: 145. BRINKHURST, 1963 a: 143; 1967: 112. NAIDU, 1966: 210. CORIC AND CORIC, 1965: 233. BUNKE, 1967: 193, Fig. 21.

Aeolosoma viridae Stephenson. NAIDU, 1961: 643.

Aeolosoma viridis Stephenson. DALES, 1963: 157.

Large, first zooid about 2 mm × 200-300 μ; chains up to 8 zooids, up to 8 mm. Fission-zone after VII or VIII (or IX); complicated paratomy, with simultaneous strobilation, budding and segmentation. Prostomium rounded, wider than following segments; tactile hairs up to 25 μ; ciliated field with large latero-dorsal continuations; sensory pits large, oval, situated posterior to level of brain, not connected with ventral field. Epidermal glands globular or of irregular form, green, or yellowish or brownish green, with satellite cells. Hair setae unequal, 2-8 per bundle; short setae 80-100 μ, 0-5 per bundle; long setae 190-250 μ, 1-4 per bundle. Posterior end two-lobed. Intestine dilated from anterior part of IV to VI or VII. Nephridia regularly distributed, 4-7 pairs in first zooid; first pair in II/III.

Pakistan, Jugoslavia (?), Italy, Belgium, Brazil.

Aeolosoma litorale Bunke, 1967
Fig. 13. 2B

Aeolosoma litorale BUNKE, 1967: 190, Figs. 9-10, 33, 50-61, 70-73.

Aeolosoma litorale Bunke. AX AND BUNKE, 1967: 222-224, Figs. 1, 2a, 6.

First zooid 1·6-1·8 mm × up to 175 μ; chains up to 7 zooids, up to 4 mm. Fission-zone after (VII)VIII to XII, after X in most specimens. Prostomium slightly triangular, evidently wider than following segments; short tactile hairs; ciliated field not reaching anterior tip, with small dorso-lateral continuations; evident sensory pits lacking. Epidermal glands globular or of irregular form, red, diameter 10-15 μ (but also numerous smaller glands: 3-8 μ). 7-10 hair setae per bundle; long setae 110-150 μ, 3-5 per bundle; short setae 70-95 μ, 4-6 per bundle. Intestine dilated from posterior part of III to posterior part of VII. First pair of nephridia in II/III; up to 9 pairs in the first zooid. Copulatory gland V to VI. 3 pairs of spermathecae in II, III, and IV.

Germany, River Weser near Bremen and outlets of some small rivers in the Baltic Sea, up to a salinity of 5%. Finnland, brackish water along southwest coast.

Aeolosoma olivaceum Bunke, 1967

Aeolosoma olivaceum BUNKE, 1967: 190, fig. 1g.

First zooid 1 mm×90 μ; chains up to 6 zooids, up to 3 mm. Fission-zone after VII, rarely after VI. Prostomium rounded, slightly wider than following segments; ciliated field restricted to ventral surface, which is nearly entirely covered; sensory pits separated from ventral field, diameter 12-15 μ. Epidermal glands globular or of somewhat irregular form, faintly yellowish green; no satellite cells. 3-6 (7) hair setae per bundle; long setae (110) 120-140 (160) μ, 1-3 per bundle; short setae 50-85 μ, 2-4 per bundle. Intestine dilated from posterior part of III to anterior part of VII. First pair of nephridia in II/III or III/IV.

Among vegetable debris in ditch, Camargue, France.

Aeolosoma psammophilum Bunke, 1967
Fig. 13. 1C

Aeolosoma psammophilum BUNKE, 1967: 190, Figs. 6, 8, 36-49.

Aeolosoma psammophilum Bunke. AX AND BUNKE, 1967: 222, Figs. 4-5.

First zooid 1·2 mm×75-100 μ; chains of 2 or 3 zooids, up to 2 mm. Fission-zone after VIII or IX. Prostomium slightly wider than following segments; tactile hairs 5-7 μ; ciliated field with small dorso-lateral continuations; sensory pits round, often not separated from ventral field. Epidermal glands globular, diameter 2-7 μ, red; no satellite cell. Hair setae, sigmoid setae, and intermediate forms present; total number 4-6 (7) per bundle. Typical hair setae 60-80 μ, 1-5 per bundle. Sigmoid setae with 5-10 fine teeth on convex side of distal end, 36-42 μ, 1-4 per bundle, lacking in anterior segments. Intestine dilated from posterior part of III to anterior part of VII. First nephridia in II/III; irregularly distributed.

Mesopsammon, River Weser near Bremen, Germany.

Aeolosoma tenuidorsum Baskin, 1928

Aeolosoma tenuidorsum BASKIN, 1928: 229, Figs. 1-9.

Aeolosoma tenuidorsum Baskin. MARCUS, 1944: 13. HERLANT-MEEWIS, 1954: 91. BUNKE, 1967: 266.

Aeolosoma variegatum Vejdovský. CEKANOVSKAYA, 1962: 144.

Chains 2-4 zooids, 2·5-4 mm, 11-16 segments. Fission-zone after (VI) VIII-X. Prostomium pointed, much wider than following segments; sensory pits present. Epidermal glands large, of irregular form, yellow or yellow and green. 5-9 hair setae per bundle; long dorsal setae 175-250 μ; short dorsal setae 90-135 μ; ventral setae slightly longer. Intestine dilated from posterior part of III-IX. Commissural vessel present. Pairs of nephridia from II/III to second last segment.

On waterplants, River Charkow near Charkow, U.S.S.R.

The original description was very detailed, but obscure in several aspects. Therefore Marcus and Bunke considered this a dubious species. However, this is the only large species with yellow epidermal glands, so, if the original description is correct, this species can certainly be recognized. The great variation in the location of the fission-zone (from VI to X) is somewhat unusual.

Aeolosoma evelinae Marcus, 1944
Figs. 13. 1F, 13. 2C

Aeolosoma evelinae MARCUS, 1944: 14, Figs. 3-4.

Aeolosoma evelinae Marcus. MARCUS, E., 1944: 2. HERLANT-MEEWIS, 1951 b: 232; 1954: 89. YAMAGUCHI, 1957: 162. DIONI, 1961: 108, Pl. 2, fig. 11. NAIDU, 1961: 647. BUNKE, 1967: 190, Figs. 5, 7.

First zooid 0·5-1 mm × 50-100 μ; chains up to 6 zooids, up to 3 mm. Fission-zone after VI-VIII. Prostomium much wider than following segments; tactile hairs up to 8 μ; ciliated field with latero-dorsal continuations, which are pointed anteriorly; sensory pits deep, round and relatively large, connected with the ventral field. Epidermal glands red, globular, diameter up to 6-10 μ, without satellite cells. Hair setae unequal, 0-7 per bundle; long setae 70-110 μ, 0-2 per bundle; short setae 45-70 μ, 0-5 per bundle. Sigmoid setae sickle shaped, smooth or with a number of small teeth on concave side of distal end, 28-40 μ, 0-2 per bundle, only in the ventral bundles, where they are mostly restricted to the posterior segments, occasionally even absent in the first zooid. Nephridia irregularly distributed, first pair in III/IV. Intestine dilated from the posterior part of III to the anterior part of VI.

Common in S. America (Uruguay, Brazil and Surinam). Once found in Germany.

The specimens from Germany differ from the specimens from S. America in that the sigmoid setae bear a row of small teeth in the former and are smooth in the latter. The S. American specimens are generally somewhat smaller.

Aelosoma leidyi Cragin, 1887
Figs. 13. 1G, 13. 2A

Aeolosoma leidyi CRAGIN, 1887: 31.

Aeolosoma leidyi Cragin. BEDDARD, 1892 b: 353; 1895: 177. MICHAELSEN, 1900: 15. KENK, 1941: 1, Figs. 1-2. MARCUS, 1944: 11. PENNAK, 1953: 289, Figs. 175, 183B, 183D. GOODNIGHT, 1959: 528. NAIDU, 1961: 648. VAN DER LAND, 1965: 238. JUGET, 1967: 218. BUNKE, 1967: 193, Fig. 22.

Chains of 2 or 3 zooids 1·5-3 mm × 120-170 μ. Fission-zone after IX or X. Prostomium slightly wider than following segments; tactile hairs 5-14 μ; ciliated field not reaching anterior tip, restricted to ventral surface, but lateral parts visible in dorsal view; sensory pits round, connected with ventral field. Epidermal glands globular or of irregular form, with satellite cell, pale olive to bright green, diameter up to 25 μ. Hair setae unequal, 0-6 (8) per bundle; long setae 90-180 μ; short setae 30-80 μ. Sigmoid setae with one or more small teeth on concave side of distal end or smooth (?), only slightly sickle shaped, 45-70 μ, 0-4 per bundle, often lacking in II. First pair of nephridia in II/III, usually in all segments. Intestine dilated from anterior part of IV to posterior part of VII.

U.S.A., Kansas, Shawnee County, in creek; Michigan, Ann Arbor, dry bottom of temporary pool. France, Lac Léman. The Netherlands, several localities, in watermains and sandfilters of waterworks.

These animals move rather slowly and they tend to attach to detritus particles. Mucous secretion is considerable, so this probably is a tube-building species.

Aeolosoma sawayai Marcus, 1944
Fig. 13. 1D

Aeolosoma sawayai MARCUS, 1944: 10, Figs. 9-10.

Aeolosoma sawayai Marcus. MARCUS, E., 1944: 2. YAMAGUCHI, 1953: 282. DIONI, 1961: 108, Pl. 2, fig. 10. NAIDU, 1961: 648. BUNKE, 1967: 193, Fig. 24.

Very small, first zooid 0·5-0·6 mm × 50-60 μ; chains up to 4 zooids, up to 2 mm, but generally smaller, about 1 mm. Fission-zone after VI or VII (or VIII). Prostomium slightly wider than following segments; a few tactile hairs; ciliated field restricted to ventral surface, not reaching anterior tip; sensory pits relatively large, diameter up to 15 μ, most often connected with ventral field. Epidermal glands globular, lemon. Hair setae unequal, 0-3 (4) per bundle; short setae 40-60 μ; long setae 80-105 μ. Sigmoid setae with some very small teeth on concave side of distal end or smooth (?), slightly sickle shaped, 28-35 μ, 0-2 per bundle, most often absent in II. First nephridium in II/III, usually one per segment. Intestine dilated from posterior part of III to posterior part of V.

S. America, common in Brazil and Surinam, recorded once from Uruguay.

The sigmoid setae of the specimens from Brazil are stated to be bifurcate with 2 or 3 teeth near distal end (Marcus, 1944), but Dioni (1961) emphasizes that they are smooth in the specimens from Uruguay. In specimens from Surinam the sigmoid setae are certainly not bifurcate, but teeth can often be observed. However, it should be stressed that the teeth are very small indeed and that they cannot be made visible under all circumstances and with every microscope.

The animals are agile and move very rapidly.

Aeolosoma travancorense Aiyer, 1926
Fig. 13. 1E

Aeolosoma travancorense AIYER, 1926: 136, Fig. 4.

Aeolosoma travancorense Aiyer. AIYER, 1929: 15, Fig. 1, Pl. 1, fig. 1. STEPHENSON, 1930: 723. STOLTE, 1933-55: 126, 546. CHEN, 1940: 26. SCHAEFER, 1941: 170. MARCUS, 1944: 10, Figs. 11-12, 15, 75; 1945: 25. MARCUS, E., 1944: 2. HERLANT-MEEWIS, 1951 a: 430; 1951 b: 232; 1954: 80. YAMAGUCHI, 1953: 282. STOUT, 1956: 99. NAIDU, 1961: 643, Fig. 3. DIONI, 1961: 108, Pl. 2, figs. 7-9. FOMENKO, 1962: 543. CEKANOV-SKAYA, 1962: 97, Fig. 73. NAIDU, 1963: 224; 1966: 210. BUNKE, 1967: 190, Figs. 27-28. BRINKHURST, 1967: 112.

Aeolosoma tracanvorense Aiyer. DIONI, 1961: 114.

First zooid 0·6-1 mm × 60-150 μ; chains up to 4 zooids, 1-2·5 mm. Fission-zone after VII or VIII. Prostomium slightly wider than following segments; tactile hairs long, up to 14 μ; ciliated field restricted to ventral surface, not visible in dorsal view; sensory pits round, widely separated from ventral field. Epidermal glands globular, without satellite cells, colourless, often not strongly refractive; besides cells with granular contents are common. Hair setae unequal or subequal, 50-100 μ, 0-5 per bundle. Sigmoid setae slightly sickle shaped, with very thin distal point and two rows of 5-10 rather large teeth near distal end, 30-45 μ, 0-4 per bundle, not in II. Nephridia irregularly distributed; first pair in II/III. Intestine dilated from anterior part of IV to anterior part of VI.

In all kinds of fresh-water habitats in Europe (U.S.S.R.; Germany; The Netherlands), India, and S. America (Uruguay; Brazil; Surinam). In Surinam it is by far the most common species of the genus.

This is a tube-building species. Movements are awkward, with much peristalsis.

Aeolosoma aureum Marcus, 1944

Aeolosoma aureum MARCUS, 1944: 15, Fig. la-b.

Aeolosoma aureum Marcus. HERLANT-MEEWIS, 1951 b: 277; 1954: 89. STOUT, 1956: 99. NAIDU, 1961: 648. BUNKE, 1967: 193, Fig. 23.

Chains up to 4 zooids, up to 3 mm × 60-100 μ. Fission zone after XI (or X?). Prostomium with short, inconspicuous tactile hairs; ciliary field restricted to ventral surface; sensory pits present, approximately round. Epidermal glands brilliant golden-green, diameter 10 μ. Only hair setae; (2) 4-6 (8) per bundle; short setae about 60 μ; long setae 100-150 μ, ventrally not over 120 μ. Intestine dilated from posterior part of III to anterior part of IX. Pairs of nephridia from III/IV to VII/VIII.

São Paulo, Brazil. In mountain stream, among Hepaticae on stones.

This species is perhaps identical with *Aeolosoma flavum*, from which it mainly differs in the more posteriorly located fission zone.

Aeolosoma corderoi Marcus, E., 1944.
Fig. 13. 1A

Aeolosoma corderoi MARCUS, E., 1944: 1, Figs. 1-3.

Aeolosoma corderoi Marcus, E. NAIDU, 1961: 647. BUNKE, 1967: 192, Fig. 13.

First zooid 0·4-0·6 mm × 60-80 μ; two zooids up to 1 mm. Fission zone after VII or VIII. Prostomium narrower than following segment; long tactile hairs (20 μ); small ventral ciliated field, not reaching the edges; round sensory pits, not contiguous with ventral field. Epidermal glands red. Hair setae 60-85 μ, 1-5 per bundle, sometimes lacking in posterior bundles. Sigmoid setae strongly sickle-shaped, with 6 or 7 teeth on concave side of distal end, 40-60 μ, 1 or 2 per bundle, lacking in II and sometimes in III. Intestine dilated from posterior part of III to middle of VI. First nephridia in III.

São Paulo, Brazil. In river, among *Utricularia* and *Eichhornia*. Tube-building species.

Aeolosoma flavum Stolc, 1903

Aeolosoma flavum STOLC, 1903 a: 3, Fig. 2.

Aeolosoma flavum Stolc. STOLC, 1903 c: 76. MARCUS, 1944: 12. NAIDU, 1961: 648. BUNKE, 1967: 190.

Aeolosoma tlavum STOLC, 1903 c: 77.

Chains of 3 zooids 1·5-2 mm × 110 μ. Fission zone after VIII, sometimes VII or IX. Prostomium wider than following segments; ciliated field restricted to ventral surface; sensory pits present. Epidermal glands golden-yellow, of irregular form, diameter 10 μ. Only hair setae, 2-5 per bundle; short setae 50-60 μ; intermediate setae 70-85 μ; long setae ventrally 85-100 μ, dorsally 110-140 μ. Intestine dilated from III-V. First nephridia in III/IV.

Pond near Prague (Czechoslovakia). Psammon of River Weser near Bremen (Germany).

SPECIES INQUIRENDAE

Aeolosoma aurigenum (Eichwald, 1847)

Nais aurigena EICHWALD, 1847: 359, Pl. 9, fig. 15.

Nais aurigena Eichwald. VEJDOVSKY, 1885 a: 22. BEDDARD, 1889 b: 51.

Aeolosoma aurigenum (Eichwald). CZERNIAVSKY, 1880: 302. VAILLANT,

1890: 463. BEDDARD, 1895: 182. MICHAELSEN, 1900: 15. MARCUS, 1944: 12. BUNKE, 1967: 265.

About 5 mm. Epidermal glands golden-yellow. Prostomium broad. 3-4 short setae per bundle.
Kangern near Riga (USSR).

The original description lacks detail. The epidermal glands are said to be arranged in longitudinal rows, but this is not shown in the figure. The depicted specimen has very short setae and 25 segments, but fission zones are not shown.

Aeolosoma balsamoi Maggi, 1865

Aeolosoma balsamo MAGGI, 1865: 6, Pl. 1, figs. 2a, 3a, 4a.

Aeolosoma balsamo Maggi. VAILLANT, 1890: 463, 469-470 .

Chaetodemus balsamoi Maggi. CZERNIAVSKY, 1880: 307.

Aeolosoma bolsamo Maggi. VEJDOVSKY, 1885 a : 17.

A large species. In original figure 7 segments. Epidermal glands orange-red. Two bundles of setae on either side of each segment, one with short and one with long setae.
Valcuvia, Lombardia, Italy.

A great difference between the dorsal and ventral bundles of setae would be quite characteristic, but it is doubtful whether the observations are reliable.

Aeolosoma crassum Moszynski, 1938

Aeolosoma crassum MOSZYNSKI, 1938 b: 271, Figs. 1-6.

Aeolosoma crassum Moszynski. MARCUS, 1944: 13. MOSZYNSKI AND MOSZYNSKA, 1957: 338. BRINKHURST, 1967: 112. BUNKE, 1967: 266.

Chains up to 850 μ; first zooid 450-750 μ. Fission-zone after VI to VIII. Only hair setae; (3) 4-5 (7) per bundle, generally including two long setae.
Altamira, Spain. Numerous in pools in cave.

Original description based on fixed material.

Aeolosoma decorum Ehrenberg, 1831

Aeolosoma decorum EHRENBERG, 1831: 60.

Aeolosoma decorum Ehrenberg. OKEN, 1832: 1286. DESMAREST, (?1845): 374. GRUBE, 1851: 105. LEYDIG, 1865: 360. MAGGI, 1865: 4. CLAUS, 1876: 420. GRIMM, 1877: 116.

Aeolonais decorum (Ehrenberg). GERVAIS, 1838: 14.

Aeolosoma ehrenbergi p.p. See under *Aeolosoma hemprichi*.

1·2 and 3·7 mm; 9-20 segments. Epidermal glands red. Setae shorter than width of body. Intestine dilated in IV.
Berlin, Germany.

An unidentifiable species.

Aeolosoma fiedleri Bretscher, 1896

Aeolosoma fiedleri BRETSCHER, 1896: 500.

Aeolosoma fiedleri Bretscher. BRETSCHER, 1899: 370; 1903: 31.

1-2 mm; 12-13 segments. Prostomium pointed, narrrower than body width; with long tactile hairs. Epidermal glands colourless. 2-4 hair setae per bundle; rather long, at least reaching the base of the next bundle; all of equal length. First nephridia in II/III.
Bogpools, Katzensee near Zürich, Switzerland.

This species has characters in common with both *Aeolosoma hyalinum* (length, pointed prostomium, long hair setae) and *Aeolosoma niveum* (setae of equal length). It is undoubtedly identical with one of these species, but the original description is insufficiently detailed to permit a definite conclusion.

Aeolosoma gracile Stolc, 1903

Aeolosoma gracile STOLC, 1903 a: 3.

Aeolosoma gracile Stolc. STOLC, 1903 c: 77. MARCUS, 1944: 13. BUNKE, 1967: 265.

A small species. First zooid 8 segments. Prostomium and hair setae as in *A. hemprichi*. Epidermal glands small, brilliant yellow.
Czechoslovakia, in ponds.

An unidentifiable species, probably identical with *A. flavum* or *A. variegatum*.

Aeolosoma italicum Maggi, 1865

Aeolosoma italicum MAGGI, 1865: 6, Pl. 1, figs. 1a, 5a, 6a, Pl. 2, figs. 7-10.

Aeolosoma italicum Maggi. VEJDOVSKY, 1885 a: 17. VAILLANT, 1890: 463. ISSEL, 1901: 8. CHINAGLIA, 1910: 2.

A small species. Epidermal glands red. Hair setae only, 2 per bundle, shorter than body-width. Intestine dilated in IV.
Valcuvia, Lombardia, Italy.

Aeolosoma josephi Stolc, 1903

Aeolosoma josephi STOLC, 1903 a: 2, Fig. 4.

Aeolosoma josephi Stolc. STOLC, 1903 c: 76. MARCUS, 1944: 13. BUNKE, 1967: 265.

A small species. Fission-zone after VI or VII, sometimes VIII. Prostomium as in *A. hemprichi*. Epidermal glands red(?). Hair setae only, length intermediate between those of *A. hemprichi* and *A. quaternarium*.
Moldau near Prague and some other localities in Czechoslovakia.

Aeolosoma laticeps (Dugès, 1830)

Derostoma laticeps DUGÈS, 1830: 77, Pl. 2, fig. 9.

Derostoma laticeps Dugès. MICHAELSEN, 1900: 14.

Nais? laticeps (Dugès). DUGÈS, 1837: 30.

Chaetogaster laticeps (Dugès). VAILLANT, 1890: 452.

2 mm. Epidermal glands red.
France.

Aeolosoma macrogaster Scmarda, 1861

Aeolosoma macrogaster SCHMARDA, 1861: 10, Pl. 17, fig. 154.

Aeolosoma macrogaster Schmarda. VEJDOVSKY, 1885 a: 18; 1885 b: 279. BEDDARD, 1895: 182. JANDA, 1901: 4. STEPHENSON, 1923: 40; 1930: 726. MOSZYNSKI, 1938 b: 273. MARCUS, 1944: 13. BUNKE, 1967: 265.

Pleurophleps macrogaster (Schmarda). VAILLANT, 1890: 459.

Pleurophleps macrogaster (Schmarda). MICHAELSEN, 1900: 16; 1903 a: 41.

2 mm×330 μ, 23 segments. Indistinct intersegmental furrows. Prostomium narrower than following segments; conspicuous tactile hairs. Coloured epidermal glands lacking. 4 hair setae per bundle. Intestine dilated from VII to XI.
San Juan del Norte, Nicaragua.

The description lacks detail, but would suffice for recognition. However, it is probably unreliable.

Aeolosoma maggii Cognetti, 1901.

Aeolosoma maggii COGNETTI, 1901: 2.

Aeolosoma maggii Cognetti. CHINAGLIA: 1910: 2. MARCUS, 1944: 13. BUNKE, 1967: 265.

2 mm; 17 segments; first zooid 11 segments. Prostomium not wider than following segments. Epidermal glands orange-red. Hair setae of equal length, 5-7 per bundle in anterior segments, less in posterior segments. Intestine dilated from IV. First pair of nephridia in III.
Regione Rizzeddu near Sassari, Sardinia, Italy; 1 specimen in well among algae.

Aeolosoma panduratum (Leidy, 1851)

Chaetodemus panduratus LEIDY, 1851 : 286.

Chaetodemus panduratus Leidy. CZERNIAVSKY, 1880 : 307. VAILLANT, 1890 : 463. MICHAELSEN, 1900 : 14. CHEN, 1940 : 26.

Aeolosoma panduratum (Leidy). VEJDOVSKY, 1885 a : 17.

Chains of 3 zooids 2 mm × 45 μ. First zooid 7 segments, 2nd. zooid 5 segments, 3rd. zooid 9 segments. Prostomium panduriform, wider than following segments, 50 μ. Coloured epidermal glands absent. No hair setae (?). Sigmoid setae aristate, 38 μ, 4 per bundle.
Probably Philadelphia, U.S.A.

Aeolosoma stokesi Cragin, 1887

Aeolosoma stokesi CRAGIN, 1887 : 31.

Aeolosoma stokesi Cragin. BEDDARD, 1889 a : 262.

Chain of two zooids 1·1×0·1 mm. Fission-zone after VII. Epidermal glands salmon-red. 4 or 5 unequal hair setae per bundle.
Common in Shunganunga creek, Shawnee County, Kansas, U.S.A.

An unidentifiable species.

Aeolosoma ternarium Schmarda, 1861

Aeolosoma ternarium SCHMARDA, 1861 : 10, Pl. 17, fig. 153.

Aeolosoma ternarium Schmarda. SCHMARDA, 1871 : Fig. 258. VEJDOVSKY, 1885 a : 18; 1885 b : 279. BEDDARD, 1895 : 182. JANDA, 1901 : 4. STEPHENSON, 1923 : 23; 1930 : 726. MOSZYNSKI, 1938 b : 273. MARCUS, 1944 : 12. NAIDU, 1961 : 643. MENDIS AND FERNANDO, 1962 : 45. NAIDU, 1966 : 210. BUNKE, 1967 : 265.

Pleurophlebs ternarius (Schmarda). VAILLANT, 1890 : 472.

Pleurophleps ternaria (Schmarda). MICHAELSEN, 1900 : 16; 1903 a : 41: 1909 b : 104.

2·5 mm × 0·5 mm; 11 segments. Prostomium rounded; with 10-12 tactile hairs. No coloured epidermal glands. Hair setae shorter than width of body, 3 per bundle. Intestine dilated in III and IV.
In stagnant water, near Galle, Ceylon.

Aeolosoma thermophilum Vejdovsky, 1905

Aeolosoma headleyi (Beddard). MRAZEK, 1903 : 3.

Aeolosoma thermophilum VEJDOVSKY, 1905 : 2, Fig. 1-3.

Aeolosoma thermophilum Vejdovsky. MEYER, 1926 : 343. STEPHENSON, 1930 : 216. STOLTE, 1933-55 : 496, Fig. 423. SCHAEFER, 1941 : 198. MARCUS, 1944 : 13. SPERBER, 1948 : 264. BUNKE, 1967 : 265.

Aeolosoma termophilum Vejdovsky. HERLANT-MEEWIS, 1954: 91. ERCOLINI, 1960: 9.

9 segments. Epidermal glands green.
Aquaria in greenhouse, botanical gardens, Prague, Czechoslovakia.

Vejdovský only described the nephridia in detail. It was his intention to give a complete description later, and his manuscript or notes may still be available.

Aeolosoma venustum Leidy, 1857

Aeolosoma venustum LEIDY, 1857: 46, Pl. 2, figs. 8-12.

Aeolosoma venustum Leidy. MAGGI, 1865: 5. VEJDOVSKY, 1885 a: 17. CRAGIN, 1887: 31. VAILLANT, 1890: 463.

Oeolosoma venustum Leidy. D'UDEKEM, 1858: 23.

0·6 mm; 8 segments. Prostomium large, rounded. Epidermal glands red. 4 hair setae per bundle; long setae longer than body width; short setae shorter than body width. Intestine dilated in IV.
Philadelphia, U.S.A.

An unidentifiable species.

GENUS **Hystricosoma** MICHAELSEN, 1926

Type species: Hystricosoma chappuisi MICHAELSEN

Very small species (chains about 1 mm). ?Epidermal glands. Intersegmental furrows present. Only hair setae; dorsal setae of each bundle arranged in two transverse rows or in an ellipse. Commissural vessels absent. ?Septa. ?Nephridia. ?Reproductive organs. Asexual reproduction by budding.
Europe. Epizoic on crayfishes (*Astacus*).

Hystricosoma chappuisi Michaelsen, 1926
Fig. 13. 3D

Hystricosoma chappuisi MICHAELSEN, 1926: 93. Figs. A-B.

Hystricosoma chappuisi Michaelsen. MICHAELSEN, 1928: Fig. 18. STEPHEN-SON, 1930: 642. WESENBERG-LUND, 1937: 332. MOSZYNSKI, 1938 c: 69. MARCUS, 1944: 8. GEORGEVITCH, 1957: 93, Figs. 1-7. CEKANOVSKAYA, 1962: 82, Fig. 76. BRINKHURST, 1967: 112.

Hystricosoma chapuisi Michaelsen. MICHAELSEN, 1933: 328.

Chains 0·5-1 mm×0·1-0·14 mm. Fission zone after X-XII. Dorsal bundles with (8)9-11(12) setae, up to 80 μ long; ventral bundles with 5-7 setae; slightly smaller than dorsal ones; setae within each bundle of approximately equal length.

Epizoic on *Astacus astacus* (L.) and perhaps other Astacidae. Germany, Czechoslovakia, Rumania, Jugoslavia.

Detailed descriptions of living specimens are still lacking.

SPECIES INQUIRENDAE

Hystricosoma insularum Michaelsen, 1933

Hystricosoma insularum MICHAELSEN, 1933: 238, Pl. 1, fig. 1.

Hystricosoma insularum Michaelsen. LASTOČKIN, 1935 a: 642. MARCUS, 1944: 8.

Chains of two zooids 0·7 mm × 140-150 μ. Fission zone after VII-X. Only hair setae, 4-7 per bundle, up to 150 μ.

In ponds on the Caribbean islands Bonaire, Klein Bonaire and Aruba.

The original description, based on fixed material, lacks detail. Michaelsen placed the species in the genus *Hystricosoma* but neither the original description nor the type material provide any reason for this. The species may be identical to several known *Aeolosoma* species. The various samples probably included more than one species.

Hystricosoma pictum (Schmarda, 1861)

Aeolosoma pictum SCHMARDA, 1861: 10, Pl. 17, fig. 155.

Aeolosoma pictum Schmarda. VEJDOVSKY, 1885 a: 18; 1885 b: 279. BEDDARD, 1889 a: 262; 1895: 182. VAILLANT, 1890: 460. MICHAELSEN, 1900: 15. JANDA, 1901: 3. STEPHENSON, 1931: 297. MOSZYNSKI, 1938 b: 273.

Hystricosoma pictum (Schmarda). MICHAELSEN, 1933: 328. MARCUS, 1944: 8.

1 mm × 0·2 mm; 11 segments. Intersegmental furrows present. Prostomium small. Epidermal glands purple. 4 very short hair setae per bundle. Intestine not dilated.

In stagnant water, Cauca-valley near Cali, Columbia.

GENUS **Potamodrilus** LASTOČKIN, 1935

Type species: Stephensoniella fluviatilis LASTOČKIN

Very small species (about 1 mm); 7 segments. Refractive or coloured epidermal glands absent. Intersegmental furrows present. With postanal adhesive tail. Hair setae present. Mouth with a pharyngeal pocket with a muscular tongue and two lateral swellings ventral to the pharynx. Incomplete septae. Commissural vessels absent. One pair of normal nephridia in anterior part of body. Copulatory gland in VI/VII. One median spermatheca just anterior to copulatory gland; consisting of several cells. Testes in IV and V. Two differentiated nephridia in V and VI act as male ducts, both opening to the outside in a midventral longi-

tudinal furrow in VI. A pair of ovaries in V-VI. Perhaps asexual reproduction by fragmentation.
Europe. Freshwater meiopsammon.

Potamidrilus fluviatilis Lastočkin, 1935
Figs. 13. 3C, E

Stephensoniella fluviatilus LASTOČKIN, 1935 a : 637, Figs. 1-5.

Potamodrilus fluviatilus (Lastočkin). LASTOČKIN, 1935 b : 488. MARCUS, 1944: 7. BUNKE, 1967: 189, Figs. 76-97.

Potamodrilus rivularis NEISWESTNOWA-SHADINA, 1935 : 568, Fig. 14A.

Potamodrilus stephensoni LASTOČKIN, 1937: 233.

Potamodrilus fluviatolis (Lastočkin). MOSZYNSKI, 1938 b : 271.

Potamodrilus rivularis Neiswestnowa-Shadina. WISZNIEWSKI, 1947 : 11. SCHULZ, 1952 : 9. DELAMARE-DEBOUTTEVILLE, 1960 : 185.

Potamodrilus stephensoni Lastočkin. SILLO, 1953 : 271. CEKANOVSKAYA, 1962 : 151, Fig. 75; 1964 : 113. BRINKHURST, 1967 : 112.

Stephensoniella fluviatilis Lastočkin. AX, 1963 : 59, Fig. 7B.

1-1·3 mm×0·1 mm, 7 segments. Many adhesive gland-cells in all segments. Prostomium with two transverse furrows and three relatively small mid-ventral fields of cilia; bundles of tactile hairs on the dorsal and lateral sides; sensory pits absent. Intersegmental furrows strongly marked, especially in posterior part of body. Short cylindrical tail, about 50 μ long. Hair setae unilaterally feathered; usually two per bundle, a long (125-150 μ) and a short one (50-70μ), sometimes 3 or 4; dorsal and ventral bundles close to each other. Anterior nephridia in II and III; nephropore near ventral setae in III. Copulatory gland consisting of a single circle of large gland-cells around the female pore, which is situated in the intersegmental furrow VI/VII. Spermatheca two-lobed; sperm stored in 3 or 4 cells of each lobe. Mid-ventral seminal groove in anterior half of VI; male pores in posterior end of this groove.

Meiopsammon of several rivers in Russia (Don, Moskva, Oka, Volga), Poland (Prypec), and Germany (Weser, Elbe). Beach of Baltic Sea in Russia (Kurskiy Zaliv).

GENUS **Rheomorpha** RUTTNER-KOLISKO, 1955

Type species: Aeolosoma neizvestnovae LASTOČKIN

Very small species (chains about 1 mm). Refractive or coloured epidermal glands mostly present. Intersegmental furrows absent. Two posterior terminal adhesive papillae. Setae absent, replaced by adhesive papillae. Septa absent. Commissural vessels absent. Two pairs of nephridia, one in anterior and one in posterior half of body. Spermathecae consisting of one cell. Functional testes in several segments. The posterior nephridia act as male ducts. One functional

ovary in the mid-body. Asexual reproduction by budding.
Europe. Freshwater meiopsammon.

Rheomorpha neizvestnovae Lastočkin, 1935.
Fig. 13. 3B, F

Aeolosoma neizvestnovi LASTOČKIN, 1935 a: 643.

Aeolosoma neiswestnovi Lastočkin. NEISWESTNOWA-SHADINA, 1935: 568,
Fig. 14B.

Aeolosoma neizwestnovae LASTOČKIN, 1937: 233.

Aeolosoma neistvestnovi Lastočkin. MOSZYNSKI, 1938 a: 275, Figs. 1-3;
1938 b: 271.

Aeolosoma neizvestnovae Lastočkin. MARCUS, 1944: 7. MOSZYNSKA, 1962:
8.

Aeolosoma neisvestnovae Lastočkin. WISZNIEWSKI, 1947: 11.

Aeolosoma neiswestnovae Lastočkin. LASTOČKIN, 1949: 123.

Rheomorpha neiswestnovae (Lastočkin). RUTTNER-KOLISKO, 1955: 56, Figs.
1-7; 1956: 384. AX, 1957: 430; 1963: 59, Fig. 7A. DELAMARE-
DEBOUTTEVILLE, 1960: 142, Figs. 54, 70. BRINKHURST, 1963 a: 139.
JUGET, 1967: 217. BUNKE, 1967: 189, Figs. 31-35.

Rheomorpha neiswestnowae (Lastočkin). RUTTNER-KOLISKO, 1957: 425, Fig.
3B. CEKANOVSKAYA, 1962: 82.

Aeolosoma neisvestnovi Lastočkin. MOSKYNSKI AND MOSZYNSKA, 1957: 341.

Aeolosoma neisviestnovi Lastočkin. MOSZYNSKI AND MOSZYNSKA, 1957: 390.

Rheomorpha neiwestnovae (Lastočkin). DELAMARE-DEBOUTTEVILLE, 1960:
Fig. 70.

Lastockinia nieznestnovae (Lastočkin). NAIDU, 1961: 639.

Rheomorpha neisvestnovae (Lastočkin). CEKANOVSKAYA, 1962: 85, Fig. 74.

Rheomorpha neisvestnovae (Lastočkin). NAIDU, 1963: 222.

Rheomorpha neizvestnovi (Lastočkin). BRINKHURST, 1967: 112.

First zooid 0.5-0.7×65-75 μ; chains up to 5 zooids, up to 1.3 mm. Prostomium
with a pair of lateral sensory pits; ventral cilia covering nearly the whole
ventral surface. Segmentation extremely obscure; first zooid probably 6 segments. Refractive epidermal glands colourless, yellowish, yellow-green or
yellow-orange, sometimes absent. Adhesive papillae most evident ventrally in
III-VI. Intestinal dilatation in III and IV. Nephridia in III and V. Clitellum
in V.

Meiopsammon of lakes and rivers in Russia, Poland, Germany and the
Alps. Beach of Baltic Sea in Russia (Kurskiy Zaliv).

NOMINA DUBIA

(non) *Aeolosoma gineti* Juget, 1959

Aeolosoma gineti JUGET, 1959: 391, Fig. 3.

Aeolosoma gineti Juget. BRINKHURST, 1967: 112. BUNKE, 1967: 266.

The description of this species is based on two fixed specimens from a cave in France. Nothing in this description indicates that the specimens belong to *Aeolosoma*; they are most probably Naididae.

(non) *Aeolosoma setosa* Georgévitch, 1957

Aeolosoma setosa GEORGÉVITCH, 1957: 95, Figs. 8-9.

Aeolosoma setosa Georgévitch. BRINKHURST, 1967: 112.

0·4 mm × 220 μ; 10-11 segments. Prostomium small, with two rows of conical teeth at its base. Intersegmental furrows mostly evident. No coloured epidermal glands. Hair setae only; exact arrangement obscure (there are said to be normal bundles and bundles with 8-10 shorter setae with their bases united). Septa present.

Epizoic on crayfishes (Astacidae), Cetina, Jugoslavia.

This species has been described poorly. If pharyngeal teeth are indeed present, it is a quite remarkable animal, certainly generically different from *Aeolosoma*. A more detailed study is necessary.

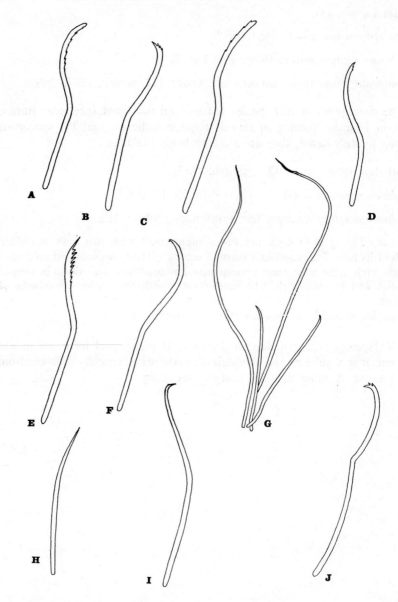

Fig. 13.1. *Aeolosoma* species, setae. (A-F, I,J—sigmoid setae).
A. *A. corderoi*; B. *A. japonicum*; C. *A. psammophilum;* D. *A. sawayi*; E. *A. travancorense*; F. *A. evelinae*; G. *A. leidyi*, setal bundle; H. *A. quaternarium*, short hair seta; I. *A. beddardi*; J. *A. tenebraum*.
(B,G,H,J from specimens from Netherlands, D-F,I from Surinam).

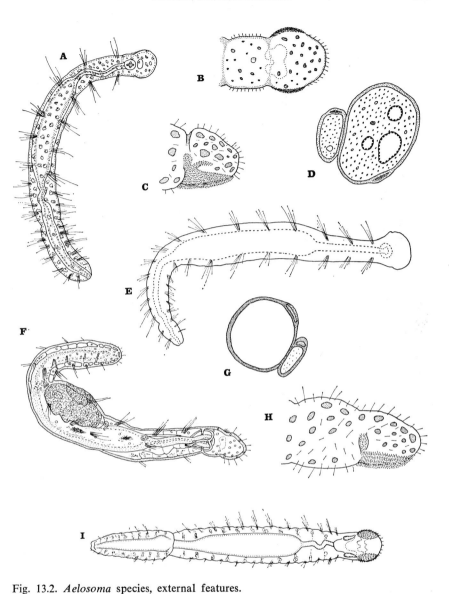

Fig. 13.2. *Aelosoma* species, external features.
A. *A. leidyi*, whole worm, slightly flattened. B. *A. litorale*, prostomium and first segment, dorsal view. C. *A. evelinae*, prostomium, lateral view. D. *A. variegatum*, epidermal gland. E. *A. gertae*, whole animal, from life. F. *A. quaternarium*, sexually mature individual. G. *A. tenebrarum*, epidermal gland. H. *A. quaternarium*, prostomium and first segment, lateral view (contracted). I. *A. marcusi*, whole worm. (A,D-H) from specimens from Netherlands, C from a specimen from Surinam).

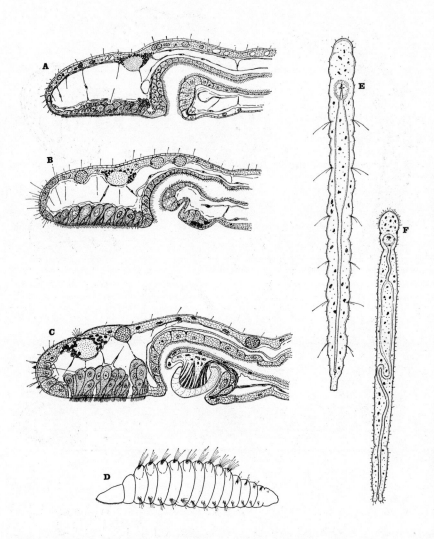

Fig. 13.3. *Potamodrilus, Hystricosoma, Rheomorpha,* and *Aeolosoma.*

A-C. Cross section of pharynx showing ventral pharyngeal bulb and lack of dorsal
  pharyngeal pad in *Aeolosoma* (A), *Rheomorpha* (B), and *Potamodrilus* (C). No
  illustration of this structure in *Hystricosoma* available.
D. *H. chappuisi,* whole animal, lateral view.
E. *P. fluviatilis,* whole animal, dorsal view.
F. *R. neizvestnovae,* whole animal, dorsal view.

REFERENCES

AIYER, K. S. P. 1926. Notes on the aquatic Oligochaeta of Travancore II. *Ann. Mag. nat. Hist.*, (9) **18**, 131

— 1929. An account of the Oligochaeta of Travancore. *Rec. Indian Mus.*, **31**, 13.

ALTMAN, L. C. 1936. Oligochaeta of Washington. *Univ. Wash. Publs Biol.*, **4** (1), 1.

ANONYMOUS, 1949. *Illustrated Encyclopedia of the fauna of Japan*. Tokyo.

AUCLAIR, J. L., HERLANT-MEEWIS, H., and DEMERS, M. 1951. Analyse qualitative des acides aminés de deux oligochaetes: *Aeolosoma hemprichi* et *Aeolosoma variegatum. Revue can. Biol.*, **10**, 162.

AVEL, M. 1959. Classe des Annélides Oligochaetes. In Grassé, P. P., *Traité de Zoologie*, **5** (1), 224. Paris.

AX, P. 1957. Die Einwanderung mariner Elemente der Mikrofauna in das limnische Mesopsammal der Elbe. *Zool. Anz. Suppl.*, **20**, 428.

— 1963. Die Ausbildung eines Schwanzfadens in der interstitiellen Sandfauna und die Verwertbarkeit von Lebensform-charakteren für die Verwandschaftsforschung. *Zool. Anz.*, **171**, 51.

— and BUNKE, D. 1967. Das Genitalsystem der Aeolosomatidae mit phylogenetisch ursprünglichen Organisationszügen für die Oligochaeten. *Naturwissenschaften*, **54**, 222.

BASKIN, B. 1928. Über eine neue Art der Gattung *Aeolosoma. Zool. Anz.*, **78**, 229.

BEDDARD, F. E. 1888. Observations upon an Annelid of the genus *Aeolosoma. Proc. zool. Soc. Lond.*, **1888**, 213.

— 1889a. Notes upon certain species of *Aeolosoma. Ann. Mag. nat. Hist.* (6), **4**, 262.

— 1889b. Note upon the green cells in the integument of *Aeolosoma tenebrarum. Proc. zool. Soc. Lond.*, **1889**, 51.

— 1892a. Note upon the encystment of *Aeolosoma. Ann. Mag. nat. Hist.*, (6) **9**, 12.

— 1892b. On some aquatic Oligochaetous worms. *Proc. zool. Soc. Lond.*, **1892**, 349.

— 1895. *A monograph of the order of Oligochaeta*. Clarendon Press, Oxford.

BEGER, H. 1966. *Leitfaden der Trink-und Brauchwasserbiologie*. 2nd ed. by J. Gerloff and D. Lüdemann. Stuttgart.

BERG, K. 1938. Studies on the bottom animals of Esrom Lake. *K. danske vidensk. Selsk.*, (9) **8**, 1.

— 1948. Biological studies on the river Susaa. *Folia limnol. scand.*, **4**, 1.

BOTEA, F. 1962. Contributii la studiul râspîndirii oligochetelor limicole din R. P. R. (valea Motrului) IV. *Studii Cerc. Biol.* (*Ser. Biol. animal.*), **14**, 401.

BOVERI-BONER, Y. 1920. Beiträge zur vergleichenden Anatomie der Nephridien niederer Oligochaeten. *Vischr. naturf. Ges. Zürich.* **65**, 506.

BRACE, E. M. 1898. Notes on *Aeolosoma tenebrarum Proc. Am. Ass. Advmt Sci.*, **47**, 363.

— 1901. Notes on *Aeolosoma tenebrarum. J. Morph.*, **17**, 177.

BRETSCHER, K. 1896. Die Oligochaeten von Zürich in systematischer u. biologischer Hinsicht. *Revue suisse Zool.*, **3**, 499.

— 1899. Beitrag zur Kenntnis der Oligochaeten-Fauna der Schweiz. *Revue suisse Zool.*, **6**, 369.

— 1900. Mitteilungen über die Oligochaetenfauna der Schweiz. *Revue suisse Zool.*, **8**, 1.

— 1903. Zur Biologie und Faunistik der wasserbewohnenden Oligochäten der Schweiz. *Biol. Zbl.*, **23**, 31, 119.

BRINKHURST, R. O. 1962. A checklist of British Oligochaeta. *Proc. zool. Soc. Lond.*, **138**, 317.

— 1963a. The aquatic Oligochaeta recorded from Lake Maggiore with notes on the species known from Italy. *Memorie Ist. ital. Idrobiol.*, **16**, 137.

BRINKHURST, R. O. 1963b. A guide for the identification of British aquatic Oligochaeta. *Scient. Publs Freshwat. biol. Ass. Br. Emp.*, **22**, 1.

— 1966. A contribution towards a revision of the aquatic Oligochaeta of Africa. *Zool. afric.*, **2**, 131.

— 1967. Oligochaeta. In Illies, J., *Limnofauna Europeae*. Stuttgart.

BUCHER, E. 1965. Der Wurm *Aeolosoma*. *Mikrokosmos*, **54**, 97.

BUISÁN, E. G. 1946. Contribución al estudio de los Elosómidos (Oligoquetos Naidimorfos). *Publnes Inst. Biol. apl., Barcelona*, **2**, 149.

BÜLOW, T. von. 1955. Oligochaeten aus den Endgebieten der Schlei. *Kieler Meeresforsch.*, **11**, 253.

— 1957. Systematisch—autökologische Studien an eulitoralen Oligochaeten der Kimbrischen Halbinsel. *Kieler Meeresforsch.*, **13**, 69.

BUNKE, D. 1967. Zur Morphologie und Systematik der Aeolosomatidae Beddard 1895 und Potamodrilidae nov. fam. (Oligochaeta). *Zool. Jb. (Syst.)*, **94**, 187.

CARUS, J. V. 1863. Vermes. In Peters, Carus and Gerstaecker, *Handbuch der Zoologie*, II: 422. Leipzig.

CEKANOVSKAYA, O. V. 1962. *Vodnye maloščetinkovye cervi fauny SSSR*. Moskva.

— 1964. Maloščetinkovye cervi reki Oki. *Trudy zool. Inst. Akad. Nauk*, **32**, 113.

CERNOSVITOV, L. 1930. Oligochaeten aus Turkestan. *Zool. Anz.*, **91**, 7.

— 1931. Příspevky k poznání fauny tatranských Oligochaetu. *Mém. Soc. r. Sci. Bohème*, **1930** (9), 1.

CHEN, Y. 1940. Taxonomy and faunal relations of the limnitic Oligochaeta of China. *Contr. biol. Lab. Sci. Soc. China (zool. Ser.)*, **14**, 1.

CHINAGLIA, L. 1910. Catalogo sinonimico degli Oligocheti d'Italia *Boll. Musei Zool. Anat. comp. R. Univ. Torino*, **27** (655), 1.

CLAUS, C. 1876. *Grundzüge der Zoologie*, 3e Aufl. Marburg and Leipzig.

— and GROBBEN, K. 1917. *Lehrbuch der Zoologie*, 9e Aufl. Marburg.

COGNETTI, L. 1901. Gli Oligocheti della Sardegna. *Boll. Musei Zool. Anat. comp. R. Univ. Torino*, **16** (404), 1.

COLE, G. A. 1955. An ecological study of the microbenthic fauna of two Minnesota lakes. *Am. Midl. Nat.*, **53**, 213.

COLLINS, D. S. 1937. The aquatic earthworms (Microdrili) of Reelfoot Lake. *J. Tenn. Acad. Sci.*, **12**, 188.

CORBELLA, C., DELLA CROCE, N., and RAVERA, O. 1956. Plancton, bentos e chimismo delle acque e dei sedimenti in un lago profondo (Lago Maggiore). *Memorie Ist. ital. Idrobiol.*, **9**, 125.

CORDERO, E. H. 1931. Notas sobre los Oligoquetos del Uruguay I. *An. Mus. nac. Hist. nat. B. Aires*, **36**, 343.

— 1951. Sobre algunos oligoquetos limicolas de Sud America. *Publnes Inst. Invest. Cienc. biol. Montev.*, **1**, 231.

CORIC, L. A., and CORIC, D. 1965. Activité et localisation de la cholinestérase chez certaines espèces d'Aeolosomatidae au cours de la reproduction asexuée. *Bull. Soc. zool. Fr.*, **89**, 232.

CRAGIN, F. W. 1887. First contribution to a knowledge of the lower Invertebrata of Kansas. *Bull. Washburn Coll. Lab. nat. Hist.*, **2**, 27.

CZERNIAVSKY, V. 1880. Materialia ad zoographiam ponticam comparatam. fasc. III. *Vermes. Byull. mosk. Obshch. Ispyt. Přír*, **55** (4), 213.

DALES, R. P. 1963. *Annelids*. London.

DEHORNE, L. 1916a. Contribution à l'étude du genre *Aeolosoma*. *Bull. Mus. Hist. nat., Paris*, **22**, 122.

— 1916b. Les Naidomorphes et leur reproduction asexuée. *Archs. Zool. exp. gén.*, **56**, 25.

DELAMARE DEBOUTTEVILLE, C. 1960. *Biologie des eaux souterraines littorales et continentales*. Paris.

DESMAREST, E. (no date; about 1845). *Nais*. In d'Orbigny, *Dictionnaire universel d'histoire naturelle*, **9**, 373.

DIONI, W. 1961. El género *Aeolosoma* en el Uruguay. *Actas Trab. Primer Congr. sudam. Zool.*, **2**, 107.

DITLEVSEN, A. 1904. Studien an Oligochäten. *Z. wiss. Zool.*, **77**, 398.

DUGÈS, A. 1830. Aperçu de quelques observations nouvelles sur les Planaires et plusieurs genres voisins. *Annls. Sci. nat.*, (1) **21**, 72.

— 1837. Nouvelles observations sur la zoologie et l'anatomie des Annelides abranches sétigères. *Annls Sci. nat.* (*zool.*), (2) **8**, 15.

EHRENBERG, C. G. 1828. Symbolae Physicae, seu Icones et descriptiones Corporum Naturalium novorum aut minus cognitorum, quae ex itineribus per Libyam, Aegyptum, Nubiam, Dongalam, Syriam, Arabiam et Habessiniam. *Pars. Zoologia*; *Animalia evertebrata, exclusis insectis.* Plates. Berolini.

— 1831. Text of the same (Pages not numbered).

EICHWALD, (E). 1847. Erster Nachtrag zur Infusorienkunde Russlands. *Bull. Soc. imp. nat. Moscou*, **20** (4), 285.

ERCOLINI, A. 1960. Nota sistematica sopra la specie *Aeolosoma viride* Steph. (Aeolosomatidae, Oligocheti, Microdili) rinvenuta in Piemonte. *Boll. Ist. zool. Univ. Torino*, **6**, 145.

FERRONNIÈRE, G. 1899. IIIᵉ Contribution à l'étude de la faune de la Loire-inférieure. Oligochètes littoraux et supra-littoraux (Annélides oligochètes). *Bull. Soc. Sci. nat. Ouest France*, **9**, 229.

FINOGENOVA, N. P. 1968. Malošcetinkovye cervi bassejna reki Nevy. *Trudy zool. Inst. Akad. Nauk*, **45**, 233.

FOMENKO, N. V. 1962. Novi vidi malošcetinkovih červiv. (Oligochaeta) dlja r. Dnipra. *Dopov. Akad. Nauk ukr. RSR*, **4**, 542.

FRENZEL, J. 1891. Untersuchungen über die mikroskopische Fauna Argentiniens. Vorläufiger Bericht. *Arch. mikrosk. Anat.*, **38**, 1.

GARBINI, A. 1895. Appunti per una limnobiotica Italiana. *Zool. Anz.*, **18**, 105.

GEORGÉVITCH, J. 1957. Contribution à la connaissance des aeolosomatidés de la Yougoslavie. *Bull. Acad. serbe Sci. Cl. Sci. math. nat.* (*N.S.* 18 *Sci. nat.*), **5**, 93.

GERVAIS, P. 1838. Sur la disposition systématique des Annélides chétopodes de la famille des Nais. *Bull. Acad. r. Belg.*, **5**, 13.

GOODNIGHT, C. J. 1959. Oligochaeta. In H. B. Ward and G. C. Whipple, *Freshwater Biology*, 2nd ed, 522. New York.

GREBNICKIJ, N. 1873. Materialy dija funy Novorossijskago kraja III. K. faunê otkrytyh limanov. *Zap. Novoross. Obshch. Estest.*, **2** (2), 267.

GRIFFITHS, A. B. 1898. Sur la composition de l'aeolosomine. *C. r. Acad. Sci.*, *Paris*, **127**, 448.

GRIMM, O. A. 1877. K poznaniju fauni Baltijskago morja i istorii eja vozniknovenija *Trudy petrogr. Obshch. Estest.*, **1878**, 116.

GRUBE, A. E. 1851. *Die Familien der Anneliden mit Angabe ihrer Gattungen und Arten. Ein systematischer Versuch.* Berlin.

HÄMMERLING, J. 1924. Die ungeschlechtliche Fortpflanzung und Regeneration bei *Aeolosoma hemprichii.* Histologische und experimentelle Untersuchungen. *Zool. Jb.* (*Phys.*), **41**, 581.

— 1930. Vergleichende Untersuchungen über Regeneration, Wachstum und Embryonalentwicklung bei Tubifex. *Zool. Jb.* (*Phys.*), **48**, 349.

HARMAN, W. J. 1966. Some aquatic Oligochaetes from Mississippi. *Am. Middl. Nat.*, **76**, 239.

— and PLATT, J. H. 1961. Notes on some aquatic oligochaetes from Louisiana. *Proc. La Acad. Sci.*, **24**, 90.

HERLANT-MEEWIS, H. 1950a. Les lois de la scissiparité chez *Aeolosoma Hemprichi* (Ehr.) *Annls ACFAS*, **16**, 140.

— 1950b. Réproduction asexuée et enkystement chez *Aeolosoma hemprichi* (Ehrenberg). *Revue can. Biol.*, **9**, 211.

— 1950c. Les lois de la scissiparité chez *Aeolosoma hemprichi* (Ehrenberg). *Revue can. Biol.*, **9**, 123.

— 1950d. Cyst-formation in *Aeolosoma hemprichi* (Ehr.). *Biol. Bull. mar. biol. Lab., Woods Hole*, **99**, 173.

— 1951a. Enkystement chez les oligochètes Aeolosomatidae. *Revue can. Biol.*, **9**, 429.

— 1951b. Les lois de la scissiparité chez les Aeolosomatidae. *Aelosoma viride. Annls Soc. r. zool. Belg.*, **82**, 231.

— 1953. Contribution a l'étude de la régéneration chez les oligochètes Aeolosomatidae. *Annls. Soc. r. zool. Belg.*, **84**, 117.

HERLANT-MEEWIS, H. 1954. Étude histologique des Aeolosomatidae au cours de la réproduction asexuée *Archs Biol., Liège*, **65**, 73.
— and BOULANGER, J. P. 1950. Le genre *Aeolosoma. Annls ACFAS*, **16**, 137.
ISSEL, R. 1900. Saggio sulla fauna termale italiana. *Atti Acad. Torino*, **36**, 53.
— 1901. Osservazioni sopra alcuni animali della fauna termale italiana. *Boll. Musei Lab. Zool. Anat. comp. R. Univ. Genova*, **106**, 1.
ISTVÁN, A. 1955. *Gyürüsférgek* I. *Annelida* I. *Magyarország Allatvilága*, **3** (10). Budapest.
JANDA, V. 1901. Příspěvky ku poznání rodu Aeolosoma. *Sber K. böhm Ges. Wiss. math. nat Classe*, **1900** (31), 1.
— 1902. Bemerkungen zu M. Brace's Arbeit "Notes on *Aeolosoma tenebrarum*". *Zool. Anz.*, **25**, 172.
JÄRNEFELT, H. 1956. Materialien zur Hydrobiologie des Sees Tuusulanjärvi. *Acta Soc. Faun. Flor. fenn.*, **71** (5), 1.
JUGET, J. 1959. Recherches sur la faune aquatique de deux grottes du Jura méridional francais: La grotte de la Balme (Isère) et la grotte de Corveissiat (Ain). *Annls Spéléol.*, **14**, 391.
— 1967. Quelques données nouvelles sur les Oligochètes du Léman: Composition et origine du peuplement. *Annls Limnol.*, **3**, 217.
KAESTNER, A. 1955. *Lehrbuch der speziellen Zoologie*, **1** (1). Wirbellose. Jena.
— 1960. *Lehrbuch der speziellen Zoologie*, **1** (1). Wirbellose. Jena.
KAMEMOTO, F. I., and GOODNIGHT, C. J. 1956. The effects of various concentrations of ions on the asexual reproduction of the oligochaete *Aeolosoma hemprichi. Trans. Am. microsc. Soc.*, **75**, 219.
KENK, R. 1941. Notes on three species of *Aeolosoma* (Oligochaeta) from Michigan. *Occ. Pap. Mus. Zool. Univ. Mich.*, **435**, 1.
KESSELYÁK, A. 1942. Über einige für die Tierwelt Ungarns neue Wasser-Oligochaeten. *Fragm. faun. hung.*, **5**, 47.
KNÖLLNER, F. H. 1935. Ökologische und systematische Untersuchungen über litorale und marine Oligochäten der Kieler Bucht. *Zool. Jb. (Syst.)*, **66**, 425.
KONDO, M. 1936. A list of naidiform Oligochaeta from the waterworks plant of the city of Osaka. *Annotnes zool. jap.*, **15**, 382.
KORSCHELT, E. 1914. Über Transplantationsversuche, Ruhezustände und Lebensdauer der Lumbriciden. *Zool. Anz.*, **43**, 537.
KOWALEWSKI, M. 1914. Materyaly do fauny polskich skaposzczetów wodnych (Oligochaeta aquatica) II. *Spraw. Kom. fizyogr., Kraków (Sect. bot. zool.)*, **48**, 107.
KRAEPELIN, K. 1886. Die Fauna der Hamburger Wasserleitung. *Abh. naturw. Ver. Hamburg*, **9**, 3 (reprint).
KRIBS, H. G. 1910. The reactions of Aeolosoma to chemical stimuli. *J. exp. Zool.*, **8**, 43.
LAAKSO, M. 1967. Records of acquatic Oligochaeta from Finland. *Annls zool. fenn.*, **4**, 560.
LAMEERE, A. no date. *Précis de zoologie*, II. *Caractères fondamentaux des coelomates. Les vers*. Liège.
— 1895. *Manuel de la faune de Belgique*, I. *Animaux non Insectes*. Bruxelles.
LAND, J. VAN DER, 1965. Notes on microturbellaria from freshwater habitats in The Netherlands. *Zool. Meded.*, **40**, 235.
LASTOCKIN, D. A. 1935a. Two new river Aeolosomatidae (Oligochaeta limicola). *Ann. Mag. nat. Hist.*, (10) **15**, 636.
— 1935b. New name for the genus *Stephtensoniella* Lastočkin (Oligochaeta). *Ann. Mag. nat. Hist.*, **16**, 488.
— 1937. Novye vidy Oligochaeta limicola v faune evropejskoj casti SSSR (New species of Oligochaeta limicola in the European part of the USSR). *Dokl. Akad. Nauk SSSR*, **42**, 233 (*Dokl. Akad. Nauk SSSR for. Lang. Edn*, **42**, 233).
— 1949. Kolcatye ščetinkovye červi (Chaetopoda). In V. I. Zadina, *Zhizn presnyh bod SSSR*, **2**, 111.
LEENTVAAR, P. 1946. Biologie en mogelijke bestrijdingswijze van enkele wormachtige organismen, voorkomende in waterleidingnetten. *Water Den Haag*, **1946** (12, 13), 2 (reprint).

LEIDY, J. 1851. Corrections and additions to former papers on helminthology published in the Proceedings of the Academy. *Proc. Acad. nat. Sci. Philad.*, **1851**, 285.

— 1857. Descriptions of some American Annelida abranchia. *J. Acad. nat. Sci. Philad.*, (2) **2** (1), 43.

LEVINSEN, G. M. R. 1884. Systematisk-geografisk-Oversigt over de nordiske Annulata, Gephyrea, Chaetognathi og Balanoglossi II. *Vidensk. Meddr. dansk naturh. Foren.*, **1883**, 92.

LEYDIG, F. 1865. Ueber die Annelidengattung *Aeolosoma. Arch. Anat. Physiol.*, **1865**, 360.

LEUNIS, J .1886. *Synopsis der Thierkunde*, II. 3e Aufl. von H. Ludwig. Hannover.

MAGGI, L. 1865. Intorno al genere *Aeolosoma. Memorie Soc. ital. Sci. nat.*, **1** (9), 3.

MALEVIC, I. I. 1956. Maloščetinkovye červi (Oligochaeta) Moskovskoj oblasti. *Uchen. Zap. mosk. gor. ped. Inst.*, **61**, 403.

— 1957. K. faune maloščetinkhovyh cervej (Oligochaeta) reki Hopra i ego pojmy. *Uchen. Zap. mosk. gor. ped. Inst.*, **65**, 109.

MARCUS, E. 1944. Sôbre Oligochaeta límnicos do Brasil. *Bolm Fac. Filos. Ciênc. Univ. S. Paulo (Zool.)*, **8**, 5.

— 1945. Sôbre Microturbelários do Brasil. *Comun. zool. Mus. Hist. nat. Montev.* **1**, 1.

MARCUS, EVELINE DU BOIS-REYMOND, 1944. Notes on freshwater Oligochaeta from Brazil. *Bolm Fac. Filos. Ciênc. Univ. S. Paulo (Zool.)*, **1** (20), 1.

MAUPAS, E. 1919. Expériences sur la reproduction asexuelle des Oligochètes *Bull. biol. Fr. Belg.*, **53**, 150.

MENDIS, A. S., and FERNANDO, C. H. 1962. A guide to the fresh-water fauna of Ceylon. *Bull. Fish. Res. Stn Ceylon*, **12**, 1.

MEYER, A. 1926. Die Segmentalorgane von *Tomopteris catharina* (Gosse) nebst Bemerkungen über das Nervensystem, die rosettenförmigen Organe und die Cölombewimperung. Ein Beitrag zur Theorie der Segmentalorgane. *Z. wiss. Zool.*, **127**, 297.

MICHAELSEN, W. 1900. Oligochaeta. *Das Tierreich*, **10**. Berlin.

— 1903a. *Die geografische Verbreitung der Oligochaeten*. Berlin.

— 1903b. Oligochaeten. Hamburgische Elb-Untersuchung IV. *Mitt. naturh. Mus. Hamb.*, **19**, 169.

— 1905. Die Oligochäten Deutsch-Ostafrikas. *Z. wiss. Zool.*, **82**, 288.

— 1909a. Oligochaeta. In A. Brauer, *Die Süsswasserfauna Deutschlands*, **13**. Jena.

— 1909b. The Oligochaeta of India, Nepal, Ceylon, Burma and the Andaman Islands. *Mem. Indian Mus.*, **1**, i-ii, 103.

— 1926. Schmarotzende Oligochäten nebst Erörterungen über verwandtschaftliche Beziehungen der Archioligochäten. *Mitt. zool. StInst. Hamb.*, **42**, 91.

— 1928. Oligochaeta = Regenwürmer und Verwandte. In Kükenthal and Krumbach, *Handbuch der Zoologie*, **2** (2) 8, 1. Berlin and Leipzig.

— 1933. Süss und Brackwasser-Oligochäten von Bonaire, Curaçao und Aruba. *Zool. Jb. (Syst.)*, **64**, 327.

— and BOLDT, W. 1932. Oligochaeta der Deutschen Limnologischen Sunda-Expedition. *Arch. Hydrobiol. Suppl.*, **9**, 587.

MINKIEWICZ, S. 1914. Przeglad fauny jezior tatránskich. *Spraw. Kom. fizyogr., Kraków*, **48** (2), 114.

MOORE, J. P. 1906. Hirudinea and Oligochaeta collected in the Great Lakes region. *Bull. Bur. Fish., Wash.*, **25**, 153.

MOSZYŃSKA, M. 1962. *Oligochaeta. Catalogus faune Poloniae*, **11**, Warszawa.

MOSZYŃSKI, A. 1925a. Contribution à l'étude de la faune des Oligochètes aquatiques (Oligochaeta limicola) de la Grande Pologne. *Bull. Soc. Amis. Sci. Lett. Poznan (Ser. B)*, **1**, 27.

— 1925b. Materjaly do fauny skaposzczetów wodnych (Oligochètes limicola). *Pr. Kom. mat.-przyr., Poznan (Ser. B)*, **3**, 1.

— 1933. Skaposzczety (Oligochaeta) miasta Poznania (Les Oligochètes de la ville de Poznan). *Kosmos, Lwów (Ser. A)*, **57**, 235.

— 1935. Niektóre dane a ilościowem rozmieszczeniu skaposzczetów (Oligochaeta) jezior Wigierskich. *Archwm Hydrobiol. Ryb.*, **9**, 79.

MOSZYŃSKI, A. 1938a. *Aeolosoma neisvestnovi* Last. 1935 un Oligochète intéressant, psammique, nouveau pour la faune polonaise. *Archwm Hydrobiol. Ryb.*, **11**, 275.

— 1938b. Ein neuer Oligochaet, *Aeolosoma crassum* n. sp., aus der Tropfsteinhöhle von Altamira bei Santillana, Spanien. *Zool. Anz.*, **123**, 271.

— 1938c. Oligochètes parasites de l'écrevisse (*Potamobius astacus* L.) de la Yougoslavie, *Glasn. skops. nauči Društ.*, **18**, 69.

— and MOSZYŃSKA, M. 1957: Skaposzczety (Oligochaeta) Polsik i niektórych krajów sasiednich. Studium ekologiczno-zoogeograficzne. *Prf. Kom. biol.*, *Poznan*, **18**, 318.

— and URBAŃSKI, J. 1932. Étude sur le faune des serres de Poznan (Pologne). *Bull. biol. Fr. Belg.*, **66**, 45.

MRÁZEK, A. 1903. Ein Beitrag zur Kenntnis der Fauna der Warmhäuser. *Sber. böhm Ges. Wiss. math. naturw. Klasse*, **1902** (37), 1.

MUNSTERHJELM, E. 1904. Luettelo Hämeessä, Sääksmaen pitäjässä tavatuista vesi-oligochaeteista. *Meddn Soc. Fauna 'Flora fenn.*, **30**, 32.

— 1905 Verzeichnis der bis jetzt aus Finnland bekannten Oligochaeten. *Festschrift für J. A. Palmén*, **13**, 1. Helsingfors.

NAIDU, K. V. 1961. Studies on the freshwater Oligochaeta of South India I. Aeolosomatidae and Naididae. *J. Bombay nat. Hist. Soc.*, **58**, 639.

— 1963. Studies on fresh-water Oligochaeta of South India I. Aelosomatidae and Naididae, part 5. *J. Bombay nat. Hist. Soc.*, **60**, 201.

— 1966. Check-list of freshwater Oligochaeta of the Indian sub-continent and Tibet. *Hydrobiologia*, **27**, 208.

NEISWESTNOWA-SHADINA, E. 1935. Zur Kenntnis des rheophilen Mikrobenthos. *Arch. Hydrobiol.*, **28**, 555.

OKEN, 1832. Symbolae physicae. Animalia evertebrata exclusis Insectis. C. G. Ehrenberg, *Isis*, **1832**, 1274 (Review of Ehrenberg, 1828 and 1831).

ÖRSTED, A. S. 1842. Conspectus generum specierumque Naidum ad faunam Danicam pertinentium. *Naturh. Tidskr.*, **4**, 128.

OYE, P. VAN, 1927. Courte note au sujet du genre *Aeolosoma* au Congo Belge. *Revue zool. Afr.*, **15**, 359.

PARKER, R. A., and KAMEMOTO, F. J. 1959. The effects of various concentrations of ions on the asexual reproduction of the oligochaete *Aeolosoma hemprichi* II. *Trans. Am. microsc. Soc.*, **78**, 207.

PASQUALI, A. 1938a. Notizie sistematiche sugli Oligocheti acquicoli di Padova. *Boll. zool.*, **9**, 19.

— 1938b. Note biologiche sugli Oligocheti acquicoli di Padova. *Boll. zool.*, **9**, 25.

PENNAK, R. W. 1953. *Fresh-water Invertebrates of the United States*. New York.

PERRIER. E. 1897. *Traité de Zoologie*, IV, 1345. Paris.

PETERS, W. 1967. Untersuchungen zur Cystenbildung limikoler Oligochaeten, insbesondere der Gattung *Aeolosoma*. *Sber. Ges. naturf. Freunde Berl.* (*N.F.*), **7**, 50.

PIGUET, E. 1906. Oligochètes de la Suisse française. *Revue suisse Zool.*, **14**, 389.

— 1913. Notes sur les Oligochètes. *Revue suisse Zool.*, **21**, 111.

— and BRETSCHER, K. 1913. *Oligochètes. Catalogue des Invertébrés de la Suisse*, **7**, Genève.

POINTNER, H. 1911. Beiträge zur Kenntnis der Oligochaetenfauna der Gewässer von Graz. *Z. wiss. Zool.*, **98**, 626.

REISINGER, E. 1925. Ein landbewohnender Archiannelide (zugleich ein Beitrag zur Systematik der Archianneliden). *Z. Morph. Ökol. Tiere*, **3**, 197.

RIIKOJA, H. 1955. Eesti NSV selgrootute fauna uurimise küsimusi. *Loodusuur. Seltsi Aastar.*, **48**, 7.

ROMIJN, G. 1919. Aeolosomatidae en Naididae in Nederland. *Tijdschr. ned. dierk. Vereen.*, (2) **18**, xxxix.

RUTTNER-KOLISKO, A. 1955. *Rheomorpha neiswestnovae* und *Marinellina flagellata* zwei phylogenetisch interessante Wurmtypen aus dem Süsswasserpsammon. *Öst. zool. Z.*, **6**, 55.

— 1956. Psammonstudien III. Das Psammon des Lago Maggiore in Oberitalien. *Memorie Ist. ital. Idrobiol.*, **9**, 365.

RUTTNER-KOLISKO, A. 1957. Der Lebensraum des Limnopsammon. *Zool. Anz. Suppl.*, **20**, 421.

SCHAEFER, K. 1941. Histologische Untersuchungen über den Bau von *Aeolosoma quaternarium. Z. wiss. Zool.*, **155**, 159.

SCHMARDA, L. K. 1861. Neue Turbellarien, Rotatorien und Anneliden. In *Neue Wirbellose Thiere beobachtet und gesemmelt auf iner Reise um die Erde 1853 bis 1857*, I (2), 1. Leipzig.

— 1871. *Zoologie*, I. Wien.

SCHULZ, E. 1952. Über das Vorkommen des Oligochaeten (*Potamodrilus*) *Stephensoniella rivularis* (Lastochkin) in Deutschland. *Faun. Mitt. Norddeutschl.*, **1**, 9.

SCIACCHITANO, I. 1934. Sulla distribuzione geografica degli Oligocheti in Italia. *Archo. zool. ital.*, **20**, 1.

SEKERA, E. 1896. Über einen interessanten Turbellarienfundort. *Zool. Anz.*, **19**, 375.

SEMAL-VAN GANSEN, P. 1958. Les vers Naïdomorphes. *Naturalistes belg.*, **39**, 235.

SHRIVASTAVA, H. N. 1962. Oligochaetes as indicators of pollution. *Wat. Sewage Wks*, **109**, 40.

ŠILLO, N. V. 1953. K. izučeniju fauny bespozvonočyh Dona v predelah Gramjačenskogo rajona. Priroda hozjajstvo Gremjačinskogo rajona. *Voronežsk. goc. univ.*, **1953**, 266.

ŠMIDT, J. Je. 1896. K. poznaniju roda Aeolosoma. *Trudý Imp. S-peterb. Obshch. Estest.*, **27**, 161, 169.

SMITH, F. Notes on species of North American Oligochaeta. III. List of species found in Illinois, and descriptions of Illinois Tubificidae. *Bull. Ill. St. Lab. nat. Hist.*, **5**, 441.

SOUTHERN, R. 1909. Contributions towards a monograph of the British and Irish Oligochaeta. *Proc. R. Ir. Acad.*, **27** (*B*), 119.

SPERBER, C. 1948. A taxonomical study of the Naididae. *Zool. Bidr. Upps.*, **28**, 1.

STENROOS, K. E. 1898. Das Thierleben im Nurmijärvi-See. Eine faunistisch-biologische Studie. *Acta Soc. Fauna Flora fenn.*, **17** (1), 1.

STEPHENSON, J. 1907. Descriptions of two freshwater Oligochaete worms from the Punjab. *Rec. Indian Mus.*, **1**, 233.

— 1909. The anatomy of some aquatic Oligochaeta from the Punjab. *Mem. Ind. Mus.*, **1**, 255.

— 1911. On some aquatic Oligochaeta in the collection of the Indian Museum. *Rec. Ind. Mus.*, **6**, 203.

— 1913. On intestinal respiration in Annelids; with considerations on the origin and evolution of the vascular system in that group. *Trans. R. Soc. Edinb.*, **49**, 735.

— 1923. *Oligochaeta. The fauna of British India, including Ceylon and Burma.* London.

— 1930. *The Oligochaeta*. Oxford University Press.

— 1931. The Oligochaeta. Reports of an expedition to Brazil and Paraguay in 1926-7, supported by the Trustees of the Percy Sladen Memorial Fund and the Executive Committee of the Carnegie Trust for Scotland. *J. Linn Soc. (Zool.)*, **37**, 291.

ŠTOLC, A. 1890. O pohlavních organech rodu *Aeolosoma* a jejich pomeru ku organům exkrecním. *Sber. K. böhm. Ges. Wiss.*, **1889**, 183.

— 1903a. O životnim cyklu nejnižších sladkovodnich cervu kroužkovitých a o něketrych otazách biologických. Na základě pozorováni českých druhů rodu *Aeolosoma. Rozpr. ceské Akad.* (*Ser.* 2), **10** (17). 1.

— 1903b. Pokusy v řešení otázky o dědičnosti vlastnosti získaných mechanickým zasáhnutím neb vlivem ústředi při množeni nepohlavním. *Rozpr. českéAkad.* (*Ser.* 2), **10** (32), 1.

— 1903c. Über den Lebenscyklus der niedrigsten Süsswasserannulaten und über einige sich anschliessende biologische Fragen. Auf Grund der Beobachtungen an bömischen *Aeolosoma*-Arten. *Bull. int. Acad. tchéque Sci.* ۱*Sci. math.*), **7**, 74 (the same as Štolc, 1903a, but without plate).

— 1903d. Versuche betreffend die Frage, ob sich auf ungeschlechtlichem Wege die durch den mechanischen Eingriff oder das Milieu erworbenen Eigenschaften vererben. *Bull. int. Acad. tchéqe Sci. (Sci. math.)*, **7**, 153 (the same as Štolc, 1903b).

ŠTOLC, A. 1903e. Versuche betreffend die Frage, ob sich auf ungeschlechtlichem Wege die durch mechanischen Eingriff oder das Milieu erworbenen Eigenschaften vererben. *Arch EntwMech. Org.*, **15**, 638 (the same as Stolc, 1903b).

STOLTE, H. A. 1933-1955. Oligochaeta. In H. G. Bronn, *Klassen und Ordnungen des Tierreichs*, 4 (3, 3). Leipzig.

STOUT, J. D. 1952. The occurrence of aquatic Oligochaetes in soil. *Trans. R. Soc. N.Z.*, **80**, 97.

— 1956. Aquatic oligochaetes occurring in forest litter I. *Trans. R. Soc. N.Z.*, **84**, 97.

SZARSKI, H. 1947. Skaposzczety wodne zebrane w okolicach Krakowa w r. 1942. *Kosmos, Wroclaw (Ser. A.)*, **65**, 127.

TAUBER, P. 1879. *Annulata Danica I. En kritisk Revision af de i Danmark fundne Annulata Chaetognatha, Gephyrea, Balanoglossi, Discophoreae, Oligochaeta, Gymnocopa og Polychaeta.* Kjøbenhavn.

TIMM, R. 1883. Beobachtungen an *Phreoryctes Menkeanus* Hoffmr. und *Nais*, ein Beitrag zur Kenntnis der Fauna Unterfrankens. *Arb. zool.-zoot. Inst. Würzburg*, **6**, 109.

TIMM, T. 1959. Ülevaade Eesti Magevee-Väheharjasussidest (Oligochaeta) (A survey of the freshwater Oligochaeta of Estonia). *Faun. Märkm.*, **1**, 23.

TOIVONEN, D. 1911. Bidrag till kännedomen om södra Finlands vattenoligochaetfauna. *Meddn. Soc. Fauna Flora fenn.*, **37**, 15.

UDE, H. 1929. *Oligochaeta.* F. Dahl, *Tierwelt Deutschlands*, **15**. Jena.

UDEKEM, J. D. 1858. Nouvelle classification d'Annélides sétigères abranches. Mém. *Acad. Sci. Lett. Belg.*, **31**, 1.

— 1861. Notice sur organes génitaux des *OElosoma* et des *Chaetogaster*. *Bull. Acad. r. Belg. Cl. Sci.*, **2** (12), 243.

VAILLANT, L. 1890. Lombriciens, Hirudiniens, Bdellomorphes, Térétulariens et Planariens. In De Quatrefages and L. Vaillant, *Histoire naturelle des Annéles marins et d'eau douce*, **3** (2), 341. Paris.

VEJDOVSKY, F. 1880b. Vorläufiger Bericht über die Turbellarien der Brunnen von Prag, nebst Bemerkungen über einige einheimische Arten. *Sber. K. Böhm Ges. Wiss.*, **1879**, 501.

— 1884. Revisio Oligochaetorum Bohemiae. *Sber. K. Böhm Ges. Wiss.*, **1883**, 215.

— 1885a. *System und Morphologie der Oligochaeten.* Prag.

— 1885b. Aeolosoma variegatum Vejd. Prispevek ku *noznání nejnižších Annulatův. Sber. K. böhm. Ges. Wiss.*, **1885**, 275.

— 1892. Ueber die Encystierung von *Aeolosoma* und der Regenwürmer. *Zool. Anz.*, **15**, 171.

— 1905. Über die Nephridien von *Aeolosoma* und *Mesenchytraeus*. *Sber. K. böhm. Ges. Wiss. (Math.-nat. Kl.)*, **1905** (6), 1.

WACHS, B. 1963. Zur Kenntnis der Oligochaeten der Werra. *Arch. Hydrobiol.*, **59**, 508.

WALDVOGEL, T. 1900. Das Lützelsee und das Lautikerried 'ein Beitrag zur Landeskunde. *Vjschr. naturf. Ges. Zürich*, **45**, 277.

WERFF, A. VAN DER, 1946. Het voorkomen van wormachtige dieren in Nederlandse waterleidingsbedrijven. *Biol. Jaarb.*, **13**, 251.

WESENBERG-LUND, C. 1937. *Ferskvandsfaunaen biologisk belyst. Invertebrata*, 1, Köbenhavn.

WHITELEGGE, T. 1889. List of the marine and freshwater Invertebrate fauna of Port Jackson and the neighbourhood. *J. Proc. R. Soc. N.S.W.*, **23**, 163.

WISNIEWSKY, J. 1947. Remarques relatives aux recherches recentes sur le psammon d'eaux douces. *Archwm Hydrobiol. Ryb.*, **13**, 7.

YAMAGUCHI, H. 1953. Studies on the aquatic oligochaeta of Japan VI. A systematic report, with some remarks on the classification and phylogeny of the Oligochaeta. *J. Fac. Sci. Hokkaido Univ. (Zool.)*, **11**, 277.

— 1957. On *Aeolosoma hemprichi* Ehrenberg, obtained from subterranean water in Japan. *J. Fac. Sci. Hokkaido Univ. (Zool.)*, **13**, 161.

ZACHARIAS, O. 1885. Studien über die Fauna des grossen und kleinen Teiches im Riesengebirge. *Z. wiss. Zool.*, **41**, 483.

POSTSCRIPT—R. O. BRINKHURST

The Suborder Lumbricina will be considered in a slightly different manner from the rest of the Haplotaxida apart from the Enchytraeidae. The family Alluroididae is entirely aquatic (the habitat of one poorly known species being in doubt) and is dealt with in its entirety. Many of the genera of the Glossoscolecidae are wholly aquatic, and these are dealt with in depth, with a review and keys to the generic level for the rest included in order to make identification of the aquatic forms possible.

The remaining families (Megascolecidae, Eudrilidae, Lumbricidae) and the small order Moniligastrida (with its single family) are omitted for two reasons. Primarily, the aquatic representatives of these families are less numerous than the terrestrial forms—these are the true earthworms, some of which inhabit riverine, lacustrine or marine littoral habitats. Secondly, however, we must admit that there is need for a detailed review of much terrestrial material before the systematic position of the aquatic forms can be determined. Rather than delay the appearance of this volume further, the authors have left treatment of these families to a separate review to be published elsewhere at a later date. The junior author (B.G.J.) is undertaking the preparation of this review.

14

FAMILY ALLUROIDIDAE

Type genus: Alluroides BEDDARD, 1894

Setae single-pointed, sigmoid, 4 pairs per segment commencing on II; genital or penial setae sometimes present. Clitellum one cell thick, commencing on XI, XII or XIII and occupying 2-6 segments. Male pores one pair, intraclitellar, at the anterior border of XIII or in its setal arc. Female pores one pair, at the anterior border, or more posterior in XIV. Spermathecal pores on 1-3 of segments VI-IX, at or near their anterior margins, lateral to dorsal, paired or single. Dorsal pores (always?) absent. Alimentary canal with or without oesophageal gizzards; lacking diverticula or other appendages. Nephridia holonephridia. Testes in X or in X and XI. Seminal vesicles, when present, projecting into XI only or extending posteriorly through several segments as in the microdriles. Prostate glands either discharging separately from the male pores or through the latter, in some cases receiving the vasa deferentia entally. Ovaries anterior in XIII. Mature oocytes large and yolky, in ovisacs which extend posteriorly through several segments. Spermathecae simple, without diverticula.
Africa (S. of the Sahara) and S. America (Argentina).

The family Alluroididae was established by Michaelsen (1900) for *Alluroides pordagei* Beddard, 1894. Subsequent additions were *A. tanganyikae* Beddard, 1906, and two species erected by Brinkhurst (1964) in a revision of the family, namely *A. ruwenzoriensis* and *A. americanus*. The latter species is the only American record for an otherwise African group. A number of forms have been described (Michaelsen, 1913, 1914 a, 1914 b, 1915, 1935, 1936) which are of uncertain status but are referred in the present work to *A. pordagei*. In a taxonometric investigation, the author

708

(Jamieson, 1968) included, in the Alluroididae, the sub-family Syngeno-drilinae, a monotypic group containing only *Syngenodrilus lamuensis* Smith and Green, 1919, which those authors had placed in the Moniligastridae. In the same work the type of a new monotypic genus, *Standeria trans-vaalensis* (from the Transvaal) was described, *A. americanus* was made the type of a new, monotypic genus, *Brinkhurstia*, and *A. tanganyikae* s. Brink-hurst, 1964, (Mt. Elgon) was removed from *A. tanganyikae* Beddard and made the type of a new species, *A. brinkhursti* of which a new sub-species, *A. brinkhursti abyssinicus* was described from Ethiopia. *Alluroides*, *Brink-hurstia* and *Standeria* thus constituted the sub-family Alluroidinae while the Syngenodrilinae, as Gates (1945) had suggested it should, formed the other sub-family of the Alluroididae. The two sub-families may be dis-tinguished as follows:

1. Outer setal couples dorsal. Gizzards absent. Testes in X only. Seminal vesicles projecting into XI only, or absent. Prostate pores com-bined with the male pores   *Alluroidinae*

— Outer setal couples lateral, not dorsal. Giz-zards present. Testes in X and XI. Seminal vesicles extending posteriorly through several segments. Prostate pores separate from the male pores   *Syngenodrilinae*

SUBFAMILY ALLUROIDINAE

Type genus: Alluroides BEDDARD, 1894

A longitudinal lateral line present on each side between the ventral and dorsal setal couples. Dorsal median intersetal distances only about 0·2 of the circum-ference of the body ($dd=0.2$ u); genital setae or penial setae present or absent. Clitellum commencing in XII or XIII. Male pores ventral to lateral in the setal arc of XIII. Female pores at or near the anterior border of XIV in line with the male pores or nearly so. Spermathecal pores lateral to dorsal but never in line with the male pores, paired or dorsal median and single in VI-IX, maximally in 3 of these segments. Dorsal pores absent. Subneural blood vessel absent. Nephridia avesiculate. Prostates (atria) tubular or bulbous, receiving the male ducts, or discharging with the latter but separately from them into a terminal chamber; internal epithelium surrounded by a muscular sheath outside which prostatic cells are usually present. Gizzards absent, Proandric; testes in X only. Sperm funnels with their mouths directed anterodorsally. Seminal vesicles projecting into XI or absent. Ovisacs extending posteriorly through several segments.

Africa, S. of the Sahara. Argentina. (All aquatic.)

KEY TO GENERA, SPECIES AND SUBSPECIES

1. Spermathecal pores lateral. Ventral setal
    couples of XIV and XV replaced by enlarged
    genital setae   *Standeria transvaalensis*
—   Spermathecal pore(s) near to or in the dorsal
    midline. Ventral setal couples of XIV and XV
    unmodified.   2

2. Ventral setal couples of XIII replaced by one
    pair of elongated, bullet-shaped penial setae.
    Spermathecal pore mid-dorsal in IX. Atria
    very slender coiled tubes, each opening into a
    copulatory chamber separately from the vas
    deferens   *Brinkhurstia americanus*
—   Ventral setal couples of XIII unmodified or
    absent. Spermathecal pore(s) in one or more of
    VI to IX. Atria wide tubes or bulbous,
    receiving the vasa deferentia   3

3. Spermathecal pore(s) either in VIII or IX   4
—   Spermathecal pores unpaired in VI, VII and
    VIII or in VII and VIII only   *Alluroides ruwenzoriensis*

4. Atria wide, tortuous tubes. Spermathecal
    pores paired or singly dorsally in VIII   *Alluroides pordagei*
—   Atria massive and bulbous. Spermathecal
    pores unpaired in VIII or IX   5

5. Intestine commencing in XIX. Spermathecal
    pore at the anterior margin of IX   *A. tanganyikae*
—   Intestine commencing in XV or XVI. Sperma-
    thecal pore anterior in VIII or IX   6

6. Spermathecal pore anterior in IX   *Alluroides brinkhursti brinkhursti*
—   Spermathecal pore anterior in VIII (appearing
    posterior in VII)   *Alluroides brinkhursti abyssinicus*

GENUS **Alluroides**—BEDDARD, 1894, emend. JAMIESON, 1968

Type species: Alluroides pordagei BEDDARD

Alluroidinae in which the atria are wide, coiled tubes or are bulbous; prostate cells surround the muscular sheath of the atrium, and the vas deferens (always?) enters each atrium entally. Male pores lateral; ventral setae of XIII present or absent; genital and penial setae absent. Spermathecal pores paired or single, dorsal in VI, VII and VIII, or in VII and VIII only, or in either VIII or IX. First septum 3/4; septum 10/11 attenuated. First nephridia in XIII-XVI. Intestine commencing in XIII-XIX With or without distinct seminal vesicles but always with spermatogenic masses partly occluding XI.

Kenya; Rhodesia; Republic of the Congo; Tanzania (L. Tanganyika); Uganda (Mt. Ruwenzori; Mt. Elgon); Ethiopia.

Collectively the features listed above adequately diagnose the restricted genus from the remainder of the sub-family. An individual character of particular note which is not shared with *Brinkhurstia* and *Standeria* is the termination of the vasa deferens entally on each atrium, clearly demonstrated by Brinkhurst (1964, Fig. 4) for *A. ruwenzoriensis* and observed by the author in *pordagei*. The vasa deferentia of *brinkhursti* and *tanganyikae* have not been observed but the overall phenetic resemblance of these to the other two species points to the existence of such a connection. In the other genera entry of two ducts into each terminal male chamber is readily seen while *tanganyikae* and *brinkhursti* show no semblance of dual terminalia.

At the specific level the systematics of *Alluroides* cannot be considered settled. Much larger series, from more localities, together with ecological data, are needed before the validity of the present division into *A. pordagei*, *tanganyikae*, *ruwenzoriensis* and *brinkhursti* can be conclusively determined.

Alluroides brinkhursti brinkhursti Jamieson, 1968
Fig. 14. 3A, Frontispiece

Alluroides tanganyikae Beddard (part.) BRINKHURST, 1964: 528, Figs. 1-3, Pl. 1, Fig. 1-2, Pl. 2, Fig. 1-2.

Alluroides brinkhursti brinkhursti JAMIESON, 1968: 76, Fig. 13.

(Non) *A. tanganyikae* BEDDARD, 1906: 215. MICHAELSEN, 1913: 7; 1914 a: 89; 1914 b: 165; 1935: 36.

$l=35$-45 mm, $w=1$-$1\cdot5$ mm, $s=150$. Length of a lateral seta from XV ental to the node, $0\cdot109$ mm. in XI, $aa : ab : bc : cd : dd = 3\cdot7 : 1\cdot0 : 3\cdot1 : 1\cdot0 : 3\cdot6$; $dd : \mu = 0\cdot21$; $bc : aa = 0\cdot8$. Clitellum annular in $\frac{1}{2}$XII to XV. Male pores ventrolateral in XIII, just dorsal to b. Female pores in line with the male pores immediately behind furrow 13/14. Spermathecal pore single, mid-dorsal on IX just behind 8/9 on a prominent papilla, Septa very muscular from 4/5-9/10 inclusive; 3/4 not apparent; 10/11 dorsally partially thickened, mostly evaginated posteriad as a very thin-walled seminal vesicle; 11/12 and 12/13 only slightly thickened; 13/14 posteriad but not appreciably thickened. Intestine beginning in XVI. Chromophil septal glands in V-VIII, attached to posterior septum. Nephridia commencing in XV or XVI. Testes ventral on the anterior wall of X. No testis-sacs apparent. Seminal vesicle almost filling XI. Sperm funnels, posterior and ventral in X facing anteriorly and dorsally so as to serve both X and the seminal vesicle. Atria large and globular, almost filling XIII, 0·347 mm wide narrowing ectally to 0·204 mm. Each with lining epithelium surrounded by a thick muscular sheath outside which, but not closely adherent, are tongues of unicellular gland cells which communicate with the atrial lumen by a large number of small ductules which penetrate the muscular layer. Penis eversible and elongate. Ovaries in XIII, on its anterior wall; funnels on its posterior wall, penetrating 13/14. Ovisac extending posteriad into XVI. Mature oocytes large and yolky. Nephridia commencing in XV or XVI. Spermatheca 0·138 mm wide, single, lying recurved in 9.

Uganda, Mt. Elgon, in tributaries of the Kiriki River, which flows into Lake Okolitorom.

Alluroides brinkhursti abyssinicus Jamieson, 1968
Fig. 14. 1B, C; 14. 3B

Alluroides brinkhursti abyssinicius JAMIESON, 1968: 77, Fig. 9-11, 14.

$l=45$ mm, $w=1·6$ mm, $s=144$. Zygolobous. Body form slender, tapering uniformly to the somewhat pointed tail end; cylindrical throughout. A lateral line visible as a conspicuous white line running the length of the body on each side slightly above mid-*bc* and commencing at the lower limits of the peristomium. Ventral setal couples absent from XIII.

Clitellum annular, over XII-XV but its limits not determinable with certainty; best developed in XII-XIV. Male pores one on each side in the setal arc of XIII immediately below the lateral line and thus almost exactly lateral. Each at maturity at the depressed centre of a very protuberant, transversely elliptical porophore which extends forwards to intersegmental furrow 12/13 and is preceded by a small tumescent crescent in XII. Female pores 1 on each side, each a small but conspicuous transverse slit, with minute elliptical lips, in intersegmental furrow 13/14 which is deflected posteriorly; slightly more ventrally located than the male pores.

Spermathecal pore, a conspicuous unpaired mid-dorsal slit immediately in front of the posterior border of VII, with obvious but only slightly tumid lips which are surrounded by a transversely oval pale field. This field extending anteriorly almost to the setal arc and posteriorly impinging slightly on VIII.

Septa 4/5-12/13 fairly strong, with the exception of 10-11 which is very thin and is evaginated posteriad to form a bulbous seminal vesicle; 7/8-9/10 the thickest; 3/4 imperfect and very delicate; 13/14-15/16 thin but not as delicate as the succeeding septa. 13/14 is deflected posteriad by the atria so as almost to touch 14/15 and is prolonged backwards to XVIII as the (unpaired?) ovisac. Chromophil glands extend posteriorly to the posterior wall of VII. The first nephridia with distinguished lumina in XVI, but peritoneal masses in XV and some more anterior segments possibly represent nephridia.

The alimentary canal widens in XIII, but not until XV, which may be the segment of intestinal origin, is enlargement pronounced. Dorsal and ventral blood vessels are well developed; subneural vessel absent. Sperm morulae in X form a compact mass which suggests the presence of a testis-sac and are also contained in a fairly narrow stalked seminal vesicle formed by backward evagination of septum 10/11 into XI which it almost fills. The sperm funnel is situated in the spermatogenic mass at the neck of the seminal vesicle and is directed dorsally so as to serve both X and the seminal vesicle. Atria each a bulbous tube with narrow lumen, restricted to XIII, with lining epithelium of eosinophil glandular cells surrounded in turn by a thick sheath of predominantly circular muscle and a layer of similar width and several cells thick of unicellular gland cells; total width of muscular tube (external surface) 0·38 mm, narrowing to 0·195 mm ectally. Ovaries and funnels in XIII. The ovisac is almost occluded at the intersegmental septa but bulges dorsally to the gut in each segment and posteriorly contains large oocytes full of large eosinophil yolk granules. Spermatheca unpaired; a simple, narrow, blind tube, 0·14 mm wide, recurving at the ental third, widening ectally before opening, via a narrower section at its pore dorsally in VIII.

Ethiopia: Stream at 10,500 feet in Choké Mts (Mt. Talo). Senan District, Gojjam Province.

Alluroides pordagei Beddard, 1894
Fig. 14. 3C-E

Alluroides pordagei BEDDARD, 1894: 244, Fig. 4, 5.

Alluroides pordagei Beddard, BRINKHURST, 1964: 527. JAMIESON, 1968: 75, Fig. 15-18

? *Alluroides tanganyikae* Beddard. MICHAELSEN 1913: 7, Pl. XIX, fig. 9; 1914 a: 89; 1914 b: 165, Pl. V, fig. 13; 1915: 29; 1935: 36; 1936: 37 (part.)

(Non) *A. tanganyikae* Beddard, 1906.

l=25-48 mm, *w*=1·0 mm, *s*=202. Prolobous. Form approximately cylindrical throughout; anterior segments are at least biannulate with the transverse furrow shortly postsetal. Lateral line in *bc*, nearer to *b* than to *c*. Anus terminal. Ventral setal couples of XIII absent at maturity; all setae sigmoid, with approximately central, well developed node; none modified as genital setae. Outer setal couples dorsal, *dd* : *u*=*c*. 0·2. Clitellum annular, XII-XVI. Male pores a pair precisely lateral, or slightly ventrolateral, nearer to *b*—than to *c*— lines in the setal arc of XIII, each surrounded by a circular tumescence almost filling the segment longitudinally. Female pores in 13/14 directly in line with the male pores, inconspicuous. Spermathecal pore paired or single, dorsal and anterior in VIII.

Septa 4/5 (the first clearly developed) 9/10 exceptionally strongly thickened, 10/11 thin and attenuated, 11/12 and (though less so) 12/13 moderately thickened; the succeeding septa thin. Septal glands posterior in V-IX? Nephridia commencing in XII-XVI. Pharynx anterior to septum 4/5. Intestine commencing in XIII, internal epithelium with tall cilia; oesophagus similarly ciliated in XII in front of the oesophageal valve; ciliation not certainly present further anteriorly. Brown-granuled chlorogogen cells commencing (always?) at ½ IX. Dorsal blood vessel large and adherent to the gut in XIII posteriorly; slender and freer anteriorly to this. Hearts very tortuous extending posteriorly into XI(XII?). Septum 10/11 bulging posteriorly far into XI to form a wide-mouthed (unpaired?) seminal vesicle. Muscular tubes of the atria forming slender somewhat coiled tubes extending from XIII into XIV or as far as XVI and widening immediately before opening at the male pores. With the exception of the terminal ectal chamber each tubular atrium is surrounded by a sheath of unicellular prostate cells, which approximately doubles the width of the apparatus, the cells communicating with the atrial lumen by minute ductules. Vasa deferentia running in the glandular prostatic duct and presumably entering the atrium entally. Ovaries and funnels in XIII. Oviduct penetrating the body wall immediately behind septum 13/14. Female gonadial cells, at maturity, in XIII-XXI, probably in ovisacs. Spermatheca single or paired in VIII.

Kenya: swamp 4 miles inland from Mombasa (Type locality, Beddard, 1894); lower forest region of Mt. Kenya at 2400 m (Michaelsen, 1914 a, b); Kahawa stream, Athi River tributary, near Nairobi, collecter B. G. M. Jamieson, 11 Dec. 1967, new record. Republic of the Congo: Mulongo, Niunzu; Albertville (Michaelsen, 1935); Leopoldville (Michaelsen, 1936). Rhodesia: swampy earth, Zambezi River, near Victoria Falls (Michaelsen, 1913).

The above account omits reference to Michaelsen's material, the status of which is somewhat uncertain, but incorporates data from the new material from the Athi River, Kenya. The extreme septal thickening, and its distribution, now emerge as diagnostic features of *A. pordagei* (condition unknown, however, in *A. tanganyikae* s. Beddard and s. Michaelsen). Intraspecific variation from a paired to a single spermatheca (in VIII) is confirmed and Brinkhurst's conclusion (1964) that *A. tanganyikae* s. Michaelsen (spermatheca single in VIII), but not s. Beddard (spermatheca single in IX), should be included in *pordagei* thus gains further support.

Alluroides ruwenzoriensis Brinkhurst, 1964

Alluroides ruwenzoriensis BRINKHURST, 1964: 531, Fig. 4.

Alluroides ruwenzoriensis Brinkhurst. JAMIESON, 1968: 74.

l=100 mm, w just exceeding 1 mm, s=150. All segments, except first 2 or 3, triannulate. Setae broad, simple pointed; ventral couples absent from XIII at maturity. Spermathecal pores middorsal on VII and VIII (also on VI in 1 specimen). Sperm funnels, testes, seminal vesicles and ovaries as in *pordagei*. Atria elongate, with glandular covering, standing vertically in XIII to which they are restricted. Vas deferens thin walled, joining the median face of the atrium about midway along the length of the latter. Each atrium narrowing distally and terminating in a well-defined penis which is visible externally on a raised papilla. Spermathecae single (in one case the spermathecal duct had 2 unequal ampullae).

Uganda: Ruwenzori Mountains (lakes and stream at 12,500 and 12,900 feet, Nyamagasani Valley).

Alluroides tanganyikae Beddard, 1906

Alluroides tanganyikae BEDDARD, 1906: 215.

Alluroides tanganyikae Beddard (part.) BRINKHURST, 1964: 528.

(Non) *Alluroides tanganyikae*; MICHAELSEN, 1913: 7; 1914 a: 89; 1914 b: 165: 1915: 29; 1935: 36.

$l=30$ mm, $w=1·5$ mm, $s=60$. Prostomium long and pointed, transversely bisected by a constriction, but zygolobous. Ventral setal couples absent from XIII. Clitellum? Male pores on XIII in line with the ventral setae (of adjacent segments). Female pores in intersegment 13/14 in *ab*. Spermathecae single, pore middorsal in 8/9, with tumid periphery and very conspicuous.

Septal glands obvious. Intestine beginning in XIX, transition from moniliform oesophagus abrupt. Chloragogen cells apparently beginning in IX. Nephridia? Atria a pair ending posteriorly in oval expansions; directed posteriorly to the pores.

Lake Tanganyika (at about 10 fathoms).

The status of *A. tanganyikae* Beddard (1906) is uncertain owing to omissions and contradictions in the type description and the paucity of information yielded by the holotype (Brinkhurst, 1964). Its inadequate characterization, together with an apparent difference in location of the male pores and in intestinal origin, have made it necessary to separate *A. tanganyikae* from the better known material from Mt. Elgon referred to this species by Brinkhurst (1964), the latter being regarded (Jamieson, 1968) as a distinct species, *A. brinkhursti*. The affinity between *tanganyikae* and *brinkhursti* cannot be settled until topotypic material of the former taxon is collected from Lake Tanganyika. Lake Tanganyika is noted for the high endemicity of its fauna. Occurrence of *A. tanganyikae* in the lake at a depth of 60 feet, whereas *brinkhursti* is a montane form, lends some support to taxonomic separation. Its affinities with *pordagei* are also uncertain but in the present study *A. pordagei* has been shown to be clearly distinguished from *A. brinkhursti* (Mt. Elgon and Ethiopian sub-species) in its extreme septal thickening. If this distinction is found to exist between *pordagei* and *tanganyikae* it will considerably strengthen the grounds for specific separation.*

GENUS **Brinkhurstia** JAMIESON, 1968

Type species: Alluroides americanus BRINKHURST

Alluroidinae in which the atria are very slender, much coiled tubes, prostatic cells surround the muscular sheath of the atrium, and the opening of each atrium into the rounded terminal, non-muscular chamber is separate from that of the corresponding vas deferens. Male pores ventrolateral; ventral setal couples of XIII absent. Penial setae present. Spermathecal pore single, dorsal, in IX. First septum 3/4; 10/11 not attenuated. First nephridia in IX? Intestine commencing in XIII-XIV. No distinct seminal vesicles.

Argentina (Monotypic).

Unique features, within the Alluroidinae, of this genus are the extreme

* See Lauzanne (1968) listed in Appendix, p. 838.

attenuation of the atrial prostates, the ratio width muscular tube of the atrium: width body being only 0·036 at maturity, and the presence of penial setae. It is also the only known American species. The generic significance of the penial setae, *per se*, is questionable, however, as their presence or absence varies intragenerically in other families.

Brinkhurstia resembles *Standeria*, and differs from *Alluroides*, in possessing at each male pore a distinct, rounded terminal chamber into which two ducts, positively identified as the atrium and vas deferens of the corresponding side, discharge but the chamber is exceptional in *Brinkhurstia* in being non-muscular. A particularly noteworthy difference from *Standeria*, shared with *Alluroides*, is the presence of prostatic cells ensheathing the atria.

Brinkhurstia americanus (Brinkhurst), 1964
Fig. 14. 1E

Alluroides americanus nom. nud. CERNOSVITOV, 1936: 19.

Alluroides americanus BRINKHURST, 1964: 533.

Brinkhurstia americanus (Brinkhurst), JAMIESON, 1968: 80, Fig. 3, 12.

$l=30$ mm, $w=0·75$ mm, $s=100$-130. Prostomium short and rounded. Ventral setal couples of XIII replaced by a single penial seta on each side; in the ovarian segments $aa:ab:bc:cd:dd=4·7:1:5:1·1:6·2$; $dd=0·25$ μ; length of a lateral seta from XI, base to node$=87$ μ. Clitellum annular, slightly less developed ventrally; segments ½XII-XV. Lateral lines present. Male pores ventrolateral in XIII. Female pores anterior in XIV at about one-quarter of the distance from septum 13/14-14/15. Spermathecal pore single, middorsal, anteriorly in IX.

Septum 3/4 excessively delicate and doubtfully complete; 4/5 and 14/15 hardly appreciably, 6/7-8/9 fairly strongly thickened (two tiers of muscle fibres) the remainder intermediate; 10/11 has one tier of muscle fibres and is not backwardly deflected. Alimentary canal expanding greatly in XVI but no definite transition from oesophagus to intestine. Perienteric blood sinus in IX posteriad, not appreciable in front of this. Testes discrete elongated tongues of tissue dependent from the anterior wall of segment 10. Gonadial tissue absent from 11. Seminal vesicles absent. Atria very thin much coiled tubes extending from mid-XIII into XIV; discharging in the close proximity of the corresponding vas deferens through the posterior wall of a globular non muscular ectal chamber; a simple, straight, unornamented penial seta in a follicle approximately twice the length of the chamber, penetrates the anterior wall of the chamber. Each prostate has a cuboidal lining epithelium surrounded in turn by a very thin muscular sheath and an external layer of prostatic cells only one to a few cells thick, though considerable masses of prostatic cells are formed between adjacent coils. Width muscular tube of atrium (external) 28 μ; width ectal chamber 37 μ; length penial seta 115 μ; greatest width 18 μ. Ovaries dependent from the anterior wall of XIII and extending posteriad into ovisacs formed by the backward evagination of septum 13/14 through as many as 8 segments; diameter of the largest oocyte-nuclei 33 μ. First nephridia (from peritoneal masses) in IX. Spermatheca in IX a simple elongated pouch, 87 μ wide, with no distinct ampulla, lined internally by a short columnar epithelium invested in a single layer of circular muscle and an outer glandular? peritoneal sheath.

Argentina: Arroyo Pastora.

GENUS **Standeria** JAMIESON, 1968

Type species: Standeria transvaalensis JAMIESON

Alluroidinae in which the atria have narrow ducts and bulbous ental ends, the prostatic cells are solely internal to the muscular sheath of the atrium, and the opening of each atrium into the terminal chamber is separate from that of the corresponding vas deferens. Male pores ventral at the sites of the absent ventral setal couples of XIII. Genital setae present. Spermathecal pores paired, lateral in IX. First septum 4/5; 10/11 attenuated. First nephridia in XI. Intestine commencing in XV. No distinct seminal vesicles.

Transvaal (Monotypic).

Standeria shows some notable resemblances to *Brinkhurstia* which distinguish the two genera from *Alluroides*. In both, rounded terminal male chambers each receive an atrial duct and a second duct which in *Brinkhurstia* is known to be the vas deferens and in *Standeria* is apparently so. Furthermore in both genera the atria are significantly thinner relative to the body width than those of *Alluroides*. Possession of genital setae in *Standeria* and penial setae in *Brinkhurstia* are possibly further evidence of phylogenetic affinity. Both genera are distinctly austral.

Standeria transvaalensis Jamieson, 1968
Fig. 14. 1A; 14. 2A-F

Standeria transvaalensis JAMIESON, 1968: 82, fig. 1, 2, 4-8.

$l=25$ mm, $w=1\cdot1-1\cdot4$ mm, $s=82$. Prostomium epilobous 1/2 although lacking a separate dorsal tongue. Form cylindrical; somewhat flattened ventrally at the clitellum. Lateral lines conspicuous, in mid *bc*; the posterior third of IV and a number of more posterior segments set off as a distinct annulus. Intersetal ratios in XII; $dd=0\cdot2$ u; $bc=1\cdot7$ aa; ventral pairs absent in XIII; replaced in XIV and XV on each side by 6 enlarged genital setae located on paired papillae.

Lengths of 4 of 6 genital setae on the left side in XIV$=0\cdot33-0\cdot35$ mm. Length of lateral setae from the same segment $0\cdot19$ mm. Neither type of seta with distinct node but both sigmoid with tapering extremities. Greatest widths: 15 μ for a genital seta and 10 μ for a lateral seta; ornamentation absent. Clitellum saddle-shaped; apparently extending from 1/2 XII-*c*. XV, 1/2 XVI, 1/2 XVII ($=3$ 1/2, 4, 5 segments); ventral margins at c. 1/4 *bc*. Male pores crescentic slits on low porophores at the sites of the absent ventral setae of XIII. Each porophore continuous posteriorly with a slight tumescence which extends to approximately midway between the former and the setal zone of XIV; indentations at the posterior limits of each tumescence are the female pores, anterior in XIV. Ventral setal couples of XVI and XVII each on slight but distinct protuberances. These protuberances confluent medially to form feebly developed transverse pads of glandular appearance. Spermathecal pores: 1 pair, perforate and conspicuous in intersegmental furrow 8/9, lateral of setal lines *b*, with very large tumescent anterior and posterior lips each of which occupies a third of the adjacent segment. Nephropores in setal-lines *a* very close to the setal follicles.

Septa: 4/5 the first visible, it and 12/13 slightly thickened. 5/6, 7/8, 9/10 and 11/12 relatively strongly thickened; 6/7 and 8/9 moderately thickened; 10/11 and 13/14 unthickened and very delicate, as are 14/15 and succeeding septa. Septal glands attached dorsally to anterior faces of septa as far as 9/10 and also 11/12, 12/13? Oesophagus widening to form the intestine, which is at least twice as wide, in mid-XV; longer and more slender in XIII than in other segments. Dorsal and ventral blood vessels single. No subneural. A pair of commissural loops winding far laterally is present in XII anteriad. Nephridia at least as far anteriorly as XI; ducts expanding in the outermost region of the longitudinal muscle but apparently lacking bladders. Large lobed testes ventral and anterior in X embedded in free morulae and abundant spermatozoa which fill the remainder of the segment. Sperm funnels large convoluted areas of septum 10/11; seminal vesicles absent. No reproductive organs or cells in XI. On each side, overlying the external male pore on XIII is a rounded muscular atrial chamber into which discharges the tubular entally dilated atrium of the corresponding side. Its ectal portion is narrowed to form a slender duct which curves ventrally and medially to join the anterolateral aspect of the atrial chamber. This ectal region is obscured by a web of muscle fibres, passing from the ventral to the lateral body wall, which bind it to the latter. A second, smaller sac about the width of the atrial duct, projects from the junction of the atrial chamber with the duct. The latter sac apparently represents the thickened end of the vas deferens.

The walls of the atria and of the appended sacs are lined by cuboidal epithelium outside which is a layer of circular muscle fibres several times as thick. The wide lumen of the atrial chamber is lined by columnar epithelium outside which is a thick sheath of radial and circular muscle fibres. There is no glandular investment of these male terminalia.

Ovaries in XIII. Septum 13/14 evaginated posteriorly on each side as a broad funnel continuous into XIV as a slender oviduct. Spermathecae a pair of elongated, adiverticulate pouches in IX; their scarcely narrowed ducts being almost wholly embedded in the body wall.

S. Africa: Eastern Transvaal (Lake Chrissie).

Subfamily Syngenodrilinae

Defintion as the only genus.

genus **Syngenodrilus** SMITH AND GREEN, 1919

Type species: Syngenodrilus lamuensis SMITH AND GREEN

Lateral lines unknown. Genital and penial setae absent; the dorsal median intersetal distance = 0·6 of the circumference of the body ($dd = 0·6\ \mu$). Clitellum commencing in XI; intraclitellar tubercula pubertatis present. Male pores lateral and anterior in XIII. Prostate pores separate from the male pores on XI, XII, and XIII. Female pores in the setal arc of XIV. Spermathecal pores two pairs, posterior in VII and VIII. Dorsal pores absent. Septa 8/9-12/13 delicate and displaced. Gizzards present, one in each of VIII and IX. Nephridia with pyriform terminal bladders. Holandric, testes in X and XI enclosed in testis-sacs. Sperm funnels with their mouths directed ventrally. Seminal vesicles of the microdrile type, i.e. extending posteriad within the ovisacs through several segments. Sperm ducts simple; prostate glands entirely separate.

Kenya (Monotypic).

Syngenodrilus lamuensis Smith and Green, 1919

Syngenodrilus lamuensis Smith and Green, 1919: 145 Fig. 1-8.

Syngenodrilus lamuensis Smith and Green. GATES, 1945: 393. PICKFORD, 1945: 397. JAMIESON, 1968: 85.

$l=52$ mm, $w=4$ mm, $s=137$. Clitellum ring-shaped, 2/3 XI-XVI. A copulatory band extends from 11/12 to setal arc XIV and fills bc. Setae very small; *aa*: $ab:bc:cd:dd=17:1:17:0.9:92$; $dd=0.6$ μ; $bc=1.0$ aa; present in the clitellar region. Nephropores commencing at the anterior margin of IV; some midway between cd and the mid-dorsal line others in cd. Male pores at the anterior margin of XIII at about 3/5 b-c. Prostate pores intrasegmental in XI, XII and XIII, slightly dorsal of setal line b, presumably paired. Female pores slightly anteriad and dorsal to setae b of XIV. Spermathecal pores two pairs immediately in front of intersegmental furrows 7/8 and 8/9, slightly ventral of setal lines c.

Septa 5/6-7/8 very strong. 8/9-12/13 very thin, displaced and imperfect. Strong gizzards in VIII and IX. Intestine commencing in XII or XIII but oesophagus widened between it and the gizzard. Hearts in VI-XI (XII?), possibly absent from X. Nephridia paired; absent from I-III, XI and XII; with pyriform terminal dilation. Testes and funnels paired in X and XI and enclosed in (unpaired?) testis sacs containing hearts and nephridia. Seminal vesicles of the microdrile type, a pair extending backwards in the ovisacs to XX. Vasa deferentia lacking terminal atria. Three pairs of short tubular prostates, in XI, XII and XIII, opening separately from the male pores by very short ducts which are confined to the body wall. Ovaries extensive in XIII; funnels broad and short on the anterior face of septum 13/14. Ovisacs extending posterior of XXII. Spermathecae two pairs, in VIII and IX, each with an irregular, simple tubular ampulla, *c*. 0.35 mm wide and somewhat longer than the duct which is 0.07-0.1 mm wide; without diverticula.

Kenya: Lamu (Habitat unknown).

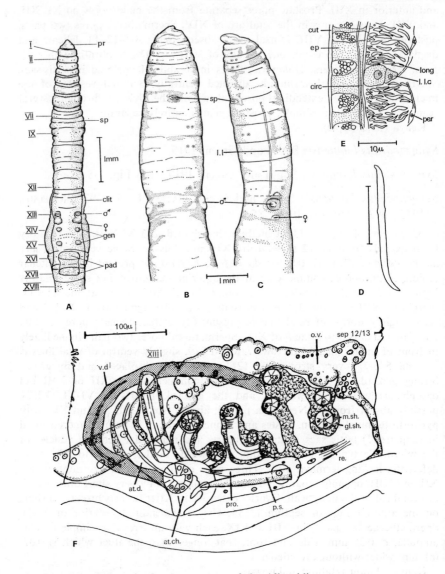

Fig. 14.1. Morphology in the three genera of the Alluroidinae.

Standeria transvaalensis. A—ventral view.

Alluroides brinkhursti abyssinicus. B—dorsal view, C—lateral view, D—a lateral seta from segment XX.

Brinkhurstia americanus. E—transverse section of the body wall showing the lateral line, F—longitudinal section through male terminalia.

at. ch, terminal chamber of atrium; at. d, atrial duct; circ, circular muscle; clit, clitellum; cut, cuticle; ep. epidermis; gen, papillae bearing genital setae; o, ovary; ♀, female pore; gl. sh, glandular sheath; l. l, lateral line l. l. c, cell of same; long, longitudinal muscle; ♂, male pore; m. sh, muscular sheath; o, ovary; pad, ventral pads; per, peritoneum; pr, prostomium; pro, protractor muscle; p. s, penial seta; re, retractor muscle of same; sep, septum; sp, spermathecal pore; v. d, vas deferens.

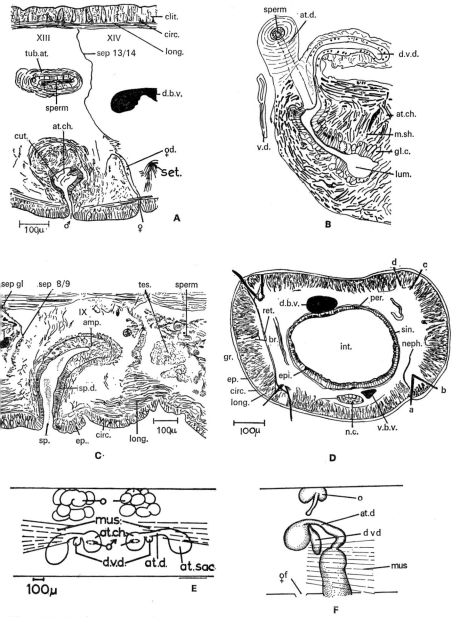

Fig. 14.2. *Standeria transvaalensis.*

A—longitudinal section through the male and female genital apertures.
B—longitudinal section through the muscular, terminal atrial chamber.
C—longitudinal section passing through a spermatheca.
D—transverse section in the intestinal region.
E—the two atria *in situ* in segment XIII.
F—atrial duct and vas deferens opening into the atrial chamber.
amp, ampulla; at. sac, saccular portion of atrium; br, hiatus in longitudinal muscle; d. b, v, dorsal blood vessel; d. v. d, dilatation of vas deferens; epi, epithelium; ♀ d, oviduct; ♀ f, oviducal funnel; gl. c, gland cells; int, intestine; lum, lumen; mus, muscle fibres; n. c, nerve cord; neph, nephridium; sep. gl, septal gland; set, seta; sin, blood sinus; sperm, spermatozoa; sp. d, spermathecal duct; t. testis; tub, at, atrium proper; v. b. v, ventral blood vessel. (Other abbreviations as Fig. 14.1.).

REFERENCES

BEDDARD, F. E. 1894. A contribution to our knowledge of the Oligochaeta of tropical eastern Africa. *Q. Jl microsc. Sci. (N.S.)*, **36**, 201.
— 1906. Zoological results of the Third Tanganyika Expedition, conducted by Dr. W. A. Cunnington, 1904-1905. Report on the Oligochaeta. *Proc. zool. Soc. Lond.*, **1906**, 206.
BRINKHURST, R. O. 1964. A taxonomic revision of the Alluroididae (Oligochaeta). *Proc. zool. Soc. Lond.*, **142**, 527.
CERNOSVITOV, L. 1936. Notes sur la Distribution mondiale de quelques oligochètes. *Mém. Soc. zool. Tchecosl.*, **3**, 16.
GATES, G. E. 1945. On the oligochaete genus *Syngenodrilus* and its taxonomic relationships. *J. Wash. Acad. Sci.*, **35**, 393.
JAMIESON, B. G. M. 1968. A taxonomic investigation of the Alluroididae (Oligochaeta). *J. Zool. Lond.*, **155**, 55.
MICHAELSEN, W. 1900. Oligochaeta. *Tierreich*, **10**, 1.
— 1913. Oligochaeten vom tropischen und südlichsubtropischen Afrika. *Zoologica, Stuttg.*, **25** (68), 1.
— 1914a. Oligochäten vom tropischen Afrika. *Mitt. naturh. Mus. Hamb.*, **31** (2), 81.
— 1914b. Oligochaeta in *Beiträge z. Kenntnis der Land u. Süsswasserfauna d.-Südwestafrika*, **1**, 137. Hamburg.
— 1915. Vers II. Oligochetes, in *Resultats scientifiques, voyage de Ch. Alluaud et R. Jeannel en Afrique orientale*, 1911-1912. Paris.
— 1935. Oligochäten von Belgisch-Kongo. I. *Rev. Zool. Bot. afr.*, **27**, 33.
— 1936. Oligochäten von Belgisch-Kongo. II. *Rev. Zool. Bot. afr.*, **28**, 213.
PICKFORD, G. E. 1945. Additional observation in the oligochaete genus Syngenodrilus. *J. Wash. Acad. Sci.*, **35**, 397.
SMITH, F., and GREEN, B. R. 1919. Descriptions of new African earthworms including a new genus of Moniligastridae. *Proc. U.S. natn. Mus.*, **55**, 145.

* See also LAUZANNE, 1968 in Appendix, p. 838.

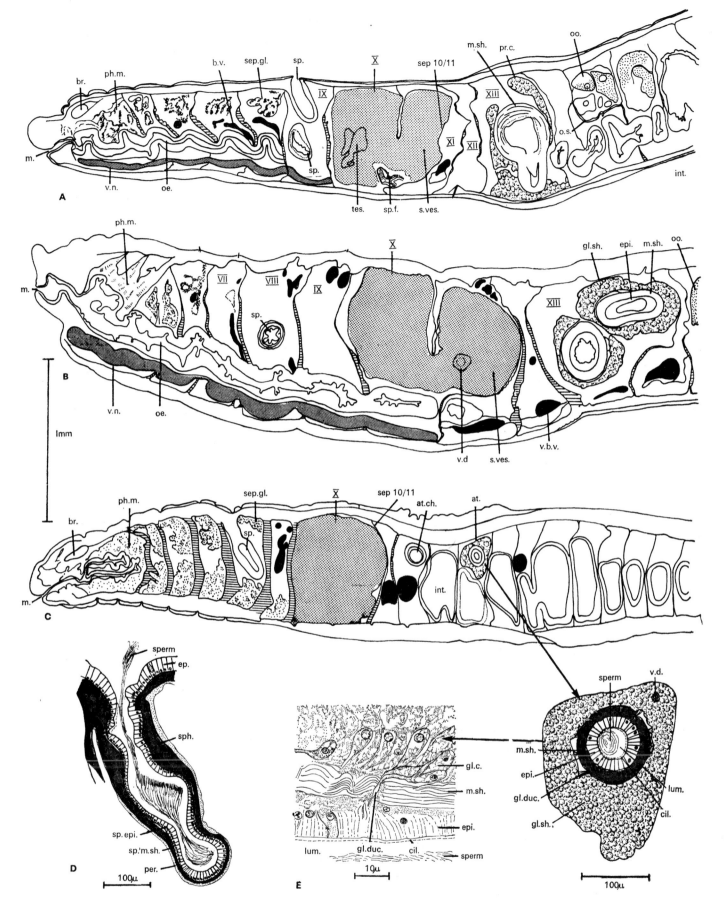

Fig. 14.3. Morphology of *Alluroides*.

A to C—longitudinal sections of the anterior end of the body of: A—*Alluroides
brinkhursti brinkhursti*, B—*Alluroides brinkhursti abyssinicus*, C—*Alluroides
pordagei*.

Alluroides pordagei. D—longitudinal section of a spermatheca, E—longitudinal section
of wall of atrium.

at, atrium; br, brain; bv, blood vessel; cil, cilia; gl. duc, glandular ductule(s); m,
mouth; oe, oesophagus; oo, oocyte; os, ovisac; ph. m, pharyngeal musculature; pr. c,
prostate cells; s, ves, seminal vesicle; sp, spermatheca; sph, sphincter; sp. epi,
spermathecal epithelium; sp. f, sperm funnel; sp. m. sh, muscular sheath of
spermatheca; v. n, ventral nerve cord. (Other abbreviations as Figs. 14.1. and 2.).

15

*FAMILY GLOSSOSCOLECIDAE**

Lateral lines present or absent. Somatic setae usually sigmoid, simple pointed, with 4 pairs per segment forming 8 longitudinal rows along the body (the lumbricine condition), rarely irregularly arranged or more numerous; occasionally with bifid tips; frequently ornamented transversely by scales, ridges, tooth-rows or striations. Penial setae absent; genital setae often present; sometimes longitudinally grooved. Clitellum multilayered, usually beginning behind segment 14, and usually extensive, frequently occupying ten or more segments. Male pores a pair, exceptionally 2 pairs, in the anterior part of the clitellar region or anteclitellar, very rarely postclitellar. Female pores in 14, or rarely, in 13 and 14.

Oesophageal gizzards 1-3 in front of the testis-segments, sometimes absent. Calciferous glands present or absent. Anterior end of the intestine with or without gizzard-like thickening of its musculature. Hearts variable in number (usually in 7-11, 12); dorsal vessel single or double; ventral vessel well developed; median subneural and supra-oesophageal vessel present or absent. Nephridia holonephridia; exonephric with, rarely, some anteriorly enteronephric; rarely 2 pairs per segment (meronephry). Holandric; proandric or metandric; testis-sacs present or absent. Ovaries in segment 13 or, rarely, in 12 and 13. Vasa deferentia (always?) concealed, in the body wall musculature; the ectal end usually simple but often with muscular copulatory sac, rarely associated with prostate glands. Prostate-like glands sometimes present in the vicinity of normal or genital setae. Spermathecae before, in, or behind the testis-segments; paired or transversely multiple; very rarely with diverticula.

Holarctic; Neotropical; Palaeotropical and Malagasian (see remarks).

(Key to sub-families, p. 728.)

* Terrestrial subfamilies and tribes reviewed to generic level only.

The least variable distinctive characters of the Glossoscolecidae are the extensive, anteriorly located clitellum, which is always multilayered (far posterior in some *Alma* species however); the intra-parietal male ducts, running deep in the body wall muscle (so far as is known invariable and diagnostic); and, with the sole exception of *Opisthodrilus*, location of the male pores shortly in front of, or on, the anterior portion of the clitellum.

Michaelsen (1918) recognized a family Lumbricidae s. lat. which he divided into the sub-families Lumbricinae, Glossoscolecinae, Spargano-philinae, Microchaetinae, Criodrilinae and Hormogastrinae. In his final monograph (1928 c), however, each of these sub-families was elevated to the rank of a family, the six families comprising a new super-family ("Familienreihe") Lumbricina. Other workers have retained the Lumbricidae as a family but opinions as to the rank of the other families has varied. Stephenson (1930) reduced them all to sub-families of the family Glossos-colecidae while Gates (1958a etc.) regards them as separate families. Omodeo (1956b) elevated the Glossoscolecinae to familial rank and placed the Hormogastrinae as a sub-family in the family Microchaetidae s. strict (including only *Microchaetus, Tritogenia, Kynotus, Callidrilus* and *Glyphid-rilus*). He considered the status of *Criodrilus* (the sole member of the Criodrilinae) and of *Alma* and *Drilocrius* (included in the Microchaetinae (-dae) by other workers) to be problematical and suggested that *Criodrilus* should probably be placed in the same (unnamed) group as *Drilocrius* and, less certainly *Alma*, comprising with them either a distinct family or a sub-family within the Glossoscolecidae. He appeared doubtful whether the Lumbricidae were to be regarded as separable at the familial level from the *Criodrilus-Alma-Drilocrius* group.

In the present work the Lumbricidae are retained as a separate family as, apart from other features of their organization which have made them a recognizable taxon, they are the only members of the Super-family Lum-bricoidea (=Familienreihe Lumbricina) in which the vasa deferentia are not, or (*Allolobophora; Bimastos*) are not deeply embedded in the body wall. The family Glossoscolecidae is retained in the broad sense of Stephen-son (1930) though internal groupings are profoundly altered. The type genus of the Microchaetinae (*Microchaetus*) is shown to be morphologically closer to members of the Glossoscolecinae than either is to other glossos-colecids; and *Microchaetus* and its close relative *Tritogenia* are included in the Glossoscolecinae as the Tribe Microchaetini. The sub-family Micro-chaetinae has thus been suppressed and its remaining genera have been placed in other sub-families. Of these genera, *Alma, Drilocrius, Glyphidri-locrius* gen. nov., *Callidrilus* and *Glyphilidrilus* have been placed in a Tribe Almini in the reconstituted sub-family Alminae, and *Criodrilus* has been included in the Alminae as the Tribe Criodrilini. *Kynotus* and *Hormogaster* have each been given the rank of a separate sub-family, the Kynotinae and Hormogastrinae respectively. The reasons for regrouping the Glossos-colecidae s. Stephenson, 1930, will best be appreciated by a discussion of each of the sub-families recognized in the present work.

SUB-FAMILY GLOSSOSCOLECINAE

The Microchaetinae are distinguished from the Glossoscolecinae in Stephenson (1930) solely by the phrase "Spermathecal pores usually altogether behind the testis segments". The difference is invalidated by the glossoscolecine *Rhinodrilus fafner* in which the spermathecae extend right through the testis segments, with pores in 6/7-14/15, and by the microchaetine *Glyphidrilus stuhlmanni* in which the spermathecae occupy 9/10-21/22 though only functional behind the testis segments, in 12/13 posteriorly.

Similarities between the two groups are marked whereas differences from other glossoscolecids are equally pronounced. Of the many points of similarity the following are found in no other glossoscolecids.

(1) Nephropores are located in or very near *cd* lines.
(2) Extramural calciferous glands are present.
(3) The nephridia of the S. African *Microchaetus* are shown in the present study to be identical with those of *Pontoscolex* and some other S. American glossoscolecines and to be of a type seen nowhere else in the Oligochaeta (see Chapter 1 and Fig. 1. 9A, B and 1. 10D, E).

Supposed distinctions between the Glossoscolecidae (s. strict, Michaelsen, 1928 c) and the Microchaetidae (s. Omodeo, including *Microchaetus, Tritogenia, Kynotus, Glyphidrilus* and *Callidrilus*) have been summarized by Omodeo (1956 b) and will now be considered in turn.

The calciferous glands of the Glossoscolecidae are considered to be more complex than those of the Microchaetidae and their complexity is believed by Omodeo to preclude considering the Glossoscolecidae to be more primitive than the Microchaetidae, contrary to the view held by Michaelsen (1918). Leaving aside the latter consideration, it is noteworthy that Michaelsen (1918) emphasized the strikingly close resemblance in the structure of the calciferous glands of *Microchaetus* and *Tritogenia* (Microchaetidae) and those of a *Holoscolex-Enantiodrilus-Glossoscolex-Fimoscolex* group (Glossoscolecidae), the glands being tubule sacs ("Schlauchtaschen") in all cases.

The spermathecae of the Glossoscolecidae are said to differ from those of the Microchaetidae in being restricted to a pair in each intersegment and in being distinctly differentiated into ampulla and duct. These differences usually hold but in *Martiodrilus* (=*Thamnodrilus*) *crassus* there are as many as six spermathecal pores in an intersegment and the spermathecae are poorly differentiated pear-shaped organs which, like those of the Microchaetidae s. Omodeo, are covered by the musculature of the body wall (Pickford, 1940). Furthermore, in *Rhinodrilus fafner* (see Michaelsen, 1918) there are variants with multiple spermathecae.

Puberty ridges (tubercula pubertatis), which are characteristic of *Micro-chaetus* are said to be absent from the Glossoscolecidae s. strict, but they are present in, for instance *Martiodrilus crassus* (Pickford, 1940).

Dorsal pores, absent from *Microchaetus* but said to be present in the Glossoscolecidae s. strict, are absent from *Martiodrilus* and are present or absent in other intrafimilial groups, e.g. the Ocnerodrilinae.

The only features of *Microchaetus* of those noted by Omodeo which constitute valid differences from the glossoscolecid genera are the absence of copulatory sacs and the relative simplicity of the circulatory system especially the absence of a supra-oesophageal, and a subneural, vessel. Copu-latory sacs are not, however, limited to the Glossoscolecidae (occurring also in *Kynotus*) and are absent from many glossoscolecids and there is wide variability in the Glossoscolecidae s. lat. with regard to presence or absence of supra-oesophageal and subneural vessels (Fig. 1. 8). To these two differ-ences may be added the location of the gizzard in segment 7 in Microchaetids as opposed to segment 6 in Glossoscolecids. Segmental variation in the position of the gizzard occurs intragenerically, however, in for instance, *Glyphidrilus* and *Callidrilus*. In view of the striking similarities noted above between *Microchaetus* (its satellite *Tritogenia*) and the genera of the Glossoscolecinae s. Stephenson 1930, and their restriction to these genera, *Microchaetus* and *Tritogenia* are here included in the Glossoscolecinae, their S. African distribution and the differences noted warranting recognition of the tribe Microchaetini.

The residue of the Microchaetinae s. Stephenson is comprised by *Kynotus, Callidrilus, Glyphidrilus, Alma,* and *Drilocrius* from which the new genus *Glyphidrilocrius* is separated in the present work. These have previously been associated with *Microchaetus* almost solely on the grounds of the post testicular location of the spermathecae.

SUB-FAMILY ALMINAE

Callidrilus, Glyphidrilus, Alma, Drilocrius and *Glyphidrilocrius*, all of which are aquatic show, as a group, a number of distinctions from Micro-chaetus and *Tritogenia* and it seems likely that these differences indicate phylogenetic unity rather than that they have been convergently acquired several times over in response to an aquatic existence. The distinctions from *Microchaetus* are: a quadrangular posterior cross section, with setae at the angles; the usual presence of a dorsal respiratory groove and a dorsal or dorso-terminal anus; dorsal location of the outer setal couples, so that *dd* is significantly less than 0·3 of the body circumference; absence of genital setae, of their glands, and of testis sacs; absence of calciferous glands; location of the nephropores in or ventral to *b* lines as against in *c* lines in *Microchaetus*; loss of the nephridia in the anterior region of the body; and absence of nephridial sphincters and caeca. A further difference is the presence of a median subneural vessel except in the highly evolved Alma and Glyphidrilocrius. The five genera here comprise the tribe Almini of the sub-family Alminae, their affinities are discussed on p. 184.

The relationships of *Criodrilus* (widespread in the Palearctic) have been the subject of much debate (e.g. Michaelsen, 1918; Stephenson, 1930 a; Omodeo, 1956b) and are discussed on pp. 183 and 186. The deeply intra-parietal sperm ducts indicate that it is a true glossoscolecid and the location of its cerebral ganglia in segments 1 and 2 suggest that it is primitive rather than derived. As it shares all the points of mutual similarity noted for the Almini it is here placed with them, but in the separate tribe Criodrilini, in the Alminae.

With regard to the overall similarities within the Alminae it must be emphasized that whether such similarities have been acquired by convergence or parallelism is of no significance in phenetic classification. Nevertheless, the similarity of the dorsal intersetal/circumferential ratio to that of the Haplotaxidae, Alluroididae and other primitive oligochaetes, taken with the constant location of the nephropores which are ventrally located in the Alluroididae (and Haplotaxidae?) in the Biwadrilinae and Sparganophilinae suggests origin from primitive Glossoscolecids rather than from forms resembling the modern Glossoscolecinae and favours regarding the quadrangular section and other similarities as ancestral characters rather than as ones acquired independently several times over in secondarily aquatic lineages.

SUB-FAMILY KYNOTINAE

The terrestrial Malagasian genus *Kynotus*, the last member of the Microchaetinae s. Stephenson remaining to be considered, was regarded by Michaelsen (1918) as a derivative of an early glossoscolecine stock (as evidenced by the persistence of copulatory sacs) and as an ancestor of *Callidrilus* on the one hand and an *Archidrilocrius* on the other, the latter hypothetical genus being the supposed ancestor of *Drilocrius* and *Alma*, of *Hormogaster*, and of *Criodrilus* and hence the Lumbricidae. Alternative origins of these genera are suggested below.

Location of the gizzard in segment 5 (in *Haplotaxis gordioides* segments 4 and 5) appears to be a primitive feature relative to locations in 6-8 (*Hormogaster*), in 6 (Glossoscolecini), and in 7 (Microchaetini). As is often the case in the fauna of Madagascar it shows a generalized anatomy, linking that of these three taxa, and suggesting that it is indeed a little modified descendent of their ancestral stock. Furthermore the location of its nephropores in *bc* is intermediate between the condition in the Alminae and the Glossoscolecinae s. mihi. The anterior location of the gizzard and absence of calciferous glands and of a typhlosole exclude it from the Glossoscolecinae and it appears closer to *Hormogaster*.

Common features of *Kynotus* and *Hormogaster*, including some found in the Glossoscolecinae also, are: the rounded cross section; male pores in the vicinity of 15-16, though anteclitellar; spermathecae sometimes in 13/14; oesophageal gizzards present; calciferous glands and testis sacs absent; supra-oesophageal and subneural vessels present; nephridia with caeca but lacking sphincters and multiple spermathecae. These features certainly indicate closer affinity of *Kynotus* with *Hormogaster* than with *Microchaetus* with which Stephenson grouped it. Nevertheless, differences from *Hormogaster* and specializations of the latter genus necessitate recognition of two separate sub-families for them.

SUB-FAMILY HORMOGASTRINAE

Appropriately for an hypothesis of separate origin of *Hormogaster* from the ancestral stock of the Glossoscolecinae and possibly the Kynotinae (see above, p. 185), its morphology is of a generalized kind with few obvious specializations beyond the triple gizzard. Its primitive status is perhaps also indicated by retention of the small dorsal intersetal/circumferential ratio seen in the Alminae to which it is also phenetically close but from which it differs significantly in the presence of nephridial caeca. Resemblance to *Kynotus* has already been discussed.

Features which *Hormogaster* shares with *Microchaetus* are presence of ridges of puberty; intraclitellar male pores; the approximate location of the spermathecal pores (10/11-15/16 in *Microchaetus*); multiple spermathecae; oesophageal gizzards (one, in 7 only, in *Microchaetus*); nephridial caeca and absence of copulatory sacs. Noteworthy differences from *Microchaetus*, collectively meriting sub-familial distinction, are absence of calciferous glands (large and extramural in *Microchaetus*); the single dorsal vessel; presence of a supra-oesophagael vessel and of a subneural; and absence of testis sacs and of nephridial sphincters.

KEY TO SUB-FAMILIES

1.	3 large oesophageal gizzards, in 6-8	*Hormogastrinae*
—	Never with 3 well defined oesophageal gizzards. Rarely with gizzard-like modification of the oesophagus in the three segments 7-9	2
2.	Dorsal median intersetal distance (*dd*) 0·2-0·3 of the circumference of the body	3
—	Dorsal median intersetal distance (*dd*) 0·4-0·5 or more of the circumference of the body	5
3.	Male pores on obvious protuberances on segment 13. Spermathecae and subneural vessel absent	*Biwadrilinae**

3. Male pores whether on obvious protuberances, or inconspicuous, located in or behind segment 15. Spermathecae and subneural vessel present or absent   4

4. Spermathecae wholly in front of segment 10 and projecting far into the coelom. Intestine beginning in segment 9   *Sparganophilinae**

— Spermathecae wholly or mostly behind segment 10 and wholly or partly concealed in the body wall; or absent. Intestine beginning in or behind segment 13   *Alminae**

5. Calciferous glands and intestinal typhlosole present. Gizzard in segment 6 or 7   *Glossoscolecinae*

— Calciferous glands and intestinal typhlosole absent. Gizzard in segment 5   *Kynotinae*

SUB-FAMILY GLOSSOSCOLECINAE EMEND.

Type genus: Glossoscolex LEUCKART

Lateral lines absent. Body rounded, not quadrangular in cross section; dorsal groove absent; anus terminal; dorsal median intersetal distance greater than 0·3 ($dd:u$$\rangle$0·3); modified genital setae present. Nephropores in or very near *cd*. Clitellum usually saddle-shaped. Male pores intraclitellar, rarely (*Opisthodrilus*) postclitellar. Female pores in 14, or, exceptionally, in 13 and 14. Spermathecal pores pretesticular or posttesticular but sometimes extending anteriorly or posteriorly into, or through, the testis segments; paired or multiple in each segment; spermathecae intraparietal or projecting well into the coelom; differentiated into ampulla and duct or not; sometimes absent. Dorsal pores present or absent.

Oesophagus with a gizzard in 6 or 7 and 1-8 pairs of extramural calciferous glands which vary in complexity from ridged to composite-tubular. Muscular thickening of the intestine (always?) absent; typhlosole present. Supra- oesophageal and subneural vessels present or absent. Nephridia holonephridia with or without terminal caecum, sphincter and spiral loops; exceptionally with 4 meronephridia per segment or with multiple funnels. Holandric or proandric; testis-sacs present or absent. Copulatory sacs present or absent at the male pores. Prostates absent but prostate-like genital seta glands sometimes developed. (Cocoons never (?) elongate).

Central and S. America, the Bermudas, and the West Indies including Barbados. S. Africa. Some peregrine in warmer latitudes.

* Primarily aquatic.

KEY TO TRIBES

1.　Gizzard in segment 6. Supraoesophageal vessel
　　present　　　　　　　　　　　　　　　　　　*Glossoscolecini*

—　　Gizzard in segment 7. Supraoesophageal
　　vessel absent　　　　　　　　　　　　　　　　*Microchaetini*

TRIBE GLOSSOSCOLECINI

Type Genus: Glossoscolex LEUCKART

Nephropores in or above *cd* lines. Male pores 1 pair, rarely 2 pairs, intraclitellar or (*Opisthodrilus*) postclitellar. Female pores in 14 or exceptionally (*Enantiodrilus*) 13 and 14. Spermathecal pores pretesticular, rarely extending into or behind the testis-segments; usually a pair, sometimes multiple in each intersegment occupied; sometimes absent. Dorsal pores present or absent.

Oesophageal gizzard in 6 and behind this, 1-8 pairs of calciferous glands. Supra-oesophageal and subneural vessels present; latero-oesophageal hearts present. Nephridia holonephridia which may have terminal caecum and sphincter and multiple spiral coils, and anteriorly are sometimes enteronephric; sometimes with multiple funnels. Holandric or proandric; testis-sacs present or absent. Copulatory sacs present or absent. Spermathecae extending freely into the coelom; well differentiated into duct and ampulla or intraparietal and poorly differentiated.

Central and S. America, the Bermudas and the W. Indies including Barbados. Some species have been distributed around the world in warmer latitudes.

The Glossoscolecini (as Glossoscolecinae) were reviewed by Michaelsen (1918), Stephenson (1930 a) and again by Cordero (1945). Only Michaelsen's review provides for identification of species. Cordero's review omits the genus *Martiodrilus* Michaelsen, 1937 b, and consequently ignores changes in the constitution of *Thamnodrilus* and supression of *Aptodrilus*. The present account gives synoptic descriptions of all known genera and indicates systematic references subsequent to Michaelsen's review.

The Glossoscolecini is the dominant group of earthworms in tropical America from Costa Rica and Panama to as far south as a line joining La Plata (on the Atlantic) and the Atacam desert (on the Pacific) and approximately following the course of the Juramento-Saledo River in Argentina, the southern portion of the continent being occupied by Acanthodrilinae.

In the generic synopses of the term "gymnorchous" denotes that the testes and sperm funnels are free in the coelom, and "cleistorchous" that they are enveloped in testis sacs which are either paired or unpaired.

Michaelsen (1918 , 1937 b) attached great significance in generic classification to the internal structure of the calciferous glands, the forms which he recognized are illustrated in Fig. 1·3 (for explanation see p. 52).

KEY TO GENERA

1.	With 4 to 8 pairs of calciferous glands	2
—	With 1 to 3 pairs of calciferous glands	7
2.	With more than 4 pairs of calciferous glands	3
—	With 4 pairs of calciferous glands (in 7-10)	*Anteoides*
3.	With 5-6 pairs of calciferous glands	4
—	With 7-8 pairs of calciferous glands	5
4.	With 5-6 pairs of calciferous glands, in 9, 10-14, panicled tubular sacs with lumen	*Thamnodrilus*
—	With 5 pairs of calciferous glands, in 11-15, honey-comb sacs	*Aymara*
5.	Calciferous glands composite tubular, with large central lumen, and appendage, 7-8 pairs, in 7-13, 14	*Martiodrilus*
—	Calciferous glands with folds of the walls (composite, partitioned, or not), 8 pairs	6
6.	Folds free in the cavity of the sac	*Inkadrilus*
—	Folds united at the centre	*Quimbaya*
7.	Calciferous glands 3 pairs (in 7-9)	8
—	Calciferous glands 1 pair	16
8.	Calciferous glands of the tubular type (simple or composite).	9
—	Calciferous glands of the ridged or lamellar type	13
9.	Tubes simple	10
—	Tubes compound, panicled	11
10.	Holandric	*Onychochaeta*
—	Metandric	*Meroscolex*
11.	Male pores intraclitellar	*Pontoscolex*
—	Male pores postclitellar	*Opisthodrilus*

13(8). Calciferous glands with internal lamellae crossing the lumen with no free edges 14

— Internal lamellae with free edges ..... *Diachaeta*

14. Holandric 15
Proandric ..... *Andiodrilus*

15. 2 pairs of male pores ..... *Eudevoscolex*
— 1 pair of male pores ..... *Andiorrhinus*

16. Calciferous glands in 7 ..... *Periscolex*
— Calciferous glands in 11-12 17

17. Holandric ..... *Holoscolex*
— Metandric 18

18. Male pores unpaired ..... *Fimoscolex*
— Male pores 1 pair 19

19. With copulatory sacs at the male pores and with spermathecae 21
— With either copulatory sacs or spermathecae 20

20. With copulatory sacs but no spermathecae ..... *Glossoscolex*
— With spermathecae but no copulatory sacs ..... *Andioscolex*

21. Cleistorchous ..... *Diaguita*
— Gymnorchous ..... *Enantiodrilus*

GENUS **Andiodrilus** MICHAELSEN, 1900

Type species: Anteus schutti MICHAELSEN

Setae usually lumbricine. Male pores, intraclitellar. Calciferous glands 3 pairs, in 7-9, stalked, lamellar sacs. Proandric; cleistorchous. Seminal vesticles, when present, very short. With spermathecae.
Central America: Costa Rica. S. America: Colombia.

Like *Andiorrhinus* but lacking the testes and funnels of segment II. For all known species, see Michaelsen (1918).

GENUS **Andiorrhinus** COGNETTI, 1908

Type species: Andiorrhinus salvadori COGNETTI

Setae lumbricine. Male pores intraclitellar. Calciferous glands 3 pairs, in 7-9, stalked lamellar sacs. Holandric; cleistorchous. Seminal vesicles not extensive. With spermathecae.

Venezuela: N. Brazil; Bolivia; Paraguay.

Distinguished from *Rhinodrilus* by the lamellar calciferous glands.

For species described after Michaelsen, 1918, see Michaelsen (1925, 1934, 1936); Cernosvitov (1939); Omodeo (1955).

GENUS **Andioscolex** MICHAELSEN, 1927

Type species: Tykonus peregrinus MICHAELSEN

Setae lumbricine. Male pores intraclitellar. Calciferous glands 1 pair, in the region of 11-12, composite-tubular sacs. Metandric; cleistorchous. Seminal vesicles short or extensive. Copulatory sacs absent. With spermathecae.

Panama. Colombia; Venezuela; Ecuador; Bolivia; Brazil.

Andioscolex was erected by Michaelsen, 1927, for those species of *Glossoscolex* possessing spermathecae (1 or 2 pairs, in 8/9 and 9/10 or one of these). Cernosvitov (1934 b) added a further species, *A. geayi* (see also 1935).

GENUS **Anteoides** COGNETTI, 1902

Type species: Anteoides rosea COGNETTI

Setae lumbricine. Male pores intraclitellar. Calciferous glands 4 pairs, in 7-10, simple ridged sacs. Metandric; gymnorchous. Seminal vesicles not extensive. Spermathecae absent. Nephridia without terminal sphincter.

Bolivia; Argentina; N. Paraguay.

The two species differ from *Diachaeta*, which they closely resemble in having 4 pairs of calciferous glands. There is a doubtful record for Venezuela (?*Anteoides* sp. Omodeo, 1955).

GENUS **Aymara** MICHAELSEN, 1935

Type species: Aymara voogdi MICHAELSEN

Setae lumbricine. Male pores intraclitellar. Calciferous glands 5 pairs, in 11-15, reticulate ("honey-comb sacs"). Metandric; cleistorchous. Seminal vesicles very extensive. Spermathecae present.

Peru. (Monotypic).

GENUS **Diachaeta** BENHAM, 1886

Type species: Diachaeta thomasi BENHAM

Setae in 8 longitudinal rows or irregularly closely and widely paired. Male pores intraclitellar. Calciferous glands 3 pairs, in 7-9, small simple ridged sacs. Metandric; gymnorchous or cleistorchous. Seminal vesicles short or extending far back. With or without spermathecae. Nephridia without terminal sphincters but with some terminal dilatation; lacking a caecum.

W. Indies, including Trinidad and Curacao; Venezuela.

Michaelsen (1918) synonymized the Jamaican seashore form, *D. littoralis* Beddard, 1892, with *D. thomasi* (W. Indies, Trinidad, Curacao) and reduced the number of species to two. Subsequent additions were *Diachaeta exul* Stephenson, 1931 a, and *D. carsevenica* Cernosvitov, 1934 b, later redescribed by Cernosvitov, (1935).

Beddard (1892, *D. littoralis*) noted cup-like sphincters like those of *Pontoscolex corethrurus* but Michaelsen (1918) re-examining a type specimen, denied their existence.

GENUS **Diaguita** CORDERO, 1942

Type species: Diaguita michaelseni CORDERO

Setae lumbricine. Male pores intraclitellar. Calciferous glands, 1 pair, composite-tubular, in 11-12. Metandric; cleistorchous. Seminal vesicles short, limited to 1 segment. A pair of copulatory sacs present. With spermathecae.

N. Argentina (Monotypic).

GENUS **Enantiodrilus** COGNETTI, 1902

Type species: Enantiodrilus borelli COGNETTI

Setae lumbricine Male pores intraclitellar. Calciferous glands 1 pair, in the region of 11-12, of the composite-tubular type, with thin walled appendage. Metandric; testis sacs and seminal vesicles absent.

Argentina: Surinam.

A second species, *E. cognetti* was erected by Michaelsen (1933 c) for a simple specimen lacking genital organs. The genus is similar to *Glossoscolex* from which it is distinguished by the appendages of the calciferous glands.

E. borelli is unique in the Glossoscolecinae in being hologynous (ovaries in 12 in addition to 13), a feature shared with *Glyphidrilus kukenthali*.

GENUS **Eudevoscolex** CORDERO, 1944

Type species: Eudevoscolex vogelsangi CORDERO

Setae lumbricine. Male pores intraclitellar. Calciferous glands 1 pair, in the in 7-9, lamellar. Hollandric; cleistorchous. Lacking seminal vesicles. With 2 pairs of copulatory sacs and with spermathecae.

Venezeula (Monotypic).

GENUS **Fimoscolex** MICHAELSEN, 1900

Type species: Fimoscolex ohansi MICHAELSEN

Setae lumbricine or some irregular. Male pores intraclitellar, unpaired. Calci-

ferous glands, 1 pair, in the region of 11-12; composite-tubular. Metandric; cleistorchous. Seminal vesicles very extensive. Without spermathecae.
Brazil.

The setae of the hind end are irregularly arranged in *F. sporadochaetus* Michaelsen, 1918.

GENUS **Glossoscolex** LEUCKART, 1835

Type species: Glossoscolex giganteaus LEUCKART (see Glossoscolex sp. Leuckart, 1835).

Setae lumbricine. Male pores paired, intraclitellar. Calciferous glands one pair, in the region of 11-12, of the composite tubular type. Metandric; cleistorchous. Seminal vesicles short or extensive. Copulatory pouches present, chambers with thick muscular walls which may occupy several segments. Spermathecae absent.
Brazil; Uruguay; Argentina; Paraguay; Ecuador.

Michaelsen (1927) restricted the genus to those species lacking spermathecae (see *Andioscolex*). Species are described by Michaelsen (1925), Cernosvitov (1934 a and b, 1935), Cordero (1942, 1943).

GENUS **Holoscolex** COGNETTI, 1904

Type species: Holoscolex nemorosus COGNETTI

Setae lumbricine. Male pores intraclitellar. Calciferous glands 1 pair, in the region of 11-12, conjecturally of the composite-tubular type. Holandric; gymnorchous. Seminal vesicles not extensive. With spermathecae.
Ecuador (Monotypic).

GENUS **Inkadrilus** MICHAELSEN, 1918 emend. MICHAELSEN, 1935

Type species: Anteus abberatus MICHAELSEN

Setae lumbricine. Male pores intraclitellar. Calciferous glands 8 pairs, in 7-14, composite ridged sacs (Saumleistentaschen). Holandric; cleistorchous. Seminal vesicles extensive. With spermathecae.
Peru.

Erected as a sub-genus of *Thamnodrilus* by Michaelsen (1918) and subsequently elavated to generic rank (Michaelsen, 1935), *T.* (*Inkadrilus*) *cameliae* being excluded and made the type of *Quimbaya*.

GENUS **Martiodrilus** MICHAELSEN, 1937

Type species: Hypogaeon heterostickon SCHMARDA

Setae lumbricine. Male pores intraclitellar. Calciferous glands generally 8 pairs, rarely (*M. savanicola*) 7 pairs; always commencing in segment 7; composite-tubular glands with a large central lumen (mostly "honey-comb glands"); (always?) with a terminal appendage with the probable exception of those in segments 13 and 14. Holandric; cleistorchous. Seminal vesicles not extensive. Spermathecae present.

Ecuador (most species) and Colombia. A few in Panama, Peru, Surinam and Guyana. 41 species.

Michaelsen (1937 b) erected *Martiodrilus* for all those species which in 1918 he had placed in *Thamnodrilus* (*Thamnodrilus*) with the exception of *T. gulielmi* Beddard, 1887, which he recognized as the type species of a restricted *Thamnodrilus*. Additions since Michaelsen (1918) are *T. cognetti* Beddard 1921 and *T. gonggrijpi* Michaelsen, 1933 c, and *T. tenkatei* (Horst) "var" *geayi* Cernosvitov, 1934 b.

Beddard (1892) showed that in the type species the nephridia, which have convoluted tufts of tubules, have a large terminal sac and a caecum, and that the nephropores are presetal in *d* lines.

GENUS **Meroscolex** CERNOSVITOV, 1934

Type species: Meroscolex guianicus CERNOSVITOV

Setae lumbricine. Male pores intraclitellar. Calciferous glands 3 pairs in 7-9, simple tubular type. Metandric. With or without testis sacs. Seminal vesicles extensive. With spermathecae. Nephropores with sphincter.

French Guiana.

The two species were redescribed by Cernosvitov (1935).

GENUS **Onychochaeta** BEDDARD, 1891

Type species: Diachaeta windlei BEDDARD

Setae not, or not throughout the whole length of the body, arranged in longitudinal rows. Calciferous glands 3 pairs in 7-9, fairly simple tubule sacs (? panicled sacs). Holandric; gymnorchous. Seminal vesicles when present short, limited to a single segment. With spermathecae. Nephridia with external sphincter.

Panama; Venezuela; Guyana; Cuba.

A key to the two species is given by Michaelsen (1918) and *O. windei* is redescribed by Cordero (1944 c). Michaelsen (1923) describes a "var. *cubana*" of *O. elegans*.

GENUS **Opisthodrilus** ROSA, 1895

Type species: Opisthodrilus borelli ROSA

Setae lumbricine. Male pores and ridges of puberty located far behind the clitellum. Calciferous glands 3 pairs, in 7-9, panicled tubular sacs arising

dorsally from the oesophagus. Metandric; gymnorchous. Seminal vesicles extending far back. With spermathecae. Nephridia with terminal sphincter.
Brazil; Paraguay; Argentina.

Of the two specis *O. borelli* has been recorded by Stephenson (1931 a and *O. rhopalopera* Cognetti, 1906 a, by Cognetti (1926), since Michaelsen's review (1918).

GENUS **Periscolex** COGNETTI, 1905

Type species: Diporochaeta profuga COGNETTI (vide COGNETTI, 1905)

Setae in 8 longitudinal rows or more numerous. Male pores intraclitellar. Calciferous glands 1 pair, in 7, ridged sacs with wide lumen. Holandric (cleistorchous) but with only 1 pair of seminal vesicles, on septum 11/12, which extend backwards a greater or lesser distance by penetration of septa. With spermathecae. Nephridia without terminal sphincter.
Panama; Colombia; Ecuador.

Four of the six species have more than 8 setae per segment, the largest number being 35-40, in *P. mirus* Cognetti, 1905. As in *Pontoscolex*, some species have forked setae.

GENUS **Pontoscolex** SCHMARDA, 1861

Type species: Lumbricus corethrurus MÜLLER

Setae in the hind region normally arranged in quincunx (*a b c* and *d* not forming 4 longitudinal rows, the two setae of a pair set somewhat widely apart and those of successive pairs alternating in position). Male pores intraclitellar. Calciferous glands 3 pairs, in 7-9, panicled tubular sacs arising dorsally from the oesophagus. Metandric; cleistorchous. Seminal vesicles very extensive. With spermathecae. Nephridia with terminal sphincter.
Bermudas; W. Indies; S. America; Venezuela; Surinam and circum-mundane in the tropics.

P. hingstoni Stephenson, 1931 a, and *P. vandersleeni* Michaelsen, 1933 c, have been added to *P. corethrurus* (Müller) and *P. lilljeborgi* Eisen, 1896 b. *P. corethrurus* is probably the most ubiquitous earthworm in the tropics. The author (unpublished) has recorded it near Nairobi, Kenya, and in many parts of Queensland, in forest soils, gardens, and along the banks of rivers. A recent description is that of Gates (1943); its reproduction is discussed by Gates (1962); the structure of its nephridia is lucidly treated by Bahl (1942) and its distribution is given by Vannucci (1953). Other references to *P. corethrurus* subsequent to Michaelsen (1918) are Cernosvitov (1942 a), Cordero (1944 c), Gates (1931, 1933, 1938, 1948 a, 1949, 1961), Omodeo (1955 b, 1956 a).

GENUS **Quimbaya** MICHAELSEN, 1935

Type species: Rhinodrilus cameliae MICHAELSEN, 1914

Setae lumbricine. Male pores intraclitellar. Calciferous glands 8 pairs in 7-14, partitioned sacs (Fachkapseltaschen), with folds meeting at the centre; Holandric, cleistorchous. Seminal vesicles short, limited to 1 segment. With spermathecae.

Colombia (Monotypic.)

GENUS **Rhinodrilus** PERRIER, 1872

Type species: Rhinodrilus paradoxus PERRIER

Setae lumbricine. Male pores intraclitellar. Calciferous glands 3 pairs, in 7-9, all, or those of the first 2 pairs, panicled tubular sacs. Holandric; cleistorchous. Seminal vesicles limited to one segment. With spermathecae.

Colombia; Venezuela; Brazil; Guyana; Argentina; Paraguay.

Species according with this description but with lamellar sacs are referred (Michaelsen, 1918) to *Andiorrhinus*. For species described after Michaelsen, 1918, see Michaelsen (1925, 1928 a, 1931 b, 1934); Cernosvitov (1934 b); Cordero (1943, 1944 a, b, c); Omodeo (1955). See also *Quimbaya*. *Rhinodrilus* includes *Urobenus* Benham 1886, *Geogenia* (part.) Vaillant, 1889, and *Anteus* (part.) Horst, 1891 a.

GENUS **Thamnodrilus** BEDDARD, 1887, emend. MICHAELSEN, 1937

Type species: Thamnodrilus gulielmi BEDDARD

Setae lumbricine. Male pores intraclitellar. Calciferous glands 5 or (*gulielmi*) 6 pairs in 9, 10-14; panicled tubular sacs without central lumen; appendages absent or rudimentary on the glands. Holandric; cleistorchous. Seminal vesicles not extensive. Spermathecae present.

Ecuador; Colombia; N. Brazil; Guyana.

The former sub-genus *Inkadrilus* was removed from *Thamnodrilus* and elavated to generic rank by Michaelsen (1935 a).

Michaelsen (1937 b) reinvestigating a type-specimen of *gulielmi* Beddard, 1887, showed that this differed from the remainder of the genus in having typical panicled-tubular calciferous glands (Rispenschlauchtaschen). As *T. gulielmi* was the type species of *Thamnodrilus* sensu Beddard, the remaining species were given a new generic name, *Martiodrilus* (q.v.). The structure of the calciferous glands in *T. guliemi* accorded with that in *Aptodrilus* Cognetti, 1904 a, and the latter genus was therefore supressed as a synonym of *Thamnodrilus*.

The erection of *T. salathei* Michaelsen, 1934, brought the number of species of *Thamnodrilus* emend. to eight.

Cordero (1945) overlooked the supression of *Aptodrilus* and erection of *Martiodrilus* and his definition of *Thamnodrilus* is therefore invalid.

TRIBE MICROCHAETINI

Type genus: Microchaetus RAPP

Nephropores of holonephidria in or near *c* lines. Male pores 1 pair; intraclitellar. Female pores in 14. Spermathecal pores immediately posttesticular or also occupying the east testis-segment; paired or multiple in each intersegment. Dorsal pores absent.

Oesophageal gizzard in 7; a pair of calciferous glands of the tubular or combined tubular lamellar type in the region of 9/10. Supra-oesophageal and subneural vessels absent. Lateroparietals 1 on each side. Nephridia holonephridia with terminal caecum and sphincter and, in the anterior segments, numerous coiled loops; the anterior one (always?) enteronephric; or secondarily (*Tritogenia*) with 4 meronephridia per segment which lack caeca and sphincters. Holandric or proandric; testis-sacs present. Copulatory sacs and prostates absent. Genital seta glands large, one to several at each genital seta, or absent. Spermathecae not projecting far into the coelom.

S. Africa.

GENUS **Microchaetus** RAPP, 1849

Type species: Lumbricus microchaetus RAPP

Setae lumbricine, closely paired; $dd:u=0.4-0.5$; some modified as genital setae. Nephropores in or near *c* lines. Clitellum saddle-shaped in the region of 9-34; ridges of puberty and puberty papillae present. Male pores intraclitellar, a variable distance behind 16, in a depressed flat or slightly raised area. Spermathecal pores inconspicuous, in the region of intersegments 10/11-15/16 (sometimes absent), one to several on each side per intersegment.

Oesophageal gizzard in 7; calciferous glands, of the tubular type, one pair in the region of 9/10, often fused medially; intestinal origin in 13, typhlosole present. Dorsal vessel frequently double; subneural and supra-oesophageal vessels (always?) absent. Nephridia one pair per segment, including the anteriormost segments; each with much coiled loops; the wide tube with a large caecum; pore with well developed sphincter. Holandric or proandric, with testis sacs. Copulatory sacs and prostates absent. Genital seta glands at each genital seta or absent. Spermathecae sometimes serpentine tubes.

S. Africa (terrestrial).

Data on the blood vascular system and nephridia are derived from an examination of material of *Microchaetus pentheri* (*=saxatilis*) B.M. (N.H.) 1934.10.11.5/8 (Fig. 1. 9B). Accounts of species, subsequent to Michaelsen (1918), are Michaelsen (1928 b, 1933 b), Brock and Dick (1935) and Sciacchitano (1960). The genus includes *Geogenia* Kinberg, 1867.

Ljungstrom (in *litt.*) states that *Geogenia*, united with *Microchaetus* by Michaelsen (1918), should be resurrected for *M. natalensis* (Kinberg, 1866) and other species which have only a single, intramural, ring- or horseshoe-shaped calciferous gland (in 9). *Microchaetus*, he considers, should be limited to paired, extramural calciferous glands connected by short, thick ducts to the oesophagus in 9.

GENUS **Tritogenia** KINBERG, 1867

Type species: Tritogenia sulcata KINBERG

Setae 2-4 pairs per segment, closely paired: $dd\!:\!u\rangle 0.5\ u$; some modified as genital setae. Male pores intraclitellar, behind 16. Spermathecal pores inconspicuous, in the region of intersegments 11/12-16/17, 2-8 on each side per intersegment.

Oesophageal gizzard in 7; stalked calciferous glands (a combination of the lamellar and tubular types) on pair, entirely separate, in the region of 9-10; typhlosole present. Dorsal vessel single or double. Nephridia 2 pairs per segment, apparently including the anteriormost segments; terminal ampulla and caecum but ducts slightly widened in the body wall. Holandric, with testis sacs. Copulatory sacs and prostates absent but body wall in the region of the male pores with a pair of glandular cushions externally. Genital seta glands one to several at each genital seta. Spermathecae sometimes serpentine tubes.

S. Africa (Terrestrial).

Tritogenia is sufficiently similar to *Microchaetus* for Michaelsen (1913 b) to have united the two genera. His recognition (Michaelsen, 1918) that *Brachydrilus benhami* Michaelsen, 1900 a, which was known to have two pairs of nephridia per segment, was virtually identical with *Michrochaetus* (=*Tritogenia*) *sulcatus* (Kinberg) led him to reinstate *Tritogenia* for forms with reduplicated nephridia. Five such species were recognized, three species being transferred from *Microchaetus*, one from *Brachydrilus*, and one being new.

SUB-FAMILY KYNOTINAE NOV.

Definition and distribution as for the type and only genus, *Kynotus* Michaelsen, 1891.

GENUS **Kynotus** MICHAELSEN, 1891

Type species: Geophagus darwini KELLER

Lateral lines absent. Body rounded, not quadrangular, in cross section. Anus terminal. Setae lumbricine, closely paired; $dd\!:\!u\rangle 0.3$; genital setae present in the region of 13-16. Nephropores in *bc*. Clitellum annular or saddle-shaped, in the region of 18-47; tubercula pubertatis absent. Male pores anteclitellar, very conspicuous, on 16, or rarely 15, on a flat area or, in erection, on everted copulatory sacs. Female pores in 14. Spermathecal pores posttesticular in the region of intersegments 13/14-16/17. more than 2, sometimes many, in at least some of the intersegments occupied. Dorsal pores absent.

Oesophageal gizzard well developed in 5; calciferous glands absent; intestine lacking muscular thickening; typhlosole (always?) absent. Supra-oesophageal and median subneural vessel present. Nephridia a pair per segment including the anteriormost segments; each with coiled loops; nephridial caeca (always?) present from the region of the male pores posteriorly. Holandric; testis sacs (always?) absent. Ovaries in 13. Tubular prostate-like glands associated with the pair of copulatory sacs and with the follicles of the genital setae. Spermathecae spherical to tubular.

Madagascar.

SUB-FAMLY HORMOGASTRINAE

Definition and distribution as for the type and only genus, *Hormogaster* Rosa, 1887.

GENUS **Hormogaster** ROSA, 1887

Type species: **Hormogaster redii** ROSA

Large pigmented or unpigmented worms (150-750 mm long and with 380-600 segments). Prostomium zygolobous or prolobous. Form cylindrical. Anus terminal. Setae lumbricine, very closely paired the lateral more closely paired than the ventral; $dd:u\langle0{\cdot}3$; the ventral couples in the clitellar region enlarged as genital setae; longitudinally grooved (always?) and ornamented distally with transverse incisions or with circlets of minute teeth. Nephropores immediately above *b* lines. Clitellum saddle-shaped, commencing on 13 or 14 and extending to 24-32. Tubercula pubertatis commencing in 17-19 and reaching 23-29. Male pores in intersegmental furrow 15/16, above *b* lines, between glandular elevations. Female pores paired, above *b* lines shortly presetal to posterior in 14. Spermathecal pores in 2-4 intersegments from 9/10-13/14 or even 14/15; paired, or multiple with 2-8 on each side in a transverse row. Dorsal pores absent.

Pharynx ending in 5; 3 oesophageal gizzards, in 6-8; typhlosole well developed. Extramural calciferous glands absent; oesophagus in 9-16; highly vascularized. Dorsal and ventral blood vessels continuous to the anterior end, single. Dorsoventral commissural vessels in 3-11, those of 5-11 forming 7 pairs of hearts; not latero-oesophageal. Supraenteric sinus forming a definite supra-oesophageal vessel in 13 running forward and capillarising on the pharynx. Suboesophageal vessel present. Subneural vessels continuous to the anterior end. Nephridia (always?) commencing in III, with funnels, the anterior nephridia each with a long bladder bearing a lateral caecum; terminal sphincter absent. Ovaries in 13. Spermathecae ovoid.

Mediterranean region: Italian Peninsula and Tuscan Archipelago; Sicily; Corsica; Sardinia; Tunisia; Algeria; S.E. Spain; France (E. Pyrenees). Terrestrial, at considerable depths.

Omodeo (1956 b) discusses the systematics and geographical distribution of the Hormogastrinae and gives synonymies of the two species. An earlier paper (Omodeo, 1948) gives an account of the morphology and biology of *H. reddii*. Both papers give extensive bibliographies of the sub-family. The fullest account of the anatomy of *H. redii* remains that of Rosa (1889). The illustrations of the nephridia of *H. praetiosa* are from Baldasseroni (1914).

SUB-FAMILY ALMINAE DUBOSCQ, 1902 EMEND.

Type genus: **Alma** GRUBE

Lateral lines absent. Body quadrangular in cross section; dorsal groove (always?)

present; anus dorsal or dorsoterminal; dorsal median intersetal distance less than 0·3 (*dd*: *u*⟨0·3⟩); genital setae, if present, ornamented with transverse serrations or cicatricing and/or with longitudinal grooves. Nephropores in or ventral to *b* lines. Clitellum annular. Male pores conspicuous or not, intraclitellar or anteclitellar. Female pores in 14 or rarely 13 and 14. Spermathecae post-testicular, rarely extending into, or in front of, the testis segments; largely intraparietal; never with clearly differentiated, intracoelomic duct; usually multiple in each intersegment occupied; rarely absent. Dorsal pores absent.

Rudimentary or well developed gizzard-like thickening of the oesophageal musculature present in 1 to 3 of segments 5 to 9, or such thickening absent. Calciferous glands absent. Intestine commencing in or behind 13; usually with some thickening of its musculature anteriorly; dorsal typhlosole present. Supra-oesophageal and subneural vessels present or absent; lateroparietals 1 only on each side. Nephridia absent from 10 or more anterior segments; all exonephric holonephridia, never with terminal sphincter or caecum nor with spiral coils. Holandric; testis-sacs absent. Glandular bursae present at the male pores or absent. Ovisacs present or absent. (Cocoons elongatedly fusiform, with several to many embryos.)

Tropical Africa and the Nile Valley. Central and S. America. India. Burma. Malaysia and Indonesia. Europe. Syria and Palestine.

KEY TO TRIBES

1.   Male pores on conspicuous oval protuberances
     which occupy segment 15 and extend into
     adjacent segments; each corresponding intern-
     ally with a large hemispheroidal bursa. Sper-
     mathecae absent   *Criodrilini*

—   Male pores not on segment 15, or if on 15
     lacking internal bursae. Spermathecae present
     (in the region of segment 10 to *c*. 200) though
     often recognizable only by stripping away the
     longitudinal muscle of the body wall   *Almini*

TRIBE ALMINI NOV.

Type genus: Alma GRUBE

Genital setae if present little if at all modified (except when on claspers). Nephropores in *ab* or *b* lines. Male pores one pair on 15-30; always inconspicuous; intraclitellar or preclitellar. Female pores in 14 or 13 and 14. Spermathecal pores post-testicular; rarely continued into and in front of the testis-segments; sometimes translocated into the hindbody, usually multiple in an intersegment. Cerebral ganglia in 2-3. Rudimentary or well developed gizzard-like thickening of the oesophageal musculature present in 1-3 of segments 5-9 or such thickening absent. Oesophagus posterior to this region highly vascularized (sinusoidal) and its posterior end with a relatively little-vascularized straining apparatus. Intestine commencing in 15-38; with or without anterior thickening of its musculature. Supra-oesophageal blood vessel present; at least some hearts latero-oesophageal. Median subneural present throughout; or absent from anterior segments; or totally lacking, in which case its function is subserved by a pair of lateroparietals. Lateroparietals, if coexisting with a median subneural communicating with the latter. Prostate-like glands rarely present.

Tropical Africa and the Nile Valley. Central and S. America north of the Juramento-Salado River. Throughout the Oriental Region.

KEY TO GENERA

1. Elongate ribbon like claspers bearing 2 or more genital setae present on or behind segment 18. Median subneural vessel absent *Alma*

— Claspers only exceptionally present and there located on segment 16 and lacking setae. Median subneural present or absent 2

2. A pair of well developed, non-muscular prostate glands present at the male pores in segment 17. Subneural vessel present behind them *Callidrilus*

— Prostates absent. Subneural vessel present or absent. Male pores not in 17 or, if (rarely) in 17 a subneural vessel absent throughout 3

3. A single gizzard, in segment 7 or 8. Low narrow, wall-like or delicately lamella- or wing-like tubercula pubertatis extending through several clitellar segments *Glyphidrilus*

— Gizzard weak in 5 or 6 absent. Tubercula pubertatis if present not narrow; rarely wing-like but then stout 4

4. Male pores on 15 or 16. Median subneural vessel present, at least in the mid and hind-body *Drilocrius*

— Male pores on 17. Median subneural absent throughout *Glyphidrilocrius*

GENUS **Callidrilus** M:CHAELSEN, 1890

Type species: Callidrilus scrobifer MICHAELSEN

Prostomium zygolobous. Body quadrangular behind the clitellum with a deep dorsal groove; anus dorsal. Setae 8 per segment, closely paired throughout, the couples narrowing posteriorly; the dorsal median intersetal distance equal to 0·2-0·3 of the circumference. Setal ornamentation oblique encircling ridges; or closely spaced scars in spirals and in longitudinal rows. Specialized genital setae absent. Clitellum wholly, or posteriorly, annular, occupying 19 to 29 segments, beginning in or behind 12 and extending as far posteriorly as 37. Wings absent but indistinct longitudinal ridges of puberty present lateral to *b* (in 17-20, 21) between which the male field protrudes, in the contracted state, as a conspicuous posteriorly U-shaped cushion. Male pores, each on a papilla, posterior in 17, slightly if at all lateral of *b*-lines; prostate pores diagonal on the papilla at or immediately behind setae *b*. Nephridiophores in front of setae *b*. Genital markings, with peripheral rim and central area, paired in several segments in the clitellar region, including those of the male field, lateral to, to median to, *ab* lines; unpaired ventral markings, singly or in rows, sometimes present. Female pores in 14, presetal in *a* lines. Spermathecal pores inconspicuous and numerous in (12/13), 13/14-14/15 (as many as 41 per furrow).

A moderate to large oesophageal gizzard in 5 and 6, or in 7-8, 9; weakly constricted intersegmentally; alimentary diverticula absent; intestine commencing in 15 or 16, lacking muscular thickening; a dorsal typhlosole (always?) present; oesophagus with less vascular region (straining apparatus?) interpolated between plexus and intestine. Paired hearts in 7-(11)12, some of them with connectives to the supra-oesophageal vessel. Subneural vessel present, divaricating anteriorly, in 15 or 16, to form a pair of lateroparietal vessels of which one may be suppressed. Holonephric; adult nephridia commencing in 12 or 13; nephrostomes large, single; ducts avesiculate, or with small bladder, and without caeca. Testes and funnels free in 10 and 11, male ducts deeply intramural; seminal vesicles 4 pairs, in 9-12. Prostate glands

Callidrilus scrobifer Michaelsen, 1890

"Intraclitellide" STUHLMANN, 1889 : 457.

Callidrilus scrobifer Michaelsen, 1890 : 20.

Callidrilus scrobifer Michaelsen. MICHAELSEN, 1891 : 210, Pl. VIII, fig. 7; 1897 b : 57; 1900 a : 458; 1913 a : 56; 1918 : 352.

Callidrilus dandaensis MICHAELSEN, 1897 b : 57.

Callidrilus scrobifer forma *dandaensis* Michaelsen. MICHAELSEN, 1913 a : 56.

Callidrilus scrobifer forma *reservationis* MICHAELSEN, 1913 a : 56.

Callidrilus scrobifer forma *typica* MICHAELSEN, 1913 a : 56.

Callidrilus scrobifer forma *dundaensis* Michaelsen. MICHAELSEN, 1913 a : 56.

Callidrilus scrobifer forma *reservationis* Michaelsen. MICHAELSEN, 1918 : 353.

Callidrilus scrobifer forma *nyassaensis* MICHAELSEN, 1937 a : 503.

l=115-200 mm, $w\langle$3-5 mm, s=200-250. Setae ornamented with closely spaced scars in irregular spirals and longitudinal rows. Clitellum 12, 15, 16, 17-(24?), 30, 32, 36, 37 (=(10?), 16-23 segments); wholly or posteriorly annular. Male genital field including a highly elevated rounded-rectangular cushion on the ventral surface of the clitellar segments which occupies 16, $1/n$16, 17, 16, to 20, $\frac{1}{2}$ 21, 23, and laterally extends over the ventral setal couples. Longitudinal ridges of puberty may or may not be evident bounding the cushion laterally. A pair of papillae situated at the anterior end of these ridges or at the anterior corners of the cushion, bears the male pores at 17/18, and just behind setae b of 17, the prostate pores. Genital markings on anterior paired series in 11, 12, 13-14, 15, 16, diverging posteriorly from postsetal in ab (segment 11) to lateral of b (segment 16). A posterior paired series and a midventral row absent, or present in some or all of 21-24. Spermathecal pores in 12/13, 13/14-14/15, numerous (e.g. 24-30) in each intersegment, the undermost in bc.

Oeosophageal gizzard in 5 and 6, not sharply demarcated; intestine beginning in 16; typhlosole not certainly present. Hearts in 7-11, 12, all, with the possible exception of those in 7, latero-oesophageal. Dorsal vessel ending with the hearts of 7. Subneural in 15 posteriorly. Nephropores presetal in b lines. Male pores encircled by intramural glands. Prostate glands free in the coelom. Ovaries compact. Spermathecae projecting in front of and behind the corresponding funnel.

Mosambique (Quilimane, the type locality); Tanzania; Malawi.

Michaelsen (1913 a) recognized 4 forms—*"typica"* (Quilimane, Mozambique); *dundaensis* (Dunda in Kingani; and Lake Chumgrum, N.W. of L. Nyassa); *reservationis* (Ngura and Mtibwa, Tanzania) and *nyassaensis* (Fort Johnston, Malawi)—by the distribution of the genital markings. No clear distinction on this basis is apparent from his accounts, however, examination of the types is required before the validity of recognizing infraspecific taxa can be certainly rejected. The statement (Michaelsen, 1897 b) that the last hearts are in XI in *dundaensis*, compared with XII in *"typica"*, requires confirmation.

Callidrilus ugandaensis (Jamieson, 1968)
Fig. 15. 1

Glyphidrilus ugandaensis JAMIESON, 1968 : 387, Fig. 1.

$l\rangle$295 mm, w=7-10 mm, $s\rangle$291. In XII aa:ab:bc:dd=6·3:1·2:6·9:1·0:8·0 (rear of 2); dd=0·20-0·28; setae ornamented by some 6 widely spaced oblique encircling ridges. Clitellum annular, 11, 12-35·38, 1/2 39 (=24-28 $\frac{1}{2}$ segments). Male genital field in 17, $\frac{1}{2}$, 17, $\frac{1}{2}$, 22, 22, 24, producing a posteriorly U-shaped cushion or contraction; longitudinal puberty ridges in 17-20, 21, lateral to b lines or unrecognizable. Male pores in 17/18 immediately behind the prostate pores in b lines, both pores of each side on a papilla. Paired genital markings in bc in 15-18 or some of these. Transverse rows of postsetal papillae in 13 and 14. Spermathecal pores in (12/13), 13/14-14/15, as many as 41 in an intersegment.

Gizzard well developed, in 7 and 8, and typically extending into 9, slightly constricted intersegmentally; oesophageal plexus in 11-13; intestine commencing in 15; dorsal typhlosole from 22. Hearts in 7-12, the last two pairs latero-oesophageal, the other hearts receiving "coronaries" from supraoesophageal-circumoesophageal vessels. Dorsal vessels ending with the anterior pair; supraoesophageal in 13 anteriorly. Subneural widely divergent to form the paired? lateroparietal, in 15 or 16; First nephridia very reduced in 12 or 13, ducts avesiculate. Ovaries plicate transverse bands. Septal pouches in 13-15, dependent from the anterior walls. Prostates racemose, at the male pores. Spermathecal ampullae largely or wholly intramural.

Uganda (Lake Victoria). Kenya (near Nairobi, collector E. Oxtoby, 1967, new record).

Exclusion of *ugandaensis* from *Callidrilus* was formerly necessitated by characteristics given in the definitions of this genus by Michaelsen (1918) and Stephenson (1930 a). These were wide pairing of anterior setae, location of male pores on a midventral puberty tubercle, absence of any reference to longitudinal tubercula pubertatis and genital markings, and location of the gizzard in segments 5 and 6. Correspondence with *Glyphidrilus* was on the whole close, even the low tubercula pubertatis in place of wings seeming to have an intermediate condition in *G. tuberosus* and *G. birmanicus*. The only departure from the definition of *Glyphidrilus* given by Michaelsen (1918) in which the male pores were stated to lie median to a pair of long "Pubertatsaumen" more or less far posterior to segment 16, lay in the intersetal ratios in *ugandaensis*. A survey of descriptions of *Callidrilus scrobifer* reveals, however, that correspondence between *ugandaensis* and *C. scrobifer* is closer than former generic definitions suggested, as is indicated in the new definition of *Callidrilus* above. A notable similarity between the two species is location of the male pores posteriorly in segment 17 in setal line *b* in association with a racemose prostate gland, whereas in *Glyphidrilus* only *weberi* reputedly has prostates (at 27/28) (p. 753). Location of the gizzard in the region of 5 and 6 in *Callidrilus* against 7 to 9 in *ugandaensis* may not be an important difference in view of variation in the location of the gizzard which occurs intragenerically in *Glyphidrilus*.

The similarity between *Callidrilus* s. lat. and the genus *Glyphidrilus* is nevertheless, striking and suggests a closer relationship than was suggested by Michaelsen (1918). It is noteworthy in view of this similarity, that the location of the gizzard in 6 in *Callidrilus scrobifer* corresponds with that in Gates hypothetical "protoglyphidrilus".

GENUS **Glyphidrilus** HORST, 1889

Type species: Glyphidrilus weberi HORST

Prostomium zygolobous (sometimes prolobous, pro-epilobous or epilobous?). Body quadrangular behind the clitellum; with a posterior dorsal groove; anus dorsal, dorsoterminal or exceptionally terminal. Setae 8 per segment, widely paired in the forebody, usually more closely paired posteriorly; the dorsal median intersetal distance equal to 0·2 of the circumference. Setal ornamentation consisting solely of one to several transverse or oblique distal ridges. Specialized genital setae unknown but those of the clitellar region at least sometimes enlarged. Nephropores in *b* lines. Clitellum annular, occupying several to many segments beginning in or behind segment 12. Body wall protuberant at maturity as a ridge which usually forms a delicate lamella (ala or "wing") dorsal to the ventral setal couples and crossing several to many clitellar segments. Male pores inconspicuous, located ventral to the wings, at or one to a few segments in front of their posterior limits. Paired and, often, median genital markings present in front of and less frequently behind the wings, exceptionally (*quadrangulus*) absent; each usually with a peripheral rim and a central area, rarely a simple papilla. Female pores paired in 14 or (*kukenthali*) in 13 and 14, presetal in front of or slightly lateral of setae *b*. Spermathecal pores inconspicuous; occupying some or all of the setal lines, or more numerous, in 2 to 6 intersegments from 12/13 to 18/19; exceptionally more extensive and including the testis-segments.

An oesophageal gizzard in 7 or 8, sometimes extending into an adjacent segment; alimentary diverticula absent; intestine commencing in (14?) 15 to 18, lacking muscular thickening; a dorsal typhlosole present; a ciliated "straining apparatus" separating the oesophageal and intestinal blood sinuses. Paired hearts in 7 to 11 or rarely to 12 or 13, some of them with connectives from the supraoesophageal vessel, those of 7 or 8 sometimes absent. Subneural vessel present, rarely reaching the anterior end, well developed only posterior to its junction with a pair of lateroparietal vessels. Holonephric; adult nephridia absent in front of segment 12; nephrostomes single ducts avesicles and without sphincters or caeca. Testes and funnels free in 10 and 11; male ducts intramural seminal vesicles usually 4 pairs, in 9-12. Prostate glands (always?) absent. Ovaries and funnels in 13 or 12 and 13; ovisacs present or absent. Spermathecae with ducts, and often the subspherical ampullae, concealed in the parietes; lacking diverticula.

Tanzania; India; Ceylon; Burma; Hainan; Malaya; Sumatra; Java; Borneo; Celebes. In rivers, lakes (sometimes at great depths) and moist soil.

The genus includes *Bilamba Rosa,* 1890, and *Annadrilus.* (Key to species, p. 753).

Ridge-like tubercula pubertatis are present in all species in *bc* and are elevated for part or, more usually, the whole of their lengths as delicate, laminate "wings" (alae). These may (abnormally?) grow out unilaterally as a foliating tumour-like mass in *G. tuberosus.* Few accounts indicate their height but *G. gangeticus* and *kukenthali* appear to display a typical development. Those of *birmanicus* are described as "rather low" as are those of *tuberosus* and especially, of *stuhlmanni.* As the wings are intraclitellar, and occupy much of the length of the clitellum, they are unusually far posterior (42, 43-66, 67) and long (*c.* 25 segments) in *stuhlmanni*; their extreme limits for other species being 17-35, occupying slightly less than 4-14 segments. Intraspecific variation in length of the wings may be marked. Those of the poorly described *weberi*, which may represent more than one species, were said by Horst to occupy 6-14 segments. In 116 specimens of a single colony of *gangeticus* the number of segments occupied varied from 6·5 to 9 (Gates, 1958 c). In contrast, 20 randomly selected specimens of *kukenthali* showed no variation in location or extent (present study). Genital markings consist of a peripheral rim and a central area (a depression or tubercle), as seen in *Callidrilus*, or in *tuberosus*, are simple rimless papillae. They have not been recorded for *quadrangulus*. Paired series are present in 14 species and unpaired, median series in all but *birmanicus*. Only in *tuberosus* is there more than one pair (basically three pairs) of markings in a segment. Extreme locations of the markings in the genus are 10-38 or (*stuhlmanni*) 50. Within these limits there is wide variation in their arrangement and, even where segmental distributions of paired and median markings are similar in apparently different species, frequency for each segment may differ significantly, as in the case of *kukenthali* and *gangeticus*. It appears true of the genus that paired (and usually also unpaired) markings do not extend into the region of the wings, but accounts are rarely explicit in this regard.

The location of the apertures of the vasa deferentia, which lie ventral to the wings, is known only for *annandalei* (anteriorly in segment 29 or 30), *ceylonensis* (intersegment 20/21), *kukenthali* (segment 24), *malayanus* (19/20), *quadrangulus* (21/22) and *weberi* (27/28). The location in numbers of segments anterior to the hind end of the wings is respectively 4, 3, 0, 1 1/3, 4 and 5. This contrasts with the location of the male pores in *Callidrilus* (including *C.* (=*Glyphidrilus*) *ugandaensis*) at the anterior end of the tubercula pubertatis, in segment 17.

Nair (1938) has given a detailed account of the gross morphology and histology of the alimentary canal of *G. annandalei* which largely accords with the author's finding for *G. kukenthali*. Other accounts permit only broad generalizations. Notably that an oesophageal gizzard is restricted or lies mainly in segment 8 in all species with the exception of *ceylonensis*, *jacobsoni*, *tuberosus* and *stuhlmanni* in which it lies in 7. In 9 of these species it occupies 2 segments. The intestine commences in 15 more frequently than in 16, and, only in *stuhlmanni* in 18. Origin in *annandalei* apparently varies from 14 [?] Cognetti, 1911) to 15 (e.g. Nair, 1938) or (reexamination of types) 16. In none is the muscular thickening of the anterior region of the intestine, so common in *Alma*, developed. A dorsal typhlosole begins in 16-19 (7 species).

From studies of *G. annandalei* and *G. kukenthali*, the following regions of the alimentary canal can be recognized (Fig. 15.2).

The pharynx: The pharyngeal mass pushes septum 4/5 far posteriorly and is preceded by a thin walled buccal cavity. It is enveloped dorsally and laterally by chromophil glands which extend from segment 4 into segment 7 (*annandalei*) or 6 (*kukenthali*). Nair gives details of secretory and other cells of the pharyngeal complex.

The oesophagus: A pre-gizzard portion appears to commence in the posterior region of the pharyngeal mass in segment 4 in *kukenthali* and in Nair's illustration for *annandalei*, and extends a little into segment 7. Its lumen is diverticulate, giving it a much folded exterior, and its inner circular and outer longitudinal muscle layers are well developed. Internally it is lined by an unciliated columnar epithelium which is bounded on its free surface by a hyaline layer which Nair identified as a thin cuticle. It is possible that this layer is a brush border. Nair does not mention the moderately well developed vascularization which is visible outside the musculature in *kukenthali*.

The gizzard (Fig. 15. 2A-F): The gizzard is lined internally by a thick cuticle which in *kukenthali* is striated at right angles to the surface. No striation was seen in *annandalei*. The cuticle is underlain by a simple, elongated columnar epithelium at the base of which (*kukenthali*) are scattered large rounded cells of unknown function. External to these is the extremely thick circular muscle layer and (*annandalei*) outside this a sparse layer of longitudinal musculature. The whole is enveloped in a clear "endothelium" (Nair) or a low, vascularized peritoneum (*kukenthali*).

Nair observed chromophil glands at the junction of gizzard and the succeeding oesophagus in *annadalei*. He comments on the uniqueness for the megadrili of this observation of chromophil cells beyond the pharynx. The author has observed groups of chromophil cells along the oesophagus of *Alma nilotica*. The *post-gizzard oesophagus* is divisible into a highly *vascularized unciliated portion*, here termed the sinusoidal region, which is the only portion of the canal, which Nair termed the oesophagus, and behind this a ciliated portion which Nair termed the "straining apparatus".

The *sinusoidal region* extends in *annandalei* from septum 8/9 to septum 13/14 (segments 9-13) and in *kukenthali* from mid-segment 9 to the posterior end of 12. The vascularization of this region in *kukenthali* corresponds with that described by Nair for *annandalei*. The oesophagus is here encircled by numerous blood vessels (lying between internal epithelium and muscular sheath) from which the supra-oesophageal blood vessel rises dorsally. In both species the supra-oesophageal commences in 13 and passes anteriad onto the pharynx but whereas in *annandalei* the oesophagus in segment 13 resembles that in the preceding segments, in *kukenthali* it is transitional in nature between the straining apparatus and the sinusoidal oesophagus, having the stiff cilia of the former and the vascularization, though less developed, of the latter.

The straining apparatus: In *annandalei* there are no intramural blood spaces in the straining apparatus, which is restricted to segment 14 or (re-examination of the types) occupies 15 also. The straining apparatus thus divides the alimentary blood system into two distinct regions, the plexus or sinusoids of the oesophagus and the sinus of the intestine. In *kukenthali* the same holds for the posterior portion of the straining apparatus lying in segment

14 but, as has been said above, the apparatus differs in extending, in transitional form, into 13 where a reduced intramural plexus occurs. In *kukenthali*, as in *annandalei*, the straining apparatus is externally covered by a very thick coating of chloragogen cells which gives it a distinctive appearance; the muscle layer is clearly developed (though thin) and the internal epithelium is thrown into longitudinal folds which considerably reduce the lumen; the epithelial cells are tall and slender, with elongated nuclei, and bear long, closely set cilia, such long cilia being found nowhere else in the alimentary canal. It would appear that Nair is correct in proposing that the cilia, projecting into the narrow lumen comprise an efficient straining apparatus. Nair rightly states that there is nothing in the histology of the straining apparatus to suggest that it is the homologue of a gizzard. The author has observed a similar apparatus in *Alma nilotica* and *A. stuhlmanni* in which it preceds a region of the intestine with muscular thickening which may be considered an intestinal gizzard.

The intestine: The straining apparatus protrudes into the intestine, forming an oesophageal valve which presumably limits regurgitation of food into the oesophagus. The wall of the intestine is extremely thin, consisting of a low epithelium underlain by a narrow uninterrupted blood sinus outside which are a very sparse muscular layer and the chloragogue. The dorsal typhlosole, beginning in segment 16 or 17 in *annandalei* and 17 in *kukenthali* has been observed by Nair to extend to the pre-anal segments in which it becomes flush with the surface, having diminished in girth by the posterior third of the worm; ciliation continued almost to the anus, there being no specialized rectum.

The dorsal blood vessel is continuous throughout the length of the body in *G. ceylonensis* (vide Gates, 1945) and in *G. weberi* but in the six other species for which its extent has been recorded it ends with the anterior pair of "hearts". According to Nair (1938), who had access to abundant fresh material of *annandalei*, the anterior "hearts", in segment 7, are not pulsatile and lack valves unlike those of the succeeding segments. They are thus strictly not hearts, though referred to as such in the species descriptions, but are merely vessels produced by bifurcation (and termination) of the dorsal vessel and connecting this, as do the true hearts, with the ventral vessel. The dorsal vessel is provided in every segment, in front of the septum with a pair of valves opening forwards (Nair, 1938: 53).

The *ventral vessel* runs beneath the gut, beginning in the third segment; from there backwards it gives off a pair of branches in every segment; just behind the points where the hearts join it paired valves opening backwards occur (Nair, 1938: 55).

The *hearts* (including the non-pulsatile anterior pair) lie in segments 7-11 in 7 of the 9 species, including *annandalei* in which their full extent has been recorded; in *gangeticus* (Fig. 15.2H), however, they occupy 9-11 (a reduction recorded by Cognetti (1911) for *annandalei*) or rarely (Gates,

1958 c: 55) 8-11; that on the left side being absent in a specimen examined by Stephenson (1920), giving paired hearts in 10 and 11 only; and in *spelaeotes* 8-11. In *weberi* and *malayanus* they have been erroneously recorded as ending in 12. In *stuhlmanni* they occupy 8-13. Two types of hearts may be distinguished: lateral hearts (sensu Gates, 1958) which simply connect the dorsal and ventral blood vessels, and latero-oesophageal hearts (s. Gates 1958) which have the same course but in addition receive a branch from the supra-oesophageal vessel. In the latter type Nair (1938: 52) has shown that both connectives (from the dorsal and the supra-oesophageal vessels) are provided with valves opening towards the heart. Recognition of the supra-oesophageal link is difficult in museum material but the observation of Nair (1938) for *annandalei* that only the last two pairs (i.e. those of 10 and 11) are latero-oesophageal, by virtue of the condition of his material and his exhaustive examination can be accepted. The same condition has been found in *birmanicus, papillatus* and *weberi* but Gates (1958 c: 57) found only the hearts of 11 to be latero-oesophageal in *gangeticus*, a fact confirmed in the author's examination, whereas those of 9-11 were latero-oesophageal in *ceylonensis*.

The *supra-oesophageal vessel* arises from the intramural plexus of the oesophagus and is therefore first apparent in 13 in *annandalei, kukenthali* and *gangeticus*, as in *weberi*, though in *gangeticus* there is some suggestion of development in 14 or even in 15. It seems probable that such vascularization as the straining apparatus possess normally drains into the supra-oesophageal trunk in the genus. This trunk continues into segment 2 where it bifurcates to join the anterior latero-parietals (*gangeticus* Gates, 1958 c) or bifurcates to supply the prostomium (*annandalei*, vide Nair, 1938). It receives in segment 2 (*gangeticus*) and in each of segments 6 and 7 (*gangeticus* and *annandalei*) a pair of vessels from the latero-parietal (latero-intestinal) vessels; in 8 it vascularizes the gizzard, and in 9-13 is fed by the plexus of the sinusoidal region of the oesophagus (*gangeticus* and *annandalei*); while in 10 and 11 (*annandalei*) or 11 only (*gangeticus*) it supplies roots to the latero-oesophageal hearts.

A supra-oesophageal vessel is apparently absent as a distinct vessel in *ceylonensis* and is weakly developed in *weberi* in both of which the dorsal vessel is exceptional in extending to the anterior end of the worm.

The latero-parietal and subneural vessels: In *annandalei* and *ceylonensis* the latero-parietals (lateral intestinals of Nair, 1938, extra-oesophageals of Gates, 1945) originate anteriorly (at the peristomium in *annandalei*) and, passing backwards one on each side, connect in 6 and 7 with the supra-oesophageal vessel (a distinct vessel and these connectives absent in *ceylonensis*) and in 8-13 (*annandalei*) or 9-13 (*ceylonensis*) with a pair of suboesophageal vessels; posteriorly, in 20 and 23 (*annandalei*) or 19-24 (*ceylonensis*) they bend medianly and join to form the subneural vessel. From the point of union of the two vessels the subneural continues forwards as a slender vessel to segment 14, where it capillarizes in *annandalei*, or to the pharyngeal region, in *ceylonensis*. A closely similar situation obtains in *gangeticus* and *kukenthali* but the latero-parietal vessels are typically (but not always in *gangeticus*) divided into two distinct sections on each side: an anterior and a posterior latero-parietal.

In *gangeticus* the anterior pair unites in one or all of segments 8-13 with the suboesophageal vessels or, less commonly they turn medially in 12, 13 or 14 and unite with the posterior latero-parietals, the united vessel on each side then joining the suboesophageals. The posterior latero-parietals more commonly open into or become the suboesophageal vessels in 12, 13 or 14 independently of the anterior pair. The transverse vessels may be in different segments instead of being paired in one segment. The subneural trunk has a pair of branches passing laterally just behind each septum. Any one, or any pair, of the branches, in the region of 23-28 may be much enlarged. An enlarged branch in 22, 27 or 28 after passing laterally for some distance turns almost at right angles and then continues forward as the posterior latero-parietal vessel which may join in any of segments 18-23 by another large branch of the subneural or, if the subneural has disappeared in the meanwhile, the two latero-parietals may be interconnected by a large commissure. The subneural disappears in *gangeticus* before reaching 17 (Gates, 1958 c) or in 16 (present study).

The condition in *kukenthali* falls within the variation seen in *gangeticus*: each latero-parietal sends a transverse vessel beneath the oesophagus (to the suboesophageals) the anterior latero-parietals doing so in 9 or 10 and the posterior latero-parietals in 12; junction to form the subneural occurs in 24, and the subneural disappears in 17.

The *suboesophageal vessels*, with which the latero-parietals connect as noted above, are part of the plexus of the sinusoidal region of the oesophagus. In *annandalei* they are two longitudinal vessels which run closely parallel to each other on either side of the midventral line of the oesophagus and are free from the gut wall at regular intervals; posteriorly they end at septum 13/14 but anteriorly they break up into small branches on the ventral surface of the gizzard, one of the branches so formed passing sideways and spreading over the parietes (Nair, 1938). In *gangeticus* the suboesophageal vessels may be paired, unpaired and median, or large on one side and small on the other, or present on one side only, or the condition may vary from one segment to another (Gates, 1958 c). Paired suboesophageals ("infra-intestinal vessels") were noted by Horst (1894) in *weberi*. *Latero-oesophageal vessels* appear to have been observed in *annandalei* only by Nair (1938). They were not seen by the author in long-preserved material of *annandalei* and poor preservation probably accounts for the absence of any record of them in other species. Shortly above its junction with the ventral vessel each heart receives two blood vessels from the parietals. The larger vessel of each side are interconnected longitudinally to form a longitudinal vessel (the latero-oesophageal) lying clear of the gut and parallel with the latero-parietal. Behind the region of the hearts the latero-oesophageal vessel is directly connected with the ventral vessel by a transverse branch in each of segments 12, 13 and 14. It also extends backwards to segment 15, 16 or even 17 (Nair, 1938). It appears to the author that the latero-oesophageals are homologous with the ventral latero-parietals of *Sparganophilus*.

Prostate glands were reported for *G. weberi* by Horst but their presence has not been confirmed in a re-examination of the types (from Java). Their presence in the genus therefore seems inacceptable unless they occurred in types of *weberi* from other localities.

KEY TO SPECIES

1.	Gizzard wholly, or for most of its length, in 7	2
—	Gizzard wholly, or for half or more of its length, in 8; or 2 gizzards, with the larger in 8	5
2.	Wings beginning in 42 or 43 (hardly protuberant). Intestine commencing in 18	*G. stuhlmanni* (Tanzania)
—	Wings beginning far forward of 42. Intestine commencing in front of 18	3
3.	Wings beginning in 26 to 28. Intestine commencing in 16.	*G. ceylonensis* (Ceylon)
—	Wings beginning, or first lamellar, in 20 or 21. Intestine commencing in 15	4
4.	Wings in 20-24, low; continuous with slight anterior sometimes and posterior, ridges. As many as 3 pairs of genital markings per segment. Spermathecal pores in 13/14 and 14/15, in *a*, *b*, *bc* or also *c*.	*G. tuberosus* (India)
—	Wings in 21-26, high lamellae; continuous with anterior and posterior ridges. Paired genital markings maximally 1 set per segment. Spermathecal pores in 12/13-16/17 in *a*, *b*, *c* and *d* lines	*G. jacobsoni* (Sumatra)
5.	Ovaries (always?) in 12 and 13 (female pores in 13 and 14). Wings beginning in 18 or 19. Median papillae, in 1 to 3 of segments 10-15, transversely doubled (i.e. 2 in a single rim)	*G. kukenthali* (Borneo)

5.   Ovaries in 13 only (female pores in 14). If wings begin on 18 or 19, median papillae not doubled, though in *papillatus* and *spelaeotes* median *paired* papillae may occur on 12, 17-20, 24-29   6

6.   Spermathecae in setal lines c and d only or with supernumeraries median to c or in d. Wings in 22 or 23   7

—   Spermathecae not limited to c and d lines. Wings rarely (*papillatus* and *birmanicus*) beginning in 23   8

7.   Intestine commencing in 15. Wings in (22?) 23- (27, 28?) $\frac{1}{2}$32.   *G. weberi* (*Java*)

—   Intestine commencing in 16. Wings in $\frac{1}{2}$23 to part of 27   *G. horsti* (Straits of Malacca)

8.   Genital markings absent. Wings beginning in 19 or 20   *G. quadrangulus* (Sumatra)

—   Genital markings present. Wings rarely beginning in 19 or 20   9

9.   First hearts or commissurals in 7   10

—   First hearts in 8 or 9   13

10.   Median genital markings present   11

—   Median genital markings absent   *G. birmanicus* (Burma)

11.   Wings beginning on 18-23 and extending to 23-26. Genital papillae unpaired, median, on 11-21 and 23-33; paired median on 12, 17-20, 24-29; paired laterals on 12-28; or in some of those segments only   *G. papillatus* (Burma)

—   Wings beginning in or behind 25. Genital papillae not as above   12

12.   Wings in 25-27 to $\frac{1}{2}$32-33 (35). Lateral markings from 13 or 15-26, and often also in 1-5 segments behind the wings. Median markings in 2 or more of segments 11-14 to 26 and occasionally on 33-38   *G. annandalei* (India)

—   Wings in $\frac{1}{2}$25, 25-30. Lateral markings in 17-24. Median markings in 12-24 and 29-31   *G. buttikoferi* (Borneo)

13. Intestine commencing in 15   *G. spelaeotes* (Assam)
—   Intestine commencing in 16   14

14. Wings in 17, 18-$\frac{1}{3}$, 21, $\frac{1}{3}$ 22 (4$\frac{1}{3}$ seg-
     ments)   *G. malayanus* (Malaya)
—   Wings beginning on 17-19 and ending on
     23-26 (=6$\frac{1}{2}$-9 segments)   *G. gangeticus* (India)

Glyphidrilus annandalei Michaelsen, 1910
Fig. 13. 3A-C

Glyphidrilus annandalei MICHAELSEN, 1910 a : 101.

Glyphidrilus annandalei Michaelsen. COGNETTI, 1911: 502, Pl. XIII fig. 11,
   12. MICHAELSEN, 1913 c: 92; 1918: 344. STEPHENSON, 1916: 349; 1921:
   767; 1922: 387; 923: 491. NAIR, 1937: 300; 1938: 39, Fig. 1-24. GATES,
   1958 c: 54.

Glyphidrilus achencoili (laps.) COGNETTI, 1911: 506.

Glyphidrilus fluviatilis and *G. elegans* and *G. rarus* and *G. saffronensis*
   RAO, 1922: 53, 62, 64, 66, Fig. 1-4.

?*Criodrilus* sp (?*lacuum*). STEPHENSON, 1925: 903.

l=90-265 mm, w=2·5-11 mm, s= 125-322. Zygolobous, sometimes prolobous.
Anus dorsal. Setae widely paired throughout but more closely posteriorly;
aa : ab : bc : cd : dd=2·2-2·3 : 1·0-1·1 : 2·3-2·4 : 1·0-1·1 . 2·5-3·0, in segment 12; dd : u=0·18-0·20;=3·0 : 1·0 : 2·5 : 1·2 : 4·8 : or 2 : 1 : 2 : 1 : 3·5
in the hindbody, dd : u=0·25-0·43. Clitellum annular, commencing in 13-18
and extending to 35-41, (=approximately 19-25 segments) usually well developed
only from 17 posteriad; its limits usually indistinct. Wavy "wings" 1·25-1·5 mm
high in 25, 26, 27 to $\frac{1}{2}$32, 32, $\frac{1}{2}$33, 33 or occasionally 35 in bc: usually continued forward as ridges, angles, or whitish lines to $\frac{1}{2}$17 or 18 and sometimes
posteriorly of the wings as far as 36. Genital markings, with rims, in 2 series,
median and lateral; the lateral markings 1 on each side between b and d
beginning in 13, or on or behind 15 and ending, in front of the wing as far
posteriorly as segment 26, with a second, lateral series often present, in 1-5
segments immediately behind the wings; the median series in 2 or more of
segments 11, 12, 13 or 14 to 26 and occasionally present behind the wings, on
some of 33-38, the total number varying from 2 to 14. Nair states that markings
are always absent in the segments occupied by the wings, but Cognetti (1911)
noted additional paired markings in 27-33 in some specimens. Male pores two
point-like depressions in or slightly in front of intersegment 28/29 or 29/30
(variation in the types), in b lines. Female pores in or immediately median or
lateral to b lines, presetal in 14 nearer the anterior margin than the setal arc.
Spermathecal pores inconspicuous or undetectable externally in 4 intersegments,
13/14-16/17 and occasionally in 17/18 also, basically in setal lines a b c and d
and in mid bc, rarely ventral to a; usually 10 or 11, in a transverse row with

the extremes of variation 2-13 per row; never present in the middorsal or mid-ventral intersetal spaces. Alimentary canal as p. 749, gizzard large, always in 8, sometimes occupying more or less of 7. Intestine commencing in 15 or (re-examination of type) 16. Dorsal typhlosole beginning in 16 or (type) 17 (or 20?). Vascular system as p. 750. Hearts in 7-11, the "hearts" in 7 formed by bifur-cation of the dorsal vessel and neither pulsatile nor valvular, unlike the others. Hearts of 7-9 dorsoventral, those of 10 and 11 latero-oesophageal. Supra oeso-phageal in 13 to the anterior extremity. Lateroparietals uniting in 20-23 to form the subneural vessels; a slender subneural continuing forward from the junction to 14. Nephridia in 12, 13 or 14 (sometimes 9) posteriad; single nephrostome with 15 marginal cells; avesiculate. Seminal vesicles in 9-12; those in 12 the largest, pushing septum 12/13 back to 13/14. Septal pouches in 13, dependent from 12/13. Vasa deferentia deeply emebedded in the longitudinal muscle of the body wall; their terminal cells unciliated and glandular. Plicate ovaries and funnels in 13; ovisacs located above the funnels, evaginations of septum 13/14 into 14. Spermathecae lying in the segment behind their pores; ovoid sacs 125×75 μ or more in size, embedded in the body wall or protuberant into the coelom; ducts short and straight.

India: Malabar Coast; Travancore; Coorg; Mysore.

The identity of the specimens from Mysore, of which only external features were described (Stephenson, 1921) is perhaps questionable. The wings originated in 25, one segment in front of the most anterior origin noted by other authors. The disposition of the genital markings was unusually constant, the median markings terminating in 21.

The above account contains data derived in the present study from two syntypes (from Quilon, Hamburg Museum V. 3600 (Fig. 15. 3B, C) and two clitellate specimens from Travancore (B.M. (N.H.) 1925.5.12. 86/87 (Fig. 15. 3A)).

Glyphidrilus birmanicus Gates, 1958

Glyphidrilus birmanicus GATES, 1958 c: 61.

Glyphidrilus papillatus (part.) (ROSA). GATES, 1933: 603.

l=95-103 mm, w (clitellum)=6 mm, elsewhere 5 mm. Zygolobous. Form? Anus dorsal or dorso-terminal. Nephridiopores in setal lines *b*. Clitellum 12, 13-43, 44. "Wings" low, 21, 22, 23-28, 29, 30 in *bc*, just lateral to the genital markings which are postsetal above setal lines *b* on 12, 13 to 21, 22, 23, 26 and 29, 30 to 31, 33 or 34 (and do not occupy *aa*).

Gizzard in 8 and apparently part of 7. Intestine beginning in 15; Typhlosole beginning in 18 thickly lamelliform. Dorso-ventral hearts in 7 to 11, those of 10 and 11 latero-oesophageal. Dorsal vessel ending with hearts of 7. Septal pouches in 13 and 14. Spermathecae concealed in the parietes, 2 to 8 on each side in a transverse row, their pores in 13/14 to 17/18.

Burma.

Distinguished from *papillatus* by the larger size, absence of genital papillae in *aa*, and location of the wings. The author has included here specimens from Mong Mong Valley and Bana, referred to *G. papillatus* by Gates (1933), in which the wings occupied 21, 22, 23-29, 30, lateral papillae occupied some of 13-26 and 29-30 and markings were absent from *aa*.

Glyphidrilus buttikoferi Michaelsen, 1922
Fig. 15. 3D-F

Glyphidrilus buttikoferi MICHAELSEN 1922: 9.

l=110-150 mm, w=4 mm, s=c. 370. Prolobous to proepilobous. Anus dorsal but not occupying several segments. Setae in the forebody widely paired, especially the dorsal couples; $aa : ab : bc : cd : dd$=1·7 : 1 : 1·7 : 1·3 : 2; $dd : u$=0·17. At the hind end the ventral pairs narrowed (aa=c. 3ab); the dorsal pair narrow or very wide owing to the entirely irregular location of seta d. Clitellum segments 12 to 30, indistinctly delimited. A pair of wavy, moderately wide "wings" extends, between setal lines b and c, from ¼25 or 25 (in which development may be very weak) to 30 (5½ or 6 segments). Numerous large, transversely oval genital markings which occupy almost the entire lengths of their segments present: 16 unpaired ventral median in 12-24 and 29-31 and 16 paired, between setal lines b and c, in 17-24; one or two of these may be absent. Spermathecal pores in intersegments 14/15 to 17/18, on each side in groups of 1 to 6 mainly in the setal lines; supernumaries median to a lines and in bc.

A small or almost vestigial gizzard, with moderately thick walls, in segment 8. Intestine commencing in 15; typhlosole apparent by 19. Hearts in 7 to 11. Supra-oesophageal vessel continued onto the pharynx. Dorsal vessel ending anteriorly with the hearts of 7. Nephridia in 15 posteriad. Sperm funnels in 10 and 11. Seminal vesicles in 9 to 12. Ovaries appearing large and bushy, but basically plicate laminae, in 13. Ovisacs somewhat smaller than the ovaries, in 14. Spermathecae simple, subspherical.

Borneo: Sintang.

The three alate syntypes (Hamburg Museum V. 9301) have been re-examined in the present study (Fig. 15. 3D-F).

Glyphidrilus ceylonensis Gates, 1945

Glyphidrilus ceylonensis GATES, 1945: 89.

l=180-280 mm, w=5-7 mm. Dorsal groove? Anus a dorsal triangular orifice, apex anterior. Zygolobous. Setae ab c.=$cd\langle bc\langle aa\langle dd$. Low ridges beginning in 16-19 and ending on 32-35, slightly dorsal to b lines posteriorly but passing gradually dorsal further anteriorly, at the front end markedly curved upwards and reaching nearly to c lines; almost twice as high over 26, 27 or 28-30, 31 or 32 than elsewhere. Post setal genital markings (with central area and peripheral rim?) just ventral to the ridges of puberty and close to b lines, on some or all of 16-25 and 31-35. Also, in one specimen, a marking just median to seta a on one side in 33 and 34. Clitellum? Spermathecal pores in 13/14 to 16/17 in a and b lines, "two in bc on each side".

Gizzard strong, confined to 7. Intestinal origin in 16; typhlosole beginning

in 18 to 19, fairly large, triangular. Dorsal blood vessel continued to region of
the brain. Subneural reaching to pharyngeal region but larger behind 19 to 24.
Hearts in (5), (6), 7 to 11 those of (9), 10 and 11 latero-oesophageal. Nephridia
absent from 11 anteriorly, small in 12-14, large thereafter. Holandric (?); seminal
vesicles in 9-12.

Ceylon: Palmadulla, 1000-1500 ft.

G. *ceylonensis* appears to be closest to the S. Indian G. *annandalei*
Michaelsen 1910, from which it is distinguished by its large size, restriction
of the gizzard to 7, and absence of spermathecal pores from *c* and *d* lines.
Gates gives a detailed account of the blood system.

Glyphidrilus gangeticus Gates, 1958
Fig. 15. 1H; 15. 3G, H

Glyphidrilus sp. ? MICHAELSEN, 1909: 244.

Glyphidrilus papillatus; STEPHENSON, 1920: 258-260; GATES, 1948 b: 175;
GATES, 1951: 17; (part.) STEPHENSON, 1923 a: 493.

Glyphidrilus sp. GATES, 1947: 121.

Glyphidrilus gangeticus GATES, 1958 c: 55.

$l=85$-200 mm, $w=2.5$-5.5 mm. $s=202$-325. Anus dorsal or dorsoterminal, a
median longitudinal slit widening to form a triangle posteriorly. Zygolobous.
Setae widely paired throughout; $aa : ab : bc : cd : dd = 1.6 : 1 : 2 : 1.1 : 2.5$
in segment $12; = 2.5 : 1 : 2 : 1.2 : 3.7$ in c. 300; $dd : u = 0.20$ and 0.25 respectively. Setae in the clitellar region bearing a few distal oblique, jagged ridges
which give them a minutely but sparsely toothed profile; this ornamentation
less developed in posterior setae; length and greatest width (at the node) 418 μ
and 26 μ (segment 25), 300 μ and 22 μ (hindbody); all sigmoid. Male pores in
intersegment 20/21 in *b*. Female pores minute white points in *b* lines, midway
between the anterior border and equator of 14. Wings beginning on 17-19,
ending on 23-26 ($=6\frac{1}{2}$-9, usually 7 or $7\frac{1}{2}$, segments) wavy lamellae 0.45 mm
high and 0.15 mm wide, in *bc*. Genital markings unpaired, median and postsetal
in (10), 11-14 (15-16), 17-18 (19, 28-32), or in some of these or absent; paired
lateral to *a* or in the dorsal half of *bc* in 10, 12-17, 18 (19, 23), 24, 25, 26, 27
(28-30); with rim and central portion. Clitellum annular, 13, $\frac{1}{2}$ 13, 14, 16(?)-26-34
(?) 37 (?) ($=11$-$23\frac{1}{2}$ segments); limits indefinite. Spermathecal pores chiefly
in setal lines *a-d*, 4 or 5 on each side in a transverse row in 12/13, 13/14-16/17;
sometimes, reduced in number and size, in 17/18.

Gizzard with its major part in 8 or 2 separate gizzards, in 7 and 8 with the
larger in 8. Intestinal origin in 16; typhlosole from 17 or 18, large, wide, not
lamelliform. Oesophagus (always?) narrow with vascular striae in 9-13; moniliform and poorly vascularized in 14 and 15. Hearts in (8), 9-11; only the last
pair latero-oesophageal; dorsal vessel ending anteriorly with the hearts of 9;
supra-oesophageal in 13 (14, 15?) anteriad; latero-parietals well developed
uniting in 22; median subneural vessel in 16 or 18 posteriad. Nephridia in 13
posteriad; rudimentary in 12. Holandric; sperm ducts not concealed in body
wall. Seminal vesicles in 9-12; the last pair deflecting septum 12/13 to 13/14.

Septal pouches on the anterior walls of 13, 14 (15, 16). Ovaries in 13, transversely extensive unfolded laminae composed of longitudinal chains of oocytes. Inseminated spermathecae large laterally adpressed ellipsoidal sessile sacs with their long axes filling their segments longitudinally and projecting freely into the body cavity, 4 or 5 on each side in a transverse row; or 2 or less if the row is anterior of the usual locations.

India: Western part of the Gangetic plain, from Saharanpur to Lucknow, Ahraura and Sohagi; including the Jumna River at Delhi (collector M. C. Balani, 28 ii 1961,—new record).

G. gangeticus is distinguishable from *G. papillatus* (Rosa) 1890, with with which it has often been confused, by the termination of the dorsal vessel with the hearts of 9 and by the intestinal origin in 16 (segments 7 and 15, respectively, in *papillatus*). The material from Lucknow, ascribed by Stephenson (1920) to *papillatus* but synonymized by Gates (1958 c) with *gangeticus*, has the first hearts in 9 and intestinal origin in 16, as in *gangeticus* from the type locality (Saharanpur) but has the spermathecal pores restricted to 13/14-16/17, as in *papillatus* (Burma). Intraspecific variation in the number of rows of spermathecae occurs in other glossoscolecoids, however, and there is therefore little reason to doubt that the Lucknow specimens have been correctly assigned to *gangeticus*. It now appears that *papillatus* does not occur in India. No obvious difference in the disposition of wings and genital markings separates the two species but the Lucknow specimens were peculiar in having the paired genital markings beginning in segment 10, whereas 10 and 11 lacked them in all 116 Saharanpur specimens of *gangeticus* which Gates examined for this character.

Glyphidrilus horsti Stephenson, 1930

Glyphidrilus horsti STEPHENSON, 1930 b: 4.

$l=35$ mm, $w=2$ mm, $s=145$-167. Anus terminal. Zygolobous (?). Setae paired, more closely posteriad; in front of the clitellum $ab=\frac{1}{2}$ aa (or more than $\frac{1}{2}$ aa at the anterior end)$=3/7$ bc (or at the anterior end $\frac{1}{2}$ bc)$=\frac{1}{4}$ cd; dd $c=1\frac{1}{3}$ aa,$=0\cdot20$ to $0\cdot25$ μ; behind the clitellum $ab=\frac{1}{3}$ $aa=\frac{2}{3}$ bc, $c=cd$, $dd=1\frac{1}{4}$ aa $=0\cdot20$ to $0\cdot25$ μ. Clitellum segments 17 to 28 ($=12$). Wings occupying $\frac{1}{2}$ 23 to $\frac{1}{3}$, $\frac{1}{2}$ or $\frac{3}{4}$ 27; not very wide, between setal lines b and c, a little nearer b; they are turned somewhat downward, may be applied to the body wall, and are sometimes slightly frilled. Genital markings with rims; a mid ventral series on the hinder part of segments 16, 17, or 18, to 20 or 22 and usually also on 27, or 27 and 28; paired markings laterally on the hinder half of 27, and on the anterior half of 23, immediately behind the last and immediately in front of the first segment of the ridge of puberty. Genital apertures?

Gizzard large and barrel shaped, nearly twice as long as broad, in 8, forcing the septa apart, with a small portion in 7. Intestinal origin 16. Nephridia commencing in 13. Seminal vesicles in 9, 12 and probably 11; sperm masses in 10 apparently free. No prostates seen. Spermathecae subspherical sacs, the ducts concealed in the body wall, usually 2 on each side per segment, in line with setae *c* and *d*, in intersegments 14/15 to 16/17; one supernumerary seen in 13/14 in *d* line in one specimen. Spermathecae when dilated fill the length of a segment and are pressed together to form a compact mass of six on each side.

Straits of Malacca: Pulau Berhala, in a freshwater streamlet.

Glyphidrilus jacobsoni Michaelsen, 1922
Fig. 15. 3I-J

Glyphidrilus jacobsoni MICHAELSEN, 1922: 10.

$l=160$ mm+, $w=4$ mm, $s=274+$. Prolobous (?). Anus small, dorsal, but extending only a very little anteriorly. Setae anteriorly widely paired, especially the dorsal couples, posteriorly fairly closely paired. In the forebody aa : ab : bc : cd : $dd=2$: 1 : 2 : 1·3 : 2·3 : dd : $u=0·17$. Clitellum segments 18-30; clitellar region widened and dorsoventrally somewhat depressed, anteriorly and posteriorly (especially distinctly anteriorly in segments 18-20, less distinctly posteriorly, in 27-30) bordered by a somewhat folded ridge. In the intervening region, in 21-26, the ridge is replaced by a rather high, ventrally hanging, delicate "wing". The ridge and wing lie in *bc*. Genital markings with rims, in *bc* or including setae *b*; 3 pairs in front of and 3 pairs behind the wings, i.e. in 18-20 and 27-29, those of 29 less well developed; in one specimen supernumerary papillae were present on the right side in 14 and 15, and on the left side in 13. A median, unpaired papilla may occur in 13 and 14. Spermathecal pores in 5 intersegments 12/13 to 16/17 in setal lines *a*, *b*, *c* and *d*; supernumeraries absent.

Last chromophil glands in segment 6. Gizzard large and firm in 7, with softer anterior portion in 6. Intestine commencing in 15. Hearts in 8 to 11; the last pair (only?) latero-oesophageal. Dorsal vessel ending anteriorly by bifurcation to form the hearts of 8. Supra-oesophageal continuous over the gizzard; commencing in 12. Subneural vessel ending anteriorly in 13, very large further posteriorly. Latero-parietals large, traceable posteriorly as far as *c*. 26. Nephridia rudimentary in 12; well developed in 14. Seminal vesicles in 9 to 12; those of 12 extending to 14. Vasa deferentia ending at 22/23 very slightly median to *b* lines; elsewhere running slightly lateral to these. Ovaries delicate laminae of linear strings of oocytes. Ovisacs absent? Spermathecae elongate-subspherical, over or behind the intersegments which contain the pores.

Sumatra: Korintji, Soengai Koenbang.

Re-examination of the single alate syntype (Kurintji; Hamburg Museum V. 9293) (Fig. 15. 31J) permits the first description of the internal anatomy of this species.

Glyphidrilus kukenthali Michaelsen, 1896
Fig. 15. 2A-F; 15. 4A-D

Glyphidrilus kukenthali MICHAELSEN, 1896 a: 195 Pl. XIII, fig. 1.

Glyphidrilus kukenthali Michaelsen. MICHAELSEN, 1900 a: 460; 1918: 344.

$l=75-90$ mm, $w=2-3$ mm, $s=153-200$. Zygolobous. Anus dorsal or dorsoterminal. Setae widely paired throughout, aa : ab : bc : cd : dd=2·5 : 1·0 : 2·8 : 1·2 : 33, with dd : u=0·21, in segment 12;=2·7 : 1·0 : 2·7 : 2·0 : 6·0, with dd : u=0·30 new segment 100. Clitellu mannular, $\frac{1}{2}$ 13-34, limits indistinct, weakly developed in front of 17 or 18. Wings high, in (18, $\frac{1}{4}$18) 19-24 (1/3, $\frac{1}{4}$25). Female pores 2 pairs on inconspicuous low papillae, midway between setae b and the anterior margins of 13 and 14 (sometimes in 14 only?). Spermathecal pores in (12/13, a single pore, in one specimen) 13/14-17/18, maximally 6, minimally 3, on each side, one in each of the setal lines and in ab and bc, that in ab and d often, and others sometimes, absent. Genital markings 3 types: (1) *pre-alar median markings*, always present, each with a well-developed rim surrounding 1 or 2 papillae, in one to three of segments 10-15, usually in 13 and 14; (2) *post-alar median markings*, rarely absent, in one to five of segments 25-32, each with a transversely elliptical postequatorial papilla surrounded by a more or less distinctly developed rim; segment 30 the most commonly occupied; (3) *paired markings*, each with single papilla and rim, invariably present in each of segments 18 and 25, i.e. in the immediate pre- and post-alar segments. An additional pair commonly present in 17, and sometimes in 16, 26 and/or 27. Rarely traces of paired papillae in 28-31. Centres of the paired papillae varying from below c lines (16 and 17) to in or above b lines (18), in b or occasionally ab (25) and slightly further median, usually in ab (26 posteriad). Alimentary canal (details p. 749 and Fig. 15.2 A-F) with highly muscular gizzard in 8; intestinal origin in 15 (16?); typhlosole beginning in 17. Hearts in 7-11; dorsal vessel ending anteriorly with the hearts of 7; supra-oesophageal vessel in 13 anteriad; subneural vessel ending anteriorly in 17; latero-parietals as p. 752. Nephridia in 13 or 14 posteriorly; avesiculate with very slender ducts discharging in front of setae b. Testes and funnels free in 10 and 11. Seminal vesicles in 9-12 (sometimes in 10 and 11 only?). Vasa deferentia visible internally throughout their lengths, diverging lateroposteriorly in 23 and disappearing (presumably ending) shortly behind septum 23/24 a little below c lines. Prostate glands absent. Ovaries flat, tongue-like laminae, and their funnels, free in 12 and 13 (rarely 13 only?); ovisacs absent. Spermathecae subspherical sacs with intramural ducts in the segments behind their pores (14-18) except those discharging in 13/14, which occur in 13 also. Inseminated spermathecae protuberant far into the coelom, those in a-b laterally adpressed, as in cd, those of bc usually standing apart.

Borneo: Barem River (Borneo and Sarawak).

The above account is drawn from the author's detailed examination of 20 clitellate specimens (B.M. (N.H.) Reg. No. 1904. 10.5 1102/1112) (Fig. 15. 4A-D) hitherto undescribed, from the Barem River (Sarawak) the type locality. Conspecificity with the single mature type specimen (only two immature, non-alate types remain) is indicated by the paired papillae between the ventralmost setae in 13 and 14 in the latter. No other genital markings are described in the brief account of the type but as those in 13 and 14 were rudimentary (Fig. 1, Michaelsen) it is probable that their absence elsewhere was due to incomplete maturity. Differences described for the type are origin of the wings in 18 (constant in segment 19 in the 20 newly examined specimens), doubtful restriction of seminal vesicles to

9 and 10, and a single pair of ovaries, in 13.

Hologyny in the species is discussed on p. 60.

Glyphidrilus malayanus Michaelsen, 1903
Fig. 15. 4E-F

Glyphidrilus malayanus MICHAELSEN, 1903 b: 35.

Glyphidrilus sp. ? GATES, 1938: 221.

(?) *Glyphidrilus malayanus* Michaelsen. GATES, 1958 c: 54.

$l=85$-90 mm, $w=2$-2·5 mm, $s=236$-256. Zygolobous. Anus dorsoterminal or dorsal and close porteriorly by a pygidial knob. Setae widely paired through out; in segment 12 *aa* : *ab* : *bc* : *cd* : *dd*=2·1 : 1 : 2·2 : 1·4 : 2·2; *dd* : *u*=0·17; at the posterior end *aa*=1·9 *ab*. Wavy wings in 17, 18-1/3 21, 1/3 22 (=4 ⅓ segments), above setal lines *b*. Clitellum annular, 15-25. Ventral median genital markings in 12-15 or 21, 22-25; paired in 21 or in 15-17 and 22 in setal lines *b*; single, asymmetrical and lateral in 14 or lateral in 15. Male pores (from vasa deferentia) in 19/20 in *b* lines. Female pores presetal in 14 immediately lateral of *b* lines. Spermathecal pores up to 2 on each side in intersegments 14/15-16/17, in setal lines *b* and *c* only.

Gizzard large and firmly muscular in 8, preceded by a shorter less muscular, gizzard-like region in 7. Oesophagus sinusoidal as far posteriorly as 13 and transitionally so in 14; in 15 not apparently vascularized, rounded and relatively thick walled (straining apparatus?). Intestine commencing at septum 15/16; typhlosole beginning anteriorly in 17. Hearts in 9-11; at least those in 10 and 11 latero-oesophageal. Dorsal vessel ending, anteriorly, at the hearts in 9. Supra-oesophageal vessel in 13 to the anterior end of the body, giving off a large pair of vessels in 7 to the latero-parietals. The latero-parietals originating from the median subneural in 24 in front of which no subneural is apparent. Nephridia rudimentary in 12; well developed in 15 posteriorly, ducts avesiculate, slender, entering the parietes in *b* lines. Tenuous lobed testes and large iridescent sperm funnels in 10 and 11. Seminal vesicles in 9, 10, 11 and 12; those in 12 deflecting septum 12/13 posteriorly. Ovaries each a large flat lobe or slenderly pear-shaped, with scattered oocytes, in 13. Small ovisacs in 14, not directly connected with the funnels. Spermathecal ampullae subspherical, in 15-17; with narrow sharply demarcated intraparietal ducts.

Malay Peninsula: Lubock Paku (Pahang River).

Michaelsen's account is augmented and emended above from the author's re-examination of a mature syntype (Hamburg Museum, V, 5875) (Fig. 15. 4E-F).

The status of Gates' material (1938, 1958 c), from Johore, is very uncertain as it differs in several respects from that of Michaelsen. Furthermore Gates was doubtful of the accuracy of his account of the internal anatomy owing to maceration of the two specimens. Reference to them has therefore been excluded from the above description. Differences from the types which are apparent from the accounts of the two authors are: Wings occupying posterior 18-24 or 18/19 to anterior 24 or 23/24; paired genital markings in 17-18, 24, 24-26; unpaired genital markings in 15-18 and 25-26 or in 13-14, 16 and 18; and 3, instead of 2, as the maximum number of spermathecae on each side in a transverse row and an intestinal origin in 18, observed in the single specimen dissected, an origin known elsewhere only in the very distinct *G. stuhlmanni*.

Glyphidrilus papillatus (Rosa), 1890

Bilimba papillata ROSA, 1890: 386, Fig. 1.

Bilimba papillata Rosa. BEDDARD, 1895: 687.

(?) *Glyphidrilus papillatus* (Rosa). MICHAELSEN 1896: 195. CHEN, 1938: 426.

Glyphidrilus papillatus (Rosa). MICHAELSEN, 1900 a: 459; 1918: 344; GATES, 1931: 431; 1958 c: 60.

Glyphidrilus papillatus (Rosa) (part.); STEPHENSON, 1923: 493. GATES, 1933: 603.

$l=100$ mm (immature type), $w=5$ mm, $s=330$. Prolobous. Dorsal groove? Anus wholly dorsal, with posterior lobe. Setae in the forebody very widely, in the hindbody fairly closely paired; *ab* in the forebody little smaller than *aa*, in the hindbody $\frac{1}{3}$ *aa*; *c* and *d* dorsal. Clitellum in at least 14-40. "Wings" begin on 18-23 and extend on to 23-26, in *bc*. Genital papillae: unpaired medians on 11-21 and 23-33, paired medians on 12, 17-20, 24-29; paired laterals on 12-18; or in some of those segments only. Spermathecal pores in 13/14-16/17.

Gizzard in 8 but apparently reaching well into 7. Intestinal origin in 15. Typhlosole in about 18 posteriad, thickly lamelliform. Dorsal blood vessel ending anteriorly with the hearts of 7. Hearts in 7-11 ($=5$ pairs), those of 7-9 lateral, 10-11 latero-oesophageal. Seminal vesicles in 9-12. Ovaries in 13, ovisacs in 14. Spermathecae in 14-17, mostly within the parietes.

Burma: Cheba or Biapo Districts; Mong Kung, Namkham; Rangoon; Maymo; Tharrawaddy and Bassein. Common in permanently moist soil, e.g. river banks, swampy grounds, buffalo wallows. Hainan?

G. papillatus was erected for a singly poorly preserved specimen and the type description is inadequate in many respects. Additional details have been included from Burmese material described by Gates (1931, 1958 c). Distribution of the spermathecae in the type is unknown and Gates does not specify the number per row; 6-9 on each side were reported for specimens from Hainan Is. which Chen (1938) referred to *papillatus*. The Hainan specimens showed an arrangement of genital markings similar to both *gangeticus* and *papillatus* and in the absence of information on the segment of origin of the intestine, cannot certainly be referred to *papillatus* s. Gates. Their geographical location, to the east of Burma favours identification with *papillatus* rather than *gangeticus*, however. The cavernicolous *G. spelaeotes* Stephenson, 1924 (Assam) is doubtfully distinct, being principally separable by commencement of hearts in 8 and absence of spermathecal pores from 16/17. Its intestinal origin (segment 15) is that of *papillatus*.

Of the specimens referred to G. *papillatus* by Gates (1933), only those (from Bassein and Tharrawaddy) with wings in 18-24 or 25 and, it is implied, possessing median genital markings are here included in the species. The remaining specimens are placed in G. *birmanicus*. Some of the Rangoon specimens of G. *papillatus* described by Gates (1931) appear to be intermediate between *birmanicus* and *papillatus*, having the posteriorly located wings of the former but, it appears, possessing the median genital markings of the latter. Further work is required to determine whether the two species are distinct or whether *birmanicus* should be included in a highly variable *papillatus*.

Glyphidrilus quadrangulus (Horst, 1892)

Annadrilus quadrangulus HORST, 1892: 44.

Annadrilus quadrangulus Horst. BEDDARD, 1895: 680.

Glyphidrilus quadrangulus (Horst). MICHAELSEN, 1896: 195; 1900 a: 460; 1918: 345.

$l=50$ mm, $s=200$. Epilobous 1/3. Dorsal groove? Anus dorsal. Setae in the forebody widely paired, in the hindbody more closely paired; setae *ab* ventrally, *cd* dorsally situated. Clitellum "characterized by a narrow, folded ridge (wings) along segment (19) 20-25", dorsal to setal lines *b*. Male pores in intersegment 21/22 in setal lines *b*; spermathecal pores in groups of 2 to 5 on each side in intersegments 13/14-15/16 (14/15-16/17?) in *a* and *b* or in *c* and *d* also,, exceptionally in *bc*. Gizzard in segment 8. Testes? Nephridia commencing in 13. Seminal vesicles in 9-12, "three pairs" (?). Prostates apparently absent. Spermathecae small.

Sumatra: Lake Danau diatas (near Alahan Pandjang).

Glyphidrilus spelaeotes Stephenson, 1924

Glyphidrilus spelaeotes STEPHENSON, 1924: 133.

Glyphidrilus papillatus (part); GATES, 1958 c: 54 (Footnote synonymizing *spelaeotes*).

$l=175$ mm, $w=2$-3 mm, $s=310$. Anus dorsal, a pear shaped opening with narrow end forward. Zygolobous. Setae, especially the lateral, widely paired. In front of the wings $ab=0.5$ $aa=0.4$ $bc=0.8$ cd, $dd=1.7$ aa; a little behind the wings $ab=0.4$ $aa=0.3$ $bc=0.8$ cd, $dd\rangle aa=3$ cd; towards the hind end dd relatively greater, $aa=or\rangle bc$, $ab=cd$; ornamentation a few fine transverse lines near the tip. Clitellum embracing 16 to 30; 14 and 15 also slightly modified. Wings in bc, extending from 18, 19 to 24, $\frac{1}{2}$ 25; on 18 if present less prominent than elsewhere and separated by an incision. Genital markings with rim; paired in (14, 15, 16, 19) 25, 27, (28), occasionally absent on one side; median in 11 or in 17 and 18; each pair occupying only the posterior two thirds of each segment. Dorsal median non-mamillate papillae may be present in 18 to 28, ? 29. Genital apertures?

Gizzard rather soft in 8 and the hinder part of 7. Intestinal origin in 15. Hearts 4 pairs, in 8 to 11. Seminal vesicles in 9-12, largest in 12. Ovaries in 13; ovisacs (?) in 14. Spermathecae small longitudinally ovoid sessile sacs, in 13/14, 14/15 and 15/16; 5 (once 4) on each side in each row, in the four setal rows, with the supernumerary in *bc*. No prostates seen.

Assam: Siju Cave, Garo Hills, 2000 and 3000 ft in from the entrance. The affinities of this species are discussed under *papillatus*.

Glyphidrilus stuhlmanni Michaelsen, 1897
Fig. 15. 4G, H

Glyphidrilus stuhlmanni MICHAELSEN, 1897 b: 62.

Glyphidrilus stuhlmanni Michaelsen. MICHAELSEN, 1900 a: 461; 1918: 346.

$l=190$ mm, $w=4$ mm, $s=540$. Zygolobous. Anus dorso terminal. Setae widely paired throughout though posteriorly more closely $ab=cd=2/3$ bc, $dd=2$ cd; distally ornamented by irregular toothed scars. Clitellum 22, 23-66, 67 ($=44$, 46 segments). "Wings" in 42, 43-66, 67, in *bc*, very low, broad ridges and scarcely protuberant. Lateral genital markings in *bc* beginning in 16 or 18 and ending in the first case in 23 and in the second case in 26 or 27 (left and right respectively); increasingly overlapping the intersegment behind the segment occupied in a posterior direction. Ventral median markings in 46/47-50/51 ($=5$) 64/65-67/68 ($=4$) or in 44/45-49/50 ($=6$) and 64/65-67/68 ($=3$). Male pores (?). Female pores anterior to setae *b* of segment 14.

Gizzard large and firm, occupying 7 and part of 8, not sharply demarcated from the oesophagus. Oesophagus in 8, 9 to 15 highly vascularized (sinusoidal); thicker walled and not apparently vascular in 16 and 17 (straining apparatus?); intestine thin walled, commencing with distinct oesophageal valve at 17/18. Typhlosole beginning in 19. Hearts in 8-13 only; dorsal vessel apparently ending, anteriorly, with those of 8; those of 12 and 13, at least, latero-oeso-phageal. Supra-oesophageal vessel from 15 to the anterior end of the pharynx. Subneural vessel present behind the spermathecae. Small testes and large iridescent sperm funnels in 10 and 11. Seminal vesicles 4 pairs, in 9-12. Sperm ducts not traceable. Ovaries plicate, with numerous oocytes in 13. Ovisacs in 14, apparently not connected with the egg funnels. Spermathecae simple pouches; (always?) embedded in the septa which are each split into two laminae; pores 1 to 5 transversely on each side, in 9/10 and 10/11 (rudimentary, seen by Michaelsen in serial sections) and 12/13 to 20/21 or 21/22 (well developed) in setal lines *a*, *b*, *c* and *d*, with or without some supernumeraries in *bc*. Of the normal spermathecae, those in the setal lines *c* and *d* commence in 13/14 and reach to 21/22 or (re-examined type) 19/20 (right) or 20/21 (left); those in the setal lines *a* and *b* start in 12/13 but are absent behind 17/18 with one exception (Michaelsen) or end in 18/19 (*a* lines) and 15/16 (*b* lines).

Tanzania: Dunda on the Kingani.

Re-examination of a previously undissected syntype (Hamburg Museum

V. 4512) (Fig. 15. 4G, H) permits the above confirmation and extension of Michaelsen's account.

Glyphidrilus tuberosus Stephenson, 1916
Fig. 15. 4I

? *Criodrilus lacuum* Hoffmeister. STEPHENSON, 1914: 256.

Glyphidrilus tuberosus STEPHENSON, 1916: 349, Pl. XXXIII, fig. 37.

Glyphidrilus tuberosus Stephenson. STEPHENSON, 1923: 494, Fig. 262. GATES, 1958 c: 59.

l=60-118 mm, w=3 mm, s=221. Anus dorsoterminal. Prolobous (or zygolo-bous?), the delimiting groove a shallow depression only. Setae behind the clitellum at the angles of the section; aa=bc=2 ab=2 cd; dd=3 cd; in front of the clitellum setae widely paired and rather irregular, ab=½ aa or less. Clitellum 14, 15, 16-28, 29 (or 30 dorsally). "Wings" low, on 20-24, continued forwards as a slight ridge to 15 or 14 and sometimes back to 26, 27 or 28; they may (abnormally?) grow out into a foliating tumourlike mass, extending nearly to a lines; similar patches may be present above the wings. Genital markings small, white, rounded papillae without rims in the posterior parts of their segments; an anterior set, on 10-12, 13, maximally a midventral and on each side two lateral, one in ab the other above b; pre-alar and post-alar sets on 16, 17, 18-19 and in 24-28, 30 respectively, in three pairs per segment, one papilla on each side median to a, one in ab and one above b but 1 to 5 papillae may be absent from a segment. Female pores minute, very slightly lateral to b lines and nearer to the setal equator than to 13/14. Spermathecal pores in intersegments 13/14 and 14/15, 2-4 on each side per segment, in a and b and in bc or also in c.

Gizzard in 7, sometimes also extending a little into 6; development variable, well developed and firm or, often, vestigial. Intestinal origin in 15. Last hearts in 11. Seminal vesicles in 9-12. Ovisacs in 14. Spermathecae in 14 and 15, small subspherical sacs with short, thin duct.

India: Orissa, Bengal, Madras Presidency.

G. tuberosus is unique in the genus in possessing rimless genital markings and in the occurrence of more than one pair in a segment.

Glyphidrilus weberi Horst, 1889 emend
Fig. 15. 4J-M

Glyphidrilus weberi (part?) HORST, 1889: LXXVII.

Glyphidrilus weberi Horst. (part?) HORST, 1891 b: 11; 1893: 37, Pl. II, figs. 15-19, Pl. III, fig. 20.

l=82 mm, w=0·3 mm, s=283. Anus dorso-terminal. Setae widely paired throughout; in segment 12 aa:ab:bc:cd:dd=2·3:1:2·7:1·3:3·3; dd:u=0·22; in the hindbody aa:ab⟨3:1. Clitellum annular, in 16-32. Wings wavy delicate laminae in bc of 23-½32 (=10½ segments). Rimmed genital markings in mid bc in 16, 17 and 19; in c in 18 and immediately above b in the pre-alar segments

(22); those of 18 and 22 present, respectively on the right and left sides only. Female pores very slightly median of *b* lines and presetal in 14. Spermathecal pores not recognizable; from internal examination at the anterior borders of the spermathecal segments, in 13/14-17/18, 1 in each of *c* and *d* lines, with a supernumerary on each side in 15/16 and 16/17 median to *a* lines.

Last chromophil glands in 7, attached to the anterior septum. Septal pouches absent. Gizzard long and moderately wide, with muscular sheen but readily compressible, in ½7-½8, septum 7/8 appearing to be attached to it equatorially. Oesophagus sinusoidal in 9-12; thicker walled and not evidently vascular in 13 to ½15; intestine commencing at ½15 and typhlosole in 18. Dorsal blood vessel continued onto the pharynx. Supra-oesophageal narrow throughout; depleted in 7 and 8 and probably not continued further forwards. Dorsoventral commissurals in 7 slender and not heart-like. Large hearts, increasing slightly in size posteriorwards, in 8-11; only those of 10 and 11 latero-oesophageal. Lateroparietals present, especially thick in 9 anteriorly in which they connect, as aparently in adjacent segments, with the slender subneural vessel. Nephridia beginning in 12; with preseptal funnels; ducts slender, avesiculate, entering the parietes in front of setae *b*. Slender, long testes and large funnels in 10 and 11. Seminal vesicles 4 pairs, in 9-12, largest in 12, next largest in 9. Vasa deferentia (intraparietal) ending at 27/28 in *bc* slightly nearer *b* than *c*; prostate glands absent. Ovaries long transverse sinuous bands extending from the midventral line to *d* lines in 13. Oviducal funnels fairly small, sessile. Large oocyte-containing ovisacs in 14. Spermathecal ampullae laterally adpressed; with spermatozoal iridescence; sessile subspherical sacs projecting freely into the coelom.

Java: Buitenzorg. (Also Sumatra, Celebes and Flores?)

The above account has been restricted to a single specimen (Fig. 15. 4J-L) of six alate re-examined syntypes, all from Java, in the collections of the Zoologisches Museum, Hamburg, and the Rijksmuseum, Leiden. This specimen (Hamburg Museum, V. 5097) conforms with Horst's account in location of the wings in segments 22-32 and in termination of the vasa deferentia in 27/28. It differs, among other respects, in lacking prostate glands and in the absence of hearts from segment 12. Prostate glands are absent from four Javanese syntypes examined for this character and, as types from the other localities, which might have possessed them, are no longer traceable, the species must be redefined as aprostatic on the basis of the Javanese specimens. It seems likely, from the wide distribution and morphological variation attributed by Horst to *G. weberi*, that he was dealing with more than one species. Variation within the material from Java is also considerable, but is probably intraspecific, that for the wings and genital markings in six specimens (including that described above) being as follows.

Alar segments	Lateral genital markings	Median genital markings
(1) 22-½ 32 =10½	16, 17, 18, 19	Absent.
(2) 22-28 = 7	16, 17	Absent.

Alar segments		Lateral genital markings	Median genital markings
(3) 22-27	=6	18, 21	Absent.
(4) 23-28	=6	17	Absent.
(5) 23-28	=6	17	19, 20.
(6) 23-29	=7	16, 17	Absent.

Variation in origin of the wings from 22-23 coincides with maintenance of a normal location of the female pores, on segment 14. Examination of specimen 4) above reveals no differences from specimen described which would merit specific distinction of the specimens with shorter alae (6-7 segments) but location of the male pores was indeterminable. The gizzard was stronger and filled segment 8 longitudinally but the oesophagus also had a muscular sheen in 7. These individuals are separable from *G. horsti* Stephenson, 1930 b, almost solely by the segment of origin of the intestine.

GENUS **Alma** GRUBE, 1855

Type species: Alma nilotica GRUBE

Prostomium zygolobous, rarely prolobous. Body quadrangular behind the clitellum; with a posterior dorsal groove; anus dorsal or, exceptionally, terminal; 2 rows of gills sometimes present posterodorsally. Setae 8 per segment, widely paired in the forebody, usually more closely paired posteriorly, the dorsal median intersetal distance equal to 0.2 of the circumference. Ornamentation of the somatic setae consisting solely of transverse annulations which are (always?) minutely spinose. Clitellum annular, occupying 20 to 69 segments between segments 35 and 295. A pair of ribbon-like claspers, which may be longer than the preceeding portion of the body, occupying 2 or 3 segments and centred on segment 19 or, much less commonly, on intersegment 18/19 or (abnormally?) 23/24 or segment 25. A male pore located distally on each clasper. Genital setae, smaller than the somatic setae, 2 or more than 200 on a clasper, sigmoid, or pointed rod-like to lancet-like; ornamentation, if present, jagged annuli, or circlets of spines. Nephropores in *b* lines. Rimmed or other genital markings absent. Female pores paired, presetal in or lateral to *b* lines (rarely in *a* lines?) of segment 14. Spermathecal pores inconspicuous; occupying some or all of the setal lines, or more numerous, in 7 to 37 intersegments (or exceptionally 138 intersegments) from 18/19 to 253/254, thus always well behind the testis segments.

Oesophageal gizzard and alimentary diverticula totally absent; oesophagus with unciliated region with an intramual blood plexus and, behind this, a ciliated and more strongly muscularized portion. Intestine unciliated, beginning in 16 to 28, its musculature in the first 2 to 4 segments augmented to form a rudimentary gizzard; a dorsal typhlosole present. Paired hearts in 7, or 8, to 12 or rarely 13; some or all of them with connectives to the supra-oesophageal vessel. Subneural vessel absent. Latero-parietal vessels running close to the nerve cord as neuro-parietals in the mid- and hind-body. Holonephric; adult nephridia absent in front of 13; lacking terminal caeca; avesiculate; nephrostomes single. Testes and funnels free in 10 and 11; seminal vesicles 4 pairs, in 9-12; rarely 2 pairs, in 11 and 12, when the posterior pair extends through many segments;

rarely, all male organs 1 segment further posteriorly. Prostate and setal glands absent. Ovaries and funnels in 13 or rarely, 14. Ovisacs absent; septal pouches present. Spermathecae with their ducts, and often the simple ampullae, concealed in the parietes; lacking diverticula.

Tropical Africa and the Nile as far as Cairo.

Synonyms of *Alma* are *Siphonogaster* and *Digitibranchus*, both of Levinsen (1890). A key to species is given on p. 773. Genital markings and tubercula pubertatis, or alae, so characteristic of *Callidrilus* and *Glyphidrilus*, are totally lacking but a pair of ribbon-like appendages, which function as claspers in copulation, is present. The claspers may exceed the preceding portion of the body in length. They occur in no other oligochaetes, although structures approaching them in form occur in the S. American *Drilocrius alfari*. Khalaf (1950 a) has given a detailed account of their anatomy in *A. nilotica* and (1950 b) their mode of action in copulation. In this process one individual, acting solely as a male, applies the claspers to the spermathecal region of a concopulant which acts as a female, there being no simultaneous exchange of sperm. No other attachment exists between the partners. The anterior ends of the two worms are directed the same way in contrast with the familiar lumbricid copulation where the concopulants, mutually exchanging sperms, lie in opposite directions one to the other. A vas deferens discharges on the inner surface of each clasper, at the distal third in *A. nilotica* or near the free end in *emini*, *stuhlmanni* and *tazelaari;* the orifice is always incospicuous and its location is unknown in most species. Each clasper bears genital setae, usually on small papillae or in sucker-like depressions, there being one or two setae at each of these loci. The number and arrangement of the loci is of great taxonomic value but examination of the external and internal morphology of the worms is needed for certain specific identification. The numbers of genital seta loci on each clasper varies from 2, located in a distal depression or cupule, in *A. stuhlmanni* to 220 in *A. multisetosa*. The existence of only 2 genital setae in *stuhlmanni* suggests a primitive condition and this is apparently endorsed by the fact that they are sigmoid, like the body setae, whereas those of other *Alma* species (Fig. 15. 5) are either approximately peg-shaped (i.e. tapering cylinders), as in *eubranchiata*, *multisetosa*, *pooliana* and *ubangiana*, or somewhat spear-shaped (i.e. distally expanded and flattened), as in *basongonis*, *emini*, *kamerunensis*, *nasuta*, *nilotica*, and *togoensis* or intermediate between these two often ill-defined forms as in *millsoni*, and *tazelaari*. Against too ready an acceptance of the primitiveness of the condition in *stuhlmanni* it must be noted that the 2 genital setae are not simply the 2 ventral setae of the segment bearing the clasper, carried on to the clasper during outgrowth of its rudiment, as the *b* setae remain on the body wall in this and the two adjacent segments, which are also occupied by the base of the clasper and similarly lack the *a* setae. Primitiveness of *A. stuhlmanni* is indicated, nevertheless, in other features, notably in the morphology of the alimentary canal.

In addition to the genital setae loci, the clasper frequently bears proximally what in the present work are termed asetigerous papillae ("Stielpapillen", Michaelsen, 1918; "basal suckers", Khalaf 1950 a). These are sucker-like, with peripheral rim and a central region which may be more or less deeply sunken or protruberant to the extent of forming a hemispheroid knob. Asetigerous papillae are present in 10 species though they are rare in sub-species *millsoni*, while normally present in sub-species *zebangui* of *A. millsoni*. In *millsoni millsoni* and in *A. basongonis*, the asetigerous papilla is scarcely larger than a genital seta locus. The number of asetigerous papillae on a clasper is 1 in 6 species (including *millsoni*), with abnormal variation to 2 in *stuhlmanni*; 1 or 2 in *A. millsoni zebangui* and 2 or 3 in *kamerunensis* and in *pooliana* which sometimes (Gates, 1958 a) apparently may possess a further pair. They are absent from *emini*, *nasuta*, and *togoensis* Khalaf (1950 b: 35) states that the muscular and glandular "basal sucker" in *A. nilotica* augments the action of the muscles, glands and genital setae of the clasper in effecting adhesion of concopulants.

A notable difference from *Glyphidrilus* is the total absence of an oesophageal gizzard. Despite this, the organization of the gut is found in a close examination of *emini*, *nilotica*, *stuhlmanni* (Fig. 15.5) and *tazelaari* in the present study to be strikingly similar to that in *Glyphidrilus*, a far greater resemblance existing than is displayed by another Glossoscolecid, *Pontoscolex corethrurus* or by the lumbricid *Allolobophora caliginosa trapezoides* or, especially, the Ocnerodrilid *Pygmaeodrilus affinis*, which were examined for comparison.

As in *Glyphidrilus* a portion of the oesophagus is unciliated and has within its walls a plexus of encircling vascular striae, which feeds the supra-oesophageal blood vessel. This sinusoidal region, as the author has termed it, is followed by a ciliated portion which appears to be homologous with the "straining apparatus" (Nair 1938) of *Glyphidrilus*. Following this, as in *Glyphidrilus*, is the intestine which lacks cilia but has a conspicuous brush border. The term "straining apparatus" is here retained for the ciliated region in *Alma*, but only in *stuhlmanni*, of the four species mentioned, is the lumen as markedly occluded by folding of the walls as is the case in *Glyphidrilus*. Of the two genera, very considerable thickening of the musculature of the gut is restricted to *Alma* and commences with the ciliated region. This muscular strengthening is especially well developed at the beginning of the intestine where it may form a rudimentary gizzard. There may be no obvious constriction at the junction of oesophagus and intestine but in such forms (e.g. *A. emini*) the intestine is taken to commence with cessation of ciliation as it does where the valve is present. Records of muscular thickening and of the segment of origin of the intestine in the literature are in some cases unreliable owing to failure to distinguish between muscularization of the ciliated portion of the oesophagus and that of the intestine as here defined. Interruption of the alimentary sinus by the "straining apparatus" which Nair (1938) found to occur in *Glyphidrilus annandalei* is seen in *A. stuhlmanni* but does not occur in all species. Thus in *A. tazelaari* there are sinuses at the oesophageal valve.

The degree of development of the intramural sinuses in the sinusoidal region of the oesophagus also varies significantly. In *emini* and *nilotica* the sinuses are very conspicuous and are reminiscent of rudimentary calciferous glands; in *stuhlmanni* they are less apparent and in *tazelaari* they are only weakly developed.

The segment of origin of the intestine and its muscular thickening in *Alma*, as far as is known is as follows (numbers of segments in which thickening occurs, and origin of the typhlosole, in parentheses in that order): 16, *togoensis* (2); 17, *stuhlmanni* (3) (20-22), *tazelaari* (4) (24); 18, *basongonis* (4) (44), *millsoni* (2-3) (22-24); 20, *emini* (4) (23-27); 21, *nilotica* (3) (25); 25 (?), *pooliana* (3?) (29); 28 *eubranchiata* (3) (32-33) and (?) *nasuta* (?).

In conclusion, the histological differentiation of the gut of *Alma*, but not the metamerism, is constant throughout the genus and resembles that seen in *Glyphidrilus*. Variation in the segmental location of the regions is very marked, and is of great taxonomic value, as intraspecific variation seems slight.

The chief elements of the vascular system in *Alma* are (*i*) the dorsal vessel, (*ii*) the ventral vessel, (*iii*) the dorso-ventral hearts, (*iv*) the supra-oesophageal vessel and (*v*) the latero-parietal vessels. There is no subneural vessel (*pooliana*; Gates, 1958 a: 7—*stuhlmanni* and *nilotica*; Jamieson and Ghabbour, 1969—*millsoni* and *tazelaari*, present study), a notable contrast with *Glyphidrilus*.

(*i*) The *dorsal vessel* has its anterior termination in the segments of the first hearts in all species in which its course has been described (*eubranchiata, emini, millsoni, nilotica, tazelaari*). In *Glyphidrilus annandalei* it terminates by bifurcation to form a pair of heart-like but non-pulsatile dorso-ventral commissural vessels (Nair, 1938) but in *Alma* it continues in front of the first hearts for the length of the segment containing them, though vestigial in this segment (*A. stuhlmanni*; Jamieson and Ghabbour, 1969; *A. eubranchiata, A. millsoni* and *A. tazelaari*, present study) or may be absent in front of them (*nilotica*; Jamieson and Ghabbour; alternative condition in *tazelaari*, present study). The dorsal vessel supplies the hearts and in *A. stuhlmanni* (v. Jamieson and Ghabbour) connects in segment 13 with the latero-parietal vessels.

(*ii*) The ventral vessel runs beneath the gut and is persistent throughout the body.

(*iii*) The *hearts* commence in segment 7 (*eubranchiata nilotica*, some *A. emini*) or in 8 (*millsoni, pooliana, stuhlmanni, tazelaari, togoensis*, some *A. emini*). The last pair lies in the pre-ovarian segment, i.e. in 12, in all of these species but in 13 in *eubranchiata*, the number of pairs being

5 to 7. All the pairs are latero-oesophageal, i.e. connect both with the dorsal and with the supra-oesophageal vessels, in *stuhlmanni* and *nilotica*, though supra-oesophageal connectives were not certainly demonstrated for the anterior pair in the latter. In *pooliana* those in 8 and 9 lack these connectives while those in 10 to 12 possess them (Gates, 1958 a: 7). In *tazelaari*, also, only the last three pairs appear to be latero-oesophageal.

(*iv*) The *supra-oesophageal vessel* arises from the intramural plexus of the oesophagus and is therefore first apparent, posteriorly, at or within a segment's length of the posterior end of the plexus, e.g. in 16 (*emini*), 18 (*nilotica*), 15 (*stuhlmanni*) or 13 (*tazelaari*). Gates (1958 a) states that this trunk does not extend in front of segment 8 in *pooliana* but in *stuhlmanni* it continues on to the pharynx (present study). The supra-oesophageal receives in each of segments 13 to 15 (*stuhlmanni*) 6 to 10 (*nilotica*), 8 (*pooliana*; Gates) and 14 (*tazelaari*) a pair of vessels from the latero-parietal vessels; it is fed by the plexus of the sinusoidal region of the oesophagus, and supplies connectives to the latero-oesophageal hearts as noted above.

(*v*) The *latero-parietal vessels* (neuro-parietals of Gates, 1958 c) run closely adjacent to the nerve cord from the posterior end of the body throughout most of its length but anteriorly diverge and then run forwards far to each side of the cord, divergence occurring in 15 (*stuhlmanni*), 14 (*nilotica*) or 24 (*pooliana*). In *Millsoni* a neuro-parietal lies slightly to the right of the nerve cord and is straight as far forward as 18 but is tortuous anteriorly; pairing is not apparent, presumably because of poor preservation. In *tazelaari* the latero-parietal of the right side is enlarged relative to that of the left side behind segment 49, the reverse being true in front of this segment and the transverse subneural connective is enlarged at the point of "cross over" the impression is thus given of a thick latero-parietal trunk running first on the left side of the nerve cord and then, more posteriorly, on the right side.

The latero-parietals send connectives to the oesophageal plexus in 10, 13, 14 and 15 and possibly do so in other segments in *stuhlmanni*, and in 13 to 18, and (always?) 4 and 5, in *nilotica* (Jamieson and Ghabbour, 1969).

The oesophageal plexus

The exact extent of the intramural plexus and the sinusoidal region of the oesophagus is not always precisely determinable and probably varies intraspecifically, the observed extent in *emini* being 11-17 (Stephenson, 1930 c), 14-16 (Stephenson, 1928) and $9-\frac{1}{2}17$ (present study). It clearly must extend at least as far posteriorly as the posterior limit of the supra-oesophageal vessel as this originates from the plexus.

Alma

1. Each clasper with only 2, sigmoid, genital setae (rarely duplicated), located near the free end. 1, basal, asetigerous papilla   *A. stuhlmanni*

— Each clasper with more than 2 genital setae which are never sigmoid. With or without asetigerous papillae   2

2. Each clasper spatulate, with long narrow stalk. Lacking asetigerous papillae. Genital setae on the stalk and 10-20 distal in an anterior and a shorter posterior marginal row. (The tip of each with a conspicuous rhomboidal expansion.)   *A. emini*

— Genital setae not arranged as in *emini* though clasper may be spatulate. Asetigerous papillae present or absent. (Tip of genital seta if expanded not extremely.)   3

3. Each clasper with a short stalk which bears a large asetigerous papilla distally. 19-80 irregularly arranged genital setae on the roughly parallel sided lamina. (A row of conspicuous branched gills may be present on each side of the dorsal groove posteriorly.)   4

— Claspers not as above. (Gills never present.)   6

4. Intestine commencing in 28. (28-80 genital seta loci per clasper.)   5

— Intestine commencing in 21. (19-43 genital setae per clasper.)   *A. nilotica*

5. Female pores in segment 15. Asetigerous papilla *c.* 1·4 mm wide. (Gills present.)   *A. eubranchiata eubranchiata*

— Female pores in segment 14. Asetigerous papilla more than 2 mm wide. (Gills absent?)   *A. eubranchiata catarrhactae*

6. Genital setae very numerous, more than 200 on each clasper   *A. multisetosa*

6. Genital setae less than 60 per clasper   7

7. 2 short, non-marginal distally convergent
    rows each of 3-8 genital seta loci near the
    rounded free end of the clasper and proxi-
    mally 1-4 genital seta loci. Asetigerous
    copulatory papillae absent. (Spermathecal
    pores in the region of 36/37-51/52.)   *A. togoensis*

—   Genital setae not distributed as in *togoensis*.
    Asetigerous papillae present or absent   8

8. Each clasper with a distal distinctly circum-
    scribed area or depression (cupule) containing
    2-5 genital setae loci which are relatively
    isolated from those on the remainder of the
    lamina   9

—   Distal setae, if present, not isolated from the
    remainder and not in a cupule   11

9. Cupule or terminal area containing only 2
    setae   10

—   Cupule or terminal area containing 4 or 5
    setae   *A. tazelaari*

10. 6-15 genital seta loci proximal to the cupule
    in 2 longitudinal rows. Asetigerous papilla
    absent or, rarely, rudimentary. Spermathecal
    pores in the region of 18/19-42/43   *A. millsoni millsoni*

—   2-5 genital seta loci proximal to the cupule
    in 2 longitudinal rows. 1 or 2 small basal
    asetigerous papillae. Spermathecal pores in
    the region of 23/24-46/47   *A. millsoni zebangui*

11. Stalk one third or less of the total length of
    the clasper. A weak asetigerous papilla or
    none developed   12

—   Stalk (slender and distinctly demarcated)
    approximately half the total length of the
    clasper and bearing 2 or 3 well developed
    asetigerous papillae   13

12. Stalk broad, not obviously demarcated; aseti-
    gerous papillae absent. Genital seta markings
    numbering *c.* 35, irregularly arranged but
    tending to longitudinal single and double

rows, throughout the clasper and most of the stalk

A. nasuta

12. Stalk distinctly demarcated, flattened, about one third total length of the clasper. Asetigerous papilla hardly protuberant, proximal on the lamina. Genital seta markings 13 in number, in 3 longitudinal rows

A. basongonis

13. Spermathecal pores (in the region of 56/57-70/71) between setal lines *b* and *c* and with a few between *a* and *b*; none in the setal lines. 2 or 3 asetigerous papillae on the stalk. Genital seta markings in 2 longitudinal submarginal rows (e.g. a posterior row of 15 and an anterior row of 8 or 9) totally absent from the region of the free end

A. kamerunensis

— Spermathecal pores in setal lines *a*, *b*, *c* and *d* with supernumeraries above *d* or in *bc*. Genital seta markings present or absent at the free end

14

14. Genital seta loci 15-30 in 2 longitudinal rows, absent from the free end. Spermathecal pores in the region of 75/76-95/96 in setal lines *a*, *b*, *c* and *d*, with supernumeraries above *d*. 2 asetigerous papillae on the stalk of the clasper

A. ubangiana

— Genital seta loci 17-35 on the lamina, forming distally an irregular group of 10 or 11 or short marginal rows and more proximally 2 or 3 more or less distinct longitudinal rows; 3-7 markings also present on the stalk, some or all of them proximal to the distal asetigerous papilla. Spermathecal pores in the region of 41/42 to 75/76 (12-18 intersegments) in setal lines *a* to *d* and with supernumeraries in *bc*

A. pooliana

Alma basongonis Michaelsen, 1935
Fig. 15. 7A

Alma basongonis MICHAELSEN, 1935 b: 43, Fig. 5, Pl. I, fig. 2.

In front of claspers $l=21$ mm, $w=5$-6 mm. Dorsal groove? Anus? Gills ?, hind end missing). Prolobous? $aa:ab:bc:cd:dd=2.5$-$3:1:2.5$-$3:1:2.5$-3. Clitellum embracing 42 (?), 43 to 77 (?) ($=35$ segments?). Claspers arising from 19 and the adjacent halves of 18 and 20, in setal lines *a*, about 18 mm long and reaching almost to the anterior end of the body; stalk, thin, somewhat flattened, 1·5 mm wide and about one third the entire length of the clasper; the remainder spoon-shaped, reaching 2·7 mm wide, its hollow narrowing entally and sharply delimited from the stalk. Asetigerous papilla a hardly protuberant structure, only slightly larger than a genital seta marking, situated near one edge in the ental region of the hollow. Genital seta markings 13 in each clasper arranged in a median row of 5 and 8 markings distributed more or less equally in two lateral rows; each marking bearing a single genital seta which is unornamented, straight and spear-shaped, 600 μ long, of which one sixth is the head; the cylindrical shaft c. 38 μ wide, narrowing to 31 μ at the ental and 18 μ at the neck of the spear head; the head flattened and somewhat hollowed, 25 μ wide in the middle, tapering to the sharp, slender, three-sided tip. Spermathecal pores occupying 19 intersegmental furrows, 34/35 to 52/53, and therefore including less than half of the clitellar region; 4 in each furrow, the upper in setal lines *c*, the lower in *bc* slightly below *c*. The first pair of pores on the right side were fused.

Four small intestinal gizzards, in segments 18-21; intestine widening suddenly behind them (in 22?); wide circular typhlosole commencing in 44. Spermathecae with small ampulla and very narrow short duct concealed in the body wall.

Republic of the Congo: Basongo.

Alma emini (Michaelsen, 1892)
Fig. 15. 5A; 15. 6A; 15. 7B, C

Siphonogaster emini MICHAELSEN, 1892: 8, Pl. fig. 4, 5.

Siphonogaster emini Michaelsen. MICHAELSEN, 1896 b: 6, Pl. II, fig. 27.

Alma emini (Michaelsen). MICHAELSEN, 1897 b: 68; 1900 a: 467; 1914 b: 126; 1915: 280. GROVE, 1931: 224. STEPHENSON, 1928: 13. WASAWO, 1962: 113.

Alma aloysii-sabaudiae COGNETTI, 1906 b: 1.

Alma aloysii-sabaudiae Cognetti. COGNETTI, 1908 b: 693, Pl. fig. 1-16; 1909: 44, Pl. IV, fig. 59-63.

?*Alma* sp. MICHAELSEN, 1913 d: 57.

Alma emini Michaelsen var. *aloysii-sabaudiae* Cognetti. MICHAELSEN, 1918: 369.

Alma emini Michaelsen forma *typica* MICHAELSEN, 1918: 369.

Alma emini forma *typica* Michaelsen. STEPHENSON, 1930 c: 505.

Alma worthingtoni STEPHENSON, 1930 c: 501, Fig. 9, 10.

Alma worthingtoni Stephenson. WASAWO, 1962: 113.

(Non) *Alma emini* Michaelsen. GATES, 1958 a: 5.

l=160-430 mm, *w*=3-6 mm, *s*=270-407. Prolobous or zygolobous. *aa*:*ab*:*bc*:

$cd:dd=2\cdot8:1\cdot0:2\cdot8-3\cdot0:1\cdot2-1\cdot3:3\cdot0-3\cdot5$ in front of the claspers (e.g. segment 12); $=3\cdot8-3\cdot9:1\cdot0-1\cdot1:3\cdot1-4\cdot0:1\cdot0:4\cdot3-4\cdot6$ in the hind body; $dd:u=0\cdot18-0\cdot22$ and $0\cdot21-0\cdot22$ respectively, and within these limits throughout the body. Clitellum annular, commencing in 47-61 and extending to 87-100 ($=31$-50 segments), some clitellar modification seen in one specimen as far forward as segment 30 and involving 67 segments. Claspers spoon-, racquet- or lancet-shaped, with long slender stalks and expanded distal lamina; their bases in setal lines a, and laterally extending to b, and occupying $\frac{1}{2}$ to $\frac{1}{3}$ of 18, the entire length of 19, and $\frac{1}{3}$ to $\frac{2}{3}$ of 20; setae a absent but setae b present in these segments; total length in sexual specimens 20-40 mm, the lamina 3·5-4·5 mm wide, the stalk comprising 0·7 of the total length and reaching 21 mm, with a width of 1·0-1·25 mm; the margins of the stalk, and sometimes the lamina, may be inrolled. Genital seta loci 10-20 in a submarginal row around the distal end of the lamina and divided by the tip into an anterior and a posterior row, the anterior row numerically superior, with 6-17 loci, the posterior with 3-7 loci; a row of 2-10 loci present on the stalk and extending on to the anterior or posterior edge of the proximal region of the lamina, sometimes absent. Each locus consisting of a sucker-like pit containing 1 or 2 genital setae and with or without raised margins; the proximal loci, especially those on the stalk more protuberant and often not appreciably depressed centrally. Asetigerous genital markings absent. Genital setae distinctly lancet-like, with the distal fourth to almost a half expanded and flattened to form a lamina (always? slightly hollowed), 16-40 μ wide, with rhomboidal dorsal outline; joining the shaft by a slender neck 11 μ wide; the shaft straight or entally curved, 20-21 μ wide; ornamentation absent; the genital setae from the stalk longer, at 490-550 μ, than the distal setae which are 270-360 μ long. Somatic setae sigmoid, 600-700 μ long and 50 μ wide, distally ornamented by numerous (10 or more) transverse undulating striations which in some individuals at least are interrupted circlets of fine points. Nephropores at anterior margins of their segments in 14 posteriad, in b lines. Male pore near the distal end of each clasper. Female pores externally invisible, near the anterior margin and 14 in a lines (or in b lines?). Spermathecal pores not externally apparent, in intersegmental furrows 50/51, 51/52, 52/53, 55/56, 65/66-70/71, 74/75, 77/78, 78/79, 84/85, ($=10$-33 intersegments), in transverse rows maximally at 18-34 little or not at all interrupted midventrally and dorsally extending to bc or above d with no relation to the setal lines; alternatively only 4 or 5, in setal lines a and b. Total number of spermathecae 42 to several hundreds.

Oesophageal plexus in 9-$\frac{1}{2}$17; oesophagus ciliated in $\frac{1}{2}$17-$\frac{1}{2}$20; intestine beginning at $\frac{1}{2}$20, with muscular thickening (not always externally apparent) in $\frac{1}{2}$20 to $\frac{1}{2}$23 dorsal typhlosole beginning in 23-27. Hearts in 7 or 8 to 12, the dorsal vessel ending anteriorly in the segment of the first pair; at least the last 3 pairs latero-oesophageal. Supra-oesophageal vessel arising from the oesophageal plexus, as far back as 16 or 17, ending anteriorly in 8. Latero-parietal vessels present; median subneural vessel absent. Nephridia beginning on 14 or 15. Testes and funnels free in 10 and 11, seminal vesicles in 9-12, those of 12 the largest and displacing septum 12/13 posteriorly. Ovaries undulating but not plicate transverse bands anteriorly in 13. Septal pouches on the anterior walls of 13, 14-18, 20 ($=6$-7 pairs). Spermathecae elongated parallel to the body axis, mostly concealed in the parietes.

Tanzania: L. Victoria (Bukoba, the type locality; Ulambwe Bay) L. Mohasi; L. Tanganyika (Kigoma Harbour) *Uganda:* L. Victoria (Bugalla Is., Sesse), Mutoma in Ankole; L. Edward; Kazinga Channel between Lakes Edward and George); near Kampala (Kabanyolo: Bukasa Swamp; Kabaka's Lake); Masaka road; Kigezi (Kiseisero); West Nile (Dufile); Ruwenzori; Nabieso on L. Kwania. *Kenya:* Kahawa stream, tributary of Athi River, near Nairobi, collector E. Oxtoby 11 December 1967, new record. *Republic of the Congo:* Lufuko River at Ngansa; Katompe; Ituri River (Ituri Forest, collector Lord Howard der-Waldon, B.M. (N.H.) 1930. 7-30. 77/89, new record), (?) Semiliki River, N. of L. Albert; Avakubi in Aruwimi. *Sudan:* White Nile (Tongo). Immature material from the Victoria Falls, Zambesi, referred to *emini* by Michaelsen (1913 d) cannot, on the basis of the inadequate description, be assigned to a species.

The above synonymy excludes material from Stanleyville which was referred to *A. emini*, by Gates, 1958 a.

The Stanleyville material is here referred to *A. pooliana* Michaelsen, 1913 d, for which Stanleyville is the type locality. Significant differences between Stanleyville specimens and *emini* are: presence of asetigerous papillae on the claspers; location of genital seta loci on the lamina in several alternative arrangements but never solely in an anterior and a posterior distal row, the invariable arrangement in *emini* as so far reported; the absence of a narrow-necked, lancet-like distal expansion of the genital setae recorded for *emini* (Cognetti, 1909; Stephenson, 1928; Michaelsen, 1935 b); and the presence of distal ornamentation of the setae. The morphology of the alimentary canal confirms the discreteness of the two entities for, while in *emini*, the sinusoidal region of the oesophagus and the associated supra-oesophageal vessel do not extend behind 16 or 17 (Stephenson, 1928, 1930 c, personal observation; Fig. 15. 6A, Kigoma specimen, B.M. (N.H.) 1927.10.13.96/110, the supra-oesophageal in Gates' material ends in 18. Furthermore, origin of the intestine and/or of muscular thickening of this in 20 in *emini* (Cognetti, 1909; Stephenson, 1928; present study) contrasts with origin in 25 which is apparent from Gates' account.

The location of the spermathecae in the two Stanleyville collections agrees very closely (57/58, 58/59 to 73/74-75/76, Gates, 1958 a; 59/60 to 75/76, Michaelsen, 1915) though the clitellar extent differs considerably (79-121, Gates, 1958 a; (50)52-112(114) in a Bipindihof specimen). Agreement in the form of the claspers and genital setae is close. Unfortunately, only Gates has provided information on internal features other than spermathecae.

Illustrations of claspers and of a genital seta are from an Ituri Forest specimen (Fig. 15. 7B, B.M. (N.H.) 1930.7.30/77-89 and a syntype of *A. aloysiisabaudiae* (Hamburg Museum V. 3284).

Alma eubranchiata Michaelsen, 1910

Alma eubranchiata eubranchiata Michaelsen, 1910

Fig. 15. 7D

Alma eubranchiata MICHAELSEN, 1910 b: 162, Pl. fig. 14-16.

Alma eubranchiata Michaelsen. MICHAELSEN, 1915: 307; 1918: 372.

$l=420$ mm, $w=8$ mm, $s=450$. Zygolobous. Anus? In the forebody $aa:ab:bc:cd:dd=2\cdot7:1:2\cdot3:1:3$; elsewhere$=5:1:4:1:6$; $dd:u=0\cdot21$ and $0\cdot26$. A pair of gills present in each of the last c. 100 segments, median to d lines, on each side of the dorsal groove; simple, digitiform processes anteriorly; dendritic posteriorly, with a conical stem and c. 6 median branches which increase in size and complexity of branching upwards. Clitellum very indistinctly delimited; well devolped in 225, 230, 236, 247 to 268, 275, 278, 295 ($=44$, 46, 43, 49 segments respectively). Claspers originating in $\frac{1}{2}$. $\frac{1}{3}$ 19 to $\frac{2}{3}$, $\frac{4}{5}$ 20 or, exceptionally, $\frac{1}{4}$ 20 to 4/5 21, almost as long as the anterior part of the body; medianly in contact or separated by a more or less raised transverse ridge; setae b of the two segments of origin unaltered; setae a, the sites of which are occupied by the bases of the claspers, supressed. Each clasper c. 26 mm long and maximally 1·1 mm wide, with a narrow stalk; elsewhere slenderly tongue-like with medianly inrolled lateral margins; bearing a single asetigerous papilla medianly 1·4 mm wide near the distal limit of the stalk; its distal half bearing medianly 3-6 irregular longitudinal rows of minute papillae (28 and 47 in number in two claspers), each of which carries a genital seta. Genital setae c. 460 μ long and maximally, in the middle, c. 25 μ thick, mostly straight, seldom (especially proximally) somewhat curved; the mid-region forming an indistinct node; the ectal end cylindrical, slightly thinner, terminally sharply pointed, with many closely spaced, not always regular, annuli of fine spines. Male pores not apparent. Female pores small but distinct in front of setae b of segment 15 or, in an exceptional specimen, 16. Spermathecal pores extending considerably into the clitellar region; in the 37 intersegmental furrows 217/218 to 253/254; up to 6 pores on each side per intersegment, 4 in setal rows a b c and d and 1 or 2 supernumeraries in bc. Spermathecal count in one specimen 288.

Oesophagus moniliform and with high internal vascular folds in 15-24 (sinusoidal region); in 25-27 moniliform, thicker walled and neither visibly vascularized nor internally greatly folded (straining apparatus). Intestine not moniliform, commencing in 28 with abrupt dilatation to at least twice oesophageal width; its musculature quite strongly thickened in 28 and less so in 29 and 30; thin walled in 31 posteriorly. Typhlosole commencing in 33; not well developed until 35. Hearts in 7-13, at least the last pair latero-oesophageal. Dorsal vessel apparently terminating with the hearts of 7; supra-oesophageal vessel commencing in 24 and traceable anteriorly as far as 6. Latero-parietals very large in 7 anteriorly. Nephridia beginning in 16; duct avesiculate, entering the parietes in a lines. Testes, slender, sinuous transverse bands, and their very large, convoluted funnels in 11 and 12. Seminal vesicles 4 pairs, in 10-13; those in 13 the largest, and filling 14. Ovaries and funnels in 14; the ovaries long, sinuous transverse bands, very short in an anterior-posterior direction, with only 1 or 2, marginal, rows of oocytes; funnels very large and delicate. Ovisacs absent. Septal pouches dependent posteriorly from septa 14-28; each consisting of several elongate. digitiform lobes.

Republic of the Congo: Kuka Muna on the River Luburi (Chiloango District); ? Duma, Ubangi District.

The illustration of a clasper is from a syntype (Fig. 15. 7D; Hamburg Museum V. 3654).

Possession of gills and the form of the claspers suggest that *A. eubranchiata*, in the Congo Basin, is perhaps a geographical replacement of *nilotica*, more specialized in the posterior shifting of the clitellum and spermathecae and elongation of the sinusoidal region of the oesophagus. Its resemblance to *nilotica* is however, less close than was suggested by Michaelsen (1910 b). It differs from *nilotica* in a number of significant respects (character state in *nilotica* in parentheses): termination of the sinusoidal oesophagus in 24 ($\frac{1}{4}$18) and of the straining apparatus in 27 ($\frac{1}{4}$21); intestinal origin in 28 ($\frac{1}{2}$21) and, correlated with this, "gizzards" in 28-30 ($\frac{1}{2}$21-23); typhlosolar origin in 33 (25); location of spermathecal pores between 217/218 and 253/254 (123/124 and 165/166); and a correspondingly more posterior location of the clitellum.

A. catarrhactae, on the other hand, is shown in the present study to be morphologically closely similar to *A. eubranchiata*, notably in the detailed morphology of the alimentary canal and is here given only subspecific status. It would appear that *eubranchiata* has arisen from a *catarrhactae*-like progenitor by development of a genetical predisposition to interpolation of a supernumerary heart-containing metamere (as evidenced by the additional pair of hearts) with consequent shift in the location of the gonads and their ducts one segment posterior to the normal location in megadriles (location of female pores in 15 verified for 7 type specimens taken at random). Congruence of metamerism of the gut in the two taxa indicates that adjustment of segmentation of the sinusoidal and more posterior regions of the gut occurs in the ontogeny of *eubranchiata* bringing it into agreement with the specific norm. The reported absence of gills in *catarrhactae* still requires confirmation as the types are posterior regenerates. It is especially noteworthy that an undescribed specimen (in the Hamburg Museum), from Leopoldville, which was identified by Michaelsen as *A. catarrhactae*, has gills of the *eubranchiata-type*. The specimen is abnormal in having the female pores in 17 and claspers in 21 and 22.

Alma eubranchiata catarrhactae Michaelsen, 1935
Fig. 15. 7E

Alma catarrhactae MICHAELSEN, 1935 b: 50, Fig. 6.

Alma schoutedeni MICHAELSEN, 1935 b: 47, Pl. I, fig. 3.

l=325-500+mm, w=4·5-7 mm, s=386-455+. Gills absent (but posterior regenerates). Prolobous. Intersetal ratios $aa:ab:bc:cd:dd$=2·5:1:2·5:1:2·5, in the forebody; 3·5:1:3·5:1:3·5, in the mid body. Claspers arising in segment 19 and the adjacent parts of 18 and 20 in setal lines a; extending forward as far as segment 4; setae a absent, setae b present in these segments. Asetigerous papilla large and almost circular, 2·2-2·8 mm wide, ental on the flattened spoon-shaped clasper immediately ectal to the short stalk. Genital seta loci 55 to at least 80, irregularly disposed, especially dense in the middle of the ectal half of the clasper; each locus bearing a single genital seta which is straight, spike-shaped, hardly appreciably flattened, 430-500 μ long, the shaft 21-25 μ wide, narrowing towards the ectal end which forms a slender, sharp point and is ornamented by very fine hair-like spines arranged in closely set annuli. Clitellum purple, embracing segments 186, 200, 206, 225-252, 273 (=47, 53, or 88 segments). Female pores on well developed papillae anteriorly in 14, in b lines. Spermathecal pores in 176/177, 187/188-214/215 to 204/205, 215/216-234/235 (=20, 28 or 29 intersegments); or in a re-examined syntype, inseminated spermathecae in 135/136 to 167/168 and empty, terminally rudimentary sperma-thecae extending to segment 88 anteriorly and 225 posteriorly; pores in the 8 setal rows but supernumeraries in bc and sometimes median to a lines.

Oesophagus swollen and not visibly vascularized, in 25-27; chloragogenous in 10-24 and distinctly vascularized as far anteriorly as segment 12. Intestine commencing in 28; its walls much thickened in 28, 29 and 30; typhlosole beginning in 32. Hearts in 7-12; those in 12, and apparently the other pairs, latero-oesophageal. Supra-oesophageal vessel ending, posteriorly, in 23. Nephri-dia rudimentary in 13 (and further anteriorly?); well developed in 14 posteriorly. Testes, slender, sinuous transverse bands, and their very large, convoluted funnels in 10 and 11. Seminal vesicles 4 pairs, in 9-12; the posterior pair the largest and projecting into 14. Ovaries, very long sinuous transverse bands and their delicate very large funnels in 13. Palmate septal pouches dependent posteriorly from septa 13/14-26/27 and apparent homologues on posterior and anterior faces of 28/29 and on the anterior face of 30/31. Spermathecae small, with globular ampulla and narrow almost entirely intraparietal duct.

Republic of the Congo: Tschoppo Falls; Mongende.

A. eubranchiata catarrhactae appears to differ from *A. eubranchiata eubranchiata* in the larger diameter of its copulatory papillae and the greater number of genital setae. The spermathecal pores are also more anterior in the types of *catarrhactae*. Those in *A. schoutedeni*, which is here synony-mized with *catarrhactae*, correspond in location with those of *eubranchiata eubranchiata*, but the larger number of setae (80), the large asetigerous papillae, the absence of gills (though the latter may be due to posterior amputation), and lack of information on the location of the female pores preclude union with the nominal sub-species. Location of the claspers in 23 and 24 in the single type specimen of *schoutedeni* is here regarded as an abnormality (c.f. location in 21 and 22 in branchiata material from Leopoldville identified as *catarrhactae* by Michaelsen). The illustration of a clasper (Fig. 15. 7E) is from a syntype of *catarrhactae* (Hamburg Museum, V. 12175).

Alma kamerunensis Michaelsen, 1915
Fig. 15. 7F

Alma kamerunensis MICHAELSEN, 1915: 298.

Alma kamerunensis Michaelsen. MICHAELSEN, 1918: 370.

$w=12$ mm. Dorsal groove?; Anus?; gills (?; posterior amputee). Zygolobous (?). $aa:ab:bc:cd:dd=2:1:2:1:2$ and $5:1:4\cdot5:1:7\cdot5$, with $dd:u:=0\cdot17$ and $0\cdot29$ anteriorly and in the midbody respectively. Claspers unusually large, 42-47 mm long, almost twice the length of the forebody; the base of each an elongated longitudinal field, occupying the whole length of 18, 19 and 20 in setal lines *a*, and displacing furrows 17/18 and 20/21 respectively anteriorly and posteriorly in its vicinity. Setae *a* of these segments supressed, setae *b* unmodified and separated from the base of the claspers by a distinct intervening strip. Clasper stalk comprising the proximal half, about 4 mm wide but marginally inrolled except at its base; distal half of the clasper elongate-elliptical, maximally $6\cdot5$ mm wide, concave, marginally inrolled. Stalk with 2 or 3 wide asetigerous papillae spaced along it almost as far proximally as the beginning of the distal expansion, about 20 mm from the base. Genital seta markings circular, $0\cdot7$ to $1\cdot1$ mm wide, in 2 submarginal rows a posterior row of 15 and an anterior row of 8 or 9. Genital setae $0\cdot5$ mm long, proximally 25 μ wide, basally slightly curved otherwise straight, the basal portion cylindrical, the remainder slightly expanded but somewhat narrowed at its middle; distal tip sharp, slightly curved; the distal two fifths of the seta appearing lancet-like but not greatly flattened, being 50 μ wide and 33 μ deep. The wide distal part ornamented with circlets, 3 μ apart, of fine spines. Clitellum 54 to 112 ($=69$ segments), but limits indistinct. Spermathecal pores in intersegmental furrows 56/57 to 70/71 ($=15$ intersegments), 1-7 on each side in each intersegment, mainly between setal lines *b* and *c*, 1 or 2 between *a* and *b*; thus not associated with the setal lines.

Cameroun: Bipindihof.

Only a single specimen of this species, which lacked the hind end, has ever been seen and of this only a single clasper remains (Fig. 15. 7F; Hamburg Museum, V. 8413). Specimens in Hamburg Museum from Fernando Poo referred to *A. kamerunensis* by Michaelsen (unpublished) are immature, lacking claspers, and cannot be identified with certainty.

Until the internal anatomy, especially that of the alimentary canal, is known, the status of *A. kamerunensis* cannot be determined with any certainty. The possibility of synonymy with *A. pooliana*, from the same locality must be considered. The restriction of genital seta markings to two marginal rows, and the wider genital setae appear to distinguish it from *pooliana* but the arrangement of the markings in the latter species is highly variable and sometimes includes marginal rows. Location of the spermathecae between rather than in the setal rows at present precludes synonymy, however. The morphology and unusually large size of the claspers suggests relationship with *A. ubangiana*.

Alma millsoni (Beddard, 1891)

Alma millsoni millsoni (Beddard, 1891)
Fig. 15. 5E-J; 15. 7G-I

Siphonogaster millsoni BEDDARD, 1891 b: 48, Fig. 1-3.

Siphonogaster millsoni Beddard. BEDDARD, 1893: 264, Fig. 17-21; 1895: 685.

Alma millsoni (Beddard). MICHAELSEN, 1895: 12; 1900 a: 467. BEDDARD, 1901: 219.

Alma schultzei MICHAELSEN, 1915: 280.

Alma schultzei Michaelsen. MICHAELSEN, 1918: 368.

(Non) *Alma millsoni*; MICHAELSEN, 1913 d: 59.

l=150-350 mm, w=2-4, s=240-289. Zygolobous. Anus dorsoterminal; gills absent. $aa:ab:bc:cd:dd$=2·2:1·0:2·0:1·5:2·5, in segment 12,=3·8:1·0:4·0: 1·0:5·0 in segment 200; $dd:u$=0·18 and 0·24 respectively. Claspers shorter than the forebody; their bases occupying the posterior half of 18 and half to the whole of 19, centred in a lines, separated medially by about 1/3 aa; replacing setae a and b of these segments but b follicles sometimes recognizable. Each clasper about 10-15 mm long; its width increasing slightly from the base, which lacks a distinct stalk, to the semicircular distal end; apposable surface slightly tumid, with numerous transverse furrows, and 6 to 15 genital seta loci, each 0·5-0·8 mm wide, in two longitudinal rows, of which a distal pair lies in a more or less distinct depression or cupule; margins not, or only slightly inrolled; an asetigerous copulatory sucker rarely present at the proximal fourth. Genital setae 410-600 μ long with a large node, 32-44 μ wide, in the proximal half; distally only slightly widened but conspicuously flattened; notched and encircled (except ventrally) by irregular transverse lines; the underside concave or longitudinally grooved. Clitellum annular, commencing in 47-65 and extending to 82-89 (=20-42 segments). Female pores inconspicuous in 13/14 lateral to b. Spermathecal pores in 18/19, 20/21, 35/36, 36/37-42/43 (=16, 17, 25 intersegments), 2 on each side per intersegment, in setal lines a and b with some omissions. Nephridiopores in setal lines b near the anterior segmental margins.

Oesophagus widened and with encircling vascular striae in 10-15; narrow and non-vascular in 16 and 17; intestine beginning in 18; a crop or gizzard in 18 to 19 or 20, with or without appreciably increased musclarization; typhlosole commencing in 22-24. Heart in 8-12 (5 pairs); dorsal vessel rudimentary in front of them; supraoesophageal vessel in 15 anteriad; no median subneural. Nephridia well developed in 15 posteriad, rudimentary in 14 (or 13 also?). Holandric; 4 pairs of seminal vesicles, in 9-12; those of 12 deflecting septum 13/14 posteriad; vasa deferentia of each side uniting anteriorly, running in the parietes. Ovaries paired in 13, ribbon-like, transverse bands; not plicate. Septal pouches (always?)in 13 and 14. Spermathecae when inseminated clearly visible projecting into the coelom but not sufficiently large to be adpressed, longitudinally ovoid; apparently sessile.

Nigeria: Lagos. Cameroun: forest by Yukaduma.

Alma millsoni, of which 5 type specimens have been re-examined by the author, is very similar to *A. tazelaari* n.sp. (Ghana). It differs from *tazelaari* in the closer pairing of its setae posteriorly, in restriction of the origin of the claspers to 18 and 19, in the presence of only 2 setae in the distal cupule, in the lesser extent of the clitellum and the spermathecae; and in the location of the "gizzard"; furthermore, its genital setae appear to differ in being hollowed out ventrally at the distal end.

Examination of the internal characters of *A. schultzei*, which have not previously been described, necessitates its inclusion in *millsoni* as a junior synonym. The minute gills reported by Michaelsen in *schultzei* are not observable in the specimen for which he illustrated them.

Alma zebangui Duboscq, 1902, is included in *millsoni* as a distinct subspecies. Its internal anatomy, previously little known, has been found on re-examination of a type specimen (Hamburg Museum V. 6104) to differ in no significant respects from that of the types of *millsoni*. It appears to be a Congo form of *millsoni* differing from the nominal sub-species in the apparently constant (rather than rare) occurrence of asetigerous papillae on the claspers. Illustrations of claspers are from syntypes of *A. millsoni* (B.M.N.H. 1904.10.15.1031-41; Fig. 15. 7G), *A. schultzei* (Hamburg Museum, V. 8432; Fig. 15. 7I); those of genital setae are from a syntype of *millsoni* (Fig. 15. 5E, original; 15. 5G-J, after Beddard) and from a syntype of *A. schultzei* (Fig. 15.5F, after Michaelsen).

Alma millsoni zebangui Duboscq, 1902
Fig. 15. 7J

Alma zebangui DUBOSCQ, 1902: 97; Fig. 1-3.

Alma zebangui Duboscq. MICHAELSEN, 1915: 280; 1918: 369.

l=110-130 mm, w=4-5 mm, s=271-288. Zygolobous. Setae $aa=dd\rangle bc$; $ab=cd$. Lengths of setae 530 μ behind the clitellum; 430 μ in the hindbody; 320 μ in the last segments; all crochets with fine annulations proximal to the terminal curvature. Clitellum segments 47, 50-77, 80 (=31 segments). Claspers originating in $\frac{1}{3}$ $\frac{1}{2}$18-19, $\frac{2}{3}$ 20; setae a supressed; setae b present or absent. Each clasper 10 mm long, 3 mm wide with one or two asetigerous papillae near its insertion; 2-5 genital seta loci (papillae or sucker like structures) arranged in two longitudinal rows, a little distal of the middle of the clasper; and invariably 2 distal loci situated one on each side at the bottom of a terminal cup-like depression (cupule), which may be elevated. Intermediate loci bearing 2 or 1 setae the distal locus 1 seta each. Number of loci differing on the two claspers of an individual. Genital setae approximately equal in length at 480 μ, to the somatic setae of the midbody; cylindrical with the extremity flattened in the form of a spear head hollowed out into a groove or spoon-shape. Vas deferens discharging between the two loci in the cupule?

Sinusoidal oesophagus ending, posteriorly, in 14; oesophagus not visibly vascularized in 15-17. Intestine commencing in 18; thick-walled in 18 and 19 (and $\frac{1}{2}$20?); typhlosole beginning in 22. Hearts in 8-12. Supra-oesophageal commencing in 14. Nephridia beginning in 14. Seminal vesicles in 9-12. Ovaries plicate bands larger transversely than longitudinally. Last septal pouches dependent posteriorly from 16-17; commencing in 13.

Spermathecae commencing a short distance behind the claspers and reaching into the clitelar region, e.g. in intersegments 23/24, 25/26 to 45/46, 46/47

($=22$-23 intersegments) in setal lines a and b, total number 80; form broadly bulbous, projecting relatively freely into the coelom. No supernumeraries. Spermatophore suender and elongated.

Central African Republic (border with The Congo); Bangi, affluent of the Ubangi River.

Duboscq examined 10 clitellate specimens with claspers and therefore the few anatomical features which he described may be taken to be characteristic of the species as a whole or, at least, not aberrant. His limited description revealed no significant differences from *A. millsoni* as revised by the author beyond the normal occurrence of asetigerous papillae on the claspers. Examination of a syntype in the present study necessitates inclusion in *A. millsoni* as a sub-species (Fig. 15. 7J; Hamburg Museum, V. 6104).

Alma multisetosa Michaelsen, 1915
Fig. 15. 5K; 15. 8A

Alma multisetosa MICHAELSEN, 1915: 302.

Alma multisetosa Michaelsen. MICHAELSEN, 1918: 371.

$w=6$-8 mm. Zygolobous. Dorsal groove? Anus? Gills? (posterior amputees). $aa:ab:bc:cd:dd$ in the fore and midbody respectively $2\cdot5:1:2\cdot5:1:2\cdot5$ and $7:1:7:1:12$ with $dd:u=0\cdot17$ and $0\cdot33$. Segments 18, 19 and 20 shortened; the base of each clasper forming an elongated field bordered by intersegmental furrows 18/19 and 19/20 and stretching from mid 18 to mid 20 between setal lines a and b. Claspers longer than the anterior portion of the body, 40-52 mm long; the stalk 30 mm long and $2\cdot5$ mm wide owing to inrolling of the margins. Distal part widened into a medially concave elongate-tongue shaped lobe, maximally 7 mm wide; 2 copulatory papillae (sucker-like) on the stalk of each clasper, one about 5 mm, the other about 25 mm from the proximal end. Genital setae markings $c.$ 220 in number, irregularly scattered on the distal expansion and in a row on the stalk well proximal to the distal sucker; each a rounded depression $c.$ $0\cdot4$-$0\cdot7$ mm wide, with a diaphragm-like centre bearing a single genital seta. Genital setae $c.$ $0\cdot5$-$0\cdot6$ mm long; maximally, somewhat proximal of the middle, $c.$ 35 μ wide; gradually tapering to the end; only basally slightly curved; distally sharply pointed, not lancet-like or flattened; the distal half with very closely situated annuli of slender spines. Male pores at the middle of the ladle-like enlarged part of the clasper, some 11 mm from the distal end. Clitellum annular, (80), 84 to 136 ($=53$, (57) segments); indistinct in 80 to 83 but sharply delimited posteriorly. Spermathecal pores in 65/66 to 81/82 ($=17$ intersegments). Usually 1 in each setal line a-d, i.e. 8 per intersegment; omissions very occasional: supernumeraries (above d or in bc) rare. Internal structure characteristic of the genus.

Cameroun: Nyui River at Yukaduma and Lau (Nigeria?)

A. multisetosa is morphologically near *A. pooliana* but differs notably in its larger size and the large number of genital setae, compared with the 24-34 in the latter species. The remaining type specimen (Hamburg Museum V. 12165) is too severely macerated for examination of internal features (clasper, Fig. 15. 8A). The illustration of a genital seta (Fig. 15. 5K) is after Michaelsen (1915).

Alma nasuta Michaelsen, 1935

Alma nasuta MICHAELSEN, 1935 b: 37.

$w=5\cdot3$ mm. Zygolobous; nasute. Anus? Gills? (posterior amputees). Setae in the forebody fairly closely, in the hindbody, closely paired; in the forebody $aa=3\text{-}4$ $ab,=bc,=3\text{-}4$ $cd,=c.$ $5/6$ $dd.$ Clitellum? Genital pores? The base of each clasper occupying ab from about mid 18 to the posterior margin of 20; segments 18-20 shortened and narrowed. Claspers spoon-shaped with short, broad not obviously demarcated stalks; about 6 mm long (about one fifth to one fourth the length of the forebody); maximally, at the distal third $2\cdot5$ mm wide; the apposable surface segmented and concave, the other surface smooth and arched. Asetigerous papillae absent. Genital seta markings numbering 35 (1 clasper); irregularly arranged but tending to longitudinal single and double rows; only the basal part of the stalk lacking them; each with 1, rarely 2, setae. Genital setae straight, 320 μ long, with a stalk 150 μ long and 25 μ wide, widening and flattening imperceptibly into a lancet-shaped distal portion which is maximally, at the middle, $c.$ 50 μ wide and is slightly concave-convex; tip sharp and three sided; transverse lines on the distal stalk and the expansion with the exception of the tip, probably representing minutely spinose annuli.

Intestinal musculature apparently slightly thickened in 28; rectangular typhlosole present. Holandric but testes not recognizable; 2 pairs of large much folded sperm funnels free in 10 and 11; seminal vesicles 2 pairs, in 11 and 12; from the basal part of each vesicle arises a tube which meets that of the other side above and below the gut. The tubes associated with the seminal vesicles of 11 fill segment 11 and project into 10; those of the seminal vesicles of segment 12 project far posteriorly, into segment 35; the tubes are deeply constricted at the intersegmental septa.

Republic of the Congo: Koteli (Congo basin).

The absence of asetigerous suckers and the shortness of the claspers may have been due to the immaturity of the specimens. Michaelsen considered *A. nasuta* to be near *A. millsoni* as indicated by the small claspers, absence of copulatory papillae and the form of the genital setae. Distinguishing features mentioned were the barely visible ornamentation of these setae, their number and arrangement and the nasute prostomium. The location of the intestinal gizzard also differs. The types (Hamburg Museum, V. 12165) are fragmentary, lack claspers, and yield no useful information.

Alma nilotica Grube, 1855
Fig. 15. 5B-D; 15. 6B; 15. 8B, C.

Alma nilotica GRUBE, 1855: 129.

?*Siphonogaster aegyptiacus* LEVINSEN, 1890: 319, Fig. 1-6, 1895: 684.

Digitibranchus niloticus (Grube). LEVINSEN, 1890: 321, Fig. 7-8.

Alma nilotica Grube. MICHAELSEN, 1895: 7, Fig. 14; 1897 b: 67; 1899 b: 119; 1900 a: 466; 1915: 280; 1918: 370; REA, 1901: 174. KHALAF EL DUWEINI, 1950 a: 94, Fig. 1-18, Pl. I-V; 1950 b: 29, Fig. 5, Pl. I; 1951: 52, Fig. 1-4, Pl. I; 1954: 102, Fig. 6; 1957: 833, Fig. 1-5; 1965: 31, Fig. 1-7; KHALAF ET GHABBOUR 1963: 23; JAMIESON ET GHABBOUR, 1969:

Alma nilotica Grube (part.). KHALAF EL DUWEINI, 1940: 113, Pl. IV, fig. 2, Pl. V, fig. 1, 2 (large immature branchiate specimens without "penes").

l=280-550 mm, w=4-6 mm (Khalaf. in litt.), s=472-480. Anus? Dorsal groove bordered by branched gills. Zygolobous or indistinctly prolobous. $aa:ab:bc:cd:dd$=2·8:1·0:2·7:1·1:2·9 (mean of 3) in segment 12;=5·4:1·0:4·8:1·3:4·3 posteriorly, (mean of 2); $dd:u$=0·19 and 0·18 respectively. Clitellum annular, commencing in 133-155 and extending to 173-202 (=40-60 completely modified segments). Claspers appearing many-segmented; in preservation 0·9-1·8 cm long wide the mode at 1·6 cm for fully mature specimens and 1·7-4 mm wide;〈4·0 cm long in life; much shorter than to somewhat longer than forebody; their bases occupying 19 and adjacent portions of 18 and 20; setae a absent, setae b retained and not translocated. Male pores one on each clasper, median and shortly distal of the middle. Basal sucker oval, 1·4-1·8 mm at its greatest width; its centre located at the proximal seventh; the short stalk half the general width, the remainder parallel sided with inrolled margins; genital seta loci low or knob-like papillae, ony about 0·15 mm wide, or lacking a recognizable field frequently in transverse furrows; 19-43 per clasper (mean 29); in no specific pattern, although 2-4 irregular rows may be apparent, but more densely clustered in the distal fourth; very few in the proximal half. Genital setae 302 μ long and 22 μ wide; straight or basally slightly curved; appreciably expanded in the pointed distal fourth and expanding again just proximal to the middle; the intervening waist and the distal expansion bearing closely spaced irregular circlets of spines which are so minute as to be unrecognizable except under oil immersion; the expanded regions appreciably flattened. Spermathecal pores commencing in 123/124-144/145 and extending to 145/146-165/166 (=22-27 intersegments); one in each of the 8 setal lines though there are frequently omissions in each transverse row; single supernumeraries infrequent or common in bc; less frequent or absent median to setal lines a on each side.

Nephridia well developed in 14 or 15 posteriad, sometimes rudimentary in 13 and 14, discharging at b. Oesophageal plexus in $\frac{1}{2}$13-$\frac{1}{2}$18, ciliation in $\frac{1}{2}$18-$\frac{1}{2}$21. Intestine unciliated commencing in $\frac{1}{2}$21 forming a rudimentary gizzard in $\frac{1}{2}$21-23; typhlosole commencing in 25. Hearts 6 pairs, in 7-12; with the possible exception of those in 7, latero-oesophageal. Dorsal vessel absent in front of the hearts of 7. Supra-oesophageal in 18 anteriad sometimes recognizable only from 17 or 16; communicating in 6-10 with longitudinal latero-parietal trunks. Median subneural absent. 4 pairs of seminal vesicles in 9-12; those of 12 deflecting septum 12/13 posteriad. Ovaries, ruffled transverse bands, and their funnels in 13. Septal pouches in 13-22. Inseminated spermathecae intramural or exposed.

Egypt: Mansura (Type locality); Giza; Bedrashin; Zamalek (Cairo); El Khatatba (Nile Delta). Sudan (Nile): Shimbat (near Khartoum). Wadi Haifa (collector Loat, B.M.N.H. 1907.11.12.6, labelled *A. stuhlmanni*, new record; Fig. 15. 5).

The illustrations of genital setae (Fig. 15. 5B, C) and of claspers (Fig. 15. 8B, C) are from Jamieson and Ghabbour (1969). That of the alimentary canal (15. 6B) is from a specimen from Abu Rawwash.

Alma pooliana Michaelsen, 1913
Fig. 15. 8D, E

Alma pooliana MICHAELSEN, 1913 d: 57, Pl. II, fig. 19, 20.

Alma pooliana Michaelsen. MICHAELSEN, 1915: 300.

Alma emini (Michaelsen). GATES, 1958 a: 5. (Non) *Alma emini* (Michaelsen 1892)).

l=65-290 mm, w=2-8 mm, s=298-410. Gills absent. Zygolobous. *Setae* anteriorly widely paired, with the exception of those of the peristomium which are closely paired, in the mid and hindbody (always?) closely paired; in the forebody $aa:ab:bc:cd:dd$=3:1:3:1·5:4 or 2:1:2:1:2, $dd:u$=0·22 or 0·17 respectively; ab=cd=4-5 aa in the midbody. Nephropores at or close to b lines. Clitellum annular (50), 52, 79-112, (114), 121 (=43 or 61 (65) segments). Claspers with narrow bases in the ventral setal lines, occupying segment 19 and a little of 18 and 20; each about 26-30 mm long with a long narrow stalk; the spoon-shaped lamina 10-13 mm long and 4-6 mm wide. Asetigerous papillae situated on the stalk, circular with central knob, one near the base, a second near the lamina, and occasionally a third between them, or an additional pair distally. Genital seta markings 17-35, on the lamina, which form distally an irregular group (e.g. of 10 or 11) or short marginal rows and more proximally 2 or 3 more or less distinct longitudinal rows, of which the middle row is particularly regular; 3-7 (or more?) genital seta markings also present on the stalk, some or all of them proximal to the distal papilla; each marking circular, c. 0·7 mm wide, and bearing a single genital seta which is 600 μ long, maximally 35 μ wide, varying in form from straight to sigmoid; the shaft ornamented ectally with closely situated circlets of very fine points and tapering to a point which is bent towards the concave face. Female pores presetal in b lines of 14, each on a slight tumescence. Spermathecal pores in 41/42, 57/58, 58/59, 59/60 to 52/53, 72/73-75/76 (=12, 16, 17, or 19 intersegments?) located in a-d, 1-11 on each side; in the middle of the spermathecal region they lie in the setal lines with 4-7 supernumeraries in bc; at the extremities present only in c and d; total number of spermathecae per individual c. 120 to more than 200.

Gut with transverse vascular striae in 8-17, widened and white in 19-24, in 25-27 still wider and nearly spheroidal, the muscular layer thickest in 25 in which the intestine apparently commences. Dorsal typhlosole commencing in 29. Dorsal and ventral blood vessels not reaching the ends of the body; supra-oesophageal vessel in 8 (and further anteriorly?) to 18 with posterior bifurcations ramifying in 19, connecting with a pair of lateroparietals in 8. Subneural absent. Hearts in 8-12, those of 10-12 latero-oesophageal. Holandric, seminal vesicles in 9-12. Ovaries plicate. Spermathecae sessile or concealed in the parietes; ampullae when inseminated filling a segment longitudinally. Septal pouches (?) in 13-25 or 26.

Republic of the Congo: Stanleyville (River Congo);
Cameroun: Ukaduma (Nyui River); Bipindihof.

Reasons for synonymizing material from Stanleyville, previously referred to *A. emini*, are given under that species. A close relationship between *pooliana* and *emini* is indicated by the markedly spatulate shape of the claspers, their segmental location, the similar location of the clitellum and spermathecae, and the very large number of the latter. The distinctive form and arrangement of the genital setae, and the absence of asetigerous papillae in *A. emini* are, however, only two of several clear indications of specific distinction of the two taxa. Although both species are now sympatric in headwaters of the Congo, it seems possible that they have speciated from an ancestral species, populations of which were isolated on each side of the Albert-Tanganyika rift, that on the Congo side giving rise to *A. pooliana*, and that on the east to *A. emini*, which is now so ubiquitous in the great lakes.

Further collections from Bipindihof should be examined thoroughly as the distinctive spermathecal distribution suggests the possibility of subspecific or even specific distinction from Yukaduma and Stanleyville populations.

The illustrations of a clasper are from the type specimen (Hamburg Museum, V. 7655; Fig. 15. 8D is original; 15. 8E, after Michaelsen, 1913d).

Alma stuhlmanni (Michaelsen, 1892)
Fig. 15. 6C; 15. 8F, G

Siphonogaster stuhlmanni MICHAELSEN, 1892: 10, Fig. 7-9.

Siphonogaster stuhlmanni Michaelsen. MICHAELSEN, 1896 b: 4, Fig. 28; BEDDARD, 1895: 686.

Alma stuhlmanni (Michaelsen) MICHAELSEN, 1895: 8; 1905: 363; 1910 c: 88; 1915: 280; 1918: 366. JAMIESON et GHABBOUR, 1969:

Alma sp. BEDDARD, 1901: 215.

Alma budgetti BEDDARD, 1903: 222.

Alma nilotica (Michaelsen) (part.); KHALAF EL DUWEINI, 1940: 113 (small form with "penes").

Alma sp. KHALAF et GHABBOUR, 1963: 24; 1964: 89; 1965: 9.

?*Alma sp.* OMODEO, 1958: 97.

$l=73$-170 mm, $w=1.5$-3 mm, $s=225$-258. Gills absent; Zygolobous or indistinctly prolobous. Setae widely paired throughout; $aa:ab:bc:cd:dd=3.9:1.0:2.8:1.0:4.0$ in segment 12; $=2.4:1.0:2.0:1.1:2.3$ in the hindbody; $dd:u=0.23$ and 0.18 respectively; ornamented with fine teeth. Clitellum annular, commencing on 37-55 and extending to 69-92 (129?) ($=26$-45 (76?) segments). Claspers 0.6-1.2 cm long, more than half as long and longer than the forebody, and 1 mm wide; originating in 19, usually part of 18 and less frequently part of 20, in all of which setae *a* are absent and *b* present; abnormally originating one segment anteriad; many-segmented, narrow and parallel-sided, only slightly widening in the vicinity of the semicircular free end; the apposable surface flat or slightly concave but not marginally inrolled; the other surface transversely convex; basal sucker circular, 0·7 mm wide, located at the proximal seventh. An abrupt eversible depression or cupule near the distal end of the clasper houses the male pore, on each side of which is an unornamented sigmoid genital seta, 100-170 μ long, and, at the node, 10-13 μ wide, i.e. much shorter than the somatic setae which are 314-500 μ long and 21-25 μ wide. Female pores each on a minute papilla in 13/14 lateral to b. Nephridiophores in front of setae b. Spermathecal pores commencing in 20/21, 24/25, 27/28 and extending to 26/27, 33/34, 37/38, 50/51 (=7-27 intersegments), in a and b, maximally 4 per intersegment; omissions frequent, especially in the a rows from which spermathecae are often completely absent; supernumeraries occasionally present in bc.

Nephridia commencing in 14. Oesophageal plexus in $\frac{1}{3}$ 12-14; oesophagus ciliated in $\frac{1}{2}$16-$\frac{1}{2}$17. Intestine commencing posteriorly in 17; thickened in 17-19 to form a rudimentary gizzard; typhlosole commencing in 20, 21 or 22. Hearts 5 pairs, all latero-oesophageal, in 8 to 12; supra-oesophageal located in 15 anteriad and connected with a pair of lateroparietal trunks in 14 and 15 and to the dorsal vessel in addition to these in 13. Testes and funnels in 10 and 11, free; seminal vesicles 4 pairs, in 9-12; not extending their segments or, sometimes, reaching into 14. Acinous ovaries, and funnels, in 13; septal pouches projecting posteriorly from 13/14 only or 13/14 to 16/17. Spermathecae concealed in the body wall.

Tanzania: Bukoba (Lake Victoria) (Type locality). Gambia: McCarthy Island. Cameroun: Bamenda. Republic of the Congo: Leopoldville; Kilo; Kinyawanga; Kassenge=Kasenyi?; Beni. Uganda: Entebbe. Egypt: Abu Rawwash near Cairo. Guinea: Mt. Nimba.

The most widely distributed species of *Alma*. The illustrations of claspers (Fig. 15. 8F, G) are from Jamieson and Ghabbour (1968).

Alma tazelaari n.sp.
Fig. 15. 5; 15. 8H, I.

l=265 mm, w=4 mm, s=378. Zygolobous. Gills absent. Setae anteriorly very widely, in the hindbody fairly widely paired; $aa:ab:bc:cd:dd$=2:1:2·1:1·2: 3·4 in segment 12;=2·4:1:2:4:1:2·9 in the hindbody; $dd:\mu$=0·24 and 0·21 respectively. Claspers about half the length of the anterior part of the body, their bases in $\frac{1}{2}$ 18 to $\frac{3}{4}$ 20, in ab, the sites of setae a occupied, those of setae b immediately lateral to the bases; setae a and b nevertheless absent in 18 to 20. Each clasper has a many-segmented stalk $\frac{1}{5}$ to $\frac{1}{3}$ the total length of the clasper; distal to the stalk the clasper widens slightly but uniformly to the distal end; near the distal end there is an abrupt, deep, cuplike median depression (cupule) the floor of which may be partly or wholly everted to form a more or less convex cushion which is white and glandular in appearance and bears on each side of the midline 2 (or in one specimen 3 or 2) round protuberant genital seta markings, each with one or rarely two visible genital setae. These markings are darkly translucent and flank a longitudinal median strip of similar appearance which continues to the proximal rim of the cupule, the male pores apparently lying at the junction. At the junction of distal lamina and stalk there is a large, convex basal asetigerous sucker, maximally 1 mm wide. Between this and the cupule there are two rows of round genital seta markings (maximally 13 in a row), each with a single visible genital seta. Some transverse duplication of the rows may occur; the number of genital seta markings per clasper totals 15-23 (including these in the cupule); means for 7 claspers are 4 (cupule), 9 (median row), 6 (lateral row), total 19. Genital setae 330 μ long, maximally, at the node which is proximal of the middle, 16-27 μ wide, proximally curved slightly or through a right angle; distally only slightly expanded and ornamented with several circlets of jagged, irregularly interrupted ridges which in profile have the appearance of shingles slightly divergent from the surface of the seta at their free distal margins; at the extreme tip the ridges sometimes replaced by a few scattered scales. Clitellum annular, 52, 58 to 98, 103 (46-47 segments). Female pores each on a minute papilla concealed in intersegmental furrow 13/14 at 1/3 *bc*. Spermathecal pores (externally invisible) in *a* and *b* lines, with a single one displaced from *a* to *ab*, in 22/23 to 52/53 (31 intersegments); 2 per side in each transverse row.

Oesophageal plexus in 11-13; ciliation in 15-$\frac{1}{2}$ 17, in which the muscularization is increased. Intestine (unciliated) beginning at $\frac{1}{2}$ 17, its muscularization increased in $\frac{1}{2}$ 17-20 to form a single easily compressible gizzard; dorsal typhlosole beginning in 24.

Hearts in 8-12 (5 pairs), the last 2 or 3, pairs latero-oesophageal; dorsal vessel ending in 8 in which it is thin or absent in front of the hearts. Supra-oesophageal vessel in 13 anteriad, in 14 divaricating around the gut to form two latero-parietal trunks. Subneural absent. Nephridia well developed in 15 posteriad; rudimentary in 13 or 14. Seminal vesicles large and lobed, in 9-12; those of 12 displacing septum 12/13 to the succeeding septum. Ovaries plicate transverse bands. Ovisacs absent. Racemose septal pouches in 14-17, the last pair minute. Spermathecae, even when inseminated, completely intramural.

Ghana: Ho and unspecified locality, 4 sexual specimens, collector M. Tazelaar, B.M.N.H. 1964. 2. 189/190 and 203/204.

Alma tazelaari appears to have close affinities with *A. togoensis* (Togo) and especially with *A. millsoni* (Nigeria) which have a terminal cupule on each clasper. It shows clear morphological distinctions from these, however, and the four or five large genital seta papillae in the cupule permit ready identification. Differences from *A. millsoni* are discussed under that species.

The illustrations of genital setae (Fig. 15. 5LM) are from a paratype and those of claspers are from the holotype (15. 8H) and from specimen B.M.N.H. 1964.2.203 (Fig. 15. 8I).

Alma togoensis Michaelsen, 1915
Fig. 15. 5N, O; 15. 8J, K.

Alma togoensis MICHAELSEN, 1915: 289, Pl. XV, fig. 35 (for *A. millsoni* (Beddard); MICHAELSEN, 1913: 59).

Alma togoensis Michaelsen. MICHAELSEN, 1918: 363.

(Non) *A. millsoni* (Beddard, 1891).

$l=115+$mm, $w=4$ mm, $s=217+$. Zygolobous. Cylindrical in front of the claspers; quadrangular behind; gills apparently absent. Anus? $aa:ab:bc:cd:dd=2\cdot8:1:3:1\cdot6:3\cdot1$ in segment 12; $5:1\cdot2:3\cdot5:1:5\cdot5$ in segment 200; $dd:u=0\cdot18$ and $0\cdot25$ respectively. Claspers with bases occupying the adjacent halves or two thirds of segments 18 and 19, setae a and b of which they replace; 12-13 mm long; the proximal $\frac{1}{3}$-$\frac{2}{3}$ straplike, 1-1·3 mm wide, with medianly inrolled edges; the remainder widening to a maximum width of 2-2·5 mm; its rounded free end bearing 2 distally convergent rows, each of 3-8 closely spaced genital seta loci, the numbers of setae per row being less owing to omissions, a variation of 2-7 having been recorded. The area around these setae may form a heart-shaped depression. Elsewhere 1-4 genital seta between the base and the distal region of the clasper, each on a round, sometimes suckerlike genital marking, 0·4-1 mm wide, submarginal but extending over the midline; asetigerous papillae absent.

Genital setae almost straight, only basally slightly bent, 480-500 μ long and maximally (in proximal and middle regions) 18 or 28 μ wide; the distal fourth flattened, hollowed out on one surface and bearing relatively large slender or broadly triangular spines closely pressed to the surfaces, which are scattered or are arranged in irregular and often incomplete transverse rows which may give the margins a serrated appearance; distal tip blunt. Clitellum annular distinctly developed in 35, 36-49, 60, 63 ($=15$, 26, 28 segments) but weakly developed in 25, 33, 34-54, 65, 72 ($=22$, 32, 48 segments). Female pores small but fairly conspicuous, without tips, in intersegment 13/14 at $\frac{1}{3}$ bc. Nephridiopores in b lines in 15 posteriad. Spermathecal pores clitellar, in 36/37, 41/52-46/47, 50/51, 51/52 ($=10$ or 11 intersegments), in setal lines a and b, seldom in a or b; supernumeraries absent.

Hearts in 8-12 (5 pairs); latero-parietal vessels large. Intestine commencing in 16; with very slight muscular strengthening in 16 and 17. Nephridia in 15 posteriad. Very large sperm funnels in 10 and 11; 4 pairs of seminal vesicles in 9-12; the posterior pair projecting into 13. Spermathecae bulbous, projecting freely into the coelom; with short, narrow stalks.

Togo: Sokode, Ghana (collector Dr. Thomas, B.M.N.H. 1960. 4.7.1/3, new record).

The two convergent rows of genital setae at the distal end of each clasper, and the few proximal setae on very large papillae, are distinctive of this species. The illustration of genital setae are after Michaelsen, 1915 (Fig. 15. 5N) and original from a specimen from Ghana (Fig. 15. 5O); those of claspers are from the type (Hamburg Museum. V. 7368; Fig. 15. 8J) and a specimen from Ghana (Fig. 15. 8K).

Alma ubangiana Michaelsen, 1915
Fig. 15. 8L

Alma ubangiana MICHAELSEN, 1915: 305.

Alma ubangiana Michaelsen. MICHAELSEN, 1918: 372.

$l=380$ mm, $w=7$ mm, $s=600$. Gills absent. Anus terminal, small. Zygolobous. $aa:ab:bc:cd:dd=2:1:2:1:2$ and $4:1:3:1:6$, with $dd:u=0\cdot17$ and $0\cdot30$ in the forebody and in 26 posteriad respectively. Clitellum annular, (100) 101 to

159 (162) (=59 (63) segments). Claspers each with its base in and a little above *a* line and occupying the entire lengths of 24-26; the ventral setal couples of these segments absent; length of a clasper 85 mm compared with 30 mm for the anterior part of the body; its proximal half a slender stalk, 2 mm wide basally; the distal half a slender lamina maximally 5 mm wide. 2 asetigerous papillae, one at 20 mm, the other at 40 mm from the base, the latter slightly the larger with a diameter of 1·5 mm. Lamina with 2 longitudinal rows of mostly small genital markings, maximally 1 mm wide, midway between the midline and margins of the clasper; each row with about 15 to as many as 30 markings. Genital setae 400 μ long and maximally 30 μ wide; straight, very slightly flattened, fairly sharply pointed; not expanded distally; finely annulated, probably with spines. Spermathecal pores in intersegments 75/76 to 95/96 (21 intersegments); one in each setal line, i.e. 8 per intersegment; a single supernumerary above *d* line. Spermathecae almost entirely concealed in the body wall.

Republic of the Congo: Banzyville on the Ubangi.

The far posterior position of the claspers is unique but is possibly an abnormality as only a single mature specimen which lacked one clasper was described. Only the clasper remains for re-examination (Hamburg Museum, V. 8448, Fig. 15. 8L). The three immature types apparently lacked claspers.

GENUS **Drilocrius** MICHAELSEN, 1918

Type species: Criodrilus iheringi MICHAELSEN

Prostomium prolobous. zygolobous or epilobous. Body quadrangular behind the forebody ; (always?) with a dorsal groove; anus dorsal or dorsoterminal. Setae 8 per segment, closely paired; $dd:u\langle0\cdot3$; sigmoid, single pointed; unornamented or with minute scars irregularly strewn or in circumferential circlets; distinctly modified genital setae absent though some setae in the forebody may be elongated and lie on glandular papillae. Nephropores presetal in *b* lines. Clitellum annular, occupying 22-30 segments; immediately behind or anteriorly including the male pores. Male pores in 15 or 16 in or immediately lateral to the ventral setal lines and median to paired genital lobes in *bc* which are moderate to very prominent and may be digitiform or may form clasper-like laminae sometimes (*buchwaldi*) on the lobes. Female pores a pair in 14 median to *a* lines or exceptionally (*alfari*) in *b* lines. Spermathecal pores posttesticular, inconspicuous in 1-7 intersegmental furrows, in front of or behind the male pores; 6-20 in a transverse row, from *b* lines to as far as the middorsal line.

Oesophagus with or without a rudimentary gizzard in 5 and/or 6; lacking calciferous glands. Intestine beginning in 17-20, usually with appreciable muscular thickening anteriorly; dorsal typhlosole present. Hearts in 7-11, some at least, latero-oesophageal. Dorsal vessel ending with those of 7 or continuous onto the pharynx; median subneural vessel well developed throughout or reduced or absent in anterior segments. Nephridia absent from 10 or more anterior segments; lacking terminal bladder or caecum. Testes free in 10 and 11; seminal vesicles 2 or 3 pairs, in 10, 11-12, those of 12 enlarged in 13 and (always?) deeply constricted by septum 12/13. Vasa deferentia intraparietal; copulatory sacs and prostate glands absent. Ovaries with more than one string of mature oocytes. Ovisacs (always?) present. Paired septal pouches present or absent. Spermathecae largely or wholly concealed in the body wall.

Central and S. America: Costa Rica to Argentina. In the mud of swamps, streams and rivers.

Drilocrius was erected by Michaelsen (1918) for the Central and S. American species of *Criodrilus*, which possess spermathecae. The sole species then remaining in *Criodrilus*, the palaearctic *C. lacuum*, was distinguished from *Drilocrius* by lacking spermathecae, by possessing grooved genital setae and by the absence of gizzard-like thickening of the alimentary musculature. Thickening in 5-7 has since been shown to exist in *C. lacuum*, however.

D. alfari and *D. buchwaldi* show a relatively posterior location of the spermathecae and each is clearly a valid species. *D. ehrhardti*, also with spermathecae behind the male pores, is here made the type of a new genus, *Glyphidrilocrius*. Of the remaining 5 species, all of which have spermathecae in 13/14 or 13/14 and 14/15, *D. dreheri* appears from its description to be inseparable from *D. iheringi*; *D. hummelincki* is not certainly distinguishable from *D. breymanni* and *D. burgeri* is inadequately known. Type material of most of these species has been re-examined but longer series are needed before their interrelationships can be satisfactorily elucidated.

Drilocrius

1.	Genital lobes long tongue-shaped laminae (claspers) with their bases restricted to segment 16	*D. alfari*
—	Genital lobes with their bases not restricted to 16. Not tongue shaped laminae	2
2.	Spermathecal pores in 17/18-20/21 (vasa deferentia entering the genital lobes)	*D. buchwaldi*
—	Spermathecal pores in 13/14 or 13/14 and 14/15. (Vasa deferentia not entering the genital lobes.)	3
3.	Two conspicuous genital protuberances on each side, in 15 and 16 respectively, of which at least the anterior is digitiform or penis-like. (Spermathecal pores in 13/14 only.)	4

3. Genital lobes stout lateral ridges, low (if present) in 15, highest in 16 (spermathecal pores in 13/14 and 14/15.)   5

4. Intestine beginning in 17   *D. iheringi*

— Intestine beginning in 18   *D. burgeri*

5. Ventral setae in 12-19 on conspicuous protuberances. Subneural vessel weakly developed or absent in pregonadial segments   *D. hummelinki*

— Setae of 12-19 not on conspicuous protuberances. Subneural vessel well developed throughout   *D. breymanni*

Drilocrius alfari (Cognetti, 1904)
Fig. 15. 10A, B

Criodrilus alfari COGNETTI, 1904 b: 4; 1906 a: 62, Pl. fig. 34-38.

Drilocrius alfari MICHAELSEN, 1918: 362 .

l=100-120 mm, w=3-4 mm, s=310. Epilobous. Cross section quadrangular in 7 posteriorly; a wide dorsal gutter filling dd; anus dorsal, immediately subterminal. In the mid-body $aa:ab:bc:cd:dd$=6·8:1·0:7·6:1·4:8·4; $dd:u$=0·24 Ventral setal couples of 13, 15-21, 22 on white protuberances; seta a (of segment 14) with several closely spaced circumferential circlets of minute denticulations; 560 μ long by 43 μ wide. Seta b (of 14) with scattered lunate scars, some tending to form tranverse, others longitudinal rows; 530 μ long by 39 μ wide. Setae in the hinder region of the clitellum ornamented distally with irregularly strewn but numerous minute transverse lunate scars; node recognizable by its greater width but not distinctly demarcated; a seta a and b, respectively, 520 μ and 500 μ long by 39 μ and 35 μ wide. Nephropores presetal in b lines (recognizable on the clitellum). Clitellum annular, 14-40, 42, 43; pigmented in 22, 24 posteriorly. Tongue-like claspers present, lacking setae or papillae; maximally 5-6 mm long and 3 mm wide; their bases filling 16 longitudinally and extending laterally as far as ⅓ bc and mesially including b lines; setae a and usually also b, persistent in this segment. Male pores minute, lateral to setae b, near the bases of the antero-ventral face of the claspers. Female pores inconspicuous, presetal in b lines of 14. Spermathecal pores, externally invisible except where sperm ejected, from internal distribution of spermathecae numbering 3 to 11 on each side in each transverse row in intersegments 16/17-21/22, 22/23; a full row extending from c lines to slightly median of a line; all rows extending to c line but the first 3 rows not reaching a lines.

Cerebral ganglia in 3. Oesophagus very thin walled and vascular in 7-17; thicker walled and not evidently vascular in 18 and 19; gizzard small, soft, but distinct in 6; intestine beginning in 20, with muscular crop-like dilatation in 20-22, 23; typhlosole beginning (very low) in 24, well developed by 27 and full-sized by 29, zigzag, deep and cylindrical at each end of the intestine. Hearts large and moniliform, in 7-11 (connection with dorsal vessel only, verified); dorsal blood vessel apparently continuous onto the pharynx. Slender supra-oesophageal vessel in 17 to 9. Median subneural vessel thick in 27 posteriorly; anteriorly slender; not traceable in front of 17; suboesophageal present above the ventral vessel in 24 anteriorly and probably derived from the subneural. 1 pair of latero-parietals present, in 7 anteriorly. Nephridia commencing in (13?), 14, 15; slender coiled tubules with avesiculate ducts. Large iridescent sperm funnels, (testes?), and free sperm masses in 10 and 11. Seminal vesicles 2 pairs only, in 11 and 12; those of 12 greatly enlarged in 13 the posterior septum of which is displaced posteriorly (to 14/15). Fairly small bushy ovaries and their delicate pleated funnels in 13; ovisacs larger than the ovaries and with numerous large oocytes in 14, connected with the funnels. Very small septal pouches in 15. Spermathecal ampullae visible internally when inseminated; subspherical; located in the segment in front of the pore.

Costa Rica : San Jose.

Cognetti's account has been augmented and emended above from a re-examination of the type specimens (Hamburg Museum V. 3286; Fig. 15. 10A, B). Location of the male pores was not determined in the re-examination. *D. alfari* is distinct from the remainder of the genus in its Central American locality, in possessing claspers and in location of the male pores on 16. There is no evidence against regarding it as congeneric with the other *Drilocrius* species, however.

Drilocrius breymanni (Michaelsen, 1897)
Fig. 15. 9A-C

Criodrilus breymanni MICHAELSEN, 1897 a : 383, Pl. 33, fig. 13, 14.

Drilocrius breymanni MICHAELSEN, 1918: 357; 1933 a : 345. GATES, 1958 a : 3.

l>70 mm, w=3·5-5 mm, s>310. Zygolobous. Anus dorsal. aa=bc<dd. Somatic setae simple pointed, sigmoid, with nodulus distal from the middle and a fine ornamentation consisting of irregularly strewn minute scars which are some-what longer than wide; a mid-body seta 500 μ long and 20 μ wide. Ventral setae of 16 more slender, 680μ long and 28 μ wide, with more pronounced and more extensive ornamentation possibly to be considered genital setae. Nephropores presetal in b lines. Clitellum annular, 14-35 (indistinct). Dome-shaped or thick wing-like or ridge-like protuberances located in bc of 16 and extending into 15 and 17, the ventral surface and sides of these segments to shortly above d and the dorsal surface of 16 with clitellum-like modification. A strongly raised tubercle present on each side in 15 and bearing the male pore at the bottom of a deep transverse slit at the anterior border of which are situated setae ab. Female pores inconspicuous, immediately presetal, median to a lines of 14. Spermathecal pores in 13/14 and 14/15, 3 or 4 on each side in setal lines cd and dorsal-median of this, inconspicuous, nipple-like.

Oesophagus with weak to strong muscular thickening in (5 and) 6, sometimes widened in 6 but not externally recognizable as a gizzard; narrow, with folded epithelium in 7-14 or 12-13; in 14, 15 to 17 somewhat expanding; valve posteriorly in 17; in 18 and 19 with strongly thickened masculature (intestinal gizzard); a large dorsal typhlosole from 21-22 (23?). Dorsal blood vessel simple; hearts (lateral?) in 7-11; median subneural larger than the ventral vessel anteriorly. Vasa deferentia of a side uniting shortly before their common pore; deeply intraparietal. Seminal vesicles of 12 filling 13 also and constricted by septum 12/13. Septal pouches in 13 and (ovisacs?) 14. Fan shaped ovaries, and funnels, in 13. Spermathecal ampullae partly or wholly projecting into the coelom.

Columbia: Palmyra (Type locality); Bitaco, in muddy bottom and banks of stream entering Rio Bitaco.

Location of spermathecae in *dd* in 13/14 and 14/15 presumably necessitated referring the Bitaco material (Gates, 1958) to this species. Origin of the intestine in 18 agreed with the type description of *breymanni* but not with an emendation based on serial sections of one specimen (Michaelsen, 1933a) giving the origin in *breymanni* as 17. The validity of Michaelsen's emendation for the species as a whole is questionable, however, as it suggested also a posterior heart location in segment 10 which is unknown in the genus and related genera. Intestinal origin is nevertheless in 17 in four other species, and the possibility exists that intraspecific variation from 17 to 18 occurs in *breymanni*.

The internal organs of the remaining type specimen (Hamburg Museum, V. 3844; Fig. 15. 9A) have been previously removed but it is possible to confirm Gates' observation that the subneural vessel is very large even at the anterior end of the body; the recorded location of the spermathecae; and that the male pores each lie in segment 15 in a deep depression on the anterior wall of which are situated the corresponding setae *ab*. Two vasa deferentia are traceable to but not posterior to the depression (left side) and what appears to be a seminal groove runs from it to the base of the dome-shaped tuberculum in 16.

Drilocrius buchwaldi Michaelsen, 1918
Fig. 15. 10C, D

Drilocrius buchwaldi MICHAELSEN, 1918: 359, Pl. II, fig. 36, 37.

$l=120$ mm, $w=4$ mm, $s=264$. Zygolobous. Anus dorso-terminal. Body quadrangular but not dorsally grooved. In segment 12 $aa:ab:bc:cd:dd=5:1:8:1:9.4$; $dd:u=0.27$. Genital setae apparently absent; setae (segment 14) sigmoid, 650 μ long, 40 μ wide. Clitellar tumescence in 20-47 but intersegments as far forward as 16/17 partly obscured dorsally. A pair of very protuberant, roundly pyramidal genital lobes located just above *b* lines in 15-19; with a span of twice the body width; most protuberant in 15 and 16; continued in 17-19 as wide ridges which are low in 19. Intersegmental furrows 16/17-18/19 forming deep transverse furrows incising the posterior region of the lobes, whereas 15/16 is present only dorsally on the lobes. Segments 15-19 3- or 5-annulate. Each lobe is internally widely open to the coelom and the two vasa deferentia of each side penetrate deeply into the internal cavity of the corresponding lobe. Female pores (from parietal penetration of oviducts) presetal in 14; median to *a* lines. Spermathecal pores minute beadlike papillae in 17/18 to 20/21, 5 to 8 on each side in each intersegment, in *cd* to just above *b*, with no apparent correspondence with the setal lines. From segment 8-47 the setal couples are on distinct papillae of which the ventral are largest in 14 and 17, with the exception of 15 and 16 in which the ventral papillae and apparently the setae are absent.

Gizzard distinct, if rudimentary, in 5. Intestine with a small typhlosole. Dorsal blood vessel single. Hearts, in 7-11, those of 9-11 particularly large. Subneural vessel well developed in and behind 16; very attenuated and soon disappearing in front of it. Seminal vesicles of 12 constricted by septum 12/13 so as to appear bipartite. Female organs in the usual location; without ovisacs. Spermathecae minute, narrowly ovoid, completely intraparietal.

Equador: Guayaquil (in Sabonne swamp).

Except for their larger size, the genital lobes of *buchwaldi* are closely similar to those of *breymanni* but a careful study of the course of the vasa deferentia reveals that in *breymanni* these open to the surface in front of the lobes while in *buchwaldi* they penetrate the lobes. Further differences from *breymanni* are the location of the spermathecal pores posterior to the lobes and the attenuation of the subneural vessel in the forebody. Little else of significance is discernible in the remaining type specimen as the internal organs have previously been removed.

Fig. 15. 9C and D are redrawn from Michaelsen, 1918.

Drilocrius burgeri (Michaelsen, 1900)
Fig. 15. 9D, E

Criodrilus burgeri MICHAELSEN, 1900 c: 236.

Drilocrius burgeri (Michaelsen). MICHAELSEN, 1918: 356.

l=120 mm, w=5·5 mm, s=220. Prolobous. In 12 $aa:ab:bc:cd:dd$=6:1:6·6: 1:6·4; $dd:u$=0·22. Clitellum, at least centrally, annular, limits indistinct, possibly 21-50 (=30 segments). Two ear-like male genital lobes on each side separated transversely by a furrow; the anterior lobe, in 15, projecting forward over 14 and 15 when well developed; the posterior, in 16 and 17; the lobes bordered medially by a (seminal?) groove in *ab* lines. Male pores (?). Female pores inconspicuous minute slits anteriorly in 14 median to *a* lines. Spermathecal pores inconspicuous white points in 13/14, 7-10 on each side from *b*, or slightly below, to *c* or almost to the middorsal line.

Gizzard rudimentary in 6. Intestine originating in 18; either without or with slight muscular thickening in 18 and 19. Hearts in 7-11; the dorsal vessel ending anteriorly with those of 7. Supra-oesophageal vessel originating posteriorly in 16, continued onto the pharynx. Median subneural vessel in 18 posteriad, continued as an unpaired (?) suboesophageal vessel closely adherent to the oesophagus, but separable, in 17 forward. Nephridia commencing in 10, small in 10 and 11 but all with thick peritoneal investment; discharging by slender adiverticulate ducts in front of setae *c;* funnel single, preseptal. Seminal vesicles of 11 and 12, or only those of 12 extending into 13. Tufted, or subspherical slightly incised ovaries and funnels in 13; ovisacs in 14 opening into 13 immediately dorsomedian of the funnels by wide pores. Septal pouches in 15,

16 and 17. Spermathecae subspherical, entirely intraparietal.

Columbia: Bogota (in a stream and in moist earth). Argentina: Arroyo Pastora (collector L. Cernosvitov, 13. iii. 1932, 1 specimen—new record).

Though conspecificity of Columbian and Argentinian material seems questionable, the material from Argentina (Fig. 15. 9E) agrees closely with *D. burgeri*, the only noteworthy departure from the types being absence of spermathecae above the dorsal setae and the anterior thickening of the intestinal musculature, and it must at present be included in the Columbian species.

Drilocrius hummelincki Michaelsen, 1933
Fig. 15. 9G

Drilocrius hummelincki MICHAELSEN, 1933 a: 344, Pl. I, fig. 11.

$l=110$ mm, $w=2.7$ mm, $s=200$. Zygolobous. Cross section quadrangular in 7 posteriorly, with a shallow dorsal gutter. Anus a long, dorsal, subterminal slit through several short segments. Setae: in segment 12 $aa:ab:bc:cd:dd=3.8:$ $1:5.8:1:8.8$; $dd:u=0.3$; in the midbody $c.$ 500 μ long by 22 μ wide, slenderly sigmoid with indistinct node somewhat ectal of the middle; ornamentation (ventral couple of 21) $c.$ 8 circumferential rows of shaggy scales near the distal extremity; appearing notched in profile. Nephropores minute, presetal in b lines. Clitellum annular, $\frac{1}{2}$ 13-38, reddish; intersegmental furrows slightly obscured. Male pores behind the ventral setal couples of 15, immediately in front of intersegment 15/16; wide transverse slits extending a little over a and b lines. Female pores inconspicuous, antero-medial to setae a of 14. Spermathecal pores in intersegments 13/14 and 14/15, 3 or 5 on each side in an intersegment, in and dorsal to cd or in bc and ab lines also. A stout ridge-like superficially puckered puberty tubercle on each side in $\frac{1}{2}$ 15-17, 18 or 19 between b and c lines, highest in 16. Ventral setal couples of 12-19 on clearly raised, transverse cushions; possibly modified as genital setae but not appearing distinct in balsam sections.

Oesophagus with slight muscular thickening in 5 and 6, hardly sufficient to be termed rudimentary gizzards; sinusoidal region ending at 13; thicker walled in 14, 15 and especially 16. Intestine commencing in 17, with a definite oeso-phageal valve at 16/17. Intestinal musculature slightly thickened in 17 and 18, the intestine here widened and hardly constricted by septum 17/18 so as to appear like a gizzard. Typhlosole commencing at 19, $\frac{1}{2}$ 21, well developed by 22. Hearts in 7-11; the last 3 pairs at least, latero-oesophageal. Dorsal blood vessel ending anteriorly at septum 7/8 with the hearts of 7. Supra-oesophageal vessel ending posteriorly in 13, extending as a large vessel to the anterior end of the body; receiving a pair of large transverse parietal vessels in each segment from 9 anteriorly. Subneural vessel large in 16 posteriorly, receiving a pair of large transverse parietal vessels in 16 and lesser ones in anterior and posterior seg-ments; thinner but still considerable by 12; no longer visible in 9. Nephridia rudimentary in 14; well developed in 15; ducts avesiculate and without caeca. Testes and large iridescent funnels in 10 and 11; seminal vesicles 3 pairs, in 10-12; those in 12 enlarging to an equal size in 13 and narrowly constricted by septum 12/13. Ovaries each a broad, flattened lobe with numerous large but indistinctly visible oocytes and elongated, minutely ruffled oval funnels in 13. Ovisacs racemose in 14 dependent from the funnels.

Aruba, Fontein. (In slime at the bottom of an outlet from a pool with more than 100, less than 1000 mg Cl/litre.)

Re-examination of a mature type specimen (Hamburg Museum V. 11711; Fig. 15. 9G) permits the above extension of Michaelsen's account. A notable departure is the occurrence of spermathecae ventral to setal lines *cd* and their absence dorsal to these, in the re-examined specimen.

D. hummelincki is morphologically exceedingly close to *D. breymanni* from which it differs in the lesser development (distension?) of the sub-neural vessel anteriorly and intestinal origin in 17 (reported for *breymanni* by Michaelsen, 1933 a, however). Observed differences in the form of the genital lobes and the more obvious development of protuberances around the ventral setae in the genital field in *hummelincki* are probably not significant. Synonymy must be deferred, however, at least until larger series are obtained.

Drilocrius iheringi (Michaelsen, 1895)
Fig. 15. 9F, H-I

Criodrilus iheringi MICHAELSEN, 1895: 5, Fig. 15.

Drilocrius iheringi (Michaelsen). MICHAELSEN, 1918: 353; 1926 b: 300.

Drilocrius dreheri MICHAELSEN, 1926 b: 301, Fig. E.

$l=120$ mm, $w=6$ mm, $s=125$. Zygolobous. Mid and hindbody with deep dorsal groove. In segment 12 $aa:ab:bc:cd:dd=6·5:1·0:6·7:1·2:3·3$; $dd:u=0·12$. A seta of the midbody 500 μ long, the node located slightly distal of the middle, 35 μ wide; distally ornamented with minute scars, some irregularly strewn, others arranged in several jagged or wavy transverse rows. Setae *a* of segment 15 similar. Setae *ab* of 13 and 14 (always?) at maturity on papillae; not apparently modified as genital setae although 650 μ long. Nephropores fine white points midway between the setal arc and the anterior border of their segments, in *b* lines Clitellum indistinct, 16-42 ($=27$ segments). The region of setal lines *a* in the entire length of 15 and 16 tumid, the tumid region bounded laterally by a deep groove or cleft from which a rounded strongly protuberant penis-like lobe projects in 15; this lobe separated by a deep transverse furrow from a corresponding irregular moundlike protuberance in 16. Male pores somewhat lateral of setae *b*, in the grooves in 15, each probably discharging into a (seminal?) groove which longitudinally incises the median border of each male lobe. Female pores (from sections) apparently in 13/14 above *b* lines. Spermathecal pores externally unrecognizable, 7-9 on each side in intersegmental furrow 13/14, in *bc*. Spermatophores external, flat, irregularly incised.

Oesophagus without distinct muscular thickening; with at most, a very rudimentary gizzard in 5 and 6; narrow in 5 to 13; especially posteriorly strongly folded and vascularized; in 14 to 16 losing its folding and vascularization but with thickened musculature. Intestine beginning with abrupt expansion in 17; its musculature anteriorly unthickened or thickened in 17-20 and especially in 18-½ 20. Last hearts in 11. Subneural vessel present (re-examination of mid-and hind-body). Seminal vesicles of 12 extending into 13 and deeply constricted by septum 12/13. Spermathecae wholly intraparietal.

Brazil: Sao Paulo (Pericicaba River, the type locality) and Franca in the Rio Grande Basin.

Drilocrius iheringi, the type species of the genus, was erected for three aclitellate specimens. *D. dreheri* is here synonymized with it because Michaelsen (1926 b: 301; see Fig. 15. 9F) concluded that its genital lobes might represent a mature condition of those observed in *iheringi* and because the only distinctive features of *dreheri* given were the distribution of septal thickening and greater development of the posterior seminal vesicles. The anterior end of the intestine was described as unthickened in the types but estimation of thickening is subjective and increased muscularization has elsewhere been reported as present or absent intraspecifically (see *D. burgeri*).

Of the two available type specimens of *iheringi* (Hamburg Museum V. 170) one (Fig. 15. 9H) has previously been dissected and has had the internal organs removed. The other lacks external male genital lobes and is not therefore ascribable with certainty to this species. The types of *dreheri* are not traceable.

GENUS **Glyphidrilocrius** NOV. GEN.

Type and only species: Drilocrius ehrhardti MICHAELSEN

Zygolobous. Body quadrangular in cross section; grooved dorsally. Setae closely paired; $dd : u\langle0·3$; sigmoid with distinct node and distal circumferential sculpturing. Nephropores slightly median to b lines. Clitellum annular, embracing more than 30 segments and extending shortly anterior to the male pores. Stout wing like lateral ridges in bc extending through several segments behind the anterior border of the clitellum and circumscribing the male pores in 17. Female pores median to setae a of 14. Spermathecal pores multiple in each intersegment occupied; post testicular.

Oesophagus lacking anterior muscular thickening but with thickened walls (straining apparatus?) posteriorly; elongated. Intestine with well developed gizzard-like thickening of its musculature anteriorly. A large dorsal typhlosole present. Hearts in 8-13; all latero-oesophageal; dorsal vessel terminating anteriorly with those of 8. Supra-oesophageal vessel continuous onto the pharynx. A pair of latero-parietals present; communicating with a pair of suboesophageal vessels. Subneural vessel totally absent. Nephridia stomate, without bladders or caeca; absent from the preclitellar region. Testes and funnels in 10 and 11; seminal vesicles restricted to 11 and 12. Ovaries lamina-like, in 13. Ovisacs in 14. Vasa deferentia and spermathecae intraparietal.

Brazil.

The internal anatomy of the type-species has been examined for the first time in the present study (holotype Hamburg Museum, V. 9810). The examination reveals sufficient overall phenetic divergence from the morphology of the type species of *Drilocrius* to warrant recognition of a new genus. Notable differences from *Drilocrius* are: absence of a subneural vessel, of posterior enlargements of the seminal vesicles of 12 in segment 13; presence of hearts in 13; location of the male pores in 17 and the presence of lateral alae. The great extent of the oesophagus, which is longer than that known in any other species of the Oligochaeta, though remarkable, may be a purely specific character. It is variable interspecifically in, for instance, *Glyphidrilus* and *Alma,* in each of which there appears to be an evolutionary tendency to elongation of the oesophagus.

Glyphidrilocrius ehrhardti (Michaelsen, 1926)
Fig. 15. 10E-G

Drilocrius ehrhardti MICHAELSEN, 1926: 256, Fig. F.

$l=230+$mm (posterior amputee), $w=5$ mm, $s=245+$. Segments anteriorly triannulate (weakly hexannulate). Lateral lines absent. Forebody very narrowly and uniformly tapering to the small zygolobous prostomium. Cross section quadrangular from segment 4 posteriorly, soon becoming concave on all sides; the dorsal surface markedly wider than the ventral. A broad flat bottomed dorsal gutter in *c.* 16-40, its lateral walls, bearing the dorsal setal couples, meeting over it behind the male field. Setae closely paired throughout; in segment 12 $aa:ab:bc:cd:dd=5\cdot6:1\cdot0:7\cdot8:1\cdot0:8\cdot5$, $dd:u=0\cdot25$. Ventral setal couples of 14 and 15 each on a transversely oval papilla within a larger rim-like protuberance; dorsal setal couples of 15 and 16 on similar rimmed papillae. Setae sigmoid, with distinct node; distally ornamented by transverse jagged ridges which form about 8 circumferential but (especially "laterally") interrupted circlets which give the outline of the seta a notched appearance. (Genital (?) setae broken off, form and sculpturing not determinable.) Nephropores visible on the clitellum, minute points in *ab*, slightly median of *b*, anteriorly in their segments. Stout wing-like longitudinal lateral lobes in *bc*, extending from $\frac{1}{2}$15-$\frac{1}{2}$21; highest in 17; each circumscribing a large suboval depression which fills 17 and part of 18 longitudinally and contains the ventral setal couples (only *b* detected) of 17. Male pores in the depressions, equatorial (?) in 17 lateral of *b* lines but not externally recognizable. Clitellum annular, 15-46; intersegmental furrows less distinct than elsewhere. Female pores on well marked, rounded papillae, median to setae *a* of 14. Spermathecal pores 4 per intersegment, in *cd* lines, in 17/18-21/22.

Oesophagus lacking anterior muscular thickening but thickened and not evidently vascularized in 35-37; transversely vascularized and narrowly sacculated to as for posteriorly as 34. Intestine commencing in 38 (only slightly more constricted at 37/38 than at adjacent intersegments); moniliform in 38-40 in which, especially in 39, its walls are strongly muscularized; thin walled in 41 posteriorly; typhlosole rounded, beginning in 43; at first low; in the hindbody almost filling the intestinal lumen. Hearts elongate moniliform, in 8-13, all laterooesophageal, increasing in size posteriorwards. Dorsal vessel ending anteriorly with the hearts of 8. Supra-oesophageal vessel ending posteriorly, in 34, where it appears to join the dorsal vessel; anteriorly continuous onto the pharynx; as large as or larger than the dorsal vessel and lying to one side of the latter;

receiving vessels from the oesophagus in 24-34. Subneural vessel absent through-out. A pair of posterior latero-parietals (distant from the nerve cord) uniting in 36 to form the large left suboesophageal vessel; the right suboesophageal vessel small and lacking this connection. Latero-parietals also continuing, but small, anterior to their junction. Suboesophageal vessels each receiving a trans-verse parietal vessel posteriorly in each segment.

Nephridia commencing in 15; preseptal funnels large and three-lobed; ducts avesiculate and lacking caeca. Sperm funnels large and vase-shaped, with whitish iridescence in 10 and 11; seminal vesicles restricted to 11 and 12. Vasa deferentia intra-parietal, penetrating the parietes at mid 17 well lateral of b, anterior to the middle of a glandular mass corresponding to the external depression. Ovaries in 13, extensive laminae of parallel chains of oocytes. Ovisacs large, morula-like, in 14. Bunches of glandular cells located in the follicular pits but scarcely pro-tuberant into the coelom around the ventral setal couples of segments 14, 15, 17 and 18 and also around setae cd of 15 and 16. Spermathecae intra-parietal but pushing the parietal peritoneum into the coelom; clavate or three-lobed or irregular, sometimes diverticulate.

Brazil: Manacapuru on the Amazon (Solimoes), above Manaos (in a stream).

TRIBE CRIODRILINI NOV.

Definition and distribution as for the type and only genus, *Criodrilus* Hoffmeister, 1845.

GENUS **Criodrilus** HOFFMEISTER, 1845

Type species: Criodrilus lacuum HOFFMEISTER

Zygolobous. Body quadrangular in cross section. Anus dorsal, subterminal. Lateral lines absent. Somatic setae ornamented by circumferential ridges or by rows of crescentic scars; closely paired; $dd:u\langle 0.3$. Genital setae elongated; distally with 4 longitudinal grooves which may coexist proximally with sculptur-ing similar to that of somatic setae. Nephropores (always?) in b lines. Clitellum extensive, just including the male pores anteriorly. Male pores posterior in 15 on conspicuous porophores which extend into adjacent segments. Female pores in ab lines. Spermathecae absent; horn-shaped external spermatophores developed. Dorsal pores absent.

Suprapharyngeal ganglia in 1 and 2. Oesophageal musculature slightly thickened in the region of 5-7. Intestine commencing in 13; with some thicken-ing of the musculature at or near its anterior end. Dorsal typhlosole present. Hearts in 7-11. Supra-oesophageal vessels absent. A single pair of latero-parietals present, originating from the dorsal vessel. Median subneural present through-out. Nephridia absent from anterior segments almost to the clitellum. Holandric; testis-sacs absent. Seminal vesicles in 9-12. Vasa deferentia intra-parietal; large bursae or prostates present at the male pores. Metagynous with diffuse oocytes. Ovisacs present. Cocoons elongate-fusiform, with several to many embryos..

Holarctic to as far East as Syria and Palestine.

Michaelsen (1895, 1897 a, 1900 c) and Cognetti (1904 b) added a number of S. and Central American species to the genus but the additional species were later (Michaelson, 1918) transferred to a new genus, *Drilocrius*, as they differed from *lacuum* in possessing spermathecae. Additional grounds for their separation are advanced in the present work. *Criodrilus* remained monotypic until Georgevitch (1950) added the dubious species *C. ochridensis*. Data from the latter species have not been included in the generic description as even its generic status must remain uncertain until the species is adequately redescribed. The affinities of *Criodrilus* are discussed on p. 186.

Criodrilus lacuum Hoffmeister, 1845
Fig. 15. 11A-H

Criodrilus lacuum HOFFMEISTER, 1845: 41, Pl. III, fig. 9a-c.

Criodrilus lacuum Hoffmeister: PANCERI, 1875: 238. HATSCHEK, 1878: 1. ÖRLEY, 1881: 602, Fig. 10a-f; 1887: 552, Pl. XXXVIII, fig. 1-8. VEJDOVSKY, 1884: 57, Pl. X, fig. 21, Pl. XIII, fig. 12-24, Pl. XIV, fig. 1-5. ROSA, 1886 a: 681; 1886 b: 1. BENHAM, 1887: 561, Pl. XXXVIII, fig. 9-19. COLLIN, 1888: 471, Pl. XXIII. BEDDARD, 1895: 667. MICHAELSEN, 1900 a: 468; 1909 b: 59, Fig. 108, 109; 1918: 373; 1926: 351. JANDA, 1926: 1. UDE, 1929: 111, Fig. 155-157. CERNOSVITOV, 1938: 549. GEORGEVITCH, 1950: 79. MCKEY-FENDER AND MACNAB, 1953: 373. GRAFF, 1962: 369, Fig. 1.

Criodrilus. VEJDOVSKY, 1879: 358.

?*Criodrilus lacuum* Hoffmeister. STEPHENSON, 1913: 53.

Criodrilus lacuum var. *macedonica* UDE, 1922: 161.

Criodrilus lacuum Hoffmeister. (part) STEPHENSON, 1923: 495.

[non] *Criodrilus lacuum* Hoffmeister. STEPHENSON, 1914: 256.

$l=$85-400 mm, $w=$3-10 mm, $s=$150-450. Zygolobous. Form quadrangular in cross section from approximately segment 6-9 posteriorly; grooved dorsally in and behind the clitellar region. Anus dorsal, subterminal. Body wall musculature very thick, fasciculate. Lateral lines absent. Setae fairly closely paired throughout; in segment 12 $aa:ab:bc:cd:dd=$6·2:1·2:8·3:1·0:10·5, $dd:u=$0·28. Nephropores in b (or a?) lines. Ventral setal couples of segments, 9, 10, 11-12, 13 and 17, 18 to 19, 20 or 23 each on an ovoid prominence; whitish glandular fields largely or wholly limited to setae a, however. Setae a of at least segments 12, 13, 16-18 modified as genital setae: approximately the terminal fourth, distal to the node, bearing 4 deep longitudinal grooves the proximal ends of which grade into irregular transverse jagged ridges which in profile have the appearance of long, distally directed teeth; the distal-most of these tooth-rows each forming a crescent (with proximal convexity) limited to the corresponding groove, the more proximal rows progressively less crescentic and finally forming complete, if irregular, circumferential circlets. Somatic setae unornamented in Macedonian populations. Setae b in the segments bearing the genital setae resembling those of other segments, being ornamented over approximately the distal half of the portion distal to the node by as many as 13 complete serrated circular ridges or with 2-4 alternating rows of crescentic scars. Genital setae 842-910 μ long and 26-28 μ wide at the node; unmodified (somatic) setae 560-790 μ long and 38-45 μ wide. Clitellum indistinctly delimited anteriorly and posteriorly; setae and intersegmental furrows not obscured; annular, embracing 14, 15, 16 to 45, 47 (=30, 32, 34 segments), but the epidermis hypertrophied even at the anterior end of the body. Consisting histologically of an outer columnar epidermis continuous with that of the general body surface and 3 or 4 layers of club-shaped glandular cells with basal nucleus and filled with highly refractive small spherical globules. Male porophores very strongly protuberant, transversely placed, ellipsoidal mounds the rim-like bases of which fill segments 15 and 16 longitudinally and push intersegments 14/15 and 16/17 apart (the pores sometimes translocated 1 segment further posteriorly); the base of each porophore medianly including seta a of segment 15 which it may translocate appreciably towards the ventral midline and laterally extending to about mid bc. The male pore a transverse cleft lateral to line, deeply bisecting the summit of the male porophore and in line with intersegment 15/16 which is obscured ventrally. The ventral body wall arched upwards and puckered between the porophores. Spermatophores curved, horn-shaped, hard but flexible structures approximately 1 mm long and maximally about 0·4 mm wide, at the expanded base; one to several attached in the vicinity of the genital field in or above b lines. Female pores each a small transverse slit in intersegmental furrow 14/15 in a lines or between a and b lines; sometimes on swellings which are smaller than those bearing the male pores. Dorsal and spermathecal pores absent.

Oesophageal musculature appreciably thickened in 5-6, 7, in which the oesophagus may appear widened and glossy. Intestine commencing abruptly, with distinct oesophageal valve, anteriorly in 13; usually thicker walled in 15, 16, 18, 19-20, 21 (=5 or 6 segments), sometimes, however, appearing thicker in 13 and 14 than further posteriorly. Dorsal typhlosole beginning, but very low, in 17 or in 20, 20/21; well developed by 18 or 21-23; disappearing gradually in the vicinity of 75. Hearts in (6) 7-11; none latero-oesophageal (supra-oesophageal vessel absent). Dorsal vessel continuous onto the pharynx. Latero-parietals a pair arising from the dorsal vessel in 12. Median subneural present throughout; adherent to the nerve cord; lateroneural vessels absent. Holonephric; nephridia commencing in 13 though rudiments may be present in 12. Ducts avesiculate but wide, entering the parietes immediately in front of setae ab. Testes free; digitate, or delicate, transversely slightly plicate lobes in 10 and 11. Large, much convoluted sperm funnels in 10 and 11. Seminal vesicles 4 pairs, in 9-12. Vasa deferentia concealed deeply in the unusually thick body wall musculature, emerging in the coelom of 15 where that of each side of the body joins the antero-dorsal aspect of a large hemispherical male bursa or prostate gland which is restricted to 15. The gland consisting of cells similar to and continuous with those forming the epidermis of the clitellum; the muscular layers of the body wall here thin and covering the inner surface of the gland; the vas deferens continuous through the substance of the gland to the male pore. Ovaries each a solid, tongue-like or paddle-shaped lobe showing few external indications of oocytes, almost filling the length of segment 13. Oviducal funnels rosette-like, rather thick but small. Ovisacs in 14, at maturity at least as large as the ovaries; containing large oocytes; projecting into 14 from septum 13/14 and closely associated with but apparently not directly connected with the funnels. Supra-pharyngeal ganglion in segments 1 and 2.

Europe: Germany (Type-locality, "Tegel-See"); Austria; France; Yugoslavia; Hungary; Italy; Poland; S. Russia. Syria. Palestine. U.S.A.: Baltimore (in plant pots). Usually in fresh or brackish water.

Örley (1887) stated that there were two pairs of spermathecae which appeared to open on the ventral surface between segments 9 and 10, and 10 and 11 but no other worker has observed these organs and Benham's explanation that Örley mistook testes for spermathecae is accepted here.

Conspecificity of material from Lake Tiberias, Palestine, described by Stephenson (1913) is not entirely certain. The different form of the sperma-tophores (compressed ovals) is probably not significant as it would seem that the horn-shaped spermatophores of *lacuum* must break off on passage of the cocoon over them, a view supported by observation of portions of spermatophores within the cocoons by Örley (1887). Hearts in 6-11 may be due to interpolation of a supernumerary segment as evidenced by trans-location of the male pores to 16 in some specimens. Apparent differences in the extent of the clitellum (maximum extent 15-37) are probably un-important.

Specimens from Lake Chilka (Pakistan) referred to *C. lacuum* by Stephen-son (1914) though immature, are here excluded. The wide pairing of the setae necessitates exclusion and suggests that they may have been immature specimens of a *Glyphidrilus* species.

It is possible that a sub-species may have to be recognized for Syrian specimens. In those examined by Michaelsen (1918), and in de Kerville's material examined by the author (Fig. 15. 11G, H), whitish glandular fields surround setae *a* as far as 23 posteriorly and contrast conspicuously with the reddish brown clitellum whereas in European material no such colour differentiation has been seen and glandular protuberances have not been observed behind segment 20. Furthermore, in the Syrian material the typhlosole begins in 20 (3 re-examined specimens) while in the single type (?) specimen dissected (B.M.N.H. 1923.12.31.200-202 Fig. 15. 11A-C, E-F) it begins in 17. Larger series must be examined before the validity of erect-ing subspecies can be established, however.

Fig. 15. 11D is redrawn from Vejdovský (1884).

SPECIES INQUIRENDAE

Criodrilus? ochridensis Georgevitch, 1950

Criodrilus ochridensis GEORGEVITCH, 1950: 77-79, Fig.

l=50-150 mm, s=100-150. Dorsal surface more or less concave; ventral surface similar or flattened; hindbody almost square in cross section; anterior end almost circular. Anus terminal. Prolobous or less frequently zygolobous. Male pores very protuberant on 15 which is the widest segment. Puberty protuberances, usually a pair, in each of several segments, notably 12-14. Body wall thickened and furrowed ventrally in 14-17; elsewhere smooth and thin (immature).

Septa thin. "Pharynx" large and muscular extending as far as 7; narrow oesophagus in 8-12; muscular stomach in 13-20, followed by an intestine of the same width. Hearts 5 pairs, in 9-13. Testes free in 10 and 11. 4 pairs of simple or lobed seminal vesicles, the first and last pairs the largest. "Seminal receptacles" two pairs, close together or distant, the first in 12, the last as far as 15. Ovaries small, in 13.

Yugoslavia: Lake Ochrid (1-4 m depth).

The variable position of the "receptacles seminaux" and the fact that Georgevitch included *ochridensis* in a genus defined as lacking spermathecae suggest that the receptacles were external spermatophores. If they were spermathecae *ochridensis* must be excluded from *Criodrilus* s. Michaelsen. Presence of spermathecae behind the testis segments would indicate Glossoscolecid affinities but apart from this the generic status of *ochridensis* is indeterminable from Georgevitch's description. Derivation by sympatric speciation (Stankovic, 1960) from *Criodrilus lacuum* is clearly insupportable.

SUB-FAMILY BIWADRILINAE NOV.

Type and only genus: Biwadrilus GEN. NOV.

Lateral lines present between the ventral and dorsal setal couples. Grooved dorsally but cross section not quadrangular. Anus dorsal, subterminal. Dorsal median intersetal distance only about 0·2 of the circumference of the body (*dd* $c. = 0·2\ u$); genital setae present. Clitellum extensive, immediately behind the male pores. Male pores on conspicuous porophores in 13. Female pores in 14. Spermathecal and dorsal pores absent.

Cerebral ganglia in segments 1 and 2. Calciferous glands, typhlosoles and gizzard like thickening of the alimentary musculature lacking. Supra-oesophageal blood vessel present. Subneural vessels absent. Latero-parietal vessels 2 on each side, one originating from the dorsal the other from the suboesophageal vessel. Nephridia holonephridia; absent from the forebody; lacking terminal bladder, caecum or sphincter; all exonephric. Holandric; testis-sacs absent. Vasa deferentia intra-parietal. Lobular masses of prostatic glands discharging near the male pores together with a pair of prostate-like setal glands. Ovisacs restricted to 14 but an extensive subenteric septal pouch, arising further anteriorly, may be present.

Holarctic: Japan.

As only a single species is known in the sub-family it is not possible to determine with any certainty which characters have familial, generic or purely specific significance.

The single known species has been very fully described by Nagase and Nomura (1937) under the name *Criodrilus miyashitai*. They distinguished *C. miyashitai* from *C. bathybates* Stephenson by the presence of genital setae in segment 13 from which ventral setae were absent in Stephenson's immature specimens. Yamaguchi (1953) demonstrated genital setae in specimens from Lake Biwa, the type locality of *bathybates*, but only unmodified setae in immature specimens. As the mature specimens were identical with *miyashitai* but had cocoons identical with those of *bathybates*, he concluded that *miyashitai* was a synonym of *bathybates*. Nothing further in Stephenson's account of *bathybates* militates against inclusion of *miyashitai*; restriction of seminal vesicles to segment 12 would seem, in view of the holandric condition, to be an expression of the immaturity of the types.

Criodrilus bathybates is here made the type of the new genus *Biwadrilus* because of the following differences from *Criodrilus lacuum*: absence of ornamentation of the somatic setae; absence of grooving of the genital setae (the sole diagnostic character of the Criodrilinae sensu Stephenson (1930: 903) being, supposedly, genital setae of the grooved type); presence of a supra-oesophageal vessel and of two pairs of latero-parietal vessels; possession of a sub-enteric septal pouch; presence of copulatory (genital seta) glands; and presence of lateral lines. The location of the male pores on segment 13 is a further difference and is presumed to be a primitive condition, in view of the unspecialized anatomy of the species, and the relict-like geographical distribution. The same location of the male pores is seen in the Alluroididae, in which *Brinkhurstia americanus* has remarkably similar male terminalia. This accords with the view (Michaelsen, 1919; Omodeo, 1956), which is supported in the present work, that the origins of the Glossoscolecidae are probably to be sought in a Haplotaxid stock of which the Alluroididae are little modified proandric descendants.

Glossoscolecid affinities are indicated by a group of characters: the extensive multilayered clitellum; the location of the male pores near its anterior limit; the copulatory glands, in association with the genital setae, which resemble the setal glands of *Microchaetus*; and the intraparietal sperm ducts. The subenteric septal pouch has been observed also in *Microchaetus saxatilis* (present study). Inclusion in the Criodrilinae by Stephenson presumably was prompted by the absence of gizzards and of spermathecae and the appearance of the male porophores but these negative characters cannot be taken to indicate especially close phylogenetic proximity to *Criodrilus*.

Lumbricid affinities, hinted at by Stephenson (1930: 911) are contraindicated by a number of characters, including the anterior location of the clitellum and the intraparietal male ducts. The affinities of the genus are further discussed in Chapter 2.

Phenetically *Biwadrilus* appears to be closer to *Sparganophilus* than to other Glossoscolecids, notable similarities of the two genera being presence of two pairs of latero-parietal vessels, their origin in segment 14, and the absence of a subneural vessel. Reported presence of lateral lines, absence of spermathecae, the anterior location of the brain and other peculiarities of *Biwadrilus* at present necessitate placing it in a separate and apparently very primitive sub-family, however.

GENUS **Biwadrilus** NOV.

Type and only species: Criodrilus bathybates STEPHENSON

As for the Biwadrilinae. Also: setae closely paired; genital setae situated at the male pores, bifid, not grooved. Somatic setae unornamented. Latero-parietals joining the suboesophageal and dorsal vessels in 14. Testis sacs absent; seminal vesicles in 11 and 12. A copulatory (setal?) gland at each male pore.

Japan: in lakes and rivers.

Biwadrilus bathybates (Stephenson, 1917) n. comb.
Fig. 15. 12A-H

Criodrilus bathybates STEPHENSON, 1917 e: 96, Pl. IV, fig. 8.

Criodrilus miyashitai NAGASE et NOMURA, 1937: 361, Fig. 1-43.

Criodrilus bathybates Stephenson. YAMAGUCHI, 1953: 309.

l=200-300 mm, w=2-4 mm, s=250. Zygolobous. Cross section at the male porophores crescentic with concavity ventral; in front of them rounded; at the hind end elliptical with the short axis vertical or flattened with dorsal groove. Clitellum annular, 14 (ventrally), $\frac{1}{2}$14, 15 (dorsally) to 31, 34. Male genital porophores each a quadrangular pyramid, its apex above b in 13, its base in 12-14; each when fully developed divided apically into an anterior and a posterior papilla between which is a transverse slit, 0·3 mm or more deep, in the dorsal part of which lies the dorsally directed male pore. Female pores immediately presetal in ab of segment 14. Setae closely paired; $aa=bc$; dd=1·3 aa in front of segment 13; in the midbody $aa:ab:bc:cd:dd$=6·1:1·3:6·5:1·0:7·0; $dd:u$=0·23 (approximate from a photograph). Somatic setae without ornamentation, the node at the distal third; dorsal setae 500-550 μ long; ventral setae 370-450 μ long; node of largest setae 40 μ wide; ventral setae of 13 replaced by genital setae; the latter lying in the slit between the two male papillae and just inside the opening of the copulatory gland; each genital seta 500-570 μ long and only 13 μ wide; an inconspicuous smooth node at its proximal third; the distal end abruptly bent and two pronged; (2), 4, (6) per bundle (Grooving not mentioned and presumably absent; the bifurcate form certainly non-lumbricine).

Buccal cavity in 1, pharynx in 2-$\frac{1}{2}$4, pharyngeal glands in 4-8; oesophagus beginning at $\frac{1}{2}$4 with a dorsal diverticulum similar to that in *Criodrilus lacuum*; chloragogen usually increasing in 11 or 12; intestinal origin indefinite, in the region of 11-13; typhlosole absent. Dorsal and ventral blood vessels present throughout; supra-oesophageal vessel distinct, and suboesophageal indistinct, anterior to 9; subneural absent. Latero-parietals 2 on each side, running along the lateral lines; the dorsal thicker, originating from the dorsal vessel, the ventral from the suboesophageal; both originating posteriorly in 14 and reaching the prostomium. Hearts in 7-11 (5 pairs), forming tortuous and extensive lateral loops, dorsoventral commissures in 5 and 6 similarly valvular and therefore heart-like. Mature nephridia in 15 posteriorly; nephrostome meganephridiostomal; duct with minute terminal ampulla discharging in front of setae *ab*. Testes and funnels free in 10 and 11; seminal vesicles 2 pairs, in 11 and 12, the two sperm ducts of a side intra-parietal for much of their lengths, uniting only at the base of the conical male pore. Prostate glands consisting of numerous lobules with branching ducts, bundles of ducts and common ducts opening into the male slit just ventral to the common male pore. A large cylindrical "copulation gland" (setal gland of the *Microchaetus*-type?) on either side in 13 opening into the male slit ventrally to the prostate orifices; and just externally to the genital setae; each gland with a terminal duct and a glandular portion consisting of outer peritoneum, a middle muscular-vascular layer, an inner glandular layer with 3 types of gland cells and a simple lumen. Lobed ovaries, and funnels, free in 13; ovisacs in 14. Septal pouch single, median, below the gut, in 6-13 (-16), containing coelomic corpuscles. Spermathecae absent. Spermatophores deposited ventrally near the male pores.

Japan: bottom of Lake Biwa, to 180 feet depth (type locality); and in streams and rice paddies.

Fig. 15. 12 is redrawn from Nagase and Nomura (1937).

SUB-FAMILY SPARGANOPHILINAE

Type and only genus: Sparganophilus BENHAM, 1892

Lateral lines absent. Body rounded, not quadrangular in cross section but capable of forming a dorsal groove. Anus dorsal (or terminal?). Dorsal median intersetal distance less than 0·3 of the circumference ($dd: \mu \langle 0\cdot 3$); setae 8 per segment, single pointed, with transverse grooves but otherwise all unmodified. Male pores 1 pair, intraclitellar, very inconspicuous in intersegmental furrow 18/19 or anteriorly in 19. Clitellum saddle-shaped and bearing tubercula pubertatis. Dorsal pores sporadically present.

Cerebral ganglia in segments 2 and 3. Calciferous glands, typhlosoles and gizzard-like thickening of the alimentary musculature lacking. Intestine commencing in 9. Supra-oesophageal and subneural vessels absent. Latero-parietal vessels 2 on each side, one originating from the dorsal, the other from the ventral vessel. Nephridia absent from the forebody. Holandric; testis-sacs absent. Prostate-like glands may be present in few to many segments. Vasa deferentia intraparietal. Ovaries with a single terminal row of oocytes. Spermathecae intracoelomic, pretisticular; paired or multiple.

Nearctic. Palearctic (by transportation?).

The anterior location of the oesophageal valve (behind the hearts in all other subfamilies) and the origin of the paired latero-parietals from the dorsal and ventral vessels are diagnostic.

Glossoscolecid affinities appear to be confirmed by a group of characters although most of these characters taken individually, are not restricted to or even general in the family. They are: the extensive clitellum and the histology of this; the intraclitellar male pores with longitudinal tubercula pubertatis; the transversely scarred setae; the tendency to transverse multiplication of the adiverticulate spermathecae; the intraparietal sperm ducts, the posterior dorsal groove and the dorsal anus. Prostate-like glands in association with ventral setae are seen also in *Kynotus* and in *Biwadrilus bathybates* (Biwadrilinae).

GENUS **Sparganophilus** BENHAM, 1892

Type species: Sparganophilus tamesis BENHAM

Setae closely paired: *dd*: *u* less than 0·3; sigmoid, single pointed, with or without transverse ridges; none modified as genital setae. Nephropores presetal in *ab*. Clitellum saddle-shaped (but the epidermis tumid ventrally in the clitellar region) occupying 8 to 12 segments in the region of 15-19. Tubercula pubertatis in the clitellar region, ridge-like or composed of a series of papillae on each side, lateral to *b*. Male pores very inconspicuous, intraclitellar, in intersegment 18/19 or anteriorly in 19, ventral or dorsal to the tubercula. Female pores inconspicuous, presetal in *ab* of 14. Spermathecal pores inconspicuous, intersegmental in and dorsal to *cd* lines, in 6/7-8/9 or 5/6 also; a single pair to four pairs per intersegment. Dorsal pores present or absent intraspecifically. Pores of prostate-like glands, if the latter are present, minute in the vicinity of setae *b* in several segments in the clitellar region and sometimes at the *a* follicles in a variable number of anterior segments.

Intestine commencing anteriorly in 9; typhlosole, calciferous glands and gizzard absent. Dorsal and ventral vessels confluent at the anterior end; dorso-ventral commissurals in 2-11, those of 7 or 8-11 moniliform and heart-like; latero-parietals 2 on each side, arising posteriorly in segment 14, one from the dorsal, the other from the ventral vessel. Subneural and supra-oesophageal blood vessels absent. Nephridia holonephridia, lacking a terminal ampulla and caecum; post-septale at maturity beginning in 12-16. Testes and funnels free in 10 and 11, seminal vesicles 2 pairs, in 11 and 12. Vasa deferentia intra-parietal. Ovaries (always of the lumbricid type?) and funnels in 13, the funnels with small ovisacs in 14. Spermathecae adiverticulate, 2-8 in a transverse row, projecting far into the coelom.

England. France. North and Central America from Guatemala and Mexico to Ontario. In lakes, streams and rivers.

Sparganophilus

1. Tubercula pubertatis longitudinal bands commencing in segments 15-19 and extending to 22-24, sometimes with anterior extensions to segment 12. Spermathecae paired, in 7, 8 and 9   *S. tamesis*

— Tubercula pubertatis longitudinal bands or series of papillae in segments 19 to 27 or 28, sometimes with extensions to segment 10.

Spermathecae in segments 6-9; more than 1
pair in each of 7-9   2

2.   Tubercula pubertatis strongly elevated longitudinal strips, broken or depressed intersegmentally. Spermathecae 8 per segment in
7-9; 1 pair in 6   *S. smithi smithi*
—   Tubercula pubertatis each a longitudinal series
of papillae. Spermathecae 4 per segment, in
7-9; 1 pair in 6   *S. smithi sonomae*

Sparganophilus tamesis Benham, 1892
Fig. 1. 5D-F 1. 6A; 15. 13; 15. 14A-C, H-Q, U, V-X

Sparganophilus tamesis BENHAM, 1892: 156, Pl. XIX, fig. 1-11, Pl. XX,
fig. 12-33.

Sparganophilus tamesis Benham. H. MOORE, 1895: 473. EISEN, 1896 a: 1.
MICHAELSEN, 1918: 302. ČERNOSVITOV, 1945: 531, Fig. 20-30.

Sparganophilus eiseni SMITH, 1895: 142.

Sparganophilus eiseni Smith. EISEN, 1896 a: 153. MICHAELSEN, 1900 a:
464; 1918: 304. J. MOORE, 1906: 170. HEIMBURGER, 1915: 283. SMITH,
1915: 556. HAGUE, 1923: 1, Pl. I-IV, fig. 1-19. OLSON, 1928: 8; 1940: 8.
GATES, 1935: 43; 1943: 94; 1965: 12. DAVIES, 1954: 9. MURCHIE, 1956:
67. OMODEO, 1963: 127. HARMAN, 1965: 22, Fig. 1.

Sparganophilus benhami EISEN, 1896 a: 154, Fig. 79-119.

Sparganophilus benhami Eisen. MICHAELSEN, 1900 a: 464.

Sparganophilus benhami guatemalensis EISEN, 1896 a: 155, 167.

Sparganophilus benhami var. *guatemalensis* Eisen. MICHAELSEN, 1900 a:
464; 1918: 305.

Sparganophilus benhami carneus EISEN, 1896 a: 155.

Sparganophilus benhami var. *carnea* Eisen. MICHAELSEN, 1900 a: 464.

Sparganophilus benhami Eisen forma *typica*. MICHAELSEN, 1918: 305.

Helodrilus elongatus FRIEND, 1911: 192.

Sparganophilus elongatus (Friend). FRIEND, 1919: 426; 1921: 137, Fig. 1-3;
1923: 52, Fig. 22, 24, 25. ČERNOSVITOV, 1942 b: 268.

Pelodrilus cuenoti TÉTRY, 1934: 322.

Pelodrilus cuenoti Tétry. TÉTRY, 1938: 188.

l=60-200 mm, w=1·5-5 mm, s=125-260. Zygolobous. Body cylindrical, in preservative often with a deep dorsal groove which may become shallowly V-shaped posteriorly; anus dorsoterminal, often terminal but probably only in posterior amputees. A lateral line sometimes visible at one third bc below c, coincident with an internal interruption of the longitudinal musculature of the body wall. In segment XII, $aa:ab:bc:cd:dd=4\cdot6\text{-}5\cdot1:1\cdot0:5\cdot0\text{-}6\cdot6:1\cdot0\text{-}1\cdot2:8\cdot3\text{-}10\cdot0$; $dd:u=0\cdot28\text{-}0\cdot32$; less closely paired posteriorly. Dorsal setae on the clitellum concealed at maturity. Dorsal and ventral setae similar, 350-400 μ long in the forebody; almost straight, with slightly curved ends and distinct, somewhat distal node; each seta ornamented distally by one to a few very irregular jagged ridges encircling the whole or part of the circumference. Nephropores minute, but usually visible, in front of the ventral setal couples. Clitellum saddle-shaped, commencing on $\frac{1}{2}$ 14-16 and extending to 24-27; its ventral margins considerably above b lines. Tubercula pubertatis tumid, often translucent, longitudinal bands median to the ventral borders of the clitellum, their ventral borders at or including b lines, commencing on $\frac{1}{2}$ 15-19 and extending to $\frac{1}{2}$ 22-24, narrowing at both ends, sometimes continued as ridges anteriorly, as far as $\frac{1}{2}$ 12. Ventral surface of segment 26, and sometimes also 23-25, glandular and with setae ab on indistinct papillae. Pores of prostate-like glands minute, often unrecognizable, the first on (16), 23, (24?), the last on (25), 26, (27), and not infrequently on 1 or more of segments 3-10, immediately lateral of setae b. Setae cd of segment 23 sometimes on conspicuously protuberant papillae. Male pores unrecognizable or inconspicuous, in 19, presetal in bc nearer b lines than c lines, on (or immediately lateral to?) the lateral margins of the tubercula pubertatis. Female pores midway between the setal arc and the anterior border of 14, slightly lateral of a; without bordering tumescence but often fairly conspicuous. Spermathecal pores unrecognizable or inconspicuous, a pair in each of intersegments 6/7-8/9, in cd lines, or in or just ventral to c lines. Dorsal pores absent or, rarely, present in intersegments 1/2-5/6.

Oesophagus ciliated; intestine originating with abrupt expansion in 9, with shorter cilia. Dorsal and ventral blood vessels single, uniting anteriorly, the dorsal thin in front of 9. Two longitudinal latero-parietal vessels present laterally, close together, from the prostomium to the posterior end of segment 14 in which one originates from the dorsal, the other from the ventral vessel. Hearts in 7-11, those of (7), 8-11 moniliform; dorsoventral commissurals also present in 2-6. Subneural and supra-oesophageal vessels absent. Dorso-intestinal vessels beginning in 13 which perhaps marks the beginning of the true intestine. Nephridia with preseptal funnels, their postseptal portions beginning in (12), 13, (14), 15, (16) at maturity (as far forward as segment 3 in embryos); with long, bent, avesiculate duct. Seminal vesicles lobulated, those in 12 the larger and sometimes displacing septum 12/13 posteriorly. Vasa deferentia forming slightly coiled epididymes behind the male funnels; wholly intramural further posteriorly. Prostate-like glands arced or tortuous tubular glands with muscular ducts, in one to as many as 14 segments but usually restricted to 23-26 and sometimes totally absent even at maturity. Ovaries and funnels in 13, the ovaries with a "tail" of a single row of oocytes at the free end as in lumbricids; oviducts opening on 14; ovisacs in 14 communicating with the funnels. Spermathecae 3 pairs, in 7-9, opening at the anterior margins of their segments, each when well developed with an elongate-ovoid ampulla and a more or less distinctly decarcated, poorly muscularized duct of equal or slightly greater length. Maximum recorded length of a spermatheca, 1·2 mm.

England: Goring-on-Thames (Type-locality); Windermere (lake) (Cernosvitov, 1945, and a new record, B.M.N.H. 1968); Pencarrow, Cornwall (*S. elongatus* (Friend), from a slate water-lily tank); Kew Gardens, Surrey (lily house, B.M. (N.H.) 1909.9.16. 1-5, new record). (Also found in Cheshire Meres-R.O.B.)

N. and Central America: Guatemala and Mexico northwards to Michigan (St. Mary's River, new record) and Ontario. Locally abundant in submerged mud or gravel at the margins and at the bottoms of river systems and lakes; in Windermere favouring localities with an organic substrate, e.g. *Scirpus* swamps.

Erection of *Sparganophilus eiseni* by Smith (1895) and its recognition in numerous N. American publications stemmed from an omission in Benham's type-description of *S. tamesis*. Neither Benham (1892; Goring-on-Thames specimens) nor Cernosvitov (1945; Windermere specimens) observed in *S. tamesis* the prostate-like glands which characterized *S. eiseni*. Re-examination of the type-specimens of *S. tamesis* in the British Museum has revealed, however, that, notwithstanding the incomplete immaturity of the specimens, prostate-like glands are present in segments 23-26 (Fig. 15. 14O, U, X). Furthermore, in new material of a *Sparganophilus* from Lake Windermere (Fig. 15. 14C, L-N, W) kindly supplied by the Fresh Water Biological Association, prostate-like glands are again present in (and restricted to) these segments. No grounds thus remain for separating *S. eiseni* from *S. tamesis* either as a specific or an infraspecific taxon. Hague (1923: 31) provided evidence that the prostate-like glands of *"eiseni"* contribute to the formation of, or at least adhesion of, the slime tubes of concopulants. It seems possible therefore that the Windermere material examined by Cernosvitov which was clitellate yet apparently lacked prostate-like glands, though having wart-like genital markings above *b* in segments 23-26 (precisely those segments which most commonly have prostate-like glands in *S. eiseni*), may have represented a non-copulating, uniparental morph of *tamesis*. It seems unlikely that Cernosvitov could have overlooked the presence of the glands.

Hague argued convincingly for suppression of three of Eisen's forms (1896 a): *benhami* (Mexico) (Fig. 15. 14A, H, redrawn from Eisen), *guatemalensis* (Guatemala) and *carneus* (Iowa) as synonyms of *eiseni* and these must now be included in *S. tamesis*.

The status of *S. elongatus* (Friend, 1911) is not entirely certain. Cernosvitov (1942, 1945) on the assumption that Friend's accounts, though grossly inadequate, were correct, was inclined to regard *elongatus* as a good species. An incompletely mature type specimen in the collections of the British Museum has been examined in the present study and its anatomy has been found to agree in all respects with that of *S. tamesis*, though prostate-like glands are absent. "Coelomic glands" were, however, observed by Friend in segment 3 and the type specimens examined have porelike markings in mid *bc* in segments 1-23, in two specimens, and in 1-22, in a further specimen. Though somewhat lateral of the usual location, and extending into segments 1 and 2, they are presumably homologues of the pores of prostate-like glands. Spermathecae, not described by Friend, are present as in *tamesis* in segments 7-9 (Fig. 15. 14P, Q). Clitellar origin on 15, reported by Friend, is common in *tamesis* and termination in 27 is only one segment behind the extreme limit recorded for *"eiseni"*. In the re-examined types clitellar tumescence is not externally recognizable but dorsal setal couples are absent in 15-22, 23 and dorsal incision confirms that clitellar thickening of the epidermis occurs in 15-23. Limits of 15-24 are known for *tamesis* (including *eiseni*). Tubercula pubertatis, said by Friend to occupy 19-22, but not recognizable in the re-examined types, have limits of 18-22 in some specimens of *"eiseni"*.

Presence of *S. tamesis* in the Lily House at Kew Gardens, England (B.M.N.H. 1909.9.16.1-5; Fig. 15. 14B, I-K, V) adds support to Benham's contention that the species has been introduced into England with N. American water plants.

Sparganophilus smithi Eisen, 1896

Sparganophilus smithi smithi Eisen, 1896
Fig. 1. 5G; 15. 14F, R. S

Sparganophilus smithi EISEN, 1896 a: 154, Fig. 120-122, 124, 129-139.

Sparganophilus smithi (typicus) Eisen. MICHAELSEN, 1900 a: 465.

Sparganophilus smithi Eisen f. *typica* MICHAELSEN, 1918: 304.

Sparganophilus smithi Eisen. GATES, 1942: 100.

$l=200$ mm, $w=3.5$ mm, $s=185$. Prostomium zygolobous, with dorsal pit. Anus dorsal or terminal. Setae sigmoid, slightly hooked, lacking ornamentation; none modified as genital setae; *cd* located dorsally. Clitellum annular; dorsally ½16-½28, ventrally 19-25 (=8-13 segments). Tubercula pubertatis each a very elevated, ridge, broken or depressed intersegmentally, in 19-27, usually convex laterally; anteriorly continued forward as a distinct semi-circular ridge ending at 9/10. Male pores in intersegment 18/19, well lateral to *b*, but ventral to the tubercula pubertatis. Female pores clearly visible in front of *ab* in 14. Apertures of prostate glands not externally visible, in front of setae *ab* of 22-24. Spermathecal pores: 2 in front of or slightly dorsal to *cd* of segment 6; 8 in front of and slightly dorsal of *cd* of each of segment 7, 8, and 9. Dorsal pores in 1/2-12/13.

Alimentary canal and vascular system similar in general morphology to other species; sacculated intestine beginning in 13. Hearts in 8-11; commissurals absent in 12 and 13. Nephridia beginning in 16. Seminal vesicles in 11 and, larger, in 12. Ovaries flat and unlobed; oviducts with ovisacs. Prostate-like glands 3 pairs, in 22-24, sometimes a pair in 24 only; muscular duct absent. Spermathecae very large, tall and slender, filling their segments longitudinally, with narrow muscular ducts and wider, non-muscular ampulla.

United States: Laguna Puerca pond, San Francisco, above and below water.

S. smithi is distinguished from *S. eiseni* by location of its sperm ducts and male pores ventral to the tubercula pubertatis.

Sparganophilus smithi sonomae Eisen, 1896
Fig. 15. 13G, T

Sparganophilus sonomae n. subsp. (sic.) EISEN, 1896 a : 154, Fig. 123, 126.

Sparganophilus smithi sonomae Eisen. MICHAELSEN, 1900 : 465.

Sparganophilus smithi var. *sonomae* Eisen. MICHAELSEN, 1918 : 304.

Sparganophilus sonomae Eisen. GATES, 1942 : 100.

$l=200$ mm, $w=3·5$ mm, $s=200$. Clitellum dorsally 17-28, ventrally 26-29. Tubercula pubertatis 19-28; broken up into a series of numerous tubercles, generally 2 or 3 on each segment. Male pores in 18/19 ventral to the tubercula. Spermathecal pores 2 in 5/6 in *d* lines; 2 pairs (4) in each of 6/7-8/9 in *d* lines and *cd* lines. Prostate-like glands in 23-25, without muscular ducts. Spermathecae not slender.

United States: creeks and springs around Sebastopol, Sonoma County, California.

Division of each tuberculum pubertatis into a series of little knobs and the presence of 4 spermathecae in each of segments 7-9 distinguished all the specimens for which Eisen erected *sonomae* from *smithi*.

Gates (1942) considers that *S. smithi* and *sonomae* may be specifically distinct. Neither form has been seen since its original description.

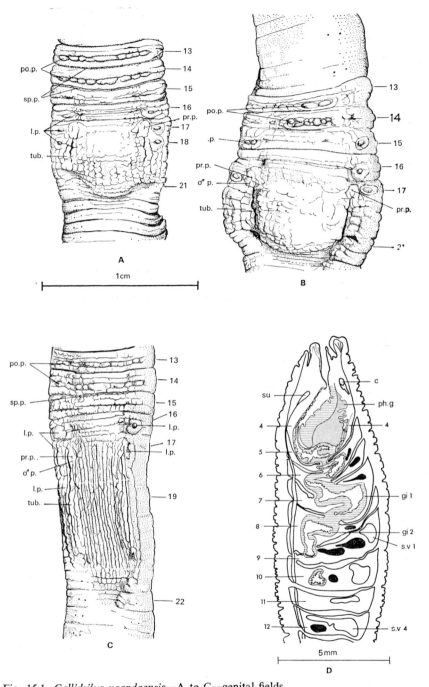

Fig. 15.1. *Callidrilus ugandaensis*—A to C—genital fields.
A—partly contracted, B—strongly contracted, C—relaxed (Types), D—longitudinal
   section of anterior end (original; specimen from Kenya), C—cerebral ganglia.
gi 1, first gizzard; gi 2, second gizzard; 1p, lateral papilla (genital marking); ♂ p,
   male pore; ph.g., pharyngeal gland; po.p., transverse post-setal papillae; pr.p., prostate
   pore; s.v. 1, anterior seminal vesicle; s.v. 2, posterior seminal vesicle; sp.p.,
   spermathecal pores; su, subpharyngeal ganglion; tub., tuberculum pubertatis.

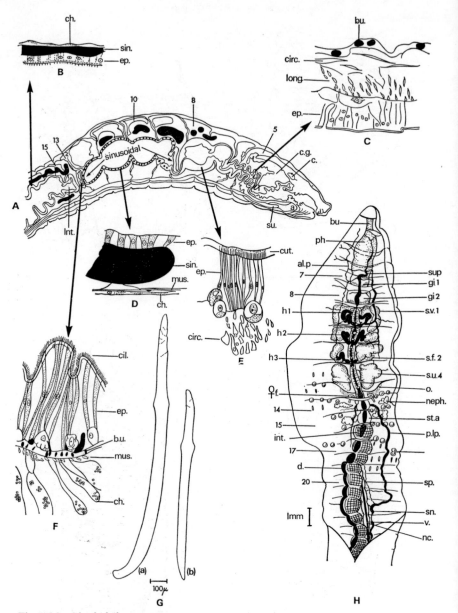

Fig. 15.2. *Glyphidrilus*—anatomy.

G. kukenthali—A to F—longitudinal sections of the alimentary canal: A—*in situ*.
B—wall of intestine. C—wall of anterior oesophagus. D—wall of sinusoidal
oesophagus. E—innermost layers of gizzard. F—wall of straining apparatus.

G. gangeticus. G—somatic setae: (a) from segment 25, (b) from hindbody. H—general
dissection (all original, camera lucida).

a.lp., anterior lateroparietal vessel; bu, buccal cavity; bv, blood vessel; c, cerebral
ganglia; ch, chloragogue; cil, cilia; circ, circular muscle; cut, cuticle; d.v, dorsal
vessel; ep, epithelium; ♀ f, oviducal funnel; gi 1, gi 2, anterior and posterior gizzards;
h 1-3, first to third pairs of hearts; int, intestine: int 1, intestinal origin; long,
longitudinal muscle; mus, muscle fibres; neph, nephridia; nc, nerve cord; o, ovary;
ph, pharynx; p.lp, posterior lateroparietal vessel; s.f. 2, posterior sperm funnel; sp,
spermatheca; st.a, straining apparatus; sup, supra-oesophageal vessel; s.v. 1 and 4,
anterior and posterior seminal vesicles; v, ventral vessel.

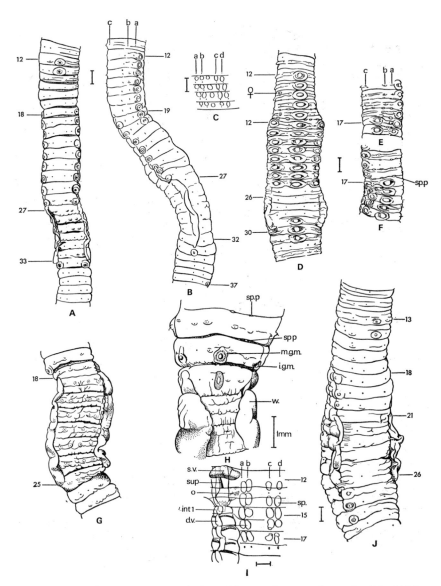

Fig. 15.3. Distribution of alae, genital markings and spermathecae in *Glyphidrilus*.
A to C, *G. annandalei* (C—spermathecae). D to F, *G. buttikoferi*—ventral and lateral
views of a single syntype. G to H, *G. gangeticus*—Jumna River specimens. I to J,
G. jacobsoni (all original, camera lucida; scale 1 mm).

d.v, dorsal vessel; int 1, intestinal origin; l.g.m, lateral genital marking; m.g.m, median
genital marking; sp, spermatheca; sp.p, spermathecal pore; sup, supra-oesophageal
vessel; s.v, seminal vesicle; w, wing (ala).

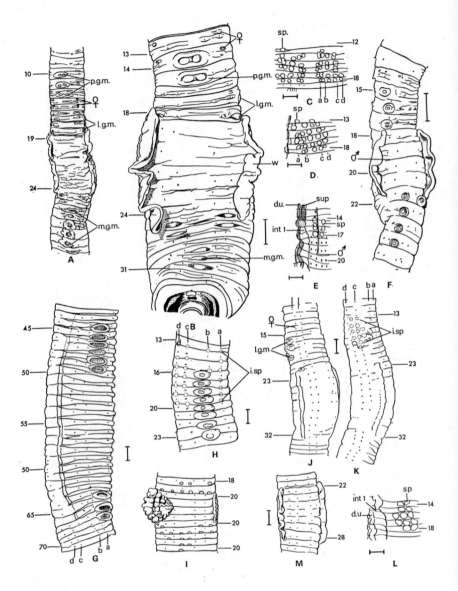

Fig. 15.4. Distribution of alae, genital markings and spermathecae in *Glyphidrilus* (continued).

A to D, *G. kukenthali*. E to F, *G. malayanus*. G to H, *G. stuhlmanni*. I, *G. tuberosus*. J to M, *G. weberi* (J-L specimen with long alae, M, with short alae; both from Hamburg Museum.) (All except (I) original, camera lucida; scale 1 mm).

i.sp, inseminated spermathecae visible through body wall; ♂, male pore; p.g.m, medium paired genital marking. Other abbreviations as Fig. 15.3.

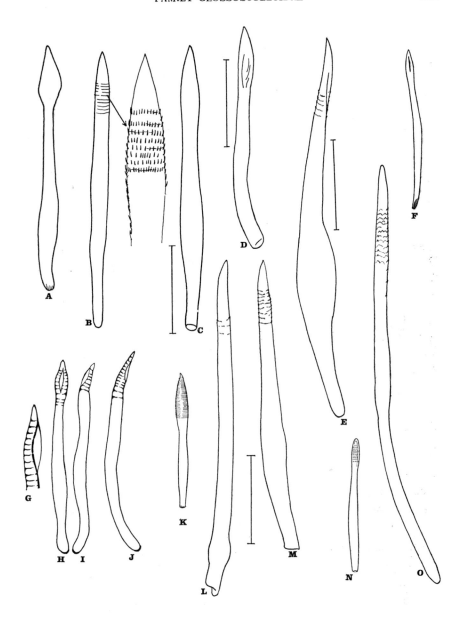

Fig. 15.5. Genital setae in *Alma*.
A—*A. emini* (from posterior row of right clasper). B,C—*A. nilotica* (Egypt). D—*A. nilotica* (Wadi Halfa). E to J—*A. millsoni*: E,G-J, from syntypes of *millsoni*, F, from syntype of *A. schultzei*. K—*A. multisetosa*. L,M—*A. tazelaari*. N,O—*A. togoensis*. (Scale 100μ).

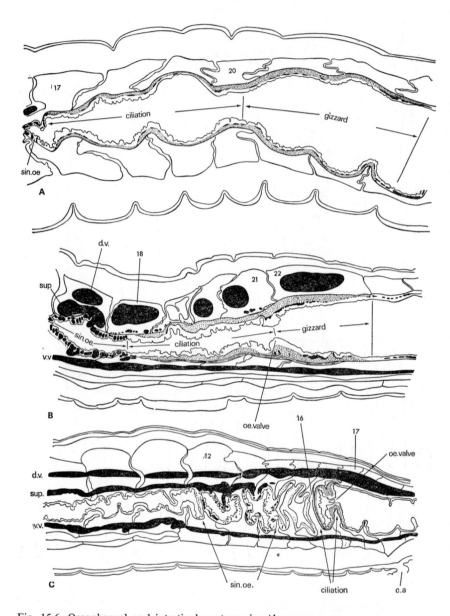

Fig. 15.6. Oesophageal and intestinal anatomy in *Alma*.
A—*A. emini*. B—*A. nilotica*. C—*A. stuhlmanni*. (All original, camera lucida).
cla, clasper; d.v, dorsal vessel; oe.valve, oesophageal valve; sin.ve, sinusoidal
oesophagus; sup, supra-oesophageal vessel; v.v, ventral vessel.

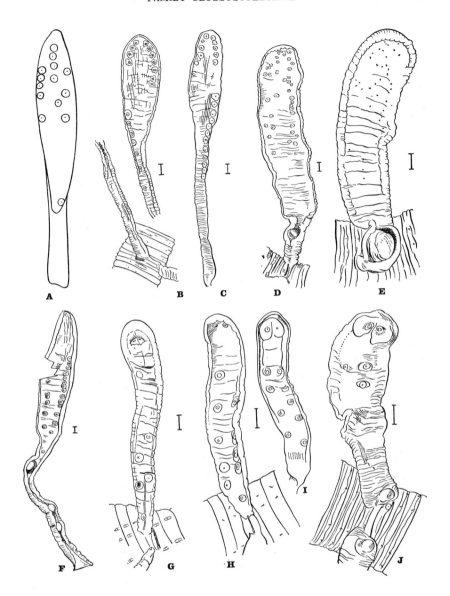

Fig. 15.7. *Alma* claspers.

A—*A. basongonis.* B—*A. emini* (Ituri Forest). C—*A. emini* (*aloysiisabaudiae*). D—*A. eubranchiata eubranchiata.* E—*A. eubranchiata catarrhactae.* F—*A. kamerunensis.* G—*A. millsoni millsoni.* H,I—*A. millsoni millsoni* (=*A. schultzei*). J—*A. millsoni zebangui.*

(All original, camera lucida; claspers with inrolled margins have been flattened by coverslip pressure. Scale 1 mm).

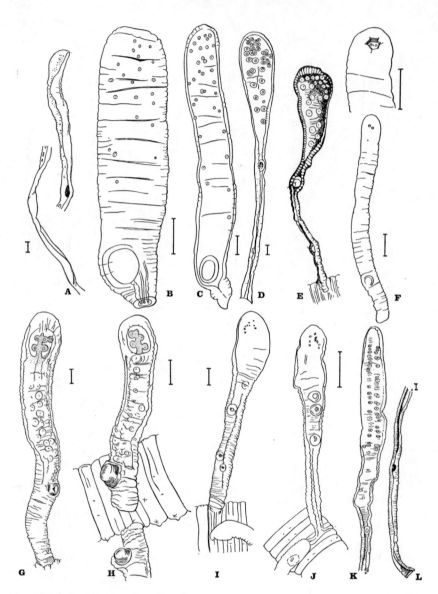

Fig. 15.8. *Alma* claspers (continued).

A—*A. multisetosa*. B—*A. nilotica* (Egypt). C—*A. nilotica* (Sudan). D,E—*A. pooliana*. F,G—*A. stuhlmanni*: F, entire clasper; G, distal end of a further clasper. H,I—*A. tazelaaro*. J,K—*A. togoensis*: J, type; K. from Ghana. L.—*A. ubangiana*.
(A-D, F-L, original or with Ghabbour, camera lucida; claspers with inrolled margins have been flattened by coverslip pressure. Scale 1 mm).

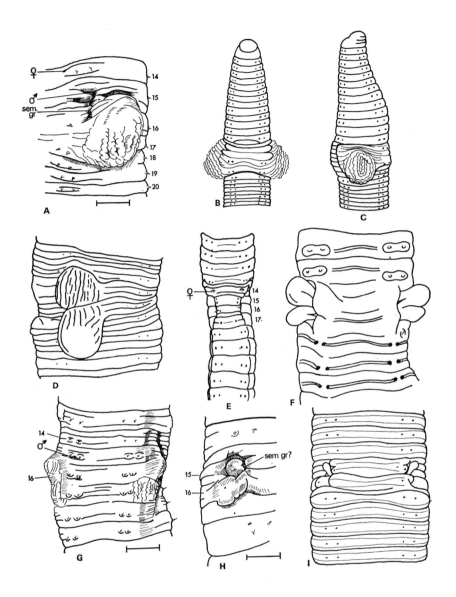

Fig. 15.9. Male genital fields in *Drilocrius*.

A to C—*D. breymanni*; D to E—*D. burgeri*: D from Colombia, E from Argentina; F—*D. iheringi* (type of *D. dreheri*); H to I—*D. iheringi* (type); G—*D. hummelincki*. ♀, female pore; ♂, male pore; sem.gr, seminal groove.

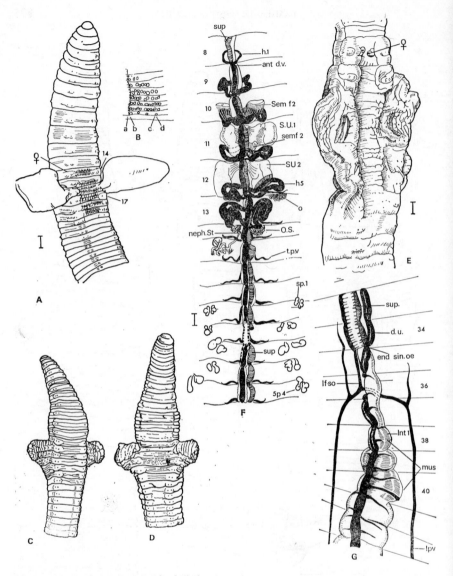

Fig. 15.10. *Drilocrius* and *Glyphidrilocrius*.

A,B—*Drilocrius alfari*: A, showing claspers; B, spermathecae of the right side.
C—*D. buchwaldi*: dorsal and ventral views.
E to G—*Glyphidrilocrius ehrhardti*: E, male genital field; F—dorsal dissection of
    cardiac and spermathecal region; G—dorsal dissection to show intestinal origin and
    associated blood vessels.
ant.d.v, anterior end of dorsal vessel; h 1, first heart; h 5, last heart; d.v, dorsal vessel;
    end.sin.ve., posterior limit of sinusoidal oesophagus; gizz oe, gizzardlike portion of
    oesophagus; int 1, intestinal origin; lf.so, left suboesophageal vessel; l.p.v, latero-
    parietal vessel; nephst., nephrostome; o.s, ovisac; sem.f. 1, sem.f. 2, seminal funnels
    of 10 and 11, respectively; sp. 1, anterior spermathecae; sp. 2, posterior spermathecae;
    sup, supraoesophageal vessel; s.v 1, s.v 2, seminal vesicles of 11 and 12, respectively;
    t.p.v, transverse parietal vessel.

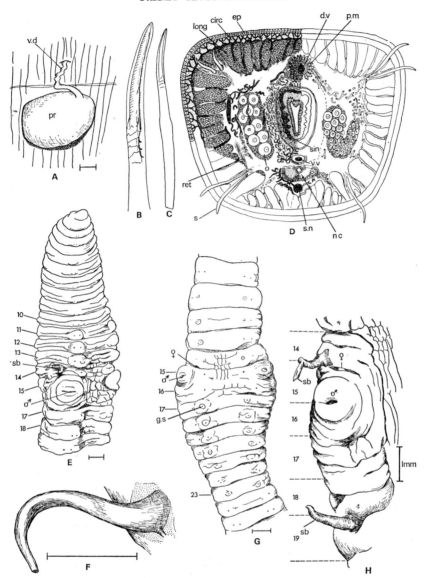

Fig.15.11. *Criodrilus lacuum.*

A—prostate gland; B,C—a single genital seta; D—transverse section in the intestinal region; E—genital field and anterior end; F—a spermatophore. (All except D original, camera lucida, from Tegel. See specimen). G,H—genital fields of two Syrian specimens. (Scale 1 mm).

Circ, circular muscle; d.v, dorsal vessel; ep, epidermis; ♀, female pore; g.s, genital seta; long, longitudinal muscle; o, male pore; n.c, nerve cord; o, ovary; p.m, peritoneal and muscular sheath; pr, "prostate" gland; ret, retractor muscle of seta; s, seta; sb, spermatophore; s.n, subneural vessel; v.d, vas deferens; v.v, ventral vessel.

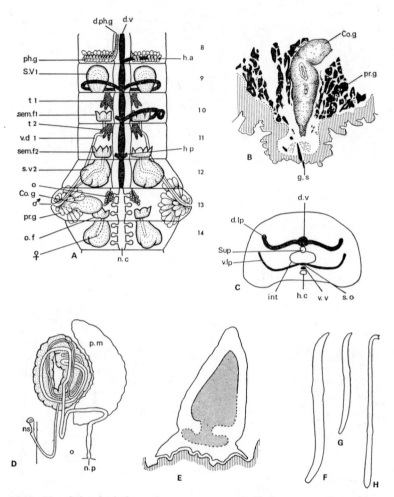

Fig. 15.12. *Biwadrilus bathybates.*

A—diagram of genital organs; B—copulatory and prostate glands in longitudinal section of body; C—diagram showing origins of dorsal and ventral lateroparietal vessels from the dorsal and suboesophageal vessels, respectively; D—nephridium; E—spermatophore attached to body wall, in section; F—largest dorsal seta from segment 4; G—seta from last segment; H—genital seta.

co, copulatory gland; d.lp, dorsal lateroparietal vessel; d.ph.g, duct of pharyngeal gland; d.v, dorsal vessel; ♀, female pore; g.s, genital seta; h.a, heart of segment 8; h.p, last heart; int. intestine; ♂, male pore; n.c, nerve cord; np, nephropore; ns. nephrostome; o.f., oviducal funnel; p.m, peritoneal mass; ph.g, pharyngeal gland; pr.g, prostate gland; s.v 1 and s.v 2, anterior and posterior seminal vesicles; s.o, suboesophageal vessel; sem.f. 1 and 2, anterior and posterior male funnels; sup, supra-oesophageal vessel; t 1 and t 2, anterior and posterior testes; v.a 1, anterior vas deferens; v.lp, ventral lateroparietal vessel; v.v, ventral vessel.

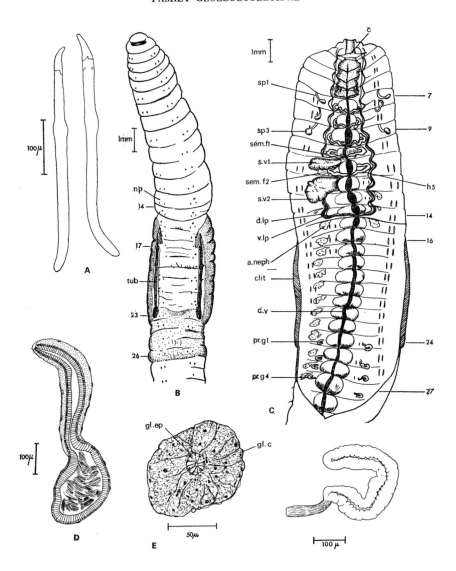

Fig. 15.13. *Sparganophilus tamesis.*
A—somatic setae from forebody; B—external view of anterior region; C—general
dissection; D—a spermatheca containing sperm; E and F—posterior prostate-like
glands: E, transverse section, F, entire gland from segment 24 of specimen C.
(All from Michigan (St. Mary's River) specimens by camera lucida).
a.neph, first well developed nephridium; c, cerebral ganglia; clit, clitellum; gl.c, gland
cells; gl.ep, glandular epithelium; pr.g 1 and pr.g 4, anterior and posterior prostate-
like glands; sp 1 and sp 3, anterior and posterior spermathecae; other abbreviations
as Fig. 15.12.).

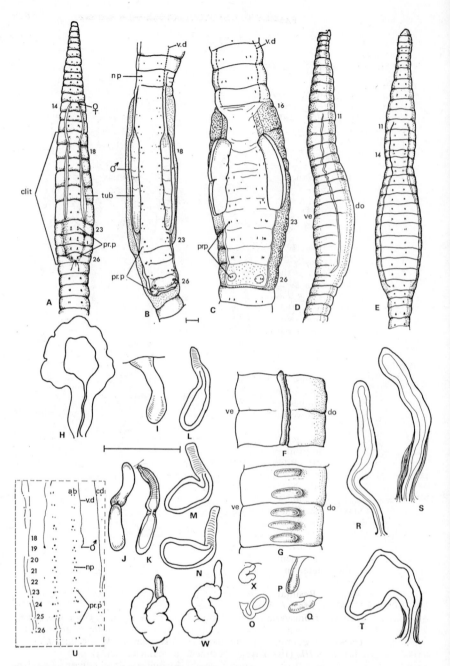

Fig. 15.14. *Sparganophilus tamesis* and *S. smithi*.

A to E—Genital fields showing tubercula pubertatis: A to C—*S. tamesis*; D to E—*S. smithi smithi*.

F to G—Tubercula pubertatis: F—*S. smithi smithi*; G—*S. smithi sonomae*.

H to T—Spermathecae: H to Q—*S. tamesis*; R,S—*S. smithi smithi*; T—*S. smithi sonomae*.

U—Male genital field of *S. tamesis* (type); V to X—Posterior prostate-like glands of *S. tamesis*.

(Scales indicated=1 mm. All spermathecae and prostate-like glands to same scale excepting H and R to T).

clit, clitellum; do, dorsal; ♀, female pore; ♂, male pore; np, nephropore; pr.p, pores of prostate-like glands; tub, tubercula pubertatis; ve, ventral.

REFERENCES

BAHL, V. N. 1942. Studies on the structure, development and physiology of the nephridia of Oligochaeta. iii. The branching and division of nephridia and Eisen's so-called safety valves in *Pontoscolex. Q. Jl miscrosc. Sci.,* **84**, 1.

BALDASSERONI, V. 1907. Contributo alla conoscenza dei Lombrichi italiani. *Monitore zool. ital.,* **18**, 48.

— 1914. Sui nefridii dell' *Hormogaster praetiosa* Mchlsn. *Monitore zool. ital.,* **25**, 160.

BEDDARD, F. E. 1887. On the structure of a new genus of Lumbricidae (*Thamnodrilus gulielmi*). *Proc. zool. Soc. Lond.,* **1887**, 154.

— 1890. On the structure of a species of earthworm belonging to the genus *Diachaeta. Q. Jl miscrosc. Sci.,* **31**, 159.

— 1891a. The classification and distribution of earthworms. *Proc. R. phys. Soc. Edinb.,* **10** (2), 235.

— 1891b. On an earthworm of the genus *Siphonogaster* from West Africa. *Proc. Zool. Soc. Lond.,* **1**, 48.

— 1892. The earthworms of the Vienna Museum. *Ann. Mag. nat. Hist.,* (6) **9**, 113.

— 1893. Two new genera and some new species of earthworms. *Q. Jl miscrosc. Sci.,* **34**, 243.

— 1895. A monograph of the order Oligochaeta. Oxford.

— 1901. On the clitellum and spermathophores of an Annelid of the genus *Alma. Proc. zool. Soc. Lond.,* **1901** (1), 215.

— 1903. On a new genus and two new species of earthworms of the family Eudrilidae with some notes upon other African Oligochaeta. *Proc. zool. Soc. Lond.,* **1903** (1), 210.

— 1921. On a new species of the oligochaete genus *Thamnodrilus* (Beddard) with notes on *Th. gulielmi. Ann. Mag. nat. Hist.,* (9) **7**, 153.

BENHAM, W. B. 1886. Studies on earthworms. II. *Q. Jl microsc. Sci.,* **27**, 77.

— 1887. Studies on earthworms. III. *Criodrilus lacuum,* Hoffmeister. *Q. Jl microsc. Sci.,* **27**, 561.

— 1892. A new English genus of aquatic Oligochaeta (*Sparganophilus*) belonging to the family Rhinodrilidae. *Q. Jl microsc. Sci.* (*N.S.*), **34**, 155.

BERGH, R. S. 1888. Zur Bildungsgeschichte der Exkretionsorgane bei *Criodrilus. Arb. Zool.-Zoot. Inst. Würzburg,* **8**, 223.

BROCK, G. T., and DICK, J. 1935. The anatomy of the Grahamstown species of *Microchaetus. Rec. Albany Museum,* **4**, 236.

CERNOSVITOV, L. 1934a. Eine neue *Glossoscolex*—Art aus den Sammlungen des Nationalmuseums in Prag. *Zool. Anz.,* **105**, 183.

— 1934b. Les oligochètes de la Guyane française et d'autres pays de l'Amerique du Sud. *Bull. Mus. Hist. nat., Paris,* (2) **6**, 47.

— 1935. Oligochaeten aus dem tropischen Süd-Amerika. *Capita zool.,* **6** (1), 1.

— 1938. The Oligochaeta. In: Studies on the Freshwater of Palestine. *Ann. Mag. nat. Hist.,* **2**, 535.

— 1939. Oligochaeta. Resultats scientifiques des crossières du Navire-école belge "Mercator". *Mém. Mus. r. Hist. nat. Belg.,* (2) **15**, 115.

— 1942a. Oligochaeta from various parts of the world. *Proc. zool. Soc. Lond. B III,* (1941), 197.

— 1942b. A revision of Friend's types and descriptions of British Oligochaeta. *Proc. zool. Soc. Lond. B III,* (1941). 237.

— 1945. Oligochaeta from Windermere and the Lake District. *Proc. zool. Soc. Lond. B (IV)* **114**, 523.

CHEN, Y. 1938. Oligochaeta from Hainan, Kwangtung. *Contr. biol. Lab. Sci. Soc. China. Zool.,* **12**, 375.

COGNETTI de MARTIIS, L. 1902a. Terricoli boliviani ad argentini. *Boll. Musei Zool. Anat. comp. R. Univ. Torino,* **17** (420), 1.

— 1902b. Un nuovo genere della Fam. "Glossoscolecidae". *Atti Accad. Sci., Torino,* **37**, 432.

COGNETTI de MARTIIS, L. 1904a. Oligochetti dell' Ecuador. *Boll. Musei Zool. Anat. comp. R. Univ. Torino*, **19** (474), 1.

— 1904b. Nuovi oligocheti di Costa Rica. *Boll. Musei Zool. Anat. comp. R. Univ. Torino*, **19** (478), 1.

— 1905. Oligocheti raccolti nel Darien dal Dr. E. Festa. *Boll. Musei Zool. Anat. comp. R. Univ. Torino*, **20** (495), 1

— 1906a. Gli oligocheti della regione neotropicale. *Memorie Accad. Sci. Torino*. (*Cl. Sci. Fis, Mat. and Nat.*), (2) **56**, 1.

— 1906b. Spedisione al Ruwenzori di S. A. Luigi di Savoia. Un nuovo oligochete criodrilino (*Alma aloysii sabaudiae*, n. sp.). *Boll. Musei Zool. Anat. comp. R. Univ. Torino*, **21** (534), 1.

— 1908a. Lombrichi di Costa Rica e del Venezuela. *Atti Accad. Sci.*, *Torino*, **43**, 913.

— 1908b. I cosi detti peni dei *Criodrilini*. Ricerche anatomoistologiche e fisiologiche *Atti Accad. Sci.*, *Torino*, **43**, 1122.

— 1909. Lombrichi del Ruwenzori e dell' Uganda. Spedizione al Ruwenzori di S.A.R. il Principe 1. Amedeo di Savoia. *Parte Scientifica* V. 1 (Milano, Hoepli), **1909**, 359.

— 1911. A contribution to our knowledge of the Oligochaeta of Travancore. *Ann. Mag. nat. Hist.*, (8) **7**, 494.

— 1914. Nota sugli oligocheti degli Abruzzi. *Boll. Musei Zool. Anat. comp. R. Univ. Torino*, **29** (689), 1.

— 1926. Eine neue *Opisthodrilus*—Art aus Brasilien. *Denkschr. Akad. Wiss., Wien* **76**, 41.

COLLIN, A. 1888. *Criodrilus lacuum* Hoffm. Ein Beitrag zur Kenntnis der Oligochaeten. *Z. wiss. Zool.*, **46**, 471.

CORDERO, E. H. 1942. Oligoquetos terricolas del museo Argentino de ciencias naturales. *An. Mus. argent. Ciene. nat.*, **40**, 269

— 1943. Oligoquetos sudamericanos de la familia Glossoscolecidae, i. El genero *Glossoscolex* en el Uruguay, con una sinopsis de las especies del grupo *truncatus*. *Comun. zool. Mus. Hist. nat. Montev.*, **1** (2), 1.

— 1944a. Oligoquetos sudamericanos de la familia Glossoscolecidae, iii. *Rhinodrilus francisci* n. sp., de Pernambuco, Brasil. *Commun. zool. Mus. Hist. nat. Montev.*, **1** (10), 1

— 1944b. Oligoquetos sudamericano de la familia Glossoscolecidae, IV. Sobre algunas especies de Venezuela. *Commun. zool. Mus. Hist. nat. Montev.*, **1** (14), 1.

— 1944c. Oligoquetos sudamericanos de la familia Glossoscolecidae, V. *Eudevoscolex vogelsangi* n.g. n. sp. de Venezuela, nueva forma con cierto numero de caracteres primitivos. *Commun. zool. Mus. Hist. nat. Montev.*, **1** (18), 1.

— 1945. Oligoquetos sudamericanos de la familia Glossoscolecidae, VI. Los generos de la subfamilia Glossoscolecinae, su probables relaciones fileticas y su distribucion geografica actual. *Commun. zool. Mus. Hist. nat. Montev.*, **1** (22), 1.

DAVIES, H. 1954. A preliminary list of the earthworms of Northern New Jersey with notes. *Breviora*, **26**, 1.

DUBOSCQ, O. 1902. *Alma zebanguii* n. sp., et les *Alminae* oligochètes de la famille des Glossoscolecidae Mich. *Archs Zool. exp. gen.*, (3), **10** (7), xcvii.

EISEN, G. 1896a. Pacific Coast Oligochaeta II. *Mem. Calif. Acad. Sci.*, **2** (5), 123.

— 1896b. *Pontoscolex lillieborgi* with notes on auditory sense cells of *Pontoscolex corethrurus*. *Festskrift Lilljeborg, Upsala*, **4**, 1.

FRIEND, H. 1911. A new earthworm. Zoologist, (IV), **15**: 192-193.

— 1919. *Sparganophilus*: a British oligochaet. *Nature, Lond.*, **103**, 426.

— 1921. Two new aquatic Annelids. *Ann. Mag. nat. Hist.*, (9) **7**, 137.

— 1923. *British earthworms and how to identify them*. London.

GATES, G. E. 1931. The earthworms of Burma. II. *Rec. Indian Mus.*, **33**, 327.

— 1933. The earthworms of Burma. IV *Rec. Indian Mus.*, **35**, 413.

— 1935. The earthworms of New England. *Proc. New Engl. zool. Club*, **15**, 41.

— 1938. Earthworms from the Malay Peninsula. *Bull. Raffles Mus.*, **14**, 206.

— 1939. Thai earthworms. *J. Thailland Res. Soc.*, **12**, 65.

— 1942. Check list and bibliography of North American earthworms. *Am. Midl. Nat.*, **27** (1), 86.

— 1943. On some American and Oriental earthworms. *Ohio J. Sci.*, **43** (2), 87.

GATES, G. E. 1945. On some earthworms from Ceylon ii. *Spolia zeylan*, **24** (2), 69

— 1948a. On some earthworms from the Buitenzorg Museum, III. Results of the 3rd Archbold expedition 1938-1939. *Treubia*, **19**, 139.

— 1948b. On segment formation in normal and regenerative growth of earthworms. *Growth*, **12** (3), 165.

— 1949. On some earthworms from Perlis and Kedah. *Bull. Raffles Mus.*, **19**, 5.

— 1951. On the earthworms of Sahranpur Dehra Dun, and some Himalayan Hill Stations. *Proc. nat. Acad. Sci. India (B)*, **21**, 16.

— 1958a. On some miscellaneous lots of earthworms belonging to the American Museum of Natural History. *Am. Mus. Novit.*, **1887**, 1.

— 1958b. On a hologynous species of the earthworm genus *Diplocardia*, with comments on oligochaete hologyny and consecutive hermaphroditism. *Am. Mus. Novit.*, **1886**, 1.

— 1958c. On Indian and Burmese earthworms of the genus *Glyphidrilus*. *Rec. Indian Mus.*, **53**, 53.

— 1961. Ecology of some earthworms with special reference to seasonal activities. *Am. Midl. Nat.*, **66** (1), 61.

— 1962. Miscellanea megadrilogica. V. On some instances of supposed variation in the night crawler, *Lumbricus terrestris* L. *Proc. biol. Soc. Wash.*, **75**, 137.

— 1963. Miscellanea megadrilogica. VII. Greenhouse earthworms. *Proc. biol. Soc. Wash.*, **76**, 9.

— 1965. Louisiana earthworms. I. A preliminary survey. *Proc. La Acad. Sci.*, **28**, 12.

GEORGÉVITCH, J. 1950. Contribution a la connaissance de la faune du lac d' Ochrid. *Criodrilus ochridensis*, nov. sp. *Bull. Acad. serb. Sci. Cl. Sci., math. nat.*, **1** (1), 75.

GRAFF, O. 1962. Ein *Criodrilus* aus Südfrankreich. *Vie Milieu*. **13**, 369.

GROVE, A. J. 1931. The structure of the clitellum of *Alma emini* Mich. *Q. Jl microsc. Sci.*, **74**, 223.

GRUBE, E. 1855. Beschreibungen neuer oder wenig bekannter Anneliden. *Arch. Naturgesch.*, **21** (1), 81.

HAGUE, F. S. 1923. Studies on *Sparganophilus eiseni* Smith. *Trans. Am. microsc. Soc.*, **42**, 1.

HARMAN, W. J. 1965. Life history studies of the earthworm *Sparganophilus eiseni* in Louisiana. *SWest. Nat.*, **10** (1), 22.

HATSCHEK, . 1878. Studien über Entwickelungsgeschichte der Anneliden. *Arb. zool. Inst. Univ. Wien*, **3**, 277.

HEIMBURGER, H. V. 1915. Notes on Indiana earthworms. *Proc. Indiana Acad. Sci.*, **1914**. 281.

HOFFMEISTER, W. 1845. *Die bis jetzt bekannten Arten aus der Familie der Regenwurmer*. Braunschweig.

HORST, R. 1889. Over eene nieuwe soort order de Lumbricinen door Prof. Max Weber uit nedenl. Indië medegebracht. *Tidjdshr. ned. dierk. Vereen.*, **2**, 77.

— 1891a. Descriptions of earthworms. *Notes Leyden Mus.*, **13**, 77.

— 1891b. Preliminary note on a new genus of earthworms. *Zool. Anz.*, **XIV**, 1.

— 1892. Earthworms from the Malay Archipelago. In Max Weber, *Zool. Ergebn. einer Reise in Niederlandisch Ost-Indien*, **2**, 28.

JAMIESON, B. G. M. 1963. A revision of the earthworm genus *Digaster* (Megascolecidae, Oligochaeta). *Rec. Aust. Mus.*, **26** (2), 83.

— 1968a. A new species of *Glyphidrilus* (Microchaetidae: Oligochaeta) from East Africa. *J. nat. Hist.*, **2**, 387.

— and GHABBOUR. 1969. The genus *Alma* (Microchaetinae: Oligochaeta) in Egypt and the Sudan. *J. nat. Hist.*

KELLER, C. 1887. *Reisebilder aus Ostafrika and Madagaskar*. Leipzig.

KHALAF EL-DUWEINI, A. 1940. The earthworms of Egypt. *Bull. Inst. Egypte*, **22** (1939-1940), 99.

— 1950a. On the so-called penes of *Alma nilotica* Grube. *Proc. Egypt Acad. Sci.*, **5**, 94.

— 1950b. On the copulation in *Alma nilotica* Grube, with a description of the spermathecae in this species. *Bull. zool. Soc. Egypt*, **9**, 29.

KHALAF EL-DUWEINI, A. 1951. On the clitellum and cocoons of *Alma nilotica* Grube. *Proc. Egypt. Acad. Sci.*, **6** (1950), 52.

— 1954. On the internal genital organs of *Alma nilotica* Grube, with a note on the septal pouches in this species. *Proc. Egypt. Acad. Sci.*, **9**, (1953), 102.

— 1957. On the gills and respiration in *Alma nilotica* Grube. *Publ. 2nd Sci. Arab Congr.*, (1955), 833.

— 1965. The excretory system of *Alma nilotica* Grube. *Bull. zool. Soc. Egypt*, **20**, 31.

— and GHABBOUR, S. I. 1963. A study of the specific distribution of megadriline oligochaetes in Egypt and its dependence on soil properties. *Bull. zool. Soc. Egypt*, **18**, 21.

KINBERG, J. G. H. 1867. Annulata nova (continuatio). *Ofvers. K. VetenskAkad. Forh.*, **23**, 97.

LEUCKART, F. S. 1835. Abbildung eines neuen Genus von Ringelwurmern *Notizen Froriep*, **46**, ? 88.

— 1836. Abbildung eines neuen Genus von Ringelwürmern. *Isis (Oken)*, **1836**, ? 764.

LEVINSEN, G. M. R. 1890. On two new earthworms from Egypt: *Siphonogaster aegyptiacus*, n.g. and sp., and *Digitibranchus niloticus*, n.g. and sp. *Vidensk. Meddrdansk naturh. Foren.*, **1889**, 318.

MCKEY-FENDER, D., and MACNAB, J. A. 1953. The aquatic earthworm *Criodrilus laccum* Hoffmeister in North America (Oligochaeta, Glossoscolecidae). *Wasmann J. biol.*, **11**, 373.

MICHAELSEN, W. 1890. Beschreibung der von Herrn Dr. Franz Stuhlmann im Mündungsgebiet des Sambesi gesammelten Terricolen. *Mitt. naturh. Mus. Hamb.*, **7**, 21.

— 1891. Terricolen der Berliner Zoologischen Sammlung I. Afrika. *Arch Naturgesch.*, **57**, 205.

— 1892. Beschreibung der von Herrn Dr. Fr. Stuhlmann am Victoria Nyanza gesammelten Terricolen *Jb. hamb. wiss. Anst.*, **9** (2), 1.

— 1895. Zur Kenntnis der Oligochaeten. *Abh. naturw. Ver. Hamburg*, **13**, 1.

— 1896a. Oligochaeten. Ergebnisse einer zoologischen Forschungsreise in den Notukken und Borneo, im Auftrage der Senckenbergischen naturforschenden Gesellschaft ausgeführt von Dr. Willy Kükenthal. *Abh. Senckenb. naturforsh. Ges.*, 23, 193.

— 1896b. Regenwürmer aus Deutsch Ost-Afrika. Die Thierwelt Ost-Afrikas. IV. Wirbellose Thiere. Berlin, *Geogr. Verlagshudl. Reimer*, **8**, 1.

— 1897a. Organisation einiger neuer oder wenigbekannter Regenwürmer von West-indien und Südamerika. *Zool. Jb. (Anat.)* **10**, 359.

— 1897b. Neue und wenig bekannte afrikanische Terricolen. *Mitt. naturh. Mus. Hamb.*, **14**, 1.

— 1899. Terricolen von verschiedenen Gebieten der Erde. *Mitt. naturh. Mus. Hamb.*, **16**, 1.

— 1900a. Oligochaeta. In *Das Tierreich*, **10**. Berlin.

— 1900b. Zur Kenntnis der Geoscoleciden Südamerikas. *Zool. Anz.*, **23**, 53.

— 1900c. Die Terricolen-Fauna Columbiens. *Arch. Naturgesch.*, **66**, 231.

— 1903a. Die Oligochäten Nordost-Afrikas, nach den Ausbeuten der Herren Oscar Neumann und Carlo Freiherr von Erlanger. *Zool. Jb. (Syst.)* **18**, 435.

— 1903b. Neue Oligochaeten und neue Fundorte altbekannter. *Mitt. Mus. Hamb.*, **19**, 1.

— 1905. Die Oligochaeten Deutsch-Ostafrikas. *Z. wiss. Zool.*, **82**, 288.

— 1909. The oligochaeta of India, Nepaul, Ceylon, Burma and the Andaman Islands. *Mem. Indian Mus.*, **1** (3), 105.

— 1910a. Die Oligochäten-fauna der vorderindisch-ceylonischen Region. *Abh. naturw. Ver. Hamb.*, **19** (5), 1.

— 1910b. Oligochäten von verschiedenen Gebieten. *Mitt. naturh. Mus. Hamb.*, **27** (2), 47.

— 1910c. Die Oligochäten des inneren Ostafrikas und ihre geographischen Bezie-hungen. *Wiss. Ergebn. dt. ZentAfr.-Exped. Leipzig, 1907-8*, **3** (1), 1.

— 1913a. Oligochäten vom tropischen und südlich subtropischen Afrika. *Zoologica, Stuttg.*, **67** (B26), 139.

— 1913b. The Oligochaeta of Natal and Zululand. *Ann. Natal Mus.*, **13**, 397.

MICHAELSEN, W. 1913c. Oligochäten von Travancore und Borneo. *Jb. hamb. wiss. Anst.*, **30** (1912) (2), 73.

— 1913d. Oligochäten vom tropischen und südlichsubtropischen Afrika, 2. *Zoologica, Stuttg.*, **68**, 1.

— 1914a. Die Oligochaeten Columbias. In Fuhrmann and Mayor. *Voyage d'Exploration scientifique en Colombie. Mém. Soc. neuchât. Sci. nat.*, 202.

— 1914b. Oligochäten vom tropischen Afrika. *Mitt. naturh. Mus. Hamb.*, **31** (2), 81.

— 1915. Zentralafrikanische Oligochäten. *Wiss. Engebn. dt. ZentAfr.-Exped. 1910-1911*, **1** (8), *Zool.*, **1** (1), 185.

— 1918. Die Lumbridicae. *Zool. Jb.*, **41**, 1.

— 1922. Oligochäten aus dem Rijks—Museum van natuurlijke Histoirie zu Leiden. *Capita zool.*, **1** (3), 1.

— 1923. Oligochäten von den wärmeren Gebieten Amerikas und des Atlantischen Ozeans sowie ihre faunistischen Beziehungen. *Mitt. naturh. Mus. Hamb.*, **41**, 71.

— 1925. Zur Kenntnis einheimischer und ausländischer Oligochäten. *Zool. Jb.*, **51**, 255.

— 1926a. Notes sur les oligochètes rapportes, par M. Henri Gadeau de Kerville de son voyage zoologique en Syrie. In *Voyage Zool. d'Henri Gadeau de Kerville en Syrie*, **1**, 351.

— 1926b. Zur Kenntnis einheimischer und ausländischer Oligochäten. *Zool. Jb. (Syst.)*, **51**, 255.

— 1927. Die Oligochätenfauna Brasiliens. *Abh. senckenb. naturforsch. Ges.*, **40**, 369.

— 1928a. Miscellanea oligochaetologica. *Ark. Zool.*, **20** (A2), 1.

— 1928b. Die Oligochäten Borneos. *Ark. Zool.*, **20**, (A3), 1.

— 1931a. Ausländische opisthopore Oligochäten. *Zool. Jb.*, **61**, 523.

— 1931b. Zwei neue aussereuropäische Oligochäten des Senckenberg-Museums. *Senckenbergiana*, **13**, 78.

— 1933a. Zoologische Ergebnisse einer Reise nach Bonaire, Curaçao und Aruba im Jahre 1930. *Zool. Jb.*, **64**, 327.

— 1933b. Opisthopore Oligochäten aus dem mittleren und dem südlichen Afrika hauptsächlich gesammelt von Dr. F. Haas während der Schomburgh-Expedition 1931-32. *Abh. senekenb. naturforsch. Ges.*, **40**, 411.

— 1933c. Die Oligochätenfauna Surinames, mit Erörterung der verwandschaftlichen und geographischen Beziehungen der Octochätinen. *Tijd schr. ned. dierk. Vereen.*, (3), 112.

— 1934. Opisthopore Oligochäten des Königlichen naturhistorischen Museums von Belgien. *Bull. Mus. r. Hist. nat. Belg.*, **10** (25), 1.

— 1935a. Oligochaeten aus Peru *Capita zool.*, **6** (2), 1

— 1935b. Oligochäten von Belgisch-Kongo. *Rev. Zool. Bot. afr.*, **27**, 33.

— 1936. African and American Oligochaeta in the American Museum of Natural History. *Am. Mus. Novit.*, **843**, 1.

— 1937a. On a collection of African Oligochaeta in the British Museum. *Proc. Zool. Soc. Lond. (B)*, **107**, 501.

— 1937b. On the genus *Thamnodrilus* Beddard. *Proc. Zool. Soc. Lond. (B)*, **1937**, 1172.

MOORE, H. F. 1895. On the structure of *Bimastus palustris*, a new oligochaete. *J. Morph.*, **10**, 473.

MOORE, J. 1906. Hirudinea and Oligochaeta of the Great Lakes Region. *Bull. Bur. Comml. Fish.*, **25** (1905). 153.

MÜLLER, F. 1857. *Lumbricus corethrurus*, Bürstenschwanz. *Arch. Naturg.*, **23**, 1, 113.

MURCHIE, W. R. 1956. Survey of the Michigan Earthworm fauna. *Pap. Mich. Acad. Sci.*, **41** (1955), 53.

NAGASE, I., and NOMURA, E. 1937. On the Japanese aquatic Oligochaeta *Criodrilus miyashitai.* n.sp. *Sci. Rep. Tohoku Univ.*, (4) **11**, 361.

NAIR, K. B. 1937. The anatomy of *Glyphidrilus annandalei* Mich *Proc. Indian Sci. Congr.*, **24**, 300.

— 1938. On some points in the anatomy of *Glyphidrilus annandalei* Mich. *Z. wiss. Zool.*, **151**, 39.

836 AQUATIC OLIGOCHAETA OF THE WORLD

ÖRLEY, L. 1881. A magyaroszági oligocheták Faunaja. *Math. Termesz. Kozlem.*, **16**, 561.
— 1887. Morphological and biological observations on *Criodrilus lacuum*, Hoffmeister. *Q. Jl microsc. Sci.*, **27**, 551.
OLSON, H. W. 1928. The earthworms of Ohio *Bull. Ohio biol. Surv.*, (17) **4** (2), 1.
— 1940. Earthworms of New York State. *Am. Mus. Novit.*, **1090**, 1.
OMODEO, P. 1948. Oligocheti della Campania. ii. Morfologia e biologia dei Terricoli: *Microscolex dubius* (Fletch.) (Megascolecidae, Acanthodrilinae); *Hormogaster redii* Rosa (Glossoscolecidae, Hormogastrimae). *Annuar. R. Mus. zool. R. Univ. Napoli (N.S.)*, **8** (2), 1
— 1952. Oligocheti della Turchia. *Ann. Ist. Mus. Zool. Univ. Napoli*, **4** (2), 1.
— 1954. Problemi faunistici riguardanti gli oligocheti terricoli della Sardegna. Atti *Soc. tosc. Sci. Nat.* **61** (*suppl.*), 1.
— 1955. Oligocheti terricoli du Venezuela raccolti del dr. Marcuzzi. *Memorie Mus. civ. Stor. nat. Verona*, **4** (1954), 199.
— 1956a. Oligocheti dell' Indocina e del mediterraneo orientale. *Memorie Mus. civ. Stor. nat. Verona*, **5**, 321.
— 1956b. Sistematica e distribuzione geografica degli Hormogastrinae (Oligocheti). *Archivio Botanico e Biogeografico Italiano* 32, (*Ser.* 4), **1** (4), 159.
— 1958. La réserve naturelle intégrale du Mont Nimba. Oligochèta. *Mem. Inst. fr. Afr. noire*, **58**, 9-109.
— 1960. Oligocheti della Sicilia (I). *Memorie Mus. civ. Stor. nat. Verona*, **8**, 69.
— 1961. Oligocheti della Francia meridionale e di localita limitrofe. *Memorie Mus. civ. Stor. nat. Verona*, **9**, 67.
— 1963. Distribution of the Terricolous Oligochaetes on the two shores of the Atlantic. In *North Atlantic Biota and their history*. Oxford.
— 1964. Oligocheti della Sicilia. II. *Boll. dell' Accad. Gioenia Sci. nat. Catania*, **4**, 8 (2), 73.
PANCERI, P. 1875. Catalogo degli Anellidi, Gefirei, e Turbellarie d'Italie. *Atti Soc. ital. Sci. nat.*, **18**, 201.
PERRIER, E. 1872. Recherches pour servir à l'histoire des Lombriciens terrestres. *Arch. Z. exp. gen.*, **1**, lxx.
— 1874. Sur les Lombriciens terrestres exotiques des genres *Urochaeta* et *Perichaeta*. *C. r. hebd. Séanc. Acad Sci. Paris*, lxxviii, 814.
RAO, N. 1922. Some new species of earthworms belonging to the genus *Glyphidrilus*. *Ann. Mag. nat. Hist.*, (9) **9**, 51.
RAPP, W. 1849. Über einen neuen Regenwurm vom Cap. (*Lumbricus microchaetus*). *Jber. Ver. Naturk. Wurttenberg*, **4**, 142.
ROSA, D. 1886a. Noti sui Lombrici del Veneto. *Boll. Musei Zool. Anat. comp. R. Univ. Torino*, **4** (3), 673
— 1886b. Nota preliminaire sul *Criodrilus lacuum*. *Boll. Musei Zool. Anat. comp. R. Univ. Torino*, **1** (15), 1.
— 1887. Hormogaster redii n.g., n.sp. *Boll. Musei Zool. Anat. comp. R Univ Torino*, **11** (32), 1.
— 1889. Sulla struttura della *Hormogaster redii*. *Mem. Acc. Tor.*, **39**, 49.
— 1890. Viaggio di Leonardo Fea in Birmania e Regioni vicine XXV, Moniligastridi, Geoscolecidi ed Eudrilidi. *Annali Mus. civ. Stor. nat. Giacomo Doria*, **9**, 368.
— 1892. *Kynotus michaelsenii* n.sp. (Contributo alla Morfologia dei Geoscolicidi). *Boll. Musei Zool. Anat. comp. R. Univ. Torino*, **7** (119), 1
— Oligocheti terrioli (inclusi quelli raccolti nel Paraguay dal Dr. Paul Jordan). In *Viaggio Borelli Argentina Paraguay. Boll. Musei Zool. Anat. comp. R. Univ. Torino*, **10** (204), 1.
SCHMARDA, L. K. 1861. *Neue wirbellose Tiere, beobachtet und gesammelt auf einer Reise um die Erde 1853-1857*, Leipzig.
Reise um die Erde 1853-1857, Leipzig.
SCIACCHITANO, I. 1960. Oligochaeta part. In *South African animal life. Results of the Lund University Expedition in 1950-1951*. Stockholm. 7, 9.
SIMPSON, G. G. 1961. *Principles of animal taxonomy*. New York.
SMITH, F. 1895. A preliminary account of two new Oligochaeta from Illinois. *Bull. Ill. St. Lab. nat. Hist.*, **4**, 138

SMITH, F. 1915. Two new varieties of earthworms, with a key to described species in Illinois. *Bull. Ill. St. Lab. nat. Hist.*, **10** (7), 551.

STANKOVIC, S. 1960. *The Balkan Lake Ohrid and its living world.* Monographie Biologicae. den Hague 1960.

STEPHENSON, J. 1913. Aquatic oligochaeta from the Lake of Tiberias. *J. Asiat. Soc. Calcutta*, **9**, 53.

— 1914. Littoral Oligochaeta from the Chilka Lake on the East Coast of India. *Rec. Indian Mus.*, **10**, 255.

— 1915. Fauna of the Chilka Lake. Oligochaeta. *Mem. Indian Mus.*, **5**, 138.

— 1916. On a collection of Oligochaeta belonging to the Indian Museum. *Rec. Indian Mus.*, **12**, 299.

— 1917. Aquatic oligochaeta from Japan and China. *Mem. Asiat. Soc. Beng.*, **6**, 83.

— 1920. On a collection of Oligochaeta from the lesser known parts of India and from Eastern Persia. *Mem. Indian Mus.*, **7** (3), 191.

— 1921. Oligochaeta from Manipur, the Laccadive Islands, Mysore and other parts of India. *Rec. Indian Mus.*, **22**, 745.

— 1922. A note on some supposed new species of earthworms of the genus *Glyphidrilus*. *Ann. Mag. nat. Hist.*, (9) **9**, 387.

— 1923. Oligochaeta. In *The fauna of British India.* London.

— 1924. Oligochaeta of the Siju Cave, Garo Hills, Assam. *Rec. Indian Mus.*, **26**, 127.

— 1925. Oligochaeta from various regions, including those collected by the Mount Everest Expedition 1924. *Proc. zool. Soc. Lond.*, **1925**, 879.

— 1928. Oligochaeta from Lake Tanganyika. (Dr. C Christy's Expedition, 1926.) *Ann. Mag. nat. Hist.*, (10) **1**, 1.

— 1930a. *The Oligochaeta.* Oxford.

— 1930b. On some oligochaeta from Berhala Island in the Straits of Malacca. *Miscnea. zool. sumatr.*, **48**, 1.

— 1930c. On some African Oligochaeta. *Archo. zool. ital.*, **14** (2-4), 485.

— 1931a. Reports of an expedition to Brazil and Paraguay in 1926-27, supported by the Trustees of the Percy Sladen Memorial Fund and the Executive Committee of the Carnegie Trust for Scotland. The Oligochaeta. *J. Linn. Soc.*, **37**, 291.

— 1931b. Oligochaeta from Burma, Kenya and other parts of the world. *Proc. Zool. Soc. Lond.*, **1931**, 33.

STUHLMANN, F. 1889. Zweiter Bericht über eine mit Unterstützung der Königlichen Akademie der Wissenschaften nach Ostrafrika unternommenen Reise. *Sber. dt. Akad. Wiss.*, **2**, 645.

TÉTRY, A. 1934. Description d'une espèce francaise du genre *Pelodrilus* (Oligochètes). *C. r. hebd. Séanc. Acad. Sci., Paris*, **199**, 322.

— 1938. *Contribution à l'étude de la faune de l'Est de la France (Lorraine).* Nancy.

UDE, H. 1922. Regenwürmer aus Mazedonien. *Arch. Naturgesch.*, **88** (A), 155.

— 1929. Oligochaeta. In Dahl, *Die Tierwelt Deutschlands*, **15** (1), 1.

VAILLANT, L. 1889. *Histoire naturelle des Annelés marins et d'eau douce. Lombriciniens, Hirudinées, Bdellomorphes, Térétularieus et Planariens.* 3.

VANNUCCI, M. 1953. Biological notes I. On the Glossoscolecidae earthworm *Pontoscolex corethrurus. Dusenia*, **4** (3-4), 287.

VEJDOVSKY, F. 1879. Ueber die Entwickelung des Herzen von *Criodrilus. Sber. K. bohm. Ges. Wiss.*, **1879**, 358.

— 1884. *System und Morphologie der Oligochaeten.* Prag.

WASAWO, D. P. S. 1962. The taxonomic position of *Alma worthingtoni* Stephenson. *Ann. Mag. nat. Hist.*, (13) **5**, 113.

YAMAGUCHI, H. 1953. Studies on the aquatic Oligochaeta of Japan. VI. *J. Fac. Sci. Hokkaido Univ. (Zool.)*, **11**, 277.

APPENDIX
RECENT ADDITIONS TO
THE LITERATURE

Since this manuscript was prepared a number of publications have appeared, and in order to update the text they will be reviewed very briefly here under the authors in alphabetical order.

BRINKHURST, R. O. 1969. Aquatic Oligochaeta of the Azores and Madeira. *Bol. Mus. Municip. Funchal.* **23**, 46. A poor fauna with cosmopolitan species.

—— 1970. A further contribution towards a study of the aquatic Oligochaeta of Africa. *Rev. Zool. Bot. Afr.* **81**, 101. A new phreodrilid from L. Tanganyika (Uvira) (*P. tanganyikae*) is described, and new records from Africa (mostly the Congo) are given. The generic name *Potamothrix* (Tubificidae) is discussed, as well as the validity of identifications of African naidids by Hrabe, 1966a.

—— 1971. The aquatic Oligochaeta known from Australia, New Zealand and the adjacent islands. *Univ. Queensland Papers.* **3**, 99.

COOK, D. G. 1969. Observations on the life history and ecology of some Lumbriculidae (Annelida, Oligochaeta). *Hydrobiologia.* **34**, 561.

—— 1970a. *Peloscolex dukei* n.sp. and *P. aculeatus* n.sp. (Oligochaeta, Tubificidae) from the north-west Atlantic, the latter being from abyssal depths. *Trans. Amer. Microsc. Soc.* **88**, 492.

—— 1970b. *Torodrilus lowryi.* New genus and species of marine tubificid oligochaete from Antarctica. *Trans. Amer. Microsc. Soc.* **89**, 282.

—— 1970c. Bathyal and abyssal Tubificidae (Annelida, Oligochaeta) from the Gay Head-Bermuda transect, with descriptions of new genera and species. *Deep-Sea Res.* **17**, 973. Describes *Bathydrilus asymmetricus, Adelodriloides voraginus*, and *Phallodrilus profundus.*

COSTA, H. H. 1967 A systematic study of freshwater Oligochaeta from Ceylon. *Ceylon J. Sci.* **7**, 37. Aeolosomatidae and Naididae—new records.

ERCOLINI, A. 1969. Su alcuni Aeolosomatidae e Naididae della Somalia (Oligochaeta, microdrili). *Ital. J. Zool.* **2** (suppl. 3), 9.

HRABE, S. 1966. On some Naididae from the Volta Lake in the Ghana. *Publ. Fac. Sci. Univ. Brno.* **477**, 373. *Aulophorus ghanensis, A. tridentatus*, and *Allonais paraguayensis ghanensis* are described as new entities, but vide Brinkhurst (1970).

—— 1969a. On certain points in the structure of *Tubifex minor* Sok. and *Tubifex amurensis* n.sp. Sok. and Hrabe. *Ibid.* **506**, 259. Incomplete serial sections are used to establish the absence of coelomocytes in *T. minor* and to describe

T. amurensis which has no cuticular penis sheath but does have penial setae. This combination of characters makes it highly unlikely to be a *Tubifex*, and it is best to regard the specimen as unidentifiable as did Sokolskaya (1961).

HRABE, S. 1969b. Some remarks to the paper on *Euilyodrilus thermalis* (Pop). *Ibid.* **506**, 265. See Pop (1968) below.

—— 1969c. *Peloscolex kozovi* n.sp. from Baikal Lake. *Ibid.* **506**, 269. Two mature specimens briefly described and compared to the inadequately described *P. pigueti* and *P. wereschtschagini*.

LAAKSO, M. 1969. New records of aquatic Oligochaeta from Finland. *Ann. Zool. Fennici.* **6**, 348.

LAUZANNE, L. 1968. Inventaire preliminaire des oligochètes du Lac Chad. *O.R.S.T.O.M. Hydrobiol.* **11**, 83. Ten species are recorded from the region. *Aulodrilus tchadensis* n.sp. is part of the *A. pigueti-remex-kashi-stephensoni-cernosvitovi-prothecatus* complex, all grouped under the single name *A. pigueti* above. A description of *Alluroides tanganyikae.*

MILBRINK, G. 1970. Records of Tubificidae (Oligochaeta) from the great lakes (L. Mälaren, L. Vättern, and L. Vänern of Sweden. *Arch. Hydrobiol.* **67**, 86.

PICKAVANCE, J. R. & COOK, D. G. 1971. *Tubifex newfei* n.sp. (Oligochaeta, Tubificidae) with a preliminary reappraisal of the genus. *Can. J. Zool.* **49**, 249. The new species is probably *Tubifex nerthus* which it closely resembles. The generic definitions used in the subsequent discussion are narrow, and recent literature is overlooked. The merging of genera as subgenera of *Tubifex* would achieve little beyond re-stating the problem of generic definition in this area discussed herein.

PODDUBNAYA, T. L. 1966. On the classification of *Chaetogaster diaphanus* Gruith. (Oligochaeta, Naididae). Proc. Acad. Sci. USSR Inst. Biol. *Interior Waterways.* **12**, 120. Re-evaluates the description of the species.

POP, V. 1968. *Ilyodrilus thermalis*, eine neue Tubificidenart (Oligochaeta). *Zool. Anz.* **181**, 136. The species belongs in *Potamothrix*, and is a synonym of *P. heuscheri* (vide Brinkhurst 1970).

SOKOLSKAYA, N. L. 1964. Material on the aquatic microdrile fauna of South Sakhalin in *Lakes of the South Sakhalin and their Fauna*, Moscow, 1964. Describes *Tubifex lastockini*, now regarded as a synonym of *Ilyodrilus templetoni* (Tubificidae) vide Sokolskaya 1969a.

—— 1968. On the aquatic oligochaete fauna of the eastern USSR. *Nauch. Doklady. Ves. Skol. Biol. Nauk.* **595**, 14.

—— 1969a. Material on the fauna of aquatic oligochaetes of Kunashire Island. *Bull. M.O-Va. Isp. Prirod. Otd. Biol.* **74**, 62. The name for the newly described species *Ilyodrilus orientalis* (Tubificidae) is preoccupied. Brinkhurst (1970) supstituted *Ilyodrilus sokolskayae* nom.nov.

—— 1969b. New species of Lumbriculidae (Oligochaeta) from Kamchatka. *Zool. Zhurnal.* **48**, 342. Describes *Styloscolex opisthothecus* and *Kurenkovia magna*, n.spp., but the latter is not only unique but would have a completely nonfunctional set of reproductive organs if the observations are accurate and if the specimen is fully mature.

TIMM, T. 1970. On the fauna of the Estonian Oligochaeta. *Pedobiologia.* **10**, 52. The paper redescribes *Potamothrix bedoti* (Tubificidae) and maintains its identity in contrast to the synonymy with *P. bavaricus* claimed herein, but the distinction in anatomy and ecology is not fully clarified. *Tubifex smirnovi* is redescribed without reference to Laakso (1969).

WESTHEIDE, W. & BUNKE, D. 1970. *Aeolosoma maritimum* nov.sp. Die erste Salzwasserart aus der Familie Aeolosomatidae (Annelida, Oligochaeta). *Helgoländer wiss. Meeresunters.* **21**, 134.

SUBJECT INDEX TO PART I

Numbers in parentheses refer to Glossoscolecidae or 'megadriles' in general.

SYSTEMATIC INDEX
TO PART II

Specific epithets appear in Roman *type with the appropriate generic names in parenthesis; generic names are also in* Roman *type.*

  e.g. Tubifex
     tubifex (Tubifex)

Synonyms are italicized

  e.g. blanchari (Tubifex)

Nomina dubia, species dubia and species inquirendae are listed in Roman *type followed by* (n.d.), (sp. d.) *or* (sp. i.).

845

DATE DUE

JE 2 4 '81			
NO 1 2 82			
			FRINTED IN U.S.A.